THE ROUTLEDGE ENCYCLOPEDIA OF CITIZEN MEDIA

This is the first authoritative reference work to map the multifaceted and vibrant site of citizen media research and practice, incorporating insights from across a wide range of scholarly areas.

Citizen media is a fast-evolving terrain that cuts across a variety of disciplines. It explores the physical artefacts, digital content, performative interventions, practices and discursive expressions of affective sociality that ordinary citizens produce as they participate in public life to effect aesthetic or socio-political change. The seventy-seven entries featured in this pioneering resource provide a rigorous overview of extant scholarship, deliver a robust critique of key research themes and anticipate new directions for research on a variety of topics. Cross-references and recommended reading suggestions are included at the end of each entry to allow scholars from different disciplinary backgrounds to identify relevant connections across diverse areas of citizen media scholarship and explore further avenues of research.

Featuring contributions by leading scholars and supported by an international panel of consultant editors, the *Encyclopedia* is essential reading for undergraduate and postgraduate students as well as researchers in media studies, social movement studies, performance studies, political science and a variety of other disciplines across the humanities and social sciences. It will also be of interest to non-academics involved in activist movements and those working to effect change in various areas of social life.

Mona Baker is Professor Emerita of Translation Studies, University of Manchester, UK and Director of the Baker Centre for Translation & Intercultural Studies, Shanghai International Studies University, China. She is co-coordinator of the Genealogies of Knowledge Research Network, author of *Translation and Conflict* (Routledge 2006/2019), editor of *Translating Dissent* (Routledge 2016) and co-editor of *Citizen Media and Public Spaces* (Routledge 2016). She posts on translation, citizen media and Palestine on her personal website, www.monabaker.org and tweets at @MonaBaker11.

Bolette B. Blaagaard is Associate Professor of Communication at Aalborg University, Copenhagen, Denmark. Her research focuses on the intersections of culture and journalism with an emphasis on citizen media and postcoloniality. Blaagaard is the author of *Citizen Journalism as Conceptual Practice* (2018) and co-editor of *Citizen Media and Public Spaces* (Routledge 2016) with Mona Baker.

Henry Jones is Lecturer in Translation and Intercultural Communication at Aston University, Birmingham, UK and co-coordinator of the Genealogies of Knowledge Research Network. He has published research on translation practices in emerging online contexts, media theory and corpus-based methodologies.

Luis Pérez-González is Professor of Translation Studies and Co-director of the Centre for Translation and Intercultural Studies at the University of Manchester, UK. He is author of *Audiovisual Translation: Theories, Methods and Issues* (Routledge 2014) and editor of *The Routledge Handbook of Audiovisual Translation* (Routledge 2019).

Critical Perspectives on Citizen Media

Series Editors:

Luis Pérez-González, University of Manchester (UK)
Bolette B. Blaagaard, Aalborg University (Denmark)
Mona Baker, University of Manchester (UK)

Critical Perspectives on Citizen Media aims to define and advance understanding of citizen media, an emerging academic field located at the interface between different disciplines, including media studies, sociology, translation studies, performance studies, political science, visual studies and journalism studies. Titles in the series are focused on high-quality and original research, in the form of monographs and edited collections, made accessible for a wide range of readers. The series explores the relationship between citizen media and various cross-disciplinary themes, including but not restricted to, participation, immaterial work, witnessing, resistance and performance. The series editors also welcome proposals for reference works, textbooks and innovative digital outputs produced by citizen engagement groups on the ground.

Translating Dissent
Voices *from* and *with* the Egyptian Revolution
Edited by Mona Baker

Citizen Media and Public Spaces
Diverse Expressions of Citizenship and Dissent
Edited by Mona Baker and Bolette B. Blaagaard

Citizen Media and Practice
Currents, Connections, Challenges
Edited by Hilde Stephensen and Emiliano Treré

The Routledge Encyclopedia of Citizen Media
Edited by Mona Baker, Bolette B. Blaagaard, Henry Jones and Luis Pérez-González

Translating the Crisis
Politics and Culture in Spain after the 15M
Fruela Fernández

For more information, visit: https://www.routledge.com/Critical-Perspectives-on-Citizen-Media/book-series/CPCM

Dedicated series website: http://citizenmediaseries.org

THE ROUTLEDGE ENCYCLOPEDIA OF CITIZEN MEDIA

Edited by Mona Baker, Bolette B. Blaagaard, Henry Jones and Luis Pérez-González

Routledge
Taylor & Francis Group
LONDON AND NEW YORK

First published 2021
by Routledge
2 Park Square, Milton Park, Abingdon, Oxon OX14 4RN

and by Routledge
52 Vanderbilt Avenue, New York, NY 10017

Routledge is an imprint of the Taylor & Francis Group, an informa business

British Library Cataloguing-in-Publication Data
A catalogue record for this book is available from the British Library

Library of Congress Cataloging-in-Publication Data
A catalog record has been requested for this book

ISBN: 978-1-138-66556-9 (hbk)
ISBN: 978-1-315-61981-1 (ebk)

Typeset in Bembo
by Newgen Publishing UK

CONTENTS

ILLUSTRATIONS

Figure

Table

CONSULTANT EDITORS

CONTRIBUTORS

Randa Aboubakr
Cairo University, Egypt

Tina Askanius
Malmö University, Sweden

Michael Atkinson
University of Toronto, Canada

Konstantinos Avramidis
University of Portsmouth, UK

Jess Baines
London College of Communication, University of the Arts (UAL), UK

Bolette B. Blaagaard
Aalborg University, Denmark

Julie Boéri
Hamad Bin Khalifa University, Qatar

Çiğdem Bozdağ
University of Groningen, the Netherlands
Bremen University, Germany

Jacob Breslow
London School of Economics, UK

Vince Carducci
College for Creative Studies, USA

Maria Chatzichristodoulou
Kingston University London, UK

Karen Cross
Roehampton University, UK

Michael S. Daubs
Victoria University of Wellington, New Zealand

Lina Dencik
Cardiff University, UK

Christian DuComb
Colgate University, USA

Jens Eder
Film University Babelsberg Konrad Wolf, Germany

Fruela Fernández
Universitat de les Illes Balears, Spain

Cristina Flesher Fominaya
Loughborough University, UK

Dorismilda Flores-Márquez
Universidad De La Salle Bajío, México

Benjamin Franks
University of Glasgow, UK

Owen Gallagher
National College of Art and Design, Ireland
Bahrain Polytechnic

Britta Hartmann
University of Bonn, Germany

David Harte
Birmingham City University, UK

Andreas Hepp
University of Bremen, Germany

Arne Hintz
Cardiff University, UK

Moira Inghilleri
University of Massachusetts Amherst, USA

Petros Iosifidis
City, University of London, UK

Engin Isin
Queen Mary University of London, UK
University of London Institute in Paris, France

Henry Jones
Aston University, UK

Sigrid Kannengießer
Bremen University, Germany

Anne Kaun
Södertörn University, Sweden

Abigail Keating
University College Cork, Ireland

Alex Khasnabish
Mount Saint Vincent University, Canada

Dorothy Kidd
University of San Francisco, USA

Marwan M. Kraidy
Northwestern University, USA

Eileen Le Han
Michigan State University, USA

Eva Cheuk-Yin Li
Lancaster University, UK

Dario Lolli
Keio University, Japan

Marianne Maeckelbergh
Ghent University, Belgium

Daniela Mansbach
University of Wisconsin-Superior, USA

Stefania Milan
University of Amsterdam, the Netherlands

Nina Grønlykke Mollerup
University of Copenhagen, Denmark

Mette Mortensen
University of Copenhagen, Denmark

Christina Neumayer
University of Copenhagen, Denmark

Aoileann Ní Mhurchú
University of Manchester, UK

Evgenia Nim
National Research University Higher School of Economics, Russia

Gwen Ottinger
Drexel University, USA

Ian Alan Paul
Stony Brook University, USA

Paul Reilly
University of Sheffield, UK

Alexandra Reynolds
Solent University, UK

Martin Johannes Riedl
The University of Texas at Austin, USA

Clemencia Rodríguez
Temple University, USA

Julia Rone
Université Libre de Bruxelles, Belgium
University of Cambridge, UK

Mette Marie Roslyng
Aalborg University, Denmark

Neil Sadler
Queens University Belfast, UK

Digdem Sezen
Teesside University, UK

Tonguc Ibrahim Sezen
Teesside University, UK

Sandra Smeltzer
Western University, Canada

Kristina Smith
University of Toronto, Canada

Hilde C. Stephansen
University of Westminster, UK

Zoetanya Sujon
London College of Communication, University of the Arts London (UAL), UK

Einar Thorsen
Bournemouth University, UK

Gavan Titley
Maynooth University, Ireland

Omid Tofighian
University of New South Wales, Australia
University of Sydney, Australia

Emiliano Treré
Cardiff University, UK

Filippo Trevisan
American University, USA

Carlie D. Trott
University of Cincinnati, USA

Myrto Tsilimpounidi
Feminist Autonomous Centre for Research, Greece

Matilda Tudor
Uppsala University, Sweden

Jerome Turner
Birmingham City University, UK

Christopher Vito
Southwestern College, USA

Karin Wahl-Jorgensen
Cardiff University, UK

Melissa Wall
California State University, Northridge, USA

Katie Warfield
Kwantlen Polytechnic University, Canada

Manès Weisskircher
TU Dresden, Germany

Mark R. Westmoreland
Leiden University, the Netherlands

Elaine Yuan
University of Illinois at Chicago, USA

Derya Yüksek
Vrije Universiteit Brussel, Belgium

ACKNOWLEDGEMENTS

The editors would like to express their gratitude to consultant editors Stuart Allan, Susan Haedicke, Engin Isin, Stefania Milan, Mette Mortensen, Vinod Pavarala, Hilde C. Stephansen, Denise Tse-Shang Tang, Emiliano Treré and Haiqing Yu for their invaluable guidance throughout the project. Further thanks are due to Guobin Yang, Anne Kaun, Marianne Maeckelbergh, Jeffrey Juris, Julie Boéri, Maeve Olohan, Julia Rone and Neera Chandhoke, all of whom provided generous assistance in identifying potential contributors for entries that we had difficulty commissioning.

We are additionally grateful to the team at Routledge for their confidence in the project and their support at every stage of the process. Our special thanks go to Natalie Foster, Louisa Semlyen and Jennifer Vennall for their expertise and patience in guiding the publication forward.

Finally, the editors would like to acknowledge the help of the administrators of the MeCCSA listserv through which a call for contributors and entry suggestions was circulated in the spring of 2018.

Funding statements

The entry on Big data by Stefania Milan is informed by the findings of a project that has received partial funding from the European Research Council (ERC) under the European Union's Horizon 2020 research and innovation programme (grant agreement No. 639379-DATACTIVE, https://data-activism.net).

The entry on Film studies and citizen media by Jens Eder and Britta Hartmann is informed by the findings of a project that has received funding from the Volkswagen Foundation (A115999: Bewegungs-Bilder 2.0:Videoaktivismus zwischen Social Media und Social Movements).

NOTES ON REFERENCING CONVENTIONS

Different editions of the same work

Where there are two or more different editions from which different authors have quoted stretches of text, these are listed separately in the bibliography to ensure reliability of citations. For example:

Baym, N. (2010) *Personal Connections in the Digital Age*, Cambridge, UK and Malden, MA: Polity Press.
Baym, N. (2015) *Personal Connections in the Digital Age*, second edition, Cambridge: Polity Press.

Particles in European names

Common particles in Dutch and Flemish names – such as *van*, *de* and *der* – are ignored for the purposes of ordering authors alphabetically in the bibliography but are otherwise included in intext references. Hence, van Dijck is listed under *D* in the bibliography but referenced as *(van Dijck 2014)* or *as van Dijck argues*, in the body of the text.

The same applies to other European names beginning with *de* or *De*: hence, De Angelis is listed under *A* in the bibliography but referenced as *De Angelis* in the text, and Boaventura de Sousa Santos appears as *Santos, B. de Sousa* in the bibliography and as *de Sousa Santos* within the relevant entries.

For further information on relevant referencing conventions, see http://eprints.bournemouth.ac.uk/11930/2/Referencing_Dutch_Flemish_names.pdf.

Note: Italian names beginning with particles such as *della* appear in alphabetical order under the relevant particle in the bibliography – for example, della Porta is listed under *d* not *p*.

INTRODUCTION

Various social, political and technological developments have redefined the role of media in social life since the beginning of the twenty-first century, as attested by a burgeoning body of literature that interrogates the scholarly and disciplinary implications of such changes. Widely endorsed conceptualizations of the function of mass media in the Habermasian public sphere until the early 2000s have been superseded by new scholarly perspectives that foreground the impact on contemporary media practices of emerging forms of creative and political agency, labour structures, institutional agendas and technological platforms. Whereas the mass media of the public sphere were seen as mediators between the systemic world of political and economic institutions and the lifeworld of ordinary people sharing their feelings, experiences and values through everyday relationships (Habermas 1964/1989), the new media ecology is predicated on a more intricate enmeshing of the institutional domain and the subjectively experienced world. In this reconfigured public sphere, the systemic and the public are bound together as an organic whole by the discourses circulating not just within each of these domains but also across the two realms (Torgerson 2010). The *Routledge Encyclopedia of Citizen Media* explores the shifting conceptual network required to gain a better understanding of this reconfigured public sphere, thematizing new means and practices of representation and expression by ordinary people who wish to exercise citizenship in every domain of today's informational society.

This is by no means the first attempt to capture and elucidate the fluid and rapidly evolving interplay between media and the social fabric, as the relatively high number of encyclopedias on media and communication studies published since the turn of the century confirms. Whether they are published as multi-volume works (Valdivia 2013; Johnston 2003; Donsbach 2008) or in more compact formats (Schaefer and Birkland 2006; Danesi 2013), these cartographies of established pathways within media studies include entries on mediated discursive practices that reflect a variety of world views. Most of the scholarly work on media content produced by ordinary people during this period has orbited around two key concepts theorized in a range of monographs: radical media (Downing 1984, 2001) and alternative media (Atton 2002, 2004). Both designate citizen-driven projects seeking to resist and transform the establishment primarily through one of three channels: participatory journalism, e.g. blogs reporting about activist issues and events; mediated mobilization facilitated by platforms that host resources or circulate information about upcoming protests; and commons knowledge transmitted, for example, through wikis that are made up of citizen threads or posts (Lievrouw 2011). But while Downing (1984:ix) uses 'radical' to characterize media aiming to effect social change and build solidarity around their activist agendas, Fuchs (2010:178) favours the label 'alternative' to designate media "that challenge the dominant capitalist forms of media production, media structures, content, distribution and reception" and whose content articulates "oppositional standpoints that question all forms of heteronomy and domination" (ibid.:179). The difficulty in disentangling radical from alternative media – one of the reasons why critics have questioned the terms' explanatory power (Atkinson 2017) – is compounded by the fact that the same term may carry different meanings for different scholars. Unlike Fuchs, for example, Atton (2002:8) extends the scope of the term 'alternative' beyond political and resistance media to encompass "artistic and literary media (video, music, mail art, creative writing)", including "newer cultural forms such as zines and hybrid forms of electronic communication".

During the second decade of this century, debates over which term best represents the remit of oppositional media projects and initiatives seeking to provide alternatives to their mainstream counterparts have given way to retheorizations of old practices/acts and explorations of new ones. Large editorial projects have ushered in revised categorizations and new designations – reflecting the extent to which the two key terms have become inflected by competing themes or disciplinary perspectives. Acknowledging the drawbacks of using the term 'alternative' to refer to media that bypass the usual channels of commercial production and distribution, Atton's (2015) *Routledge Companion to Alternative and Community Media* effectively equates alternative media with community media to foreground the fact that they "are most often organised and produced by 'ordinary' people, local communities and communities of interest" (ibid.:1). Thematizing the role of communities in today's media landscape – whether these take the form of transnational networked subjectivities or, alternatively, map onto neighbourhoods, indigenous groups or diasporic constituencies, to give but a few examples – opens up new perspectives that are as concerned with issues of power, representation, participation and citizenship as they are with the subject matter of media content produced outside mainstream media structures.

In a near contemporaneous major editorial project, on the other hand, community media are subsumed under the category of social movement media, together with a myriad of other media types. Driven by "an anthropological and social movement perspective", Downing's *Encyclopedia of Social Movement Media* (2011:xxv) embraces a broad spectrum of media, ranging from established forms of activist journalism to "murals, graffiti, popular song and dance ... video and cinema", from "[l]ow-power community radio to hitech digital networks". As Downing himself notes in the introduction to this Encyclopedia, the term 'social movement media' is not chosen here to advance the disciplinary agenda of this scholarly domain, which has so far ignored activist initiatives outside the Global North and neglected the study "of communication and media as integral dimensions of social movements" (ibid.).

Despite the differences in their size, nomenclature and guiding principles for the selection of topics and contributors, Atton's and Downing's projects deliver comparable surveys of contemporary media actors and practices that straddle the digital and physical spheres, with an emphasis on those that critique dominant power structures. By contrast, Gordon and Mihailidis take a different slant on this mapping exercise in a major collective project entitled *Civic Media: Technology, design, practice* (2016). Civic media, the core concept at the heart of this publication, is defined as "all the technologies, designs, and practices that connect people to government, institutions, and more generally, the practice and promise of contemporary democracy" (ibid.:20). Most of the examples of civic engagement in digital culture explored here focus on "people around the world harnessing the affordance of digital media to enable democratic participation, coordinate disaster relief, advocate for policy change, empower marginalized communities, or simply to strengthen local advocacy groups" (ibid.), although some of the contributions to the collection explore how organizations and institutions can use digital technologies to develop closer relationships with citizens and service users. While approached primarily from a technology-centred perspective, the concept of civic media is ultimately underpinned by an understanding of citizenship, justice, inclusion and equity as the drivers of critical and engaged sociality.

Atton's (2015) emphasis on communities; Downing's (2011) conceptualization of social movements and their repertoires of media practices as instruments of resistance against the 'antidemocratic' and 'antisocial' forces at play in today's society; and Gordon and Mihailidis' (2016) focus on the connective practices and critical discourses fostered by the ubiquity of civic tools all bring into sharp relief the challenges that the shift towards digital culture poses to media and communication scholars. Although a number of thematic threads run through these reference works, thus pointing to the existence of shared core concerns across individual perspectives, the differences in nomenclature also expose the conceptual overlap and redundancy between them. By extension, these differences also expose gaps in the remit of the three editorial projects, as illustrated by the relatively small amount of attention paid to individual media practices in Atton's companion on community media; to the potential of playfulness and affect to feed and stir resistance against social structures in Downing's *Encyclopedia of Social Movement Media*; and to the role of embodied, on-the-ground performance of dissent in Gordon and Mihailidis' volume on civic media.

The Routledge Encyclopedia of Citizen Media has been conceived as an intervention in ongoing terminological and conceptual debates. Informed by Baker and Blaagaard's (2016) extended discussion of the term, citizen media is theorized in this volume as the physical artefacts, digital content, performative interventions, practices and discursive formations of affective sociality that ordinary citizens produce as they participate in public life to effect aesthetic or socio-political change. The contribution that the concept of citizen media makes to extant scholarship is multifold, as are the manifestations of citizen media that Baker and Blaagaard's definition encompasses. Unlike other conceptualizations of non-mainstream media, their understanding of citizen media places equal emphasis on both the digital and the material, as well as on the interplay between virtual and physical forms of expression. Seen from this new perspective, digital technologies act as the technical infrastructure enabling and amplifying multi-voiced discursive practices from networked subaltern counterpublics (Fraser 1990), providing individuals and collectives with the means and opportunities to share social imageries and discourses with other members of their networked imagined communities (Anderson 1991). Counterpublics sometimes take the form of disembodied spaces of discourse (Stephansen 2016), but they can also arise from a multiplicity of material processes and a constellation of individual and collective interventions in publics through the medium of the human body that numerous strands of media studies have tended to downplay (Baker and Blaagaard 2016).

Any area of human experience is therefore seen here as subject to negotiation by ordinary individuals and groups positioned largely outside the mainstream institutions of society and committed to expressing their concerns – politically, artistically, emotionally, physically or otherwise – in their own, often novel ways and using any media at their disposal. The large, global movement for natural birth, which challenges the widespread medicalization of women's bodies, is thus a form of citizen media practice as understood in this Encyclopedia, as is the small, local Ikwe Marketing Collective, a grassroots economic development project on the White Earth Indian reservation in northern Minnesota which supports economic subsistence in the community by ensuring its crafts and wild rice are sold at a fair price (*Cultural Survival Quarterly Magazine* 1987). In resisting the conflation between citizen media and the digital, and acknowledging the potential of virtual and physical artefacts and practices to operate through and upon the other (Isin and Ruppert 2015), this Encyclopedia engages with participatory cultures across the entire spectrum of a population, irrespective of their level of access to digital media: even illiterate members of a society experience and engage in embodied forms of expression, while it is primarily the (relatively) privileged and (relatively) young who have access to and know how to use social and digital media effectively.

The inclusion of citizens' physical practices within the remit of citizen media, as conceptualized here, affords a new lens to explore how these forms of unaffiliated expression and representation, whether embodied (e.g. flash mobs, graffiti and street art or parkour) or virtual (e.g. selfies or other forms of user-generated content), performatively confront the disciplining norms of society – one of the main themes running through this Encyclopedia. Drawing on a body of literature on performance studies and theories of performativity, Baker and Blaagaard have argued that the intervention of citizens in publics, often with a view to subverting or renegotiating prevailing social norms, "is a form of performance that participates in constructing the social space in which we live and act as citizens" (2016:7), both locally and in networked spaces of cosmopolitan solidarity. Indeed, the growing amount of scholarly attention that the constructed, performative dimension of these interventions is attracting is not simply a reflection of the fact that studies of citizen media accommodate multiple ontologies and foster epistemological plurality. Ultimately, this increased attention recognizes that "by actively intervening in and transforming the established mediascape" and "contesting social codes, legitimized identities, and institutionalized social relations" (Rodríguez 2001:20) citizen media enable the enactment of citizenship.

Some of the most studied practices and interventions "through which individuals *become* citizens" (Stephansen 2016:28; emphasis in original) aim to bypass various structures and systems of control and surveillance (e.g. hacking and hacktivism, surveillance, sousveillance), voice political or ideological dissent and enact protest. As a theme, resistance features very prominently in this Encyclopedia, foregrounding the connections between our conceptualization of citizen media and some of the more established notions in the field, such as radical media and critical media. Entries gravitating around this theme focus on various categorizations of resistance (e.g. activism, civil disobedience, direct action), attempts to silence protest and prevent uprisings in public spaces (discussed, to some extent, in the entries on autonomous movements and space and place), and the threat of co-optation through which existing structures of political and corporate power try to contain the subversive (co-optation).

A third, interlocking theme in this Encyclopedia pertains to the centrality of witnessing in a wide range of citizen media practices. As befits an editorial project firmly anchored in the age of digital culture, the Encyclopedia includes a range of entries that examine how the enactment of citizenship and performance of resistance are documented (e.g. archiving, community media, digital storytelling, citizen journalism, Indymedia, documentary filmmaking, mobile technologies, photography), whether events are witnessed by citizens on the ground or by networked communities capitalizing on the affordances of digital technology. By "engaging people's potential to care" (Chouliaraki 2010:305) and turning witnessing experiences into mediated discourses, citizen media thematize moral demands on publics and audiences to act on behalf of ordinary individuals and communities who suffer in situations of political and armed conflict and through natural disasters. In these sites of struggle, citizen media contribute to bridging space and time, and to developing structures of moral and affective sociality, as discussed in the entries on solidarity and precarity. The growing need for protocols to facilitate the independent verification of citizen media content adds another dimension to the role of witnessing in the context of citizen media studies (see the entry on authenticity).

While other theorizations privilege the communal dimension of citizen media (Atton 2015; Rodríguez 2011), the conceptualization of citizen media underpinning this Encyclopedia places particular emphasis on unaffiliated citizens acting, potentially on an individual basis, "in the pursuit of a non-institutionalized agenda, and without the mediation of a third party or benefactor" (Baker and Blaagaard 2016:15). A number of entries relating to participation, the fourth of the major themes covered in this Encyclopedia, explore how citizen media practices, whether they involve individual or collective agents, facilitate the construction of physical (e.g. the entry on the World Social Forum) or virtual networks or platforms (e.g. Facebook, Twitter and hashtags, Weibo, YouTube) of like-minded individuals around a collective sense of purpose. By examining how unaffiliated citizens participate in publics, it is possible to gain a better understanding of the collaborative production and dissemination of knowledge and information (e.g. big data, citizen science, commons, wikis); of debates on roadmaps to replace vertical structures of democratic representation with horizontal models of deliberation (e.g. social movement studies, prefiguration, civil society); and of the dynamics of sites of engaged or aesthetic, often playful, affectivity (culture jamming, video games).

Participation intersects fruitfully with the fifth of the major themes running through this Encyclopedia: immaterial work. Explored in a range of general entries (e.g. immaterial labour, amateur), immaterial work is responsible for the proliferation of free media content produced by unpaid ordinary citizens – whether they act on their own impetus or their involvement is solicited by specific corporations or institutions (see the entries on crowdsourcing and crowdfunding). Amid fast-paced changes blurring traditional boundaries between the production, distribution and consumption of media content (convergence) and fostering the development of participatory sites of 'prosumption', the involvement of individuals and communities in different forms of co-creational work (fandom, remediation, social media, content moderation and volunteer participation) is disrupting traditional structures of labour and eroding the social status of established professions in the creative and media industries. The cluster of entries engaging with the logic of the gift economy raises important issues pertaining to the interplay between citizen media and digital technologies, whether this is ultimately shaped by a relationship of empowering synergy or driven by dynamics of regulative tension – in those cases where the technologization of these interventions effectively restricts the transformative power of the practices that enable the enactment of citizenship.

In addition to mapping out practices, acts and platforms whose contours continue to change at an accelerated rate, we have also sought to offer critical accounts of the core conceptual framework at the heart of this emerging interdisciplinary domain of study (entries on citizenship, media, media ecologies, media event, media practices, mediatization, process vs. event, public sphere, publics), as well as more pervasive phenomena that are important to understand in the context of how individuals and groups negotiate social and political spaces dynamically to express their aspirations (e.g. authenticity, precarity, solidarity, subjectivity, temporality).

The grouping of entries under each of the various themes outlined above is not meant to put in place a rigid compartmentalization of citizen media content and practices, insofar as most entries are connected with one or more of these themes. Indeed, as defined in this volume, the study of citizen media necessarily draws on theoretical and methodological linkages that have not been addressed so far within the scope of a single companion volume. For instance, in engaging with the practices of citizens voicing their views, articulating their shared interests and negotiating their personal and collective identities in public, some entries make excursions into domains such as social movements, activism and fandom studies. To shed light on the affective and transformative potential of citizen media practices – whether these take the form of amateur art installations or community radio, among other examples – the various entries surveying these areas necessarily bring into sharp relief the entanglements between media and relevant forms of multimodal expression. And in discussing digital practices of citizenship such as digital storytelling and remediation – which, unlike their traditional embodied counterparts, are delinking citizens' agendas from their immediate physical environment, disrupting the traditional entwining of people, culture and place and weaving transnational affective bonds across different regions of cyberspace and digital media spaces – the Encyclopedia draws on insights from globalization theories and studies on participatory and networked cultures. In light of the rapidly evolving and ever more challenging configuration of citizen media studies, which is not yet recognized as a distinct field of study, the Encyclopedia adopts a broad, cross-disciplinary approach that cuts across practically all fields of enquiry – from core areas of the humanities to social science, and to the natural sciences and the study of medicine. A number of entries are thus intended to map scholarly developments at the interface between citizen media and other disciplines or interdisciplinary domains, in order to encourage the nurturing of synergies and engage scholars in related fields of study (e.g. anthropology and citizen media, conflict & humanitarian studies and citizen media, film studies and citizen media, migration studies and citizen media, performance studies and citizen media, philosophy and citizen media, political science and citizen media, popular culture and citizen media, postcolonial studies and citizen media, race & ethnicity studies and citizen media, social movement studies and citizen media). The risks that citizens incur in producing media content and the ethical responsibilities of researchers investigating this domain are also explored in a dedicated entry (ethics of citizen media research).

A number of concepts, practices and media (such as aesthetics, affect, co-creation, blogs, vlogs) and disciplines (including gender studies, sexuality studies, legal studies and urban studies) that we hoped to engage with remain sadly absent from the volume. Some were commissioned but did not materialize, while others proved too complex at this stage of developing the concept of citizen media as understood here to attract willing contributors. We hope to be able to address these gaps, and several others, in a future edition of the Encyclopedia, when the interdisciplinary domain of citizen media we have attempted to chart here is more developed. Until that future edition materializes, we trust this reference work is strong and flexible enough to offer a theoretical anchor and methodological steer through the rugged emerging terrain of citizen media studies.

The Editors
January 2020

References

Anderson, B. (1991) *Imagined Communities: Reflections on the origin and spread of nationalism*, London: Verso.

Atkinson, J. D. (2017) *Journey into Social Activism: Qualitative approaches*, New York: Fordham University Press.

Atton, C. (2002) *Alternative Media*, Thousand Oaks, CA: Sage.

Atton, C. (2004) *An Alternative Internet: Radical media, politics and creativity*, Edinburgh: Edinburgh University Press.

Atton, C. (ed.) (2015) *Routledge Companion to Alternative and Community Media*, London and New York: Routledge.

Baker, M. and B. B. Blaagaard (2016) 'Reconceptualizing Citizen Media: A preliminary charting of a complex domain', in M. Baker and B. B. Blaagaard (eds) *Citizen Media and Public Spaces: Diverse expressions of citizenship and Dissent*, London and New York: Routledge, 1–22.

Chouliaraki, L. (2010) 'Ordinary Witnessing in Post-television News: Towards a new moral imagination', in L. Chouliaraki (ed.) *Self-mediation: Citizenship and new media.* Special issue *Critical Discourse Studies* 7(4): 305–319.

Cultural Survival Quarterly Magazine (1987) 'Native Rice, Native Hands: The Ikwe Marketing Collective'. Available online: www.culturalsurvival.org/publications/cultural-survival-quarterly/native-rice-native-hands-ikwe-marketing-collective.

Danesi, M. (ed.) (2013) *Encyclopedia of Media and Communication*, Toronto, Buffalo, NY and London: University of Toronto Press.

Donsbach, W. (ed.) (2008) *The International Encyclopedia of Communication*, 12 volumes, Malden, MA: Wiley-Blackwell.

Downing, J. (1984) *Radical Media: The political experience of alternative communication*, Boston, MA: South End Press.

Downing, J. (2001) *Radical Media: Rebellious communication and social movements*, Thousand Oaks, CA: Sage.

Downing, J. (ed.) (2011) *Encyclopedia of Social Movement Media*, Los Angeles, CA, London, New Delhi, Singapore and Washington, DC: Sage.

Fraser, N. (1990) 'Rethinking the Public Sphere: A contribution to the critique of actually existing democracy', *Social Text* 25/26: 56–80.

Fuchs, C. (2010) 'Alternative Media as Critical Media', *European Journal of Social Theory* 13(2): 173–192.

Gordon, E. and P. Mihailidis (eds) (2016) *Civic Media: Technology, design, practice*, Cambridge, MA and London: The MIT Press.

Habermas, J. (1964/1989) 'The Public Sphere: An encyclopaedia article', translated by S. Lennox and F. Lennox, in S. E. Bronner and D. Kellner (eds) *Critical Theory and Society: A reader*, New York: Routledge, 136–142.

Isin, E. and E. Ruppert (2015) *Being Digital Citizens*, London, Lanham, MD, Boulder, CO and New York: Rowman & Littlefield.

Johnston, D. (ed.) (2003) *Encyclopedia of International Media and Communications*, Amsterdam: Academic Press.

Lievrouw, L. (2011) *Alternative and Activist Media*, Cambridge: Polity Press.

Rodríguez, C. (2001) *Fissures in the Mediascape: An international study of citizens media*, Cresskill, NJ: Hampton.

Rodríguez, C. (2011) *Citizens' Media against Armed Conflict: Disrupting violence in Colombia*, Minneapolis, MN and London: University of Minnesota Press.

Schaefer, T. and T. Birkland (eds) (2006) *Encyclopedia of Media and Politics*, Washington, DC: CQ Press.

Stephansen, H. C. (2016) 'Understanding Citizen Media as Practice: Agents, processes, publics', in M. Baker and B. B. Blaagaard (eds) *Citizen Media and Public Spaces: Diverse expressions of citizenship and dissent*, London and New York: Routledge, 25–41.

Torgerson, D. (2010) 'Policy Discourse and Public Spheres: The Habermas paradox', *Critical Policy Studies* 4(1): 1–17.

Valdivia, A. N. (ed.) (2013) *The International Encyclopedia of Media Studies*, Malden, MA: Wiley-Blackwell.

ACTIVISM

Christina Neumayer

Activism is a relatively new term. While its political use can be traced back to Swedish activists' petitioning for the end of neutrality in the First World War in 1915, some argue that film activism may predate these events, with the appearance of suffragettes in film as early as 1899 (Marchetti 2016). Because the suffragettes' street protests in their battle for women's rights were largely ignored by politicians and newspapers alike, they took to other forms of action and entered private homes and cinemas through newsreel footage (Hutchinson 2015). Using film in social struggles places these activist films outside genres such as political documentaries (Marchetti 2016). The term activism generally refers to a form of citizen-led intervention that seeks to improve society through political, economic, environmental or social change, which suggests some form of direct action in public space. Demonstrations, sit-ins, strikes, riots and violent confrontation as well as various forms of civil disobedience are drivers of social change generally associated with activism, but as the early example of suffragettes in film suggests, there are numerous other strategies for pressuring governments and policymakers to effect change gradually.

The term activism has been contested and challenged, partly because of its ambiguity (Bevington and Dixon 2005; Hoofd 2012). To reduce this ambiguity, it is useful to differentiate between activism and activist. As suggested in an online article titled 'Give up Activism', "[t]he activist is a specialist or an expert in social change ... Defining ourselves as activists means defining our actions as the ones which will bring about social change, thus disregarding the activity of thousands ... of other non-activists" (Andrew X 1999). A broader understanding of activism would go beyond this definition of activist to encompass – in addition to the radical subculture composed of activists who identify as such – a broader spectrum of actions that can lead to social change alongside direct action. If the concept of a collective identity of social movements is taken as a point of departure, identifying oneself as an activist raises high expectations, but "one can 'do activism' without 'being activist'" (Bobel 2007:149). In this sense, activism may be understood as concerned with *doing* and the practice of political action rather than with *being*. As Gitlin (2003b:5) explains in his *Letters to a Young Activist*, the word activism is useful "because it reminds us that the world not only is but is made" and that "human beings make history". The active making of the world and society again refers to activism as something we *do*. Cammaerts (2007a:217) thus argues that the interaction between "agency and makeability of society" is key to any understanding of activism. In this context, the media function as a "symbolic arena" in which struggles over meanings and ideas take place (ibid.:217), and these struggles involve various types of action, ranging from lobbying through NGOs, to classic forms of direct action, to artistic performances.

Media traditionally played a minor role in studies of activism and were assigned the purely functional tasks of organizing, communicating, movement building and providing collective action frames in social movement studies. The term activism in this context has often been used solely in connection with social movements, but this has changed dramatically since the turn of the century. With the emergence of the Internet, digital media, Web 2.0, social media and big data, numerous studies have been conducted on how these technologies are used to mobilize support for particular causes and allow unaffiliated individuals to act together (Lievrouw 2011; Bennett and Segerberg 2013). What had previously been understood as media activism transformed into numerous terms that describe new technological developments accompanied by new forms of political action and civic engagement (Neumayer and Rossi 2016). Cyberactivism (McCaughey and Ayers 2013), slacktivism (Morozov 2009), clicktivism (White 2010), hacktivism (Jordan 2002a), Internet activism (Kahn and Kellner 2004), social media activism (Gerbaudo and Treré 2015), mobile activism (Liu, J. 2015), subactivism (Bakardjieva 2009) and design activism (Fuad-Luke 2013) or data activism (Milan and van der Velden 2016) are examples of terms used to refer to different forms of action associated with media-related technological developments. This use is in line with similar patterns in journalistic and public discourse about new media technologies for activism, as evident in journal article titles referencing 'Egypt's Facebook Revolution' (Smith 2011) and 'Iran's Twitter Revolution' (Keller 2010). Many of these catchphrases are based on the misconception that new media technologies deterministically foster political engagement and spark activism; this misconception has informed numerous publications about the role of digital media in the Middle East uprisings and the Occupy and Indignados movements (Castells 2012; Curran et al. 2012; Dahlgren 2013).

While the focus in social movements is often on the collective, studies on activism also engage with the individual perspective, especially in the context of digital and social media. In social media, Bennett and Segerberg (2013) have argued for connective action as an organizational principle, alongside collective action: connective action gives rise to personalized campaigns organized by loosely connected individuals. Such forms of action which connect unaffiliated individuals through social media do not quite fit into concepts of social movements or collective action, which is one reason why many have acknowledged that new forms of activism are now emerging for which we have not elaborated a stable terminology. These forms of activism are enabled by emerging media technologies; at the same time, media are appropriated by activists but also possess their own material characteristics, which shape how people engage, protest, resist and struggle. It is at this intersection between people and media that activists develop new tactics and renegotiate the meaning of established forms of communication, transforming them into activist practices in their struggles (Gerbaudo 2012). In line with the definition of citizen media adopted in this volume (Baker and Blaagaard 2016a), we can conceptualize and theorize activism across different fields and disciplines, including social movement studies, media studies, communication studies, science and technology studies (STS), political economy and critical theory (Cammaerts et al. 2013; Gerbaudo and Treré 2015; Lievrouw 2011). Activism can then be empirically and conceptually understood as a mediated form of resistance: mediated by media technologies as well as material and immaterial artefacts and devices endowed with expressive power to communicate information, emotions, values and narratives.

Artefacts, media and the digital

Many forms of political action revolve around street protests, specific events or an ephemeral spontaneous form of resistance. These include spontaneous lecture hall occupations by university students (Maireder and Schwarzenegger 2012), local protest events to protect a community-owned bathhouse (Svensson et al. 2014), and protests organized around global events such as a climate conference (Askanius and Gustafsson 2010) or against local actions of authoritarian governments (Liu 2016). They also include transnational movements of loosely connected individuals such as Occupy Wall Street (Feigenbaum 2014; Juris 2012). In these events, artefacts such as the tear gas canister, mobile phone and the human body carry expressive power in and of themselves. As Winner (1986/2009) argues, we are surrounded by architecture and objects that are inherently political. In acts of resistance, various artefacts can thus be appropriated and endowed with expressive power to become objects of contention. In the Occupy movement, for example, the tent and the protest camp became particularly prominent as architectures of resistance, their materiality endowed with symbolic, emotional and performative power that "turns them into intended or unintended elements of communication systems" (Feigenbaum 2014:16). It is these material artefacts and their appropriation as objects of resistance and tools of various acts of civil disobedience that communicate activism and connect unaffiliated individuals in acts of protest. The umbrella (initially used to protect activists from tear gas) as a symbol of resistance against authorities in Hong Kong (Chan 2014), and the human body – specifically, the gesture of raising the hands to indicate 'don't shoot' as a symbolic form of resistance against unjustified police violence in the Black Lives Matter movement (Mirzoeff 2017) – are all examples of material artefacts that are used to communicate meaning and distinguish different movements. These objects, moreover, not only possess symbolic power but also contribute to forming alliances and establishing connections among activists. Despite the burning down of tents in Tahrir Square in 2011 and the total removal of these objects of resistance, the communicative and collaborative infrastructure that they represented persisted and was replicated in other protest movements (Mollerup 2015). In other words, the material value of the tents for activist agendas and pursuits endured beyond any individual movement. Similarly, as objects of resistance such as the umbrella or the tent travel through digital media in the form of images – mediated and remediated, appropriated, memefied and personalized – they become symbols of resistance and forms of activism in their own rights.

Media representations, in their consistency and in their power, tend to delegitimize and marginalize alternative framings of different issues, and in so doing they define the asymmetries, hierarchies, presences and absences of public space. Internet and digital technologies facilitate rapid mobilization across distances as well as *ad hoc* action, but the question is, as Fenton (2008) notes, whether these *ad hoc* actions can result in a coherent oppositional ideology that can effect policy change. The hope for a better world that underlies these ideologies must be capable of overcoming fragmentation if it is to turn acts of resistance into a sustainable political programme. Building on Hannah Arendt's work, Roger Silverstone defines a mediapolis as a

> mediated space of appearance where the world appears and in which the world is constituted in its worldliness, and through which we learn about those who are and who are not like us. It is through communication conducted through the mediapolis that we are constructed as human (or not), and it is through the mediapolis that public and political life increasingly comes to emerge at all levels of the body politic (or not).
>
> *Silverstone 2007:31*

Maria Bakardjieva suggests to understand the contemporary mediapolis as a "triple helix comprising online media, traditional media and the physical spaces of the city" (Bakardjieva 2012:77). Activists who wish to make a difference struggle for appearance by visibility work and creative navigation for effective civic action across these strands of the mediapolis (ibid.). From this perspective, activism – particularly mediated forms of activism – includes direct action in public space as a means of seeking social change but also mediated forms of direct action as well as the mediation of bodies and objects endowed with expressive power to disrupt dominant representations.

Digital forms of action include expressing activist identity, coordinating, connecting individuals, mobilizing online and offline to protest or campaign, and traditional political actions such as signing petitions and reporting on events onsite or showing solidarity from a distance (Earl and Kimport 2011; Neumayer and Svensson 2016). In an overview of alternative and activist new media, Lievrouw (2011) discusses culture jamming, artistic expression, hacking, participatory journalism (such as Indymedia), mediated mobilization, and coordination of physical protest as forms of political action related to new media. Numerous other studies have been conducted on how activists use online platforms to mobilize support and organize themselves and their campaigns (Earl and Kimport 2011; Neumayer and Svensson 2016), with the interaction between control and emancipation fostering a particular strand of critical investigation into activism and digital media (Dencik and Leistert 2015; Uldam and Vestergaard 2015; Trottier and Fuchs 2014).

Activists have always built alternative media in the hope of counteracting the commercial influence of mainstream media and creating space for civil society (Atton 2004; Downing et al. 2001). There are, however, many challenges involved in building these alternative spaces using corporate media in digital space, including social media. Poell and Borra (2012:695) note that for "crowd-sourcing alternative reporting" the content of tweets is framed by mainstream news reporting to ensure wider circulation. Similarly, scholars have argued that violent action frames often dominate not only the news media but also digital media reporting by activists in order to achieve visibility (Cammaerts 2012). Leistert (2015:39) argues that corporate social media have become "algorithmic mass media", using censorship through algorithms as a normalization and standardization tool for activists' communicative action. This silencing of critical voices beyond the dominant violent action frames reinforces the neoliberal values in which both mainstream and corporate social media are embedded (Couldry 2010). The shaping of activist communication around spectacular, news-oriented reporting shifts the focus away from the actual protest (Poell and van Dijck 2015).

Media presence affects "virtually every aspect of a challenger's experience – recruitment efforts, organisation, strategy, and tactics" (Gamson 1992:147). While digital media may play an important role in creating visibility for counterinformation, this visibility could also become subject to surveillance and control by hostile authorities (Neumayer and Stald 2014). The ambiguity of being in the spotlight to create awareness and being exposed to a high level of risk is one aspect of the tension between the need to share information widely in order to enact radical forms of collective action on a large scale and the need to protect movements against surveillance and co-optation in digital space. In his critical theory of technology, Feenberg (2002) focuses on human agency, arguing that technology reinforces the prevailing political system's hierarchies and power relations. However, Feenberg argues, technological invention also provides new possibilities for subversive actors, who can challenge the system by appropriating new media technology for their cause. A critical analysis of the relationship between technology and activism requires us to open the black box of the materiality of media "as active agents shaping the symbolic and organizational processes of social actors" (Milan 2015:897).

Mirzoeff (2011:476) argues that visuality is a "discursive practice for rendering and regulating the real that has material effects". The classification of subjects through the aesthetics of their representation prevents actors from cohering as political subjects. In today's saturated media environments, riots, peaceful protests, artistic action and police and news media form a mosaic of perspectives in a struggle for attention in protest events. Various authors have suggested that activists can build powerful alternative public communication platforms using social media (Castells 2012). Their claims, however, raise questions regarding how corporate social media and their inherent logics might shape activists' communication (Fuchs 2012a; Poell and Borra 2012; Youmans and York 2012), and some have argued that self-absorbed and egocentric social media practices can counteract the aims of sustainable activist collectives (Fenton and Barassi 2011).

In media studies, it is assumed that activists use media in general and online media in particular to express political demands in their struggles, communicate alternative perspectives, organize, challenge dominant discourses and coordinate protests (della Porta and Tarrow 2004; Dunbar-Hester 2009; Rucht 2004). There are diverse perspectives on what should be considered activism in this plethora of forms of action and their relationship to media (Neumayer and Svensson 2016). Hacker groups such as Lulzsec (a name which combines Lulz – the plural of the acronym for 'Laughing Out Loud' or lol – and security) and Anonymous are examples of loosely connected groups of individuals clearly identifying as activists and possessing a clear political agenda, despite claiming to do things 'for the lulz' (popular catchphrase used as an explanation for trolling on the Internet) (Coleman 2013). Popular forms of Internet culture, such as memes, can become forms of activism and resistance. Simultaneously, images from protest events can be memefied and then decontextualized and entirely detached from the political critique that they originally expressed. A famous example of such memefication is the pepper-spray cop. During the Occupy movement demonstrations at the University of California, Davis, protesters seated on the pavement were pepper sprayed by university police for refusing to leave. A photograph taken of this incident went viral and was turned into a meme accompanied by the disclaimer 'Don't mind me, just watering my hippies', alongside hundreds of new photoshopped pictures (Bayerl and Stoynov 2016).

Individuals who would not consider themselves activists show support for and even participate in actions (for instance by producing memes) that require the involvement of a large number of people. Hackers, too, cannot automatically be labelled 'hacktivists' since not all hackers identify with this term (Taylor 2005). As Coleman (2013) argues, apart from a political message, hacking also includes an artistic component. This again shows how closely different forms of activism work together and how blurred the boundaries between them can be, especially given how often they are referred to using the same terminology. Anonymous and WikiLeaks, for example, use very different forms of participation to create visibility and resistance but are frequently included in the same category of online activism (Coleman 2013).

Activism and citizen media

By departing from a traditional understanding of media, this volume's perspective on citizen media challenges the dichotomy between online and offline action as well as between mediated activism and direct action in public space. Encompassing media technologies as well as material and immaterial artefacts and devices endowed with expressive power to communicate information, emotions, values and narratives within the same definition can provide an analytical perspective that can overcome these dichotomies. To conceptually understand contemporary activism, we must understand the interaction between objects and artefacts as well as their material and expressive power, which extends to and is being altered by its appropriation and decontextualization within the digital. Media technologies are thus themselves objects that carry expressive power to communicate and act politically in their own right. The smartphone in street protests, for example, alters activists' tactics towards more fluid organizational structures, particularly in the case of unaffiliated individuals. At the same time, eyewitnessing and the sharing of information and news both with the public and within movements of direct action in public space can itself become a form of action and activism (Neumayer and Stald 2014). Indeed, Galis and Neumayer (2016) argue that filming and photographing police violence can become a more relevant form of resistance than assisting fellow activists who are exposed to violence. Furthermore, making police violence visible online via mobile phone images and video can provide alternative perspectives that mainstream media might not be willing to broadcast.

Using media technologies can alter the ontology of activism. Activists' use of social media can result in acts of resistance as well as the building of (im)material alliances against injustice (Papadopoulos 2011). Including both material (umbrella, tent, mobile phone) and immaterial artefacts in an analysis of activism can advance our conceptual understanding of these altered ontologies of contemporary activism. It enables us, for instance, to treat citizen media as political machines (Barry 2001) which constitute an alternative to corporate media and perform technopolitical activism (Galis and Neumayer 2016). That is, a critical analysis of activism and citizen media contributes to a reweaving of the political and the material by creating spaces in which individuals and collectives act to confront injustice and establish alternatives that can realize sociomaterial justice (Papadopoulos 2011; Galis and Neumayer 2016). In understanding activism as a mediated form of resistance, we pay epistemological and ontological attention to the obstacles to and enablers of citizen media, their political economies and techno-commercial materiality, as well as their sociomaterial and performative potential for activism.

See also: autonomous movements; big data; civil disobedience; culture jamming; direct action; hacking and hacktivism; media; mobile technologies; prefiguration; social media; social movement studies and citizen media; twitter and hashtags

Recommended reading

Dunbar-Hester, C. (2009) '"Free the Spectrum!" Activist encounters with old and new media technology', *New Media & Society* 11(1–2): 221–240.

Based on a case study with activist groups, this article unpacks the role of older and emerging media technologies for activism. It gives valuable insights into the complex interactions between activism and technologies.

Feigenbaum, A. (2014) 'Resistant Matters: Tents, tear gas and the "other media" of Occupy', *Communication and Critical/Cultural Studies* 11(1): 15–24.

This article discusses objects and architecture that mediate and communicate resistance in contemporary activism. It traces the history of such material objects (such as tents and tear gas) in social movement activism.

Lievrouw, L. (2011) *Alternative and Activist New Media*, Malden, MA: Polity Press.

Bringing together alternative media, social movements and activist arts this book gives an overview of contemporary forms of media activism and traces their historical roots.

AMATEUR

Karen Cross

There is limited scholarship focusing directly on the subject of the amateur, especially on the role of the amateur in the context of citizen media. A number of media and cultural studies scholars have made significant contributions by beginning to map the elusive histories of the amateur within specific domains of cultural activity. Some have explored, for instance, the role of the amateur within the development of photography (Bloore and Seiberling 1986; Edwards 2012). Others have focused on amateur film and cinema (Zimmermann 1995), as well as the development of amateur radio (Bartlett 2007) and theatre (Dobson 2011). Sociologists have also emphasized the importance of amateur activities as productive forms of leisure, especially in sustaining health and wellbeing, and in providing people with an occupation in so-called post-work contexts, such as redundancy or retirement (Stebbins 1992, 2014). Given the problems inherent in the rise of "free labour" in today's economy (Terranova 2000:33), and the shifting patterns of contemporary work and employment that raise questions regarding social justice (Taylor and Luckman 2018), investments in the figure of the amateur are never innocent. Amateurs are arguably key drivers of the new creative economy made possible by the development of the Internet (Hunter et al. 2013). However, the challenge that we are presented with today is one of reconciling the freedom that is apparently enjoyed by the amateur with what we now know about the new networks of power and control at play in the digital age (Merrifield 2017).

Adopting a historical perspective can be very helpful when discussing the amateur. This is primarily because it allows us to understand the shifting cultural landscape within which positions of marginality have become important, and for what reasons. In the past, the notion of the amateur tended to conjure up the idea of one who is engaged in serious leisure, aspiring to the heights of the professional, whilst simultaneously suggesting a failure to meet the standards required. Another sense of the term, which has gained further traction in today's context, is that of one who acts in opposition to the professional and who resists professionalism. This can be seen, for example, in the context of contemporary art practice (Roberts 2008) and in the more recent developments of citizen news reporting and social uses of camera phones (Pantti and Andén-Papadopoulos 2012). The notion of the amateur as a lover, as one who engages in an activity not for profit or personal gain but out of passion, ties into the prevailing discourse of the free market and gift economy. Against this backdrop, the amateur has become a symbolic figure of productive political change, and there are interesting ways in which the identity of the amateur is now connected with that of the citizen, especially within online settings. This entry begins by outlining some of these correlations as they have been viewed in the context of Web 2.0. It then focuses on the conceptual and theoretical challenges associated with taking up the marginal position of the amateur, drawing on the work of Edward Said in particular, before finally considering the case of visual journalism and reporting, an area that has been of key interest to those writing on citizen media today.

The historical and political circumstances of the amateur

The recent shift in the politics of media production, which has seen the amateur coming to the fore, coincides with the rise of a new system of governance emerging through online forms of social communication. This new regime relies on amateur content and the constant upload of material to create the data traces that can be monitored and transferred to different domains of analysis, and thus used to understand public activities and opinion. In this respect, the shifting fortunes of the amateur connect to a longer history of social change and commercial activity that exploits the social use of technology as a means of retaining a hold on power. The amateur has been subsumed by a range of new technological categories, such as the *produser* (Bruns 2008:21) – a neologism conflating *producer* and *user* – as well as ideas of "doing it together" and "doing it yourself" (Jenkins et al. 2015:181) that underlie the ethos of participatory culture. These terms suggest more active subject positions that raise new challenges for governing media institutions in particular, and for the professionals who work with them.

At the same time, there has been a certain amount of backlash against the amateur, as well as an attempt to tame and control environments of amateur participation. At a certain point during the development of Web 2.0, it was argued that we seemed to be enthralled by a "cult of the amateur", the phrase used in the title of Keen's (2007) book, and that online participation demanded that we suspend our critical faculties. In the cacophony of voices that proliferated through the web, the idea was that we were facing the death of the cultural intermediaries who effectively told us what was 'good' and 'bad' in culture, and who helped us discern what we should pay attention to. Yet, it would now seem more accurate to argue that, far from a flattening of hierarchical divisions, what we are now seeing in digital culture is a redevelopment of professional activity in light of a new culture of amateur creation and participation. This is especially apparent within the field of visual news reporting. In order to understand this, we need to reflect also on the longer history of amateur and professional entanglement.

Amateur activities have long been the primary site through which institutionalized and commercial interests encroach upon the domains of public and private life. Leisure and popular forms of technological consumption during the twentieth century especially were key to sustaining the productive powers of industrial capitalism, as well as the relations of the new service and information economies that have emerged since the turn of the century. We can look to the popularization of photography in the late nineteenth and early twentieth centuries as an example here. The development of popular box cameras placed the means of production in the hands of the people, but it also resulted in a process of deskilling that was to foreclose any radical political possibilities that might arise through the broader public circulation of images (Tagg 1988). This was especially problematic for the working class and for women, who were increasingly associated with the burgeoning aspirational and promotional cultures of that period and were positioned as somewhat passive consumers. We also see certain kinds of visual practices normalized during this period, practices which encouraged a focus on family life and the memorialization of the domestic frame.

Beyond this, there are also those practices of the 'serious' amateur, perhaps a camera club member, an intensely masculinized figure seeking mastery of the camera technology. Such kinds of practice often involve commitment to historical (sometimes nationalistic) forms of representation, or inhabiting some other aspect of technologically or artistically invested identity. Unsurprisingly, these types of visual culture became the focus of critical interest in the latter part of the twentieth century, especially for feminists who sought to expose sexist and ideologically problematic forms of amateur vision (Spence and Holland 1991). The development of online sharing of images has involved a turning away from these earlier modernist visions, but somewhat at the cost of a critique of vernacular outlooks. What we now see is not a reduction of interest in photographic technique but rather a move towards a position that is associated with the artistic avant-garde, where the amateur has long held the status of an "ego ideal" (Roberts 2008:22) for the artist-professional attempting to resist institutional incorporation. This, of course, presents certain challenges because to take on the position of the amateur is also today to take on the position of the failed aspirant or artist intentionally working towards failure, and hence there is an unwillingness to reflect on political power structures that have an impact on the amateur.

The marginal position of the amateur

It is not just within the domain of contemporary art that such an attachment to the amateur has been formed. The idea of the amateur as an outsider and oppositional figure has been taken up across the wider arts and humanities, and also within the context of critical theory, where attempts have been made to construct alternative political identities. In his 1993 Reith Lectures, published later as *Representations of the Intellectual* (1994), Edward Said adopts the concept of the amateur as a means of challenging the increasing pollution of the field of higher education by a toxic form of professionalism. In his lecture, we find a damning critique of the university and its instrumental attachment to the identity of the professional, which, Said argues, works to fundamentally undermine opportunities for social and political change.

The endeavours of the professional – the one who operates as a state functionary – should, Said contends, be separated out from those of the intellectual who, released from the burdens of control, can operate more in the service of freedom and justice. According to Said (1994:9), the role of the amateur-intellectual is to "represent all those people and issues that are routinely forgotten or swept under the rug". This suggests some kind of innate or intuitive sensing of history that is activated by a sense of responsibility in the world, or what might be considered as a form of citizenship. Yet, this citizenship is one that also requires investing in a certain critical position. As Said points out, Walter Benjamin's (1940/1999:247) theses on history provide a productive way forward in this respect. The materialist historian is characterized in Benjamin's work as a kind of angelic witnessing presence offering an alternative account of history that brushes against the grain of history. The role of the angel-historian, according to Benjamin, is to "seize hold of a memory as it flashes up in a moment of danger" (ibid.). The point is not to look at the past in linear terms; not to be guided by "the storm of progress" (ibid.), or what we might today understand as technological determinism. Instead, the figure of the angel, largely equated with the amateur in Said's manifesto, is made to provide an access point to the terrain of memory, and therefore to a set of social relational experiences that are often ignored or sidestepped within history writing. The intellectual must resist professionalism in order to be able to capture stories untold and, in turn, to be able to speak truth to power – especially power that adheres to nascent processes of specialization as defined through the career trajectory of a modern-day professional historian.

Said's amateurism thus becomes a strategy for navigating the terms of institutional and state power in general, but his investment in such a position is also heavily entangled with his own experience of living as an intellectual in exile. This is interesting to consider in light of the growing significance of migrant and transitional identities today. The ideas explored by Theodor Adorno (1951/2005:38–39) in *Minima Moralia: Reflections from a Damaged Life* allow Said to argue that, for the person who no longer has a homeland, "writing becomes a place to live". Today, we may want to add that the online world becomes a place to live, and it is through the use of digital media that performances of identity in exile are now possible. Just as Said argues, however, even for those living in exile, there can be no slackening within the rigour of self-analysis. One's own subjective positioning, however inherently political, must always be interrogated and challenged. The process by which one is able to find citizenship must also be challenged.

As Said (1994:44) points out, there are certain pleasures that can be found in the marginal position of the exile: "[o]ne of course is the pleasure of being surprised, of never taking anything for granted, of learning to make do in circumstances of shaky instability that would confound or terrify most people".

One has to invent a means of living in what are quite often untried and untested ways, in ways that are suggestive of the amateur experience. The precarious state that underlies the experience of the person living in exile also offers a chance to adopt what Said refers to as a "double perspective" (ibid.) – a perspective that brings to consciousness the disjuncture and contradictions of opposing worlds, different time frames and states of governance.

The experience of exile (of operating in amateur ways) thus becomes a useful standpoint from which to understand certain rationalizations of power that play out across different domains of experience and action, and which reformulate themselves across space and time, and across different cultural contexts. However, not all experiences are equal, and when it comes to the position of the amateur, we are, by using the term, evoking a whole history of privilege and power that is associated with the rise of consumerism and state forms of governance. This needs to be taken into account as we critically reframe the study of citizen media practices and forms.

Amateur and professional media

As noted above, the amateur has accrued a new kind of value in today's world, especially within online settings. This is partly a consequence of the rise of platforms that require the creation of user-generated content, but also due to a renewed, somewhat nostalgic interest in popular analogue technologies. When reflecting on different examples of citizen media, especially citizen journalism, we can see that images circulated online are valued for their amateur qualities. This is evident, for instance, in some of the discussions of reporting around key events, such as the London Bombings (Cross 2016), where it appears that professionals were challenged by the arrival of user-generated content. Scholars have noted the various ways in which amateur images signify a form of authenticity perceived to be absent from professional forms (Pantti and Bakker 2009:486; Pantti and Andén-Papadopoulos 2012). The amateur is also thought to represent a form of ethical 'witnessing' (Allan 2013), tapping into the critical histories of looking awry and from the point of view of memory. Amateur images have thus come to represent politicized visions that are believed to work against the norms of the mainstream media. With this development also comes the assumption that the inclusion of the amateur in public spaces will inevitably shift the vision of professionals, encouraging them to adopt more ethical strategies of representation. It is additionally the case, however, that the proliferation of amateur forms of production and modes of engaging the audience have created new and interesting roles for professionals. New professional identities and subjectivities have been honed within the context of new media environments that privilege the amateur.

The pervasiveness of digital technologies and the proliferation of devices operated by citizens have brought about a "post-scarcity culture" (Hoskins 2014:238) that one would have thought would fundamentally unravel mainstream reporting. It is more the case, however, that digital mediations sustain what can be described as a "chain of memory" (ibid.:239) in which a kind of pattern repetition occurs. We still see the spectacular displays of professional reporting dominating the news, for instance. Yet, there have also been attempts by professionals to shift the focus by adopting an amateur aesthetic, especially in the context of photojournalism. The attachment to the analogue in the domain of visual culture is one way in which this has become evident, but perhaps more significant is the way in which professionals and scholars who seek to foreground the amateur draw upon quite familiar terms of mediation – especially those associated with visual art, including the categories of genre and artistic expression – in order to reestablish their role as mediators (if not creators) of citizen forms.

Ultimately, identification with the amateur forms a key part of the relational dynamics through which technological transitions become more culturally significant. Identification with the amateur has become one of the ways in which political transformation and social change are activated. Yet, it is important that we challenge recent assumptions underlying the attachment to the amateur, asking questions that are not limited to unravelling our current conceptions of positions of marginality. We may ask, for instance: in what way can the position of the amateur provide a legitimate resolution to historical absences? Can the amateur really allow access to the domain of personal experience? If so, does this necessarily allow us to understand social problems and ways in which they can be changed? And, with such questions in mind, we may also begin to decipher the ways in which amateur and professional identities, together, are beginning to configure the kind of relational experience that allows for the expression of citizen concerns and interests within the online world.

Future directions

What becomes apparent in the end is that we cannot adopt the term amateur, or its present-day technological correlates, in any simplistic sense, and that in attempting to set out a definition of the amateur and map its various historical meanings and cultural investments, it is instructive to consider how various disciplines and domains of practice enmesh the amateur and the professional within distinctive representational strategies and meaning-making processes. There is also a long history within which the amateur has formed a figure of resistance against professionalism, and therefore the codes and conventions of a given time period and institutional context must be borne in mind. It is consequently important to take a balanced and measured view when attempting to argue for the liberatory potential of the amateur and when making claims about what purpose the use of the term might serve in naming and establishing new directions within the wider context of citizen media practice and study.

Moreover, the adoption of the perspective of the amateur can be seen to stabilize a practice of seeing from the point of view of 'the other', and represents at least an attempt to see what cannot ordinarily be seen through a professional lens. Just as Said points out, the amateur also perhaps offers a contact point for those experiencing exile – a reality that relates to a growing number of the world's population. The challenge though will be to not too easily or readily conflate these different positions of marginality. It is still necessary to be able to account for differences, and, most importantly, we must be able to account for the various historical and institutional alignments that serve to naturalize new configurations of power, especially alignments that involve the lives of those who have very little power at all.

See also: authenticity; citizen journalism; citizen science; photography

Recommended reading

Hunter, D., R. Lobato, M. Richardson and J. Thomas (eds) (2013) *Amateur Media: Social, cultural and legal perspectives*, London: Routledge.

An edited collection of essays that maps the integration of the amateur across a range of media, including policy and related economic issues, in the UK, Europe and Singapore. This volume is especially useful for the way it highlights the tensions between the amateur and the professional, through case studies relating to intellectual property and media law.

Pantti, M. and K. Andén-Papadopoulos (2011) *Amateur Images and Global News*, Bristol, UK: Intellect.

Focuses on the ethical issues that arise when amateur images appear in mainstream news, raising questions about the role of amateurs in producing knowledge, especially in relation to crisis and disaster.

Said, E.W. (1994) *Representations of the Intellectual: The 1993 Reith Lecture*, London: Vintage.

A comprehensive, personal and critical reflection on the role of the intellectual in society and the problems and pitfalls of working in academic institutions where professionals have been transformed into state functionaries. Said makes an argument for taking on the position of the amateur as a form of resistance.

ANTHROPOLOGY AND CITIZEN MEDIA

Nina Grønlykke Mollerup

This entry engages with anthropological contributions to the study of citizen media that are based on long-term, ethnographic fieldwork and immersive engagement. Citizen media have not formed a core focus for anthropological research to date. Rather, anthropological attention to citizen media has mainly grown out of studies of other topics. Thus, the exploration of citizen media does not yet constitute a coherent field of inquiry within the discipline. However, important insights on citizen media have been engendered by a wide range of different anthropological subdisciplines, including media anthropology (Barassi 2015a; Mollerup and Gaber 2015; Postill 2018), visual anthropology (Razsa 2014; Stone 2015; Westmoreland 2016) and indigenous anthropology (Fisher 2009; Ginsburg 1997, 2007; Ginsburg and Myers 2006).

At its most fundamental, anthropology is the study of humans as social beings. In Tim Ingold's words, the objective of anthropology "is to seek a generous, comparative but nevertheless critical understanding of human being and knowing in the one world we all inhabit" (2007:69). The methodological foundation of this discipline is fieldwork. While anthropological fieldwork has many different forms, few will dispute that time is a crucial aspect. Anthropologists often spend from at least a few months to sometimes many years in the field. Though our fields today are often more accessible than they once were, largely thanks to advances in transport and communications technologies, the core of anthropological fieldwork in many cases remains participant observation. Anthropologists seek to know by using themselves as tools, placing themselves next to the people whose lifeworlds they are interested in learning about, taking an active part in their lives. It is not unusual that anthropologists enter the field with the intention of studying something particular, but through sustained engagement with people over time, they frequently shift their attention to other aspects of people's lives, which turn out to be more significant. Thus, while anthropologists may intentionally position themselves in certain places, they cannot always anticipate or shape what will happen in these places while they are there. This has repercussions for the way anthropologists study citizen media, as they are often able to capture processes that evolve over time. Thus, anthropological accounts can provide an in-depth understanding of processes that enable and encourage the production of citizen media, even before the citizens concerned engage in media practices.

This entry seeks to explore the implications that these methodological foundations hold for the ways in which anthropologists study citizen media. It shows that through ethnographic methods and a mode of analysis committed to richly detailed, textured and contextualized understandings of subjects' lifeworlds and cultural practices (Urla and Helepololei 2014), anthropology can draw greater attention to what are termed here quiet ruptures, that is, to everyday unspectacular changes in power relations which over time can be significant to larger processes of change. The entry next demonstrates that people involved in the production of citizen media are not always actively and deliberately aiming to do what their actions might eventually achieve, thus highlighting the emphasis anthropologists place on unplanned, unintended and unpredictable aspects of resistance. Finally, the entry seeks to illustrate the empirical depth and geographical spread of anthropological work on citizen media and consequently the discipline's capacity to provide thick description of and concrete engagement with such practices in a broad range of geographical and cultural settings.

Quiet ruptures: broadening understandings of citizen media

Through their temporal commitment and attention to the mundane, anthropological studies may often reveal a diversity of citizen media practices that would otherwise tend to pass unnoticed. An example of this is found in Tenhunen's (2018) ethnographic study of a village in rural India. Tenhunen originally went to India in 1999 to study women's political participation. Returning repeatedly to the village over the next fifteen years, Tenhunen came to explore how "the multifaceted use of mobile phones has influenced economic, political, and social relationships and how these new social constellations relate to culture, social change, and development" (2018:1). In particular, Tenhunen argues that the introduction of the cell phone has been part of a substantial reconfiguration of women's positions in their marital homes, a change which might be played out mainly by women speaking on the phone within the confines of their house, but which has implications not just for their relationships with their family, but also for the position of women in society. Essentially, "women benefit because phones have enabled the reconstruction of the meaning of the home and the outside world" (ibid.:19). Tenhunen's work thus serves to unsettle clearcut distinctions between private and public implicit in many definitions of citizen media by focusing on the ways in which acts in private spaces can serve to fundamentally challenge how social life can play out in public spaces. While Baker and Blaagaard (2016a) for example include the idea of acting in public space as a key element in their conceptualization of citizen media, Tenhunen shows how media technologies are co-constitutive of private and public space, and can be part of reconfiguring the tension between the two. By doing so, she also allows for an understanding of citizen media that does not fully reject the private sphere as a site of enactment.

Abu-Lughod (1986, 1990) similarly challenges the distinction between the private and the public in her classical study of the Awlad 'Ali Bedouins in Egypt, with whom she has carried out fieldwork over several years. Through continuous visits in which Abu-Lughod has lived as a guest and like a daughter of a Bedouin family, participating in everyday life and listening to women's storytelling in particular, she traces the transformation of power and resistance in this Bedouin society through several decades. In a society in which marriages are a communal affair, arranged between allies, friends and kin, Abu-Lughod has been part of discussions of potential marriages and heard older women's tales of their marriages and potential marriages. She ethnographically describes the ways young women resist unwanted marriages through practices which challenge neat distinctions between private and public. For instance, she describes how women sing taunting songs in private, female settings, while ensuring these are heard well beyond this setting, thus achieving a public ridiculing of an unwanted groom without compromising their modesty (1990:43–44). Abu-Lughod's participation in everyday life allows her to show such small performative acts as a significant form of resistance that reconfigures power relations, gender roles and access to spaces. Thus, through Abu-Lughod's ethnography, it becomes possible to recognize these women's practices as a form of citizen media.

Highlighting the messiness of human experience

As anthropologists are well aware, social life is messy. It is unpredictable and involves many unintended and contradictory events and activities. Rather than trying to impose order on the chaos of social life, anthropologists find the complexity informative. As Abu-Lughod shows, the complex workings of social power can be traced in the rich and sometimes contradictory details of resistance (1990:42). In seeking to interrogate contradictions and complexity, anthropologists are bound to continuously reconsider the categories they think with when entering the field. Often, seemingly clearcut categories become blurry once interrogated in their messy detail. This has implications for how citizen media practices are explored by anthropologists. Most notably, ethnographic research has forced a reconsideration of common understandings of what constitutes resistance. Armbrust's (1998) work serves to broaden this notion, questioning whether habits of media spectatorship can be seen as a form of resistance which can contribute to social transformation. Armbrust looks at moviegoing in downtown Cairo as "a secular ritual that can potentially enable kinds of social change not necessarily envisioned by the state, approved by normative society, or engineered through films themselves" (ibid.:413). He looks beyond the films and focuses on the types of encounters they enable. This point is particularly important in authoritarian settings where laughing at the same implicit parody of a ruler can create a feeling of shared resistance, even if this resistance is limited to a laugh. Thus, Armbrust's work compels us to think of spectatorship as an active site of production while he also points to the importance of unpredictability and the constant negotiations over meaning. Similarly, Mollerup and Gaber (2015) describe participation in illegal street screenings taking place during the Egyptian uprising between 2011 and 2013 as a revolutionary act. The first screenings were held in Tahrir Square during the summer sit-in in 2011 under the name of Tahrir Cinema. Later, screenings took place in neighbourhoods around the country, often organized by people from the neighbourhood under the name *'askar Kazeboon* (The military are liars). The original screenings did not have a more specific purpose than to show videos and raw footage of violent assaults on citizens at the hands of the army and the police to those in the square who had not seen them and to remind others of them. What was shown in these screenings was negotiated by participants in the screening and often a person from the crowd would volunteer a video, which would be transferred from the individual's phone to the makeshift screen by way of Bluetooth, a computer and a projector. These screenings explicitly contest distinctions between observing and acting by merging spectatorship with the occupation of public space. They thus work to position the act of witnessing as an act of defiance. Indeed, as Mollerup and Gaber's (2015:2906) ethnography showed, during the uprising in Egypt from 2011 to 2013, activists considered media less "a tool for the transmission of information" than "a tactic for producing new environments and collectivities", in many ways establishing the site of viewing as an active site of production. This insight echoes earlier calls from media anthropologists to radically rethink the divide between production and reception, and to pay attention to "the important but theoretically and empirically neglected area of distribution as a central process through which media helps constitute and reflect social difference, as power and status are signified through spatial and temporal dimensions of exhibition" (Ginsburg 2007:306). The participatory quality of street screenings in Egypt is of course also closely related to the participatory filming of revolutionary videos and filmmaking as a political action. Westmoreland (2016) has argued that the pervasive presence of cameras in the Egyptian uprising enabled a hypervisibility of the street in times of protest, which made image-making practices both threatening and powerful through their potential to cultivate new kinds of political subjectivity and collectivity.

The insight that Tahrir Cinema and *'askar Kazeboon* began with such vague and modest intentions points to how citizen media practices – in this case the sustained organization of street screenings – can often be initially unplanned and unpredictable yet come over time to have a very specific purpose and impact. Coleman's book, *Hacker, Hoaxer, Whistleblower, Spy: The many faces of Anonymous* (2014), further illustrates this tension in impressive empirical detail, based on her six years of extensive ethnographic research with the hacktivist collective Anonymous. Coleman describes how the collective has developed from making strange, sarcastic videos with a dark and deviant style of humour that was hardly comprehensible to anyone outside the group to becoming "integral to some of the most compelling political struggles of our age" (ibid.:2), exposing wrongdoings and supporting resistance to oppression across the world through their digital activism. Coleman's own prolonged participation in the movement was crucial for allowing her access to and engagement in the private discussions that shaped the development of the movement. This access enabled her to examine in particular detail the ways in which resistance can be unplanned, random and continually shifting, particularly in the context of collective action. This is an important insight for studies of citizen media more generally; though it can be easy to create a coherent historical narrative after the fact, when citizen media are interrogated as they unfold, they are often less determined and more ambiguous than they appear in retrospect.

Graffiti as acts of resistance

Ethnographic fieldwork often entails researchers moving through the same places as those their interlocutors move through, and this draws attention to things they did not imagine to be of importance before entering the field. The growing body of anthropological scholarship on graffiti, mainly written by anthropologists who did not know they were going to study graffiti until they realized they were already doing so, exemplifies this. Graffiti is particularly interesting for the way it simultaneously records and intervenes in often very unequal relationships of power. Like political demonstrations, graffiti occupies spaces and expresses its maker's existence (Khosravi 2013). Through fieldwork, anthropologists insert themselves into these spaces and participate in situations when and where graffiti is made, remade, read and erased; moreover, like the people under study, the paths and attention of researchers are directed by the graffiti they see. Indeed, in her analysis of the cultural landscape of the occupied West Bank during the 1980s and 90s, Peteet (1996:139) suggests a method for reading "the battle of the walls much the way an archaeologist reads stratigraphy – layer by layer – each layer of paint indicating a partial and temporary victory in an ongoing battle". She asserts that many examples of graffiti in the Occupied Palestinian Territories before the turn of the millennium did not merely send messages or signify defiance, but rather that "their mere appearance gave rise to arenas of contest in which they were a vehicle or agent of power" (ibid.:140). Peteet's engagement was not only with the walls, but also with the people who moved around them. She recounts a young woman from Ramallah telling her, "when I wake in the morning and see new graffiti I know that resistance continues. It tells me that people are risking their lives and that they live right here in this neighbourhood" (ibid.:151). What Peteet's study illustrates, therefore, through her engagement with people to whom the graffiti was significant, is that the content of the graffiti was less significant than the resistance it enabled.

Recognizing that graffiti records domination also brings significance to how city walls can tell an ongoing story of changing political situations. The walls of Cairo illustrate this point very clearly. Before the uprising broke out in Egypt in 2011, explicitly political graffiti was sparse, but walls in the country spoke of more mundane issues through graffiti with advertisements for mechanics or romantic declarations (Schielke and Winegar 2012). Schielke and Winegar (2012) additionally point to the importance of the emplacement of particular pieces of graffiti, for instance when the walls surrounding the presidential palace became a specific site to challenge political legitimacy during the years of the uprising when political graffiti was widespread. They hold that

> the new writings after 2011 are a testament to rich and interrelated modes of verbal and visual expression in Egypt, of the links between politics, love, death and the struggle to make a living, of the contests in many aspects of Egyptians' personal and social lives.
>
> *ibid.*

Similarly, through repeated returns to Cairo during the military dictatorship, the revolutionary uprising and the subsequent restoration of the military dictatorship, Mollerup (2015) found that the walls around the city told the strongest tale of the political climate at a given time. During the years of the uprising, new writings significantly changed the visual experience of public space in Egypt as people moving through it were constantly reminded of – and implicated in – the ongoing political struggles. As the Egyptian military started once again to succeed in repressing opposition, this time with renewed severity, the walls turned eerily silent. At one point, in 2015, a single piece of graffiti bravely decried, "your voice is not on this wall" (Mollerup 2015:66; my translation), serving as a sorely clear reminder of the many activists and journalists who were killed, imprisoned, beaten, tortured and threatened for making their voices heard.

Indigenous media and the right to be recognized

Anthropology has grown out of colonialism and was concerned at first with helping "classify non-European humanity in ways that would be consistent with Europe's story of triumph as 'progress'" (Asad 1994:314). With the breadth of European expansion, this means that anthropology has historically engaged with people from very diverse geographical and cultural backgrounds mainly outside of Europe. Since the 1960s, this geographical and cultural focus has been exchanged for a fundamental interest in human being and knowing irrespective of people's position in relation to colonial power structures. Certain regions and themes dealt with in anthropology do, however, continue to engender greater anthropological interest, including the struggles of colonized peoples to be recognized as political subjects. Yet, the violent history of the field has led anthropologists to be particularly aware of how they can create knowledge with people, rather than merely about people, and how this knowledge is made to matter to them. One strand of anthropology that has focused on citizen media is that which deals with the rights and cultural identities of indigenous groups, and which also engages in work to support these rights, including language rights (Urla 2012), the rights of incarcerated Aboriginal people (Fisher 2009) and civil rights (Budka 2015). Ginsburg (1997, 2007) places indigenous activist engagements with media in direct relation to media "being produced by a variety of other minoritized subjects who have become involved in creating their own representational framework as a counter to dominant systems" (Ginsburg 2007:303). This framework, she points out, includes work being done by people with AIDS (Juhasz 1995), Palestinians in Israel's occupied territories (Kuttab 1993) and Tibetan Buddhist activists (McLagan 1996). Ginsburg (1997, 2007) approaches the creative and self-conscious process of objectification through this media production as a form of cultural activism, drawing on George E. Marcus' (1996:6) term the activist imaginary, in which film and video are used to "pursue traditional goals of broad-based social change through a politics of identity and representation". Such practices also raise issues about "citizenship and the shape of public spheres within the frame and terms of traditional discourse on polity and civil society" (ibid.). Ginsburg (1997) maintains that the perspectives of Aboriginal activists are tied to the struggle for land and religious rights, which characterizes the concerns of First Nations or indigenous social movements in particular. She argues that

> products of indigenous expressive culture are part of self-conscious efforts to sustain and transform culture in aboriginal communities, an activity that is linked to indigenous efforts for rights to self-representation, governance, and cultural autonomy after centuries of colonial assimilationist policies by surrounding states.
>
> *ibid.:119*

These efforts, she holds, are not so much against the state, but rather express a desire to "be recognized and granted entitlements and reparations within the terms of the dominant legal code" (ibid.:120). The work of filmmakers and other cultural activists, then, is part of changing the cultural landscape of the Australian continent, which is bound by

> the paradox of the persistence, growth, and increasing circulation of such work in Indigenous cultural production, despite the alarming political turn against gains made by Indigenous Australians over the last decade, not only by right-wing politicians but intellectuals as well.
>
> *Ginsburg and Myers 2006:95*

In their joint work, Ginsburg and Myers (ibid.:97) explore how the work of indigenous filmmakers and cultural activists creates possibilities for Aboriginal futures outside the defining limits of law and policy by tracking "a history of Indigenous futures in Australia, over a period in which Indigenous people have slowly but surely been re-imagining what they might be". Ginsburg and Myers' work is not merely about this work to create futures. Through their sustained and concrete engagement with these communities, for instance in co-creating cultural productions, they are also part of the work to create these potential futures.

Future directions

As noted in the introduction to this entry, anthropological attention to citizen media has largely developed out of research concentrating on a wide range of other topics. While these disparate studies will surely continue to engender important insights into citizen media, it would nevertheless be productive in future work for anthropologists to concentrate more directly on practices of citizen media, and to bring these diverse perspectives more explicitly into conversation. With its non-media centric approach to understanding citizen media practices and its emphasis on how they are enacted by and make sense to people, anthropology has much to offer to this emerging area of interdisciplinary investigation. In particular, through its commitment to long-term and concrete engagement with communities in a broad range of geographical and cultural settings, anthropology can help unpack global connections and further our understanding of ways in which citizen media in different places can help unsettle power structures well beyond local contexts.

See also: documentary filmmaking; graffiti and street art; hacking and hacktivism; public sphere

Recommended reading

Barassi, V. (2015) *Activism on the Web: Everyday struggles against digital capitalism*, New York and London: Routledge.

This book draws on anthropological research among three very different political groups in the UK, Italy and Spain to examine the everyday tensions that political activists face as they come to terms with the increasingly commercialized nature of web technologies. It makes an important contribution to our theorization of the digital aspects of citizen media by suggesting that if we want to understand connections between digital and political participation, we should not focus merely on disruption and novelty.

Coleman, G. (2014) *Hacker, Hoaxer, Whistleblower, Spy: The many faces of Anonymous*, London: Verso.

This seminal work on Anonymous is based on extensive anthropological fieldwork with the worldwide movement. The book captures the movement's development from an obscure group of reckless geeks to an agenda-setting, government-disrupting political movement.

Tenhunen, S. (2018) *A Village Goes Mobile: Telephony, mediation, and social change in rural India*, New York: Oxford University Press.

This book is grounded in extensive ethnography carried out between 1999 and 2013. Through repeated returns to an Indian village, Tenhunen captures subtle yet significant changes in social relationships and in the ways in which these changes are entangled with the introduction of the mobile phone, allowing for an analysis which connects the private and the political.

ARCHIVING

Jess Baines

Archives and archiving have expanded from their once traditional role as restricted and institutionalized domains to include a growing host of autonomous, often community-orientated but also artistic resources and practices that act against the exclusions and notions of historical significance which permeated their formal counterparts (Schwartz and Cook 2002). These alternative archives, sometimes referred to as counter archives or community archives, can be invaluable resources for researchers of citizen media, and the alternative archive itself may be seen as an instance of citizen media in practice. Alternative archives are not simply repositories for excluded narratives. They contain documents and can inspire practices that bear witness, affirm identities, forge collective memories and new socialities, offering sites for critical engagement with both the past and the present (Baladi 2016). In that sense they may be anticipatory, even hopeful endeavours (Appadurai 2003). The affordances of digital technologies have enabled the growth of alternative archives and made them more visible to researchers, with a proliferation of online archives and digitization projects available for analysis. However, the community or counter archive has a longer history. This entry draws on examples of both pre- and post-digital initiatives to illustrate the range of objectives and forms that these types of archives can take. It also engages with some of their vulnerabilities, as well as strategies that have been used or proposed to mitigate these. It begins by addressing what is meant by the term archive and discussing the definitions attributed to alternative archives.

Defining archives and records

As the practice of archiving has expanded so have ideas of what constitutes an archive, particularly in the digital age. Historically it meant a collection of preserved documents or records, especially those of government or corporations, and the physical place those records are stored. The etymology of *archive* relates to both the ancient Greek for 'government' and the place where public records are kept. Records are conceived of as primary sources which evidence an occurrence and are created by either participants or witnesses at the time of the event. Over time the definition expanded to include the records of organizations more generally and of (usually notable) individuals. With the arrival of personal computing, and then the web, a newer concept of archive as digital storage has entered everyday consciousness and usage. While in computing terminology *archive* and *archive file* have specific meanings, "the popular imaginary of the Internet is that of an archive of archives" (Snickars 2009:292). However, Hogan makes the point that "[a]s a large unsorted store, the online archive, without assessment of its content, communities, and cultures of use, is allegorical to the dumpster" (2015:13). There are also conflicting views about whether platforms for user-generated/uploaded content can be meaningfully described as archives. For example, Tifentale and Manovich (2015:116) describe Instagram as "a giant archive", while van Dijck argues that such sites "lack even the most elementary principles of an archive's ordering and preservation system" (2017:223) and Spigel suggests that YouTube is best understood as an "unintentional archive" (2014:66).

Overlapping with these technological developments and the promotion of broader – and contested – notions of archive has been the longer, ongoing 'archival turn' within the arts, humanities and social sciences. Through this turn, the notion of archive has also been extended and interrogated as a site of memory, of loss and of power. This critical interrogation has often been directed towards the archives of nation-states, which by the nineteenth century had become increasingly rationalized systems, central to the governance of internal and colonial populations and territories (van Alphen 2014). Archives and records were not only associated with European political and scientific endeavours to comprehensively know and order the world, they also came to be understood as crucial sites of a nation's history and memory. This fed into the development of history as an academic discipline, with the archive offering up 'the facts' of the past, awaiting the historian's reconstruction (Featherstone 2006). However, as a wealth of this critical scholarship has shown, the archive, and its interpretation, are shaped by vested interests – and by silences (Spivak 1985; Trouillot 1995; Stoler 2002). Some of these theoretical re-conceptualizations and evocations have been influenced by the writings of Michel Foucault (1969/1972) and Jacques Derrida (1995/1998), who in very different ways explored the notion of the archive in non-literal terms (Manoff 2004). Conceptual engagement with the archive by artists has resulted in critical, speculative or fictive forms of the archive, which may expand and query what counts as a record, as evidence or may invent classification systems and new archival subjects (Foster 2004; van Alphen 2014).

In relation to grassroots or citizen-led archives, the definition of what counts as a record and as an archive has arguably always been more inclusive than in their official counterparts. Community or counter archives may include physical or digital materials of all kinds; publications and ephemera, artworks, placards, clothing, videos, oral history recordings, as well as the sorts of evidentiary records historically associated with archives. Furthermore, traditional distinctions between library, museum, information centre and archive are often blurred in relation to this type of citizen-led archive, as they may perform several of these functions but describe themselves by only one (Flinn et al. 2009). Indicating a shift in some sections of the archival profession, South African archivist Verne Harris offers the following contemporary definition:

> 'archives' is defined by three fundamental movements or attributes: one, a trace on, or in a surface; two, a surface with a quality of exteriority, and three an act of deeming such a trace to be worthy of protection, preservation and the other interventions which we call 'archival' … Anyone can *deem*.
>
> *Harris 2012:150; original emphasis*

The point is that an archive contains what its archivists and users (who may be the same) deem to be worthy of inclusion and preservation.

While the concept of archives has become more open, the rise and variety of citizen archiving initiatives, and particularly of scholarly awareness and interest in them, has led to a proliferation of labels to distinguish them; not only from official archives but also from each other. These terms include community archives, archives from below, participatory archives, activist archives, indigenous archives, counter archives, autonomous archives (Moore and Pell 2010; Artikişler Collective 2016), rogue archives (De Kosnik 2016), dissonant archives (Downey 2015), spontaneous archives (Heimo 2017) and living archives (Hall 2001). Some of these terms appear to be interchangeable or fairly open, others are more specific. Sometimes the same phrase is used to mean different things. Living archives, for example, may mean archives that are added to and used in the present, or it may mean a reservoir of cultural traditions maintained through bodily and social practices (Thorpe 2017). The term community archives has gained significant traction, primarily through the work of archival scholars in Britain, notably Andrew Flinn and his colleagues. They define community archives as

> collections of material gathered primarily by members of a given community and over whose use community members exercise some level of control … the defining characteristic … is the active participation of a community in documenting and making accessible the history of their particular group and/or locality *on their own* terms.
>
> *Flinn et al. 2009:73; original emphasis*

As Flinn et al. note (as have Williams 1976 and Young 1986), community is a contestable concept, often used politically as a homogenizing or exclusionary device; however, in relation to community archives it usefully describes a shared interest, attribute or experience that forms the basis of an archiving project, and as such is mainly used here along with reference to some of the other terms where relevant. More broadly, the definition allows for the inclusion of archiving projects with varying degrees of explicitly political motivation. There are nevertheless limitations; for example, community archives may not be appropriate to describe archiving initiatives which are only accessible in a restricted way, such as artists' projects which appear solely in gallery spaces. The broader term of counter archives is probably better applied here.

Counter archives and community archives in practice

The history of counter archiving and community archiving is long and diverse. Examples might include the archives created by exiled English Catholics in the sixteenth century (Corens 2016), the amassing of black history materials by Arturo Schomberg in late nineteenth- and early twentieth-century USA (Des Verney Sinnette 1989), the archive of women's organizations compiled by Dutch feminists in the 1930s (de Haan 2017) and the Yizkor books produced by Jewish survivors after the Second World War (Riedlmayer and Naron 2009). What these endeavours have in common with each other and with many citizen-led archives since is that they seek to redress absences in the official record, absences that Caswell (2014:27) has described as "symbolic annihilation". Not only this but the collecting is done by members of those archivally absent social groups and in the first instance is for those groups.

The more sustained rise of community archiving began as a corollary to the civil rights and social movements that began to mobilize from the 1960s onwards, as well as the related history from below movements (Gilliland and Flinn 2013; Bastian and Alexander 2009). The basis of many of these movements was the rights and recognition of groups that have been (and to varying degrees still are) historically excluded, marginalized and misrepresented in society, history and mainstream culture. The collection, creation and self-management of community archives, libraries and resources was usually understood as part of those struggles, and the resources themselves as tools for self-education and campaigning. As such they were orientated towards action in the present and the promise of future change. Between the 1970s and early 1990s, in different parts of the world (including the USA, Canada, Australia, New Zealand, India, South Africa, Britain, Ireland, France, the Netherlands and Italy) a wide variety of grassroots black, feminist, LGBT, working-class and radical history archives and collections were initiated. By claiming the right to create their own historical institutions and narratives, such archival endeavours implicitly or explicitly dispute the authority of their traditional counterparts. Continuing examples from this period include Lesbian Herstory Archives in New York (1974), Black Cultural Archives in London (1981) and SPARROW women's sound and picture archive in Mumbai (1988).

Although diverse in subject matter, archives of this type often share certain characteristics and ambitions beyond those mentioned above. They frequently begin in someone's home and are initially built out of the collection of one person or a small group of individuals. Along with more informal attitudes towards handling and preservation they often use their own cataloguing and ordering systems, which may have little bearing on those used in formal archives (Bosi and Reiter 2014). The physical space of the archive is felt to be significant, both symbolically as a counter institution and as a space of respite and autonomy. These archives also share challenges such as the fluctuations of volunteer labour, the changing priorities of the community they are orientated towards, precarious finance, the securing of affordable and adequate space, and maintaining regular access to collections. Survival strategies may mean foregoing original aspirations of autonomy and self-sustainment through community support. For example, many of the LGBT and feminist archival projects that started as grassroots initiatives in the 1970s and 80s have since deposited their collections into university or public sector archival provisions, in some cases maintaining a degree of independence within them. That this has been possible also demonstrates changing attitudes within parts of the mainstream archival sector, as well as wider social and cultural shifts, at least in some places. Also, as Cvetkovich (2012) and Gilliland and Flinn (2013:11) have pointed out, new relationships between community archives and institutions "may result in the changing, reimagining or queering of that archive and its practice". A notable exception to the institutionalized trajectory is the US-based Lesbian Herstory Archives. They have maintained their early principles of operating within "the community" rather than the academy, the latter "by definition closed to many women", with "archival skills … taught, one generation of Lesbians to another, breaking the elitism of traditional archives". Any funding, the website announces, "shall be sought from within the communities the Archives serves [sic], rather than from outside sources" (Lesbian Herstory Archives 2018).

Online citizen archiving

The development of digital tools and online technologies have enabled the flourishing of new types of community and counter archives. Based online, core issues of space and access that troubled their bricks and mortar counterparts are resolved. Citizens' autonomous online archives developed in tandem with the public launch of the web in the early 1990s. New kinds of participatory archives created by various fan fiction communities came into being, with hundreds of such sites appearing by the end of the decade, enabling the free publishing, sharing and reviewing of their creative works and, more broadly, entry into a community of interest (De Kosnik 2016). A very different kind of early examples includes the sites set up in response to the Bosnian War (1992–1995) by Bosnians, who used the new technologies to create "virtual memory books" (Riedlmayer and Naron 2009:159). One of the longest running was Žepa-online, which acted as an online collective commemoration site for lost relatives, friends and neighbours from Žepa, as well as a means for survivors to post information, news and questions. The creation of online archival projects by those facing the traumas of war, occupation or dislocation would continue to feature in the landscape of counter and community archiving. The site Palestine Remembered was set up "by a group of ordinary Palestinians" with the aim of "educating the world about towns and villages destroyed in 1948 and their inhabitants" (PalestineRemembered.com 2002). This site includes listings of towns, maps, images, chronologies, politically instructive quizzes and oral history recordings. It has been proposed by Appadurai (2003) that the archiving projects of diasporic populations in particular contain within them a promise to the future, and as such are often hopeful endeavours. In some cases, however, it might be argued that the drive to archive emerges in the face of diminishing hope about the future (Doumani 2009; Butler, B. 2009).

The largest wave of online archival activism relating to political turmoil has probably been that centred on the so-called movement of squares beginning in 2011. The Egyptian Revolution, the Occupy Movement (particularly OWS) and the 15M or Indignados protests all generated extensive participant archival activity. According to Baladi (2016; original emphasis), the artist who initiated Vox Populi, one such archive, "in [Tahrir] square, revolting *was* archiving". The now familiar technological advances enabling citizen uploading and sharing of real-time video and images, along with the instantaneity of social media feeds, generated an array of websites and dedicated social media pages, providing evidence of the "new archival multitude" (Abbas and Abou-Rahme 2013:351). Some of these archives have since lain dormant or have disappeared from view. Others have been partially preserved by institutional projects such as the Internet Archive or the American University in Cairo. One of the exceptions is the more sustained initiative of the Mosireen Collective, 858: An Archive of Resistance (858.ma), finally launched in 2018, in the context of a new political reality. Mosireen are an Egyptian media activist group set up in Tahrir Square during the revolution. They created the Mosireen channel on YouTube to distribute citizen video of the unfolding events. The 858 Archive makes public a much wider body of footage, indexed and date stamped, along with thousands of photographs and documents. The function of this archive is different from those more spontaneous ones that multiplied during the hopeful period of the revolution. The aim is now "to fight the narratives of the counter revolution [and] to keep building new histories for the future" (Mosireen 2018). Occupy Wall Street protesters set up an OWS Archives working group to preserve both physical and digital materials relating to the protests, again to be active agents in their own historical record. The vast amount of digital material created an "infoglut" which, as Kaun (2016b:5404) explains, could only be managed and analysed in a limited way. This was one of several OWS archiving projects. Much of the OWS archiving effort has since been deposited in the Tamiment Institute Library and Robert F. Wagner Labor Archives, chosen as the most politically acceptable and viable sites to ensure future access to the records of the movement (Interference Archive 2012).

Another type of online counter archive includes those which gather data to expose particular forms of injustice. For example, the issue of police violence is the focus of two US-based sites, Fatal Encounters (fatalencounters.org) set up in 2013 and The People's Archive of Police Violence in Cleveland (archivingpoliceviolence.org) set up two years later, one of the many initiatives of the Black Lives Matter movement. The latter is a community archiving project, albeit set up with support from archival professionals (Drake 2016). The site includes testimonials, protest posters for people killed by the police and documentation of activist events. The group also conducts oral histories. Fatal Encounters is data driven, featuring statistics and visualizations, and aims to produce a national database of people killed during contact with the police. Although reliant on volunteer contributions, it is the brainchild of an individual, Brian Burghart. A different kind of example is HeygateWasHome (heygatewashome.org), a UK-based site established to document and present counter information in relation to the demonization and subsequent demolition of a large public housing estate in London. The site contains evidence of media reporting about the estate, numerous council documents, obtained through Freedom of Information Act requests, and testimonies from residents. The project was coordinated by the Southwark Notes Archive Group, a volunteer collective who have been charting regeneration in the area since the 1990s, and who have amassed a significant 'Gentrification Archive' within a (physical) radical resource centre and archive, 56a Infoshop.

A final example of yet another kind of online community archival project is SAADA (saada.org), first established in 2008 as an online repository to document and provide access to the diverse stories of South Asian Americans. The material often comprises digitized post-custodial items, loaned for scanning by South Asian individuals and communities, other collections or organizations, and then returned. The site was set up to counter prevailing narratives and the pervasive erasure of South Asian American histories in the USA, through the co-creation of a more useable and meaningful past (Caswell et al. 2018).

Some of the vulnerabilities of digital archiving projects have been implied above. As De Kosnik (2016:29) explains, "digital technologies are not perfect archival technologies … they tend towards loss and disappearance". Websites require that domain names are renewed, plugins and links are updated, and commercial platforms may remove materials from the archive. File formats and storage devices change and become obsolete. Even if sites are captured by Internet Archive's crawlers on the Wayback Machine, this is only a snapshot. Social media archives raise other problems; tweets can be deleted, the most popular content is pushed to the top, and it is all owned by the platform. Along with other analysts of Internet-based media, Kaun (2016b:5397) has proposed that online archives are more often about "producing, gathering and processing data" than about the safeguarding of materials. Preservation has been central to traditional archival concerns, and indeed to many counter archival initiatives. Assman (2010:106) links archival preservation to "passive storing memory" as opposed to the "active working memory", typified by the living but unstable online archive. She also makes the point that materials may move between the two, but both are needed for collective memory.

Future directions

It is difficult to predict the future of citizen archiving, although the desire to gather records that can form the basis for counter-histories and struggles against injustice is unlikely to abate. Recognition of the value of participatory archives for marginalized groups and struggles has spread through much of the professional archival and heritage field, yet the maintenance of such enterprises is far from straightforward, as indicated above. More research is needed to explore how citizens are meeting these challenges, especially in online contexts. Citizen media scholars might fruitfully investigate, for example, the dynamics of emerging relationships between digital archivists and sympathetic non-profit software developers who have provided technical infrastructures such as Pan.do/ra, first adopted by Pad.ma to create an alternative archive of Indian film/video footage. Moreover, while the Internet has enabled a multitude of archiving projects, it is important to acknowledge that new offline alternative archives also continue to be set up, with similar aspirations to their historic forebears. For example, in New York, Interference Archives opened its doors in 2011, and MayDay Rooms in London, 2013. Both are volunteer run, dedicated to collecting the material culture of social and radical movements, and explicitly see their spaces as ones of activation.

See also: documentary filmmaking; temporality

Recommended reading

Bastian, J. and B. Alexander (eds) (2009) *Community Archives: The shaping of memory*, London: Facet Publishing.

Primarily a collection of diverse case studies of community archiving initiatives, with an informative historically grounded introduction that argues for the social importance of community and peoples archives. A very useful introductory text.

Downey, A. (ed.) (2015) *Dissonant Archives: Contemporary visual culture and contested narratives in the Middle East*, London: I. B. Tauris.

Essays, interviews and artworks from a mixture of academics, activists, curators, artists and filmmakers who in different ways interrogate and trouble conventional notions and practices of the archive in diverse Middle Eastern contexts. Downey's substantial introduction explores why the archive has become such a potent reference point for a heterogeneous array of contemporary cultural producers across the Middle East and among its diasporas.

Kosnik, A. De (2016) *Rogue Archives: Digital cultural memory and fandom*, Cambridge, MA: The MIT Press.

While focusing on the world of fan fiction archiving, this volume not only engages with the many debates surrounding digital archives, it also offers an original theoretical and analytical approach to amateur and citizen online archiving.

AUTHENTICITY

Lina Dencik

Authenticity in the contemporary age is a powerful symbolic construct that is key to how we make decisions about the way we live our lives. Whether with respect to the place we live in, the food we eat, the relationships we form or the people we trust to govern us, authenticity continues to acquire significant cultural value, particularly in an age characterized by cynicism (Banet-Weiser 2012). We search for authenticity in the media messages we consume: belief or otherwise in the authenticity of a message informs our moral frameworks and decides how much credibility we attribute to different narratives and actors. Historically, news media has often been associated with a decided lack of authenticity, with mediated communication perceived as pre-packaged, produced and contextualized by institutions, agendas and interests. At the same time, we continuously search for authenticity in media as a way of making sense of reality and deciding what is or is not true. This "paradox of mediated authenticity" means that "although we base nearly all our knowledge about the world and the society in which we live on mediated representations, we remain well aware that the media is constructed, manipulated, and even faked" (Enli 2015:1).

The advent of digital technologies, and social media in particular, has significantly restructured and underscored the role of authenticity in relation to media. In some respects, social media platforms are a continuation of what Aslama and Pantti (2006) refer to as the authenticity industry that found prominence with so-called reality TV, but they are also formatted with particular affordances that imbue them with "illusions of authenticity" (Enli 2015:1), these illusions being of special significance for the study of citizen media. This entry thus focuses specifically on the role of social media platforms in the construction of authenticity, and the way these platforms are themselves constructed as true representations of citizen voices, before going on to illustrate how such perceptions are used strategically by different political actors. The entry continues by considering how these illusions of authenticity might now increasingly be unravelling as questions of institutional agendas and manipulation have (re)entered public debates about social media platforms.

Social media and authenticity

In her book *Authentic*, Banet-Weiser (2012) argues that the question of authenticity is more important than ever in an age that hungers for anything which feels authentic, and as we increasingly lament our world as one of inauthenticity. The quest for authenticity in contemporary culture, for its expression of the genuine and the real, is an outcome of the multiple ways in which manipulations of reality now occur, not least as a result of the affordances of new technologies (Enli 2016). Authenticity acts as a counterweight to the ubiquitous, scripted moments of mediated representations of reality that we have become increasingly sceptical of, and serves as a guarantee for an undistorted and credible representation. As such, it has become an important currency in public life. In this context, Banet-Weiser (2012) argues, the binary links between, for example, commercial and inauthentic vs. non-commercial and authentic, are too simple. Rather, authenticity should be understood as a brand, as part of a brand culture that shapes not only consumer habits but also political, cultural and civic practices in the contemporary era.

Constructions of authenticity can be explored through many avenues, but social media is a particularly significant and interesting entry point because it appealed, from the outset, precisely as a format of communication and activity that bypasses institutional agendas and pre-packaged representations we commonly associate with a lack of authenticity. Similar to the appeal of television genres such as observational documentaries and reality TV that depend on the manufacturing of real feelings (Mestrovic 1997), social media is imbued with a prominent "symbolic authenticity" (Enli 2016:125). User-generated content is perceived as more authentic than mainstream media content simply because it is produced and posted by ordinary citizens rather than media companies. The amateur perspective and poorer image quality hold their own cultural value, which contrasts with the processed (and therefore potentially manipulated) images we know to be constructed for us. Indeed, studies with audiences of news content found a high rate of approval for the inclusion of user-generated content in news reports, as such content is perceived to be more real and less packaged than news produced solely by journalists (Williams et al. 2011). This perception of increased realism is closely linked to the idea that user-generated content is considered more immediate and adds drama and human emotion to a cultural form which might otherwise be dry and distanced from ordinary people. That is, authenticity appeals to our need to believe that there are spaces in our lives driven by genuine affect and emotions (Banet-Weiser 2012).

Whilst the perception of social media as an authentic representation of reality relates to our continued search for truthful representations, it is also one that has been actively advanced by both commentators and social media companies themselves. As Couldry (2015) has argued, the language we use to make wider sense of digital sites is not independent from but rather heavily indebted to the larger framing of social and political change in which the institutions that host digital networks, among others, have a strong vested interest. In what he terms a myth of us, Couldry outlines how new types of technology institutions that own and profit from the platforms where we now access media and interact with each other have come to claim, in some sense, to speak for us. That is, a language has emerged that suggests that these commercially owned 'spaces of appearances' are a new site of the social. They constitute the loci for contemporary sociality. In this context, a "myth of natural collectivity" (Couldry 2015:260) emerges that suggests social media platforms are where we are gathered naturally.

Significant in this myth of us is not only the disguise of the mode of economic necessity upon which social media platforms depend, but the assumption that "no longer encouraged to act out a role, we are forced to be 'ourselves' (in a form no less theatrical or artificial)" (Lovink 2012:13). A seamless alignment between audience (citizen) and commercial discourses is created on these platforms (Gillespie 2010); here, *social platform* is understood as a "socially constructed term that enables the continuous and seemingly unremarkable *interface* between everyday social interaction and commercially oriented tracking" (Couldry 2015:620; original emphasis). In this myth of us, moreover, media and other institutions are made to disappear altogether from the picture, giving us the impression that social media activity is focused entirely on what we do naturally. The platforms are presented as allowing you to "broadcast yourself" (YouTube slogan), seemingly free from the manipulations of institutional agendas and interests, and as aiming to "give people the power to build community and bring the world closer together" (Facebook mission statement).

Practices of mediated authenticity

The mediated construction of authenticity across expanding swathes of public life draws on the branding of authenticity (Banet-Weiser 2012) we are already familiar with from consumer culture. In *No Logo*, Naomi Klein (2000) famously argued that we now actively engage in the production of culture through the very processes of consumption, given that products have now attained meaning as brands associated with specific lifestyles. Social media has furthered this shift away from the product or service and towards the (communicative) relationship between producer and consumer, a relationship that engulfs the product or service (Terranova 2000). In this age of what Jones (2012) calls the social brand, what matters is that these relationships come to feel authentic in a way that can allow consumers to reimagine and further – and therefore validate – the value of the brand (Banet-Weiser 2012). Nothing is more emblematic of this shift than the growth of native advertising, which takes place mostly online and involves producing advertisements that are deliberately created to match the form and function of the platform on which they feature, often seeking to hide entirely from the equation the fact that a product is being sold.

As the role of authenticity as a central currency in the attribution of credibility and the conceptualization of truth has grown, social media's contribution to the branding of authenticity has unsurprisingly become a prominent focus for a range of social and political actors. In news reporting, for example, social media platforms can produce a "new authenticity" associated with citizen journalism that allows for new solidarities and cosmopolitan identities to emerge (Chouliaraki and Blaagaard 2013:3–5). This form of digital technology therefore impacts our ability to empathize and engage with others who are reported to be suffering around us, precisely through the immediacy and horizontality of the citizen voice. At the same time, in their interviews with reporters Williams et al. (2011) found that the perceived authenticity of user-generated content is highly valued and strategically used by journalists to enliven conventional news reports and create the impression of authenticity. In other words, journalists consciously integrate the affordances of social media into their news reporting in order to imbue their representations with greater cultural value garnered from the authenticity associated with social media content.

The unique position of social media as a perceived space of natural collectivity and authentic social life makes it equally pertinent for examining political communication, by social movements and governments alike. As Birks (2014) has pointed out, in the context of civil society groups and protests in particular audiences are continuously invited to assess the legitimacy of actors on the basis of the authenticity of their claims. Drawing on a distinction proposed by Habermas in relation to ascertaining legitimacy in the public sphere, Birks (2014:47) argues that the role of the media audience is to distinguish between established organizations that appear "before the public", and thus represent sectional interests and identities, and less formally organized actors who "emerge from the public" and are therefore understood to be more authentic because they are less influenced by vested interests. Historically, such assessments have often been based on different claims to legitimacy, to do with questions such as the use of violence or the ability of protesters to present themselves as victims with emotional responses (ibid.). However, the narratives that emerged around protest movements witnessed since the second decade of this century, including the various Arab uprisings, Occupy, 15M and Fight for 15, have often explicitly highlighted their extensive use of social media as a new way of legitimizing these movements. Here, social media are not seen merely as integral to the mobilization and amplification of citizen-led movements; the myth of us associated with these platforms has become key to disassociating protests from organized political forms and institutional agendas (Dencik 2015). These platforms are now celebrated precisely for the spontaneity and ordinariness that Enli (2015) identifies as key types of authenticity illusions in mediated authenticity. The centrality of social media as a mechanism for constructing "protest authenticity" (Dencik 2015:207) constitutes a significant strategic tool for civil society groups in this regard, one used to elevate their status and in some instances shift the focus away from movement architectures and historic power relations. Forms of political organization such as trade unions and large NGOs that have sometimes struggled to make themselves relevant and credible in news media have been able to turn to social media to place new emphasis on "socially organised" citizen voice, seemingly spontaneous and leaderless (Dencik and Wilkin 2015:147).

As a tool of effective political communication, the authenticity illusions of social media inevitably make it equally appealing to powerful political actors. Whilst initially treated by some politicians as a source of risk (Anstead and Chadwick 2010), the symbolic capital gained from mediated authenticity through social media has made these platforms central to political campaigning. Former US President Barack Obama famously integrated social media at an early stage of his 2008 campaign to facilitate a different kind of mobilization that was widely perceived as being citizen led and disassociated from the corporate and institutional funding traditionally linked to candidates of the establishment (Gibson 2015). As a direct counterweight to the scripted performances associated with professional political communication, social media also lends itself to humanizing politicians who otherwise struggle to appear genuine and sincere. In particular, social media has proven an effective arena for what Enli (2017:50) calls the "authentic outsider" in the political sphere, meaning that politicians use social media to explicitly step outside the conventions of the establishment by employing "authenticity markers" (ibid.:58) such as informal speech, personal pictures and, in the case of US President Donald Trump, expressed impoliteness, political incorrectness and grammatical inconsistency. The ability to employ social media to construct the appearance of authentic talk thus provides a pathway to securing trust, and such talk is offered as a "guarantee of truth" (Montgomery 2001:460). For politicians, it is this trust and appearance of sincerity that underpins their mandate. Yet, as Enli (2016:133) points out, "the rhetoric of authenticity might be a disguise for highly staged, pre-planned and expensive productions, and a way to make politicians seem like harmless and likeable everyday people 'like you and me'".

The unravelling of social media authenticity

Whilst narratives and myths surrounding social media often position social media platforms outside the institutional agendas and manipulations that otherwise mark our "world of inauthenticity" (Banet-Weiser 2012:8), the central position they now occupy in public life has inevitably invited deeper scrutiny, leading to a kind of unravelling of social media myths and a deeper questioning of the way in which collectivity and communication are structured on these platforms. Particularly since controversial political events such as the election of Donald Trump in the US and the Brexit vote in the UK, there has been increased public debate and focus on both the institutional architectures and commercial logics of these platforms that were so seamlessly integrated in the myth of us. The rapid escalation of concerns over fake news, for example, said to spread predominantly via social media platforms, has directly challenged the truthfulness of communication in these spaces, inviting us to question the sincerity of content shared among citizens (Connolly et al. 2016). The term fake news became particularly prominent in public discourse during the 2016 US presidential election, when it was suggested that a number of fake news sites had been created by different groups, for example Macedonian teenagers looking to make money, and had produced false stories on candidates. These stories were able to reach a large readership and potentially influence it because of the way they were unwittingly shared on social media networks (Wardle 2017). More serious attention subsequently shifted to fake news in relation to systematic disinformation campaigns. Such campaigns on social media allow for atoms of information to be directly targeted at specific users who are more likely to accept and share a particular message (boyd 2017a; Wardle 2017). These concerns undermine the narrative of social media as a counterweight to manipulated information and media representations, and expose platforms as easy targets for vested interests and agendas. As such, they have (re)introduced a prominent scepticism towards the idea that messages circulating on social media are inherently authentic.

Scepticism towards social media has been accelerated by a broader concern over computational propaganda which situates social media platforms at the centre of a new form of informational manipulation. Computational propaganda is defined as "the use of algorithms, automation, and human curation to purposefully distribute misleading information over social media networks" (Woolley and Howard 2016:3). Key features of this type of propaganda include decidedly inauthentic artefacts unique to social media, such as bots and click farms. These are automated accounts and processes that involve learning from and mimicking real people in order to manipulate public opinion across a diverse range of platforms (ibid.:6). Kollanyi et al. (2016) estimate that around 20 to 25 per cent of traffic on Twitter about the US 2016 presidential election leading up to the vote came from highly automated accounts, or bots. Click farms, on the other hand, or fake accounts designed to enhance the appearance of popularity and the influence of certain people or ideas, are part of a booming "global marketplace for social media fraud" (Confessore et al. 2018, n.p.). Indeed, in 2017 Facebook admitted to investors that it might be host to at least 60 million fake accounts (Confessore et al. 2018).

Cutting across these debates is a growing awareness of both the power and prevalence of the algorithmic design of social media platforms, steeped in corporate logics and institutional interests, that significantly undermines their ability to offer an authentic representation of actual users. These platforms lend themselves to distortion and manipulation through search and recommendation algorithms, sponsored content, dark advertising and other kinds of sociotechnical developments that underscore the politics of mediated authenticity on social media (Lewis 2018).

Future directions

Authenticity is key to how we make sense of the world and how we make decisions about our lives. In a world that can often feel decidedly inauthentic, steeped in profit logics and vested interests, the hunger for a space of sincerity and genuine emotion is more prevalent than ever. The ability to use different forms of media to cultivate perceptions of authenticity will continue to be central to strategies of social and political actors – whether corporations, journalists, activists or politicians – especially with the emergence of populism and affective politics. In this context, assessing competing claims to legitimacy that draw on ever-sophisticated authenticity markers is set to become an increasingly complex task. Social media continues to play a pertinent role in this regard, as citizens seek out assurances of trust in an emotionally charged political domain and try to navigate polarized public debate.

At the same time, the myths of apolitical and deinstitutionalized social media are increasingly unravelling in an age of bots, fake news and algorithmic filters that significantly put perceptions of authenticity into question. These developments are only taking on further significance as we move towards 'deep fakes' that manipulate video and voice recordings, and as policies on content moderation become increasingly contentious. The key issue in the future will be the extent to which social media platforms can hold on to their illusions of authenticity and claim to be an arena for users, and the extent to which social media companies decide to assume responsibility for the activity that takes place on their platforms. As users become alert to the possibilities for exploitation of the information ecosystem, and start questioning the wider structures that shape their activities and messages, the deliberate obscurity that has long surrounded the operations of social media platforms in order to uphold a myth of us may in the end be what marks their decline as a space in which an authentic citizen voice can be articulated.

See also: amateur; citizen journalism; co-optation; Facebook; hip-hop; remediation; social media; twitter and hashtags; Weibo

Recommended reading

Banet-Weiser, S. (2012) *Authentic™*, New York and London: New York University Press.

A key text on the role of authenticity in mediated branding. Whilst a lot of the focus is on advertising and consumerism, the text broadens the discussion to political culture and public debate. An important contribution to contemporary understandings of authenticity.

Enli, G. (2015) *Mediated Authenticity: How the media construct reality*, New York: Peter Lang.

Explores mediated authenticity through a number of case studies, across different media forms, and argues that solving 'authenticity puzzles' – separating the fake from the real – is a key part of how people engage with media. A comprehensive text on the relationship between authenticity and media.

Lovink, G. (2012) *Networks without a Cause: A critique of social media*, Cambridge: Polity Press.

An extensive critique of the politics of social media technologies and their impact on our self-understandings and society. Although not explicitly focused on the question of authenticity, the text traces the operations of social media and dismantles the myth of them as deinstitutionalized spaces void of agendas and interests. A significant book for contemplating the nature of infrastructures for providing a space for citizen voice.

AUTONOMOUS MOVEMENTS

Cristina Flesher Fominaya

Autonomy is not a consolidated formal framework but rather a set of guiding principles that characterize a rich tradition of social movement organizing. This tradition has taken different forms within and across various sites around the world, including Latin America, Asia, Europe and North America, and across time within the same site (Gagyi 2013; Flesher Fominaya 2016a; Dinerstein 2015; Katsiaficas 2006; Mikkelsen and Karpantschof 2001). Rather than a rigid orthodoxy, autonomy is a perpetual horizon towards which social movements strive (Flesher Eguiarte 2005; Pickerill and Chatterton 2006; Böhm et al. 2010). Castells defines it as "the capacity of people, either individually or collectively, to organize their lives in terms of their projects, desires, and needs without having to submit to whatever rules are established by institutions" (Castells and Kumar 2014:93). Autonomous social movements thus ultimately seek to create spaces in which they can act freely from structures of oppression and hierarchy, engage with these systems and structures in a way that challenges or transgresses dominant norms and forms, and attempt to provide alternative models of social, economic, political, cultural and communicative organization and practice.

Flesher Fominaya (2007) defines autonomous movements as organized in a horizontal network according to principles of self-organization, direct/participatory democracy, autonomy, diversity and direct action. Autonomous movements are not set up as permanent structures that seek their own perpetuation, but rather as (dis)organizations that make up nodes in biodegradable networks (Plows 1998, 2008; Wall 1999) that continually re-form and recombine. They create temporary autonomous zones (Bey 1985) in which they experiment with alternative social and cultural codes (Melucci 1996) and participate in prefigurative practices within a subculture that produces multiple forms of contentious and/or transgressive political engagement.

Autonomous social movements are often invisible to outsiders, scholars and political observers (Flesher Fominaya 2015a; Katsiaficas 2006), yet have been enormously influential and have engaged with a wide range of social issues that are seen as interconnected. Katsiaficas (2006:xiii–xiv) suggests that European autonomous movements acted as "a driving force of others", claiming that "their militant resistance to the arms race, nuclear power, patriarchy, and the housing shortage transformed single-issue struggles into an autonomous movement whose aspirations were to transform the society as a whole", that they were "independent of political parties", "their adherents would have nothing to do with established forms of politics", and that their goal was to "facilitate greater individual and community control over everyday life". Although today some forms of autonomous movement are more open to engagement with established forms of politics (Martínez López 2016), and indeed actively seek ways to "break out of the activist ghetto" (Flesher Fominaya 2015b:150), Katsiaficas' description captures the essential spirit of autonomous movements.

Autonomous movements and other political traditions

Autonomous movements attempt to carve a space free from all systems of constitutive power that produce hierarchies, exploitation and oppression. These systems include capitalism, state power, patriarchy, racism, colonialism and environmental exploitations – systems which themselves are connected to each other in myriad ways. Historically, this stance has manifested in a reflexive critique of political praxis, which has included a central preoccupation with questioning and rejecting institutionalized and orthodox leftist political practice and organizations.

In broad terms, autonomous movements are rooted in three core political traditions: libertarian Marxism, anarchism and feminism, although feminism is often ignored in genealogical accounts of autonomy. In one account of Italian autonomism and its evolution, for example, Casarini (in Iglesias Turrión 2003) does not use the word *feminist* once. By contrast, Katsiaficas (2006) offers a detailed and compelling discussion of the importance of feminism in the development of European autonomous movements, as do van der Steen et al. (2014). During the 1960s and 70s, many movements emerged that deeply questioned the hierarchical or vertical nature of what is often dubbed Old Left organizing, advocating for more horizontal and participatory spaces that prefigure the emancipatory politics to which these movements aspired. Against the gender blindness of even anti-hierarchical or revolutionary so-called New Left spaces, feminists began to critique the internal patriarchy that persisted – and continues to persist – in many leftist movement spaces. In some contexts, feminist understandings of autonomy also reflected a desire to reject the instrumentalization of womanhood in nationalist struggles (Abdulhadi 1998). Feminist theories of the relation between patriarchy, capitalism and other forms of exploitation such as racism, colonialism and heteronormativity have made significant and lasting contributions to autonomous thought and practice, notably through increased attention to the constructed nature of subjectivity and identity, the vulnerability of the body and the gendered nature of its exploitation under patriarchal capitalism, which harms both women and men in specific ways, and the theorization of the precarious subject. Feminist attention to affect and care as central to 'the political' has transformed many autonomous groups' understanding of their own goals and practices (Precarias a la Deriva 2004; Bartlett 2017a).

The influence of anti-colonial and anti-racist movements has also been very important, if more uneven in its resonance; for example, some groups still fail to engage deeply with these movements even as they embrace a commitment to anti-racist politics (Mudu and Chattopadhyay 2017). Hacker ethics and technopolitics also have strong synergies with (and indeed form part of) autonomous movements. The architecture of the Internet mirrors and facilitates the ideational frameworks that motivate autonomous political action: the Internet – at least in theory – is a horizontal, multinodal, participatory network that allows open-ended numbers of links between different points of contact, and has allowed for open source collective collaboration (Flesher Fominaya and Gillan 2017). Technopolitical struggles that aim to free the net from commodification and privatization and contest its use as an arena of exploitation, obfuscation, control and surveillance have strong synergies with the aspirations of autonomous political movements, and cyber-action often follows similar organizational logics (Pickard 2006; Flesher Fominaya 2017; Sampedro Blanco 2005; Haro Barba and Sampedro Blanco 2011; Jordan 2002a, 2002b; Flesher Fominaya 2016b).

Although some scholars such as Graeber (2013) emphasize the anarchist roots of movements like Occupy Wall Street, anarchism forms only a part, albeit a crucial one, of the multiple theoretical traditions that have enriched and continue to influence autonomous theory and practice. Similarly, the terms *autonomist* and *autonomous movements* are sometimes conflated. However, the former refers to a specific form of Italian Marxist autonomism, whereas the latter encompasses a broader range of movements sharing similar principles and practices. At the heart of Marxist autonomism is the idea that workers are active agents whose needs, desires and autonomies should be central to a Marxist analysis and practice (Osterweil 2013; Katsiaficas 2006; Cuninghame 2002). The Marxist roots of autonomy have produced provocative and influential works on alienation, precariousness and capitalism's colonization of human knowledge. These works include the writings of Franco Berardi, known as Bifo (2003, 2007, 2009, 2011), and of collectives like *Precarias a la Deriva*, who have produced feminist autonomous reflections on precarious subjectivity and the exploitation of feminine qualities under patriarchal capitalism (Precarias a la Deriva 2004).

Each specific movement group has integrated particular influences that are the product of movement contagion across time and the spaces of place and spaces of flows to use Castell's (2009) terminology (Gagyi 2013; Membretti and Mudu 2013). For example, Italian autonomous movements since the early 2000s (Cuninghame 2002; Lotringer and Marazzi 2007; Membretti and Mudu 2013; Osterweil 2013) have been influenced by earlier movements of Italian autonomism, including squatting movements, but also by their exposure to Mexico's *Zapatismo* (Iglesias Turrión 2003). Spanish autonomous movements have been influenced greatly by their own squatting movements but also by the Italian *Dissobedienti* and *Zapatismo* (both directly and via Italy) and other Latin American autonomous movements (Negri 1980; Lotringer and Marazzi 2007; ULEX 2008; Cattaneo and Tudela 2014; Martínez and García 2015; Tudela and Cattaneo 2016). Spanish autonomous activists in the 15-M movement then went on to influence activism in the Occupy Wall Street and other movements in the global wave (Romanos 2016; Flesher Fominaya 2017, 2020a; Roos and Oikonomakis 2014). UK autonomous groups have been influenced by US environmentalism as well as homegrown anti-militarist and feminist peace camps, and have in turn influenced global justice movements (Plows 1998; Wall 1999; Rootes and Saunders 2007; Flesher Fominaya 2013).

Key characteristics of autonomous groups

Autonomous actors distinguish themselves from the institutional left, rejecting representative democracy and majority rule and instead defending more participatory models, based on direct democracy and self-governance, horizontal (non-hierarchical) structures, decision making through consensus where possible and necessary, undertaken in the forum of a (normally open) assembly, and rarely with permanent delegations of responsibility. Flesher Fominaya (2007, 2020a), Katsiaficas (2006) and Chatterton and Pickerill (2010) offer a comprehensive analysis of the key features of autonomous movements, six of which are highlighted below.

First, autonomous political practice is prefigurative in that it attempts to create tools and practices today that foreshadow the future society that is aspired to. This means that there is a clear and strong connection between means and ends, and hence organizational forms, decision-making processes and various forms of action are not treated as simply means to an end, but often as ends in themselves. Social transformation comes through the creation of alternatives, not through the existing institutional system. Second, the ideological base of autonomous movements is heterogeneous and frequently not explicit. Ideologies are seen as frozen and prescriptive, denying the autonomy of the self to decide and experience her own path. Theory should arise from practice and should remain open to question, change and modification. Ideologies are seen to divide and exclude rather than enhance diversity and inclusivity. They are also identified with old leftist orthodoxies. Third, the autonomous subject is multifaceted, with multiple overlapping identities. Autonomous activists tend to reject fixed identities as limiting and prescriptive, and deliberately eschew even group identities and labels – including acronyms, symbols and flags. Flesher Fominaya (2015a) characterizes this stance as anti-identitarian. This feature is also tied to the rejection of representation, both in the sense that any individual can represent a group (politically, as in the case of a leader) and in the sense that any individual member of a group can represent the totality of membership in that group. Fourth, although the legitimate political actor is the autonomous individual, acting collectively, this does not translate into a rejection of collectives or affinity groups. Rather, the autonomous actor actively attempts to negate the isolationism and lack of empathy created by competitive capitalist consumer society, through the nurturing of social relations that create community. Solidarity and collective resolution of shared problems, even as they might affect an individual at a given point in time, are key. Fifth, autonomous movements adopt a DIY (do it yourself) philosophy. This encompasses various philosophical elements: collective self-sufficiency, the desire to work without intermediaries, the idea that pretty much anyone can learn how to do anything, and the commitment to the idea that people with more knowledge or expertise should share it in an open way. This has implications for breaking down the notion of expertise associated with hierarchies, but also is strongly connected to a philosophy of sharing and replicability: if participants provide practical, technical, substantive information freely, then others can take that information and set up their own camp, working group, movement or workshop, building the movement community in the process. Finally, a rejection of models of assistentialism, clientelism and charity follows from the commitment to a DIY culture and connects internal movement practice with systemic critique. Phenomena such as homelessness and unemployment are not seen as individual problems. Instead, they are resolutely framed as systemic failures which, within neoliberalism, are socially constructed as individual failures, leading to depression, self-blame and guilt. Autonomous groups therefore seek to foster collective self-empowerment through specific practices of resource sharing, including knowledge sharing, and solidarity.

Challenges of autonomous praxis

These general characteristics must be reflexively weighed against each other, and the outcome will vary across groups and community cultures. Rigid and orthodox interpretations of autonomous principles go against the spirit of autonomy in two related ways, first because autonomy must proceed from practice to theory, and rigidly applying theory inverts this central idea, and second because the emphasis on praxis rather than ideology means that if practice is to inform theory, then the concrete implications or material consequences of the practical application of any principle needs to be reflexively thought through. Means that do not lead to the group's desired ends, or that even prevent their realization, are not useful, however ideologically sound they might be. This can be illustrated with two common problems encountered in autonomous groups, problems that arise from rigid interpretations of the commitment to consensus and the commitment to openness.

A common problem, witnessed most often among inexperienced groups of activists, is the confusion of consensus with unanimity. Insisting on unanimity implies a conformity that contravenes the autonomous spirit: consensus is not a rule but a horizon, a desired foundation for action that reflects the inclusive participatory spirit of the group. Unanimity, in contrast, means that a single dissenter can block the action of a large group of people whose raison d'être is usually to take action. One relevant question – rarely asked by those advocating pure consensus, by which they mean unanimity – is this: If we would not allow a single person to impose their will on the group, why would we allow a single person to block action unilaterally? In practice, deciding on the threshold at which consensus may be considered sufficient to warrant action will vary, and is often contested, depending on the nature, goals and principles of the group. But rigidly adhering to means that make reaching the group's ends impossible – placing means over ends – is self-sabotage in pursuit of ideological purity, which contravenes the autonomous spirit. Another common problem that violates the practicality principle, but in this case places ends over means, arises in cases where people propose ideas for action without committing to carrying them through, or where the group lacks the resources or agency to make them materialize. This converts groups into talking shops, where no progress towards their own action goals is made. Here one solution is to follow what in Spanish is known as the *quien lo propone se lo come* principle, which loosely means 'she who proposes must be willing to act' (Bartlett 2017b). Even agreement (consensus) does not solve all these challenges to action, and a persistent inability to act can lead to burnout and exit, and hence group dissolution.

Like consensus, openness is a core value that needs to be reflexively applied. If by being completely open activists introduce elements that are seriously harmful to the integrity and survival of the group itself – for example in the form of consistently disruptive people who impede discussion or participation, trolls or police infiltrators – then an unreflexive adherence to the principle of prioritizing means again trumps ends to the point of self-sabotage. No group can attract and hold on to members unless participants take action to attend to the affective and instrumental needs of the group to some extent, and this may mean, for example, not allowing domineering trolls to send participants fleeing. Likewise, being committed to transparency and open access to information is not the same as declaring that all online spaces need to be openly accessed. Online secret or closed groups offer safe spaces to protect members from hate, backlash, trolling, surveillance, commodification and appropriation (Jeppesen et al. 2014). Autonomous activists should and do think reflexively about the particular configuration of their political practices, weighing the multiple principles, priorities and goals of group members. Failure to do so can incur high costs. While elastic exit requirements and biodegradable networks mean that in principle if an activist does not like one group's dynamic they can start another, in practice activists develop strong group identity and affective ties that can make leaving a particular group a difficult and personally costly decision, whereas starting a new group can also pose significant challenges and high costs.

Difficulties with implementing the principles of consensus and openness stem in part from a third crucial and often hidden problem in autonomous groups, what Freeman (1971/2013) refers to as the tyranny of structurelessness, whereby leaderless horizontal groups that eschew formal structures paradoxically create informal hierarchies that violate the explicit principles of the group. For example, in the absence of explicit leaders and division of responsibilities, *de facto* leaders can emerge whose power remains unchallenged because it is not named or acknowledged, rendering their effects even more pernicious. Flesher Fominaya (Flesher Eguiarte 2005; Flesher Fominaya 2010) has also shown how a commitment to horizontal and autonomous principles can work against their realization: even when activists can see a problem in the group they are often loathe to take action to correct it. Part of the reason is that introducing change requires effort and a high degree of investment in the space, along with either a strongly principled motivation and/or the belief that things will actually change if the activist does invest the energy. Introducing changes also implies a critique of established practice and those engaging in it, and the desire to not be seen as dictating to others or criticizing them, which is linked to the desire to be (and be seen to be) horizontal and democratic, ironically acts to perpetuate a non-democratic situation and to maintain the status quo.

Other prefigurative principles raise similar tensions between means and ends. A particularly problematic area involves the commitment to direct action and the forms it should take. While many groups are committed to non-violent direct action, others espouse a diversity of tactics approach (Conway 2003; Hurl 2009), where the periodic use of force is justified for disruptive or defensive purposes under certain circumstances. For example, as neo-fascism continues to rear its ugly head in the US and across Europe, justifications for violence associated with the Antifa autonomous movement against fascism re-emerge, unleashing a debate about the morality, legitimacy and strategic desirability of violent confrontation in autonomous movements. While often glossed or simplified – violence is bad, peaceful protest is good – such debates raise legitimate questions about whether marginalized communities under attack by militarized police forces, vigilantes and other protesters have the right to defend themselves. At the same time, legitimizing physical defence against violence raises the question of its strategic costs.

Yet any use of physical force can potentially compromise the moral high ground of protesters, even while defending themselves against other protesters who advocate politics that justify violence against target groups. For example, when anti-fascist protesters enter into physical confrontations with neo-Nazis who are marching in the streets, they open up a discursive opportunity for the media and the authorities to characterize them as violent, and to argue for an equivalence between fascist and anti-fascist ideologies. This then shifts the discussion away from the violence promoted by fascism, to a discussion of the violence of the tactics used to engage in contesting (violent) ideologies. Physical altercation also enables a media or opponent narrative that paints all anti-fascists as violent, even if only a small percentage of the movement participants might advocate, justify or engage in any form of physical confrontation. On the other hand, if some of those in the anti-fascist movement publicly denounce other anti-fascists for engaging in physical confrontations, even against police brutality or violent provocation by fascist protesters, they can then reinforce the narrative that neo-Nazi protesters are essentially the same (morally) as anti-fascist protesters. Debates on such diversity of tactics recur regularly and are never entirely resolvable. They also often fail to recognize the role of formerly militant organizations, such as the ANC in South Africa, in bringing about progressive social change (Seidman 2001).

Not all movements that consider themselves autonomous embrace all aspects of the principles discussed here, but rather practise what can be called hybrid autonomy (Martínez López 2016; Flesher Fominaya 2007, 2015b). For example, the argument that *de facto* leaders exist and that movements operate in a highly mediated environment has been used as a rationale for designating particular spokespeople in the interests of satisfying media demand in a world where communication is central to all forms of politics (Casarini 2003). The Zapatista EZLN's Subcomandante Marcos, for example, experimented with giving and denying the media access to a leader by playing with anonymity and visibility: although he gave interviews and released statements, he did not reveal his true identity or face. While autonomous movements reject leaders, they do often have spokespeople, and the line between the two can be blurred, especially by the media (Gitlin 2003a). Many autonomous activists are highly critical of the idea of exalting leaders and spokespeople and instead advocate for leaderless or leaderfull movements, where spokespeople are rotated, although few would ignore the power of media and political communication strategies (Jeppesen et al. 2014). The tension between means and ends again resurges here. For example, groups might recognize that in a patriarchal world men's voices receive more media attention and are taken more seriously as interlocutors. From a purely strategic point of view, these advantages could be exploited to achieve the movement's goals, but from a prefigurative point of view, doing so reproduces the very system they are seeking to dismantle, a patriarchal world order based on hierarchies and inequality by privileging ends over means. Instead, training women to be effective spokespeople and insisting on their right to that role would be a way to combine prefigurative and strategic considerations.

The ideological contradictions or cognitive dissonance for autonomous groups advocating horizontalism but practising leader-based or representation politics are difficult to overcome. This is one of the key tensions between autonomous grassroots movements and contemporary hybrid parties such as Podemos, which attempt to reconcile grassroots autonomy and representational institutional politics. In addition to the rejection of leaders, autonomous movements sit uncomfortably with hybrid parties because they problematize the notion of representation itself as the basis for democracy or political participation. Experimentation with alternative models of political party – such as using lotteries to designate leaders and representatives or adopting new forms of democracy such as direct or digitally enabled liquid/delegative democracy – characterizes part of the contemporary imaginary and evolution of autonomous movement politics (Romanos and Sádaba 2015; Feenstra et al. 2017) as activists continually seek to address these questions.

In conclusion, autonomous social movements have been largely ignored in the literature on social movements, in part because their subcultural nature and rejection of labels and identity markers makes them invisible, because their lack of formal organization and leaders makes it difficult to find and collect objective data about them, and because they do not fit neatly into dominant models of social movement scholarship (Katsiaficas 2006; Flesher Fominaya 2015a). This has changed somewhat with the turn of the century, thanks to a developing and rich body of scholarship. Their influence, however, has extended far beyond the subcultural autonomous movement spaces they create, as the global wave of occupations following the global financial crisis shows. Ultimately, the power of social movements lies not just in concrete political gains that can be objectively measured, but in the transformation of consciousness, and in imagining and striving for new and better worlds. In this endeavour, autonomous movements have played an important yet often unrecognized role.

See also: activism; civil disobedience; direct action; prefiguration; process vs event; social movement studies and citizen media

Recommended reading

Flesher Fominaya, C. (2007) 'Autonomous Movements and the Institutional Left: Two approaches in tension in Madrid's anti-globalization network', *South European Society & Politics* 12(3): 335–358.

A comprehensive treatment of the concept of autonomy within the context of a particular case study, the global justice movement in Madrid, and a comparison between institutional left and autonomous logics of action that enables a clear understanding of their differences and the implications of this difference for social movement dynamics. Includes a comparative table of key characteristics.

Flesher Fominaya, C. (2020) *Democracy Reloaded: Inside Spain's political laboratory from 15-M to Podemos*, Oxford: Oxford University Press.

Framed in debates about the crisis of democracy, this volume revolves around Spain's 'Indignados' or '15M', one of the most influential social movements of recent times. It argues that movements organized around autonomous network logics can build and sustain robust movements in the absence of formal organizations, strong professionalized leadership and the ability to attract external resources.

Freeman, J. (1971/2013) 'The Tyranny of Structurelessness', *WSQ: Women's Studies Quarterly* 41(3): 231–246.

A classic and highly influential text, originally written in 1971, on the paradoxical and unintended tyrannical outcomes of prefigurative horizontal politics that serves as a cautionary tale and critique of autonomous principles and practice. Drawn from observations in the feminist movement of the US in the 1970s, it highlights the perils that the unreflexive adoption of principles poses for groups hoping to operate in a non-hierarchical and participatory way.

Katsiaficas, G. (2006) *The Subversion of Politics: European autonomous social movements and the decolonization of everyday life*, second edition, Oakland, CA: AK Press.

A history and discussion of European autonomous movements that provides theoretical analysis and original empirical data, and traces the connections between feminist and autonomous movements in Europe.

BIG DATA

Stefania Milan

Datafication – or the transformation of various aspects of human life into information that can be processed in an automatized fashion and thus monetized (Mayer-Schönberger and Cukier 2013) – alters the exercise of civic participation. Our being citizens is increasingly mediated by digital platforms and the data they generate (Hintz et al. 2018). Citizen media, too, evolve under the pressure of datafication. Data, irrespective of their magnitude, harbour stories and new ways of looking at the world, and come to constitute a novel terrain of engagement for citizen media (Milan and Gutiérrez 2015). New varieties of media activism, exploiting the latest affordances of media innovation for activism purposes, are emerging on the fringes of the datafied society (Gordon and Mihailidis 2016; Pickard and Yang 2017). Data activism, for example, takes advantage of the possibilities for civic engagement offered by big data, incorporating data-based narratives in advocacy campaigns. It increasingly dialogues with more traditional forms of citizen self-expression and contributes to redefining citizen media today (Milan 2017).

This entry explores how and where citizen media meet big data by addressing two important questions: how big data contributes to empowering citizens, and what this empowerment means for the present and future of citizen media. Looking at emergent forms of engagement with data, the entry thus connects the current fascination with quantified information, apps and databases with the possibilities of participation and practices of empowerment enabled by citizen media. After illustrating four core features of citizen media in the age of datafication, it moves on to explore the rise of two novel configurations of citizen media: proactive data activism and civic hacking. The entry concludes by reflecting on the challenges that big data pose to citizen media and, in particular, the emergence of new divides.

From typewriters to big data: five key features of citizen media

From low-power radio transmitters to photocopier machines, from the static websites of the 1990s to today's interactive blogs, from neighbourhood television stations to social media platforms, the evolution of citizen media has closely followed the path of technological innovation. Irrespective of the supporting technology, these varieties of citizen media share four features: they empower people through their engagement with technology; present low entry barriers to participation; are collective and communitarian by nature; and allow individuals and groups to exercise democratic agency through media practice. Each of these features is elaborated below.

Empowerment, which lies at the core of citizen media, can be understood as the process through which individuals and groups come to exercise control over both messages and technology by playing an active role in reshaping the communicative processes in which they are involved (Milan 2013). Empowerment thus derives from firsthand engagement with media production – in terms of both content creation and the making of technology and infrastructure (Dunbar-Hester 2014). Learning to programme and broadcast, articulating and disseminating alternative narratives for social change, or assembling a radio transmitter from a kit bought online are some of the processes that enable empowerment.

Characterized by low entry barriers to participation, citizen media projects are accessible to virtually everyone and do not typically require advanced technical skills (Atton 2002). Instead, they capitalize on the orality of radio broadcasting, the spur-of-the-moment reporting of citizen journalists, or the immediacy of blogging and live broadcasting through a smartphone. When more specialized skills are required – for example to operate switchboards or develop software programs – experienced practitioners tend to share their knowledge through dedicated sessions that are open to newcomers.

In their capacity as media "of the people, by the people, and for the people" (Gillmor 2006:195), citizen media are by nature collective projects with a civic orientation (Downing 2001). They seek to interpret and give voice to the needs and imaginaries of their communities of reference – be they geographically bounded or interest-based (Hollander et al. 2002). In addition, citizen media emphasize the collective dimension of freedom of expression, as opposed to an individualistic approach. In the words of a practitioner, the radio "gave me the power of letting the others speak, speaking myself just a little" (Milan 2013:54).

Finally, citizen media expand and multiply spaces for political action. By making media, "a collectivity is enacting its citizenship by actively intervening and transforming the established mediascape ... contesting social codes, legitimized identities, and institutionalized social relations ... to the point where these transformations and changes are possible" (Rodríguez 2001:20). In other words, citizen media practice enables people to exercise their democratic agency beyond traditional means such as voting and institutional spaces like local administrations. By acquiring and exercising voice (Hemer and Tufte 2016), people change their perception of self, alongside their reality.

The engagement with big data and data infrastructure – the databases, data portals, apps and platforms – increasingly funnels citizen participation and alters some of the core dynamics of citizen media, with both positive and negative consequences. But we first need to understand how engaging with data can contribute to citizen empowerment.

The availability of large quantities of data and the numerical and visual forms information takes today offer novel opportunities for the exercise of democratic participation (Couldry and Powell 2014; Gutiérrez 2018). People can engage with data in a variety of ways: they can find stories in existing datasets, engage with innovative storytelling techniques, request data when they are not available, and create their own datasets.

Firstly, citizens can put existing data to new uses, promoting transformative experiments that repurpose information. For example, the *Illegal Fishing* project used publicly available satellite-tracking datasets to map and denounce prohibited fishing activities in Western Africa (Overseas Development Institute, n.d.). Similarly, the *Left-to-die Boat* project leveraged data generated by existing surveillance technologies to reconstruct the events leading to the death of sixty-three migrants in the Mediterranean Sea. The report supported a series of legal petitions against NATO member states (Forensic Architecture, n.d.).

Moreover, as data visualization software allows users to represent information with minimal effort, low-skill individuals and organizations can also use data for storytelling and to increase the effectiveness of their messages (Kennedy et al. 2016). The nonprofit Tactical Tech Collective, for example, argued that "as activists, we can't sit and wait for people to wade through sixty-page reports. To influence people we must make strong arguments and communicate them using strong evidence" (Visualising Information for Advocacy, n.d.). By way of illustration, *Out of Sight, Out of Mind* is an interactive digital visualization of US drone killings in Pakistan, which allows the public to explore the details and scope of the strikes by themselves (Pitch Interactive, n.d.).

Citizens can also request information from public administrations when this is not already made available. By 2018, over 100 countries across the world had implemented some form of freedom of information legislation, designed to facilitate the general public access to data held by national and local institutions. Data portals providing this sort of information are becoming increasingly common. *OpenSpending*, for example, gathers publicly available data to enable citizens to check and compare fiscal information from seventy-nine countries (Openspending, n.d.). Websites and entities supporting data requests are also proliferating. *ControlaTuGobierno*, for instance, is a Mexican non-governmental organization promoting the right to access information in marginalized communities (ControlaTuGobierno, n.d.).

Finally, citizens can create their own datasets to contribute to alternative narratives of a given problem. For instance, the *Syrian Archive* project (Syrian Archive, n.d.) adopts open source intelligence techniques to curate a body of visual evidence of human rights abuse in the Syrian conflict that can be used for reporting, advocacy and accountability purposes (Deutch and Habal 2018). In the same way, the *Bell¿ngcat* project, founded by a citizen journalist, engages in "open source and social media investigation" to scrutinize conflicts and the criminal underground, and offers guides "so others may learn to do the same" (Bell¿ngcat, n.d.).

What these data activism projects have in common is the belief that data can be effectively used for social change. As is also the case with citizen media, even non-experts can engage with information and technology for activist purposes. However, big data have not only introduced innovation in the tactics, but also fostered novel cooperation. In the absence of the necessary skills, unprecedented alliances have emerged: for instance, *Hacks/Hackers* connects journalists (hacks, who can tell stories) and technologists (hackers, who understand data).

The rise of proactive data activism and civic hacking

The importance of proactive data activism is borne out by a number of recent initiatives. *InfoAmazonia*, an example of proactive data activism launched in 2012, has produced investigative reports on the status of the biggest tropical forest in the world. By promoting data transparency, InfoAmazonia contributes to blocking the illegal deforestation and wildfires endangering the Amazon area (Infoamazonia, n.d.). At the core of this citizen media initiative are self-collected data about carbon monoxide, forest fires, water quality and level, and deforestation. Data are compiled through a self-organized data infrastructure that consists of sensors and mobile applications generating crowdsourced information. Combining satellite imagery with data gathered by activists, InfoAmazonia publishes interactive maps and alternative cartographies, and trains journalists, campaigners and communities in data collection and storytelling (Milan and Gutiérrez 2017). It illustrates the synergies that may obtain when the values and practices of citizen media capitalize on the novel opportunities offered by big data.

Combining complex information with innovative storytelling and collective organizing, proactive data activism embraces tactics of engagement with data seeking to take advantage of the wealth of information of big data for social change purposes. As illustrated in Figure 1, this form of activism emerges at the intersection between citizen media, journalism, big data and activism/advocacy. From journalism it takes the investigative ethos and the meticulous search for the truth; from the activism and advocacy realm the desire to make of the world a better place; from the big data domain a taste for verified information and strong evidence, such as statistics. From the citizen media domain proactive data activism derives its collective orientation, firsthand engagement with technology, storytelling with a purpose, and the possibility of exercising democratic agency through media practice. Finally, proactive data activism is also inspired by advocacy journalism and data journalism, a specialty combining traditional journalistic storytelling with advanced data analysis to produce investigations (Anderson 2018).

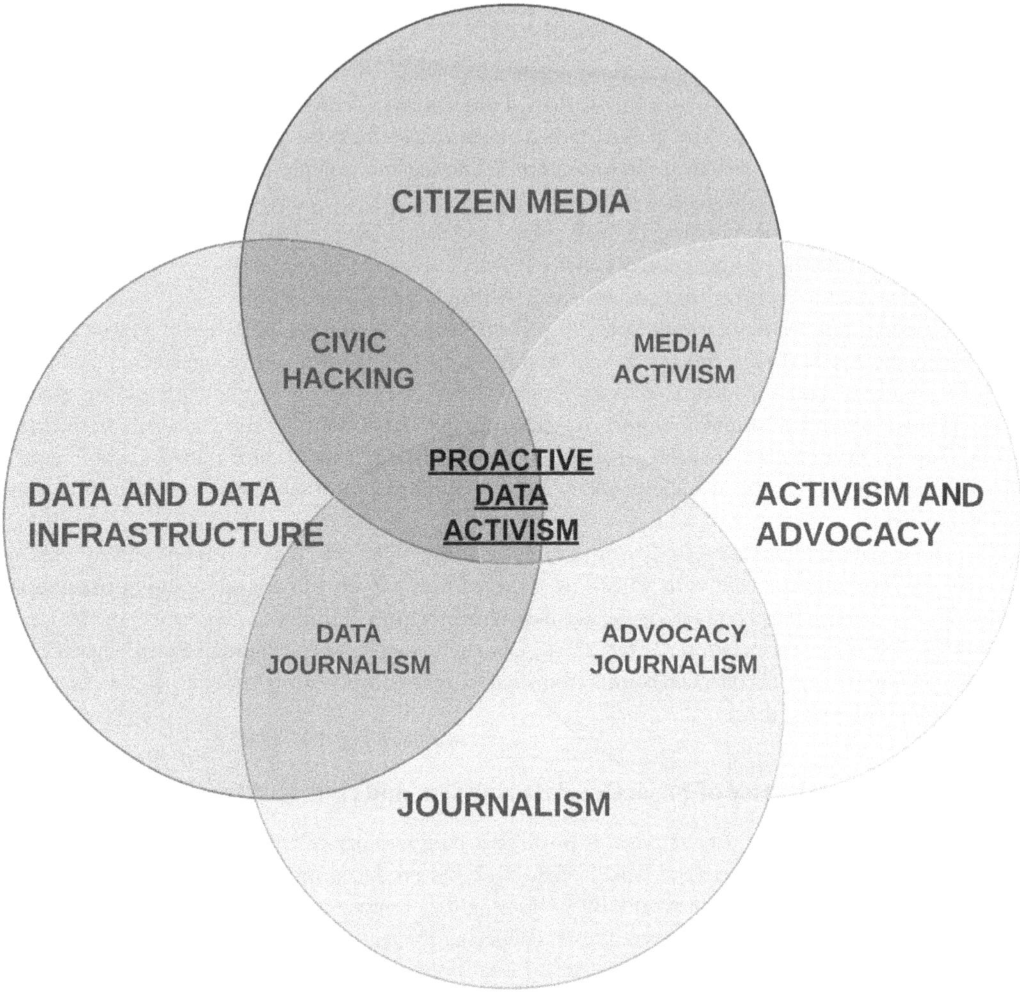

Figure 1 The emergence of two new configurations of citizen media (adapted from Milan and Gutiérrez 2015)

Civic hacking is another important phenomenon emerging at the intersection of citizen media and the big data domain. Civic hackers take advantage of publicly available information, usually government data, to improve institutional output and democratic governance. Organizing hackatons (i.e. marathons to hack data) that oddly combine the do-it-yourself culture of grassroots movements and Silicon Valley technocratic approaches, civil hackers exercise the right to information in a bid to identify solutions to social, economic and environmental challenges (Schrock 2016). Civic hacking shares with citizen media the premise that democratic agency can be exercised daily though engagement with information. They also share an interest in creating bottom-up infrastructure to match the values of citizen media. In this vein, the Open Knowledge Foundation launched the *Comprehensive Knowledge Archive Network* (CKAN, n.d.), a web-based management system for the storage and distribution of open data, used by both nonprofit and governmental organizations.

The emergence of novel divides

Proactive data activism and civic hacking signal a change in attitude towards big data that has emerged from within citizen media and storytelling projects. However, while the availability of data and the accessibility of data infrastructure clearly provide unparalleled opportunities for citizen media practitioners, they also present new challenges. In particular, novel divides emerge on a number of fronts: skills and data access capabilities vary among citizen media practitioners; the culture of individualism typical of hacking and other forms of tech activism is at odds with the collective nature of citizen media as understood in this entry; the prevalence of positivism associated with big data might hamper value-based advocacy activities; and the high concentration in the data brokers market might in the long run debilitate the efforts of citizen media to control their own infrastructure.

The skills required to engage with data and data analysis are not readily available within the citizen media realm. While software increasingly simplifies tasks, for example, by allowing even non-experts to produce appealing infographics, the entry barriers remain relatively high, especially at the cultural level. For most citizen media practitioners, handling big data is not yet part of their core skills set. In addition, software is usually available only in English. Language, literacy and income remain important obstacles for practitioners in developing countries in particular. On the whole, big data have introduced a novel divide, between the 'haves' – individuals and organizations who have access to data, state-of-the-art hardware and software, and the knowledge to meaningfully make sense of data – and the 'have nots', who are excluded from the "data revolution".

On the other hand, the power of attraction of data and data infrastructure goes hand in hand with a culture of individualism and expertise typical of the hacker scene and, more widely, of Silicon Valley. Although data projects require team work, they are markedly rooted in personal capabilities and individual practices. Big data have therefore introduced a novel tension between the individual and the collective dimensions of citizen media, which risks sidelining the cooperative nature of existing projects. Unfortunately, data-related skills are not easily transferable – at least not at the speed and the scale at which citizen media practitioners have traditionally involved newcomers in their activities.

The positivism that surrounds big data (Mosco 2014) represents another important divide, as it results in a tendency towards technocratic solutionism that is particularly visible in civic hacking. In this field of action, quantified data tend to be regarded as irrefutable information and receive higher consideration than norm- and value-based reasoning. Although data activists seek to promote alternative epistemologies countering the positivist ethos of big data (Milan and van der Velden 2016), there remains a fundamental tension between different ways of understanding and communicating the world around us – and the ways to change it. For example, the use of claims based on quantified data to justify collective demands – e.g. "we are the 99 percent", the slogan of the Occupy Wall Street mobilization – is increasingly popular, as are strategies aimed at gathering evidence as a fundamental step towards the resolution of problems (Bonechi 2012). This tension between value- and data-based reasoning deeply affects advocacy organizations and their strategies; it is indirectly reproduced by funding organizations, who prefer to support data-based projects over value-driven ones, thus reinforcing the divide mentioned above.

Finally, with the increase in magnitude and importance of big data, new data brokers such as Google and Facebook have become central to the functioning of citizen media. Although InfoAmazonia, for example, was able to produce its own data-reporting app, most citizen media are not able to do so and have to rely on commercially available products. As a result, big data contributes to weakening the self-organized infrastructure of citizen media, accelerating the move towards commercial services and promoting an even higher level of centralization of power in the hands of those who produce software, hardware and content. Unfortunately, most of these data brokers have values and agendas that are radically different from those of citizen media, and expose citizen media to monitoring and surveillance (Dencik and Leistert 2015).

Future directions

Big data have come to constitute the new frontline of citizen media practice. Two novel types of citizen media initiatives emerge, namely proactive data activism and civic hacking. The former seeks to incorporate data-based reasoning in citizen storytelling, while the latter takes advantage of available data to ameliorate democratic governance. Both share the hopeful belief that data and technology can contribute to social change. Datafication, however, presents citizen media with both opportunities and challenges. On the positive side, citizens can use data for advocacy purposes by supporting stories with data, engaging in innovative persuasive storytelling, requesting data when not readily available, and creating their own datasets to monitor phenomena of concern. Conversely, big data contribute to generating new divides within the domain of citizen media, reflecting varying levels of skills, income and access. New tensions arise between the individual capabilities of data experts and the collective nature of citizen media, and between the positivism of big data and the more traditional value- and norm-based advocacy. Finally, the self-organized infrastructure of citizen media is finding it increasingly difficult to compete with novel data brokers offering complex services that would otherwise be out of reach for grassroots practitioners.

See also: activism; hacking and hacktivism; sousveillance; surveillance

Recommended reading

Baack, S. (2018) 'Practically Engaged: The entanglements between data journalism and civic tech', *Digital Journalism* 6(6): 673–692.

Positions a new development in citizen media, namely open data, in relation to civic hacking and journalism, but also democratic participation.

Gutiérrez, M. (2018) *Data Activism and Social Change*, Cham, Switzerland: Palgrave Macmillan.

Explores proactive data activism in a number of successful projects at the intersection of advocacy journalism, data journalism and civic technologies.

Gutiérrez, M. and S. Milan (2019) 'Playing with Data and Its Consequences', *First Monday* 24(1). Available online: https://journals.uic.edu/ojs/index.php/fm/article/view/9554.

Investigates how tactics, identities and worldviews of citizen media practitioners evolve as a consequence of their exposure to data and data infrastructure.

Milan, S. and L. van der Velden (eds) (2018) *Data Activism*. Special issue of Krisis: Journal for Contemporary Philosophy 1. Available online: https://krisis.eu/issue-1-2018-data-activism/.

Reviews key aspects of data activism at the intersection with citizen media, including alternative epistemologies and the everyday experience of practitioners in engaging with data.

CITIZEN JOURNALISM

Einar Thorsen

There are a plethora of different terms that broadly describe a similar dynamic to citizen journalism, where citizens take an active role in news work or perform communicative acts that are considered by themselves or others to resemble acts of journalism or to perform a journalistic function. Citizen journalism as a term is at times seemingly all-encompassing and permeable in scope, and is therefore entangled in the boundary struggles of contemporary journalism studies, concerning "who counts as a journalist, what counts as journalism, and what is appropriate journalistic behavior" (Carlson 2015:2). The term has nevertheless retained a central position in both scholarly debates and in contemporary professional practice. Wall (2015:11) argues that

> citizen journalism has become the touchstone term for the last decade precisely because it reflects an ongoing normative belief that news is connected to a potentially positive form of civic behavior, which in turn harkens toward the long-standing idea that journalism is intimately tied to democracy.

Rather than seeking to define citizen journalism in conclusive or normative terms, this entry explores diverse forms of practice that contribute to our understanding of what is meant by citizen journalism. Examples abound where ordinary citizens have acted in journalistic ways – for example to document unfolding natural disasters, wars and conflicts, to map human rights abuses, or to challenge misuse of corporate or political power. Temporality and intentionality play an integral role in understanding different forms of citizen journalism; hence, the discussion that follows is organized under two broad categories: firstly, eyewitnessing and spontaneous forms of citizen journalism, and secondly, alternative and activist forms of citizen journalism.

Eyewitnessing and spontaneity during crises

Citizen journalists are often victims or bystanders, caught up in events beyond their control, who feel compelled to document what they are witnessing and disseminate it to friends, public audiences or professional journalists (Allan and Thorsen 2009; Thorsen and Allan 2014b). This practice is frequently referred to as accidental or amateur journalism, acts of journalism, citizen witnessing (Allan 2013) or user-generated content. Citizen journalism in this context relies significantly on the pervasive nature of mobile devices, portable cameras and Internet access, and as such these accounts are often highly visual, making extensive use of photographs or videos recorded on mobile devices. Professional news organizations have responded to the rise in this form of citizen journalism by developing sophisticated operations to source and verify authentic eyewitness reporting within a highly compressed time frame – where the citizens concerned are often themselves publishing directly to diverse audiences online, and their content in turn interpreted, appropriated and amplified by professional news organizations (Sienkiewicz 2014; Usher 2017).

The wealth of eyewitness material emerging in the aftermath of the Indian Ocean earthquake and tsunami in 2004 signalled a shift in news reporting of natural disasters, propelling citizens from victims to first-respondent correspondents who capture compelling audiovisual material in the absence of journalists (Allan 2006). This trend has continued apace and citizen journalists have since played a significant role in witnessing and documenting a range of large-scale disasters that caused massive destruction to infrastructure and loss of life – including earthquakes, superstorms, landslides and wildfires. Dramatic citizen footage has emerged in the aftermath of such disasters, documenting not only the devastation left behind from a victim's perspective, but frequently also the moment(s) of impact. Citizen journalists publish dramatic visual evidence of events as they unfold, and in so doing provide an important epistemological contribution to our collective memories and understanding of what took place.

Just as the Indian Ocean tsunami signalled a shift in the reporting of natural disasters, the iconic image of the January 2009 Hudson River plane crash captured by Jānis Krūms has become a symbol of the ability of eyewitnesses to capture and break news about transport accidents or disasters. The availability of impact footage of such accidents has in other words become increasingly normalized and constitutes an important part of fast-moving news landscapes. In April 2015, the German daily *Bild* and the French magazine *Paris Match* reported they had been shown mobile phone footage of the final moments of a Germanwings plane as it was crashed, purportedly in an act of suicide by co-pilot Andreas Lubitz; this footage was not shot by a bystander, but by one of the victims filming from the rear of the plane moments before impact. While official investigators denied the existence of such a video, the event indicates the willingness of news organizations to seek out even the most lurid and traumatic of eyewitness accounts.

The captivating nature of eyewitness material also extends to the reporting of violence, terror or criminal acts, often inciting debate about the role of perpetrators or the vivid nature of imagery. Video and still images of Gaddafi's killing in October 2011, for example, featured in news reports and on front pages across the world, sparking controversy about their graphic nature and the role of citizen imagery in showing the brutal reality of war. Arguably those capturing the images were witnesses, but questions also arose about definitional boundaries and whether the people recording would be better described as perpetrators or propagandists (Thorsen 2016). Other prominent examples include citizen eyewitness images circulating online in the aftermath of the April 2013 Boston Marathon bombing in the US and the dispute surrounding Reddit discussions that claimed to identify suspects (Mortensen 2015a), as well as the May 2013 murder of Lee Rigby in Woolwich, UK, where the attackers were filmed by eyewitnesses in the immediate aftermath boasting about their act (Allan 2014). Such is the immediacy of citizens sharing eyewitness material, that in many cases the reports circulate in the public domain even while events are still unfolding. Most harrowing of these, perhaps, are the examples of terrorist attacks, where victims have used mobile phones to report on events while their own safety remained uncertain. Examples here include the July 2011 attack on Oslo and Utøya in Norway, where young people tweeted from the island while the gunman was still hunting them down (Thorsen 2014); citizen footage from the September 2013 Westgate shopping mall attack and hostage siege in Nairobi, Kenya (Ogola 2015); and the January 2015 attack on the editorial offices of the satirical magazine *Charlie Hebdo* and the subsequent Porte de Vincennes shopping centre siege (Cooper, G. 2017).

Citizen journalism has also played a significant role in highlighting acts of police brutality (Antony and Thomas 2010; Allan 2013). Citizen videos of police brutality were particularly important triggers and focal points for the *#BlackLivesMatter* campaign against police violence in the US, with social media activism helping to document and sustain the street protests that followed (Clark 2016; Richardson 2016). Eric Garner died after being detained by a NYPD police officer using a prohibited chokehold in July 2014. His friend Ramsey Orta recorded the incident on his mobile phone, as did other bystanders, showing Garner repeatedly saying to the officers "I can't breathe" – a phrase adopted by many protesters who took to the streets in the aftermath. In April 2015, Kevin Moore and others filmed different points of the brutal arrest of Freddie Gray, also in the US; Gray would later die of injuries sustained in custody. Gray's death triggered violent protests and civil disorder in Baltimore, leading to a state of emergency being declared. The following year, in July 2016, Philando Castile was shot in his car in Minnesota after being pulled over by a police officer, while his girlfriend Diamond Reynolds was live streaming the immediate aftermath of the shooting from the passenger seat on Facebook. Eyewitness footage was in each instance crucial in documenting police behaviour, providing visual evidence shared widely on social media and remediated by professional news organizations. These witnesses felt compelled to film what they perceived to be unfair or abusive behaviour from police officers. However, they did not all share the material immediately, and some later expressed regret about sharing the footage, because they in turn became victims of harassment for their role as witnesses (Richardson 2016). While initial protests often emerge spontaneously in the aftermath of such bystander videos being published, citizen journalism also plays an ongoing role in sustaining the attention of the public, mobilizing further support and fuelling protests over longer periods of time.

Alternative, activist and counter-hegemonic citizen journalism

As in the case of *#BlackLivesMatter* and other instances of individuals documenting police brutality through the use of media, citizen journalism linked with social movements is sometimes performed by eyewitnesses and bystanders. Nevertheless, activist citizen journalism typically extends beyond spontaneous or accidental acts of journalism to encompass longer-term activities organized by collectives involved in documenting events in order to challenge established forms of power. One of the earliest examples in this regard was Indymedia, or Independent Media Center (IMC), which emerged from the protests against the World Trade Organization in 1999 in Seattle. While Indymedia had its roots in the anti-globalization and alterglobalization movements, it remains a collective of fairly autonomous media outlets whose contributors are volunteers or activists who also participate in defining and enforcing local editorial policy. In many ways Indymedia was the precursor to practices of citizen journalism that later took different forms of activist or alternative media.

We are increasingly accustomed to citizen journalists reporting from mass protests in countries with limited press freedom to counter state propaganda and lack of access for foreign correspondents. Protesters often seek to mobilize international support for opposition movements and those advocating for social change, thus blurring the boundaries between reporting as an eyewitness and propagandizing a particular cause. The disputed presidential elections in Iran and Moldova in 2009 are well known examples here, as are the Arab uprisings from December 2010 onwards, which led to the fall of governments in Tunisia, Egypt, Libya and Yemen, to civil war in Syria, and to further protests and calls for political change in Algeria, Bahrain, Jordan, Lebanon, Morocco, Sudan and Western Sahara. The rise and spread of these uprisings were closely linked with trigger events on which information was shared online. In Tunisia, images of Mohammed Bouazizi who set himself on fire in protest against government corruption were circulated by local activist bloggers and eventually picked up by international news media. The Egyptian uprising followed a similar pattern: the murder of Khaled Said, who had posted a video clip online demonstrating police corruption, prompted the creation of a Facebook page titled 'We are all Khaled Said' that acted as a focal point for mass protests against Mubarak's regime starting on 25 January 2011.

In Turkey, likewise, the May–July 2013 protests culminating in Taksim Square and Gezi Park were extensively reported on by activists who attracted global attention to the events through Twitter hashtags such as #OccupyGezi and #DirenGeziParki. In Brazil, citizen reporting drew attention to the mass protests over poverty and abuse of human rights against a background of extravagant expenditure on the FIFA Football World Cup (June–July 2013). Citizen reporting also plays an important part in drawing public attention to mass protests in liberal democracies, where state and private media often turn a blind eye to injustices in order to appease established forms of power, whether they represent state or private interests. Examples here include the riots in London and other parts of the UK in August 2011; the Occupy Wall Street movement in the US in 2011; and the mass protests against unemployment and government cuts in Spain in 2013. In all these cases, it was citizen reporting via social media that attracted international attention to the events.

Citizen journalism acted as a focal point and amplifier in such contexts, exposing underlying social tensions and enabling existing social movements and unions to mobilize. It served a dual purpose, in transnational advocacy as well as local mobilization aimed at re-appropriating public spaces (Cottle and Lester 2011; Gerbaudo 2012; Lievrouw 2011). In other words, while citizen journalism (and its intermediation) is predominantly discussed in terms of online communicative networks, it is important not to lose sight of its connection to action, or mobilization in material spaces (transmediation) – from hyperlocal to global contexts (Chouliaraki 2013a; Thorsen 2016). Ordinary citizens also contribute to relief efforts – mapping impact areas, crowdsourcing information about survivors, republishing emergency information and so forth – in what Norris (2017) has termed a form of digital humanitarianism. Here different forms of (typically online) collaboration enable people in disparate locations to form temporal networks that support disaster relief efforts through processing, curating and publishing information pertinent to distant emergencies. Bespoke platforms have also emerged to help facilitate the crowdsourcing of news and dissemination of eyewitness material. One example is *Ushahidi*, a platform originally designed to capture citizen reports of violence via text message and email in the aftermath of Kenya's disputed 2007 presidential election. The company has since evolved into a software platform with a reported 125,000 deployments. Bailard and Livingston (2014) found that citizen reporting of electoral abuse or success in the 2011 Nigerian election using the *Ushahidi* platform was significantly correlated with increased voter turnout, while Barack Obama's campaign team used *Ushahidi* during the 2012 US election to manage information collected from its election observers. *Ushahidi* further processed 40,000 reports in the first month of the Haiti earthquake in 2010, while the volunteer organization Kathmandu Living Labs (KLL) used it to provide the Nepalese army with vital emergency information to help triage rescue efforts and disaster relief following the 2015 earthquake.

Such distributed or crowdsourced investigations can also be undertaken in a more spontaneous way, with a temporary interconnection of citizens emerging around a specific event or issue. The Trafigura scandal in 2009, for example, is an often neglected case where the reporting and online mobilization by ordinary citizens managed to overturn a UK super injunction against the British newspaper *The Guardian*, drawing international attention to oil company Trafigura's culpability in dumping toxic waste off the Ivory Coast in 2006 (Thorsen 2016). Professional journalists in the UK, Netherlands and Norway were starting to uncover a trail of evidence suggesting the company had covered up the vast scale of the pollution and that it had known in advance that its waste dumping that caused several deaths was hazardous. Scant public attention was given to these reports, however, in part due to a pugnacious campaign by Trafigura's lawyers, Carter-Ruck, to sue media organizations reporting on the case for libel. *The Guardian* published a scathing editorial after it was banned from reporting parliamentary proceedings in a manner that it claimed undermined the very principles of free speech. Citizen journalists were quick to investigate what *The Guardian* had been prevented from reporting, with political blogs and social media rapidly saturated with information about the Trafigura scandal. Within 24 hours the super injunction had been overturned and the investigative contribution of ordinary citizens had helped cast light on corporate wrongdoing more effectively and with greater reach than traditional media had been able to achieve until then.

On the periphery of the journalistic field we find what Eldridge (2018) terms interlopers, people and organizations who position their work as journalism but reject the means by which mainstream media engenders journalistic values. Assuming a watchdog role they often overtly present their work as a form of journalism in order to claim credibility or even a protected legal status for their activities. Perhaps the most prominent example of such interloper media is WikiLeaks, an organization grounded in a quest for radical transparency as a means to disrupt the conspiratorial mechanisms by which elite power is sustained. It attracted worldwide attention after the publication in April 2010 of *Collateral Murder*, a video shot from a helicopter gunship hovering over Baghdad and showing a group of men, including two employees of Reuters, being slaughtered. Reuters had lobbied the US military to release the video since July 2007, but WikiLeaks demonstrated again how citizens can succeed in holding power to account where the traditional Fourth Estate is proving ineffective.

Challenging hegemony and elite power is not without risk. Chelsea Manning, the person responsible for leaking much of the material published by WikiLeaks in the early years, was sentenced in 2013 to 35 years in prison and only pardoned by President Obama in 2017 as he left office. The WikiLeaks editor-in-chief, Julian Assange, sought refuge in the Ecuadorian Embassy in London in 2012, evading extradition to Sweden on rape allegations, for fear of Sweden then extraditing him to the US. In April 2019, Ecuador withdrew its asylum and Assange was arrested and convicted of breaching bail. The US government quickly demanded his extradition to face charges of conspiracy to commit computer intrusion and assisting Chelsea Manning; these were later superseded by charges under the Espionage Act of 1917 for publication of diplomatic cables. The UK government accepted the charges, and at the time of writing there is an extradition hearing scheduled for February 2020. While these are prominent examples, the persecution and murder of citizen journalists (as with professional journalists) is not uncommon. In February 2015, Avijit Roy, a US-Bangladesh blogger advocating secularism, was hacked to death by unknown attackers; in March 2015, Raif Badawi, a Saudi Arabian blogger already sentenced to 10 years in prison and 1,000 lashes, was threatened with the death penalty for criticizing the country's clerics; and in May 2015, Evany José Metzker was found decapitated in Brazil, after blogging about politics and corruption – his latest investigation was purportedly into a child prostitution ring. These bloggers are examples of dissident or activist writers who all challenged restrictions on media freedom in their respective contexts, using digital media to report news and render visible voices and points of view that would otherwise be suppressed.

Citizen journalism has moreover helped marginalized communities gain a public voice, one which frequently transcends geographic or social boundaries. Sometimes these forms of citizen journalism are geographically defined, demographically specific or linked with social movements, but central to their purpose is to give voice to and empower otherwise marginalized peoples. Examples include blogging among racial minorities (Gabriel 2016), use of listserv by feminist movements to devise digital communication strategies (Valle 2014), news websites run by repressed indigenous people (Davies 2014) and video or photo journalism produced by disabled and homeless people (Luce et al. 2017). These forms of citizen journalism are documenting the challenges faced by groups who live at the margins of society and their everyday lifeworlds. In so doing they diverge significantly from the professional and institutional norms and ideals associated with journalism, but their reporting nevertheless provides a significant addition (and sometimes corrective) to the public information landscape. They demonstrate how citizen journalism might facilitate participatory forms of communication aimed at transformative social change, while at the same time energizing those marginalized groupings and communities.

Future directions

There has been a growing emphasis since the turn of the century on citizen journalism in the form of hyperlocal media initiatives, where the spatial parameters of reportage are confined to small villages, postcode areas or even streets. Hyperlocal journalism often arises, according to Metzgar et al. (2011), from the "public's dissatisfaction with legacy media" and as an "attempt to fill the perceived gap in public affairs coverage" (ibid.:782). While there is a diverse array of approaches to hyperlocal news, what they have in common is a focus on citizens reporting on issues in their immediate locality or community. As with other forms of citizen journalism, many commercial media organizations have sought to appropriate the hyperlocal principles by setting up partnerships between professionals and amateurs, but no model for ensuring (commercially) sustainable hyperlocal news has yet emerged (Thurman et al. 2011). This is problematic on a number of levels – local journalism is itself under attack from aggressive staff cuts and title closures (often blamed on falling advertising revenues, but also the consequence of profiteering), and if citizen journalism initiatives that might have been able to fill that vacuum are not financially sustainable in the long term, then communities will suffer.

Citizen journalists are ubiquitous across the world and their activities have had a transformative effect on news cultures at all levels. Indeed, citizen journalism in its different guises has become a normal part of the public information landscape involving crisis events, even as they unfold. Established professional values are being recast in this rapidly evolving relationship between journalists, elite sources and citizens. The rapid rise of websites facilitating photo and video sharing as well as social networking, coupled with widespread access to the Internet, including on mobile phones, have made it easier for ordinary citizens across the world to acquire and use various publishing tools. These developments have precipitated a disruptive shift in how, and by whom, global information flow is controlled (Bruns 2018). However, while citizens play an important part in content creation and dissemination, they do so primarily on platforms controlled by a small number of giant corporations that absorb any emerging threat through adoption or acquisition. Such platforms are paradoxically the reason citizen journalists transcend communities and borders, interconnecting with audiences in new and innovative ways, and at the same time these platforms provide a single point of failure should those citizen journalists fall out of favour with corporate or authoritarian power. Regulatory intervention and political leadership in the interests of citizens' rights is needed to diversify platform ownership and influence, in order to protect fundamental freedoms of speech in the future.

See also: amateur; authenticity; community media; convergence; hyperlocal media; Indymedia; journalism studies and citizen media; surveillance; wikis; witnessing/testimony; World Social Forum

Recommended reading

Allan, S. (2013) *Citizen Witnessing*, Cambridge: Polity Press.

Develops the idea of citizen witnessing as a way of describing spontaneous acts from bystanders or people caught up in events, who feel compelled to document and disseminate information about what they witness. Conceptually rich and important for anyone concerned with understanding the journalistic nature of citizen journalism.

Allan, S. and E. Thorsen (eds) (2009) *Citizen Journalism: Global perspectives*, Volume 1, New York: Peter Lang.
Thorsen, E. and S. Allan (eds) (2014) *Citizen Journalism: Global perspectives*, Volume 2, New York: Peter Lang.

These two volumes provide detailed and informative case studies of citizen journalism from every continent in the world, with an emphasis on crises, disasters and conflict. They constitute key texts for anyone seeking to understand citizen journalism in a wide range of national contexts, and where and how it has emerged in response to crisis events.

Bruns, A. (2018) *Gatewatching and News Curation: Journalism, social media, and the public sphere*, New York: Peter Lang.

Explores the current social media environment and how the normalization of networked news sharing is influencing both curation and our understanding of gatekeeping. As such, the book argues we are seeing a shift from citizen journalism in networked spaces influencing professional journalism to the latter extending its domain into social media.

CITIZEN SCIENCE

Gwen Ottinger

Citizen science refers to a phenomenon whereby individuals without formal credentials in science participate in activities typically understood as integral to the creation of scientific knowledge. These include conducting observations of the natural world to contribute to a data set to be analysed by credentialled scientists, as in the Cornell Lab of Ornithology's Project Feederwatch (Bonter 2012); using scientific instruments to monitor air pollution, as do 'bucket brigades' common in the environmental justice movement (Ottinger 2010); and detecting patterns or classifying objects in images, as Galaxy Zoo volunteers do to help scientists understand how galaxies formed (Clery 2011).

Understanding citizen science as citizen media requires, first, understanding science as a medium of expression and representation and, second, distinguishing citizen science activities which merely harness volunteer labour from those which express a critical consciousness about science and scientific institutions. This entry begins with a brief overview of the social constructivist account of science and how it permits viewing science as media. It then traces one genealogy of the term citizen science, showing how citizen science as defined by Irwin (1995) exemplifies citizen media as conceptualized in Baker and Blaagaard (2016a). Using the example of environmental justice-oriented bucket brigades, key characteristics of citizen science projects that can be meaningfully understood as citizen media are highlighted. An alternative genealogy, which sees citizen science as an extension of amateur naturalism (Miller-Rushing et al. 2012), is then discussed. The latter account is embraced by natural science-led efforts to institutionalize citizen science, and it contributes to the misrecognition of the independent, oppositional and expressive nature of grassroots-led citizen science. The entry concludes by suggesting that citizen science is subject to the same forces of appropriation as other forms of citizen media.

Science, citizen science and citizen media

Science and media are typically thought of as belonging to separate spheres. In common parlance, science is a set of facts and a method for uncovering them, and the media are implicated only insofar as scientists hope that reporters will accurately convey these facts to their audiences. However, social scientists in the field of science and technology studies (STS) have advanced a rather different picture of science. Science is best understood, they demonstrate, as a diverse set of practices through which people, situated in specific cultural contexts, aim to represent the natural world. Facts are an accomplishment that requires scientists to make value-laden judgements and engage in social and political work (Elliott, K. C. 2011; Latour 1987).

To say that science is socially constructed, as STS scholars do, is not to say that there is no underlying reality to which science corresponds (Latour 1999). The point is rather that science is not a transparent window on nature, but a more-or-less faithful representation of it that also expresses the values and concerns of its place and time. Scientific practices and their resultant facts can thus be considered media in their own right, in that they represent the natural world and express human relationships to it.

One source cited as the origin of the term citizen science is sociologist Alan Irwin's book of the same title. Irwin chose the phrase, he writes, to evoke "a science that assists the needs and concerns of citizens" as well as "a form of science developed and enacted by citizens themselves", often "outside of formal scientific institutions" (1995:xi). This independence with respect to mainstream institutions is one way in which his characterization of citizen science resembles Baker and Blaagaard's (2016a) account of citizen media. Like citizen media, Irwin's citizen science critiques mainstream scientific institutions, expresses knowledge and questions not represented by those institutions, and often seeks structural change. Specifically, Irwin argues that citizen science is a response to the contradictory position of science in the risk society. On the one hand, mainstream science and technology have contributed to contemporary environmental crises by failing to fully consider the impact of, for example, new chemicals or modified organisms on complex systems of human and more-than-human life; at the same time, dealing with such crises requires scientific knowledge and technological intervention. The route to environmental sustainability, then, involves reforming science and scientific institutions to ensure they become responsive to social and environmental concerns, to include non-scientists as participants in inquiry, and to be reflexive about science's limitations.

Irwin's book and the new term, citizen science, comprised a diverse set of phenomena that exemplified this call for reform of scientific institutions. It acknowledged and capitalized on residential communities' sophisticated but unrecognized contextual knowledge of everyday environmental hazards. It included broad participation in public discourse about risk and in deliberations on science policy questions. And it enabled active, direct involvement of non-scientists in scientific inquiry on issues of particular concern. In the last category, Irwin included science shops – university-based entities that worked with communities on projects that others might have called participatory action research or community-based participatory research – and health studies initiated by concerned citizens, dubbed popular epidemiology by Phil Brown (1992). Others would come to add participatory mapping (Allen 2000; Corburn 2005) and participatory environmental monitoring (Hemmi and Graham 2014; Ottinger 2010) to the catalogue of community-led scientific activities with a critical stance towards mainstream science and technology. Alternatives to the umbrella term citizen science, including street science (Corburn 2005) and civic technoscience (Wylie et al. 2014), have also been advanced. While none have been taken up as widely as citizen science, these names tend to underscore the independent, populist, change-oriented nature of the activities they describe – all features they share with citizen media as understood here.

One classic example of citizen science in the spirit of citizen media is the bucket brigade, a strategy for quantifying air pollution that has been used in grassroots environmental campaigns all over the world. Buckets are homemade air samplers that people living near toxic gas-emitting industries such as chemical plants and fracking sites use to collect ambient air in a non-reactive plastic bag. They then send the bag to an analytical laboratory, which determines from the sample what chemicals were in the air at the time and place the sample was taken. The *brigade* in bucket brigade refers to the mobilized collectives of community members that surround the air monitoring. Buckets are used in areas where residents have already noticed foul odours from industrial emissions, and where they have organized to demand that the local facility reduce its emissions and/or that environmental regulators take action against the facility. A few residents of the area will have buckets; others will be encouraged to keep pollution logs and alert those with buckets to unusually bad smells.

Bucket sampling is *ad hoc*. Residents only take samples when odours reach a seven on a scale of one to ten, and even very active bucket brigades seldom take more than a couple of samples per month. For instance, at the height of a campaign for relocation away from an oil refinery in southeastern Louisiana, in the US Gulf Coast region, Concerned Citizens of New Sarpy took six samples over the course of a year (August 2000 to August 2001). Similarly, a multi-sited, community-led study of local impacts of unconventional oil and gas (UOG) drilling in six US states included only thirty-five grab samples (Breech et al. 2014; Macey et al. 2014).

Even with relatively few data points, bucket brigades use their results to simultaneously represent environmental conditions and make claims about environmental responsibility. In a report titled *Land Sharks* (Louisiana Bucket Brigade 2001), Concerned Citizens of New Sarpy organized data from their six samples by chemical: under headings that included Hydrogen Sulphide and Benzene, they listed the health effects associated with the chemical, the Texas and Louisiana screening levels for the chemical (in three of the four cases, it was noted that Louisiana had none), and the dates and levels at which the chemical was detected. This arrangement of the table made two important points: residents of New Sarpy were repeatedly subjected to high levels of chemicals known to have health effects, and they were not being adequately protected from chemical intrusions by their government, which had failed to set limits on most of the chemicals. The accompanying text made the additional point that residents could not count on refinery officials to provide accurate information about the chemical concentrations to which residents might be exposed. *Warning Signs*, the report on the six-state study of UOG operations (Breech et al. 2014), made similar points. It used sampling data to highlight the presence of chemicals and the dangers to human health associated with them, then featured residents' firsthand accounts of lives changed and illnesses precipitated by new fracking sites. It underscored the paucity of monitoring and enforcement around UOG development. And it advanced an alternative vision for environmental regulation geared toward precaution and conservation.

Bucket brigades thus exemplify citizen science, and citizen media, in that they simultaneously represent the natural world by documenting levels of chemicals in the air and express critiques of mainstream institutions, which are implied to be lax about monitoring air quality *and* lax about protecting public health. Taking bucket samples is an act of cultural production in its own right: the legitimacy and the determination of marginalized communities to have their knowledge represented is expressed through photos of people one would not assume to be scientists – often people of colour, often women, often past retirement age – wielding air samplers. Indeed, *Warning Signs* explicitly calls for citizens to be included as partners in environmental protection, asserting that "[c]ommunity monitoring can be a powerful tool for assessing potential risks" and that "it should inform the action of regulators to better protect public health" (Breech et al. 2014:28).

Characteristics of citizen science as citizen media

Bucket brigades are but one example of the community-led projects that produce and engage scientific data to advance not only alternative representations of nature but also alternative accounts of government responsibility, institutional failures and sustainable futures. Although these citizen science projects differ in their particulars – indeed, they are usually highly tailored to local contexts – they share several key characteristics that resonate with those of citizen media.

First, most of these citizen science projects originate in social movements, especially environmental and health social movements. They share the movements' interest in challenging the authority structure of science (Brown and Zavestocki 2004) and see change as coming from collective action, rather than from the production of better facts (Ottinger 2017). Accordingly, Ottinger (ibid.) refers to citizen science initiatives like the bucket brigades and others that might be considered forms of citizen media as social movement-based citizen science.

Like other forms of citizen media, social movement-based citizen science is typically conducted outside mainstream institutions, although it takes place in a variety of other social formations (Baker and Blaagaard 2016a). Most citizen scientists are members of grassroots groups, ranging in formality from loose collectivities, such as Concerned Citizens of New Sarpy, to neighbourhood associations that have formally incorporated as nonprofit organizations, such as the West End Revitalization Association (Heaney et al. 2007). Further, grassroots groups usually find support for their citizen science projects from larger nonprofits like the Louisiana Bucket Brigade and WE ACT, which itself began as a neighbourhood organization. They may also find support from university-based scientists (Frickel 2004). Social movement-based citizen scientists often struggle to create equitable partnerships between grassroots groups and credentialled professionals (Cable et al. 2005), but some groups have developed models to keep the citizen or community perspective at the forefront (Heaney et al. 2007).

Finally, social movement-based citizen science often does not replicate mainstream scientific methods. Instead, it appropriates and adapts scientists' methods, instruments and interpretive techniques to better represent communities' environmental concerns and change the terms on which responsibility for them is assigned. Bucket brigades, for example, use a sampling technique that is well understood by scientists, and samples are analysed using a technique that is widely recognized and accepted by regulators. But whereas regulators typically take 24-hour samples, bucket brigades take 3–5 minute samples and then compare them to long-term health screening levels. While many scientists argue that this constitutes an error in that it compares incommensurate quantities, citizen scientists make the comparison strategically, to render visible the spikes in pollution that get averaged out in 24-hour samples, assert that the spikes have consequences for public health, and demonstrate a more precautionary way of approaching ambient air data (Ottinger 2010). Similarly, in popular epidemiology, communities adopt health survey instruments modelled on those used by professional epidemiologists but allow trained community members to administer them to fellow residents, challenging the notion that reliable information can only be collected by so-called unbiased surveyors; they also reject scientists' standards for statistical significance, arguing for alternatives that are less accepting of false negatives – for instance, not seeing a correlation between pollution and ill health that does in fact exist – and better suited to small, vulnerable populations (Brown 1992; Allen 2000).

Using alternative methods, social movement-based citizen scientists take pains to establish the credibility of their data to sceptical audiences. Organizations that use buckets, for example, frequently claim it to be a method approved by the US Environmental Protection Agency, on the basis of a quality assurance study early in the device's history. But they simultaneously tailor their methods to represent the issues of most concern to them, often leading to deviations from scientists' norms.

Institutionalizing citizen science

At the same time the term citizen science was coming to be used to describe grassroots activities that challenged institutionalized science, scientific institutions were beginning to use it to characterize the contributions that volunteers made to their research. Cooper and Lewenstein (2016) credit the earliest such uses to the Audubon Society in 1989 and to Rick Bonney of the Cornell Lab of Ornithology in the mid-1990s. Both referred to ordinary people making observations – of rainfall and of birds, respectively – and sending them in to the relevant organizations, where they became part of scientists' data sets. In this use of the term, citizen science builds on a long history of amateur naturalism (Miller-Rushing et al. 2012), and proponents argue for its legitimacy in terms of its adherence to the norms of scientific institutions. Dickinson and Bonney (2012) stress that the goal of citizen science, thus understood, is to contribute to peer-reviewed scientific papers that advance the state of knowledge. Others have demonstrated the prevalence of citizen science contributions in published studies (Cooper et al. 2014) and defended the ability of volunteers to produce data of the same quality as trained scientists (Cohn 2008; Cox et al. 2012).

Researchers characterizing the phenomenon of citizen science in the natural sciences have created a number of taxonomies to capture the diversity of citizen science projects. The most prominent of these focus on the extent to which non-scientists are involved in the research process (Shirk et al. 2012): those to which they merely submit data are called contributory, while those in which they play a role in the design of the research are dubbed co-created. These taxonomies, especially the co-created designation, have come to be used to encompass grassroots projects like the bucket brigades without acknowledging their critical stance on science-as-usual (Ottinger 2017).

This model of citizen science has been institutionalized. There are now several national and international citizen science associations, as well as at least one peer-reviewed journal devoted to citizen science in this institutionalized sense. Citizen science has also been embraced by existing institutions, including the European Commission, which sponsored an effort to increase public participation in scientific research, and the United States Environmental Protection Agency, which charged one of its major advisory boards with laying out a "Vision for Citizen Science at EPA" (National Advisory Council for Environmental Policy and Technology 2016). The taxonomy described above has been influential in structuring these efforts, and projects ranging from contributory to co-created are expressly included.

Efforts to institutionalize citizen science arise from developments within the sciences, yet they subsume a tradition of citizen science that has historically operated outside scientific institutions and called for their reform. Social movement-based citizen science has thus been appropriated by mainstream scientific institutions, echoing dynamics of appropriation in other citizen media realms. In this case, scientific institutions acknowledge the potential of social movement-based citizen science to contribute to the advancement of knowledge but (wilfully, perhaps) misunderstand its critique of science or its orientation to social change (Ottinger 2017). Some have argued that other forms of appropriation attend contributory citizen science, namely the appropriation of volunteer labour and public-spiritedness to enhance corporate profits and power (Mirowski 2017; Woolley et al. 2016).

Future directions

For citizen scientists engaged in creating alternative representations of the environment and expressing alternative visions of responsibility – that is, for citizen scientists practising citizen media – the appropriation of their endeavour by mainstream scientific institutions is double-edged. On the one hand, residents of marginalized communities are increasingly recognized as potential contributors to knowledge, after many years of seeing their local knowledge dismissed (Wynne 1996). On the other hand, this recognition may come at the cost of scientists reasserting authority over community investigations, as in the case of bucket brigades that have been persuaded to trade their buckets for the summa canister samplers that scientists favour, or that have invested extensive time and energy into creating formal quality assurance plans to satisfy regulatory audiences. It also creates the threat that social movement groups will be diverted onto a data treadmill that obscures their transformative visions for the future (Shapiro et al. 2017; Clapp et al. 2016; Kimura and Kinchy 2016). It thus remains to be seen whether social movement groups will continue to find citizen science valuable or whether they will increasingly turn to other forms of citizen media in pursuing their social and environmental justice goals.

See also: amateur; co-optation

Recommended reading

Irwin, A. (1995) *Citizen Science: A study of people, expertise, and sustainable development*, London: Routledge.

A seminal text on the limitations of mainstream science and ordinary people's responses to them.

Kimura, A. H. and A. Kinchy (2016) 'Citizen Science: Probing the virtues and contexts of participatory research', *Engaging Science, Technology, and Society* 2: 331–361.

Reviews the rationales given for citizen science and identifies several major dilemmas facing citizen scientists.

Ottinger, G. (2017) 'Reconstructing or Reproducing? Scientific authority and models of change in two traditions of citizen science', in D. Tyfield, R. Lave, S. Randalls and C. Thorpe (eds) *The Routledge Handbook of the Political Economy of Science*, London: Routledge, 351–363.

Distinguishes between social movement-based citizen science and scientific authority-driven citizen science on the basis of their embedded models of how social and political change are made.

Shapiro, N., N. Zakariya and J. A. Roberts (2017) 'A Wary Alliance: From enumerating the environment to inviting apprehension', *Engaging Science, Technology, and Society* 3: 575–602.

Theorizes several ways in which citizen science may be entrenching the very problems it tries to solve, and highlights alternative modes of inquiry into environmental degradation and change that escape the strictures of quantification.

CITIZENSHIP

Engin Isin

The concept of citizenship has become a major site of struggle since 1989, reflecting many of the profound social and political transformations that have taken place across the world following the fall of the Berlin Wall. The concept has of course been a site of struggle during earlier periods in history, such as between the two world wars when the rise of ethno-nationalism rendered millions of people without citizenship and significantly limited the civil and political rights of countless others. Yet, since 1989, many of the rights that constitute democratic citizenship have come under pressure from neoliberalism (Turner 2001). Many scholars have studied the relationships between these transformations and changes in the concept of citizenship itself. Some, like Smith (2003) and Joppke (2010), have insisted on a narrow conception of citizenship as strictly membership of a state (a legal status), while others such as Kymlicka (1995) and Young (1989) have widened the concept to include practices of belonging, identification and struggle. Thus, variations of the concept such as multicultural citizenship, sexual citizenship, transnational citizenship and digital citizenship have entered circulation. Other variations such as media citizenship and cultural citizenship are also now used. These variations create tensions when legal citizenship can have a precise meaning, dictating who may or may not act under certain capacities within a certain state, while in other senses seemingly anyone can act like a citizen. This might well point to an inherent if productive tension in the concept itself. This entry thus outlines conventional, critical and performative approaches to understanding citizenship, acknowledging that different senses of the term need to be continually encouraged, while also recognizing that it may need a relatively durable description. This entry suggests that citizenship as a concept functions under four different senses: citizenship (in) theory, citizenship (in) practice, citizenship (in) law, and citizenship (in) acts. There are always gaps and tensions between these senses, and the performativity of citizenship takes place in these gaps and through these tensions. For understanding citizen media, we especially focus on the study of performative citizenship (in) acts.

Approaching citizenship

Contestation over the meanings and functions of citizenship has already indelibly marked the twenty-first century. Questions of civil rights, political rights or social rights have repeatedly been raised during struggles over class, race, gender, sexuality, access, ecology and culture. The rights of both documented and undocumented migrants and refugees to move across international borders have also been increasingly claimed, as have the rights of indigenous peoples to land, labour and language. Whether the rights of citizens and non-citizens are protected or guaranteed by national, international, supranational or human rights laws, and indeed whether these rights should be conceived as citizenship rights at all, are among the basic political questions for our century. For example, the UK's membership of the European Union since 1973 enabled freedom of movement for European citizens both in and out of the UK at a historically unprecedented scale. By 2019, many citizens of other European countries had come to reside in the UK and many UK citizens had chosen to reside in other European countries by performing their right to free movement. When the UK notified the EU of its intention to leave the EU in March 2017, the rights of these European citizens became the focus of heated political debate. Would there be a mutual recognition of the rights of citizens between the EU and UK? Would these citizens who exercised their freedom of movement as European citizens maintain their rights to work and residence in their host country? Would the meaning of European citizenship change? Still, these questions barely engaged with the complexity of the issues when we consider the lived experience of supranational integration that freedom of movement had brought about in the lives of European citizens by that point in time: cross-border working arrangements, geographically spread generations, homes where multiple languages are spoken, dispersed property and other inheritances, fragmented health and other public services, and radically altered conceptions of self and home. Whatever the legal arrangements made as part of the UK's withdrawal agreement, these complex entanglements would continue to shape both UK and EU citizenship for decades to come.

Perhaps this is not exceptional. Tully (2014:3) does not exaggerate when he writes that "many of the central and most enduring struggles in the history of politics have taken place in and over the language of citizenship and the activities and institutions into which it is woven". Yet, it is difficult to avoid the conclusion that the twenty-first century has so far been a particularly intense period of struggles over citizenship and that this must surely be linked to the ongoing reconfiguration of the empires, states, nations, territories, sovereignties and arrangements that have long played a key role in determining the meanings and functions of citizenship around the world.

Considering these developments, a diverse body of scholarship on citizenship has emerged, and it would be a folly to seek to confine this research within all too rigid a categorization. Nonetheless, as we witness and participate in these debates, it is indeed possible to identify, in broad strokes, two approaches to understanding citizenship that have emerged. This entry will designate these as conventional and critical approaches. This is not to argue that the latter displaces the former but to bring into sharper relief some differences that have appeared in studies on citizenship, partly because of the complexities of citizenship in this century and partly because of developments in theorizing citizenship.

Conventional approaches typically describe citizenship in terms of rights, obligations and membership of the nation-state, and the nature of these state–citizen relationships remain the central focus of study. Three rights (civil, political and social) and three obligations (conscription, taxation and franchise) often characterize this understanding of citizenship (Marshall, T. H. 1992). Civil rights include the right to free speech, to conscience and to dignity; political rights include voting and standing for office; and social rights include unemployment insurance, universal health care and welfare provisions. All these rights are predicated on a fundamental right of equality before the law regardless of belief, background or origin that governs the relationship between citizens and their states (Tilly 1997). Conventional approaches to citizenship generally draw our attention to the fact that the strength of these rights has been declining in Euro-American states since 1989 and that the future of civil, political and social rights remains contested and uncertain (Marback and Kruman 2015). As for a citizen's obligations, while compulsory conscription into a national service is rapidly disappearing around the world, taxation is still fundamental; the sense of duty associated with voting, although also declining, especially among the youth, is still a vital aspect of citizenship politics.

Critical approaches to citizenship, on the other hand, make three basic interventions which distinguish them from conventional approaches. First, critical approaches typically recognize a much wider collection of rights, such as sexual rights, cultural rights and environmental rights, and emphasize struggles over their institutionalization, as in ongoing efforts to legalize same-sex marriage in many countries around the world (Somers 2008; Turner 2001). The focus on struggles over rights maintains a more dynamic sense of citizenship than in conventional approaches, and critical approaches emphasize the extent to which rights may be expanded and eroded as a result of such struggles. Second, critical approaches also recognize that increasingly – whether conventional (civil, political, social) or expanded (cultural, economic, regional, environmental, sexual, transnational and urban) – rights and obligations are negotiated through supranational and international institutions such as the United Nations (for example, through the Universal Declaration of Human Rights), the Council of Europe (through the European Court of Human Rights) and the European Union (through the European Court of Justice). This critical understanding of citizenship similarly acknowledges the importance of devolved institutions such as regional parliaments (e.g. the Quebec, Scottish or Catalan parliaments) and minority or indigenous peoples (as in calls for the application of Sharia law) that question the assumption that citizenship is membership in only a nation-state (McNevin 2011; Ní Mhurchú 2014; Nyers 2003, 2008; Rygiel 2010; Squire 2009). Critical approaches thus emphasize broader repertoires of rights that people draw upon in their efforts to expand or restrict citizenship rights (Soysal 1994). These struggles over rights are no longer contained (or, for that matter, are containable) in isolated sovereign jurisdictions (Simpson 2014). Critical approaches recognize that such struggles are fought out through a multiplicity of legal and political orders and borders, and that state–citizen relations are mediated through these multiple legal and political forms. For critical approaches, citizenship is a matter as much for international law as national law, and in fact the tensions between the two are often objects of struggle.

Third, critical approaches take a broader historical perspective on the emergence and transformation of citizenship as an institution of population management and political subjectivity. For example, critical approaches have examined imperial and colonial histories, and investigated how making up peoples as racialized, gendered, classed subjects created hierarchies of belonging and membership between citizens and non-citizens (Bloom 2017). Critical approaches have also explored the attempts made by indigenous and minoritized peoples to seek an expanded range of rights and protections beyond those they have been ascribed by imperial and colonial authorities (Banerjee 2010; Cooper 2014; Gorman 2006). These approaches emphasize the tensions between making and becoming a citizen, as well as between colonizing and decolonizing citizens (Balibar 2012/2015; Tully 2009). They thus have more affinity with the concept of citizen media adopted here than traditional approaches, principally for the way they enable a conception of the citizen as an agent of change and contestation.

Taken together, these three forms of intervention on the part of critical approaches widen the focus of debates on citizenship from discussions of an already instituted arrangement towards explorations of minoritized, subaltern and indigenous peoples as political subjects at the centre of struggles for rights, and as groups who often question, if not break, the conventions imposed on them. Nevertheless, both conventional and critical approaches have taken for granted a sovereign conception of citizenship in the sense that citizens – as opposed to non-citizens – appear as subjects of politics already authorized by the state to act as people making rights claims. Yet, this taken-for-granted assumption creates an intractable problem that is evident in any attempt to provide a presumably incontestable descriptive statement such as "citizenship is …". The trouble with such descriptive statements is that they already embody citizenship's contested histories and that any descriptive statement will always fall short of the complexities of that history. Combined with the complexity of the history of many postcolonial states that inherited various regimes from their imperial rulers, this demonstrates how difficult it is to begin with a descriptive statement when it turns out to be an ascriptive or prescriptive statement. This is, of course, the classical problem that J. L. Austin identified with all performative statements (Austin 1962, 1970). It was Austin who questioned whether it was possible to speak (or write) at all without being ascriptive and prescriptive, even if one is being descriptive (Loxley 2007). And it was Judith Butler who worked through a politics of the performative speech (or writing) that brings its objects into being (1997) and performs different senses of the subject that it produces (2015b).

For these reasons, rather than aiming to describe or prescribe citizenship, scholars now increasingly characterize citizenship as a contested institution mediating rights between the subjects of politics and the polities to which these subjects belong or with which they identify or affiliate. This minimalist characterization creates a space for thinking critically and performatively about the struggles over the meanings and functions of citizenship (Isin 2017).

The four senses of citizenship

Questions remain as to how we might best approach citizenship as it is performed both inside and outside Euro-America, and often across borders. Tully (2014:8–9) suggests that approaching citizenship critically means recognizing citizenship as a multiplicity of practices anywhere in the world. This is because, for Tully (2014:9), "citizenship is not a status given by the institutions of the modern constitutional state and international law, but negotiated practices in which one becomes a citizen through participation". In other words, we need to identify the multiplicity of negotiated practices through which people perform or enact themselves as citizens. This suggests that approaching citizenship critically involves recognizing that citizenship (as an institution mediating rights between the subjects of politics and the polities to which these subjects belong or with which they identify or affiliate) exists in several distinct, irreducible yet overlapping senses. We must additionally recognize that, rather than approaching citizenship as always already given, we need to approach it through its performative senses. Four such senses can be distinguished: (1) citizenship (in) theory, where citizenship is imagined, expressed and contested; (2) citizenship (in) practice, where citizenship is inhabited through rituals, routines and norms in everyday lives; (3) citizenship (in) law, where citizenship is codified, organized, collated and arranged; (4) citizenship (in) acts, where citizenship is cited, iterated, signified and interrupted by people taking liberties. With these four senses we might expand from conventional and critical approaches to citizenship to performative approaches that focus on the senses of the political subject as a citizen. The importance of recognizing the multiplicity of the senses of citizenship through these forms is borne out by the fact that these senses are often discordant and incongruous: there are inevitable tensions between them. But it is these tensions that reveal the urgent and pressing issues for citizenship as an object of struggle.

To begin with citizenship (in) theory, we need to recognize that citizenship exists in statements (ideas, interpretations, explanations, contentions) about what citizenship is, has been, and ought to be; in statements that produce different ideals and images of citizenship (in) theory. Citizenship (in) theory also sometimes falls short of and sometimes exceeds what transpires in citizenship (in) practice and citizenship (in) law. Yet, citizenship (in) theory nevertheless performs a significant function in maintaining a relatively enduring sense of citizenship, even in the case of contentious theories such as liberal, republican or communitarian accounts of citizenship. These descriptions are performative in the sense that they participate in shaping the ascriptive and prescriptive senses in which citizenship exists. The scholarly literature on citizenship is only part of citizenship (in) theory: there are also broader political discourses that involve statements by activists, artists, authorities, bureaucrats, intellectuals, journalists, politicians, professionals and all opinion makers who theorize citizenship in multiple ways. Citizenship (in) theory is neither reducible to nor can exhaust citizenship (in) law or citizenship (in) practice. Approaching citizenship (in) theory by developing conceptual, analytical and methodological tools with which to understand citizenship (in) law and citizenship (in) practice also means simultaneously performing citizenship. There will always be disagreement among scholars as to the precise formulations of citizenship in both national and international laws and practices, and this is largely due to the contentious and contested nature of citizenship (Brandzel 2016). Yet citizenship (in) law is always contested by citizenship (in) theory and citizenship (in) practice. Indeed, the gaps between law, theory and practices are sites of struggle over citizenship for determining who are regular or irregular, who are legal or illegal subjects of law. There are also sites of struggle over enforcement of existing laws and over changing or reforming them. Therefore citizenship (in) acts can be considered a key aspect of citizenship.

To say that citizenship also exists in acts may be stating the obvious. After all, citizenship (in) law means something transacted in a deliberative assembly and signifies the name of a decree or statute passed by a legislative body, such as a court of justice. Often, citizenship (in) law comes into being through such legislative acts concerning citizenship and nationality. Moreover, these legislative acts do not cover all acts of citizenship (in) law because there are other related laws and regulations that affect citizenship rights and obligations, such as migration law, labour law and family law. It is therefore possible that when we use the term acts of citizenship we may well be referring to citizenship (in) law. Yet, acts also have a performative meaning to signify a deed done or an action taken. Through such deeds or actions people as collectives enact their political subjectivity and constitute themselves as claimants of rights. The repertoires are those familiar to scholars of citizen media, and can be as diverse as singing, dramatizing, occupying, contesting, mocking, striking, satirizing and many more activities through which people could be claiming a right – such as that of residence, work or asylum – that they do not have, campaigning for a right that they have lost (such as the right to social care or education) or protesting a right accorded to others and that should not exist, as in the case of corporate offshore tax havens. Citizenship (in) acts symbolizes the performative politics of citizenship: in and by making rights claims people constitute themselves as citizens, sometimes without regard for any convention that might exist to authorize their act. Thus, citizenship (in) acts introduces a rupture into both citizenship (in) law, by transgressing or subverting it, and citizenship (in) practice, by breaking instituted conventions.

This key aspect of citizenship has been the subject of a growing literature that illustrates how people perform as citizens regardless of whether they are recognized by law as citizens or non-citizens. Aradau et al. (2010), for example, study sex workers as European citizens affirming their rights to mobility and free movement in Europe. Their argument is that a political subject called sex worker is brought into being by calling into question how both law and language treats members of this group as victims. By acting as political subjects sex workers constituted themselves as both subjects of law and subjects of politics – a subjectivity that is denied to them as victims. Barbero (2012) shows how people without papers have performed themselves as citizens by claiming rights to remain in Spain. Barbero does not claim that that people without papers are legal citizens but that, by refusing to be treated as illegal subjects and constituting themselves as political subjects, they acted as citizens with claims to rights. Bassel (2008) also shows how refugee women assert their rights to remain in France as French citizens (see also Bassel and Lloyd 2011). The argument here is not that refugee women are legal citizens but that by performing political subjectivity they constitute themselves in public space as visible and recognizable claimants of rights and justice. In the US, Lee (2008) shows how undocumented non-citizens collaborate with citizens to make what he calls subversive rights claims. Like Lee, Darling and Squire (2013) illustrate how citizens and non-citizens through acts of solidarity effectively defend the right of asylum seekers in sanctuary cities in the UK. These examples illustrate how so-called irregular migrants have been transformed from subjects of law to subjects of politics and to political subjects, thereby transforming categories imposed on them by state administrative practices into categories of affirmation. Many indigenous peoples find themselves subject to similar administrative practices although, unlike migrants and refugees, their displacement comes from borders that move around them. Marino (2015) shows, for example, how the rights of Adivasi peoples in India were brought into being through literature or acts of writing, effectively resisting their dispossession and displacement.

There are also examples of acts of citizenship undertaken by those who have the legal status of citizens but who occupy a subaltern status within a political order, as in the case of Roma citizens who already have (or should have) the rights of European citizenship. Yet, as Aradau et al. (2013), Çağlar (2016) and Vermeersch (2014) illustrate, it takes massive collective action to make that simple point to authorities who regard Roma citizens as strangers or outsiders. As Wasserman (2015) reminds us, sometimes a tragic event such the massacre of protesting miners in 2012 in Marikana, South Africa, can be interpreted as an act of citizenship for a democracy to come. Wasserman argues that the mainstream media's response to the massacre ignored the fact that miners were responding to democratic deficit in their everyday lives and demanding democratic rights to shape the future of their workplace (ibid.:375). McCosker and Johns (2014) similarly show how using social media can transform people from bystanders into witnesses who expose racism in everyday situations. Performing citizenship can also literally mean enacting political theatre on streets to draw attention to injustices of border regimes. Lewicki (2017:275) exemplifies this by documenting an emancipatory theatre collective in Germany, the Centre for Political Beauty, which staged various performances by digging symbolic graveyards and announcing that "the dead are coming".

This growing literature on performing or enacting citizenship suggests that citizenship (in) acts always occurs in the gap between citizenship (in) law and citizenship (in) practice, or between citizenship (in) theory and citizenship (in) practice. Citizenship (in) acts is irreducible to citizenship (in) practice because acts involve rupturing normalized social routines. It is significant to recognize this element of rupture precisely because it gives us a glimpse of a gap between how citizenship (in) law regulates subjects of rights and how citizenship (in) practice shapes and transforms citizenship (in) law. In a way, citizenship (in) acts creates fissures, conflicts, contradictions or dilemmas which enable citizens (in) practice and (in) theory to become instituted through citizenship (in) law. This is important because, as Janoschka and Mateos (2015) and Lewicki and O'Toole (2017) have argued, what may appear as inconsequential acts of citizenship can force a rupture in seemingly impenetrable organizational arrangements or ideologies.

These senses of citizenship enable us to recognize that citizenship (in) law, citizenship (in) practice, citizenship (in) theory and citizenship (in) acts are essentially sites of political and social struggle through which people make claims involving belonging, identity, inclusion, exclusion, participation, influence and engagement. To put it differently, acts of citizenship involving non-citizens – unauthorized, marginalized and excluded – are not only about their rights but also about the rights through which both citizens and non-citizens negotiate living together (Bloemraad 2017). When Soysal (1994), for example, demonstrated that non-citizens in Europe were drawing on a much wider repertoire of rights, deriving from national and international laws to human rights laws, thus ushering in a new era of postnational citizenship, she was anticipating how performing citizenship was becoming a rights-claiming practice (Soysal 1997; Zivi 2012). We now recognize that these claims not only cover rights ranging from civil, political and social rights to labour, sexual, cultural, environmental, consumer and other rights but also involve wide collections of national, international and human rights laws.

Citizenship (in) theory is both a participant and observer in these struggles, and it is difficult to draw a sharp distinction between these two types of involvement. When we discuss citizenship from within the academy as scholars we are therefore presented with the paradoxes of being both observers and participants in these struggles over the meanings and functions of an institution that organizes, regulates and distributes rights in and across contemporary polities. To recognize this is especially important as it obligates us to ask ourselves as scholars what senses of citizenship we are performing.

Approaching citizenship through these four interrelated and interwoven senses of its existence enables us to interpret how people inhabit and transform citizenship in their lives, negotiating it through various practices. It enables us to analytically distinguish its multiple senses and identify the ways in which we inhabit them. There is no reason to assume that one or the other sense of citizenship should be primary. Instead, we recognize that the tensions between these senses are the sites of struggles over the meanings and functions of citizenship. We can approach citizenship from the perspective of any of its senses and address various tensions that they generate, while acknowledging that there is no fundamental difference among the various political subjects making rights claims, whether they are scholars, activists, refugees, migrants, experts or legislators. Performative conceptualizations of citizenship are thus especially productive when considering "the physical artefacts, digital content, performative interventions, practices and discursive formations of affective sociality that ordinary citizens produce as they participate in public life to effect aesthetic or socio-political change" (Baker and Blaagaard 2016a:16). They enable us to view such examples of citizen media as interventions that shape our understanding of ourselves and others as citizens – not only as members of a nation-state but as multiply situated political subjects making claims to rights as ways of creating liveable lives.

See also: migration studies and citizen media; performance studies and citizen media; political science and citizen media; postcolonial studies and citizen media; precarity

Recommended reading

Clarke, J., K. M. Coll, E. Dagnino and C. Neveu (2014) *Disputing Citizenship*, Bristol, UK: Policy Press.

Features sociological and anthropological studies of struggles and contestations over the meanings and functions of citizenship.

Hildebrandt, P., K. Evert, S. Peters, M. Schaub, K. Wildner and G. Ziemer (eds) (2018) *Performing Citizenship: Bodies, agencies, limitations*, New York: Springer.

Features numerous analyses of performative acts of citizenship conducted from a range of disciplinary perspectives.

Zivi, K. (2012) *Making Rights Claims: A practice of democratic citizenship*, Oxford: Oxford University Press.

Illustrates how making rights claims forms an indispensable part of the performative repertoire of democratic citizenship.

CIVIL DISOBEDIENCE

Ian Alan Paul

When attempting to delineate the conceptual contours of civil disobedience, its diverse and often inconsistent genealogies and applications present themselves as obstacles. While it is clear that the concept first coherently emerged in the middle of the nineteenth century, the fact that it has subsequently been mobilized across a wide range of geographic and historical contexts brings a great deal of complexity to any comprehensive understanding of the ways in which it has been understood and deployed. Rather than attempt to produce an exhaustive account of the rich history of civil disobedience as a political practice, this entry will focus on three aspects of the tradition that, for our purposes, are most generative: the role of media and mediation in civil disobedience; the relationship between civil disobedience, violence and the law; and the ways in which sociotechnological infrastructures continue to shape practices of civil disobedience today.

The advent of civil disobedience

Before turning to contemporary examples and issues, civil disobedience must be explored historically in order to better frame and contextualize the particular ways in which it is practised today. The concept first emerged in the nineteenth century in the short essay 'Resistance to Civil Government' written by Henry David Thoreau (1849), where he contemplates the responsibilities that citizens have to assume in order to hold their governments to account. Having been jailed for refusing to pay taxes in protest against American slavery and the imperialism of the Mexican–American War, Thoreau (ibid.:198) urged his reader: "[l]et your life be a counter-friction to stop the machine". His vision of civil disobedience, which has deeply influenced its enactment since, was defined principally by radical non-cooperation and non-participation in the "machine of government" (ibid.:197) as a means of putting an end to the injustices of the state. Informed by a libertarian perspective, Thoreau states at the very beginning of the essay that "government is best which governs not at all" (ibid.:189), positing an individual's conscience and agency as necessary counters to the excesses and abuses of state power.

In terms of historical examples, perhaps the best-known is the civil rights movement in the mid-twentieth century. Martin Luther King Jr., one of the movement's central and most influential figures, advocated for a definitively non-violent use of civil disobedience as a means of achieving racial justice in the United States, writing that people of conscience must "prepare for direct action" (1963:12) and "present our very bodies as a means of laying our case before the conscience of the local and the national community" (ibid.:12). King's framing of civil disobedience was largely influenced by his Christian faith, evoking

St Augustine and Christian resistance against the Romans to justify the breaking of unjust laws. Decades earlier, in the movement against the British colonial rule of India, Mahatma Gandhi incorporated aspects of civil disobedience into the collective practice of what he called *satyagraha* (insistence on truth), itself a distinct tradition of non-cooperation and protest. Inspired by contemporary resistance to colonialism in South Africa, Gandhi advocated non-violent resistance as a means of countering state violence; "so long as there is yet life in these our bones", he wrote, "we will never comply with your arbitrary laws" (1987:42). Both King and Gandhi can be credited with theorizing and advocating for versions of civil disobedience which were understood as tactics to be taken up by the collective masses who define social movements, representing a substantial departure from Thoreau's earlier individualist and libertarian framing.

The Egyptian Revolution of 1919 against British colonial rule is often cited as one of the first popular manifestations of civil disobedience, as a series of mass demonstrations and general strikes brought the country to a standstill (Botman 1991). Resistance against the Apartheid regime in South Africa made use of civil disobedience as one of its primary tactics (Kuper 1957); civil disobedience occurred against the Nazi occupation of Europe (Buettner 2016); and since the turn of the century, during the first Palestinian Intifada, civil disobedience was widely practised in the form of boycotts, strikes and refusal to pay taxes (Lochman and Beinin 1989). All of these examples shared a commitment to non-violence, non-cooperation and disruption as forms of resistance, although what precisely was meant by any of these terms varied considerably according to the historical and geographic context in each case. As a result, any theorization of civil disobedience requires an attentiveness to the various specificities it involves, including but not limited to the way various movements have come to particular stances in relation to resistance, the law and the role of violence.

Civil disobedience, violence and the law

While the central tenets of civil disobedience remain clear and consistent in the abstract, in practice they take on a considerable deal of complexity. In order to parse the different interpretations of the principles that constitute civil disobedience and set it apart from other forms of resistance, the relationship between civil disobedience and three related concepts must be addressed in turn: non-violence, the law and, as a consequence of these first two, the practice of direct action. While non-violence has been mobilized in a variety of contexts, its definitions vary as widely as definitions of violence (Milligan 2013). For example, Gandhi differentiated between *satyagraha* and *duragraha* (prejudgement): the latter occasionally involved property destruction, which Gandhi himself did not consider to be a form of violence. In contrast, when King (1989) reflected on the urban riots of the 1960s in the United States, he considered the destruction of property to be a form of violence, although qualitatively distinct from violence practised against people. Without venturing too deeply into debates about the different ontologies of violence, it should be noted that the distinction between making political appeals and violent political coercion becomes important when differentiating between civil disobedience and other practices of resistance, which is why the notion of non-violence remains so central to the concept.

As outlined in Thoreau's original definition of the concept, civil disobedience also defines itself prominently in relation to the law. While at times acts of civil disobedience cite an appeal to a higher law, as was the case with King and divine law, what all acts of civil disobedience have in common is their wilful violation of political laws which are considered unjust, even when they rely upon abstract appeals to ethical laws for their sense of justice. Arendt (1972:75), reflecting on the difference between revolutionary practice and civil disobedience, writes that "civil disobedience can be tuned to necessary and desirable change or to necessary and desirable preservation or restoration of the status quo" but that, in both cases, an appeal is made to the law and consequently to the state as the presuppositional condition of possibility for civil disobedience. This acceptance of the legitimacy of established authority stands in marked contrast, she suggests, with revolutionary or militant practices which refuse to recognize the law as such (1972:77). However, it must also be noted that Arendt (ibid.) cites Gandhi as a figure who complicates this distinction, noting that it cannot be said that he accepted British colonial law in India. Rawls (1999:322) also elaborates on this relationship between civil disobedience and the law in his work, noting that civil disobedience understands that "law is broken, but fidelity to law is expressed by the public and nonviolent nature of the act, by the willingness to accept the legal consequences of one's conduct". From this perspective, it is the act of knowingly violating the law while also submitting oneself to the law's judgement which defines the practice, again affirming the law as the space of politics and justice.

Finally, it is helpful to distinguish here between civil disobedience and what is often considered a related practice: direct action. While direct action and civil disobedience have often been used interchangeably by King and other central figures, Graeber (2009) argues that the two can be distinguished with reference to their publicity and their contrasting relationships with the law. First, he notes that while there can be a secret direct action, civil disobedience must always, by definition, be public. Second, he proposes that while civil disobedience appeals to the political system even while disobeying its laws, direct action operates under the premise that the law is an illegitimate framework for justifying or determining the ethical value of a political act. Singer has drawn a similar distinction, describing civil disobedience as illegal but as nonetheless an act that appeals to the majority, which in democratic political systems is conceptualized as an analogue of the state: "[c]ivil disobedience is an appropriate means to these ends when legal means have failed, because, although it is illegal, it does not threaten the majority or attempt to coerce them" (2011:269). In this sense, the civil disobedient's violation of the law comes to be understood as an act seeking to preserve or correct the law of the majority, a means of asserting the democratic will on laws which are undemocratic; direct action, on the other hand, violates the law not because of any investment in that system, but because the law has already been deemed wholly illegitimate.

In sum, the relationship between violence, the law and civil disobedience can be seen as hinging around two central axes: justice and politics. While civil disobedience remains defined by its commitment to non-violence, the exact nature of what is considered to be outside of violence is historical and contextual. Nonetheless, civil disobedience's commitment to non-violence can be likened to its commitment to politics as such or, in other words, to its belief in engaging non-coercively and democratically with other members of society as a means of transforming that society and its laws. Similarly, civil disobedience's commitment to violating the law in the name of justice, which at times has been thought of as a higher form of law, ultimately reinvests in the notion that law is the exclusive space of justice. Unlike direct action which does not recognize the law, and perhaps even rejects democratic politics as conventionally defined, civil disobedience approaches the breaking of laws as a corrective to the law itself, reasserting and shoring up the centrality of democratic majoritarian sovereignty.

Disobedient media and the mediation of disobedience

In order to better grasp the practice of civil disobedience in its contemporary forms, we must in some way also contend with the fact of its mediation, that is, with the ways in which the practice of civil disobedience relies upon and at times even explicitly takes place within various media. At stake in such a conceptualization is how the space of politics, and as a result how the public itself, is conceived of. In Thoreau's (1849) work, the act of civil disobedience was theorized as something that an individual citizen must undertake in relation to the state; only later did it become understood as a mass collective act. While the subject(s) of any act of civil disobedience may differ, at least equal attention must be paid to the intended audience of that disobedience, as well as to how different media shape the corresponding forms of interaction and address made in relation to those audiences. If the mode of address is defined by an appeal to the public majority of a democracy, then the various ways in which the public itself and publicity as such come to be mediated is unavoidably of paramount concern.

While Thoreau's, King's and Gandhi's theorizations of civil disobedience had unjust laws as the object of their disobedience, it was the public that was their intended audience. In the contemporary moment, increasingly it is media which more often than not defines the contest over who speaks to and who claims to speak for the public. Butler (2015a:19) argues that popular protests, as well as the repressive governments that they oppose, both seek to establish a claim to the "popular will"; this is where "the struggle over legitimation invariably takes place in the play between public enactments and media images, where state controlled spectacles do battle with cell phone and social networks to cover an event and its significance". Who is considered part of the civil disobedience movement, and to whom this civil disobedience is addressed, comes to be shaped by the forms of media that are mobilized around and within any given act.

While radio played a significant role in igniting the 1919 Revolution in Egypt (Fahmy 2011), it was television and social media which dominated the 2011 Revolution almost a full century later. Among the most famous and widely circulated images from the eighteen days of revolt were those of large crowds gathering on Cairo's bridges chanting *silmiya* (peaceful) while being sprayed with water cannons (Feigenbaum et al. 2013). This highly visual and spectacular form of civil disobedience, arguably more spectacular for the television cameras than for those on the bridge, took place during a revolutionary period that also saw hundreds of police stations burnt and intense street fighting between security forces and those who had occupied Tahrir Square. The chanting of *silmiya* in such a setting was less a description of the situation (which was anything but peaceful) than it was a claim being directed towards global publics, as well as against the violent actors of the Egyptian state. In another instance from the same period, a video of a man who has come to be known as the 'Egyptian Tank Man' after he blocked a police water cannon was circulated widely online as a prominently publicized act of civil disobedience (Westmoreland 2016). In both cases, the imagery was used to draw a distinction between the non-violence of the uprising and the violence of the state, while also laying claim to the popular will of the people.

These kinds of images emerging from the streets of Cairo also evoked comparisons with the civil rights movement, where television images of protesters being sprayed by fire hoses in Birmingham, Alabama generated widespread public outrage against the South's segregation policies (Johnson 2007). Connections were further made with the Tank Man image from the Tiananmen Square protests in Beijing, China two decades earlier, which drew international attention to the repression of what were perceived by Western audiences as pro-democracy protests (Westmoreland 2016). In the cases of Cairo, Beijing and Birmingham, the forms of civil disobedience that were unfolding cannot be separated from the cameras that surrounded them, nor from the broader media infrastructures that carried their images around the world. In such thoroughly mediated environments, where all actors knowingly participate in relation to their own mediation, the spaces of politics and the spaces of media become increasingly imbricated.

In another case, in the Palestinian village of Bil'in, weekly protests and acts of civil disobedience against the construction of the Israeli separation wall have taken place since 2005 (Jawad 2011). These acts involve residents of the village as well as international activists, and include highly performative and visual displays which are regularly met with tear gas from the Israeli side. What makes the Bil'in protests distinct is the way in which they are strategically staged for easy consumption as media events, seeking to highlight the injustice and brutality of the ongoing Israeli occupation while displaying the non-violent discipline of the protesters, in order to attract greater sympathy from global audiences. As Jawad (2011:131) notes, the use of visually appealing non-violent protest redirects "the categorization of terror and terrorist away from the Palestinian villagers and toward the Israeli soldiers, thus reversing a dominant discourse that is used against them". In this instance, the public again extends beyond the physical site where the civil disobedience is unfolding to the diffuse and dispersed global audiences that encounter the protests always-already as media.

The networked spaces and infrastructures of disobedience

Contemporary practices have at times shifted the act of civil disobedience to forms of media themselves, essentially inverting the relation between the space of media and the space of politics. At the end of the twentieth century, as networked digital media became increasingly widespread and integral to society, activists began theorizing how protest might not only make use of, but also fundamentally engage with and contest the networked media form itself (Critical Art Ensemble 1996). While various forms of media intervention have been part of many different political traditions, it was not until 1997 that a group of activists and artists going under the collective name of the Electronic Disturbance Theater (EDT) explicitly called for civil disobedience to be taken online against the forms of state power that had also migrated there. Reconceptualized as Electronic Civil Disobedience (ECD), the EDT programmed an application they named FloodNet that would allow groups of people to collectively barrage websites and servers with automated Internet traffic, a practice which has since become known as a Distributed Denial of Service Attack (Rhizome 2017). In solidarity with the Zapatista struggle in southern Mexico, FloodNet was first deployed in 1998 against several servers of the Mexican government, effectively making them inaccessible for the duration of the action (Dominguez 2009).

Distinct from more traditional forms of civil disobedience, which are understood as occurring at particular locations while engaging broader audiences by deploying different types of media, ECD disrupts the form of mediation itself, in this case the network, as an act of protest. Raley (2009:43–44) proposes that ECD is an evolution of civil disobedience that responds to novel forms of power, where "power is no longer centralized but has become networked and nomadic; the site of resistance has in turn shifted from the street to the network; the object of electronic civil disobedience (ECD) is disturbance and obstruction". For the EDT, if power has moved online in the form of the network then the form of protest must similarly transform itself. ECD, while spectacular and demonstrative in its own way, invests itself politically in the ability to disrupt directly, absent of the mediation of the state. This is a departure from the conventional understanding of civil disobedience as a principally discursive or performative activity meant to address, convince and ultimately transform publics; it instead adopts a more interventionist pose, potentially leading some to even consider it a form of coercive violence in its willingness to interrupt and damage networked infrastructures. In this respect, ECD can at times more closely resemble direct action in its refusal to appeal to state-centric politics, although its insistence on acting in a spectacular fashion in relation to networked participants and publics can allow one to read it within the tradition of civil disobedience.

There are other meaningful aspects of ECD which differentiate it from the tradition of civil disobedience as broadly understood. Notably, civil disobedience is conceived of as a practice which occurs in the presence of or in appeal to public(s), and there is debate concerning whether the network can be considered public. Züger (2013) writes that "[i]n online activism people are not necessarily present with their physical body, they are also not as visible … they remain anonymous in front of other citizens during their action". This secrecy-contra-publicity changes the way that we might understand civil disobedience's mode of address, and represents a significant departure from the majoritarian and democratic nature of the tradition.

Lastly, it is important to consider the role of networked technologies in the organization of novel forms of civil disobedience. Castells (2011) posits that in the age of networked media defined by the Internet, power is determined by the contestation that occurs within the modes of communication. He suggests that "[i]n the network society, discourses are generated, diffused, fought over, internalized, and ultimately embodied in human action, in the socialized communication realm constructed around local–global networks of multimodal, digital communication" (2011:53). This relation between human action on the one hand and socialized communication on the other requires that we understand not merely how civil disobedience is mediated, but also how it perhaps only occurs at all as a consequence of the various forms of communication media that make it possible. This dynamic was illustrated especially clearly in the protests against Donald Trump's Executive Order 13769 that instituted a travel ban on a list of seven countries with predominantly Muslim populations. Following the order's announcement, major international airports across the United States were occupied in a mass coordinated act of civil disobedience (AP 2017). These spontaneously organized occupations, which disrupted many of the central transportation hubs in North America for several days, were planned primarily online without a single organization coordinating the various actions that were unfolding simultaneously across the United States (Rosenberg 2017). The airports – themselves part of another sociotechnological infrastructure – became one coordinate in a larger networked action in opposition against the ban. Using networked media to share information about how to get to various airport terminals, and then livestreaming the protests on platforms such as Facebook and Twitter, the demonstrations took shape in the ways that they did as a consequence of the media that was used to organize them.

The digital networked technologies which allowed for the airport occupations to occur blur the boundaries between who is participating in an act of civil disobedience and who is witnessing it, as many different forms and intensities of engagement take place. These horizontally organized protest actions were composed of people online who primarily coordinated the protest remotely, people who travelled to the airports and then shared media online and livestreamed the events, and people who did not attend but followed the actions online while sharing and commenting upon the media as they were uploaded to various social networks. In this sense, the action itself, its networked mediation and the forms of discourse that surround it all become much harder to distinguish, and as a result many more people consider themselves participants in the actions than were physically present at the airports themselves.

Tufekci (2017a) writes about how digital technologies shape the forms that protests take today, presenting distinct possibilities and challenges for those who wish to contest power. She notes that "[d]igital technologies of connectivity affect how we experience space and time; they alter the architecture of the World – connecting people who are not physically near, preserving words and pictures that would otherwise have been ephemeral and lost to time" (Tufekci 2017a:122). In such circumstances, it no longer makes sense to speak of the space of the protest and the space of the network as qualitatively distinct; they must instead be thought together as co-constitutive spaces that could not exist in the way they do without the other. In contemporary protest movements which make use of civil disobedience, the act of disobedience itself and its mediation, expressed in the networked sharing, commenting and archiving of protest media, should all be considered part of the same distributed political gesture.

See also: activism; direct action; hacking and hacktivism; social movement studies and citizen media

Recommended reading

Arendt, H. (1972) *Crises of the Republic: Lying in politics; Civil disobedience; On Violence; Thoughts on politics and revolution*, Orlando, FL: Mariner Books.

A collection of essays which provides an overview of Arendt's theorization of political action in relation to publicity and privacy, power and violence, and democracy and the state. Through an examination of American politics, Arendt articulates a critique of democratic statehood as it existed at the end of the twentieth century.

Butler, J. (2015) *Notes towards a Performative Theory of Assembly*, Cambridge, MA: Harvard University Press.

This volume elaborates on Butler's earlier work on performativity by examining the inherent sociality that is at the centre of politics, as well as the various ways in which that sociality is constituted and mediated by bodies, infrastructures, spaces and communications technologies.

King, M. L., Jr. (1963) 'Letter from Birmingham Jail', *Liberation: An Independent Monthly* 8(4): 10–16.

A significant exploration of the concept of civil disobedience which set the terms of debates concerning the practice in the decades following its publication, particularly in the civil rights movement. In the letter, King critiques liberal notions of progress, legalistic and juridical approaches to social justice, and also argues for an understanding of resistance centred on its social and historical specificities.

Thoreau, H. D. (1849) 'The Resistance to Civil Government', *Æsthetic Papers*: 189–211.

This is the foundational text that defined civil disobedience and rationalized it as a political action in relation to Thoreau's individualist philosophy. The text is notable not only for its conceptualization of a novel mode of protest, but also for its critique of state power more generally.

CIVIL SOCIETY

Manès Weisskircher

Civil society is an elastic and elusive term, typically used in reference to a wide variety of political actors from grassroots associations and social movements, to development NGOs and the nonprofit sector (Edwards 2011). It has nevertheless become a "master category in the human sciences and a key phrase often used by politicians, corporate executives, journalists, charitable foundations, human rights organizations, and citizens" (Keane 2009:461). When tracing the history of the concept, scholars tend to start with ancient Greece. Aristotle has been credited with discussing civil society first: his ancient Greek *politike koinonia* (political community) was later translated into the Latin *societas civilis*, from which the modern English term is derived. In Aristotle's *Politics*, the *politike koinonia* overlapped with the *polis* or city-state, which referred to the politically organized community: free citizens, heads of their households (*oikoi*), pursued their goals through the formalized setting of the *polis* (Ehrenberg 1999:9). This close association of civil society with state structures shaped political thinking in Europe for several centuries. Only following the decline of absolutism in the wake of democratic revolutions across Western Europe and North America did dominant theorizations of the relationship between the state and society change. In the eighteenth and nineteenth centuries, a growing number of thinkers, including key protagonists of the Scottish Enlightenment such as Adam Smith and Adam Ferguson, as well as Georg Wilhelm Friedrich Hegel in Germany, conceptualized civil society as a sphere beyond the state, the sphere of voluntary public associations that included, or even manifested itself primarily in, markets (Keane 1988).

Alexis de Tocqueville (1805–1859) is typically regarded as the founding father of the empirical study of civil society as voluntary associational life. In the two volumes that comprise his *Democracy in America* (1835–1840/2003), he studied the functioning of democracy in the United States, in his view the most advanced expression of secular development towards equality in any human society, and identified rich associational life as a crucial element in the construction of a successful democracy. Specifically, Tocqueville highlights the importance of political associations for limiting the power of the central government, viewing "freedom of association" as "a vital safeguard against the tyranny of the majority" (ibid.:223). Thus, the spread of political associations is not linked with factionalism and political instability, but rather with the strengthening of the state (ibid.:607). Tocqueville additionally emphasizes the role of non-political associations in the United States, comprising not only "commercial and industrial associations to which all belong but also a thousand other kinds, religious, moral, serious, futile, very general and very specialized, large and small" (ibid.:596). These "intellectual and moral associations" are seen as crucial components in the social structure of democracies for the way they help develop and disseminate knowledge of how to form and maintain all forms of association (ibid.:600).

Following Tocqueville, many different conceptions of civil society have been put forward. Edwards (2014) distinguishes between three broad, and overlapping, understandings of the concept: first, civil society can be regarded as "a kind of society" (ibid.:10) and specifically as a normative vision of a "good society" (ibid.:10) such as that pursued by many activists of the 1989 revolutions in Central and Eastern Europe or suggested in Kaldor's (2003) *Global Civil Society*, which she hoped would contribute to international peace. Second, civil society may be understood as the public sphere, a view which is mainly associated with the ideas of Habermas (1962/1989) and which considers civil society as a level playing field that allows everyone to engage in reason-guided discussion in search for consensus over public matters. The third approach emphasizes Tocqueville's understanding of "civil society as a part of society" (Edwards 2014:10) and focuses on the empirical reality of associational life. It is this perspective that "dominates the debate", informing the majority of contemporary discussions of the concept (ibid.:10). Such an empirical perspective is also foregrounded in this contribution given that it allows us to analyse the key dimensions of real-existing civil society and its citizen media. Going beyond normativity permits us "to see what civil society actually does or does not do for different people who inhabit the sphere" (Chandhoke 2007:613).

Within contemporary scholarship on this topic, civil society has largely been promoted as a supportive structure for electoral democratization and economic liberalization, rather than as radical-democratic potential for self-governance (Baker 1999). Explicit definitions of the concept have also tended to stress the voluntary element of participation: for Walzer (1991:293), civil society is "the space of uncoerced human association and also the set of relational networks – formed for the sake of family, faith, interest, and ideology – that fill this space". Others have focused on the notion of social capital as a key mechanism within civil society, emphasizing how social ties improve human cooperation and lead to more efficient societal output (Putnam 2000; Fukuyama 2001). At the same time, the Marxist Antonio Gramsci's vision of civil society as the sphere of contestation over cultural hegemony (Schwarzmantel 2015:199–205) has generally been marginalized in the empirical study of civil society (for an important exception, see Greskovits 2019).

The popularity of the concept of civil society peaked in the last decades of the twentieth century, especially in the context of the political transformations that reconfigured Central and Eastern Europe and Latin America during this period. Since the turn of the millennium, fascination with the term has somewhat declined, although the substantive issues at stake continue to attract attention (Edwards 2014). The study of social movements has become a popular research strand, developing largely separately from research on civil society. Given the overlap between the two concepts, however, and the conceptual difficulties involved in defining social movements, some scholars have called for bridging the two fields (della Porta and Diani 2011). According to Burstein (1999:8), "all ... attempts to distinguish ... political organizations share the same flaw. Each focuses on a continuum – of institutionalization or unconventionality of tactics, for example – and in effect declares that it can be divided at an identifiable point"; however, "none of the continua are defined precisely; how particular organizations could be placed on the continua is never spelled out; and the rationale for locating the dividing point in one place rather than another is never made clear". These problems are also impossible to overcome in any attempt to demarcate civil society from other political players. It is therefore more fruitful to discuss various perspectives that might lead us to a more critical understanding of civil society as a concept and that are also important for analysing the production of citizen media.

First, civil society overlaps and interacts with other political actors in society. It is not independent from arenas such as the market or the institutionalized political system. Aware of this complexity, Walzer (1991:293) includes trade unions and political parties as important actors in civil society. The fact that NGOs operating in developing countries are dependent on external state funding further underlines the overlap among different arenas (Steinberg and Wertman 2018). At the same time, civil society actors may have highly contentious relationships with other players, such as states, for example when activists try to prevent the deportation of a rejected asylum seeker. In extreme cases, civil society actors may even engage in revolutionary activity, such as trying to bring down an authoritarian government. It is also the case that states – even those considered liberal and democratic – may adopt extensive measures to police some of forms of action in which civil society groups engage (della Porta and Reiter 1998). Ultimately, however, states play a crucial role in providing the environment in which civil society can flourish. After all, civil society's "essential conditions ... – for instance, the rule of law, which regulates the public sphere and guarantees the rights of its inhabitants – are institutionalised by the state" (Chandhoke 2007:608f).

Second, conflict is inherent to civil society. While some interpretations of the concept mistakenly "[refuse] to acknowledge the antagonistic dimension constitutive of the 'political'" (Mouffe 2005:2), civil society must not be understood as a homogenous or even unified collective of political players who live in harmony with each other. As Bob (2011:214) puts it, "civil society itself has no interests precisely because its cacophonous components have so many different views". The French *Mouvement des gilets jaunes* provides a prime example of a very heterogeneous collective, bringing together a diverse ensemble of protestors who support a wide variety of political objectives and ideologies. And even within ostensibly clearcut elements of civil society – for example, within one of the classical social movements, such as *the* environmental movement – conflicting collective identities may separate individual organizations from each other (Saunders 2008). The 2009 United Nations Climate Change Conference in Copenhagen made conflicts within the environmental movement particularly visible, suggesting that these activist groups should be understood as "networks in contention" (Hadden 2015).

Third, civil society does not only consist of 'progressive' players necessarily in favour of left-wing or liberal values. To be sure, those who stage protest events, at least in Western Europe, are mainly left-wing political players (Hutter 2014). Still, conservative and far-right players also mobilize beyond the institutions of the political system (Blee and Creasap 2010), developing their own, sometimes flourishing, associational life. Scholars agree that their collective action needs to be analysed with the same theories used for studying other civil society players: inventing a new analytical framework to understand far-right players, drawing on a concept such as uncivil society, would run the risk of emphasizing normative positions over analytic rigour (Bob 2011).

Fourth, the consequences of a strong or weak civil society are still disputed. While Tocqueville and neo-Tocquevillian followers such as Putnam (2000) are optimistic about civil society's ability to improve democratic quality or even challenge authoritarian governments, others disagree. Notably, Berman (1997) points out that a strong civil society can also endanger democratic institutions: this was the case for example in the German Weimar Republic, where weak political institutions were undermined by powerful civil society players, contributing to the fragmentation and polarization of German society. The consequences of weak or strong civil society depend on the players who dominate civil society and on their interactions with others, including state institutions (Kopecký and Mudde 2003).

Civil society and citizen media

The development of the Internet and other digital networked communication technologies has fundamentally reshaped the media sphere. Evoking Tocqueville (1835–1840/2003:347), who made a similar claim in relation to newspapers, we might say that nothing but a viral Tweet or Facebook posting can drop the same thought into a million minds at the same moment (Fitzpatrick 2018:19). Some observers have hoped, or even predicted, that this technological development would shift the balance of power in favour of civil society actors and their media products, suggesting that the advantages for 'digital civil society' would go far beyond the reduction of transaction costs in the production, distribution and consumption of information (Castells 2012; Shirky 2008). Many of these expectations have, however, proven overly optimistic: dominant economic and political structures have not yielded to digital disruption but have continued to exert a powerful shaping force over the production and use of technology (Lovink and Rasch 2013; Morozov 2011).

In this context, and drawing on the reflections outlined in the previous section, a more critical perspective on the contribution of citizen media to the construction and maintenance of civil society in the new media environment is needed. To begin with, it is important to recognize that citizen media operates both "in the shadow of the state" and "in the shadow of the market" (Couldry and Curran 2003). On the one hand, the relationship between citizen media and corporations is complex, and goes beyond targeting private businesses in various ways, for example by performing street theatre "outside the headquarters of animal abusing companies, supermarkets, animal laboratories and zoos", as one British animal rights activist recommends (Isacat 2015). Co-optation has remained a key issue: huge media corporations have initiated platforms for soliciting independent contributions from their reading public, such as CNN's iReport.com. While these platforms can provide space for critical views, they have also been described as "aggressive cooptation … by corporate media hegemons" that financially benefit from user-created content, without significantly altering coverage (Kperogi 2011:315). In a similar vein, seemingly independent grassroots citizen media projects may turn out to be closely aligned with established economic, and political, institutions. Rone (2016) discusses one such case, showing how the local branches of Anonymous, active during the 2013 protest wave in Bulgaria, opposed popular protests against corruption. Instead, Anonymous promoted the rhetoric and interests of powerful oligarchs, closely aligned with the national government.

Similarly, state structures shape citizen media production to a significant extent. In Venezuela, the Bolivarian Revolution led by President Hugo Chávez has promoted community and alternative media through investment, leading to a situation where some citizen radio and TV producers have had to negotiate the conflicting pressures created by their reliance on state funding and their attempts to maintain critical distance from government policy (Fuentes-Bautista and Gil-Egui 2011). State repression may impose other constraints on citizen media, as in the case of *samizdat* (self-publishing) in Central and Eastern Europe during the communist era, when activists were forced to revert technologically "to a less efficient form of printing", i.e. typewriting, to evade government censors (Skilling 1989:10). The dissident singer-songwriter Wolf Biermann was expatriated from the German Democratic Republic while touring through West Germany. Even in liberal democracies, citizen media projects cannot operate in total isolation from state surveillance and interventions: the radical left Indymedia online platform, for example, has had its servers seized on multiple occasions by UK police (Sullivan et al. 2011). However, citizen media producers have at times also overcome severe repression despite targeting a dictatorship, as when the filmmaker Miguel Littín secretly entered Chile to shoot a documentary on life under Pinochet (García Márquez 1986).

Critical perspectives on civil society also alert us to the fact that the production of citizen media is rarely a fully harmonious undertaking, free of hierarchies and providing equal participation and access to everyone. Gerbaudo (2012:135) shows how informal hierarchies shaped social media use in the so-called movements of the squares, noting that, despite conceptions collectively held among members, groups such as Occupy and ¡Democracia Real YA! were not leaderless, but that usually "a handful of people control[led] most of the communication flow". Still, such soft leadership is regarded as "effective in giving collective action a certain degree of coherence and a sense of direction" (ibid.:157). Beyond such informal hierarchies, civil society can also become heavily split over the production and use of citizen media (Rone 2019). In the Mexican #YoSoy132 movement, the use of digital media led to permanent internal conflicts among activists, revolving around the control of passports, the creation of competing social media profiles, and inclusion in and exclusion from closed Facebook groups or mailing lists (Treré 2015). Bottom-up media production can further lead to the exclusion of activists from their own political organizations: at the end of the 1960s, Italian Communists, among them well-known activist Lucio Magri, established the journal *Il Manifesto*, without the consent of the Partito Comunista Italiano. They were subsequently expelled from the party after publishing a critical editorial on the intervention of the Warsaw Pact in Czechoslovakia (Anderson, P. 2011).

Finally, a growing body of research confirms the point made above about civil society and political leanings. It alerts us to the fact that citizen media is not only produced by 'progressive' players in favour of left-wing or liberal values. Far-right activists also engage in citizen media production – certainly also because mainstream media have often chosen to marginalize far-right actors (de Jonge 2018). Their activities have been particularly vibrant in one of the contemporary strongholds of the far right, eastern Germany, where the Institut für Staatspolitik has its own publishing house in the village of Schnellroda, with the bimonthly *Sezession* as its flagship publication. In Berlin, the monthly *COMPACT – Magazine for Sovereignty* has a circulation of around 40,000; it focuses on typical far-right themes surrounding ethnic homogeneity, law and order, and conspiracy theory (Schilk 2017). The popularity of these magazines underlines the ongoing relevance of print media for an interested, loyal far-right audience. Nevertheless, it is through their use of digital media technologies that far-right activists have been most successful in circulating their messages among greater audiences (Caiani and Kröll 2015; Simpson and Druxes 2015). Far-right civil society actors have created a vast online universe of blogs and discussion forums since the establishment of the Internet in the 1990s (Fielitz and Thurston 2019): the Islamophobic Gates of Vienna blog, with Norwegian far-right anti-Islamic activist Fjordman as key contributor, remains among the best-known (Berntzen 2019: 169).

Many far-right groups are also active within social media platforms, which are a key arena of contemporary (anti-)migration discourses, especially in receiving countries (Heidenreich et al. 2019). A key instance of far-right street mobilization in Western Europe – the Dresden-based Patriotic Europeans against the Islamization of the Occident (PEGIDA) (Weisskircher and Berntzen 2019) – was even first founded as a Facebook group. When activists have tried to start PEGIDA groups outside Dresden, they have often begun by establishing a Facebook presence, spreading news about far-right protests, immigration, as well as law and order, often ending up with many more online followers than protestors in the streets (Berntzen and Weisskircher 2016). For PEGIDA, live streams of their events have also remained an important propaganda tool. However, switching to social media platforms also carries risks for far-right groups, as the case of the Identitarians shows. This group, which advocates protecting European culture against the 'Islamization' of Europe through immigration, originated in France but later became particularly active in Germany and Austria. While the Identitarians have been able to gain mass media attention (Castelli and Froio 2018), social media has been crucial for initially spreading information about their protest events through pictures and videos. In 2018, however, the Identitarians were banned from Facebook and Instagram, showing how not only states, but also private corporations may inhibit the media production of such civil society groups.

Assessing the impact of citizen media in the digital age

Many observers have had high hopes for citizen media, even more so in the digital age. Some scholars even maintain that citizen media is crucial for mobilizing civil society actors and sparking revolution, pointing to the importance of social media during the Arab uprisings in the early 2010s, when authoritarian regimes in the Middle East, most prominently in Egypt, were temporarily brought down (Lim 2012). Others have found only little evidence for such an impact: according to Aday et al. (2013), Twitter content posted during these protests mainly found an audience outside the region, especially in Western Europe and North America. People in the Middle East and North Africa mainly consumed news from traditional media, although it must be noted that this too relied on content created by citizens to some extent. Generally, however, the correlation between social media penetration and the level of protest during the Arab uprisings has been found to be negative, the growth of social media usage having increased only after the onset of protest (Wolfsfeld et al. 2013).

Switching from the macro- to the meso- and micro-levels makes the impact of citizen media in the context of civil society initiatives more visible. In Hong Kong, for example, Lee and Chan (2016) have underlined how social media platforms helped activists of the Umbrella Movement in 2014 to dispel rumours, while many supporters used social media to feel part of the occupation when they were not able to be physically present. Occupy activists have elsewhere managed to create an inclusive collective identity through social media, in line with their ideological convictions (Kavada 2015). Animal rights activists have succeeded in raising awareness of animal exploitation in farm factories and laboratories by producing undercover videos, even if their targets have subsequently learned how to shield themselves from such intrusions (Jasper and Poulsen 1993). Similarly, producers of online hate speech are often quite effective in their production of content: for example, it has been shown that content containing hate speech reaches almost half of US Americans aged between 15 and 30 (Hawdon et al. 2014).

Some scholars have argued that civil society actors who understand themselves as self-sufficient and not engaged in the struggle over institutional power might have difficulty achieving wider impact (Lentz 2011). Citizen media producers who only focus on the great variety of ways in which they might articulate their ideas outside the mainstream, but who ignore the struggle over who owns and what is reported by mainstream media platforms, risk only preaching to the converted. While it is crucial for civil society actors to create their own media environment in order to disseminate their messages, it thus remains essential for them to contest what mass media players with far bigger audiences cover, and how governments and corporations may constrain independent media production – understood in the wider sense of media adopted in this volume.

See also: activism; co-optation; social movement studies and citizen media

Recommended reading

Edwards, M. (2014) *Civil Society*, Cambridge: Polity Press.

Provides an excellent and accessible introduction to key debates on civil society. The author discusses the writings of an impressive wealth of thinkers and explores diverse contexts and cases to illustrate his argument. Critical perspectives are developed on civil society as a concept and civil society activism as a practice.

Gerbaudo, P. (2012) *Tweets and the Streets: Social media and contemporary activism*, London: Pluto.

Analyses the role of social media usage during the Arab Spring, Indignados and Occupy protests – perhaps the most prominent protest movements to have emerged in the wake of the 2008 Great Recession. Based on extensive field work, the author highlights the continuing relevance of informal hierarchies and leadership structures among protestors, challenging widely held assumptions concerning social media as inclusive and egalitarian technologies.

Tocqueville, A. de (1835–1840/2003) *Democracy in America and Two Essays on America*, translated by G. Bevan, London: Penguin Books.

The classic empirical study of associational life that continues to be a key point of reference in modern discussions of civil society. Its focus on the relationship between civil society and democracy has remained an important research area until today. This edition also contains two further essays written by Tocqueville following his visit to America: *Two Weeks in the Wilderness* and *Excursion to Lake Oneida*.

Wolfsfeld, G., E. Segev and T. Sheafer (2013) 'Social Media and the Arab Spring: Politics comes first', *The International Journal of Press/Politics* 18: 115–137.

This contribution critically examines the importance of social media for the emergence of the Arab Spring protests. Adopting a methodologically sophisticated approach, the authors offer an important counterpoint to any initial enthusiasm concerning the impact of social media on regime change in the Middle East.

COMMONS

Fruela Fernández

Since the 1990s, discussions of the commons have gained considerable prominence within the conceptual frameworks of activists, economists, social scientists and politicians across the world (Caffentzis 2016:96; Subirats and Rendueles 2016:9–11). Despite this increased attention, the commons remains a problematic and ambiguous concept, one "that is both contested and innately political in nature" (Wall 2014:73) and for which multiple understandings coexist (Hudson et al. 2019). This has led some researchers to claim that it might be best seen as an "umbrella" term (Subirats and Rendueles 2016:9), used to refer to a wide variety of ideas, resources and practices. This entry will trace the historical evolution and political relevance of the concept, as well as highlighting the different tensions and debates that have shaped its development, before discussing a series of contemporary applications within the realm of citizen media.

A history of the commons: from feudalism to globalization

Historically, the commons can be defined as a wealth or resource that is shared, used and protected by a community. A key characteristic of this shared resource is the fluidity of its ownership: it can belong to the community as a whole, to a selection of members of the community with non-exclusive rights, or even to a third party that allows various forms of communal usage. According to Caffentzis (2016:96), the term commons derives from medieval-era English property law, in which it encompassed "a set of legally recognized 'assets' – including meadows, fisheries, forests, and peat bogs –" that either belonged to the king, the Church or the nobility, and upon which a community had "customary and collectively managed usage". As other scholars have shown (Federici 2004:25–75; Linebaugh 2008:21–45), these commons were fundamental in the sustainability of life during the Middle Ages, ensuring that every member of the community – regardless of age, income or gender – could have egalitarian access to basic resources.

Over the centuries, however, most British commons were dismantled through a process of privatization. This involved the creation of what became known as enclosures, where common land was fenced in to allow large landowners to gain greater profit, for instance by using the land "as pasture for increasingly lucrative sheep farming" (Meiksins Wood 2017:108). Subsequently, enclosures brought with them "the extinction of common and customary use rights on which many people depended for their livelihood" (ibid.), as commoners were no longer able to pick wood for fire or to fish in rivers and lakes. The destruction of the commons can thus be seen as a fundamental step in the birth of capitalism: between the sixteenth and eighteenth centuries, "the idea of overlapping use rights to the same piece of land" that had prevailed in previous centuries gave way "to *exclusive* ownership" (Meiksins Wood 2017:114; emphasis in the original), placing profitability over tradition and communal rights. This ideological change was also paramount in the evolution of the colonial process, which is tightly bound up with the expansion and consolidation of capitalism. When European settlers arrived in territories such as America and India, they carried with them an understanding of exclusive property rights that clashed with the communitarian, non-exclusive notions prevalent among many indigenous peoples (Cheyfitz 1997:41–58; Linebaugh 2008:144–169). In this sense, colonialism can be understood as a process of enclosure and destruction of the commons of global dimensions.

Despite this long history, neither commons nor enclosures are realities that are limited to the past. Indeed, there are many examples of commons that have been preserved (Ostrom 1990), such as the self-managed systems of irrigation that are still used in areas of Europe, Asia and Latin America. At the beginning of the twenty-first century, it was estimated that 232 million hectares of forests were still managed under some type of collective tenure regime in Latin America alone (Monterroso et al. 2019:376–377). Equally, the many cases in which common resources around the world today continue to be privatized can be understood as processes of enclosure. These include the ongoing sale of publicly owned land in Britain (Christophers 2018), the major projects carried out by the Indian state to extract profit from its rivers and forests (Roy 1999, 2011), and the continuing deforestation of the Amazon for soy plantations, cattle ranching and energy projects (WWF 2019).

Key perspectives on the commons

As Bollier and Helfrich have noted (2012), it would not be feasible to offer a "unitary perspective on the commons", given that the concept encompasses multiple material realities, social attitudes and worldviews, as well as diverse intellectual, philosophical and political arguments. Nevertheless, it is both possible and productive to identify a number of key perspectives that have emerged in the debates surrounding current uses of the concept.

Firstly, while traditionally associated with material elements such as land, water and vegetation, commons should not be understood merely as sets of things or resources, but as social relations (Federici and Caffentzis 2019:94; Harvey 2012:73). In other words, it is not the specific element enjoyed by a group of people that constitutes a common, but the whole institutional framework – in terms of rights but also of obligations (Subirats and Rendueles 2016:32–36) – developed by the community around a given resource. This perspective is vital when differentiating between common and public, though these categories do sometimes overlap. A given resource or space might be owned by the state and freely open to any citizen (which makes it public); yet, in order to be common, it must be managed and protected by its own users. For instance, citizens visiting a public park might be requested to follow certain rules (such as collecting their own rubbish or avoiding noises at certain times), but most of the activities that enable the very existence of the park – from establishing its boundaries to gardening and watering – are generally conducted by officials and workers, not by the users themselves. In the case of a common, however, all of these activities – and many others that might be necessary for its preservation and evolution over time – would be carried out by the community, emphasizing the centrality of the human bonds that enable this reality. In fact, certain thinkers prefer to use the neologism *commoning* in order to stress "not the material wealth shared but the sharing itself and the solidarity bonds produced in the process" (Federici and Caffentzis 2019:94).

A second perspective, which ties in with the first, focuses on the boundaries of a common and the question whether it should be open to everyone within a community, or limited to a restricted number of members. In her seminal study, Ostrom highlighted "clearly defined boundaries" (1990:91) as one of the main characteristics of successful, sustainable commons: the longest running commons were those which were clearly distinguishable from other social spaces and where the people who had the right to benefit from the common were explicitly defined. Moreover, these commons also had effective mechanisms in place to impose "graduate sanctions" (ibid.:94–96) for misuse and trespassing. However, other thinkers argue against these restricted commons; Federici and Caffentzis (2019:90–91), for instance, call them gated commons, and criticize the fact that, despite the fair and democratic access upon which they are based, they seem "indifferent to or even hostile to the interests of 'outsiders'". In their view, any cooperation generated by these commons "remains on the instrumental plane and rarely takes on a transformative character" (ibid.:91). While the importance of this issue is evident in the case of traditional commons based upon material resources – for instance, a housing community or a fishing area – it may be less salient in the case of contemporary, immaterial commons, such as websites offering materials that have been generated by a group of active core members to any users.

The question of access to commons also leads to another thorny matter: their ambiguous relation with capitalism. Since commons are based on relations of cooperation and regulations regarding their appropriate use, they constitute spaces where the logic of competition that characterizes capitalism is not a leading principle. However, this does not necessarily imply that commons oppose themselves to capitalism. Many successful examples of traditional commons – such as the Alpine meadows in Törbel, Switzerland (Ostrom 1990:61–65) or lobster fishing in the Gulf of Maine (Coombs 2011) – are certainly market oriented and fully embedded within capitalism, even if they do provide alternative and more humane economic spaces. In fact, growing interest in the commons at a geopolitical and institutional level seems to express a willingness on behalf of the system to accommodate popular demands for self-organization and governance from below without ultimately seeking to address economic inequality (De Angelis 2012). Nevertheless, many thinkers believe that the commons are an opportunity to start building realities that go "against and beyond capitalism", in which users would rely on "self-provisioning", while contributing to the development of "an alternative mode of production" (Federici and Caffentzis 2019:89).

Finally, it is important to remember that, even if the concept of commons has today been disseminated throughout the world due to capitalist expansion and the global domination of English in the intellectual field, understandings of the term are indebted to its historical evolution in specific Anglophone contexts. This has two important consequences: on the one hand, it means that translational and linguistic issues frequently appear in debates surrounding the concept. For instance, Das Commons Institut, a German-speaking network of researchers and activists, have retained the Anglicism *Commons* in their texts and articles, since they argue that the standard German translation (*Gemeingüter*, 'common goods') places too much emphasis on "products and resources" (Das Commons Institut 2018; author translation). For their part, however, Spanish-speaking activists have developed a variety of concepts to translate the commons (Fernández forthcoming): *procomún*, which refers to the ideological principle behind these realities; *comunes*, which denotes the specific cases shaped by collective action; and *comunal*, applied to the traditional resources belonging to a community. On the other hand, it would be worth considering whether the Anglophone model of the commons could potentially limit our ability to think these social realities, which have been originally shaped in a variety of languages across the world. Does, for instance, a member of the Quechua *ayllu* (the traditional form of community in the Andes) understand their collective action along different lines to someone involved in the Basque *auzolan* (a type of communal work based around the neighbourhood or district)? Unfortunately, no research seems to be available on the topic for the time being.

Contemporary examples and applications within citizen media

As previously noted, we can identify examples of commons not only in the past, but also in the twenty-first century. This section discusses three categories of contemporary commons that can additionally be understood as citizen media, that is, as spaces, content and practices that bring together unaffiliated citizens for a given social purpose. These are Internet-related and computer-based commons; housing and banking cooperatives; and the set of social expressions now known as "the movement of the squares" (Gerbaudo 2017b). Each of these categories presents a different perspective on the commons, as well as highlighting potential limitations and ambiguities of the concept.

Information technology has played a major role in the contemporary resurgence of the commons. Free and open source software (FOSS), peer-to-peer exchange systems and the Internet are frequently referred to as "knowledge commons" (Madison et al. 2019:77). Indeed, it was in the context of networked media that the concept was first extended beyond traditional understandings of 'natural' commons. The Creative Commons in particular has contributed to the popularization of the concept within digital culture (Linksvayer 2012). Founded in 2001, this not-for-profit organization has developed licences that allow creators to disseminate their creative work while controlling which rights they grant to users (e.g. recognition of authorship, non-modification or non-commercial use, among others).

Another well-known, albeit contested, example of a digital common would be the online multilingual encyclopedia Wikipedia, also created in 2001. In many respects, Wikipedia fits very well with the definition of a common. It is based upon collective and voluntary work; any registered user can edit content (with the exception of certain 'sensitive' articles); users with special editing and revising responsibilities are chosen by the community itself (Wikipedia 2019); and user access is unrestricted, regardless of their income, academic qualifications or institutional affiliation. At the same time, however, Wikipedia is not entirely autonomous in financial matters, and over recent years has resorted to fundraising campaigns, which have been considered controversial in light of its healthy financial situation (Dewey 2015). This signals a pertinent ambiguity: while a part of Wikipedia (the generation of content) is provided collectively, a significant part of it (salaries, management costs or information storage) is dependent on financial resources, which are subject to traditional capitalistic relations and which, most importantly, are managed by a board of trustees, falling beyond the control of the community users. This issue casts doubts on the truly communal character of Wikipedia, and also connects with wider debates on the difficulty of democratically managing large-scale commons (Subirats and Rendueles 2016:78–83). A similar debate has been raised in the case of free software (Hill 2012), as freedom of access might not automatically qualify them as commons: many among the most popular software systems and applications that can be used for free (such as Gmail or Facebook) cannot be freely altered or improved by users, thus restricting their collective potential. All of these issues are certainly relevant when exploring the overlaps between digital commons and citizen media, since a lack of accountability and democratic management, as well as potential contradictions between collective work and private benefit, must necessarily be taken into account. Leading media subscription platforms such as Netflix and Movistar+, for instance, have been exposed for using subtitles generated by members of not-for-profit online communities without permission and for commercial purposes (Ernesto 2012; Zavia 2017).

Another category of commons of special relevance to the political sphere are cooperatives, which are considered key developments in building alternatives to capitalism (Wright 2010). Along with worker cooperatives, banking and housing cooperatives are paramount in the transformation of the economic system, as both touch upon highly sensitive matters for the reproduction of life. For instance, the recently created Bank of the Commons, established through collaboration between Italian and Catalan cooperatives (Bank of the Commons 2019b), aims to rethink "money and finances as commons" in order to free society "from the control of the current banking system" (Bank of the Commons 2019a). This involves using an alternative currency (FairCoin) and a combination of open assemblies and professional work.

For their part, housing cooperatives are widespread in many countries around the world and prominent examples include Radical Routes (UK), Mietshäuser Syndikat (Germany), Habicoop (France) and FUCVAM (Uruguay). As noted above, a housing cooperative is considered by some thinkers as a type of gated commons, since access is restricted. They can also be subject to important market pressures (e.g. the cost of the land or the building process) and may even be used as a commercial asset. Nevertheless, the variety of existing cooperatives offers valuable insights into potential methods of opposing capitalist logic. A recent development in this sense has been housing cooperatives granting right to use only. According to this model, the cooperative keeps full and exclusive ownership of housing over time (Lacol and La Ciutat Invisible 2018:14–15). The property is owned collectively and members cannot sell or make profit from it, thus preventing it from becoming a commodity as has happened with social housing in countries such as the UK (Jones and Murie 2006). Moreover, members of the cooperative are involved in all stages of the project and can decide how communal life should take place within the property (Lacol and La Ciutat Invisible 2018:42).

Finally, a key example in understanding recent applications of the commons paradigm has been the so-called movement of the squares. This "array of popular and anti-establishment protest movements", which took place between 2011 and 2016 (Gerbaudo 2017b:32), involved the occupation of public spaces across a variety of countries (Egypt, Libya, United States, Greece, Turkey and Spain, among others) to protest against the economic and political system that led to the 2008 financial crisis. These occupations can be understood as forms of citizen media given that protesters gathered in response to calls and manifestoes 'from below' that were not promoted by political parties or institutions. They have also been characterized as examples of urban commons, "as people assembled there to express their political views and make demands" (Harvey 2012:73), effectively reclaiming a public good to put it at the service of the community. Most importantly, many of these improvised sites started offering basic services – such as cleaning, medical and legal services, food provision, security or sign interpreting (Kadet 2011; Castells 2012:58–60; Bravo 2014; 15Mpedia 2016) – in line with the principles of the commons: these services were carried out by the protesters themselves and freely available to any citizen. In this way, the principles of solidarity and autonomy that underpin the commons became pivotal for the practice of prefiguration (Maeckelbergh 2011), another cornerstone for contemporary social movements; instead of deferring the construction of their envisaged society to the future, protesters aim to construct it here and now, in the temporary setting of the squares.

Equally relevant is how the spirit of the commons that led to and was revitalized by these protests has inspired subsequent actions in these same cities. A notable example among many would be La Ingobernable, self-defined as a "social centre for urban commons" (La Ingobernable 2019; my translation). La Ingobernable was located in an affluent area of Madrid, in a building that formerly belonged to a public university but that was later donated to a private foundation by the local conservative government. In 2016, a group of activists occupied the building, which had been completely abandoned following its concession to the foundation. During its existence, this self-managed centre offered a variety of services to its local area – such as evening classes, sports facilities, discussion groups, workshops and conferences – and was based upon solidary and voluntary work on behalf of members. Despite its social relevance, the centre had a tense relationship with the left-wing platform – Ahora Madrid, connected to the movement of the squares, but characterized by inner strife – that governed the city until May 2019, showing once again how the concept of the commons frequently collides with institutional understandings of public ownership. After failing to secure the building's concession to activists, La Ingobernable was evicted in November 2019 by the new city council, led by a coalition of right-wing parties.

Future directions

As this entry has shown, the commons are a vibrant but also highly complex social and political reality. Combining a long historical heritage with contemporary innovation, they are likely to continue to resist an overarching definition. A number of key questions for future investigations into the commons as citizen media can nevertheless be discerned. Firstly, it is essential to consider how each common interacts both with public resources (i.e. the state) and with the structures of capitalism, as these relationships can have a negative impact on the future life of the common. For instance, if commoners have not secured a concession for the given resource from the relevant administration, it will remain at risk from enclosure. Equally, if the common is reliant on private resources that can be subject to financial pressures (e.g. electricity, Internet access, the market), it will not be autonomous and might therefore face serious challenges to its independence in the future. A second point concerns access: does the common need to be access-restricted, or is there potential for open access? As it has been shown, certain types of commons seem to be more prone to access restrictions (a housing cooperative or a natural resource) than others (knowledge commons). Either way, it is important to highlight the implications of each option, and whether a given choice could have undesired consequences or even compromise the survival of the common. Finally we must explore the model of management that is implemented in the common. By definition, a true common should be democratically managed by the members of the community; however, democratic models can differ greatly and adopt a variety of procedures. Equally, the implementation of democracy can present a number of challenges depending on the specific features of the common in question. We might consider, for example, whether and how a large common might be managed democratically: whether every step of the process must be decided collectively, whether the organizational structures of the common must be fully horizontal, and whether there is room for delegation at certain stages. Clearly, thorough reflection on democratic management, its potential and its limits should go hand in hand with the future development of the common.

See also: wikis

Recommended reading

Federici, S. (2019) *Re-enchanting the World: Feminism and the politics of the commons*, Oakland, CA and New York: PM Press/Autonomedia.

A compilation of several decades of writings by a major theoretician and activist. It includes a piece co-written by Federici and Caffentzis ('Commons against and beyond Capitalism') that offers a clear and insightful introduction to anticapitalist debates on the commons.

Linebaugh, P. (2008) *The Magna Carta Manifesto: Liberties and commons for all*, Berkeley, CA: University of California Press.

An original, innovative and beautifully written exploration of historical struggles for the preservation of the commons across various geographical locations.

Ostrom, E. (1990) *Governing the Commons: The evolution of institutions for collective action*, Cambridge: Cambridge University Press.

A classic book on the commons and arguably the best known monograph on the topic. The author, who was awarded the Nobel Prize in Economics in 2009, studies the commons as a potential tool against the excesses of the free market.

Wall, D. (2014) *The Commons in History: Culture, conflict and ecology*, Cambridge, MA: The MIT Press.

An accessible introduction covering the historical, ecological and political aspects of the commons. Particularly suitable for readers who need an overview of the main debates in this field of research, while also being useful for those who are looking for a fresh take on the topic.

COMMUNITY MEDIA

Clemencia Rodríguez

The term community media can be difficult to define because it attempts to capture the complex relationships between communities and media technologies. Although the term eludes precise definition, those involved with community media research and/or practice generally agree on a few more or less stable parameters that delineate the category of community media (Downing 2011; Gumucio Dagron 2001; Howley 2005, 2010; Kidd 1999; Rennie 2006; Rodríguez 2001, 2011). According to these scholars, the term refers to media content produced by members of a community for other fellow members. These are grassroots media, controlled and produced outside of commercial and public structures that are sometimes referred to as 'dominant media'. The main goal of community media is not to generate profit or disseminate government agendas, but to meet various local information and communication needs. Because community media originate in civil society, they are able to remain autonomous of political and government agendas. Autonomy is thus one of the central features of community media.

Community media are linked to either geographic communities or communities of interest, such as the LBGTQ community, and hence generally produce local content for specific publics. Achieving high ratings and attracting large audiences are not their main objectives. Instead, the ultimate goal of community media is to foster and promote a public sphere where a plurality of identities can have a voice. For community media practitioners, this public sphere populated by the self-expression of every unique identity that forms the social and cultural fabric of the community should be mirrored in their programming grids. Community media are thus linked to notions of voice, political agency and empowerment, and to dynamic participatory democracies.

Community media tend to promote horizontal communication and participatory media production. While dominant commercial and public media establish one-way communication processes in which the medium speaks and audiences listen, community media promote two-way or dialogic communication between content producers and receivers. Ultimately, this contributes to blurring the boundaries between different categories of people involved with community media, who are simultaneously producers and receivers of media content. As participatory media, community media generally adopt an open-door policy, allowing anyone in the community to become a media producer. They assume that everyone – including but not limited to children, youth and elders – has something to say and that their perspectives should contribute to invigorating local public spheres. Community media also frequently operate as media production training schools, where community members become equipped with the necessary skills to produce media content. It is not uncommon to find a seventeen-year-old girl directing her own radio news programme in a community radio station, for instance. If she joined the community radio station at the age of ten, seven years of media training and access to media technologies will have transformed her into a savvy citizen journalist with strong radio production skills by the age of seventeen.

The participatory and horizontal nature of community media is also reflected in the governance and day-to-day operations of these entities. While commercial and public media are generally controlled by hierarchical structures of media professionals, community media are governed by civil society collectives engaged in participatory and inclusive decision-making processes.

As each community appropriates media technology in a unique and context-dependent manner, community media take on different forms depending on the conditions from which they emerge. This entry explores three community media contexts: the Latin American context, the indigenous/Aboriginal context and the European context, while acknowledging that other regions of the world, such as Africa and the US, have produced their own unique community media. In each geopolitical context, community media emerge at the intersection of regulatory regimes, social movements for more democratic media and corporate attempts to take over media infrastructures.

The Latin American context

The roots of Latin American community media lie in the work of the Brazilian philosopher and pedagogue Paulo Freire, whose *Pedagogía do Oprimido* was published in Brazil in 1968, at a time when the Cold War was fostering the establishment of military dictatorships and brutal regimes around the world and making Latin America one of its main battle grounds. Freire's book marked the emergence of a new way of thinking about communication, language and power. According to Freire, the impact of poverty, injustice and oppression extends well beyond issues of access to material resources, plunging individuals and communities into states of isolation and what he called a culture of silence (Freire 1968). People without access to power lose their own voices and learn to mimic the voice of the powerful; internalize negative notions of themselves and their environment; and adopt stigmatized versions of their neighbourhoods and communities.

Freire proposed communication and dialogue as critical tools to break through the culture of silence; he argued that people can overcome processes of alienation, isolation and silence by appropriating their own languages and using this new fluency to speak the world in their own terms. Freire's ideal communication – known in Latin America as *comunicación popular* – transforms people into subjects who acquire their own languages, use those languages to re-signify their reality, and develop ways to move those interpretations of reality into the public sphere. According to Freire, popular communication can transform people from passive objects of others into self-determining subjects, individuals with agency. The ultimate goal of *comunicación popular* is conscientization, a consciousness-raising process that begins with language and ends in action.

Media technologies are particularly interesting because they facilitate processes of language appropriation. Learning to operate a video camera and capturing the images and sounds in one's environment is an exercise in appropriating audiovisual languages. Learning to edit sound and image and push one's own narratives into the public sphere is an act of language appropriation and dissemination of one's own voice. Media technologies exist precisely to capture raw experience and transform it into narratives, and in that capacity they offer exceptional potential for facilitating the development of voice, agency and empowerment – the main components of *comunicación popular* and conscientization.

Freire's groundbreaking understanding of communication as *comunicación popular* found fertile ground in a region where intense debates around media concentration were taking place. In the late 1970s Latin American delegates had exposed a scenario of unbalanced communication flows and global communication inequities on the floor of UNESCO. As a result, UNESCO later released its MacBride Report (UNESCO MacBride Report 1980), demonstrating that most global media traffic was controlled by a few transnational communication corporations located in wealthy industrialized nations. The MacBride Report also showed that South-to-South communication was practically non-existent. One of the Report's recommendations thus stressed the importance of equitable access to media infrastructures and a plural mediascape made up of many voices rather than just a few corporate media makers. Informed by Freire's groundbreaking ideas and the MacBride Report's call for a New World Information and Communication Order (NWICO), Latin American media activists, social justice movements and critical communication scholars began a long struggle for community media. The Latin American community media agenda includes providing democratic access to the media, offering media production training for marginalized communities and instating regulations that promote community media. Latin American social movements – including indigenous, labour and *campesino* movements – feature media democratization as central to their own agendas.

In addition to Freire's influence, the Second Vatican Council's (1962–1965) mandate to re-centre the Catholic mission around social justice gave rise to Latin American Liberation Theology, which emerged in the late 1960s with a strong commitment to marginalized communities and their resistance to oppression. Since then, Liberation Theology has played an essential role in strengthening and promoting community media in the region. Liberation Theology inspired priests, nuns and Catholic grassroots collectives to develop their own community media initiatives. The radical Church placed many of its local radio stations in the hands of the communities they served and helped develop thousands of popular communication and participatory media initiatives. The influence of Liberation Theology explains, in part, why the community mediascape in Latin America is so large (Rodríguez 2003).

Latin America has produced a significant number of world-renowned community media scholars. Argentinean communication researcher Mario Kaplún (1983, 1986) studied and analysed the unique nature of community media as dialogic communication, emphasizing its great emancipatory potential as a tool for empowerment and conscientization. In Peru, Rosa María Alfaro (1985, 1986, 1988) and Rafael Roncagliolo (1986) examined and documented the use of radio, loudspeakers, photography and video by residents in marginalized communities. Gumucio Dagron and Cajías (1989), Gumucio Dagron (2001) and Robert Huesca (1995, 1996; Huesca and Dervin 1994) studied the tin miners' radio stations in Bolivia – the earliest example of Latin American community media. In 2001, Colombian scholar Clemencia Rodríguez developed her citizens' media theory, an approach which has since proved central to understanding the role of community media in society (2001, 2010a, 2011).

Indigenous/Aboriginal communities

Indigenous, Aboriginal and First Peoples communities have also developed their own interpretation of community media. Indigenous community media respond to unique cultural understandings of time, space, communication, media message and media audience. Today's indigenous community media are the result of decades-long processes of appropriation of media technologies by indigenous communities in Australia, New Zealand, Canada, Latin America and the United States. In Canada and the United States, First Peoples' media and Native American media originated from the media that anthropologists brought into native communities. Native peoples resisted being objectified by the anthropologists' cameras and insisted on access to training and technologies (Worth and Adair 1972). Today, indigenous community media is a robust field of media practice, innovation and research.

Cultural appropriation and adaptation of media technologies is at the centre of indigenous and Aboriginal community media, which are regulated through complex systems of rules that involve kinship, land tenure, respect for the dead and sacred elements such as sites, landscapes, artefacts and designs (Michaels 1994; Rodríguez 2001). Indigenous community radio stations in northern Canada and Aboriginal community media in Australia shape programming schedules around daily indigenous life and ceremonial events. Browne documents issues that must be negotiated as a result of these processes of appropriation. Examples include a community radio station's transmission of ritual songs that are not meant to be heard by anyone;

a station trying to determine how to include 100-hour-long songs in their programming; and stations figuring out what to do about songs "meant for specific purposes, such as finding one's way across unfamiliar terrain" (Browne 1990:114). In Canada, Inuit community media privilege the privacy of individuals and family groups over the public's right to know. Similarly, camera angles are determined by indigenous rules governing eye contact and notions of respect (Madden 1992). Australian Aboriginals understand storytelling always as non-fiction, and fictional storytelling is therefore non-existent in Walpiri media (Michaels 1994). Even media spaces are created according to indigenous ways. In Australia, Aboriginal community radio leaders designed their stations' studios to fit Aboriginal lifestyles. Lounge areas built adjacent to the studios fit the Aboriginal understanding of broadcasting as deeply rooted in a sense of community, with people wandering in and out of the studios (Molnar 1994); radio stations also include outdoor areas and breezeways in an architectural design sensitive to Aboriginal camp life (Michaels 1994).

Aboriginal communities in Australia, Inuit communities in Canada and indigenous peoples in Latin America share a focus on the natural environment, which determines camera shots, scripts and many other elements of indigenous community media. A boulder, a tree or any other element of the natural world can be immensely meaningful in indigenous communities, and as a result, can take central place in a camera shot or script. The late Eric Michaels, one of the first ethnographers of indigenous media, documented how Aboriginal worldviews interpret the natural world through mnemonic clues, sacred symbolism and ancient tracks left by one's ancestors (Michaels 1994). In what Michaels (ibid.) calls the characteristic Aboriginal mise-en-scène, a camera pan can be interrupted without apparent reason on a "tree where spirits live or a flower with symbolic power" (ibid.:114).

Indigenous community media play an essential role in maintaining and strengthening indigenous languages. Browne (1996) documents how Native American community media aided the survival of Navajo and Lakota languages in the United States. In Ecuador and Chile, community media scholars O'Connor and Salazar have researched how community media help strengthen Aymara, Quechua and Mapuche languages in Latin America (O'Connor 1990; Salazar 2007). One of the best known indigenous media initiatives is Brazil's Video in the Villages project (Aufderheide 1992, 1993, 1995) through which the Kayapo Indians of the rain forest harness the power of video technologies and digital platforms. They use these tools to attract the attention of government authorities in support of land claims, intimidate opponents during disputes, capture media attention, and document and archive ceremonies and develop internal communication networks between villages (ibid.).

In Canada, Roth (2005) has documented the Inuit's decades-long struggle to develop their own media. Before the establishment of broadcasting operations, Inuit communities were receiving more than a hundred hours of English television programming per week via satellites. To counter this trend, the Inuit demanded access to their own broadcasting infrastructure and forced the central government to include native representatives in regulatory processes. The result has been that the "Inuit have designated communications services as a critical area for negotiation as part of any land claims agreement" (Valaskakis 1983:245).

Ginsburg (1991, 2003, 2004), Meadows (1992), Molnar and Meadows (2001) and Forde et al. (2009) have all made important contributions to our understanding of the complex rules that shape the production, distribution and reception of media in Aboriginal community media. In the process of appropriating media technologies, indigenous peoples adapt media to the needs and circumstances of their cultures. Vargas (1995), Ramos Rodríguez (2005), Castells-Talens et al. (2009), Castells-Talens (2010, 2011, 2012), Castells-Talens and Ramos Rodríguez (2013) and Magallanes Blanco (2008, 2014) have similarly developed a robust body of research about indigenous community media in Mexico. Magallanes Blanco and Rodríguez Medina (2016) later explored the process through which an indigenous community in southern Mexico designed *Rhizomatica Administration Interface* (RAI), a graphic interface for a local mobile phone network that is responsive to indigenous needs, resources and languages.

The European context

In Europe, community media experienced accelerated growth during the 1980s. Countering the trend toward integration and globalization, various European communities turned toward local lifestyles and cultural practices. In this framework, community radio and television emerged as forces that could help foreground the position of local life in a community. The 1980s witnessed a mushrooming of diverse community radio stations throughout Europe; by 1980 more than 2,000 community radio stations operated in Italy alone (Scifo 2014). Describing them as "green, red or black, pirate or legitimate, public or private, communal, convivial or commercial, *libre* or libertine, decentralized, devolved or deregulated, slickly professional or painfully amateurish, parochial, provincial or regional", Vittet-Philippe and Crookes (1983:7; emphasis in the original) asserted that "local radio stations in Europe make up a kaleidoscope of spectacular diversity".

At the time, European community media conceptualized themselves as promoters of local cultures, worldviews and lifestyles; their goals included reviving *communitas* or *polity*, fostering local and regional identities, increasing local autonomy, generating popular participation and promoting local cultural activities and small businesses. Coyer and Hintz (2010:275) state that, in the European context, the term community media "generally refers to self-organized, participatory, not-for-profit media that address local geographic communities and/or communities of interest". Europe's community media understood their mission as promoting diversity and pluralism and moving marginalized voices into local public spheres. In the United Kingdom, community radio originated with *Radio Caroline*, an off-shore pirate station operating on a ship in the 1960s. During the 1970s–80s, land radio pirates developed their own stations with programming that targeted ethnic minorities. In 1985, after many unsuccessful efforts on the part of the British government to interrupt pirate stations (*London Greek Radio* was raided more than 100 times in less than three years), low-power non-profit radio stations were granted licences under the rubric of 'community radio' (Keirstead and Keirstead 1987). According to Lewis, in Italy, confusing regulations or the lack of regulations created a chaotic scene in which community radio mushroomed out of control and, as a result, Italy ended up with more community radio stations than any other country in the world (Lewis 1978, 1984a, 1984b). This diversity is also reflected in the multiplicity of terms used to refer to community media. While in Britain the most common term is community media, Italy, France and Germany describe this type of media as *frei*, *libere* and *libre* (free). According to Scifo (2014:164), Spain uses the term *comunitarias*, the Netherlands uses *lokale omroep* (local broadcasting), while Scandinavian countries have opted for *near* (neighbourhood). In 1994, community broadcasters adopted the Community Radio Charter for Europe, a document that clearly links community media to communication as a human right (Scifo 2014).

Despite the growing power of corporate media and digital technologies, grassroots communities continue to find ways to appropriate communication technologies, but the process is never easy. Historically, communities have struggled against regulatory regimes that privilege corporate media and have been faced with insufficient funding, technologies designed for wealthy consumers in the Global North, media concentration and media exclusion. Since the turn of the century, the negative impact of media surveillance, data mining and manipulative use of algorithms has constituted a further challenge. And yet, communities still find ways to appropriate, adapt, hybridize, re-invent and converge media technologies to make them serve multiple and context-specific information and communication needs.

Future directions

There is overwhelming evidence that community media play a key role in maintaining healthy democracies, empowering questioning publics and creating strong local public spheres. And yet, community media are often neglected by research institutions and higher education media programmes. Indeed, most students graduate without having taken a course on community media. Donors and foundations also lack a basic understanding of community media and expect them to function like mainstream media. The interest that even world-renowned media experts and journalists often show, for example, in the audience size of community radio stations reveals a total lack of understanding of the function and value of community media and the reasons why they are valuable to our societies.

In terms of research activity in the field of communication and media studies, the vast majority of academic publications, research projects and doctoral dissertations focus on mainstream commercial and public media, as well as on digital platforms. Only a tiny fraction of academic work sets out to explore other types of media, which has contributed towards a tendency to equate very different media practices which might more productively be categorized under separate labels. As the body of knowledge generated in this area remains very small, our understanding of the distinct types of non-mainstream media continues to be muddled and incomplete. The distinctions between community media, citizens' media, militant media, indigenous media, pirate media, social movement media and radical media in particular are all too rarely made explicit. Future research should attempt to examine in greater depth such differences in order to emphasize the diversity of motivations and priorities involved, and to better understand the contributions of each to the public sphere.

See also: citizen journalism; conflict & humanitarian studies and citizen media; hyperlocal media; Indymedia; journalism studies and citizen media; race & ethnicity studies and citizen media

Recommended reading

Downing, J. D. H. (2011) *Encyclopedia of Social Movement Media*, Thousand Oaks, CA: Sage.

Explores a wide variety of community media initiatives in different world regions in the twentieth and twenty-first centuries. Includes a wide assortment of communication formats, from traditional media such as dance and graffiti, to broadcasting media and digital platforms.

Michaels, E. (1994) *Bad Aboriginal Art: Traditions, media, and technological horizons*, Minneapolis, MN: University of Minnesota Press.

Based on ethnographic studies of the Walpiri Aborigines of Western Central Australia, the book explores how Aboriginal communities appropriate and adapt media technologies according to their cultural idiosyncrasies.

Rodríguez, C. (2011) *Citizens' Media against Armed Conflict: Disrupting violence in Colombia*, Minneapolis, MN: University of Minnesota Press.

Based on fieldwork undertaken in various Colombian regions, this book investigates how unarmed civilians living in the shadow of armed groups – including leftist guerrillas, right-wing paramilitary groups, the army and drug traffickers – use community radio, television, video, digital photography and the Internet to shield their communities from armed violence.

CONFLICT & HUMANITARIAN STUDIES AND CITIZEN MEDIA

Derya Yüksek

Conflict studies has long been dominated by war-centric approaches, which have skewed our understanding of conflict by overemphasizing its negative and violent forms and influenced the way peace is conceptualized. These approaches have led to negative articulations of peace, at times associated with actions that involve overt violence, such as military interventions or covert coercion in the form of economic sanctions (Mitchell 2002:1). Academic interventions have further primed positivist perspectives, focusing primarily on prevention and elite peace-making. They have often neglected the structural and cultural drivers of conflict – in particular, power inequalities and concomitant injustices (Gleditsch et al. 2014; Diehl 2016; Richmond 2008). As a result, the agency and peacebuilding capacities of people living in conflict-torn communities have been largely overlooked.

In *Newer Ideals of Peace* (1907), Addams provides an early critique of the understanding of peace as an end to, or absence of war, which results from the prevalent militarization of society, not only through the adoption of local government practices that cherish the protection of the state and treat weak citizens as valuable resources, but also through the conceptualization of war in terms of culturally desirable values such as patriotism, courage and self-sacrifice. Instead, Addams proposes a positive, progressive understanding of peace that transcends its relationship with war, one that incorporates a socially transformative dimension: it weaves fragmented societies together by building positive relationships across social structures, from families and neighbourhoods, to cities, nations and the international community (Shields 2017). This transformation is a long-term, dynamic process that fosters lateral progress by focusing on the welfare of all individuals, including those at the bottom of the social ladder, and by locating the peacebuilding potential in "social justice, social equity, cooperation, community engagement, collaboration, effective-governance and democracy" (ibid.:37). This understanding of peace allows citizens to form an inclusive community of inquiry that promotes a sympathetic knowledge of the other, and to overcome differences by thinking and acting on shared problems within a participatory democratic framework (ibid.:38). Addams' postulates – that integrate the concepts of positive peace, social transformation and peaceful development as operational principles – provided the basis for what became a discipline in its own right at the end of the twentieth century.

This new disciplinary framework has drawn attention to the pervasiveness of war-centric perspectives in conflict studies and the extent to which they have influenced the design of peacebuilding interventions. Effectively, these have taken the form of track 1 and track 2 initiatives seeking to manage or resolve conflicts through political settlements or negotiations at the level of the state and involving elite actors. A similar approach has informed humanitarian interventions to manage (post) violent conflict, including humanitarian aid, post-war relocation, recovery and reconstruction. In these contexts, humanitarian measures and actions have been largely shaped in a top-down manner, without the involvement of the people whose daily lives and needs are ultimately affected by those actions.

These elite-centred peacebuilding approaches, with their emphasis on the parties involved in conflict and the material conditions on the ground, have failed to address the discursive aspects of violence and war (Addams 1907). Jabri (1996) has drawn attention to this dimension and examined how everyday practices underpinning the relationships between actors and social structures, and between self and society, evoke, normalize and reproduce violence – by negating plurality and generating exclusion. Within the context of asymmetrical power relations in society, the discursive hegemony of such antagonisms intensifies intergroup polarization as well as in-group cohesion, which are the main drivers of violent conflict escalation. As such, the potential for peace is located in discursive social relations; consequently, realizing that potential requires building critical, emancipatory approaches to peacebuilding that recognize difference and diversity (Jabri 1996:133–61; Ramsbotham 2010:89).

A transformative approach to peacebuilding

By the end of 1990s, the ineffectiveness of track 1 and track 2 initiatives in addressing the protracted nature of contemporary violent conflicts and in creating supportive conditions for sustainable peace brought about a shift in peace research. A new transformative approach, pioneered by Galtung (1969, 1996), Curle (1971, 1990), Rupesinghe (1995) and Dugan (1996), developed into a distinct school with the works of Lederach (1996, 2003). Grounded in a positive understanding of peace with its inherent connections to social equality and justice, the conflict transformation school focuses on "engaging with and transforming the relationships, interests, discourses and, if necessary, the very constitution of society that supports the continuation of violent conflict" (Miall 2004:4), including the structures and cultures of violence. This involves de-escalation of violence, as well as a constructive escalation of conflict for social change, connecting it with various forms of non-violent activism. For interventions to be successful, they should therefore initiate processes of change at personal, interpersonal, structural and cultural levels, with a view to maximizing the potential for self-growth, dialogue, understanding and participation, and identifying cultural resources to support these processes (Lederach 2003:27). As opposed to elite-centred approaches to peacebuilding, transformative approaches place emphasis on building non-violent relationships in society and acknowledge the agency of people in situations of conflict and in processes of change.

Reich (2006) and Francis (2011) underscore the cooperative nature of this process, arguing that transformation requires moving from a culture of exclusion and domination to a culture of inclusion and collaboration through shared power and active participation at every level of society (Francis 2011:508). While the cultural-discursive dimensions of conflict and peace still remain under-theorized in the transformational approach, some scholars (Featherstone 2000; Richmond 2005; Heathershaw 2008; Carpentier 2017) have contributed to filling this gap by studying peacebuilding from a discourse-theoretical perspective. Their critical analysis of the impact of power structures on ordinary people and traditionally oppressed segments of society advocates a radically inclusive model of society that listens to and respects these marginalized voices (Paffenholz 2009:5).

This shift towards a more progressive, maximalist view of peace has turned the attention of scholars and practitioners away from the macro field of state and political resolutions towards the micro fields of the social, through which violence is enabled and sustained in everyday life. It has also brought into sharp relief various social structures and the relationships embedded in them, in the design of peacebuilding interventions, with an emphasis on communication, participation and grassroots empowerment, where media play a key role both as channels and as agents in their own right.

Citizen media and conflict transformation

While this pivotal function of media is increasingly recognized in the scholarly literature, it remains under-researched even in the field of communication science – a discipline that, after the Second World War, placed particular emphasis on the study of how mass media mobilize people for war (Laplante and Phenicie 2009; Jusić 2009; Hoffman 2013). Peacebuilding interventions by international actors, on the other hand, started to address these effects only after the disastrous cases of inter-ethnic violence in Rwanda, the former Yugoslavia and Somalia. In all these cases, media organizations played an instrumental role in conflict escalation by propelling ethnic hatred and massacres, and rendering entire communities as spectators and actors of direct violence (Bromley 2011; Hoffmann 2013). These tragic experiences led international actors operating in the contexts of (post) violent conflict to integrate a range of communication interventions in their activities, including preventive measures such as broadcast jamming and supportive measures such as media monitoring and peace broadcasting (Hoffmann 2013:4). However, communication is still regarded as a secondary element or public relations strategy, rather than an integral constituent of post-conflict reconstruction processes (Kalathil et al. 2008:7; Hoffmann 2013).

In the field of communication studies, the role of media in conflict and peace has long been narrowly framed for two reasons: the restricted focus of media effects research on the relationship between media content and audiences (Newbold 1995), which neglects the larger cultural-discursive aspects of media use in these contexts; and the fact that definitions of media remain confined to traditional, mass communication formats and technologies. Researchers in this area have therefore placed excessive emphasis on the role of media in conflict escalation, paying less attention to their potential for conflict transformation. Exceptions to this trend include scholarly works on peace journalism (Galtung 1986), which inspired various projects of media development, including training and capacity-building activities aimed at the media sector. However, these projects were criticized for their overemphasis on journalists as peacebuilding actors, and for remaining ineffective in counteracting the hegemonic ideology of war, opening the concept of war to debate and questioning its legitimacy in the public sphere (Coban and Yanikkaya 2013:18).

Another exception is the scholarly strand that argues against binary definitions of media power – grounded in an understanding of mainstream media as all-powerful and audiences as powerless – and shifts attention to the various complex ways in which power can be exercised in and through the media, with a focus on media access and participation by ordinary people (Rodríguez 2000:157). These media have been theorized under different headings, including alternative media, participatory media, community media, activist media and radical media (Berrigan 1979; Downing 2001; Atton 2002; Bailey et al. 2008; Fuchs 2009; Carpentier 2011), to designate independent, decentralized, participatory and pluralistic media forms distinguished by their alternative-critical content and by the facilities they provide for democratic citizen participation. The technological developments of the Web 2.0 era have largely aided this shift of attention from mass media through a proliferation of media formats that facilitate the production of user-generated content and increase the reach of these outputs to larger audiences via the Internet. In addition to their support for the democratization of media environments across the world, as sites for challenging neoliberal hegemony and contesting the power of mass media, these participatory media forms have received significant attention in development contexts – mainly in the fields of development communication and participatory communication (Servaes et al. 1996; Servaes 1999; Tufte and Mefalopulos 2009) – and become privileged partners in donor-led development projects, particularly in Latin America and Africa. However, these studies have often neglected the peacebuilding potential of these media in the contexts of (post) violent conflict.

In her influential works on citizens' media and Colombian armed conflict, Rodríguez (2000, 2001, 2011) was the first to propose a rapprochement of peace studies with development communication to tap into the knowledge accumulation of the latter on alternative, community-based, participatory media, as a way of strengthening attempts to involve and empower citizens in the processes of peace construction

(Rodríguez 2000:147). Rodríguez' use of the term citizens' media was inspired by Mouffe's (2000b, 2005) theories of radical-agonistic democracy and citizenship, and highlights the transformative processes these media bring about by enabling self-representation, and the accompanying processes of empowerment that transform individual media participants into active citizens.

In the contexts of violent conflict and its aftermath, citizens' media can support conflict transformation by providing spaces "in which communities can consider, experiment with, and witness other, alternative, non-violent ways of dealing with conflict; other, alternative, collective imaginaries; other, alternative ways of interacting with others; and other, alternative, ways of understanding difference" (Rodríguez 2010b:151). Instead of communicating predesigned messages and strategies about peaceful coexistence, which is often the case in peacebuilding interventions, they serve a performative function, giving people a means to express and perform identities, lifestyles, cultures and interactions that are not pervaded by a war-driven logic, thus enacting and normalizing non-violence in a variety of ways (Rodríguez 2011:82). They give voice to the voiceless, act as a platform for communication and empathetic interaction, provide alternative sources of information with their independent position towards state and market (that are often part of the war-driven logic), connect isolated communities for collective action, and foster conscientization through a participatory engagement with social realities and various social injustices (Rodríguez 2000:152–157).

Rodríguez' studies on citizens' media have paved the way for a small yet growing body of literature relating to the potential of citizen's media for supporting conflict transformation. A number of scholars working in this area have focused on organized, community-driven media forms, their structures, content and practices (Carpentier 2015, 2017; Carpentier and Doudaki 2014); others have examined the participation of conflict-torn groups in the processes of media-making through various genres such as storytelling, video, photography and theatre (Higgins 2011, 2014; Baú 2015, 2018); yet others have zoomed in on the empowering potential of such media in contexts of protracted social conflict (Custódio 2017; Salazar 2010). In all cases, the locus of attention has been on local populations and citizen collectives, with particular emphasis on traditional audiovisual media formats in contexts with poor access to media technologies.

The emerging concept of citizen media (Baker and Blaagaard 2016a) questions the centrality of media in previous research on citizen-driven media and, in particular, the extent to which local communities have been prioritized over the larger global context in earlier scholarship. It also queries a number of premises underpinning previous research, including the comparatively higher attention that the collective agency of citizens and the use of media technologies have received at the expense of the individual and various physical forms of expression, respectively. Instead, Baker and Blaagaard (2016a:16) propose a broader definition that acknowledges and incorporates "the physical artefacts, digital content, practices, performative interventions and discursive formations of affective sociality produced by unaffiliated citizens" through their acts in public space(s) and their participation in the creation of diverse publics, including the values and agendas guiding these practices and discourses. This definition connects the material and discursive potentials of these media, links the expressive with the performative, and highlights their purpose. Ultimately, this conceptualization foregrounds the autonomous democratic expressions of citizenry in all its forms: organized and non-organized, individual and collective, physical or digital. Accordingly, citizen media manifest themselves not only in the practices and outputs of media organizations, but also in various forms of advocacy, activism and resistance by individuals and groups, enabling citizens to express and organize the expression of dissent, and to act as agents of change.

This approach is particularly useful in examining the role of media in the contexts of (violent) conflict and their conflict transformation capacities. First, it raises the often-overlooked issue of individual agency in the processes of peaceful change, enacted through various forms of political and artistic expression and intervention in conflict-affected societies. Second, citizen media's attention to physical spaces and face-to-face relationships brings into focus the human connections and interactions that are at the core of peaceful coexistence, as well as the expressive power of physical symbols and artefacts in conflicts, together with their often unnoticed transformation. Third, the concept highlights various tools and means of communication that do not necessarily involve media technologies, including the human body and various forms of bodily expression which, at times, embody the message. This has particular importance in the contexts of violent conflict, where citizens are often deprived of or denied conventional forms of media access. On the other hand, Baker and Blaagaard's (2016a) definition of citizen media does not exclude organized media, which play a crucial role in structuring and supporting the participation of citizens in the media (Carpentier 2014:91), and in facilitating bottom-up peacebuilding in conflict contexts. Finally, citizen media account for the diverse alliances formed by and between individuals and groups in relation to peacebuilding, helping to envisage peace as an umbrella struggle for all citizens through the emancipatory and collaborative potentials they offer.

Nevertheless, an approach that is too celebratory of citizen media in terms of their role in conflict transformation is problematic, as they do not always constitute progressive or peaceful initiatives. Citizen media may equally take the form of repressive and reactionary initiatives that work to trigger antagonisms, fuel conflicts and in certain settings become a cause of overt violence by spreading hate speech, essentialist discourses and dehumanizing imaginaries, as in the case of their use by alt-right, ultra-nationalist or fascist movements that pursue exclusionist agendas (Downing 2001; Atton 2002, 2006; Cammaerts 2009). In the context of (violent) conflict, these forms of citizen media may be prevalent and may even be supported and legitimized through the dominant culture of war.

Citizen media and conflict transformation: the case of Cyprus

Cyprus has been a site of protracted conflict since the 1950s, when the armed struggle against British rule fuelled an outbreak of inter-communal violence among the two most-populated communities inhabiting the country, Greek Cypriots and Turkish Cypriots. The conflict escalated against the backdrop of rising waves of nationalism that ultimately led to the geographical and ethnic division of the island in 1974. Since then, the two communities have been living on separate sides of the island, divided by a United Nations Buffer Zone. For decades, the Republic of Cyprus and the self-declared, internationally unrecognized Turkish Republic of Northern Cyprus have held political talks for reunification, with occasional interruptions. Today's Cyprus is not in a state of war. Since the division, there has been no inter-communal violence – except for a number of rare incidents involving physical attacks perpetrated by nationalist groups on both sides. However, the long-term political stalemate has resulted in a physical segregation of the Greek Cypriot and Turkish Cypriot communities, whose integration is hindered by political, cultural and psychological impediments (Zembylas 2010:439). Cyprus is thus a political environment characterized by divided communities and publics, whose violent past has created severe ruptures in communication and mutual trust, and that is kept alive by antagonistic discourses that prevail in a variety of fields, including the media systems on both sides.

Nevertheless, there is a significant variety of citizen media practices in Cyprus, partly due to the long-lasting presence of the United Nations and other international organizations that have increasingly integrated civil society and grassroots organizations in their peacebuilding interventions, and partly due to the ceaseless efforts of pro-peace individuals and groups that have shaped the bicommunal peace movement. These range from community-based media organizations and journalistic practices, to various forms of non-violent activism led by individuals and collectives that support conflict transformation in multiple ways; by facilitating dialogue and contestation among the conflict-torn communities; constructing alternative narratives and imaginaries of conflict and peace; and enabling confrontation, advocacy and resistance through non-violent means.

On the organized media front, Cyprus Community Media Centre (CCMC) – an umbrella organization established in 2009 with the aim of connecting, empowering and equipping civil society organizations and community groups from both parts of Cyprus – is a leading example of citizen media. Located in the United Nations Buffer Zone, CCMC incorporates an inter-communal, multilingual online radio station, MYCYradio, that is open to any broadcaster. Scholarly work on the conflict transformation potential of

CCMC and MYCYradio (Carpentier and Doudaki 2014; Carpentier 2014, 2015, 2017) emphasizes the empowering, networking and advocating capacities of the Centre, which acts as a bridge between the Greek Cypriot and Turkish Cypriot mediascapes, and as a rare physical location, where organized citizens across the divide meet (Carpentier and Doudaki 2014:430). The findings suggest that the Centre supports conflict transformation by enabling citizen participation and self-representation: its web radio content allows for agonistic re-articulations of conflict "by describing the many ways the divide is already overcome, by decentering it, by deconstructing the homogeneous and united self, by reverting to an idyllic past before the conflict, by envisaging a future and by showing the cost of war and the divide" (Carpentier 2015:150–151).

In addition to CCMC, there are multiple examples of organized citizen media initiatives in Cyprus. A mapping study by Voniati et al. (2018) has identified twenty-six community media organizations operating across the island, the majority of which have been recently established and are run by small collectives, individuals and NGOs. These media – that enjoy a certain degree of autonomy from market and state in their content production and daily management – seek to give voice to and facilitate dialogue between and within communities, promote a culture of active citizenship and enhance the capacity of citizens to participate in the public debate. Ultimately, they aim to oppose exclusionist power structures brought about by nationalism, militarism, neoliberalism and patriarchy, and to propose alternative modes of social organization (ibid.:26–28).

Organized media are not the only form of citizen media in Cyprus, where the peacebuilding field has long been supported by the individual efforts of citizens who also engage in certain journalistic practices. Examples of these self-led initiatives include, among others, the pioneering works of Sevgul Uludag, who used investigative journalism as a medium to facilitate dialogue between communities by sharing citizens' stories of multicultural coexistence, and to open sensitive issues concerning the island's violent past to public debate through her newspaper and blog articles, published both in Turkish and Greek. These issues include the Greek Cypriots and Turkish Cypriot civilians who went missing during the period of intercommunal violence and have not yet been declared dead or alive. *Oysters with the Missing Pearls* (Uludag 2005a) and *Cyprus: The Untold Stories* (Uludag 2005b), based on the author's detailed investigations, brought to light these stories, which had been suppressed and denied by both sides, as told by citizens who lost track of their relatives and witnessed the ensuing events. The stories facilitated a process of dialogue between the two communities, emphasizing their common pain and suffering. Uludag supported this process by setting up a hotline for her article series, allowing people to call and anonymously provide information on missing persons and burial sites, leading to numerous exhumations. These efforts have also been instrumental in launching a bicommunal citizen initiative that brings together the relatives of missing persons (Association for Historical Dialogue and Research 2011:10–11).

Citizen media in Cyprus also embrace other forms of non-violent individual and collective activism. These range from symbolic acts initiated by individual citizens to express their demands and discontent (Canalp 2017) to various educational, cultural and artistic activities and interventions (Higgins 2011; Carpentier 2019); advocacy actions such as the bicommunal Peace Day March; and a multiplicity of grassroots movements, including but not limited to Occupy Buffer Zone, Unite Cyprus Now, Demilitarized Nicosia and Conscientious Objection Initiative in Cyprus – which connect local groups to global citizen networks. Although these actions and movements have distinct and at times conflicting priorities, they share a common aim: connecting citizens, confronting and resisting the culture of war and influencing the official negotiation process through increased citizen participation, particularly during periods of turmoil. In addition to face-to-face meetings and events, they use social media extensively to enable communication between their members and supporters, and to initiate discussions on important issues in which citizens have a stake. In doing so, they serve not only as a platform for uniting people, but also as a vibrant sphere of contestation (Mouffe 2005:3) driven by solidarity and respect for diversity.

As evident in the example of Cyprus, citizen media have the potential to facilitate the processes of peace-weaving (Shields 2017) by supporting citizen engagement and empowerment; promoting solidarity and respect for diversity; acting as a means to recognize, confront, debate, negotiate and contest various conflict materialities and discourses; and providing a space for non-violent expressions of conflict and peaceful coexistence. This aligns with the long-term, maximalist conceptualization of peace that positions conflict and peace along a continuum, with peace conceptualized as a praxis, as a continuous, society-wide struggle for democracy, equality, human rights and justice.

See also: community media

Recommended reading

Carpentier, N. (2017) *The Discursive-Material Knot: Cyprus in conflict and community media participation*, New York: Peter Lang.

Presents discursive-material analyses of participation, community media organizations and conflict transformation, with a focus on the divided island of Cyprus. Blending theory and collaborative research, it explores the complexity of community media and examines their potential for transforming antagonism to agonism in conflict-affected contexts.

Lederach, J. P. (2003) *The Little Book of Conflict Transformation*, New York: Good Books.

A guide to conflict transformation with emphasis on building non-violent relationships, structures and cultures. Presents a framework for creating constructive transformation processes that reduce violence and increase justice, by attending to personal, interpersonal, structural and cultural dimensions of conflict and change.

Rodríguez, C., D. Kidd and L. Stein (eds) (2010) *Making Our Media: Global initiatives toward a democratic public sphere*, Volume 1, *Creating New Communication Spaces*, Cresskill, NJ: Hampton Press.

A critical examination of alternative and citizens' media across the world and their role in building participatory democracies. Contributors consider the relationship between these media and the people they serve, reevaluate established theoretical frameworks and present new ones in light of contemporary local and global realities.

CONTENT MODERATION AND VOLUNTEER PARTICIPATION

Martin Johannes Riedl

The work of content moderation plays an essential part in the maintenance and sustainability of many online communication platforms. Social media sites such as Facebook, Twitter, WeChat, Sina Weibo and VKontakte have assumed powerful roles in society and citizens have come to use such platforms for a variety of ends: to participate politically (Papacharissi 2010), to assemble around specific interests or concerns, as in the case of self-help communities (Myrick et al. 2015), as well as to engage in citizen journalism (Robinson and Schwartz 2014). As businesses, these sites depend on volunteer contributions of user-generated content, provided as a form of free online labour (Andrejevic 2010). It is thus in the best interest of these platforms to create inviting online environments in which people will enjoy spending time. This often involves weeding out divisive content as a means of preempting confrontation, conflict or incivility between users. Modes of governance may include top-down approaches wherein platforms regulate and moderate citizen-generated media by explicitly forbidding and policing the publication of nudity, gore or other obscene material (Roberts 2014, 2016, 2018, 2019). Peer production sites such as Wikipedia, on the other hand, operate from the bottom-up, allowing citizens themselves to self-organize to create site-specific regimes of content creation and moderation (Reagle 2010; van Dijck 2009). In this entry, the social news aggregation website Reddit serves as an exemplary case study for showing how content is moderated and managed online, both through the voluntary labour of citizen moderators and through the work performed by general users in flagging inappropriate content. The entry additionally provides a discussion of the broader social significance of content moderation as an activity which not only shapes but is also shaped by the relationship between social media platforms and their users (Gillespie 2018a; Suzor et al. 2018).

Content moderation

Within sociotechnical systems, content moderation is the editorial capacity and practice of deciding on the publication or rejection of specific content through computational means and/or human judgement. In this sense, it operates in much the same ways as journalistic editing and other forms of quality assessment and screening. However, the term is now almost exclusively applied in the context of free online user-generated content (van Dijck 2009), typically created through unpaid, voluntary labour (Terranova 2000). Content moderation can thus be defined more specifically as the "organized practice of screening user-generated content (UGC) posted to Internet sites, social media and other online outlets" (Roberts 2014:12). This process of looking at content and making a determination as to its admissibility can in some cases be undertaken through the voluntary participation of platform users who become involved in policing or flagging materials posted online (Crawford and Gillespie 2016). Alternatively, it can take the form of paid work, assumed by trust and safety teams assembled by the platform itself (Gillespie 2018b; Kaye 2018), via crowdsourcing platforms such as oDesk (Klonick 2017), or via third-party outsourcing companies (Roberts 2014, 2017). Caplan (2018) distinguishes between artisanal, community-reliant and industrial approaches to content moderation, depending on the nature of the business models, staff resources, intent and mission of the media platform concerned. Artisanal moderation follows a system based on case-by-case governance performed by small teams, as is the situation within the video-hosting site Vimeo. Community-reliant moderation, on the other hand, involves volunteer moderators enforcing rules put forth by a company (such as Reddit). Finally, industrial-style solutions (as adopted by social media giant Facebook) bring together legions of professional moderators who police content following rules created by policy teams (Caplan 2018). These trust and safety teams operate from the platform headquarters and are tasked with policy creation and the adjustment of rules. However, it is often workers not directly employed by the platform who have to enact these policies.

Machine learning and artificial intelligence solutions, as well as catalogues of inadmissible words, are some of the technological solutions offered in relation to this issue (Myers West 2018). So far, however, content moderation has not yet been satisfactorily resolved without the involvement of human judgement. Content moderation is thus frequently outsourced to third-party companies, often located in parts of the world where labour costs are lower, such as India or the Philippines (Breslow 2018; Chen, A. 2017). The Philippines, and its capital of Manila in particular, have served as a productive market for Western social media companies seeking content moderation labour, largely thanks to the good command of the English language possessed by many Filipinos, and for the fact that they share similar norms and values with the populations of many Anglophone countries such as the United States because the histories of these states have been influenced by the same religion and corresponding moral value systems (Block 2018). While it is an inescapable fact that content moderation relies on humans, the human cost of content moderation has garnered considerable criticism. The first accounts of content moderation, chiefly by Roberts (2014, 2019) and Chen (2012a, 2012b, 2014, 2017), have pointed to "the human toll of protecting the internet from the worst of humanity" (Chen, A. 2017). The policy of outsourcing moderation work to third-party companies outside the United States has also been critiqued, among others, by Breslow (2018:2), who, in the context of moderation in the Philippines, asks whether "the fact that this regulation of American sexual content is done by people in a former US colony … open[s] up the ways in which sexuality must be understood in relation to colonial power". The answer is provided in the framing of the question: colonialism is indeed implicit in the structure of Western technology companies' outsourcing of menial labour to a former colony such as the Philippines. Workers in Manila are carrying out emotional labour based on criteria imposed by US mores on what type of sexual (and other) conduct is acceptable. In the United States, lawsuits filed against Facebook in California (Reuters 2018) and against Microsoft in the state of Washington (Ghoshal 2017) speak of the psychological toll of content moderation and the possibility of moderators developing post-traumatic stress disorder (PTSD) due to their work.

In addition to paid content moderators, platforms may also co-opt the power of their user bases to maintain community guidelines. Flagging mechanisms are built into most social media platforms because they enable platforms to manage the vast scale of the task of moderation by employing the millions – if not billions – of citizen users interacting with and via their site. Wikipedia – not a social media platform but a platform with meticulous governance nonetheless – is one such example of a site in which voluntary, unpaid labour upholds the integrity of the system. Similarly, Reddit makes use of super-users who voluntarily moderate discussions within the site in exchange for status. Many platforms additionally employ upvoting or downvoting mechanisms, through which general users can influence the visibility of content posted on platforms.

The labour provided in such cases is, in Terranova's (2000:33) terms, "simultaneously voluntarily given and unwaged, enjoyed and exploited". This complexity derives from the fact that the civic labour of content moderation serves three distinct stakeholders whose interests are necessarily in tension with one another: platform participants, other moderators and the platform itself (Matias 2016). Moreover, while many platforms start out as fairly democratic in the way that they deal with content moderation, governance structures have frequently been shown to become rapidly more oligarchical as the platform expands (Shaw and Hill 2014): in larger and more complex sites, a relatively small number of paid or unpaid moderators are typically endowed with greater privileges and these may come to exercise a monopoly of power over content. In such contexts, platform moderation policies and censorship mechanisms in both commercial and volunteer platforms can clash with user interests. Since content moderation – a process which involves human judgement and is thus prone to error – can lead to false positives, this means that the voices of minorities of all kinds could be wrongfully suppressed. Therefore, due process (Suzor 2018), or the protection of the rights of contributors in cases where speech rights are curtailed by takedowns, is an important avenue for research, activism and policy development (Anderson et al. 2016).

Reddit

As social media platforms encroach upon more and more aspects of social life (from communicating with friends to commerce to dating), it is important to highlight the vast economic and political power that they have come to hold, both through the creation of pervasive advertising ecosystems and through data collection mechanisms mining information about their users. While platforms may attempt to cast themselves as neutral, this term can itself serve as a definitional and rhetorical safe house for technology companies seeking to avoid regulation. Critics have argued that social media platforms should be conceptualized as media companies rather than as neutral arbiters of information (Napoli and Caplan 2017). This is because, in Braun's (2015:1) terms, platforms are "mechanical editors, deciding algorithmically which posts and topics warrant inclusion in the continuous and often overwhelming feed of information". Moreover, as discussed above, platforms do outsource many decisions not only to algorithms but also to end-users and moderators (Matamoros-Fernández 2017), and human judgement certainly does play a vital role in a variety of their governance mechanisms.

In the early developmental stages of any social media platform it might still be possible to refrain from content moderation altogether since the norms and ideologies shared among people interacting through the platform align. Early research on moderation questioned "whether shared norms can emerge about what constitutes a good or bad post" (Lampe and Resnick 2004:543). As soon as platforms start to scale up and interest groups diversify, conflict tends to increase. As platforms grow, they have to develop governance structures and need to identify what culture of interaction they want to nurture. Freedom of speech cultures vary drastically, and in certain legislations such as the United States even hate speech falls under First Amendment protections. In Germany, on the other hand, legislation known as the Network Enforcement Act, passed in 2017, requires platforms above a specific size threshold to react to complaints about hate speech within a set time frame. Noncompliance can lead to fines of up to 50 million Euros, and the prospect of such fines is assumed to carry the potential for the over-regulation of online communication (Gollatz et al. 2018; Heldt 2019). In the case of the United States, content moderation enforced by platforms is, first and foremost, a voluntary act (Myers West 2018). Section 230 of the Communications Decency Act of 1996 (CDA 230) regulates – for the United States of America – that platforms are not to be held liable for illicit content posted on their sites by users (Electronic Frontier Foundation 2018; Medeiros 2017). Platforms are not more liable whether they moderate content or not (Matamoros-Fernández 2017). Rules for participation are commonly codified in terms of service and community guidelines. These are performative texts to the extent that, as much as they articulate what types of communications platforms might wish to enable, they also serve as a means for platforms to assert their value systems (Gillespie 2018a). They are both a normative rallying cry and a signal to user bases indicating appropriate models of behaviour. They additionally provide policy guidelines which may be used to justify content takedowns.

Reddit was founded in 2005 by Alexis Ohanian, and later sold to Condé Nast in 2006 (Massanari 2015). Predominantly an aggregator of various kinds of content published on the Internet, it also serves as a social news site (Leavitt and Clark 2014; Suran and Kilgo 2017), meaning a site through which citizens contribute their own news content. A Pew study found that in 2018, 73 per cent of Reddit users in the United States got news through the site (Shearer and Matsa 2018). Reddit thus provides avenues for large numbers of people to engage with and discuss topics that range from trivial to the arcane. Reddit has been called the "front page of the internet" (Bond 2018) and was the seventeenth most popular website in the world (fifth in the United States) in September 2018, based on its one-month Alexa traffic rank (Alexa 2018a, 2018b).

Reddit is a valuable platform in which to explore online content moderation for a number of reasons: first, the site allows users to interact within the site under a cloak of pseudonymity (van der Nagel and Frith 2015) and Reddit thus distinguishes itself from other social media platforms such as Facebook. Pseudonymity lends the platform a "sense of play and candor" (Massanari 2017:331). Reddit has become an important purveyor of free speech as a result of its largely technolibertarian culture and, in comparison with other platforms, its relatively open policy as far as the admissibility of content is concerned (Centivany 2016; Centivany and Glushko 2016; Marwick 2017). Moreover, Grimmelmann (2014:243) described Reddit as offering a "gold-plated exit option to preserve user freedom": if someone is unhappy with the content moderation policies applied within one subreddit, they may simply start their own new subforum on the platform, and thus take charge of the rules of engagement within that subforum. Marwick (2017) has argued that Reddit's stance concerning user expression has emerged primarily out of the strong hacker ethics of the platform's early community members, the legal ramifications of CDA 230 and a strong belief in the possibility of self-regulation. While liberties of this kind may certainly be seen as part of a positive, utopian attempt to enable citizens to participate, deliberate, interact and share experiences and content without limit or restriction, these same freedoms also generate greater potential for abuse, exploitation and harassment. As a purveyor of freedom of expression, Reddit has become what Massanari (2017:333) calls a "nexus for various toxic technocultures to thrive". One prominent illustration of this was #Gamergate, a full-scale harassment campaign targeted at female journalists and computer game developers (Massanari 2017). The controversy started out with the ex-boyfriend of a video game developer accusing his former girlfriend of sleeping with a journalist (Perreault and Vos 2016), but quickly evolved into a concerted wave of attacks against women in the gaming industry, organized through Reddit and the imageboard 4chan, and grouping around the hashtag #Gamergate. Purportedly, it was the intent of the campaign to convey concerns about ethical and political correctness in gaming journalism (Wingfield 2014), though the nature of the discourse speaks to widespread misogyny.

Reddit may be understood as a "community of communities" (Massanari 2017:331), with various subreddits serving as a topical structure for the discussion forum as a whole. Subreddits are threads dedicated to particular topics and themes. In 2018, Reddit held 1.16 million different subreddits (Misera 2018). These subreddits are somewhat autonomous, and display governance structures that employ Reddit's upvoting and downvoting system, through which participants can push specific posts up or down to increase or decrease their visibility (Centivany and Glushko 2016; Massanari 2017; Momeni Roochi 2014), as well as volunteer moderators in charge of the subreddits. The moderation system in place within Reddit, with voluntary moderators in charge and a liberal set of norms applied, appears to have transformed the site into a fertile breeding ground for the highly corrosive communicative behaviours described above. At the same time, however, such a governance structure can be seen as "embedded democracy", given its "voting mechanisms and distributed authority vis-à-vis subreddits" (Centivany 2016:5). Reddit's citizen moderators maintain so-called 'Reddiquette', and themselves police the platform and the norms that govern particular subreddits (Massanari 2017).

The task of moderation brings its own risks, as Grimmelmann (2014:147) has noted: "[t]he necessity of change creates the possibility of oppression". Being a moderator on Reddit can equate to a position of considerable power, depending on the size and scope of the subreddit. Reddit users may of course simply leave a particular subreddit if and when moderators begin to govern their "mini-fiefdoms" like dictators (Massanari 2017:340). Yet, "the design … doesn't prevent the moderators of a subreddit from behaving atrociously" in the first place (Grimmelmann 2014:148). In other cases, citizen moderators may choose to exercise their power against the platform itself, aware that they contribute a great deal to its success and sustainability: if moderators do not feel they are being treated fairly by the platform, they might organize a protest. This happened, for example, in 2015 during the so-called Reddit Blackout in which a large number of Reddit moderators went on strike, effectively disabling their subreddits to leverage their calls to "improve moderator software and increase its [Reddit's] coordination with moderators" (Matias 2016:8).

Future directions

Moderation, as Gillespie (2018a:13) has argued, is the "crucial commodity that platforms offer". Yet, in the case of many online environments such as Reddit and Wikipedia, this service is overwhelmingly supplied from the bottom-up, by volunteer users rather than the private corporations who own and manage these sites. Investigations into content moderation therefore provide a rich source of insight into the role of ordinary citizens within ongoing debates surrounding freedom of expression on social media platforms. Future research could usefully examine in further detail the ideological implications and power relations of volunteer moderation practices, as well as the nature of the evolving relationship between citizen moderators and the platforms for which they work. For example, following Duguay et al. (2018), studies might examine how flagging mechanisms can be used not merely to help mitigate the impact of hate speech but also misused as tools of censorship, aimed at suppressing minorities such as queer women. Such research will be instrumental in illuminating methods of creating more democratic moderation regimes and appeals processes, by way of involving citizens in decision-making about the acceptability of content at every stage in the process, as well as in reviewing wrongful content takedowns. Citizen media scholars might additionally explore content moderation not just as a hegemonic practice for platforms to control content, but also as an expressive form through which civil society might organize and claim agency in digital spaces through collective action. Examples such as the German social movement #ichbinhier, in which citizens have collaborated through social media in an attempt to drown out hateful and uncivil comments point in this direction (Ziegele et al. 2019). Content moderation brings the potential for citizens to shape communicative norms within different communities, and to engage in corrective actions. Consequently, future studies may fruitfully investigate the ways in which content moderation may function as a form of productive civic engagement.

See also: Facebook; social media; Twitter and hashtags; Weibo; wikis; YouTube

Recommended reading

Caplan, R. (2018) 'Content or Context Moderation? Artisanal, community-reliant, and industrial approaches', *Data and Society*. Available online: https://datasociety.net/wp-content/uploads/2018/11/DS_Content_or_Context_Moderation.pdf.

Emphasizes the notion that there is no one-size-fits-all approach to content moderation. This report by the Data and Society Research Institute outlines how different kinds of moderation are employed within different types of digital platforms, which impacts what realistic expectations toward moderation can be held.

Gillespie, T. (2018a) *Custodians of the Internet: Platforms, content moderation, and the hidden decisions that shape social media*, New Haven, CT: Yale University Press.

Articulates the argument that content moderation is a core commodity that social media platforms provide, and grounds the topic in a discussion of power and politics. Several case studies illustrate the discussion throughout.

Roberts, S. T. (2019) *Behind the Screen: Content moderation in the shadows of social media*, New Haven, CT: Yale University Press.

Critically examines content moderation as an often precarious form of digital labour that takes place under opaque circumstances and that often exacts an emotional toll on the workers involved.

CONVERGENCE

Henry Jones

Convergence is one of the most pervasive concepts in contemporary media studies. Although chiefly associated with the work of Henry Jenkins (2001, 2004, 2008), the term is repeatedly referenced in the titles of many publications in the field, and a search for *convergence* and *media studies* in the Google Scholar database returns many thousands of results. That said, *convergence* is most often found modified by one of a panoply of adjectives – *media, social, technological, temporal, global, visual convergence* – or alternatively working as a modifier itself, denoting a particular kind of journalism, culture, environment, regulatory framework, corporate strategy, set of technologies, texts, economies and/or industries. This diversity of collocations does not merely highlight the depth of scholarship on this topic; it is also symptomatic of the degree to which convergence has come to mean different things to different scholars (Balbi 2017:32). To cite but two examples, Dwyer (2010:2) sees convergence as "the process whereby new technologies are accommodated by existing media and communication industries and cultures", whereas Jenkins (2008:2) defines it as "the flow of content across multiple media platforms, the cooperation between multiple media industries and the migratory behaviour of media audiences who will go almost anywhere in search of the kinds of entertainment they want". Many such definitions contain the idea of certain communication tools, agents, roles and/or practices coming together and entering into new combinations but, as Ytreberg (2011:507) suggests, the differences of interpretation are often so stark that the theorists concerned "seem almost to live in different worlds, each one seemingly unaware of the others' approach and traditions".

Largely for this reason, convergence has been singled out by some as "a dangerous word" (Silverstone 1995:11). Its application to so many varieties of media-related phenomena has been accused of leading to widespread confusion as to the term's true meaning and a growing scepticism as to its value as a conceptual lens: so nebulous has convergence become, the argument goes, that it serves very little actual analytical use, describing everything and therefore, ultimately, nothing (Grant 2009:15; Noll 2003:12). Critics worry the term has become no more than a fashionable techno-buzzword, casually thrown around in all too many discussions of media, technology and contemporary culture, without contributing anything of substantial significance (Gordon 2003:57; Ytreberg 2011:502).

Advocates of the concept, on the other hand, embrace its broad applicability, arguing this constitutes one of the model's core strengths: quite simply, Meikle and Young (2012:4) suggest, "being able to explain many different kinds of media phenomenon with a single concept is a useful thing". Specifically, and as this entry shows, this is because it provides a means of emphasizing the profoundly interconnected nature of current developments in communications technologies, the emerging digital culture, capitalist economics and global society at large, as well as of understanding the complex dynamics of such changes (Miller, V. 2011:72). Moreover, despite the sceptics' fears, the term can and has been productively deployed in the analysis of concrete practices of media production, consumption and dissemination, and especially of activities that fit squarely with the definition of citizen media promoted in this Encyclopedia. Consequently, this entry highlights what are most generally referred to as technological, global, textual, industrial and cultural processes of convergence and illustrates their importance for research into citizen-led interventions in the media sphere.

Interconnected convergences

While the intellectual origins of the concept of convergence lie in the natural sciences and mathematics in the late seventeenth century, its popularization in media studies is widely attributed to the work of political scientist Ithiel de Sola Pool in the early 1980s (Gordon 2003:57; Jenkins 2008:10; Balbi 2017:31). In his book *Technologies of Freedom*, Pool (1983:23) used the word to describe the emerging technological context in which the dividing lines between different media were becoming increasingly blurred. The one-to-one associations that had previously held between specific technologies and particular kinds of content, he argued, were fast losing all meaning in an environment where, for example, the telephone network was being used not only for spoken person-to-person communication, but also the transmission of data between computer terminals, the circulation of news bulletins and the distribution of printed materials via facsimile machines (ibid.:23). Electronic technology was, Pool (ibid.:27–28) observed, "bringing all modes of communications into one grand system".

This historic shift was of course partly triggered by the wave of neoliberal deregulation that swept the West in the 1980s, and the series of pro-market government policies that have allowed for the expansion and conglomeration of a diverse range of telecommunications, news and computer companies into a small handful of vast multimedia corporations (Castells 2009:56). Indeed, as Castells (ibid.:79) argues, this process of industrial convergence has greatly facilitated the closer integration of previously distinct communication channels by offering greater economies of synergy for the businesses concerned (Meikle and Young 2012:35). However, given the unique affordances of new media tools to permit such combinations, convergence has been primarily understood as a technological process involving the "bringing together [of] multiple media functions within the same devices" (Jenkins 2008:3; Bolter and Grusin 1999:45). As such, the term is associated especially closely today with the logic of digitization, that is, with the role of the digital computer as a meta-medium, able to assimilate, combine and recreate all other modes of media production and consumption within single, often interchangeable devices, while simultaneously allowing for the fluid propagation of content across whole networks of devices (Jensen 2016; Peil and Sparviero 2017:4; Miller, V. 2011:73; Balbi 2017:35).

As Peil and Sparviero (2017:4) note, this technological paradigm shift is widely understood as the underlying, prerequisite feature of today's media environment, which has allowed for the emergence and development of all other phenomena discussed as manifestations of convergence. For example, and while acknowledging the need to avoid the trap of technological determinism (Castells 2000:5), Jenkins (2001) argues that it is largely a result of the ability of computer tools to act as a meta-medium for all forms of human communication that we are now witnessing an acceleration in what he describes as a process of global convergence, i.e. the "cultural hybridity that results from the international circulation of media content". This is because the advent of digitization has contributed powerfully to the spreadable nature of (citizen) media products by enabling all content to be converted into a standardized binary code and thus separated from its physical form, infinitely reproduced, compressed into easily deliverable packets, injected with hyperlinks and sent anywhere in the world via the web at the click of a button (Jenkins et al. 2013:3; Miller, V. 2011:74). As a consequence, networked local communities in the twenty-first century are afforded significantly greater access to cultural forms originating beyond their national borders, leading to the creation of novel hybrid practices and modes of expression. As both Tarifa (2012) and Dangl (2007) have explored, this is clearly exemplified in the music of groups such as Ukamau y Ké in El Alto, Bolivia, whose songs meld indigenous Andean folk instruments and lyrics with the aesthetics of Bronx-born hip-hop as a means of articulating shared affinities across languages and continents, and promoting social and political change within their own public sphere.

Technological convergence is also linked to experimentation in what Meikle and Young (2012:80) discuss in terms of textual convergence. As Jenkins (2001) writes, "[w]hen words, images and sounds are transformed into digital information, we expand the potential relationships between them". That is, now that single devices are able to receive all forms of media content and simulate all kinds of media experiences, the technical barriers that previously forced these different kinds of content to rely on different modes of delivery and reception have largely ceased to matter (Meikle and Young 2012:80). Thus, thanks to the affordances of networked digital media tools, an article written for the online citizen news outlet *Global Voices*, for example, can be transformed into a truly multimodal and multimedial experience: citizen journalists are now able to combine text, graphics, photography, video footage and audio files in fluid and often interactive configurations, exploiting the potential of any or all of these formats when telling their stories online (Kagumire 2018). Such practices develop ever more complex and diverse webs of intertextuality across media systems (Meikle and Young 2012:80; Jenkins 2008:93–134), as well as encouraging further creativity in the form of remediation, remix and bricolage (Deuze 2006b:70; Gallagher, this volume). These new forms of cultural expression are by no means only connected to the convergence and omnipresence of digital devices, but there can be no doubt that technology has greatly facilitated and accelerated their rapid proliferation in society (Deuze 2006b:66).

Most importantly, however, technological convergence has opened up considerable potential for a new era of cultural convergence, that is, the blurring of lines between the producers and consumers of media content (Jenkins 2001). For instance, whereas in the analogue age formidable technical, industrial and economic barriers prevented general access to the hardware required to record, process and share video footage relevant to a particular news event, the widespread availability of relatively inexpensive plug-and-play devices such as smartphones – which combine a camera, a microphone, easy-to-use software and cellular connectivity – means that today most inhabitants of developed (and many developing) states carry in their pockets all the tools required to capture, archive, edit and disseminate such content to millions of people worldwide (Murdock 2000:36). Consequently, as Allan (2006:146–147) has pointed out in a case-study focused on the reporting of the 2005 London bombings, ordinary citizens – otherwise unaffiliated with news institutions such as BBC News – are now empowered by convergent media to commit "random acts of journalism" (Lasica 2003), contributing their voices and experiences to the unfolding story on an unprecedented scale (Day 2005).

Contestation and continuity

The implications of this change in producer–consumer dynamics are often explored with reference to ideas of participatory culture (Jenkins et al. 2005), the rise of the prosumer (Toffler 1980) and/or produsage (Bruns 2008). These terms were coined to emphasize the much more active role "the people formerly known as the audience" (Rosen 2006) now play in media production and distribution, and thus focus attention on novel acts of democratic self-mediation and collaborative co-creation in which citizens now engage (Bird 2011:506). As Deuze (2006b:67) writes, however, such discussions of the emerging digital culture tend to overlook the fact that much of this citizen-led media-making takes place in an environment whose fundamental conditions, freedoms and restrictions are still largely determined by mainstream media industries and government organizations. In other words, while hyping up the democratizing potential of digital networked technologies, scholars have downplayed the continued dominance of the traditional broadcast model and a small handful of increasingly powerful media conglomerates in the contemporary media sphere (Bird 2011:508–509; Jenkins 2008:3). Viewing these ongoing developments through the lens of convergence, by contrast, provides a means of adopting a wider perspective and stressing the extent to which the social effects of media change are often characterized by the continuity of existing power structures, just as much as their contestation (Meikle and Young 2012:7–11). In the words of Jenkins (2004:37), it reveals the coexistence of and tensions between "a top-down corporate-driven process and a bottom-up consumer-driven process" in the production, distribution and consumption of media content today.

Meikle and Young's (2012:7) analysis of the activist whistleblower site WikiLeaks illustrates the explanatory potential of this expanded perspective especially well. On the one hand, by sharing hundreds of thousands of classified documents as well as leaked video footage from the frontline of the wars in Iraq and Afghanistan, WikiLeaks would seem to exemplify the world-changing possibilities provided by convergent technologies for networked citizens to uncover government corruption and radically reshape the established political order. Most notably, when in November 2010 the WikiLeaks group posted on their website 251,287 cables obtained from more than 250 American embassies and its supporters circulated the links on Twitter, Facebook, YouTube and Wikipedia, they successfully sparked a worldwide diplomatic crisis (now known as Cablegate), implicating the US government and many of its allies in a dark web of conspiracy, espionage and subterfuge (*The Guardian* 2010). The leak was also seen by many as a key factor in triggering the Tunisian revolution and the overthrow of the authoritarian president Ben Ali in January 2011, exposing as it did the sheer extent of the criminality of the nation's ruling elite (Ben Hassine 2011; Dickinson 2011). Through this action, WikiLeaks would thus appear to have harnessed the power of the web to challenge the state and mainstream media's broad monopoly over the production of news and contested the fourth-estate role of professional journalists as gatekeepers for its dissemination. Indeed, following these events, Rosen (2010, cited in Meikle and Young 2012:8) described the activist group as "the world's first stateless news organization".

Yet, when viewed from the perspective of convergence, we see that the success of Cablegate additionally relied extensively on practices and participants associated with the old broadcast media industries. As Meikle and Young (2012:7) demonstrate, at the same time as WikiLeaks was "a YouTube sensation, a Facebook sensation, a Twitter sensation", it was also fundamentally "a newspaper phenomenon". It was professional journalists employed by *The Guardian*, *The New York Times*, *Der Spiegel* and other long-established news brands who brought "credibility and authority ... to what could otherwise be dismissed as a niche website with a weird name" (ibid.:9). It was professional journalists who gave the site the visibility required to attract mainstream public attention around the world, and who were able to filter, contextualize and explain the raw materials of the leak in such a way as to make them accessible to audiences lacking "the time and expertise to process these hundreds of thousands of specialized documents for themselves" (ibid.). Convergent technologies may have allowed WikiLeaks to provide public access to the secret documents on an unprecedented scale, but it was the combination of these new technologically empowered actors with the old media industries that catalysed the explosive reaction and made possible the subsequent fallout. In sum, the lens of convergence helps researchers to focus not simply on what is new when discussing an emerging citizen media phenomenon such as WikiLeaks, but also the complex variety of ways in which new actors, new tools and new modes of distribution interact with the established structures of the status quo (Bird 2011:509).

Future directions

As the contributions to Hay and Couldry's (2011) special issue of *Cultural Studies* on 'Rethinking Convergence/Culture' make especially clear, much work remains to be done before the concept of convergence might win over its critics and more fully achieve its potential. Maxwell and Miller (2011:585), for instance, signal the "fetish of innovation" that informs much current discussion of convergence, and thus highlight the need to explore how these ideas might be applied to the investigation of pre-digital examples of citizen media production. Indeed, as Thorburn and Jenkins (2004:3) note, if we consider convergence as a process rather than an endpoint, a set of ongoing phenomena without the possibility of final termination, we find it is not too much of a stretch of the imagination to see that other convergences will likely have occurred regularly throughout the history of communications, at other moments "when an emerging technology has temporarily destabilized the relations among existing media". Analysis of these earlier convergences would, it is argued, both enrich our understanding of past citizen media practices and help better situate today's developments in their proper historical context (Balbi 2010, 2017).

Second, Bird (2011:509) advocates the necessity of examining processes of convergence beyond the limited context of the Western world. Most notably, she points to the work of scholars such as Ugor (2009), Abah (2009) and McCall (2002) whose investigations into the evolving Nigerian grassroots film industry (nicknamed Nollywood) help expose the importance of recognizing that the effects of convergence are not experienced in the same way everywhere. For example, while the proliferation of relatively affordable, small and efficient digital video technologies has enabled young Nigerian filmmakers to engage in creative and often subversive media practices that directly challenge the monopoly of state and commercial broadcasting institutions, the Internet and online platforms such as YouTube have not emerged as the primary medium of distribution for Nollywood videos. This is primarily due to the particularities of the local media environment: Internet access remains by no means as widespread among the general population of Nigeria as it is in the West and therefore real-world market places and the influence of other offline factors have continued to play a much greater shaping role in this context. The disruption of convergence is thus seen to manifest itself in different ways in different contexts, combining different channels at different scales and different paces.

Finally, Couldry (2011) stresses the necessity of deconstructing the over-bearing generality of Jenkins' (2008) presentation of convergence culture. While acknowledging that the concept, as demonstrated above, does provide a hugely valuable framework on which to build an understanding of many of the emerging trends of the twenty-first century, Couldry (2011:497) argues that the tendency to see convergence as the defining characteristic of contemporary culture risks obscuring from critical attention phenomena that fail to fit this general model. To be sure, as Sparviero et al.'s (2017) edited volume aims to show, current transformations in the media environment may additionally be described in terms of deconvergence, i.e. the "disintegration, multiplication and increased complexity" (Peil and Sparviero 2017:6) of media technologies, markets, uses, users and content: under the combined influence of digital networked technologies and neoliberalism, for example, our consumption of media is now increasingly individualized, audiences are more and more fragmented, public debate takes place in smaller and smaller echo chambers, and politics is becoming dangerously polarized. Consequently, future research must endeavour to develop perspectives on citizen media that take into account processes of both convergence and deconvergence, as well as the interplay between them.

See also: citizen journalism; immaterial labour; media ecologies; remediation; wikis

Recommended reading

Jenkins, H. (2008) *Convergence Culture: Where old and new media collide*, second edition, New York: New York University Press.

A landmark title in the study of convergence, Jenkins' book provides a series of fascinating case studies, most of which focus on the interactions between digitally empowered fan communities and the mainstream cultural industries.

Meikle, G. and S. Young (2012) *Media Convergence: Networked digital media in everyday life*, Basingstoke, UK: Palgrave Macmillan.

Deploys the concept of convergence to develop a detailed analysis of the everyday practices of media production, consumption and distribution performed today by networked citizens. The introduction additionally offers a clear and persuasive critique of previous discussions of the concept in the field of media studies.

Sparviero, S., C. Peil and G. Balbi (eds) (2017) *Media Convergence and Deconvergence*, Cham, Switzerland: Palgrave Macmillan.

The essays and case studies in this edited collection provide a useful overview of the variety of research questions currently being tackled through the lens of convergence, as well as discussion of the need to balance this perspective through attention to processes of deconvergence.

CO-OPTATION

Julia Rone

Co-optation in the context of citizen media involves powerful political or commercial actors appropriating specific media produced through citizen-led initiatives, or the very rhetoric of citizen media in general, and using them to further their own goals. In such cases, the potential of citizen media to effect "aesthetic and socio-political change or express personal desires and aspirations, without the involvement of a third party or benefactor" (Baker and Blaagaard 2016a:16) is substantially undermined. Instead of providing an alternative to an increasingly unaccountable, concentrated and de-democratized media system (Fenton 2012a), citizen media become instrumentalized and integrated in the very same power structures they aimed to subvert.

While the co-optation of citizen media has long been a widespread practice, it became the focus of significant public attention only following the UK Brexit referendum of 2016 and the election of Donald Trump as US president. These two events brought to the forefront of Anglo-American public debate phenomena such as digital astroturfing – defined as "a form of manufactured, deceptive and strategic top-down activity on the Internet initiated by political actors that mimics bottom-up activity by autonomous individuals" (Kovic et al. 2018) – and fake news spread via social or alternative online media for the purpose of propaganda or monetary gain (Tandoc Jr et al. 2018; Fenton and Freedman 2017). In academic research, however, other instances of the co-optation of citizen media had already been examined prior to these developments, and the study of co-optation represents an increasingly rich area of investigation. Examples include analysis of the commodification and commercialization of revolutionary graffiti in Egypt (Abaza 2013a), research on government co-optation and the algorithmic manufacturing of consent during the Mexico #YoSoy132 movement (Treré 2015, 2016), and studies of the oligarchic capture of citizen media in Bulgaria whereby apparently alternative media have in fact been set up or supported by oligarchs in order to promote their narratives (Rone 2016). This entry draws on this growing body of research to discuss debates over the definitions of co-optation and their relevance for the study of citizen media. Specifically, it focuses on examples of the corporate, governmental and far right co-optation of citizen media and outlines directions for future research on these topics.

Defining co-optation

According to the *Cambridge English Dictionary* (2019), the verb *co-opt* means first "to make someone a member through the choice of the present members"; second "to include someone in something, often against their will"; and third "to use someone else's ideas". These multiple dimensions of co-optation have been explored most extensively in relation to social movements (Burchell and Cook 2013; Coy and Hedeen 2005; Gamson 1975; Ho 2010; Jaffee 2012; McCarthy and Wolfson 1992; Trumpy 2008; Weisskircher, 2019). This body of scholarship has examined, for instance, the appointment of movement participants to governing positions within state institutions, the adoption and diversion of the language and rhetoric of movements by powerful political actors, and the focus on developing new technologies as a co-optation by industry (Burchell and Cook 2013; King and Busa 2017; Weisskircher, 2019). In such contexts, then, co-optation can be understood as

> the ability of the established political order to respond to or accommodate new challenges and challengers without radically altering the foundation of the established political systems and processes – a de-radicalisation of the movements and a diluting of issues to accommodate them within the established political order.
>
> *Burchell and Cook 2013:742*

The term co-optation is not, however, without its critics and detractors. Some scholars, such as King and Busa (2017), have argued that the label is too broad and that it does not account for differences in the scale of the co-optation. It does not, for example, provide a means of distinguishing between cases in which a small number of protest leaders have been appointed to a local council and cases related to the multi-billion dollar marketization of the organic food movement by corporations worldwide. Nor does the term clarify the type of actor driving the co-optation process, that is, whether it is a governmental or corporate actor. For this reason, King and Busa (2017:4) have introduced the notion of 'corporatization' to describe those special cases of co-optation in which "corporate interests come to engage with ideas and practices initiated by a social movement, and ultimately, to significantly shape the discourses and practices initiated by the movement". Other scholars have argued that the term co-optation is normatively charged and overly negative in its connotations. Lapegna (2014) has suggested that it does not reflect adequately the empowerment and positive changes that co-optation can sometimes bring about. In this sense, he argues, the "etymology of the term 'cooptation' ('from the Latin cooptare, from co-"together" and optare-"choose"') actually better captures the relational and agentic elements that are lost in the common and widespread use of the term" (ibid.:9). When considering the co-optation of citizen media, therefore, we must recognize that this is an agentic mutual process, in which the actors being co-opted sometimes readily cooperate with the state and/or commercial interests, as has been the case with many influential bloggers, as well as with volunteer or paid trolls (Woolley and Howard 2017).

Co-optation by corporations, governments and far-right actors

Co-optation by commercial interests has been a major topic of concern within citizen media studies. Scholars have noted, for instance, how individual bloggers – writing on a diversity of topics, from food to technology and culture – have since the early 2000s been targeted as 'influencers' who could be effectively engaged in brand communication and marketing (Booth and Matic 2011; Li et al. 2011; Magno 2017; Uzunoğlu and Kip 2014). In such cases, it has been the social standing and popularity of bloggers that different companies have sought to instrumentalize as part of their search for a more immediate connection with consumers and a way to bypass traditional advertising. Mass media corporations have also tried to co-opt citizen media by integrating citizen journalists in the news production process. For example, Kperogi (2011) has explored CNN's launch of iReport.com in 2006, a user-generated citizen news site that allowed the corporation not only to use unpaid journalistic labour, but also to tame and appropriate for its own profit emerging counter-hegemonic voices. Cross (2016:226) examined another example of co-optation by media hegemons by focusing on the way the BBC used citizens' amateur photos of the 2005 London bombing to drive a particular "politics of sentiment" that ultimately "was important for the justification of the War on Terror".

Non-digital practices of citizen media may also be co-opted, as Abaza (2013a) has shown through her analysis of the commodification of revolutionary graffiti from the Arab Spring. Incendiary political graffiti, whose creation involved great risk for their authors, were subsequently decontextualized and distributed as images on bags, mugs and in art galleries. Finally, we might cite Frank's (1997) classic book *The Conquest of Cool* which demonstrates how the advertising industry used and at the same time fuelled the rebel culture of the 1960s, glorifying the aesthetic of rebellion, alternative media and a newly discovered notion of cool to boost consumerism and product sales. Today, thanks to the all-pervasive logic of commodification, once counter-cultural practices of youth expression such as skateboarding, parkour, hip-hop and rap have similarly been co-opted and have ended up as integral parts of mainstream popular culture (Lorr 2005; Stapleton and Terrio 2010; Blair 2004; Swedenburg 2004; Watts 2004).

The commercialization of citizen media additionally has systemic dimensions which relate to the transformation of the very essence of the social. Online platforms such as Facebook and Twitter have co-opted the language and rhetoric associated with citizen media. The claim that they promote bottom-up collaboration, sharing and emancipation has been a crucial part of their branding strategy, aptly described as wikiwashing (Fuster Morell 2011). The problem is that no matter how radical social networks claim to be, they are ultimately designed to facilitate data mining: "[t]hey are designed to be exploited" (Ippolita et al. 2009). Community on these platforms is structured by marketing interests and citizens are reduced to a bundle of data points: "[t]he fact that online sociability is facilitated by separating users from the means of socializing and thereby creating an external, storable, and sortable collection of data about their social lives, renders the product of their online activity further alienable" (Andrejevic 2010:88).

The privatization of communication structures, the datafication of the self and the key role of algorithms in information provision not only turn citizens into transparent objects of surveillance (Couldry 2017), but also modify what they can see (Pariser 2011). The algorithmic structuring of relevance online mediates appearances and creates regimes of visibility that favour particular types of content over others on the basis of predominantly commercial criteria (Couldry and Hepp 2017). Many new citizen-led media projects claim to restore the authenticity lost in previous platforms, but commercialization soon gets hold of these self-proclaimed radical media. This is because of what has been labelled "authenticity's reactive dynamism" (Salisbury and Pooley 2017:6): as soon as a new initiative proclaims itself as a source of real authentic media, it begins to attract marketing attention, until it too becomes tainted by corporate interests, and the cycle repeats itself. The alternative, as Indymedia have found to their cost, is marginality: without developing sustainable funding mechanisms and finding other ways to attract attention in an increasingly competitive online environment, citizen media projects will all too often remain invisible to the vast majority of Internet users.

But it is not only corporations that have tried (and succeeded) in taming citizen media. The potential of citizen media to unsettle the status quo and to provide alternative values and visions of society makes them a serious threat to any established power. Thus, it is not surprising that governments across the world have experimented with different strategies for co-opting citizen media, including funding a wide range of actors who post online as unaffiliated citizens, creating government-sponsored accounts, web pages or applications, and creating content on blogging platforms in order to manufacture consent and hinder dissidence (Kovic et al., 2018; Walker 2015; Woolley and Howard 2017). Governments may additionally make use of authentic citizen media in order to collect activists' data and engage in surveillance and pre-emptive action (Morozov 2011; Treré 2015). During the #YoSoy132 movement in Mexico, for example, the government actively monitored social media profiles and the communications of activists. What is more, it turned out that one of the main websites of the movement had been created by an undercover government agent who had managed in this way to collect the data of numerous activists (Treré 2015).

Finally, while citizen media have traditionally been envisaged as vehicles of left-wing politics, inclusion and emancipation, we must recognize that far-right movements too produce citizen media to promote exclusion, discrimination and hatred (Simpson and Druxes 2015). Precisely because they are perceived as freer and less regulated than mainstream media, citizen media have proved instrumental in the diffusion of hate speech and fake news by far-right groups (ibid.). Considering the long tradition of far-right citizen media, including fanzines, magazines and early uses of computer communication networks (Berlet and Mason 2015), one can speak of co-optation not so much in chronological as in ideological terms. Adopting some of the counter-culture strategies of the 1968 leftist movements, far-right activists have presented themselves as grassroots community organizers working closely with and for the people (Castelli Gattinara and Bouron 2019). Most notably, by espousing the language and rhetoric of citizen media, groups such as the alt right movement in the US have presented themselves as active citizens, fighting against the power of corporations and governments, and against what they perceive as excessive political correctness (Berntzen and Weisskircher 2016; Bessant 2017; Love 2017; Luke 2017; Nagle 2017; Silva et al. 2017; Simpson and Druxes 2015; Weigel 2016).

Future directions

It must be noted that in an increasingly converged media environment, where the concentration of media ownership and the entanglements between media and politics have reached unprecedented levels (Fenton and Freedman 2017; Jones 2016), it is difficult to separate different types of co-optation and to treat government, political parties and corporations as completely separate actors with separate interests. On the contrary, their agendas and actions often overlap. In Bulgaria, for instance, the ruling elite of politicians and businesspersons have together managed to co-opt both individual citizen media initiatives such as the Anonymous blog online and the very language of citizen media in order to promote narratives that serve their economic *and* political agenda. Thus, instead of providing a corrective to the lack of accountability, unclear ownership and scandalous content that mark mainstream media in Bulgaria, much of the so-called citizen media in the Bulgarian context clearly serves the interests of these oligarchical groups, leading to the de-democratization of Bulgarian journalism, as well as the widespread promotion of nationalist narratives (Rone 2016).

Thus, future research should focus more on the symbiotic mechanisms through which governments, corporations and the far right co-opt citizen media. We might fruitfully explore, for example, how governments contract private firms to create content online, how media hegemons may lobby for government (de)regulation in order to take larger shares of the market, and how the far right may lobby against hate speech laws in order to protect the freedom of far-right media. In other words, the co-optation of citizen media needs to be viewed as a complex process with multiple initiators and interaction effects. Second, we need to explore the funding and business models of citizen media projects that aim to persist over extended periods of time. That is, we need to explore existing and potential strategies through which citizen media activists might support themselves financially, without being relegated to marginality or conforming to the imperatives of online advertising, data extraction and the attention economy. Third, while governments, corporations and the far right have all tried to co-opt citizen media, we might recognize that there have also been processes of "reverse appropriation" by citizen groups (Baker and Blaagaard 2016a:18–19). A good example of reverse appropriation is culture jamming (Bordwell 2002; Dery 1993b; Harold 2004; Nomai 2008), which includes practices such as billboard hacking, the appropriation of advertisements and other tactical media interventions (García and Lovink 1997; Raley 2009). While many of these practices have turned out to be less disruptive than intended, future discussion of the co-optation of citizen media should also take into account such acts of resistance to co-optation and/or proactive attempts to co-opt elements of the dominant culture in order to subvert it.

See also: authenticity; citizen science; civil society; culture jamming; diversity; Facebook; flash mobs; graffiti and street art; hip-hop

Recommended reading

Abaza, M. (2013) 'Mourning, Narratives and Interactions with the Martyrs through Cairo's Graffiti', E-International Relations, 7 October.

Shows how martyrs and martyrdom during the Egyptian revolution have been represented in graffiti as powerful instances of citizen media. These subversive forms of expression and commemoration, however, have been increasingly commodified by the needs of the international art market and turned into book albums, posters, T-shirts, handbags and a variety of other monetizable objects.

King, L. and J. Busa (2017) 'When Corporate Actors Take Over the Game: The corporatization of organic, recycling and breast cancer activism', *Social Movement Studies* 16(5): 549–563.

Explores corporatization as a process whereby large corporate entities take over ideas and practices initiated by social change advocates, dominate the field and spread corporate-friendly versions of resistance.

Kperogi, F. A. (2011) 'Cooperation with the Corporation? CNN and the hegemonic co-optation of citizen journalism through iReport.com', *New Media & Society* 13(2): 314–329.

Problematizes the way in which user-generated media is incorporated by corporate media hegemons, blurring the distinction between citizen and mainstream journalism.

Woolley, S. and P. Howard (2017) 'Computational Propaganda Worldwide: Executive summary', Working Paper No. 2017.11, Computational Propaganda Research Project.

Computational propaganda campaigns take place in both democratic and autocratic countries. State governments, political parties and foreign actors use social bots, fake accounts and automation across a variety of platforms in efforts to silence opponents and to push misinformation and propaganda.

CROWDSOURCING AND CROWDFUNDING

Alexandra Reynolds

The term crowdsourcing is derived from the term outsourcing and was coined in 2006 by *Wired* editor Jeff Howe. Crowdsourcing was initially defined in commercial terms: as a means of accessing a source of inexpensive labour supplied by members of the public and a way of helping organizations solve problems, develop content and undertake research (Howe 2006). However, since 2006, the term has been applied in a wide range of corporate, cultural and social contexts, and has been defined in numerous ways (Hossain and Kauranen 2015). In recognition of this multiplicity, Estellés-Arolas and González Ladrón-de-Guevara (2012) developed an integrated definition of crowdsourcing drawing on forty different interpretations of the term. This definition expands the concept of crowdsourcing and reframes its parameters beyond purely commercial ends. For Estellés-Arolas and González Ladrón-de-Guevara (ibid.:9), crowdsourcing can be defined as

> a type of participative online activity in which an individual, an institution, a non-profit organization, or company proposes to a group of individuals of varying knowledge, heterogeneity, and number, via a flexible open call, the voluntary undertaking of a task. The undertaking of the task, of variable complexity and modularity … always entails mutual benefit.

As a parallel phenomenon to crowdsourcing, crowdfunding has also been the focus of many research articles seeking to define and describe the concept (Moritz and Block 2016). Existing characterizations can be drawn together through Brabham's (2013a:37) succinct definition, which describes crowdfunding as a "funding model whereby individuals use the internet to contribute relatively small amounts of money to support the creation of a specific product or the investment in a specific business idea". For Brabham, crowdfunding is separate from but related to the phenomenon of crowdsourcing, relying on a markedly different mode of participation both in terms of management and creativity. A key defining characteristic of crowdsourcing is its status as a creative process operating between an organization and a given public: it is "a shared process of bottom-up, open creation by the crowd and top-down management by those charged with serving an organization's strategic interests" (ibid.:xxi). Conversely, crowdfunding is considered a form of distributed financing which does not involve creative intervention by the crowd in the development of a product. In these cases, an idea is suggested for funding, and the crowd's only involvement is in choosing a product to support (ibid.:39).

Literature reviews by scholars such as Hossain and Kauranen (2015) demonstrate that crowdsourcing has been investigated academically in relation to a wide variety of sectors, including business, planning, government, relief work, teaching, citizen journalism, citizen science and policy-making. Many studies have also discussed instances of crowdsourcing in galleries, libraries, archives and museums (Ridge 2014; Terras 2016; Andro 2018), as well as crowdsourcing in activist contexts (Heikka 2015; Massung et al. 2013). Within this broad body of scholarship, there are recurrent research themes. Many papers explore specific crowdsourced projects, creating typologies (Brabham 2013b; Oomen and Aroyo 2011), analysing motivations for participant involvement (Baruch et al. 2016; Brabham 2012), leadership models (Byrd Phillips 2014; Pan and Blevis 2011), questions of data quality and reliability (Causer et al. 2018; Goodchild and Glennon 2010), successful platform design (Koch et al. 2011; Estellés-Arolas 2018) and the impact of crowdsourcing on existing professions and practices (Eveleigh 2014; Seltzer and Mahmoudi 2012). Increasingly, the literature also includes critical analyses of crowdsourcing in the context of neoliberalism and capitalist economics (Fuchs 2013; Ettlinger 2016; Reynolds 2017). A mixture of empirical quantitative, qualitative interview and survey, critical theoretical and interpretive methods are used in this body of research.

Crowdfunding projects have in turn been explored using a similar variety of quantitative, qualitative and interpretive research methods (Gleasure 2016). Literature reviews demonstrate a focus on structural and economic models for successful crowdfunding and the role of intermediaries, fundraisers and investors within platforms and projects (Bouncken et al. 2015; Moritz and Block 2016). Analysts have additionally paid significant attention to the motivations for participation and the determinants of success in crowdfunding projects, with a further interest in the legal restrictions shaping crowdfunding initiatives. Research has discussed a diverse range of crowdfunded initiatives, from charitable and activist causes, to start ups and other commercial uses of crowdfunding, to artistic, cultural and civic crowdfunding initiatives. More critical theoretical research has been undertaken on topics such as the relation between power, value and crowdfunding (Gehring and Wittkower 2015), and the relation between crowdfunding and the construction of social norms and marginal identities (Farnel 2015; Scott 2015).

Crowdsourcing and citizen media

Examples of crowdsourced projects abound in activist, creative and alternative media contexts, where unaffiliated citizens work "in public spaces to effect aesthetic or socio-political change or express personal desires or aspirations" (Baker and Blaagaard 2016a:16). However, for a crowdsourced project to qualify as citizen media, it must also operate "without the involvement of a third party or benefactor" (ibid.). Therefore, not all socially engaged or creative crowdsourced projects qualify as citizen media. For instance, Open Ideo is a non-commercial platform launched in 2010 which uses crowdsourcing and open innovation to create design solutions to global challenges for social good. Anyone who registers an account can join the onsite community in contributing ideas to an open design process. However, this platform was developed by the global commercial design consultancy Ideo, and challenges are set, managed and funded by Open Ideo along with its external partner organizations, which include government departments, charities and foundations. Therefore, although it involves unaffiliated citizens working together for social good, Open Ideo is not a crowdsourced citizen media project. Whereas Open Ideo is developed and regulated by formal networks of third-party organizations, crowdsourced citizen media projects form organically within communities in response to collectively held beliefs. Crowdsourced citizen media projects are also funded autonomously through the collective with whom they work rather than relying on third-party benefactors.

An example of crowdsourcing in a citizen media context is Indymedia, an alternative news platform developed and launched in 1999 as part of the World Trade Protests in Seattle (Anderson, C. W. 2011:82). Part of the participatory journalism movement, this site was developed autonomously and is managed by members of the Indymedia collective. It presents itself as "a network of individuals, independent and alternative media activists and organizations, offering grassroots, non-corporate, non-commercial coverage of important social and political issues" (Indymedia 2017a). Indymedia operates through an open-source and open-publishing model which "permits everyone with internet access to post reports, images, audio or video on any Indymedia site without prior clearance by the editors" (Lievrouw 2011:133). Therefore, the definition of journalist on this site is de-professionalized, and the platform can be understood as being crowdsourced.

Another example of a crowdsourced citizen media platform is Actipedia, described as "an open access, user-generated database of creative activism" (Actipedia 2017). Launched in 2012, this project is a collaboration between the Centre for Artistic Activism – a collective founded by academic and organizer Stephen Duncombe and artist Steve Lambert – and the Yes Lab, a support initiative for creative activism headed by activist duo the Yes Men. Actipedia is a crowdsourced participatory network where anybody can freely post, and operates in a simple database form, enabling users to upload and share ideas, projects and strategies. Actipedia also uses a wiki format, which means users with an account can flag content or edit posts wherever site contributors have enabled this feature.

A third crowdsourced citizen media intervention is Sukey, a project launched in London in 2011 by Sam Gaus and Samuel Carlisle, both activists and computer engineers. This project was developed during the G20 protests and aimed to prevent protestors being kettled by the police by providing up-to-date crowdsourced information on police movements (O'Rourke 2011:50). Gaus and Carlisle coded a map of the protests which was updated in real time, with information sent to the team's headquarters by protestors on the ground. Teams of activists would then validate the data to ensure accuracy before transmission (ibid.). Sukey included an SMS warning service and Twitter feed, as well as a location feature on GPS enabled smartphones allowing protestors to see which roads were blocked, passable or difficult to access via a colour-coded system (Kingsley 2011). The application was superimposed onto a Google map, and used GPS functionality on Google Latitude enabling activists to navigate their environment in real time.

Crowdsourced citizen media projects also function through the appropriation of existing platforms by citizen groups. An example can be found on the Tumblr thread 'We are the 99%', an initiative started by members of the Occupy movement in August 2011 to render visible the challenges faced by ordinary citizens living under neoliberal capitalism. In this project, contributors often uploaded photographs of themselves holding homemade signs, describing an issue they had encountered in neoliberal society. Images regularly generated hundreds of onsite responses and heightened awareness around neoliberal policy-making and its daily impact on citizens' lives. Other entries on this thread included calls to action or questions about Occupy which were answered by the site's project team. This project team developed and managed the 'We are the 99%' thread, but states that it aims to act solely as a facilitator for other voices, explaining "we don't claim to speak for anyone, we merely present stories" (We are the 99%, 2013).

Crowdsourced projects are not solely digital. Examples of non-digital crowdsourced citizen media projects include Subway Therapy, a 2016 installation which was created in New York City's subway from Donald Trump's election on 8 November 2016 until 16 December 2016. During this time, over 10,000 post-it notes were affixed to a wall in Union Square station, expressing the feelings of New Yorkers in the wake of Trump's presidency (Rosenberg 2016). Chavez provided post-its and pens for the piece, and wrote "express yourself" as a prompt for passers-by. Chavez also produced performances in the space, where he played the part of a therapist listening to members of the public.

Crowdfunding and citizen media

Crowdfunding initiatives span a range of different forms, some of which lie closer to the definition of citizen media promoted in this volume than others. Crowdfunding platforms such as Indiegogo are not specifically focused on social or political causes but host a variety of projects aligned with citizen media. An example is the Greek Bailout Fund by Thom Feeney, which was undertaken on Indiegogo in response to the 2015 Greek financial crisis. This one-man campaign raised €1.5 million of its fixed €1.6 billion goal, offering small but incremental awards for donors of specific fixed amounts from €3 to €10,000,000. The initial project failed to reach its target and refunded all donations. However, a second related petition by Feeney raised €300,000, which was used to employ fifteen young Greek people in charities, itself helping a further 50,000 people across the country. The Greek Bailout Fund is certainly an example of a citizen mobilizing autonomously for the good of other citizens, and is listed as a scheme developed "by the people, for the people" (Feeney 2015) to which Feeney himself contributed only €10. However, Indiegogo is a commercial company founded by a group of bankers, where any successful project pays 5 per cent in platform fees to Indiegogo and 3 per cent plus 30 cents of every donation to the external credit card company Stripe (Indiegogo 2017). Indiegogo is therefore a third-party hosting organization in this sense, something which throws the Greek Bailout Fund into potential opposition with most definitions of citizen media.

A similar tension can be found in a range of other crowdfunding sites, including Change.org. This platform crowdfunds activism on sociocultural and environmental issues but operates as a profit-making certified social and environmental B Corporation. It has been the recipient of $50 million from external benefactors, including Richard Branson and Bill Gates (Change.org 2017). Therefore, although Change.org facilitates the funding of causes aligned with citizen media objectives, its status as a commercial company and its reliance on external benefactors place this platform in tension with citizen media as defined by Baker and Blaagaard (2016a).

Certain crowdfunding platforms are more aligned with citizen media in terms of the leadership and financial models adopted. For instance, Avaaz.org develops both petition-based and crowdfunding projects around a range of social, environmental and political causes worldwide. This site was cofounded by Res Publica, a "community of public sector professionals dedicated to promoting good governance, civic virtue and deliberative democracy" (Res Publica, n.d.) and MoveOn.org, an American nonprofit progressive public policy advocacy group. Avaaz.org was also supported by the North American Service Employees International Union and Getups!, an Australian nonprofit campaigning organization (Jethwany 2016:128). Avaaz.org operates like Indiegogo and Change.org in that users can donate to various causes through an online interface. However, it is a nonprofit organization wholly funded by donations from its community. Avaaz.org is also fundamentally shaped by its community, with overall priorities being set annually through all-member polls. Therefore, citizens who set up crowdfunding projects on Avaaz.org join an existing activist collective and operate within parameters set by members, rather than using a generic profit-making platform to achieve social change.

That said, perhaps the most radical uses of crowdfunding in citizen media contexts are those projects which look to their own internal networks to self-fund particular initiatives. One such example can be seen in Cowbird, a citizen storytelling platform developed between 2011 and 2017 by digital artist Jonathan Harris, which aims to build a "public library of human experience" online (Harris 2011). This site was freely available for any account holder to use. However, site members also had the option of paying a small fee to access additional writing tools and become Cowbird citizens, a strategy which helped fund ongoing platform maintenance.

Limitations of the definition

As we have seen, a broad spectrum of crowdsourcing and crowdfunding projects pursuing activist goals and social change can be identified. Some local, small-scale projects such as Sukey, Subway Therapy and Cowbird are born out of grassroots struggles by individual activists, autonomous collectives or individual artists. These tend to be funded internally by these groups and individuals, and by the communities they represent. Conversely, sites such as Indiegogo are developed by profit-making companies, but they house projects such as the Greek Bailout Fund which self-represent as citizen media. There are also organizations such as the Centre for Artistic Activism and the Yes Lab which are in fact small activist collectives utilizing crowdsourcing projects such as Actipedia as tools of intervention and learning. Similarly, crowdfunding platforms such as Avaaz.org, though run and managed by an organization, can be understood to operate very much like a citizen collective, with shared non-commercial aims and internal funding. Meanwhile, projects such as Indymedia might begin as small, autonomous activities led by activists, but become substantial organizations over time as they grow and develop. Finally, projects such as 'We are the 99%' on Tumblr can be seen to occupy profit-making, publicly available social media sites.

This diversity and complexity highlights a number of pressing questions about the relationship between crowdsourcing or crowdfunding and citizen media. While crowdsourcing and crowdfunding can be defined as the work of unaffiliated citizens managed by a leading individual or organization, citizen media is defined as the work of unaffiliated citizens operating without the involvement of a third party or benefactor. In turn, this distinction raises considerations around the definition of what constitutes third-party or benefactor involvement in a crowdsourced or crowdfunded site. We might question at what point an organizing individual or collective stops being a collaborator and starts to become a third-party organization which orchestrates or manages the work of citizens. Similarly, we might ask when a project manager becomes a benefactor or a third party for the citizens they organize, and whether an intervention by citizens within a publicly available, yet profit-making site such as Indiegogo or Tumblr can still be considered an act of citizen media under the current definition.

In this entry, sites which stem from an autonomous individual or organized collective aligned with project-specific ends are considered to meet most clearly the definition of crowdsourced or crowdfunded citizen media initiatives. Conversely, crowdsourced social change projects developed, managed or funded by large, formal organizations or companies such as Open Ideo would not be understood as citizen media projects. However, this definition is complicated by the fact that certain citizen media initiatives might utilize large, commercial, publicly available sites such as Tumblr or Indiegogo to carry out crowdfunding and crowdsourcing initiatives. In these cases, crowdsourced and crowdfunded citizen media projects can perhaps be understood to occupy sites as publicly available spaces, in much the same way that activists might occupy government-controlled public spaces in the physical world.

It is also relevant to note that, even in crowdsourced or crowdfunded projects developed by and for citizens, a hierarchy of sorts commonly exists which divides project leader from project user. There is a centralization of leadership in many crowdsourced and crowdfunded projects which means those who develop a project also design its territory, regulations, terms and conditions, and retain ultimate editorial control about who can be involved, what can be said and what the terms of inclusion within a given platform or initiative might be. Sites thus commonly operate through a striated framework where project leaders define and mediate the terms through which project users participate.

For instance, Sukey's design and motivations were developed by Gaus and Carlisle as project leaders, and any crowdsourced information collected from activists on the ground was validated by a project team before being disseminated. Similarly, Cowbird includes a code of conduct which site members must follow and operates via a centralized project team. Project leaders do collaborate with project users onsite, but ultimately retain the sole legal right to block users or delete accounts (Reynolds 2017). Actipedia operates through a wiki format, but even this site reserves some specific privileges for the organizing project team. Actipedia relies on account holders to create site content and enables users to take an active role in developing this content in cases where the original authors have allowed this. However, project leaders build the site map and site administrators retain an overarching ability to edit site posts, as stated on the project submission page (Actipedia, n.d.).

Leadership structures in Indymedia and Avaaz.org are more horizontal as both sites invite direct participation from their audiences. However, even these sites maintain some centralized power and control. A team of editors in Indymedia retain the sovereign right to highlight, hide or remove content which does not conform to the aims and principles of the initiative. These shared principles are based on fairly traditional journalistic concerns with quality and originality, although they are guided by ten Global Principles of Unity, which were collaboratively agreed by members of seventy-one International Media Centers in 2001 (Lievrouw 2011:137). Similarly, although Avaaz.org has developed a collaborative approach to negotiating its organizational priorities, the terms and conditions on this site state that Avaaz.org retains the power to take down the site or block users at their sole discretion (Avaaz.org 2017).

The reality of crowdsourced and crowdfunded sites is that there is almost always a certain striation of power, where project leaders define the terms and conditions of work for participants. This is most clearly visible in crowdfunded projects, but it is also inherent to crowdsourcing as an activity which makes use of the crowd to serve the strategic interests of an organizing group or individual. As we have seen, certain projects such as Indymedia and Avaaz.org have attempted to move away from this centralized leadership model towards a more collaborative and decentralized leadership or management style. However, in so doing such projects move towards a truly decentralized, collaborative mode of project leadership and development which can perhaps be defined more accurately as co-creation.

The hierarchical structuration of crowdsourced projects is in fact fundamental to the history of crowdsourcing as a business term referring to the cheap outsourcing of labour. More broadly, these projects can be understood to reflect the networked dynamics of power within so-called New or Network Capitalism (Sennett 2006; Castells 2009), as well as the neoliberal tendency to retain centralized power while outsourcing responsibility (Gilbert 2013:19; Bishop 2012:14). If, as Baker and Blaagaard (2016a:16) state, citizen media is regularly co-opted by capitalism and commercialism, crowdsourcing already has a fundamental relationship with capitalism. However, appropriation can work both ways. As this entry has shown, crowdsourcing projects can subvert their capitalist roots and function as citizen media. Projects cannot perhaps ever fully escape the tension between management and collaboration inherent to this cultural form. However, crowdsourcing and crowdfunding can still be interpreted and used for radical ends if developed self-reflexively.

See also: content moderation and volunteer participation; disability media; immaterial labour; user-generated content

Recommended reading

Brabham, D. C. (2013) *Crowdsourcing*, Cambridge, MA: The MIT Press.

A concise and accessible introduction to crowdsourcing, grounded in empirical research. This text defines crowdsourcing and gives an overview of its uses in corporate, government and charitable contexts, as well as considering concerns related to free labour in crowdsourced projects.

Lievrouw, L. (2011) *Alternative and Activist New Media*, Cambridge: Polity Press.

Offers an overview of five contemporary forms of activist and alternative new media, analysing the history, context and efficacy of key case studies from each formulation. Case studies include projects such as Indymedia that use collaborative and crowdsourced techniques.

Reynolds, A. (2017) 'From Crowdsourcing to the Commons: Towards critical and meaningful digital collaboration in museums', *The International Journal of the Inclusive Museum* 10(2): 49–68.

This article analyses cultural crowdsourcing projects in the broader cultural context of New Capitalism and Inclusive Neoliberalism, exploring the power relations in individual projects and their symmetrical relationship to broader hegemonic structures in Western society.

CULTURE JAMMING

Vince Carducci

Culture jamming is a form of activism that raises awareness of social and political concerns, primarily by deconstructing or otherwise interrupting communications, text and other mediated representations that emanate from authoritative sources such as corporations and government. The term cultural jamming is generally agreed to have been coined in 1984 by the San Francisco-based experimental music collective Negativland (Dery 1993a). It derives from the technique of electronically interfering with – i.e. jamming – radio or broadcast communications for military or political purposes.

Culture jamming embraces a variety of strategies and tactics, including modifying or parodying the words or images of official communications such as advertisements or websites in order to expose alternative, often critical points of view that undercut the message's original intention. The term and its practices have gained broader public recognition through the writings of Kalle Lasn and the Vancouver, Canada-based journal he co-founded, *Adbusters*, as well as Naomi Klein's first book *No Logo* (2000). As a form of citizen media, culture jamming is an assertion of individual or collective agency within the public sphere, especially in opposition to political, economic and cultural hegemony.

Origins, antecedents, affinities

Lievrouw (2011) cites early examples of culture jamming in the late 1970s, though its antecedents can be found much earlier in European culture. Culture jamming's most-widely recognized technique, parody, extends back at least to the ancient Greeks and the writer Hegemon of Thasos, whom Aristotle cites in *Poetics* as the inventor of the form by virtue of his practice of slightly altering the words of well-known poems. Its method of intervening in official messages, particularly with subversive intent, is also related to graffiti, which similarly dates back to ancient times. More modern antecedents of culture jamming can be found in the first quarter of the twentieth century, with the adoption of Cubist collage and montage by avant-garde art movements such as Dada and Surrealism. Marcel Duchamp's *L.H.O.O.Q.* (1919) – the letters of which when pronounced in French translate as 'she's got a hot ass' – deflates the pretentions of high art and bourgeois good taste by adding a moustache and goatee to a reproduction of Leonardo da Vinci's iconic *Mona Lisa*. Surrealist Meret Oppenheim's *Object* (1936) covers a teacup, saucer and spoon in fur, transforming an object typically associated with heteronormative female domesticity and respectability into a highly charged article of lesbian eroticism.

A more widely recognized precursor of culture jamming can be found in the principles of the postwar group of mainly French artists, writers and political theorists known as the Situationist International.

Particularly relevant is their technique of *détournement* (roughly translated as deflection, diversion or rerouting), which consists of appropriating and reconfiguring existing cultural ephemera in order to expose contradictions of everyday life, especially social alienation under advanced capitalism and its spectacle of commodity fetishism, and to awaken those who encounter such interventions from their complacency and encourage them to embrace alternative ways of being (Debord and Wolman 1956/2006). The influence of the Situationists can be seen clearly in the emergence of other movements in the 1970s which exhibit an affinity to culture jamming: the punk music and fashion subculture and postmodernist image appropriation and public sphere interventions such as Jenny Holzer's 'Truisms' (1978–1987), deconstructions of the language of power posted on the streets of Manhattan; Krzysztof Wodiczko's projections on corporate buildings in New York City and elsewhere starting in 1980, criticizing capitalist hegemony; and Ron English's 'POPaganda', parodies of consumer imagery appearing on billboards and in magazines.

Tactics

Hirschman's (1970) concepts of exit and voice are useful for considering the strategies and tactics of culture jamming as a form of citizen media, along with the social and political movements it can inform. According to Hirschman, constituencies, be they in a market, of the state or part of another form of social organization, have two ways of demonstrating dissatisfaction: withdrawal, which Hirschmann terms exit, or the communication of grievance or protest, which he terms voice. In the economic sphere, exit can take the form of refusal to buy or to purchase an alternative product or service. *Adbusters*' Buy Nothing Day, which takes place every year on the Friday after Thanksgiving, the biggest shopping day of the year in the United States, is a form of exit, as is the Black Spot Sneaker, also known as the Unswoosher, that *Adbusters* markets in rejection of the Nike brand. Voluntary simplicity, green and blue consumerism, boycotts, buycotts and other types of ethical and political consumption are further examples of exit that are more politically oriented; these methods are used by those outside conventional power structures as an alternate means of mobilization in the public sphere. The Sugar Boycott in protest of the slave trade in the late eighteenth and early nineteenth centuries in England is an early example, with women using their purchasing power at a time when they lacked the right to vote. Voice is more complex and often more visible, as it seeks to consciously register dissatisfaction in the hope of effecting change either in the marketplace or public sphere.

An example of culture jamming as voice is the international activist collective Brandalism's use of *détourned* advertisements and other images critical of corporate sponsors to protest the United Nations twenty-first Conference of Parties (COP21) talks on climate change, which took place in Paris in 2015. Brandalism installed 600 fake ads at bus stops, kiosks and other spaces around the city to draw attention to the alleged environmental abuses perpetrated by COP21's sponsors. One ad featuring the Mobil brand depicted a floating oil platform and carried the headline 'We knew about the impact of fossil fuels but publicly denied it'. Another branded with Volkswagen and featuring one of its sedans carried the headline 'We're sorry that we got caught', a reference to the automobile manufacturer's use of software to circumvent clean air standards in their diesel vehicles (Mufson 2015). Exit and voice can comingle, as when a consumer boycott is coupled with a culture jam, public demonstration or other tactic meant to garner media attention or regulatory action.

Brandalism's subversion of the mediated voice of authority at COP21 builds on tactics used by the early practitioners of culture jamming. The Billboard Liberation Front started in 1977 in San Francisco with a mission to improve corporate outdoor advertisements installed around the city through late-night excursions during which they altered the text and images of marketing campaigns they deemed in need of correction. The first of these, in December 1977, was a series of six billboards for the cigarette brand Fact, in which the headline copy 'I'm realistic. I only smoke Facts' was changed to 'I'm real sick. I only smoke Facts', with a thick white arrow added to the image directing the viewer's attention from the headline to the Surgeon General's disclosure of the health hazards of smoking at the bottom of the ad. Other targets for such improvement included advertisements for cosmetics, alcohol and gasoline.

Culture jamming significantly augmented its media presence with the emergence in the 1990s of the networked public sphere of the Internet and the World Wide Web, enabling the interventions to engage broader audiences with much greater immediacy. In 2001, then MIT Media Lab postgraduate student Jonah Peretti, subsequently co-founder of *The Huffington Post* and later founder of the social and entertainment site *Buzzfeed*, attracted widespread media attention as a result of an email exchange with the athletic apparel company Nike, whose NikeiD online footwear marketplace offered customers the ability to customize their sneakers by choosing fabrics, colours and other details. Nike had cancelled Peretti's order for a pair of Zoom VC USA running shoes after he requested the word *sweatshop* be embroidered on them. In the exchange, Peretti noted his desire to recognize the low level of compensation earned by the shoes' outsourced labour, for which the company had often been criticized, going back to its founding in the 1970s. The exchange soon went viral, bringing attention to the critique of Nike's labour practices and the company's dissemblance at being confronted through coverage in print, broadcast and online media.

Another high-profile culture jamming project that demonstrates similar tactics is the Yes Men, founded in the wake of the 1999 protests of the World Trade Organization conference in Seattle. The Yes Men is a collaboration between Jacques Servin (aka Andy Bichlbaum) and Igor Vamos (aka Mike Bonanno), who use brand-identity spoofs and impersonations of corporate, state and NGO officials to raise awareness of social, economic and political issues. Among their noted projects are spoofing websites of the World Trade Organization, Dow Chemical Corporation and Shell Oil Corporation to expose alleged misdeeds and ethical lapses, as well as public appearances in which they pose as official spokespersons to make pronouncements that deconstruct the media representations of their targets. The Dow Chemical culture jam famously included a television interview on the BBC in which Bichlbaum, posing as a company representative, announced that the Union Carbide subsidiary would be liquidated and the anticipated USD 12 billion proceeds used to pay for medical care and site clean-up from the 1984 Bhopal chemical disaster, in which an estimated 3,800 people immediately died and thousands more suffered long-term health effects from a toxic plume accidently released by its local pesticide plant. As news of the announcement circulated around the media, Dow's stock price fell on the Frankfurt Exchange, resulting in a temporary USD 2 billion loss in value, until the hoax was exposed and apologies offered by the BBC for its error. Three feature-length films – *The Yes Men* (Ollman et al. 2003), *The Yes Men Fix the World* (Bichlbaum et al. 2009) and *The Yes Men Are Revolting* (Nix et al. 2014) – have extended the media reach of the Yes Men in the public sphere, as has circulation of their activities via social media.

Culture jamming goals are not limited to undermining consumerism and offering social critique. The tactics discussed here have been taken up by activists working outside or in direct confrontation with institutionalized political power structures, such as organized parties, elections and the state. Waning trust in institutionalized political structures and the widespread availability of digital communications technologies have opened up the field of action for unaffiliated publics to engage in alternative forms of political expression, using new repertoires of communication, organization, mobilization and fundraising.

In 2011, a journalist, a sociologist, a graphic designer and other activists in the Czech city of Brno came together in response to the launch of a marketing campaign by the local government to promote the municipality (Macková and Macek 2014). The municipal marketing strategy espoused the key values of safety, development, creativity and openness as encapsulated under the tag line *Žít Brno* (To live Brno). The public and media immediately criticized the campaign, but the local government ignored their objections. Realizing that the city had neglected to claim the Internet domain zitbrno.cz, the journalist registered it and set about using it to criticize the government through the culture-jamming tactic of *détournement*. The *Žít Brno* website, launched in August 2011, claimed to be the official portal for the city's identity campaign, appropriating its official communications style to issue parody press releases, interviews and other statements to satirize the government. When the city later replaced the campaign's values with new ones, the Žít Brno jammers issued a statement changing the city's name to *Krno*, the *K* standing for the Czech word *koncepce* (concept), which was meant to call attention to an apparent lack of thoughtfulness on the government's part, a somewhat oblique jab that some of the public did not seem to understand, but that the group embraced as the term went viral online.

Over the next two years, Žít Brno emerged as a site of political protest, adding a Facebook page, Twitter account and individual social media profiles to enable the group's activities to go viral. The group began to use the platform as a mechanism for exposing and discussing issues that were overlooked in the media, thus playing the role of a citizen journalist. It became an oppositional voice, challenging the coalition formed by two strong political parties that had marginalized other elected representatives. Žít Brno addressed topics such as the problems of widespread gambling, the lack of transparency in government communications and decision making, and cronyism. In all of its work, Žít Brno maintained a satirical edge, using parody and other culture-jamming techniques. In one instance, it appropriated the mayor's identity and positioning statement 'Brno is not just a city, it is a lifestyle', replacing the background image of the picturesque city with an image of the mayor in front of an overflowing trash receptacle. Žít Brno's activities garnered national media attention and eventually led to it adopting a formal structure, entering the political arena by officially registering and being certified as a political movement.

A Dublin-based communications collective mounted a similar action in response to the 2010 austerity budget workout plan imposed on Ireland by the European Commission, International Monetary Fund and the European Central Bank to resolve an impending €44 billion debt crisis (Titley 2012). The group used the identity Budgetjam to organize efforts to counter mainstream media, government and civil society organization narratives that the crisis was of the nation's own making and that there was no alternative to austerity as the solution. Budgetjam's objective was not to propose a specific counter solution but to contest the dominant narrative that presented cuts to the social welfare segments of the national budget as economically necessary and even morally correct. To this end, the collective established a website and social media platforms through which it disseminated its critique of the dominant narrative, specifically targeting the International Monetary Fund, against which Budgetjam documented numerous examples of failed austerity programmes and their destructive effects in other parts of the world. Another target was Ireland's low corporate tax rate, a key element of its 'Celtic tiger' strategy, projected as sacrosanct, even though evidence suggested that raising corporate tax would not seriously impact foreign direct investment and would rather serve to augment revenue and thus help mitigate the crisis. Budgetjam maintained a live blog to monitor and contest the dominant narrative in real time; the blog also served as a platform for organizing protests and publicizing counter proposals to austerity. The hash tag #budgetjam furthered the collective's reach. Many of the perspectives developed by Budgetjam found their way into the mainstream media, on both a credited and uncredited basis, thus contributing to an emerging discourse within the public sphere that there were indeed alternatives to austerity.

Processes of co-optation

According to the Situationists, the deconstructive tactic of *détournement* can and has been countered by the reconstructing forces of recuperation, allowing powerful actors to co-opt and reabsorb radical ideas back into the dominant system (Wark 2008). Heath and Potter (2004) argue that the disruptive strategies of counter-cultural activism have ultimately served to reinforce many of the systems they mean to defy. Berger (2000) specifically cites the appropriation of the tactics of culture jamming by advertisers as a marketing trend. "Peeling away the brand veneer", as culture jamming has effectively done, opens up new avenues of consumption and thus continued adherence to capitalist hegemony in the form of green and blue consumption (eco-friendly and fair trade, respectively), new market segments that commercial interests have emerged to serve (Holt 2002:86).

The techniques of culture jamming are similarly co-opted in the practices of guerrilla and viral marketing, which use street and social media campaigns to reach consumers outside traditional print and broadcast advertising channels. For example, on 7 March 2017 a bronze figure of a Latina girl mysteriously appeared in front of the Charging Bull statue in Manhattan's Financial District. *Fearless Girl*, as it became known, appeared without warning on the day before International Women's Day, seemingly in defiance of finance capital and its white male-dominated power structure. It turned out, however, that the sculpture was part of a guerrilla marketing campaign commissioned by the New York investment firm State Street Global Advisors to mark the first anniversary of its Gender Diversity Index, which tracks the performance of large-capitalized firms in relation to their representation of gender diversity among the ranks of senior leadership. Curiously, State Street Global had recently been forced by the US Securities and Exchange Commission to pay a USD 35 million settlement for billing clients for illegitimate commissions and hidden fees (United States Securities and Exchange Commission 2017). The company also later paid some USD 5 million in back compensation and interest for equal pay violations in connection with its female and minority employees (Stevens 2017).

Future directions

As Lievrouw (2011:84) notes, despite the co-optive movements of recuperation by the hegemonic forces it has endeavoured to counter, culture jamming still retains at least some of its power to galvanize constituencies within the public sphere. Brandalism's interventions at COP21, for example, projected an oppositional perspective on environmental concerns, exposing vested corporate interests, insufficient political responses and rampant consumerism that contribute to the impending eco-apocalypse of climate change. These interventions further recorded the effects of unsustainable practices on the planet and foregrounded a political commitment to mobilize publics to work for systemic change (Lekakis 2017). They received international news coverage in print, broadcast and online and contributed to the ongoing debate surrounding the future of Earth's biosphere and humanity itself.

Adbusters continues to publish online and in print. Its 2019 campaign Milkshake Moonday, which calls for dousing corporate executives, politicians and other representatives of the powers that be with dairy products, is a *détournement* of the chicken sandwich outlet Chick-fil-A's Milkshake Monday promotion. In July 2011, *Adbusters* famously issued a call to Occupy Wall Street, inspired by the earlier uprisings in Tunisia and Egypt, and in particular activists' use of social media to galvanize and mobilize unaffiliated publics. Occupy Wall Street additionally took its cue from culture jamming's Situationist forebears, whose May 1968 actions in Paris inspired similar protests around the world. The online version of the July 2011 *Adbusters* featured a downloadable graphic file announcing the Occupy Wall Street action, which could be printed and distributed or forwarded via email or social media. Beginning on 17 September of that year, thousands of people converged on Zuccatti Park in lower Manhattan to protest growing worldwide social and economic inequality. Occupy actions soon appeared in other US cities, as well as cities around the world.

Culture jamming continues to garner academic interest, with hundreds of entries on the subject dated 2017 or later on Google Scholar. Topics cover a wide range of concerns: the assertion of indigenous identity by the Saami people of northeast Europe through posters distributed online that decolonize dominant images of 'the other' (Junka-Aikio 2018), the critique of Victoria's Secret's representations of gender via social media by the feminist activist group FORCE (Madden et al. 2018), and the reappropriation of culture jamming by the public relations industry (Weaver 2018). Other studies examine the practice theoretically through the lens of semiotics (Li 2018), the visual aesthetics of parafiction (Perucci 2018) and other forms of media analysis (DeLaure et al. 2017).

See also: activism; co-optation

Recommended reading

DeLaure, M., M. Fink and M. Dery (eds) (2017) *Culture Jamming: Activism and the art of cultural resistance*, New York: NYU Press.

A collection of essays that includes Mark Dery's 1993 treatise, the first to popularize the term culture jamming, along with entries by media scholars on culture jamming theory and practice as exemplified by artists such as Banksy, Shepard Fairey, the Guerrilla Girls, Pussy Riot and Reverend Billy and the Church of Stop Shopping.

Dery, M. (1993) 'Culture Jamming: Hacking, slashing, and sniping in the empire of the signs'. Available online: http://project.cyberpunk.ru/idb/culture_jamming.html.

First published by *Open Magazine*, Dery's essay is generally acknowledged as having popularized the term culture jamming for the mainstream media and for media scholarship.

Klein, N. (2000) *No Logo: Taking aim at the brand bullies*, New York: Picador.

One of the founding texts of the alterglobalization movement, written in the wake of the 1999 protests in Seattle against the World Trade Organization. The book describes the negative effects of global corporate brands and discusses ways in which people have fought back, including the culture jamming movement as championed by *Adbusters*.

Lievrouw, L. A. (2011) *Alternative and Activist New Media*, Cambridge: Polity Press.

A study of several major genres of alternative new media in the service of various forms of activism. In addition to a chapter on culture jamming, the book covers computer hacking, participatory journalism and independent media, the use of media as a mobilizing tool, and social media.

DIGITAL STORYTELLING

Çiğdem Bozdağ and Sigrid Kannengießer

Digital stories can be broadly defined as narratives available in different forms that are produced with and distributed through digital media (Alexander 2011). This definition encompasses almost all types of content produced using digital media. Examples include short films on YouTube and other video platforms, as well as stories in weblogs, online forums or on social networking sites. Depending on the digital media chosen to tell a story, narratives may consist exclusively of written text or combine text and (moving) pictures.

In this entry, the term digital storytelling is coterminous with "classic digital storytelling" (Lundby 2008a:5). This specific approach was originally developed in 1994 by the Center for Digital Storytelling (Center for Digital Storytelling, n.d.) in California, an initiative that today is known as StoryCenter and collaborates with organizations around the world (Hartley and McWilliam 2009a). Understood from this perspective, digital storytelling designates the workshop-based production and sharing of autobiographical stories using audiovisual digital media, providing individuals and collectives with the possibility to present and position themselves within and in relation to society.

Digital storytelling helps people from different backgrounds to express their perspectives and share their experiences on different Internet platforms. Hence, it has the potential to empower individuals by giving them a voice (Hartley and McWilliam 2009a:3; Tacchi 2009:167) and transform social structures by disrupting preestablished cultural codes and traditional power relations (Rodríguez 2001:3). Digital stories can thus be regarded as citizen media, understood as those practices through which (unaffiliated) individuals express themselves and participate in the creation of diverse publics (Baker and Blaagaard 2016a:7).

This entry explores the structural features of digital storytelling. It then moves on to examine a range of examples of storytelling from the three fields where this practice is most prevalent, i.e. education and mediated communication, development communication and health communication. Finally the entry discusses the potential of digital storytelling as a form of citizen media for democratizing existing communication structures, focusing on the issues of voice, self-representation and empowerment.

Structural features of digital storytelling

A digital story consists of a "short, first-person video-narrative created by combining recorded voice, still and moving images, and music or other sounds" (Center for Digital Storytelling, n.d.). Digital stories are produced and shared in workshops driven by different agendas and attended by participants from different backgrounds. Workshops typically start with a face-to-face story circle, in which participants sit together and tell each other their stories. Participants then proceed to produce their digital stories with non-professional media equipment, drawing on still or moving images and voice-over narrations of their stories. Although these workshops are often initiated by non-governmental organizations or public institutions, the stories are ultimately created and shared by unaffiliated workshop participants – thus falling within the remit of citizen media practices.

The main aim of digital storytelling workshops is to empower citizens who tend to be socially marginalized (Hartley and McWilliam 2009a:13). Participants in these safe spaces "feel that their ideas will be valued" and their stories will have resonance (Lambert 2009:86). In this context, the storytelling process facilitates the participants' understanding of their own "self", which is constructed socially "through the stories we tell about who we are" (Lundby 2008a:5). Significantly, the term digital story*telling* places emphasis on the storytelling process, rather than the product itself. The empowering aspect of the storytelling workshops lies in the story sharing process, during which participants learn to use digital media equipment to express themselves and develop their own voice.

The narratives and the relationship between the authors and the audiences of digital stories are changing, as digital media provide new platforms for interaction (Lundby 2008a:6; Bratteteig 2008:278–284). This transformation, which can be seen as a process of mediatization, relates to reciprocal influences between media practices and media affordances (Lundby 2008b) – thus influencing both the production and consumption of stories. In the case of digital stories, the consumption of storytelling becomes a much more interactive practice. Audiences have the chance to publish their own interpretations of the stories through comments and also interact with other audience members and authors.

Couldry (2008b:49) argues that there are four main differences in the narrative configuration of digital and oral stories. First, in digital stories there is a "pressure to mix texts with other materials" (ibid.). Within classical digital storytelling, storytellers tell their story through voice-over and by using still or moving images of themselves – whether from the present or the past – to create their story. While the visual is normally subordinated to the spoken word within classical digital stories, in some digital narratives written text is superimposed onto the visual. In these stories, written texts comment on images or sound – which in digital storytelling may take the form of music and/or other acoustic elements. Second, in digital storytelling, there is "a pressure to limit the length of narrative" (ibid.) that comes with the digitalization of stories. This means that the films produced are short, often only a few minutes long, as the availability of workshop time is limited. Third, there is "a pressure towards standardization" (ibid.) which, in the context of digital storytelling, means that the format of the films produced is very similar, with the storyteller narrating the story off screen and his or her voice dominating the visual. Despite this pressure towards standardization, these autobiographical stories have an authentic character (Hertzberg and Lundby 2008:108–109) and are often "deeply felt, poignant and gently humorous" (Burgess 2006:209). Fourth, there is always the risk that stories shared with broader publics online may have "unintended and undesired audiences" (Couldry 2008b:49). On a more positive note, authors have the possibility to remain anonymous and this gives citizens the chance to raise their voices about taboo topics that are otherwise difficult to discuss in public (Kannengießer 2012:244). Ultimately, the need for anonymous representation depends on the topic of the digital stories and the context in which these stories are produced.

Digital storytelling, as characterized in this section, provides new possibilities for the empowerment of storytellers through workshops, production of stories and the sharing of self-representation. For this reason diverse organizations and groups employ the practice of digital storytelling; examples of their different practices will be sketched in the following section.

Digital storytelling practices

The popularity of digital storytelling has increased tremendously since the term was first coined by the Center for Digital Storytelling. Today, workshops target different groups of people, including but not limited to children and youth, women or ethnic minorities (Lundby 2008b; Hartley and McWilliam

2009a). The fields in which classic digital storytelling practices are most commonly used are education and mediated communication, development communication and health communication. This section will discuss how digital storytelling is being practised in these three fields and examine concrete examples in more detail.

In the field of education and mediated communication, digital storytelling is used as a tool to empower students and to help them develop their own voice. Digital communication has the potential to foster a more participatory and student-centred approach to education and digital storytelling is considered one of the most fruitful methods to support this purpose, in both formal and informal education settings (Robin 2008; Sadik 2008; Ohler 2013). Digital storytelling is seen as "a powerful technology tool for the 21st century classroom" (Robin 2008:220) because it is based on affordable hardware and software and can help students to improve skills such as digital literacy, global literacy, technological literacy, visual literacy and information literacy, which are necessary in increasingly mediated environments (ibid.:222–224).

Digital storytelling is seen as an empowering tool for working with young people, especially those who are marginalized or silenced in society. It is used, for example, to engage indigenous youth in arts-inspired projects (Eglinton et al. 2017), young asylum seekers in Europe (López-Bech and Zúñiga 2017) and young people with migration backgrounds (Ranieri and Bruni 2012); to work with students who have learning disabilities (Manning 2010); or to improve students' intercultural awareness (Ribeiro 2016). In the context of action research, Ranieri and Bruni (2012) conducted a digital storytelling project with 11–15-year-old students from immigrant families with lower socioeconomic status in an urban suburb in Italy. Their aims were to increase the media literacy skills of the participants and to improve both their own self-representation and that of their community (Ranieri and Bruni 2012:222). The project involved several meetings with the students instead of a single workshop. In these meetings they critically discussed the storylines and the media that they were using. Through these reflective and interactive meetings, the storytelling process not only contributed to strengthen the media literacy skills of the students, but also fostered better interaction and collaboration among them. However, the stories produced were quite short and full of mistakes caused by the students' limited language skills. This shows that although digital storytelling can create a participatory learning environment and develop the media literacy of participants, it cannot redress the inequalities in the education system – which should be dealt with in the broader context of public education (ibid.:223).

A second field in which digital storytelling is increasingly used is development communication (Tacchi 2009; Reed 2010; Kannengießer 2012). In this field, digital storytelling has become an important instrument for non-governmental organizations to empower the marginalized through digital storytelling workshops. One example of this is an initiative organized by the non-governmental organizations Sonke Gender Justice Network and the Center for Digital Storytelling Speaks Initiative in South Africa. The organizations conducted workshops for young people in the Eastern Cape province (Reed 2010). The workshops functioned as computer training courses and encouraged the young participants to tell their stories, represent their realities from their own perspectives, and talk about their concerns in the films (ibid.).

Marginalized women are another target group of digital storytelling workshops in the context of development communication (Marsden et al. 2010; Simsek 2012). The aim of women-oriented workshops is to give them a voice (Kannengießer 2012) and enable them to network (Simsek 2012). One concrete example is a digital storytelling workshop with sex workers organized by the South African feminist organization Women'sNet (Kannengießer 2012). Participants report that the workshop setting functioned as a safe place that allowed them to talk about their experiences among like-minded people. Further, they described the practice of storytelling itself as a relieving and liberating experience. Many of these women participated in the workshop because they wanted to learn how to use computers and the Internet, and hoped that with this knowledge they would become qualified for jobs outside prostitution. Even though this proved to be an unrealistic aspiration, given that the five-day event did not provide comprehensive computer training, the workshop still created a 'first contact' with different digital media (Kannengießer 2012).

Health communication is the third field where the use of digital storytelling is particularly prominent. One of the key purposes of employing digital storytelling in the field of health communication is to raise awareness about a healthy lifestyle (Gubrium 2009), e.g. to increase HIV prevention (Nelson et al. 2016). The process of digital storytelling can be a therapeutic experience for participants, so digital storytelling is also used for working with adults and children who have had traumatic experiences. Anderson and Wallace (2015) adopted the workshop-based model of digital storytelling for therapeutic purposes as part of their work with children who suffered from post-traumatic stress disorder, with a view to help them cope with their experiences. Their study showed that the practice of sharing the story of their traumatic experience and creating a trauma narrative using digital media reduced their post-traumatic symptoms (ibid.:102). The children themselves rated their experience as highly helpful (ibid.).

While the specific aims and groups of the workshops depend on the organizers, individuals and the contexts where they are run, digital storytelling workshops generally aim to give participants a chance to present their perspectives through digital media production, find and raise their voices and explore self-representation and dialogue in public. These three aspects connect the practice of digital storytelling to the field of citizen media.

Classic digital storytelling as citizen media

Citizen media production enables a broader public – located outside the logic of the media industry – to exercise its self-expression through media production; it also provides an opportunity for equal access to production resources and enables people to demand recognition through self-representation in the public sphere. Therefore, citizen media production is a key practice of cultural citizenship (Burgess 2006:201; Couldry 2008a:386). Digital storytelling has now emerged as a widely used digital form of citizen media production. As is also the case with other forms of citizen media, digital storytelling has a potential for democratizing existing communication structures and challenging existing power relations by giving a voice to those who are excluded from institutionalized forms of media production. Digital storytelling offers ordinary citizens a bottom-up platform for communication through which they can create self-representations about their everyday life, raise their voices and create publics (Hartley and McWilliam 2009a:5; Couldry 2008a:384; Pettit et al. 2009:445).

The workshop-based practice of classic digital storytelling is not only important for self-representation. As is apparent in some of the examples mentioned above, participants who "find their voice" in classic digital storytelling workshops tend to find these are a therapeutic and empowering experience (Thumim 2012:8–9). Since the storytellers are telling a story to others, their stories are "becoming real" on the basis "of shared experience and affective resonances" (Burgess 2006:211). Furthermore, workshops create a space in which oppressed groups can share their experiences (Kannengießer 2012:240). This sharing is not only empowering for individuals, but can also be important for creating shared values and for the purposes of community building (Couldry 2008a:386). Although the digital stories themselves cannot revolutionize the existing exclusionary structures of mass media production, they can transform these structures by creating "the social and political conditions for making these expressions legitimate" and "reshaping boundaries" (Pettit et al. 2009:445). Through a process of remediation, digital storytelling transforms the everyday experience of ordinary citizens into shared public culture (Burgess 2006:210). By empowering individual participants, societies might become more democratic as (political) discourses become more diverse.

Digital storytelling encourages marginalized groups, who are often on the less resourceful side of the digital divide, to use digital tools for telling their stories. In this sense, workshop-based digital storytelling can also improve the digital literacies of their participants (Lambert 2009). However, the digital storytelling workshops do not easily solve the problem of the digital divide, i.e. inequalities in terms of language and technological skills or the problem of unequal access to media power (Burgess 2006:209). They can rather work as a "first contact" with computers or Internet media for workshop participants (Kannengießer

2012:248) – who, in many cases, will never use a computer again after the workshops have come to an end (Burgess 2006:209).

Digital storytelling does not only help participants to create autobiographical stories, but also provides ways of creating conversations around them. Democratic communication is not only about who gets to speak, but also about "who is being heard" and "to what end" (Burgess 2006:203; Dreher 2009:446). Digital storytelling has a potential for allowing acts of listening across differences and communities (ibid.). However, despite the wealth of good examples of digital storytelling projects that have reached broader audiences, many stories created within the workshops do not circulate outside the circles of the workshop organizers. The problem that digital storytellers have to find an audience and be heard – partly due to the high number of digital stories in circulation – is one of the main reasons why digital storytelling cannot reach its full democratic potential. But even when they are being listened to, digital storytellers may be confronted with intolerant reactions to their narratives. On the whole, however, conflictual debates prompted by digital storytelling are important to ensure an optimal functioning of democracy (ibid.:448).

This entry has focused on the practice and potentials of digital storytelling, a common communication practice today typically actualized through workshops run by non-governmental organizations or public institutions. Despite the limitations discussed above, digital storytelling has the potential to contribute to a more democratic communication environment by creating a new form of participatory communication that is accessible to broader publics. As shown by the rising popularity of digital storytelling and the growing volume of scholarly attention that it is attracting, digital storytelling is an increasingly important and relevant form of citizen media.

See also: disability media

Recommended reading

Alexander, B. (2011) The New Digital Storytelling: Creating narratives with new media, Santa Barbara, CA: ABC-CLIO.

Documents the emergence of digital storytelling, explores different formats of digital stories and reflects on the evolution of the digital storytelling movement.

Hartley, J. and K. McWilliam (eds) (2009) *Story Circle: Digital storytelling around the world*, Malden, MA: Wiley-Blackwell.

Consists of a number of empirical case studies whose authors examine the history of digital storytelling and engage with the diversity of the digital storytelling scene.

Lundby, K. (ed.) (2008) *Digital Storytelling, Mediatized Stories: Self-representation in new media*, New York: Peter Lang.

Lundy's introduction to this collection explores the concept of digital storytelling and delineates the scope of empirical research undertaken on this topic.

DIRECT ACTION

Benjamin Franks

Use of the term direct action can be traced back to the labour activism of the syndicalist movements operating in the US and UK in the late nineteenth and early twentieth centuries (Trautmann 1912/2014:30–40). It has most commonly been used to refer to the methods applied by workers to achieve their immediate goals, such as when employees deliberately sabotage production in an attempt to win better pay and working conditions. However, direct action has been applied to more than just the activities of workers at the point of production: for instance, the revolutionary syndicalist Elizabeth Gurley Flynn (1916/2014:112–113) considered women controlling their own fertility to be a form of direct action when they sought to place their interests above the needs of capitalists awaiting future supplies of workers. McKay (2012:1018) in *The Anarchist FAQ* explains the breadth of the notion: "basically direct action means that instead of getting someone else to act for you (e.g. a politician), you act for yourself". As such, direct action, as Kauffman (2017) argues, was a prominent feature of 1960s radicalism and has become an increasingly dominant feature of contemporary social movements, from Occupy to Black Lives Matter.

Direct action is an important category, not least for the way it helps demarcate specific types of political practice such as sabotage, strike action, occupation, selective vandalism and boycott, from other, more standard types of political behaviour like voting in elections, electoral campaigning, party political fundraising and lobbying. The use of direct action is also a strong indicator of ideological orientation. Its emphasis on unmediated action makes it attractive to groups opposed to hierarchies who reject the dominant political structures and norms for bringing about social change. It thus has strong affinities with the ideals of many of the more horizontally organized movements engaged in alternative and citizen media.

A veteran theorist of direct action, April Carter (1983), in a tract published by the Campaign for Nuclear Disarmament, demarcates the practice from both symbolic action and constitutional action. Symbolic actions are aimed at raising consciousness and mobilizing support rather than providing a direct solution for the problem. Examples would include the activities of the British housing charity Shelter, or citizens' actions on social media, that raise the issue of homelessness in society at large. Constitutional actions are the legitimated means for bringing about reform in accordance with the regulations stabilizing a particular social order. Advocates of direct action tend to reject constitutional forms of action as being inadequate or inherently reactionary, while opponents consider direct actionists to be ill-disciplined and damaging to the social order. For both revolutionary syndicalists and more contemporary anarchists and radical ecologists, however, the methods of direct action provide the main means of achieving effective change.

Direct action versus civil disobedience and radical interventions

Direct action can be distinguished from standard political behaviours in two respects: first, it is carried out without recourse to intermediaries, and second, it provides an immediate practical response, albeit often only partial or temporary, to a specific political issue. Direct action immediately prefigures, embodies or foreshadows the desired end goals. It is a synecdoche, representing through its action a major or minor part of a wider social vision. By contrast, other political methods might involve no relationship between method and goal. Symbolic and constitutional actions are more metaphorical: there need be no direct connection between the signifier (for example, a torch-lit parade for the homeless) and the signified (the provision of housing). While such symbolic actions are a necessary feature of any campaign, raising consciousness does not in itself house a person who was previously vulnerable. Constitutional action relies on intermediaries, usually legitimate agents of the state, to carry out the desired change. Electoral activity can have positive consequences, but the act of marking a ballot or putting in nomination forms does not by itself ameliorate the problem. By contrast, a strategy such as the homeless choosing to squat in and make habitable empty buildings does provide, albeit provisionally, a solution to their problem and thus constitutes direct action.

Freeden (1996) outlines a morphological approach to understanding political concepts that can be usefully applied here as a means of disentangling civil disobedience from direct action, as well as of exploring how they overlap. Freeden explains that contested concepts have their meaning stabilized by their priority (their importance or centrality) and their proximity to other concepts. Civil disobedience, on the one hand, although itself a disputed term, is usually characterized with reference to three interrelated concepts (Lefkowitz 2007; Rawls 1972:363–365). First, it involves the breaking of a law or strongly enforced custom (civil disobedients need to be noncompliant with a regulation). Second, it entails a commitment to nonviolence and is typically framed in terms of a withdrawal of cooperation rather than of overt confrontation. Third, acts of civil disobedience are performed with political intent in that they seek to change laws, customs, practices or significant social relationships. Rawls (1972) adds an additional characteristic, suggesting that, although civil disobedience necessarily breaks a law, it is still faithful to the legal system as perpetrators will acquiesce and even give themselves up for punishment. However, while this was a significant feature of civil disobedience in American civil rights campaigns of the 1960s and while it has continued to be prioritized in the activities of, for example, Extinction Rebellion climate activists, it should be acknowledged that many civilly disobedient militants have been less committed to handing themselves over to the authorities than Rawls indicates.

Direct action, by contrast, need not be illegal. Squatting in an empty residence was not a criminal action in England until the *Legal Aid, Sentencing and Punishment Act 2012*, but it was still a form of direct action. Anti-pollution campaigners who organize beach clear-ups are engaged in direct action, though there is no breech of legislation involved. Moreover, civil disobedience is not usually prefigurative. A protest march that refuses to move and thus blocks a busy London street to protest a war is an act of civil disobedience, but it does not itself interfere with the military action it is opposing; a similar blockade outside a military establishment that prevents the movement of the materials of war would be an instance of both direct action and civil disobedience.

These anti-hierarchical features of direct action discussed above make it particularly attractive to radical movements that distrust intermediary structures, especially those of the state or capital. As Boggs (1977a, 1977b) notes, methods such as direct action, which prioritize prefigurative responses, are typically associated with social anarchisms and heterodox Marxisms (like council communism), while competing revolutionary traditions, such as Leninism and other orthodox Marxisms, are associated with more Jacobin instrumental methods, where the ends justify the means. Social democracy, on the other hand, is aligned with the use of constitutional actions, the development of a separate political class and the diminution of emancipatory politics.

Many writers have specifically identified anarchism as being the political ideology most sympathetic to direct action, regarding it as a form of praxis that applies anarchist principles in the here-and-now and that can bring about immediate change more effectively and consistently than processes of reform (Goldman 1923; Graeber 2009; Gordon 2018; Vysotsky 2015). Such is the proximity of direct action to anarchism that many groups feature the term in their self-descriptions and publications. The anarchosyndicalist Solidarity Federation used to be called the Direct Action Movement, for instance, and still use the phrase for the title of their magazine; Earth First! (2015) issued a *Direct Action Manual* and Green Anarchist distributed *Urban Attack: A Primer for Direct Action* (1990). The early 2000s saw the rise of the Direct Action Network focused on local democracy and anti-capitalism. Trautmann (1912/2014), a militant in the revolutionary syndicalist movement Industrial Workers of the World (IWW), was aware that their preference for direct action over more constitutional methods distinguished this labour grouping from the social democratic traditions of the American Federation of Labor.

Direct action is not the only mode of prefigurative politics. The adoption of constitutional methods can foreshadow the types of representative and managerial structures sought within liberal democracies. Similarly, racist groups might prefigure their xenophobic goals by launching direct assaults on minority communities in order to drive them out (Cai and Landon 2019). However, different forms of direct action can be distinguished, such that the advocacy of unmediated political action by one ideological movement should not be confused as being equivalent to or in sympathy with other advocates of direct action, as some mistakenly infer (see for instance RAND Corporation 2005). Clear differences are apparent in the use of direct action by different groups: anarchism and its aligned social movements typically give far higher priority to prefiguration in comparison with constitutional democratic or authoritarian movements. For the latter, direct action is often a subsidiary or marginal tactical choice. In addition, the anarchist concept of prefiguration is proximate to a completely different set of principles and generates different identities to those of fascism or liberal democracy. Anarchism places prefiguration next to principles of anti-hierarchy, inclusion and social solidarity, such that prefiguration means generating direct action that embodies cooperative, egalitarian and accessible social relationships. In authoritarian or representative democratic movements, prefiguration is understood in relation to other concepts, shifting its meaning and thus the interpretation of direct action within that movement. For instance, a student Labour Party group in the early 2000s issued a poster with the slogan "Take direct action: vote". While this appears contradictory to those who regard direct action in terms of immediacy and anti-hierarchy, it becomes marginally less so if what is being prefigured are the representative functions and institutions of constitutional democracy.

Although emphasis on direct action is a strong indicator of anti-state ideologies, and although direct action is often defined as being against constitutional action, some groupings, especially in green politics, do advocate both. Influential British environmentalist Jonathan Porritt (1997:62–66), for instance, considers it a strength of the green movement to have embraced a diversity of tactics, utilizing direct action (though Porritt favours the term civil disobedience, even if the examples he cites can fit either category) and constitutional methods. Green Party members are often involved in protests, blocking airport expansion or preventing fracking, at the same time as they attempt to win votes in parliamentary and local elections. This combination of methods is, however, far from stable. A prolonged environmental occupation during an election will mean the Green Party supporter has to either abandon the site to canvass for votes, or forsake the election to maintain direct action. Either ultimate legitimacy lies with the legislature or, alternatively, it is based on those directly impacted and their methods of response. As Porritt recognizes, the result of embracing constitutional over direct action is that distinctive institutions, norms and communicative behaviours are required that are aimed at influencing existing elites rather than the myriad, fluid activist groups. Constitutional action also redefines problems and solutions in ways that prioritize the values of these dominant institutions, rather than those of other political agents and ecological subjects.

Other environmental groups that try to combine direct action with constitutional action reconceive direct action in a more symbolic manner. Greenpeace's website (2010) highlights that they run a "political unit" whose goal is to influence elected representatives to enact environmental legislation and business. "While this work rarely hits the public eye", Greenpeace (ibid.) argue that " it can be enormously influential – especially when it's combined with Greenpeace's other strands of work like direct actions or investigations". Greenpeace's statement sees direct action as a supplement to its lobbying role. As the level of engagement with business and state institutions grows, what appears to be direct action can become increasingly symbolic, leading to what Greenpeace (2019) call "direct communication". This involves spectacular acts, like dumping a pile of coal outside a senior politician's house, which are designed to influence opinion makers rather than enact immediate, prefigurative environmental change. Many groups involved in the production of citizen media are engaged in the more immediate, bottom-up and practical forms of direct action. However, as Rone (2016) notes in her study of citizen media in Bulgaria, others interact with powerful economic entities and centralized political power, and as a result become co-opted into, at best, symbolic enactments of dissent and, at worse, a further institutional resource for oligarchical power.

In sum, civil disobedience and direct action, while both militant forms of political action that contest the priority given to constitutional institutions and procedures, are nonetheless distinguished by the latter's privileging of unmediated, immediate and prefigurative action. These characteristics make direct action particularly attractive to anti-hierarchical, anti-state and anti-capitalist ideologies. Civil disobedience, by contrast, is often positioned in terms of reforming rather than replacing constitutional norms and structures. As a result, direct action is more likely to be pejoratively framed as a form of irrationalism and terror.

Direct action versus symbolic action

As noted above, direct action is often contrasted with symbolic action, and indeed some direct actionists are critical of methods that solely concentrate on the figurative. The longstanding British anarchist group Class War (2017), who have been involved in grassroots housing campaigns, advocate for squatting as they consider it as something real rather than symbolic. On their website they juxtapose this material action against the "[m]arching, demonstrating, protesting, and all the other out-of-date activities of the Left [which] have become a purely formal, symbolic activity" (ibid.). However, as Class War and other advocates of direct action are aware, symbolic action is also necessary to motivate, promote, design, coordinate and evaluate more immediate material tactics.

In addition, it is important to recognize that some communicative or symbolic activities are not just supportive of (and foundational for) direct action, but can also constitute a form of direct action themselves. As Baker and Blaagaard (2016a) indicate, graffiti is a type of citizen media as it performs communicative functions, but graffiti is also a form of direct action that interrupts and challenges state power's control of material space and its authority to impose and maintain a linguistic order that supports its dominance. Culture jammers highlight the convergence of symbolic and direct action by disrupting the production, distribution and interpretation of signs that have particular cultural and economic value. Most major corporations have carefully constructed images that constitute a brand which has identifiable marketable value. Advocates of direct action as far back as Flynn (1916/2014) have been aware of the material impact of symbolic interruptions, as in the case of aggrieved hospitality workers informing potential and current customers of the actual insanitary conditions in the kitchens (Sprouse 1992). As market managers are aware, there can be significant negative economic outcomes as a result of effective symbolic sabotage of their brands. Further, as Ordóñez (2017) argues, direct action disrupts the symbolic order of capitalism, generating new meanings and experimental, exciting and fulfilling ways to interact with one another.

The division between the merely symbolic and direct action becomes even more fluid in the case of hacktivism. Hacktivism involves the use of electronic media to disrupt the stream of images and messages on corporate and state websites that legitimize their power. The subversion of images has a material impact on the operation of the organization as it undermines the brand image and limits its commercial effectiveness (Gloeckle and Royal 2017). Hacktivism also includes persistent, organized mass postings onto targeted websites in a way that prevents traffic to and from these sites and causes significant commercial and administrative disruption (Kelion 2012).

There is often a gap between, on the one hand, how activists view their activities and how they imagine non-participants perceive them and, on the other hand, how they are actually viewed by non-participants. Activist-geographer Anthony Ince (2010) provides an example of the distinction. Ince was active in the anti-capitalist protests against the G8 meeting in Geneva in 2003. These protests were intended to directly disrupt the meeting and prevent the G8 leaders from further reorganizing the social and political order in their interests, and instead to enact an alternative in which the interests of the economically oppressed and the ecologically vulnerable were prominent. However, as Ince reports, many local people, whom the activists hoped would be supportive of the protests, simply viewed the actions as confusing, self-indulgent and paternalistic. Similarly, the more the audience for direct action becomes separated from the participants, the more it becomes symbolic and thus open to reinterpretation through intermediaries with their own institutional biases.

The strengths and limitations of direct action

Direct action has faced a number of criticisms. Kellner (1975) describes how liberal constitutionalists such as Plamenatz and Hook tend to reject political action that lacks democratic legitimacy because it has violated liberal principles of consent. Following Kant (1793/1991), they argue that free, equal citizens have the ability to influence legislation through public discussion and democratic voting, and are thus obligated to live under the laws of the land, even those against their immediate interests, because they would expect others to do the same if they had been successful in drafting legislation to meet their interests. However, as Kellner points out, the Kantian conditions of such a binding obligation to law are rarely met. Governments often deliberately misinform citizens, citizens lack full equality in the decision making process, and some laws infringe on their future ability to make democratic decisions. Furthermore, unlike civil disobedience, not all direct action necessarily breaks the law.

Although a supporter of selective direct action, Singer (2011) provides some strong arguments against such practices on largely utilitarian grounds. He argues that democratic procedures provide reliable forms of decision making that overall produce the best social outcomes. Direct action, which necessarily places legitimacy outside of the constitutional realm, undermines these benevolent political institutions. Singer is, however, aware that selective direct action might be justified on utilitarian grounds, in cases where democratic decision making is too slow to act effectively, or where the interests of some sentient beings are ignored or marginalized by majority rule. That said, in these instances, it is also possible that other forms of social organization, foreshadowed by direct action, might provide better general outcomes and avoid minority discrimination in comparison with Singer's preferred form of reformed representative democracy.

De-Shalit (2001) offers a further range of criticisms of direct action. Among the most compelling are that direct action produces elitism or exclusion as it unintentionally prioritizes specific agents and insular group identities. Given that direct action has a symbolic power which participants want to amplify, actions are chosen on the basis of their perceived effectiveness in gaining media attention. As mainstream media narratives tend to concentrate on particular media-friendly personalities rather than the complexities of the causes they are engaged in, direct action can create hierarchies of prominent activists and key spokespeople at the expense of less well-positioned practitioners, in addition to marginalizing less media-friendly but more prefigurative, anti-hierarchical activities. However, this seems less a criticism of direct action than of symbolic action. As de-Shalit acknowledges, drawing on Doherty et al. (2000), direct action is primarily about immediate impacts and is not principally concerned with attempting to shift public opinion through the media.

Direct action can appear elitist because it privileges certain groups in particular locations, such as the radical ecologists who are not burdened by immediate responsibilities and who are thus able to dedicate themselves to a protest camp for months, or the workers with the unity to maintain a workplace occupation. It can also lead to group chauvinism, where one form of direct action by one particular formal or informal collective is seen to be the primary form. Critics of direct action have suggested that constitutional democracy, by contrast, is much more egalitarian and accessible as it is far easier to vote than to participate in direct action. However, this risks misrepresenting direct action. While some groups and tactics are in a privileged position to make effective change, no type of action and no single agent or organization is supposed to provide the primary and universal moment of liberation. Oppressive power is diverse, and so are the responses to it. Unlike constitutional activity that is predicated on the single core identity of the liberal citizen, direct action's pluralism allows for different and more flexible degrees of engagement to come to the fore.

See also: activism; civil disobedience; prefiguration; social movement studies and citizen media

Recommended reading

Kauffman, L. (2017) *Direct Action: Protest and the reinvention of American radicalism*, London: Verso.

Detailed description and analysis of direct action-centred social movements and the responses they have generated. The case studies presented explore diverse contexts from the 1960s New Left to AIDS activism, anti-capitalism and Black Lives Matter.

Ordóñez, V. (2018) 'Direct Action', in B. Franks, N. Jun and L. Williams (eds) *Anarchism: A conceptual approach*, London: Routledge, 74–85.

A concise essay that identifies the core features of direct action as unmediated and prefigurative activity. Ordóñez describes and assesses its multiple manifestations and explains its particular affinity with contemporary and historical anti-hierarchical, anti-political movements.

Salerno, S. (ed.) (2014) *Direct Action and Sabotage: Three classic IWW pamphlets from the 1910s*, Chicago, IL: H. Kerr and Company.

A collection of Industrial Workers of the World pamphlets, including those by Elizabeth Gurley Flynn (1916/2014) and William Trautmann (1912/2014), that locate direct action initially within the industrial setting of the factory and mark it out from more conventional, constitutional political strategies and economic behaviours. Flynn's contribution in particular then extends the scope of direct action to encompass communicative actions and personal forms of resistance to oppression.

DISABILITY MEDIA

Filippo Trevisan

There are more than one billion people with disabilities worldwide (World Health Organization 2011). Yet, people with disabilities and disability-related issues are often ignored in mainstream news and popular media. When they are represented, coverage tends to be dominated by stereotypes that reproduce stigma and negatively affect the individual and collective identities of people with disabilities. To address these issues, people with disabilities and their organizations have created a vibrant disability media sector that is constantly in flux and open to innovation.

This entry offers a definition of disability media and provides a critical review of some of the most important initiatives in this field in recent decades. Specific examples of disability media from multiple countries are discussed to illustrate the evolution and current state of disability media by focusing on key processes and actors. There is a long tradition of disability media that predates digital platforms but that is constantly changing and being renewed with technological advancements. The most innovative, influential and thought-provoking examples of disability media have tended to come from institutionally unaffiliated writers, filmmakers and bloggers with disabilities. Their independence from corporate media structures, as well as from the fundraising goals typical of nonprofit organizations, have enabled them to contravene traditional news-making practices and challenge dominant representations that marginalize disability within the collective imagery, which characterizes their work as quintessentially citizen media (Baker and Blaagaard 2016a:15).

The overall picture that emerges from this review is one in which the creators of contemporary disability media have gone beyond the alternative media space and have established grassroots efforts that aim to influence mainstream representations and public discourse more broadly through collaboration with legacy media organizations, as well as innovative web-based outlets that are rooted in participatory culture.

Defining disability media

Disability media are media created by people with disabilities with a view to presenting distinctive disability viewpoints on key issues and experiences relevant to the disability community. Disability media encompass a broad range of artefacts and initiatives that are independent of corporate media conglomerates and, crucially, are created by editors, journalists, commentators, producers, writers, bloggers, photographers, filmmakers, actors, as well as ordinary citizens with disabilities. Some forms of disability media such as dial-in newspaper services for blind and visually impaired people, and independent YouTube sign language news programmes for Deaf and hard-of-hearing people such as The Daily Moth (The Daily Moth, n.d.) are designed to serve specific groups. However, many others seek to challenge dominant narratives and influence public discourses around disability more broadly. In doing so, they embody the ideal of alternative media because they "are produced by the same people whose concerns they represent, from a position of engagement and direct participation" (Atton 2002:16), which enables them to "offer the most thorough version of alternative news values" (ibid.). By projecting "the voices of self-advocacy" (Riley 2005:132), disability media simultaneously seek to challenge and infiltrate mainstream media and culture with a view to "effect[ing] … socio-political change" (Baker and Blaagaard 2016a:16).

It is useful to reflect briefly on the circumstances that have historically underpinned the creation of disability media and that continue to support them to this day. There are three main separate but intertwined factors that have sparked the need for disability media. These include: lack of accessibility in mainstream media; the marginalization of people with disabilities and disability issues in news, popular and entertainment media; and the use of stereotypical representations of people with disabilities in mainstream media and popular culture. Indeed, the fight for accessible media is paramount to the emancipation of people with disabilities. As technological development continues, disability rights advocates continue to push tirelessly for the application of universal design principles so that new media technologies are "born accessible … without the need for a retrofit" (Ellis and Kent 2017:5). That said, given one of the key goals of citizen media is to situate the groups and communities that stand behind them vis-à-vis society and project their grievances, needs and aspirations, it is particularly useful to focus here on the other two motivating factors behind the creation of disability media.

Historically, there has been a lack of representation of people with disabilities and disability issues in news coverage (Clogston 1994). At the same time, disability is rarely represented in entertainment media. For example, a study that looked at the most popular movies released between 2007 and 2015 found that only 2.4 per cent of speaking characters had a disability (Smith et al. 2016). Of these, only 2 per cent were shown in animated movies, virtually depriving young viewers of any opportunity to come in contact with any notion of disability (ibid.). This lack of coverage reflects the marginalization of people with disabilities in society more broadly. Part of the reason for this marginalization lies in the ongoing influence of the medical model of disability in dominant discourses about disability, which sees individual impairments – not environmental barriers – as the primary source of disability and which encourages the individualization of disability issues.

Key disability policy debates since the late 1980s have provided important opportunities for increased coverage of disability. Yet, several factors conjured against the diffusion of constructive portrayals of people with disabilities. For example, when the Americans with Disabilities Act (ADA) was negotiated and eventually passed in 1990, US disability activists made a "conscious decision … not to engage with the news media" (Burns and Haller 2015:273) and instead focused their attention on policy makers. Yet, this was a mistake as "the news media filled the void created by 'disability voices' with … those who saw it [the ADA] as a cost" (ibid.). In the UK, disability coverage has increased in conjunction with a contentious welfare reform since the turn of the century. However, British coverage has also become more ideologically driven and has often framed people with disabilities in a negative light through "the constant repetition of words like cheat, scrounger or workshy" (Briant et al. 2013:886–887). In Australia, coverage of the National Insurance Disability Scheme introduced in 2013 was dominated by a traditional mode of representing disability as a social pathology for which people with disabilities deserve state assistance as a gift, not a right (Burns and Haller 2015:272).

At the same time, positive representations of disability in mainstream media tend to be limited to simplistic stereotypes that respond to commercial imperatives and simultaneously reflect and inform widespread misconceptions about disabilities (Ellis and Goggin 2015a; Haller 2010). One popular stereotype is the super-crip, often a Paralympic athlete who is celebrated for possessing super-human abilities. Yet, these individuals are no more representative of the disability community than Olympic champions are of average non-disabled people (Riley 2005:151). A second common set of stereotypical representations are those that can be described as inspiration porn. This is an expression made popular by the late Australian activist Stella Young to signify those instances in which individuals with disabilities – often children – are celebrated for engaging in ordinary everyday activities with the aim to boost motivation for non-disabled people (Grue 2016; Haller and Preston 2017).

These representations weigh negatively not only on non-disabled audiences but also on the disability community itself. As Nelson (2000) noted, the media have an important role to play in fostering a sense of community among people with disabilities. However, in audience research people with disabilities have said that many film, television and news representations of disabled people are problematic and disempowering (Haller and Zhang 2014). Representations that foster stigma by perpetuating the stereotype of disability as deviant – whether positively or negatively – discourage individuals with disabilities from identifying with relevant stories, which "limits the ability to construct one's own identity as well as having any impact on social attitudes toward a disability" (Worrell 2018:51). This is a key obstacle to the creation of a strong sense of community among people with disabilities.

A typology of disability media

In light of this, disability media seek to project the voices of people with disabilities and offer a grassroots viewpoint while also fostering a stronger sense of community among people with disabilities themselves. Although disability media have never been systematically and comprehensively mapped, at the end of the 1990s Winston estimated that there were "at least 1,200 mass media resources for the disability community in the form of magazines, newspapers, newsletters, radio/TV programs, and recurring newspaper columns" (1999, quoted in Haller 2006:994). Since then, the range of disability media has expanded significantly with the addition of new web-based and hybrid formats that, in line with the citizen media tradition, "can be injected with specific meanings, values and expressive power" (Baker and Blaagaard 2016a:15). Dozens of new disability news websites, blogs, podcasts, sign language video services over YouTube, and other media are created every year, which can reach ever expanding audiences, both disabled and otherwise (Ellis and Goggin 2015b).

Ransom (1996) has proposed a classification with three main groups of disability media depending on their organizational nature and primary goals. These include: activist/political publications; mainstreaming/assimilationist publications; and special interest publications. An important example of activist/political publication is *Ragged Edge*, which first appeared in 1980 (then known as *The Disability Rag*) and quickly became "a small but powerful magazine" (Shaw 1994:x) rooted in the disabled people's movement. *Ragged Edge* moved to an online-only publishing model in 2004 and ceased to update its website in 2007. This trajectory is illustrative of the challenges experienced by activist disability publications that operate on a shoestring budget and rely mainly on volunteer contributors and philanthropic funding. The early 2000s constituted an especially challenging moment for this type of publication because disability discourses lost much of their urgency after the adoption of the Americans with Disabilities Act (ADA) in 1990, which was followed by comprehensive disability discrimination legislation in several other countries (Riley 2005:133). These challenges reoriented disability media toward the other two models featured in Ransom's (1996) typology (mainstreaming/assimilationist publications and special interest publications).

On the one hand, the disability media outlets that were best able to withstand the financial storms of the early 2000s were those published by large disability nonprofit organizations. Many of these fit into the mainstreaming/assimilationist category because they seek to support the acceptance of people with disabilities in society. Although there are virtually as many of these publications as there are disability non-profit organizations, good examples include the US-based National Multiple Sclerosis Society's magazine *Momentum* and the *Braille Monitor*, which the National Foundation of the Blind makes available in various formats for both blind and sighted readers. Another interesting example is *New Mobility* magazine. *New*

Mobility began as an independent publication for wheelchair users in 1989 but was acquired by the United Spinal Association in 2010, which distributes the magazine free of charge to its members and makes it available to non-members for a fee. As 'benefit of membership' publications, these magazines enjoy relative financial stability. However, their membership nature also limits their circulation, and their inherent focus on promoting services and stimulating giving has been criticized for favouring stereotypical narratives of therapy that align with the medical rather than the social model of disability (Riley 2005:137).

On the other hand, another set of publications that seems to enjoy greater financial stability are special interest publications, which have carved out loyal niche audiences with specific needs or interests. One notable example is Disability Scoop, which is a news website about developmental disabilities launched in 2008. Nevertheless, while special interest publications have a good market, they lack the reach and scope to support broader debates about topical issues on disability and society.

Disability media in the digital age

As of 2019, publications associated with disability nonprofit organizations remain a large portion of disability media. Yet, thanks in part to the affordances of digital and web-based media, a range of new initiatives have emerged, which eschew the typology illustrated above and seek to address both the need for empowering representations of disability, and the imperative of financial viability. Some of the most innovative efforts in this area have been spearheaded by unaffiliated individuals who, having sensed the need for new types of disability media, have taken advantage of digital technologies to build communities around specific initiatives. In what follows, three key trends are discussed with a view to providing a useful overview of an area of citizen media that is characterized by growing complexity, creativity and innovation.

The disability community has an ambivalent relationship with new media technologies. On the one hand, ongoing access and accessibility problems make people with disabilities less likely to regularly use the Internet than non-disabled people even in the richest and most connected countries (Anderson and Perrin 2017). On the other hand, certain forms of digital media have some important affordances for people with specific disabilities. For example, Internet radio and podcasts are increasingly important for people who are blind, YouTube and other video sharing platforms for Deaf and hard-of-hearing people, text blogging for people with autism, and Twitter for people with speech impediments (Hemsley et al. 2015). That said, one form of online media that has become particularly popular across the entire disability community are blogs, a term that is used here to include text blogs as well as photo, audio and video blogs.

Goggin and Noonan (2006:164–165) identify two main types of disability blogs including: first, blogs that are "explicitly articulated from a disability activism or disability studies perspective"; and second "activist and information blogs, often with … a diaristic or journal nature". The extended format of blog entries lends itself well to discussing disability issues, which are complex by nature, and enables new voices to "provide alternative narrations that are not necessarily in accordance with the dominant paradigms, including those proposed by social model [of disability] theorists" (Goggin and Noonan 2006:165). Blogs have introduced a new way of representing disability that goes beyond both the medical frames often pursued by established mass media and the more militant approach of publications such as *Ragged Edge*. Disability blogs offer new grassroots perspectives on how individuals with disabilities navigate the complex relationship between impairment and environmental barriers. Some of the most successful disability bloggers have since expanded well beyond their personal blogs and set up initiatives that have redefined the relationship between disability media and activism.

British disability bloggers have been pioneers in this area. For example, in 2010 a small group of British disability bloggers including Sue Marsh (n.d.) and Kaliya Franklin (n.d.), who became popular in the disability community with journal-like blogs about their experiences with the welfare system, seized on their visibility to launch a virtual advocacy campaign called The Broken of Britain that protested government plans for a disability benefits overhaul. These bloggers were able to project themselves into a position of leadership and quickly attracted thousands of supporters dissatisfied with the opposition efforts of established disability rights organizations and who recognized themselves in the personal struggles described by these bloggers (Pearson and Trevisan 2015; Trevisan 2016).

These and other similar online campaigns have been permeated by a participatory media logic, using blog sites to project representations of disability crowdsourced from hundreds of disabled contributors that form counter narratives to the stereotypes reproduced by some legacy media (Trevisan 2017, 2018). A similar initiative in the US, if somewhat more curated than the crowdsourced blogs mentioned above, is Alice Wong's (n.d.) Disability Visibility Project. Launched in 2014, this project includes a website and several social media channels (Twitter, Facebook, Instagram) that project disability narratives in various formats collected through multiple channels including the StoryCorps oral histories project, images, essays and blog posts. By using blogs and social media platforms to crowdsource and disseminate individual narratives from people in the disability community who would most likely not have been able to contribute to more traditional disability media such as magazines, these initiatives broaden the grassroots character of disability media.

Other important innovations in disability media have come from collaborations with legacy media outlets, which in recent years have opened up to a more participatory approach in covering disability. This has generated new publications that throw the brand and visibility of large news organizations behind disability grassroots voices. For example, the British Broadcasting Corporation (BBC) and the Australian Broadcasting Corporation (ABC) launched content aggregators about disability, named Ouch! and Ramp Up respectively, which also featured original commentary by writers with disabilities. While these operated only for a few years and although the former was criticized for arguably not giving sufficient prominence to the opinions of people with disabilities (Thoreau 2006; Ellis and Goggin 2015b), both these initiatives signalled renewed interest among legacy media in covering disability from a more direct and unmediated angle. In August 2016, the *New York Times* launched a regular disability column penned by a variety of writers with disabilities external to the paper. This was possibly the highest profile publication for writers with direct experience of disability at the time this entry went to print. Although most of these NYT columns include references to a personal story or experience, these are intertwined with and seek to advance the debate on key social issues that surround disability.

Capacity-building initiatives that have trained persons with disabilities in basic journalistic skills have been one of the most important drivers behind this enhanced collaboration between the disability community and legacy media organizations. One such example is the training provided by Bournemouth University in partnership with the nonprofit Access Dorset in the UK, which has led to stories about disability discrimination being disseminated through the BBC and other British media outlets (Thorsen et al. 2015). Others have gone beyond capacity building and are experimenting with a new model of communication services agency, staffed entirely by communication professionals with disabilities. Most notably, this is the case of Rooted in Rights (n.d.), which operates simultaneously as a training organization for aspiring disability storytellers and as a provider of media products and consultancy on a range of disability-related media projects, from advocacy campaigns to feature films. Disability bloggers such as Emily Ladau and Vilissa Thompson form the backbone of this operation, assisted by digital media specialists and producers with disabilities. The group offers support and advice that combine media expertise with direct experience of disability and strong ties with the community in a way that seeks to influence positively both legacy media and nonprofit communication.

Finally, a third set of actors within contemporary disability media are defined by their creative efforts to address the perennial financial precariousness in this sector, which also influences their structure in some important ways. On the one hand, some have turned to crowdfunding platforms. One notable example is that of My Gimpy Life, which is a YouTube web series created by and starring the disabled actress Teal Sherer, which uses comedy to chronicle her life in Hollywood. Following a successful first series in 2012, Sherer launched a campaign through the crowdfunding portal Kickstarter to fund a second series, which debuted in 2014. Ellcessor (2017:38) compares this approach "to that seen in the tech industry, as it rests upon the ability of individuals or groups to create a visible identity (or brand) that aligns with expectations" of supporters and potential investors. Crowdfunding promotes a special relationship between media creators and viewers/donors, in that the former often connect directly with the latter in an effort to ensure ongoing support for their projects. This can have a profound influence on the content produced.

While crowdfunding constitutes a promising model for supporting special projects, others have begun to experiment with a hybrid approach that combines commercial advertising sales with the social change orientation typical of B-Corporations. Arguably the most prominent initiative of this type within disability media has been The Mighty (n.d.). Part news site, part social media site, The Mighty is a specialized platform that publishes health and disability content written and edited by individuals impacted directly by a wide range of conditions and disabilities. Founded in 2014 by former editor-in-chief of AOL Mike Porath, The Mighty (n.d.) claims to have over ninety million unique visitors a month and more than two million registered users, while its specific disability section had 232,000 followers and 187 contributors as of the end of 2019.

The Mighty crucially offers free partnerships to nonprofit organizations with a view to accelerating their exposure to interested audiences. This is a support model similar to the one the digital advocacy platform MoveOn.org offers to its grassroots partners, who are afforded opportunities to exploit MoveOn.org's network to generate more visibility for their initiatives. The Mighty's unique set up, which blends an independent but commercially funded platform with a social change mission and community orientation, as well as its large readership, position it as an ambitious project in the current landscape of disability media.

Key trends in disability media

This entry can only scratch the surface of the vast and diverse area of disability media. Yet, in this final section, it is useful to point out three key trends that have characterized this sector as disability media creators have embraced new information and communication technologies to develop more sustainable and influential models of media production and dissemination, which in turn could provide inspiration for expanding and strengthening citizen media in other areas.

First, while disability media continue to project independent representations of disability and disability-related issues, new partnerships with legacy media organizations have been launched that have the potential to redefine how disability is conveyed in mainstream outlets. Although one may argue that this type of collaboration could expose disability media specialists to the influence of mainstream journalistic norms and the financial imperatives of legacy media, it should also be noted that, as people who are simultaneously members of the disability community and trained in production, reporting and editing skills, they are uniquely positioned to influence legacy media in positive and unprecedented ways. Another crucial benefit of these collaborations is the opportunity for disability voices to reach mass audiences that are usually precluded to specialist publications.

A second key trend is the emergence of personal storytelling as a communicative practice deployed across a variety of media forms, from personal blogs to crowdsourced activist productions that challenge dominant representations, and from the work of disability public relations agencies such as Rooted in Rights to for-profit operations such as The Mighty. These developing forms of disability media have seized on the ability of compelling personal stories to attract large audiences and promote positive identification for readers/viewers with disabilities. Crucially, by rooting the discussion of key issues in the concrete specifics of real-life experiences, these personal stories tend to offer perspectives that work to undermine the stereotypes of disability that are all too widely presented elsewhere.

Finally, a third trend that is closely connected to the previous two is that all of these new disability media initiatives have adopted – to various degrees – a participatory media logic. Indeed, disability media have always been citizen media. Yet, digital tools have greatly expanded the number of potential contributors to both crowdsourced efforts and community-oriented online platforms, which now include a much larger range of voices than could ever be represented in print magazines. That said, it is important also to remember that people with disabilities continue to be negatively affected by Internet access and accessibility issues, which exclude some of the most marginalized members of the disability community from participating effectively in these media efforts. As disability media continue to evolve, those who create them ought to be mindful of these issues and make sure there are ways for those excluded from digital media also to contribute their perspective and to enrich the disability media sector.

See also: crowdsourcing and crowdfunding; digital storytelling; media ecologies

Recommended reading

Ellcessor, E. and B. Kirkpatrick (eds) (2017) *Disability Media Studies*, New York: New York University Press.

Presents a series of case studies focused on topics as diverse as *Iron Man 3*, Lady Gaga and Oscar Pistorius, as well as analyses of historical media, independent disability media, reality television and media technologies, to consider how media form cultural assumptions about ability, the construction of disability via media technologies, and how disabled audiences respond to particular media artefacts.

Ellis, K. and G. Goggin (2015) 'Disability Media Participation: Opportunities, obstacles and politics', *Media International Australia* 154(1): 78–88.

Proceeds from an analysis of disability media spaces such as ABC's Ramp Up website and crowdfunding platforms to argue for the inclusion of a disability perspective in critiques of community, citizen and alternative media.

Ellis, K., G. Goggin, B. Haller and R. Curtis (eds) (2019) *The Routledge Companion to Disability and Media*, New York: Routledge.

Provides an overview of the main trends in key debates exploring the relationship between disability and media through contributions from leading scholars and disability media activists based around the world.

DIVERSITY

Julie Boéri

Diversity may appear as a highly consensual political principle. Indeed, the vast majority of institutions, political parties and communities would not challenge the desirability of promoting diversity in contemporary societies. Far-right populist and neo-fascist groups, who explicitly call for the hegemony of one ethnic group, race, religion, language and/or nationality over others, remain an exception, despite alarming signs of normalization of such views in mainstream political discourse. Yet, diversity is also a highly instrumentalized political principle, as evidenced in the unity in diversity political motto, which has become a staple of so-called multicultural societies at the same time as these same societies are taking decisive steps to restrict migration and political asylum. The instrumentalization of the concept of diversity is also evident in so-called multicultural marketing campaigns, which have been on the rise since the 1990s. The Benetton United Colors campaigns launched in 1984, for example, demonstrate the extent to which diversity has become a highly profitable form of branding (Hoechsmann 1997). They initially featured young people representing a wide range of ethnicities, all dressed in Benetton merchandise, but later became less focused on showing the merchandise than on displaying different ethnicities (for instance a black woman breastfeeding a white baby) living together in harmony and peace. These ads continue to receive criticism for their specious display of diversity and tolerance, and for their "insensitive and exploitative" use of images – for example, of migrants being rescued in the Mediterranean (Yeginsu 2018) – in order to promote Benetton's own corporate interests.

Diversity in a commodified media space

Not only is diversity a highly instrumentalized political principle, its application in practice is also fraught with contradictions. For example, theorists of cultural globalization have tended to see in the emergence of a transnational media and its corporate system an opportunity for cultures to reconfigure themselves and for the world to enhance cultural heterogeneity (Harvey 1989:156; Hannerz 1991:120–124; Appadurai 1990:295; Giddens 1991:199). However, the capacity of international fluxes to generate cultural diversity has been overestimated, whereas the commodification logic which facilitates these fluxes has been underestimated (Mattelart 2008:21–22). The practical consequences of the latter can be seen in the rapid extension of first world cultural hegemony, "the unchecked spread of market-ideologies, the global economic and political influence of transnational corporations, the emergence of international tourism, the dominance of Western scientific and technical paradigms, and the global spread of Western popular culture" (Cronin 2003:72). This logic is sustained by global news agencies (Bielsa and Bassnett 2009).

The diversification of channels in mainstream media has not led to a diversification of content or the inclusion of diverse voices. Instead, information abundance has been accompanied by information uniformity. In a case study of *Jornal Nacional*'s coverage of the 2002 presidential election in Brazil, Porto (2005) argues that the non-inclusion of the diversity of voices articulated around the controversies of the electoral process restricted the range of interpretations to which voters were exposed, thus serving the interests of the state and the market. Drawing on a review of the models of media regulation – ranging from the trusteeship model in a private-driven media landscape like the USA to the public broadcasting service in Europe – he concludes that public interest is under assault in the information society. In view of the dominance of such a deregulated media landscape and of its negative impact on political deliberation, Porto advocates a model which stresses the "right to knowledge", rather than the "right to information", given that media are "key sources of interpretive frameworks that allow audiences to make sense of the information that reaches them" (2005:141). In arguing for the right to knowledge to be safeguarded, for political deliberation to become a reality and for public interest to be protected, Porto (ibid.:142) draws on McQuail (1992) and Murdock (1999), who advocate two types of diversity: diversity as reflection and diversity as access. "Reflective diversity" is achieved when the media "represent or reflect the prevailing differences of culture, opinion and social conditions of the population as a whole" (Porto 2005:142); "access diversity", on the other hand, requires the media "to publicize the interpretive frameworks that are sponsored by relevant social groups or organizations" (ibid.). This twofold diversity is generally pursued by minority and alternative media, which have become increasingly visible in the participatory environment of Web 2.0.

The emergence of social media, electronic forums and blogs has often been considered as empowering a multiplicity of voices in the media space. At the same time, mainstream media are now equipped with a complex apparatus for encouraging citizens to participate as produsers, that is, as both producers and users of news as they witness various events (Bruns 2010b). A number of case studies have, however, revealed that the apparent diversity of content generated by fans and citizens in this context is misleading: it is subject to control and filtering by the traditional players of cultural industries in all sectors, including entertainment, information and infotainment (Jenkins 2006a). In their case study of Internet-user participation in *Le Monde*'s Live Blog on the scandal of Dominique Strauss-Khan, the French politician accused of attempting to rape a hotel maid in 2011, Marty et al. (2017) compared the comments submitted by Live Bloggers and those finally published on the Live Blog. They found that an "editorial filtering process remains most of the time invisible to the audience" (ibid.:16), and that this "'enunciative gatekeeping' [is] geared towards moderating the content of exchanges in accordance with the ethics and the deontology of *Le Monde* journalism" (ibid.:15). In her study of the BBC's convergent news streams from Syria and Libya, Chouliaraki (2016:205) similarly concludes that "[i]t is … the journalistic voice that ultimately dominates convergent news". Thus, the apparent multiplicity of contributions may appear to give voice to lay persons, cutting across gender, generation, political and professional backgrounds, but this voice is ultimately subsumed within patterns of homogenization that shape the way information and knowledge are circulated and shared.

Alternative practices of media production and consumption existed before the advent of Web 2.0, in the form of low-tech minority media (Rigoni and Saitta 2012) and of grassroots, independent, community-based media projects in which the same people managed the media and reported on the events they witness and experience. These initiatives, before and after the emergence of Web 2.0, tend to be more inclusive of voices which are usually marginalized or filtered out by mainstream media, and thus make available a more diverse range of interpretive frameworks on various areas of public concern. The Indymedia network created in 1999 for and by activists involved in the demonstrations against the World Trade Organization summit in Seattle is an interesting case in point. Thanks to Indymedia coverage, the Seattle demonstrations could not be ignored by mainstream media, and eventually made the headlines and TV news at the time (Cockburn et al. 2000).

Chesters and Welsh's (2004) ethnographic study of the protests against the International Monetary Fund in September 2000 in Prague describes a productive strategy for highlighting diversity and resisting mainstream media's attempts at homogenizing various groups. Activists involved in the protest split the march into three distinctive colours – silver, yellow and pink – in order to visually frame the protest as multiple, innovative and united in its diversity. Their strategy reveals the extent to which the threat of being portrayed as uniform by mainstream media leads communities to self-mediatize as diverse, through the use of technologies that may encompass not only social media and ICT devices but also other media such as flags, colour and their own bodies.

Diversity in transnational movements

In the literature on transnational movements, diversity is approached as a political principle fraught with contradiction and as characteristic of citizens' linguistic, discursive, bodily and visual performance within digital and non-digital environments.

For example, diversity is identified as constitutive of the collective identity and action of the 1960s women, black and peace movements (Polletta 1998), the alterglobalization movement in the 90s (Maeckelbergh 2007), the World Social Forum at the turn of the century (Böhm et al. 2005; della Porta 2005; Doerr 2008, 2012; Boéri 2009, 2012) and of the wave of protests that characterized the second decade of the current century, including the Arab uprisings, Occupy and the movements of the squares (Maeckelbergh 2014). These movements cut across time, space and struggles and contribute to a large and loose "movement for global justice" in which diversity is considered "an act of resistance to the homogenization of 500 years of colonial history, contemporary democracy, the mass media and consumerism" (Maeckelbergh 2007:92).

The value of diversity in this context is epistemological in nature: it is intended to challenge the homogenizing narratives of (neo-)colonialism, progress and expertise and advocate for social change as diverse, multi-layered, undefined and under construction. It is thus constitutive of an alternative approach to knowledge and power, often associated with the principle of horizontality in global movements (Böhm et al. 2005). Horizontality refers to "a radical decentralization of power, and to the creation of new types of power that allow people to 'take control' not of others, but collectively of themselves" (Maeckelbergh 2007:88). It functions as an overarching political principle which encompasses and is contingent upon participation, deliberation and prefiguration (Boéri 2009, 2012) as well as diversity (Maeckelbergh 2007:88). Social movements pursuing these principles strive to resist their opposites: verticality, representation, struggle and uniformity (Maeckelbergh 2007; Boéri 2012). The tensions between these two sets of principles remain at the heart of citizens' attempts to create networks of solidarity across struggles, as evident in the case of Babels, the international network of volunteer translators and interpreters which emerged to cover the translation and interpreting needs of the Social Forum, and which considers linguistic diversity a condition of horizontality. Case studies have shown that in practice the network is under pressure to deliver interpreting efficiently on the day of the event despite a general lack of funds, and thus to restrict interpreting to the central events of the programme rather than extend it to the autonomous spaces; to reduce the number of languages to those most commonly spoken – paradoxically the colonial languages; and to adopt top-down decision-making processes (Boéri and Hodkinson 2004; Boéri 2013, 2015).

Tensions, pressures and conflict in mass-based movements are not necessarily adversarial and destructive. A number of studies suggest that when articulated within a horizontal politics of organization, they can also result in innovation: in the case of Babels, they led to the development of an *ad hoc* training initiative (Boéri 2010) and communication protocols designed to ensure efficient and transparent consensus decision making (Boéri 2009, 2014, 2015). These initiatives demonstrate that diversity is not understood purely as a political agenda to be inscribed within the Forum itself through volunteer interpreting, but also a principle that Babels is committed to implementing in its own politics of organization (Boéri 2009, 2012). Change ultimately occurs through a process of negotiation among a diversity of viewpoints, stakes and grievances.

Linguistic diversity has been highly visible in the wave of protests that emerged across the globe in the second decade of the current century. In their case study of the Tunisian revolution, Laroussi and Lienard (2013) discuss what they refer to as *écrilectes*, that is, intensely hybrid sublanguages through which citizens express themselves using a mix of foreign and native, oral and written languages (French, English, Tunisian and standard Arabic, sometimes semi-phonologically Latinized), low and high register, in graphic and pictorial formats (words, pictures, emoticons). Drawing on El Zein and Ortiz' (2011) collection of photographs from Tahrir Square protests, Huguet's (2013) study of Tahrir scripts explores citizens' construction of a new graphic order in the wake of the Egyptian revolution, arguing that these scripts are circulated through two types of walls: the physical walls of Tahrir Square where activists write, draw and paint, and the virtual walls of the web where these inscriptions are circulated across a complex chain of communication and contextualized through different acts of framing: informing, denouncing, convincing, honouring, comforting, giving hope or conveying other emotions. This chain of communication acts and the semiotic diversity it displays to both *in situ* and remote publics contribute to the renewal of the Arabic language and the emergence of a cosmopolitan political community which has its roots in the Arab world (Huguet 2013).

Thus, despite an apparent shift from transnationalism to locally rooted struggles since 2010, citizens engaged in radical attempts to effect change are showing continued commitment to diversity as opposed to uniformity and integration (Maeckelbergh 2014). This diversity is embodied in the cultural and linguistic repertoires they draw on, the audiences they interact with and the communication tools they use. At the same time, this multi-layered diversity constructs a unique subject – the city of Cairo, or the community of solidarity with the revolution – and the very conditions of citizenship.

Diversity, intersectionality, social movement theory and technology

Introduced to feminist theory at the end of the 1980s (Crenshaw 1989), intersectionality has since functioned as a paradigm that addresses experiences of marginalization and exclusion at the intersection between gender, race, age, class and ability. Intersectionality rejects what Crenshaw refers to as "the single-axis framework" (ibid.:139), which involves "think[ing] about subordination as disadvantage occurring along a single categorical axis" (ibid.:140) such as gender or colour, but not both. This reductive prism limits "inquiry to the experiences of otherwise-privileged members of the group" (ibid.) – white, able, upper-class, heterosexual. Intersectionality thus attempts to both account for multi-layered forms of marginalization – colourism, ageism, gender discrimination or any manifestation of othering – and inform and sustain a more radical social transformation that addresses cross-cutting concerns and grievances. From the perspective of intersectionality, marginalized and oppressed individuals are understood as intrinsically diverse, and as such as bearers of a more radical transformation that is to be unlocked in practice and in theory.

Work on intersectionality has led to a rethinking of diversity in more critical terms (Longman and de Graeve 2014). It has alerted scholars to the dangers of the instrumentalization of this concept (Pinxten and Longman 2008; Longman and de Graeve 2014) and to how the dominance of a politically correct discourse on diversity has been accompanied by the loss of its critical and emancipatory potential (Ahmed 2011). As Bell and Hartmann (2007:910) put it, this institutionalized discourse "conflates, confuses and obscures the deeper sociocultural roots and consequences of diversity". It also tends to reproduce an unmarked normativity (white, Western, secular, able-bodied) that remains unquestioned under the guise of promoting diversity (Lentin and Titley 2011:183). At the micro-level, this means that minority groups are often reduced to specific identity categories to which they are perceived to belong (race, gender, culture, religion), despite the fact that they may themselves aspire to transcend these categories in an effort to contribute to radical social change. Ahmed (2011) thus calls for a critical approach to diversity which addresses the complex relationship between power and difference underpinning processes of social transformation.

This research agenda has also been pursued by social movement studies, but its application here has been restricted by the discipline's traditional focus on the mobilization strategies of what were assumed to be fairly structured groups. Social movement studies has faced the challenge of accounting for Social Movements Organizations (SMOs) which have increasingly stressed expressive goals and self-realization rather than traditional issues of labour and production (Pizzorno 1978; Touraine 1985; Melucci 1986). By the end of the 1990s, scholars had begun to formulate a theoretical response to a wide range of factors such as the collapse of communism, the explosion of feminist consciousness and the politics of multiculturalism, as well as the emergence of so-called new social movements characterized by diversity and networking rather than centralizing strategies of collective action. They thus began to reorient their analysis towards exploring bottom-up processes of identity making, self-realization and the articulation of expressive goals. The prevailing perception of SMOs as "media junkies" (Gamson 1995:85) gave way to the idea that "movements are media" (Melucci 1996:36); this has become a cornerstone of contemporary social movements, whose highly diverse deliberative spaces are intended to embody inclusivity and pluralism (Doerr 2018).

In sociology, communication and media studies, interest in diversity is evident in the study of the mediation and mediatization of ICTs in increasingly technologized environments, specifically the diversity of appropriations of technological innovations and their effect on people's experience of the world. Moving away from Everett Rogers' diffusion model, which reduces the role of users to that of mere acceptance or refusal of innovation (Boullier 1989), scholars in the sociology of (technological) uses (*Sociologie des usages* in the French-speaking tradition), in a post-1968 context, began to acknowledge the autonomy of users of technologies, and their capacity to adapt these technologies to their needs in ways that may circumvent the uses prescribed in their original design (Jouët 2000:496). This reconceptualization of individuals as actors in innovation processes rather than mere users (Chambat 1994) has fallen short, however, of studying identity making, community belonging, habitus and sociocultural aspects which structure the relationship between social actors and the world, and the ways in which they adapt and reshape the tools and artefacts they use (Granjon et al. 2009). Subsequent calls for a critical sociology attempt to account for diversity in making sense of and creatively contributing to technological innovation (Latzko-Toth and Millerand 2012). The French-speaking interdiscipline of communication and information sciences (*Sciences de l'Information et de la Communication*) has pioneered this strand of scholarship: its main focus has largely been on actual practices of communication, on the interaction between social actors and artefacts and among social actors via technological devices (Appel et al. 2010; Thiéblemont-Dollet and Koukoutsaki-Monnier 2010). Examining online and offline interactions at the micro-level can cast analytical light on converging and diverging appropriations of technologies in online communities and networks characterized by heterogeneity and can allow us to understand the dynamics of power and counter-power at the macro-level of communities, organizations, institutions and social spaces (Proulx 2005; Proulx et al. 2006; Boéri 2015).

These various strands of scholarship attempt to address the challenge of accounting for both the potential and the limits of diversity in spaces of emerging publics that are increasingly ICT-mediated. They highlight the need for a more open conceptualization of technologies, not as mere technical devices ushering social change but as apparatuses of power (Foucault 1977) that encompass tools, actors, discourses and practices whose emergence and development are interdependent with citizens' uses and appropriations. This renewed approach to technology is particularly important for understanding the complex dynamics of diversification and homogenization in citizen media initiatives, which are shaped by tensions between the local and the global, the physical and the virtual, the individual and the collective.

See also: autonomous movements; co-optation; prefiguration; race & ethnicity studies and citizen media; social movement studies and citizen media; solidarity; wikis; World Social Forum

Recommended reading

Doerr, N. (2018) *Political Translation: How social movement democracies survive*, Cambridge: Cambridge University Press.

An empirical study of the ways in which social movements address structural inequalities and patterns of marginalization in their deliberative arenas, carried out over a decade across cities in Europe and in the US. It focuses on four initiatives in which intersectional diversity typically poses a challenge to inclusion and internal democracy: the European Social Forum, the United States Social Forum, an urban development project led by a progressive city hall in California and a community forum organized by a non-governmental coalition for affordable housing in the same town. Doerr uses the term political translation to conceptualize citizens' ways of addressing "positional misunderstandings" (p. 122), which emerge in the communication encounter because of subtle and informal exclusionary practices. Political translation both includes and transcends interlingual translation; it foregrounds the power and responsibility of what Doerr refers to as 'thirds' (translators, interpreters, cultural intermediaries, bilingual facilitators) in order to construct an ethos of egalitarianism and inclusion in three ways: disruption (for example, by interrupting the meeting to effect a change in the dominant party's problematic communication behaviour towards a minority), persuasion (by "directing attention to power imbalances and drawing on the egalitarian commitments of those who otherwise would be unlikely to recognize their own structured privilege"; p. 16), and intentional organizing (as in using logistics as a political tool to make initiatives more inclusive).

Maeckelbergh, M. (2009) *The Will of the Many: How the alterglobalisation movement is changing the face of democracy*, London: Pluto Press.

An eye-opening journey through the trajectory of the alterglobalization movement, which engages extensively with the concept of diversity and connects it to related notions and practices of agency, decentralization, democracy, inclusion, constructive conflict, horizontality, identity, networks, pluralism and power. Maeckelbergh suggests reframing the tensions between diversity and unity in terms of "overlapping unities" (pp. 199–200), that is, as a dynamic flow of convergence and divergence across actors connected to each other through loose networking – a process that has the potential to open up a space for constructive and productive conflict.

Rigoni, I. and E. Saitta (2012) *Mediating Cultural Diversity in a Globalized Public Space*, Basingstoke, UK: Palgrave Macmillan.

A collected volume which addresses cultural diversity by exploring some of the ways in which the communication practices of ethnic minority groups are reshaped by three interconnected phenomena: developments in ICTs, new conceptions of space, and mobility. Three chapters engage specifically with the notion of diversity: Titley examines the positioning and reception of diversity-oriented migrant media in relation to the politics of migration, with specific reference to Ireland; Suárez Navaz and Ferrández Ferrer focus on Latin American minority media journalists in the Spanish media landscape, and the conflict they experience between uniformity, censorship and conservatism on the one hand, and diversity, critical knowledge and contestation on the other, coupled with a conflict between their socio-professional positioning and their sense of community belonging, at the interface between business and citizen journalism; Bozdag, Hepp and Suna explore audiences' appropriation of diasporic media and the extent to which these media construct an alternative, diverse public space.

DOCUMENTARY FILMMAKING

Mark R. Westmoreland

Documentary filmmaking is characterized by a civic impulse that is evident throughout its history. Aided by the proliferation of digital technology and user accessibility, documentary film has experienced a resurgence in the early twenty-first century. In its ideal form, citizen documentary filmmaking aspires to the principles of self-initiated, collectively made and civically minded productions and processes, in which unaffiliated citizens can effectively address key civic issues – whether as a means of enacting social change, shifting public narratives, claiming representational autonomy or expressing personal ambitions. At the same time, the documentary media ecology is dominated by resource intensive production processes and structurally entrenched dissemination paradigms that can heavily influence the message and impact of documentary films.

By navigating a variety of obstacles in order to make documentary films that can operate outside "the context of corporate structure, a political party, a media organization, an NGO, or similar institution" (Baker and Blaagaard 2016a:15), unaffiliated documentary filmmaking practices illuminate critical fissures in the edifice of the documentary modality. This entry thus attempts to outline the constraints and opportunities for citizen-led documentary film, which play a strong role in determining the ability to produce, share and preserve documentary productions that are, to a greater or lesser degree, unaffiliated. Infrastructural constraints specifically concern how material conditions like finance, distribution, aesthetics and technology determine the documentary film ecology. Opportunities for enacting unaffiliated documentary practices are created through various forms of institutional support, external collaborations and activist movements. The documentary imperative to leave a record for posterity is illustrated with several examples where unaffiliated filmmakers have struggled to find temporal linkages with future and past generations.

Constraints, opportunities and questions of affiliation

With increased accessibility to digital production tools and the proliferation of content platforms, documentary filmmaking now appears more open to ordinary people than ever. And yet, this proliferation of technological tools and participation in film culture has also fostered an imbalance in which an excess of filmmakers are competing for the attention of programmers and distributors, who themselves are working under both financial and formal constraints. With shrinking pools of funding, this means that filmmakers competing in this media economy must bear the initial financial risk, in the hope of recouping costs from distribution and exhibition deals at a later stage. Global film festivals are a case in point. These have become major players in the curation and cultivation of award-winning and globally distributed documentary cinema (Wong 2011; de Valck 2007); the most important festivals have implemented an assembly-chain production process from pitch session to award ceremony.

These models, in which filmmakers are now expected to adopt corporately designed impact campaigns, tend to prioritize return on investment over creative, experimental and community-based work (McLagan 2012). Rather than fostering a diversity of documentary approaches, this marketplace competition tends to favour a standardized feature-length format characterized by "emotionally manipulative soundtracks, gorgeous photography, larger-than-life characters, spiced-up scenes, and stylized reenactments" (Zimmermann and de Michiel 2017:19). Such formulaic films may pull at the heartstrings of global cineastes, but the core social issues tend to serve as a backdrop to character-driven real-life stories. In effect, such documentaries delocalize the affective responses to these social issues by targeting cosmopolitan film audiences and generating pity for victims elsewhere (ibid.:ix). While privileged global audiences who consume these documentaries ideally sympathize with the core universalizing themes of war, poverty, inequality and injustice, rarely do these film projects have an afterlife in collaboration with the communities represented.

In order to avoid these constraints, filmmakers attempt to find alternative strategies for producing civic-oriented documentary films. As Chanan points out, "even in countries where access to official screens is strictly controlled, digital video and alternative dissemination make the new documentary possible" (2008:17). For instance, digital platforms and online distribution enable documentary filmmakers to bypass the cinema to find audiences on portable screens, thus liberating documentary from more exclusive venues like film festivals and from restrictions imposed by distributors. In addition to the diversification of distribution platforms, documentaries now also adopt a wider range of modalities, "including algorithm-based work, citizen journalism, critical cartography, database and interactive documentary, locative media, live performance, social media, YouTube, and augmented and virtual reality" (Zimmermann and de Michiel 2017:ix). While digital tools have enabled citizen-led documentary initiatives to gain audiences in distant locations, technology does not guarantee a message's success. The affordances of these technologies also rely on filmmakers creating innovative ways to produce and self-distribute their films. Bootleg distribution schemes help in turn to establish alternative media networks. But in spite of technological accessibility, unaffiliated citizens must actively and creatively seek alternative opportunities to reach their intended audiences (Hight 2014). They thus need to transform their skills and practices in order to organize "an ongoing process of community mobilization and renewal" (Zimmermann and de Michiel 2017:vii).

Zimmermann and de Michiel (ibid.) have characterized such filmmaking as open documentary, explaining that it foregrounds the urgent concerns of ordinary people responding to local issues in a way that seeks solutions rather than merely diagnosis. The goal of these documentaries is not solely representational; they are attentive to the participatory processes of collaboration through their models of production. Although they may address global concerns, these citizen-oriented documentaries are not necessarily produced with global audiences in mind.

While they may openly embrace collaborative community-based projects, participatory production modalities and the affordances of new digital platforms, unaffiliated filmmakers may not necessarily reject the involvement of third parties and official infrastructures in every aspect of production. Avoiding affiliation challenges the infrastructural constraints of documentary filmmaking in compelling and crucial ways, but documentary filmmaking nevertheless often relies on varying degrees of affiliation during some phase of the production process. As an already precarious modality (financially, politically and epistemically) that typically requires a protracted production timeline, ambitions for autonomy in such projects must be situated within a spectrum of dependency enacted at different stages of documentary production and distribution, particularly where affiliations are unavoidable and perhaps advantageous for citizens.

The collaborative imperative and participatory gesture of open documentary projects are rarely about filmmaking alone. Under the banner of participatory video, such projects offer important examples of collaborative forms of expression that cede control to unaffiliated individuals and communities. These may be collective experiments with artists, designers, coders, hackers, activists, students, teachers and scholars. By foregrounding the sensibilities of citizen groups, participatory projects may challenge the professional conventions of genre, form and aesthetics in compelling ways. Indeed, such collaborations may not even result in a documentary film as the process becomes more important than the product for these kinds of community-based projects.

Such participatory and collaborative approaches to documentary filmmaking have a variety of important historical precedents. Jean Rouch, the renowned French ethnographic filmmaker, gained inspiration for his notion of shared anthropology from Robert Flaherty, who developed themes and scenes for *Nanook of the North* (Flaherty 1922) in collaboration with his Inuit hosts and then screened footage back to them for further feedback. Rouch used similar approaches to develop a series of ethnofictions with his collaborators in West Africa and France in ways that directly challenged colonial logics (ten Brink 2007; Henley 2010). Such projects have since inspired countless other initiatives, including Appalshop and the Center for Digital Storytelling, both of which enable local communities and ordinary people to produce their own films (Gubrium and Harper 2016).

These participatory and anthropological projects also helped engender various forms of indigenous media-making, which provides an important global perspective on citizen-led documentary filmmaking and demonstrates how such communities use documentary modalities to reclaim representational sovereignty (Wilson and Stewart 2008; Deger 2011). During the earth movements of the 1990s, Amazonian tribes used video to document their claims to cultural authenticity and land rights by aligning their causes with global environmental concerns (Conklin 1997). The increasing accessibility of digital cameras and editing software has helped to democratize the production of documentaries, but many citizen-led documentary projects nevertheless rely upon some form of institutional help or affiliation. Video in the Villages, founded by Vincent Carelli, provides one of the most enduring examples: it initially trained Amazonian Indians to make their own videos, but has since expanded to provide an important context for different indigenous communities to present issues of local concern and share information both within the region and beyond. Nevertheless, given the unequal power relations and appropriation of indigenous knowledge historically, indigenous communities might not be receptive to the ideas and opportunities offered by outsiders, at least initially (Elder 2016).

Citizen-led documentary filmmaking may be supported by institutional bodies. In the late 1960s, George Stoney and others helped lay the groundwork for public access television in the United States (Hazard 2015). These non-commercial stations provided the general public with free access to equipment and designated channels to narrowcast productions of local interest and concern. While dependent on institutional structures to ensure accessibility to the means of production, the establishment of public access television and nonprofit organizations has played an important role in creating production and distribution opportunities. For instance, Paper Tiger Television grew out of public access television in New York to become both a beacon for alternative citizen media as well as a platform for challenging mainstream corporate media (Halleck 2002). Similarly, the Scribe Video Center in Philadelphia offers training and tools for citizens to enact progressive social change.

Given the challenges of citizen-led filmmaking, such initiatives often rely on the mobilizing forces created around social justice issues, in which documentary filmmaking practices may only provide one among other strategies. As Sam Gregory of Witness asserts, "we have generally seen video work best in campaigns when embraced by the advocacy and communications strategy of a group, collective or movement" (Gregory 2005:xiii). Organizations like Witness, who provide various forms of support for people to document human rights abuses, prioritize the effectiveness of social justice video activism, rather than issues of affiliation, to reach their intended audiences and compel them to act accordingly. "A video" is ultimately considered "only as powerful as its ability to touch the people that watch it, to connect them to the experience of the people portrayed in the film, and to motivate them to get involved to make a difference" (Caldwell 2005:2–3). To effectively achieve this aim requires a distribution strategy with a particular purpose, which may include providing source material for broadcast news, making a public service announcement, supporting online advocacy campaigns, providing evidence for judicial proceedings, mobilizing key constituents and motivating decision makers (Gregory et al. 2005).

Organizations like Witness, who have emerged as powerful brokers of human rights video, provide tried and tested guidelines and training that help people maximize their evidentiary effectiveness. While these may help novice video activists think about best practices to effectively document human rights violations and disseminate footage to the media or other key political actors, they privilege communication models that beseech foreign audiences to act on behalf of the disempowered locals. The way they intentionally shape the form and content of unaffiliated filmmaking is not unlike the way major film festivals translate cultural policies into industry practices, which may not sit comfortably with more autonomous initiatives. Hence, while the dissemination of many citizen documentary projects benefits from the guidance and support of external organizations, activist filmmakers do not always privilege a communications model. Many activist filmmakers highlight the more visceral act of filming as a form of political agency over conventional representational aims (Hinegardner 2009; Razsa 2014). This embodied act of filming corresponds to recent theoretical interest in affective politics and calls for further consideration in citizen media projects where the shared experience may play a direct role in enacting collectivities. Nevertheless, like advocacy video and citizen journalism, activist-led documentary projects typically respond to particular situations or events in order to provide sources of information both alternative to mainstream media and unaffiliated with bureaucratic institutions.

The origins and development of activist documentary filmmaking

Most accounts of the emergence of contemporary video activism begin with the creation of the Independent Media Center (IMC) during the 1999 mass protests in Seattle. Wolfson's (2014) account, however, begins earlier, with the organizing philosophy of the Zapatistas indigenous movement. During their protests against the North American Free Trade Agreement (NAFTA) in 1994, the Zapatista Army of National Liberation (EZLN) successfully combined local political mobilization with global outreach through the nascent World Wide Web, thus signalling the emergence of the Cyber Left (ibid.). As the Zapatistas urged others to create a global communication network, media activists developed open publishing platforms specifically for the Seattle protests that allowed daily news feeds in print, radio and video from hundreds of participants. IMC media activists distributed cameras, audio equipment and computers for people to post reports on their website, along with photos, audio and video footage, and consolidated the various reports produced in different locations, thus serving as a clearinghouse for mainstream media.

Using the collectively made footage, IMC produced a series of five documentaries, including the critically acclaimed *This Is What Democracy Looks Like* (Freidberg and Rowley 2000), which combines footage from over 100 activists with a stirring soundtrack and celebrity narration. More recent examples of collective media activism elaborate the potential of crowdsourced material. For instance, *99%: The Occupy Wall Street Collaborative Film* (Aites et al. 2013) features contributions from over 100 participants from across the United States (Edgerton 2013). Despite the unconventional approach to production and their status as citizen media initiatives, these feature-length compilations became festival favourites.

Following the Seattle protests, activists from around the world, many of whom participated in the protests, opened new media centres that gave Indymedia a global scope to support the alterglobalization movement and inspire other activist media collectives around the world. With the rise of multiple mass protests a decade later – including the Green Revolution in Iran (2007), the anti-austerity movement in Greece (2010–2013), the so-called Arab Spring that spread across the region (2010–2011), the 15M/Indignados Movement in Spain (2011–2012), Occupy Wall Street in the United States (2011), Gezi Park in Turkey (2013) and the Umbrella Movement in Hong Kong (2014) – the cameraphone extended above the crowds became emblematic of both the enactment of collective politics and a means of bearing witness to this mass subjectivity (Khatib 2012). The desire to participate by filming signals a central element of these protests and a key means of enacting what Snowdon (2014:401) refers to as "the people" and defines as "essentially a performative event, rather than a pre-existing entity".

The uprisings and downfalls across the Arab world since 2010 in particular have provided fertile territory for citizen journalists and activist filmmakers to document both the assembling of mass protests and the atrocities of state violence. Despite being uncoordinated acts, the process of shooting video and uploading it online provides an important context for thinking through unaffiliated filmmaking practices (Mollerup 2015). Most of the people drawn to these street protests with their cameras had been neither politically active nor skilled in newsmaking. While many of the resulting video clips remain private images, most people uploaded their footage as part of a mass phenomenon in which thousands of videos provided a multiplicity of perspectives on the unfolding events. These networked videos offered viral, albeit momentary, glimpses of mass movements and the energy of political possibility (Westmoreland 2016). They constituted the main source of information for people across the globe before the major news outlets arrived on the scene. In order to navigate these crumbling media ecologies and generate alternative possibilities for documentary coverage of the events, a variety of collectives formed, including Mosireen, 18 Days in Egypt, Abunadara, Aleppo Media Center, Bidayyat and Raqqa is Being Slaughtered Silently, among others. Some of these organizations produced long-form documentaries. In this context, cellphone filmmaking transformed both the documentary mediascape and the documentary form, not only because of the mobile, networked and ubiquitous qualities of the technology itself, but also due to the emergence of cellphone film festivals and programmes directly related to these events, like the Syria Mobile Film Festival. Furthermore, many people who took to the streets or filmed from the security of their homes have gone on to produce feature-length films from the video fragments they collected, often with significant international backing (Appelt 2017). Unfortunately, many people who produced footage during this period lost their lives in order to offer clips and soundbites for big media (della Ratta 2018).

Future directions

As the politics in various countries continue to shift, videos of political protest take on new meanings and reveal a different range of consequences and opportunities. The transition also marks a shift from the unexpected spontaneity and political immediacy associated with documentary filmmaking to a more protracted consideration of the archival value and accessibility of these materials, posing a major challenge to unaffiliated filmmakers striving to ensure that their films remain relevant and available to future audiences.

The collaborative documentary initiative known as *18 Days in Egypt*, for instance, allows people to upload their videos to an interactive web platform. As the title indicates, the project privileges the initial phase of the Egyptian revolution that resulted in the ouster of President Hosni Mubarak. This period of mass uprisings produced a wealth of visual material (Westmoreland 2016; Lebow 2018). Unlike a general clearinghouse like YouTube, *18 Days* provides a context specific to the Egyptian revolution. The founders imagined using the material crowdsourced from unaffiliated individuals to create a documentary, but this ambition has not been fulfilled. *18 Days* relied on support from grant foundations and film festivals to sustain the efforts of the filmmakers. A similar initiative, *858: An Archive of Resistance*, was launched by the Mosireen media collective, which started life in the media tent during the occupation of Tahrir Square in 2011 (Mollerup and Gaber 2015). Mosireen made a specific choice not to seek or accept official support for the *858* project. Rather than focus only on the initial phase of the Egyptian revolution, Mosireen physically collected and produced a massive video collection over a two-and-half-year span. From this crowdsourced material, the collective produced dozens of short videos covering different issues and events specifically from the perspective of the street, many of which featured in mainstream news. As the revolutionary period came to an end, Mosireen activists became burdened with the responsibility of what is ostensibly the largest video archive of the revolution from the perspective of the street protests. After struggling to make the archive accessible, the collective released 858 hours of footage online seven years after the uprisings, using an interface that encourages crowdsourced annotation and indexing. The fate of these and other archives remains uncertain. It is unlikely that these documents will be used to incriminate leaders guilty of human rights abuses, nor compel protesters to return to the streets, but they may serve as an important resource for future filmmakers.

The history of activist documentary filmmaking remains to be written. While some can trace the emergence of the Cyber Left to the Zapatista movement (Wolfson 2014), many earlier precedents in radical filmmaking remain largely inaccessible to contemporary activist filmmakers. Despite the commonalities between different historical cases, contemporary activist filmmakers generally do not have intellectual familiarity with nor material access to earlier precedents in political filmmaking. This lack of awareness speaks to the challenges in assembling, accessing and activating archives of contingent political movements as well as other forms of documentary. For instance, Chanan (2008) notes how the *cine piquetero* movement of activist filmmaking in Argentina during the 2001 financial collapse used many of the same strategies of Solanas and Getino's *The Hour of the Furnaces*, an exemplar of Third Cinema filmed in Argentina a generation earlier. And yet, when Fernando Solanas spoke to a gathering of *cine piquetero* filmmakers, most seemed unaware of the earlier movement. Speaking of the same film, Stam stressed the importance of combining revolutionary form and content, suggesting that "[w]hile revolutionary aesthetics without revolutionary politics is often futile, revolutionary politics without revolutionary aesthetics is equally retrograde" (1998:266). At the same time, Juhasz (1999) notes how such dogmatism has resulted in a critical bias in feminist film theory, which effectively erased the realist feminist filmmaking tradition from collective knowledge in favour of formal experimentation. As a theorist and practitioner, Juhasz argues that when organized around causes like women's rights, political realism is often more effective than experimental approaches.

Like realist first generation feminist film, there is a certain risk that these micro histories may not fit well within the conventions of scholarship (Hudson and Zimmermann 2015). Since citizen documentary typically has very localized sets of concerns and may not always map onto networked structures, each production may never gain the attention of a film scholar focusing on the global film festival circuit. How to approach this work as a scholar thus presents the practical challenge of discovering micro citizen projects within the expanse of today's digital excess.

While radical formal aesthetics might appeal more to intellectuals, citizen documentary practices may prove most radical in their structural resistance to industry infrastructures, even if only partially. Whether we call this range of work unaffiliated filmmaking, open documentary, vernacular video or citizen media may matter less than the fact that it all signals a paradigm shift at the interface of arts, politics and science. These participatory and collaborative filmmaking processes demand new conceptualizations and compel us "to ask not what documentary means but *what documentary does*" (Aston et al. 2017:2; original emphasis).

See also: anthropology and citizen media; archiving; film studies and citizen media; Indymedia; media ecologies; witnessing/testimony

Recommended reading

Gregory, S., G. Caldwell, R. Avni and T. Harding (eds) (2005) *Video for Change: A guide for advocacy and activism*, London: Pluto Press.

A practical handbook offering insights from several members of the well-known and highly-influential human rights organization Witness. The strategies outlined provide methods to harness the power of video to convey compelling stories and ensure evidentiary credibility. With a focus on advocacy and social justice video activism, it outlines best practices for empowering filmmakers while remaining safe. Also includes several appendices for planning video productions, including sample release forms and budgets.

Lebow, A. (2018) *Filming Revolution*, a Stanford Digital Project, Stanford, CA: Stanford University Press. Available online: www.filmingrevolution.org/.

A meta-documentary that investigates revolutionary filmmaking in Egypt during the Arab uprisings. Based on interviews with various individuals who speak from firsthand experience about the revolution and its relationship to the creative process of filmmaking, Lebow uses a non-linear structural framework that makes it possible to trace different constellations of relations and to move through themes and topics in an organic manner or by following prescribed curated conversations.

Zimmermann, P. R. and H. de Michiel (2017) *Open Space New Media Documentary: A toolkit for theory and practice*, New York: Routledge.

Provides a framework for understanding and practising documentary filmmaking in ways that subvert conventional approaches and bypass mainstream venues. Examining a range of collaborative and community-based documentary projects from around the world, the authors outline a series of principles for conceptualizing and enacting open forms of filmmaking, focusing on social change rather than success in the festival circuit.

ETHICS OF CITIZEN MEDIA RESEARCH

Sandra Smeltzer

Citizens who produce, disseminate and consume their own media, or who disrupt and subvert existing media, challenge traditional mass media structures and often incur risks in the process. By unsettling "hierarchical notions of expertise and authority" (Luchies 2015:533), they engage in a citizen-oriented system that is networked, horizontal and participatory. Undertaken by individuals, groups and collectives for political purposes and/or as aesthetic expression, these practices expand the range of perspectives and voices that can be shared among diverse publics (Atton 2015; Baker and Blaagaard 2016a; Bélair-Gagnon and Anderson 2015; Cammaerts 2015; Downing 2001; Fuchs, 2010; Galis and Neumayer 2016; Kidd 2015; Poell and van Dijck 2015). An inclusive understanding of citizen media must recognize the performative function of media and the ways in which media production can empower communities (Rodríguez 2010a), but at the same time endanger them as sources of research data.

At ground level, citizen media practices are manifest in embodied contexts such as public protests and rallies; artivism (Diverlus 2016) and musical events (Ní Mhurchú 2016); culture jamming and flashmobs (Isin 2017; Peters 2016). Such practices often employ hybrid forms of digital and physical praxes (Baker and Blaagaard 2016a). For example, a citizen might film a public dance performance with their phone and upload it to YouTube, with a link to their Twitter account (Shresthova 2016; Croeser and Highfield 2014). Importantly, for research that includes Internet-based practices, scholars must bear in mind the "embodied realness of online behaviours" (Lüders 2015:80), which can have material consequences, especially for citizens engaged in politically oriented activities (Galis and Neumayer 2016:1). However, the rapidity of technological innovation, the multiplicity and context-dependent nature of citizen media practices and the multidisciplinary nature of citizen media research (Östman and Turtiainen 2016) make it difficult to establish a generalizable code of research ethics. Acknowledging these constraints, the Association of Internet Researchers Ethics Working Committee maintains that "ethical decision-making is best approached through the application of practical judgement attentive to the specific context" (2012:4). The discussion below takes this as its point of departure and stresses that traditional ethical principles relating to issues such as researcher reflexivity and the need to ensure the anonymity of subjects should guide research pertaining to both online and offline forms of citizen media.

Ethical principles and practices

The range of issues at stake in citizen media research requires one to be mindful, reflexive and deliberative throughout the process of inquiry (Beninger 2017; Berger 2015; Lüders 2015; Markham and Buchanan

2012). The starting point is engaging in ongoing reflexivity, which requires researchers to consider ethical issues in an open and honest manner (Cloke et al. 2000:133). Engaging in reflexivity includes recognizing researchers' own subjectivity vis-à-vis citizen media practitioners, as well as the potential power imbalances between themselves and their research participants (Gillan and Pickerill 2012; Liamputtong 2010; Smeltzer 2012). Reflexivity is thus essential for fulfilling a scholar's "fiduciary obligation to consider things from research participants' perspectives and to ensure participants' rights are safeguarded" (Palys and Atchison 2008:71). This obligation underpins the do no harm doctrine and is inextricably linked to beneficence, a core tenet of ethical practice that requires scholars to weigh the benefits and risks of their research carefully (Lunnay et al. 2015; Shilton and Sayles 2016). However, it is often difficult for researchers to know *a priori* the array of risks and unintended consequences that participants may face during and after a research project (Bishop and Gray 2017:164). This is especially the case in situations where the form or content of the media practice being investigated is politically sensitive – for example, an anti-austerity protest – and/or includes vulnerable participants such as youth, refugees and minority groups (Lüders 2015:81; Vitak et al. 2016). Reflexivity must also guide research about ephemeral and/or archivable citizen media practices. A protest, an art exhibit and a flashmob are fleeting experiences, and many blogs, YouTube videos and organizational websites similarly lack permanence. Research into these practices requires scholars to weigh the short- and long-term ramifications of their work, and to avoid rushing the process of capturing a short shelf-life media experience (Pesce 2016). When concrete media practices are circulated online, and when citizens use social media platforms to organize and advertise their activities, their information is stored and remains accessible to a hitherto unparalleled extent (Weller and Kinder-Kurlanda 2015:29). The archived nature of this information necessitates that researchers exercise considerable caution in ensuring citizens' anonymity and respecting their privacy.

At the same time, the line between public and private citizen media practices is quite porous. Although many citizens consciously engage in producing content meant for public consumption – such as community newsletters, art exhibits and political blogs – others disseminate information expecting it to be consumed by only a small number of people, like friends and family or those with similar political affinities. In either case, these individuals may assume that their activities are private even if they take place in a public domain (Lüders 2015:81). With Internet-based practices, citizens' data may be treated as public if they agreed to the terms and conditions of an online platform. But this does not mean they have agreed to have their activities used for research purposes. Hence, traditional ethical protocols pertaining to privacy remain essential (Association of Internet Researchers 2012:6–7; Lüders 2015; Saunders et al. 2015).

The situation becomes more complicated when considering the privacy expectations of those who respond to, or otherwise engage with, media products and practices. Citizens may, for example, comment on a YouTube video, retweet a post or sing along with a politically oriented music performance in a park. Although their actions are ostensibly public, the citizens in question may not regard them as data for research (Townsend and Wallace 2016:6; Sloan and Quan-Haase 2017). In the words of boyd and Crawford (2012:671), specifically in regards to big data, "just because it is accessible doesn't make it ethical" for scholars to use it for research purposes.

The public nature of most media practices raises questions about how citizens can remain anonymous if they so desire and how researchers can ensure that identities are not revealed. Social media and big data analytics make upholding anonymity in relation to online practices particularly challenging. Internet-based data are, and will increasingly become, more searchable than many users realize, and the power of corporations to store, aggregate, sell and make users' information public will continue to intensify (Beninger 2017; Shilton and Sayles 2016; Sloan and Quan-Haase 2017; Warr et al. 2016). Avoiding the disclosure of someone's identity often requires that verbatim quotes or overly descriptive paraphrasing be excluded from research findings, unless explicit consent is provided by the participant.

Upholding the principles of anonymity and confidentiality when researching concrete forms of offline media practices also requires reflexivity. Taking a picture of the face of a citizen participating in a women's march along a central street in their hometown may not constitute ethical research practice. As Warr et al. (2016:8) point out, "photographs and videos are capable of creating highly detailed and intimate portraits of individuals" and thus could constitute an invasion of their privacy. In this case, there is a potential disconnect between a researcher's legal right to take someone's picture in a public area, and the question of whether this action is ethically acceptable. Knowing that identifying software can make an individual searchable, with techniques used to protect their identity (such as blurring their face) not likely to suffice, researchers must obtain consent to use such images in the dissemination of their findings. This scenario is made more complex in cases where the photographed individual belongs to a vulnerable or marginalized population, or when cultural differences regarding privacy and capturing images need to be respected.

There are also scenarios where an academic might conduct research with an organization, collective or media outlet that includes among its members individuals who do and who do not want their identity to be known to the public. For many people, engaging in citizen media practices may be personally, politically and/or professionally empowering, but they may choose not to have their involvement amplified via an academic's research in case it makes them (more) visible to the government or to friends and family members. A group's dynamics can also change over time, and the trajectory of the form, content, target audience(s) and distribution methods for the media they generate may shift. Researchers must be attuned to these nuances and not treat collectives as homogeneous. They must further consider carefully how much identifiable information they can include in their publications and presentations, as well as the risks versus the scholarly benefits of including direct quotes, paraphrases and certain types of visuals, even via non-descriptive pseudonyms, especially in relatively small communities (Wiles 2013).

The complexity of the myriad scenarios in which citizen media research is undertaken means that obtaining informed consent from participants is part and parcel of conducting ethical research in this field. Obtaining informed consent involves disclosing to research subjects what their participation entails, any costs they may incur, any benefits they or others may receive, risks associated with participating, and any other considerations they need to be aware of in order to make an informed decision about whether to participate (Palys and Atchison 2008:74–75). Studying citizen media practices poses several challenges to gaining such consent and requires researchers to consider whether their work might harm citizens now or in the future.

First, researchers may not know who will be part of their study prior to commencing their work. At issue here is the ethics of seeking consent from participants after a researcher has already observed their media practices (Townsend and Wallace 2016:6). Second, a central component of informed consent entails participants being able to withdraw from the research activity at any point in the process. This is impossible to actualize if participants are unaware that research is being conducted (Vitak et al. 2016:950). Third, it is not always possible to determine the veracity of a citizen's identity or whether they have genuinely given consent for their data to be used by external parties for research purposes. In the case of online venues, research participants may prove to be lurkers, trolls or bots (Atkinson 2017; Sloan and Quan-Haase 2017; Yang et al. 2017), or they may be underage or part of another, similarly vulnerable population (Shilton and Sayles 2016:2). In addition, despite any identity assurance mechanisms that social media platforms may have in place, most users do not read or understand the terms and conditions of use that describe how data can be sold to and used by third parties (Townsend and Wallace 2016:6). While agreeing to such terms and conditions constitutes legal consent to information use by others, questions about the ethics of using such information for research purposes without obtaining prior agreement remain (Beninger 2017). Accessing large datasets produced by social media platforms is equally dubious from an ethical perspective because this information is seldom generated for scholarly purposes (Bishop and Gray 2017), and the large-scale, aggregated nature of big data does not preclude exercising the ethical due diligence expected of smaller-scale research (Kitchin 2017; Schroeder 2014). Fourth, participants may choose not to provide their consent, or not provide it in the form expected or required by the researcher. Atkinson (2017:49) describes his experience with "activists who have taken part in criminal activities, activists who refused to sign consent forms, and activists who were less than truthful when discussing their organizations and actions", and goes on to outline the difficulties he faced in obtaining informed consent via formal letters of information and consent that require signatures (ibid.:53–54). Kara (2018:105) similarly points out that "Western researchers may take it for granted that consent should be given in writing, but this can prove problematic in cultures where oral communication is dominant and few people are able to write". Finally, some individuals, especially younger people, may be capable of giving informed consent but may fail to consider the short- and long-term implications of their media practices, not weighing the ramifications of their digital activities being searchable and archived, or the impact of their concrete activities on familial and personal relationships, their personal safety and/or their professional lives.

Fluid contexts and the ethics of knowledge production

In many locales, producing, distributing and/or consuming citizen media can be risky and even perilous. Protecting participants during and after the research process is therefore contingent upon researchers having a solid understanding of the local socio-political terrain. The potential for harmful consequences arising from research activities is heightened where repressive forces such as state authorities, hate groups and religiously motivated extremists persecute individuals and groups who engage in activities that challenge political, economic and sociocultural status quos. Private sector actors can also stymie citizen media activities through a range of means, including libel, privacy laws and security legislation (Bélair-Gagnon and Anderson 2015:7). Consequently, researchers must judge the implications of drawing greater attention to citizen media practitioners, and be attentive to political situations that can quickly shift in unexpected ways (Atkinson 2017:57–60; Galis and Neumayer 2016; Wiles 2013). Reilly and Trevisan (2016) address such concerns in their discussion of the ethical tensions they experienced when conducting research on how citizens in Northern Ireland used Facebook to mobilize on-the-ground protests about the UK's union flag being flown over the Belfast City Hall for fewer days than prior practice (eighteen days rather than all year round). Through this platform, they gained "insight into the affective dimension of contemporary protest movements, and, in particular, how social media enables citizens to express solidarity and emotional connection with these campaigns" (ibid.:12). However, knowing that citizens' online activities could have tangible consequences in this politically fraught environment that saw increased tension between loyalists and unionists, Reilly and Trevisan avoided identifying specific citizens by choosing not to use verbatim quotes.

Importantly, researchers themselves are not exempt from being the targets of violence and must carefully assess the potential dangers associated with their work. In some situations, they too could be at risk for attending a rally, for photographing a politically contentious art exhibit or for publishing material that expands the readership of an online magazine (Smeltzer 2012:264–266). Notwithstanding these concerns, Juris (2008, 2012) contends that academics must proactively engage in collective, collaborative research, rather than separating themselves from their participants. Luchies (2015:529) similarly argues that "[w]e need to participate, theorize, and transform alongside the communities we work with"; Hale (2006) and Rodino-Colocino (2012) propound similar views. Conducting research on citizen media practices that align with their own interests, politics and perspectives does not, however, mean that scholars eschew their critical commitments (Galis and Neumayer 2016:2). As Routledge argues, researchers must address the thorny issue of "criticality versus censorship: how critical can one be and still continue to support rather than undermine a particular struggle?" (2004:87). This has direct implications for the process of knowledge production, as publishing work that casts an unfavourable light on organizations or media outlets may be read as unsupportive or serve to subvert practitioners' objectives. Conversely, academic self-censorship is ethically problematic (Clark 2006; Rodino-Colocino 2012, 2016; Smeltzer 2012), and scholars cannot focus solely on success stories of citizen media practices. Doing so would inadequately represent the difficulties individuals and groups often face in producing and sharing their messages, and can downplay the power of various agents to counter or control these messages (Christensen, C. 2011; Paré and Smeltzer 2013).

At issue here are also questions about the benefits accruing to citizens and communities from academic research, their role in guiding research projects and whether they are appropriately recognized and compensated for the time and energy they invest in the process (Gillan and Pickerill 2012:138; Tarlau 2014). Despite academic publications being shared with the people who are at the heart of scholarly work, "the language, findings and timeliness of our research can be of limited use to social movements" (Gillan and Pickerill 2012:137). Generating material that is genuinely useful beyond the academy necessitates the use of different types of skill sets, can be tremendously time-consuming and is typically not recognized in academic performance evaluations (Khasnabish 2015; Napoli and Aslama 2010; Smeltzer and Cantillon 2015).

Institutional ethics review bodies can play an important role in guiding researchers through the contextualized world of research ethics. Nevertheless, concerns have been raised about the extent to which they are attuned to recent technological changes, their capacity to fully understand issues pertaining to online forms of citizen media (Morrow et al. 2015; Sloan and Quan-Haase 2017:669) and their ability to effectively respond to the complexities of ethnographic (Bosk and DeVries 2004; Perry 2011), co-designed and participatory action research (Boser 2007; Fouché and Chubb 2017; Goodyear-Smith et al. 2015; Schrag 2011). Thus, the Association of Internet Researchers recommends that in addition to consulting with ethics boards, scholars should solicit opinions from "fellow researchers, people participating in or familiar with contexts/sites being studied ... published scholarship (within one's discipline but also in other disciplines), and, where applicable, legal precedent" (2012:5).

Although the ethical issues involved in citizen media research are neither easy to address nor straightforward in nature, and institutional ethics review boards can be slow and sometimes ineffective, scholars must not shy away from open and honest discussions about the integrity required for conducting citizen media research. This necessitates a willingness to be forthcoming about ethical conundrums. More established scholars have an added responsibility to mentor emerging colleagues and to support graduate students, especially if their citizen media research involves politically fraught environments, vulnerable populations and/or sensitive material. Given that much of what scholars learn in terms of ethical research comes from reflecting on their own experiences, they have a duty to share this knowledge with peers and with colleagues new to the field.

See also: surveillance

Recommended reading

Atkinson, J. D. (2017) *Journey into Social Activism: Qualitative approaches*, New York: Fordham University Press.

An important source for understanding myriad ethical issues related to conducting qualitative research on online and offline forms of social activism. The second half of the book focuses on the fluid and context-dependent nature of conducting research about and with social movements and activists, including a section on alternative, social and citizen media.

Sloan, L. and A. Quan-Haase (eds) (2017) *The Sage Handbook of Social Media Research Methods*, London: Sage.

A compendium of thirty-nine chapters that represent a range of contexts and disciplinary backgrounds and address the methodological challenges and ethical practices related to conducting qualitative and quantitative social media research. The volume is particularly useful for understanding ethical considerations pertaining to the use of large datasets, providing readers with insight into every step of the research process, from developing key questions to the interpretations of results. The theoretical work is supported by salient case studies of citizen-produced media.

Townsend, L. and C. Wallace (2016) *Social Media Research: A guide to ethics*. Available online: www.gla.ac.uk/media/media_487729_en.pdf.

A practical guide to conducting ethical social media research. The guiding principles and procedures described in this manual move beyond traditional ethical considerations to recognize the shifting nature of online forms of citizen media.

FACEBOOK

Zoetanya Sujon

Facebook is a complex site, marked by its massive size, continual expansion and ever-increasing platform power. In 2019, Facebook surpassed two billion monthly active users, making it the largest social media platform. Along with Google, Facebook dominates the global digital advertising market, effectively forming a global duopoly (Burrell 2019; Garrahan 2017; Reuters 2017). Facebook's size and network power mean that it has a considerable advantage over other social media platforms, sets the standard for the social landscape (Caplan and boyd 2018; Zittrain 2018; Tufekci 2014) and plays a significant role in shaping the practices of individuals, communities and industries.

Like other social media, Facebook simultaneously perpetuates the capitalist logic associated with corporate structures while reproducing a citizen-oriented logic that enables community building, social movements and many forms of creative and civic expression. In light of these opposing logics, Facebook is first presented in this entry as an instance of anti-citizen media that co-opts citizen action and subverts the social into monetizable forms of connection through metrics and the like economy (van Dijck 2013; Marwick 2013; Fuchs 2017). To illustrate Facebook's capitalist logic, this entry will draw on the Cambridge Analytica events, involving the harvesting of personal data across the Facebook platform to facilitate targeted political campaigning and advertising (Greenfield 2018). The entry will then move to examine Facebook as a form of citizen media – understood as those "artefacts, content, practices, performative interventions and discursive formations" which enable citizens (and non-citizens) to "position themselves within and in relation to society and participate in the creation of diverse publics" (Baker and Blaagaard 2016a:16). As a form of citizen media, Facebook provides a platform for citizens to come together, be heard, share and mobilize. Its grassroots and citizen-oriented logic can be observed in a range of protest and social movements, such as the Arab uprisings, Occupy and Black Lives Matter (Gerbaudo 2016; Tufekci 2017a). Facebook's potential to oppose the corporatization of citizen connections has also been explored in terms of its contribution to identity expression and community building (Miller, D. 2011; boyd 2014; Lingel 2017), as well as its contribution to citizen or participatory journalism (Rone 2016). In these contexts, Facebook facilitates connection – providing the means for ordinary people to self-identify, build and make communities of practice visible to insiders, outsiders and all those in-between.

The rise of the Facebook ecosystem

Facebook has come a long way from its early days as a single-purpose website for sharing Harvard student pictures and profiles, inspired by the paper-based face books – i.e. student directories that have been "common to American Universities from at least the 1960s" (Gray 2007:73). The Facebook platform, first implemented through the Facebook Login button (formerly called Facebook Connect), marked Facebook's expansion from a single web-based social network "into the rest of the web" by integrating external web data into its growing range of services and user information databases (Helmond 2015:1). Today, Facebook presents itself as a complex mobile-first site encompassing a "[p]latform; social plugins such as the Like button, the Share button and other similar offerings; and other media, brands, products, services, software … devices or networks now existing or later developed" (Facebook 2018). Scholars agree that providing a single definition of Facebook is challenging. Madianou and Miller describe it as a kind of polymedia or "multiplex of co-constituted and interconnected media spaces" which encompass a wide range of services and communication forms (2012:172). Likewise, van Dijck (2013; van Dijck et al. 2018) describes Facebook as a connective rather than social medium that makes social metrics – including likes, comments and shares – visible. From their perspective, Facebook encompasses many sites providing key cultural, social and economic infrastructures.

As Hoffman et al. (2018) have shown, founder and CEO Mark Zuckerberg's public statements on Facebook have also shifted over time. While he first referred to Facebook as a directory during his dorm room days, he later defined it as a "'core social infrastructure' for the web – and increasingly – the world" (as cited in Hoffman et al. 2018:205). Against the backdrop of these developments, critical scholars have come to conceptualize Facebook as an exploitative company that monetizes the social through the exploitation of user data, enables mass (self) surveillance, and exploits the free labour of users while also perpetuating a 'like ideology' (Fuchs 2014c:153, 160; Scholz 2012a; van Dijck, 2013; Sandoval 2014; Vaidhyanathan 2018b). One of the common themes emerging from this body of work is the idea that Facebook is a constitutive space of eco-systems (van Dijck 2013), underpinned by power relations (Fuchs 2014c; Sandoval 2014), and driven by the need to enable communication and connection (Gerbaudo 2016; boyd 2014; Miller, D. 2011; Jenkins et al. 2015). The key debate here does not question the fact that Facebook is powerful, but concerns how it is powerful and how all of its elements come together in systems that are predominantly exploitative, enabling and/or empowering. Bucher and Helmond (2018:235), for example, have explored the deep connections between Facebook content and the actions that it enables. These connections are presented as affordances that yield a better understanding of the actualities and potentialities of the dialectic between social interfaces and uses, behaviours or contexts (Plantin et al. 2016; Helmond 2015; Srnicek 2017).

Facebook as a corporate and anti-citizen platform

Scholars working on Facebook as a corporate platform do not only focus on the issue of social or citizen co-optation (Rone 2016), but also the monetization and subversion of the social by a capitalist logic. Notable cases illustrating Facebook's long history of anti-citizen behaviour include the radicalization of right-leaning Facebook users through targeted advertising directed at Jew-haters and white supremacists (Tufekci 2017b); India's rejection in 2016 of Facebook's Free Basics – a free service offered to countries with limited Internet access through an app similar to Facebook Lite that enables free access to Facebook – for monopolistic reasons (Bhatia 2016; Gurumurthy and Chami 2016; Vaidhyanathan 2018b); and Facebook's 2014 study on emotional contagion, involving Facebook's mood manipulation of negative and positive sentiment in the content of 700,000 unknowing Facebook users' newsfeeds (Kramer et al. 2014; Meyer 2014).

These tensions, however, are most clearly illustrated by the Cambridge Analytica events, which exposed the misuse of millions of users' personal data for political campaigning in March 2018 (Rosen et al. 2018). This particular case is important because it highlights Facebook's political and economic impact as a global power broker of personal data, confirming what many critical scholars have long argued (Tufekci 2014; Gehl 2015; Fuchs 2014c; Sandoval 2014; Marwick 2013; Scholz 2012a). In 2018, it was revealed that Cambridge Analytica, a data-driven political consulting and commercial marketing firm, used personal data from up to 87 million Facebook accounts and data gathered through an app called 'This is your digital life' – which scraped data from Amazon Mechanical Turk workers and all of their Facebook friends and contacts in 2015. The issues at stake in this case include massive invasions of privacy for Facebook users along with improper use and retention of data. Crucially, the threat that this improper retention represented was not confined to those individuals and citizens whose personal data was non-consensually shared. Cambridge Analytica reportedly used this personal data in the Trump and Brexit campaigns and developed sophisticated tactics targeting voters that may have been instrumental in Trump's electoral victory and the UK's decision to leave the EU (Tufekci 2017b, 2018; Rosenberg 2018; Ingram 2018; Greenfield 2018; Davies 2018 offers an opposing perspective).

In his public apology released via Facebook, Zuckerberg (2018) admitted that the platform allows third-party apps and developers to access the user data of anyone who downloads the apps, but also of anyone who is connected to the downloader. The Cambridge Analytica events thus revealed how the Facebook platform collects user data and enables its clients and developers to use or share that data in ways that suit them. Tufekci (2017b) argues that, contrary to what Zuckerberg's apology suggests, this was not a one-off failure. This breach clearly illustrated the impact of a core feature of Facebook's business model and the growing importance of platformization. In her critique of the implications of this business model, Tufekci (2018) concludes that Facebook "isn't a community; this is a regime of one-sided, highly profitable surveillance, carried out on a scale that has made Facebook one of the largest companies in the world by market capitalization".

Tufekci's views are informed by longstanding neo-Marxist critiques of mass media, where citizens are commodified as audiences to be bought and sold by media companies and advertisers (Adorno and Horkheimer 1944/1979; Smythe 2009; Fuchs 2012c, 2014c). From a similar perspective, other scholars have presented Facebook as a data-gathering platform that prioritizes advertising revenue over other more altruistic contributions to society (Vladeck 2018). Van Dijck (2013) argues that Facebook's main function is to automate sociality as well as to monetize social connections and personal information – a view shared by Marwick (2013) and Angwin et al. (2016). These attempts to profit from and monetize personal user data portray Facebook as a corporate and anti-citizen media platform. Instead of engaging citizens and enabling users to perform and constitute citizen-oriented connections, Facebook's monetization of data transforms acts of citizenship into data commodities.

Facebook as a citizen-oriented platform

As a powerful platform for self-expression of the networked citizen self (Baym 2015; Papacharissi 2011a), relationship maintenance and community formation (boyd 2014:7; Miller, D. 2011; Lambert 2013), Facebook is often understood as a mirror of and a window into the interface between the personal, interpersonal and public. In this context, Facebook emerges as a site where communities of unaffiliated citizens – who act independently from organized groups and collectives (Baker and Blaagaard 2016a) – and social movements coalesce and consolidate. Facebook embodies the heart of participatory culture, where ordinary citizens can realize greater communicative capacities with fewer "barriers to artistic and civic engagement" (Jenkins et al. 2015:3–4). By actualizing the Internet's "expansive possibilities for horizontal communication among citizens", it facilitates political action and fosters identity or community-building initiatives (Tufekci 2014) where the divisions between political action and self-expression often become sticky and tangled.

The interplay between the public and the private is central to various studies on the Arab uprisings, where Facebook provided a platform for activists and unaffiliated citizens in "Tunisia, Egypt, Bahrain, and latterly Yemen and Syria" (Axford 2011:682) to share information and take part in political mobilizations. Murthy (2018a), Zuckerman (2013) and Axford (2011) have explored how Facebook was able to circumvent government control and give voice to the voiceless against a complex backdrop characterized by severe restrictions of free expression, rife corruption and economic precarity. Focusing on the case of Egyptian activist Wael Ghonim, Alaimo (2015) argues that Ghonim's Facebook page 'We are all Khaled Said' became a key site of mobilization and political training. In 2011, it led to a series of mass protests, not only against the brutal murder of Said by Egyptian police, but also against widespread poverty, government failures and widespread misinformation (Alaimo 2015:4). Facebook thus became a site that transformed unaffiliated citizens into a citizen movement which sought to reshape Egyptian politics (Gerbaudo 2016; Alaimo 2015; Murthy 2018a).

A very different example of Facebook serving as a constitutive site of both community and political change is the New York drag queen community platform (Lingel 2017). Through interviews, Lingel documents the importance of social media, and Facebook in particular, for performers to promote themselves and enact "queer solidarity and mutual aid" within and across the drag community (2017:102). Drag tends to involve males dressing up in extravagant costumes to impersonate females while lip synching popular songs and disrupting particularly "conventional notions of femininity and masculinity" (Lingel 2017:101). By using pseudonyms or aliases, drag queens seek to develop their drag personae – which sometimes requires keeping their drag identities separate from their non-performing selves. Facebook was used as a form of citizen media in this context, as members of this community faced having their Facebook accounts frozen in violation of the platform's real-name policy. Lingel's respondents used Facebook not only as a site for self-expression and community building, but also as a site of protest against the real-name policy, that they regarded as "homophobic and prejudiced" (2017:117), and as problematic for abuse and domestic violence survivors. Lingel argues that the drag and queer community were successful "where others were not" because "they acted collectively" and "celebrated their alterity" and difference (2017:118–119).

Facebook's dual logics: Citizen power and platform dominance

Following longstanding patterns in the history of mass media (Murthy 2018a; Lievrouw 2011), Facebook is both an emancipatory and exploitative platform. Its role in contexts like the Arab uprisings and New York's drag community shows that Facebook is instrumental in connecting communities, coordinating collective action, fermenting political change and, most importantly, enabling users to participate in and develop diverse publics – all of which are important steps in appropriating corporate media as citizen media to foster acts of citizenship. But the very spaces that Facebook provides for unaffiliated citizens to become visible to themselves and to others are buried under deeply divisive power structures. Apart from prompting calls for its reorganization as a kind of public utility or "information fiduciary" responsible for protecting personal data and connections (Balkin 2016), Facebook's history of privacy invasions and data economics is widely held to jeopardize its potential to serve as a citizen-oriented platform. While alternative social media like Diaspora*, Ello, Minds and Mastodon may appear to be better options for a fairer, more citizen-oriented world, network effects mean Facebook is likely to maintain mainstream dominance (Gehl 2015; Fuchs 2014c; Lievrouw 2011).

Future directions

The Cambridge Analytica events are bound to influence significantly the way in which Facebook evolves over the next decade. This high-profile data harvesting case has prompted efforts to institute regulatory reviews of Facebook and other big technological companies, and drawn critical attention to various aspects of platform behaviour. These include the use of competition curbing practices, the influence of social media on the outcome of electoral processes, and the role Facebook and other platforms play in circulating disinformation – as discussed in the UK Parliament's Final Report on Disinformation and Fake News (UK Parliament 2019); and the Online Harms White Paper (DCMS 2019). Likewise, Facebook's role in perpetuating hate and reinforcing entrenched inequalities is also attracting scholarly attention (Benjamin

2019; Gillespie, T. 2018a) and bringing Facebook's role as a core social infrastructure under intense scrutiny. Beyond this sociopolitical dimension, the research agenda on Facebook is exploring its place at the forefront of many technological innovations, including artificial intelligence, sophisticated facial recognition tools, predictive analytics, finance and micro-payments, and emotional manipulation – among many other areas of public-facing research. Since it first emerged, Facebook's potential for mass connection has been socially, economically and politically transformative, so it is imperative to continue interrogating its role within the digital ecosystem and its future impact on public well-being, for both citizens of today and tomorrow.

See also: authenticity; big data; content moderation and volunteer participation; co-optation; social media; twitter and hashtags; Weibo; YouTube

Recommended reading

Gehl, R. W. (2015) 'The Case for Alternative Social Media', *Social Media + Society*, July–December: 1–12.

Drawing on the alternative media tradition, Gehl theorizes alternative social media as radically distinct from their corporate counterparts. This paper yields new insights into the power of corporate media, as well as the impact of social media on the nature of sociality.

Helmond, A. (2015) 'The Platformization of the Web: Making web data platform ready', *Social Media + Society*, September: 1–11.

Outlines the infrastructural process of platformization – understood as the extension of social media or other digital services, sites or apps across the web for data extraction and exchange – using Facebook as an example.

Tufekci, Z. (2017) *Twitter and Tear Gas: The power and fragility of networked protest*, New Haven, CT: Yale University Press. Available online: www.twitterandteargas.org/downloads/twitter-and-tear-gas-by-zeynep-tufekci.pdf.

Explores the links between social media and social movements, from the 1960s civic rights movement to twenty-first-century movements such as the Arab uprisings, Hong Kong's Umbrella Movement and Occupy.

FANDOM

Eva Cheuk-Yin Li

Fandom is a sociocultural phenomenon characterized by the creativity, playfulness and productivity of its participants. The term refers to a rich terrain of discursive, affective and performative practices within which individuals explore agency, express identities, share feelings and thoughts, interact with others with common interests, and join and create communities (Duffett 2013:288). While some have tried to define and categorize different types of fandom (Abercrombie and Longhurst 1998), others suggest that such attempts risk undermining the diversity that exists within this phenomenon (Busse and Hellekson 2006:6; Hills 2002:xiii–xiv). Drawing on the lowest common denominator across different communities, Sandvoss (2005:8) defines fandom as "the regular, emotionally involved consumption of a given popular narrative or text". Fans are not only consumers, but also "networkers, collectors, tourists, archivists, curators, producers and more" (Duffett 2013:21).

The agency of fans has historically been undermined and pathologized. Until the early 1990s, fans were largely seen as either obsessive loners or frenzied crowd members (Jensen 1992). Starting with the work of Bacon-Smith (1992), Jenkins (1992) and Lewis (1992), scholars have challenged these stereotypes and reclaimed fandom as a creative phenomenon by highlighting the imaginative and innovative practices involved in the production of fan talk, fanzines, fan fiction and various other physical and virtual fan artefacts. This paradigm shift has gone hand in hand with the rapid growth of fandom in the age of digital media convergence: following the development of networked digital communications technologies, content is able to flow much more effectively across media platforms and audience members have the tools required to share their own creative expressions on an unprecedented scale (Jenkins 2006a). Fandom has entered the mainstream and become a key shaping force in the construction of contemporary identities, social relations, culture, the economy and politics (Click and Scott 2018; Sandvoss et al. 2017). Although the Internet has transformed fan participation and interaction, scholars argue that the online performance of fandom is always complemented by fans' offline activities, and vice versa (Baym 2000:157; Bury 2005:205).

As this entry will discuss, fandom can be seen as an important locus for the production of citizen media, understood as the range of physical and digital artefacts, content and discursive formations produced by unaffiliated citizens in order to express personal desires and aesthetics or to seek sociopolitical change, without the involvement of a third party or benefactor (Baker and Blaagaard 2016a:16). The first two sections ('Participation, performance and pleasure' and 'Gender, sexuality and desire') illustrate how fandom functions as an alternative space in which individuals can voice their own opinions and feelings, construct and appropriate meanings, interact with others and celebrate playfulness. The second half of the entry ('Fan activism and civic participation' and 'Transcultural and intersectional fandom') then describes how fans explore and negotiate their agency as citizens to act in public space(s) and engender sociopolitical change in ways that reconfigure the power relations between media, public(s) and citizens.

Participation, performance and pleasure

The term participatory culture is used to refer to the widespread and active involvement of institutionally unaffiliated citizens in the creation of culture and content (Jenkins et al. 2006). As such, it has become a central concept in scholarship on fandom (Jenkins 2006b:41). Drawing upon de Certeau's theorization of the tactics and games through which audiences often engage with texts (de Certeau 1980/1984:175), Jenkins (1992:24–28; 2006b:37–38) conceptualizes fans as textual poachers who, although operating from a position of sociocultural marginality, actively participate in the production of culture by appropriating texts, unhinging their parts and creatively reassembling them in the form of fan talk, fanzines, fan fiction (fanfic), fan conventions (cons), music videos (vidding) and public performances.

When we trace the history of media fandom in the twentieth century from the arrival of cinema through rock'n'roll to television drama series, it can be seen that fans have always been active (Duffett 2013:5–17). For example, at the end of the 1920s, over 32 million fan letters, written to both male and female stars, were received by the Hollywood studios every year (Duffett 2013:7). Over the decades, fans have also frequently mobilized large-scale save-the-show campaigns, such as the letter-writing campaign initiated by the fans of science fiction television series *Star Trek* in the late 1960s (Jenkins 1992:28), and the offline petitions and boycotting campaign organized by fans of soap opera *Another World* (1964–1999) (Scardaville 2005). These campaigns demonstrate that fans are not merely members of an anonymous and isolated mass. Fans' social identities, feelings and participation are connected to other realms of the offline world.

Fans experience the pleasures of connection, appropriation and performance from engaging in fan practices (Duffett 2013:165–190). Vidding, for example, involves the editing, remixing and mashing up of a set of preexisting video clips to stage a new reading of a source text or to tell new stories through the interpretative lens of a musical soundtrack of the fan's own choice (Coppa 2008:1.1; Jenkins 1992:71). It thus not only creates new meanings for the source footage, but also enables fans to seek to develop their own artistic identities, connect with their fellow fans and acquire subcultural capital by disseminating their vids at fan conventions and/or online via Facebook and YouTube (Coppa et al. 2018:232). The performativity of vidding can also be observed in the playfulness of *SuperWhoLock* fans, who collate and combine clips from the television series *Supernatural*, *Doctor Who* and *Sherlock* in order to create and share pastiche GIF fics through Tumblr (Booth 2015:25–52).

Besides engaging in creative online activities, fans often participate and perform in offline physical spaces in ways that redefine publicness. For example, Duffett (2017) reconceptualizes female music fans screaming in public as a form of affective citizenship that rejects public disapproval of certain behaviours and modes of emotional expression enacted by female bodies. The embodied performance of cosplay (costume play) is the appropriation and actualization of stories in connection with the fan community. By engaging in this practice, fans enact and create new identity narratives, and hence transform the normative use of social spaces into spaces that are potentially playful and personal (Lamerichs 2011). Fans may also occupy and transform the existing meanings attached to a specific physical site. For example, in 2011, Bollywood dance fans in Mumbai performed a flash mob dance at one of the sites where a terrorist attack had taken place in 2008, in order to rekindle hope among their audience and to transform public memory of the site (Shresthova 2013).

Although it is evidenced that the relationship between audience and producer has been reshaped under participatory culture and media convergence, there are limitations in the extent of these changes. Power structures remain a strong influence on fandom, in particular at fan convention venues, where the hegemonic power of media corporations is naturalized and reinforced (Gilbert 2017). The concept of textual poaching is argued to be inadequate as a means of encapsulating contemporary fan practices and engagement, since the relationship between fans and commercial culture has become complexly entangled (Bennett 2014b). Alternative modes of fan engagement, such as crowdfunding, further complicate the boundaries between fans and media producers, as well as between fandom and the commercial world (Bennett et al. 2015). Indeed, scholars such as Noppe (2011) have observed the development of a new hybrid economy for derivative works: in the case of fansubbing (fan subtitling) in anime fandom, for example, anime fans work with copyright holders and commercial distributors to co-create a transnational participatory culture through their voluntary fannish labour (Ito 2012; Lee 2011).

Gender, sexuality and desire

Fandom offers a space in which unaffiliated individuals can explore the possibilities of gender, sexuality and desire. As Jenkins (1992:284) famously suggests, "fandom celebrates not exception texts but rather exception readings". Much work in fan studies focuses on the fannish productivity of heterosexual women and members of sexual minorities in reading popular culture. The fact that more fan practices are found among women than men is argued to be attributable to the sociohistorical gendering of leisure, in which women were confined to the domestic reading and viewing of texts, whereas men traditionally occupy public space to engage in an alternative range of fandom activities, such as those pertaining to the realms of sports and politics (Pugh 2005:5; Sandvoss 2005:16).

These expressions of gender and sexual desire are not only personal but can also be considered political. Feminist audience studies have provided examples of female fans expressing their desires and discontent with patriarchy. Radway's seminal work (1984/1991) demonstrates how heterosexual, married women in the United States escaped from their daily household routines by reading romantic fiction and identifying with the independent heroines. Bacon-Smith (1992) explores how female fans of *Star Trek* and other genre television series materialize their pleasure and desire by producing and circulating fan fiction and fan art. On the other hand, fans who belong to or identify as sexual minorities may seek to queer these media by reclaiming and extricating expressions of all aspects of non-, anti-, and counter-straight cultural production and reception (Doty 1993:3). Queer fandom is always rooted in the specific social trajectories of how a marginalized identity is constructed. For example, before the start of the gay liberation movement in the late 1960s, American gay men, who could not publicly share their depressed feelings due to social stigmas and oppression, resonated strongly with actress Judy Garland because they drew inspiration from her invincible spirit that had enabled her to survive considerable suffering and failures (Dyer 2004:3). The Boston Area Gaylaxians, a group of LGBT science fiction fans, protested and demanded more gays and lesbians represented in *Star Trek* through a letter-writing campaign and initiating dialogues with the producers (Jenkins 1995). They believed that an increased representation of LGBT characters in popular culture could help normalize non-heterosexual desires and empower marginalized communities. Furthermore, fandom offers a space in which gender boundaries become blurred and ambivalent desires are explored. Male fans of David Bowie in Stevenson's (2009) research expressed their sexual feelings towards Bowie; this suggests that fandom offers a space to unfix dominant masculinity and open up ambivalent possibilities for alternative masculinity.

A well-known example of fans negotiating gender and sexuality is the practice of slash reading and writing. Slash is a fanfic genre that puts two (presumably heterosexual) male characters into a romantic or erotic relationship (Green et al. 2006:78). The term originated from the convention of using the slash punctuation mark to signify a same-sex relationship between two male characters in a series (Jenkins 1992:192). An example would be Kirk/Spock (K/S) in *Star Trek*. Jenkins (1992:194) observed that many female fan writers, who were mainly heterosexual and white, regarded slash as an uncensored expression of female sexuality. Slash writers that Jenkins contacted have argued that slash stories are liberating, since the pervasive trope involving male characters' having to overcome obstacles rejects the normative construction of male sexuality and symbolizes the transgression of gender and sexual hierarchies (Green et al. 2006:67–68; Lamb and Veith 1986:254). Taking the homoeroticization of male characters further, the practice of shipping entails reimagining erotic and/or romantic relationships between male characters in dramatic texts or between male public figures. For example, the Alexter fandom that celebrates a fictional male same-sex romance between student leaders Alex Chow and Lester Sham of the 2014 Umbrella Movement in Hong Kong is said to have drawn young women's attention to the pro-democratic movement and offer a counterargument to masculine narratives of political participation (Lavin and Zhu 2014).

Parallel to slash in Anglo-American fan cultures, *yaoi* or BL (boys' love) fandom has begun to receive attention in English-language fan studies. Originally a Japanese manga and anime genre portraying male same-sex relationships, BL fandom has become a significant global female-oriented erotic subculture in which women reject normative femininity and mainstream heteronormative romance (Levi et al. 2010; Wood 2006). However, BL consumption requires the precarious management of fan identities. In Japan, female fans of BL, who are known as *fujoshi* (rotten girls) avoid publicizing their identities in everyday life to avoid stigmatization, which renders BL a "uniquely oppositional but self-denigrating feminine subculture" (Okabe and Ishida 2012:221–222).

As well as the original slash, there is also the genre of femslash, which refers to female writing on romantic or erotic relationships between two female characters (Russo 2014:452). In contrast to slash, the majority of participants in femslash fandom are characterized by the "presumed synchronicity between its participants (primarily queer women) and its content (queer relationships between women)" (Russo 2018:156). Examples include the fandom of Xena/Gabrielle in *Xena: Warrior Princess* (1995–2001) and the first multi-fandom femslash fan convention TGIF/F (Los Angeles, February 2016) (Jones 2000). Russo (2018:161) argues that queer women obtain particular pleasure not only from encouraging media producers to create explicit onscreen lesbian characters and relationships, but also from engaging in creative interventions themselves in ways that reshape the relationships between the characters inhabiting a particular storyworld.

Fan activism and civic participation

Besides expressing personal desires and aspirations, fandom can be seen as an important instance of citizen media because of its potential to effect sociopolitical change and to redefine the boundaries between formal and informal political engagement (Baker and Blaagaard 2016a:16). That said, it should be noted that the reasons lying behind fans' civic engagement are often ambiguous. Fans are more likely to be driven by the parasocial desires of star–fan intimacy and the affective sociality within (imagined) fan communities, instead of formal political motives or principles. Thus, the effects on social justice and institutional politics can be seen as a consequence rather than the driving force of fans' civic participation.

Fan activism refers to "fan-driven efforts to address civic or political issues through engagement with and strategic deployment of popular culture content", most often through existing fan practices and networks (Brough and Shresthova 2011:2.3; Jenkins 2012a:1.8; Jenkins 2014:65). Since the 1980s, it has been observed that the younger generation in the West has increasingly sought to influence civic and political life by engaging in social networks and cultural activism enacted through informal, non-institutionalized and non-hierarchical networks on the Internet, instead of by means of electoral politics (Bennett 2012; Brough and Shresthova 2011:3.2; Jenkins 2012a:2; Jenkins and Shresthova 2012). Highly organized fan activism, such as online petitions, boycotts and email and online letter-writing campaigns, have been found to be prevalent (Earl and Kimport 2009). Fans' use of social media to communicate across extended networks by liking Facebook pages and through hashtags and retweets on Twitter has enabled them to reach large numbers of people who share the same values (Bennett 2012:3.3)

The Harry Potter Alliance (HPA) offers an example of large-scale media text-driven fan activism in which fans' personal narratives and affective sociality has been collectively mobilized in struggles for social justice. The HPA is a nonprofit organization first established in 2005 (Jenkins 2012a; Jenkins 2014). Calling itself 'Dumbledore's Army for the real world', HPA re-appropriates and deploys the Harry Potter story as a keystone to draw fans' identification and engagement with its various local and global campaigns, such as those pertaining to disaster relief, fair trade, workers' rights, literacy and LGBT rights (Hinck 2011). For instance, the HPA donated more than USD 123,000 to nonprofit organization Partners for Health in response to the 2010 earthquake in Haiti. Fan activism driven by celebrity has also emerged with the help of social media. In 2010, American pop music diva Lady Gaga used public and social media to publicize her protests against the policy of Don't Ask Don't Tell (DADT), which dismisses or refuses openly gay or bisexual citizens from serving in the military (Bennett 2014a:144–148). Her fans, whom she names the little monsters, mobilized themselves in further collective actions, such as filming and uploading videos onto YouTube to publicize the protest and calling their senators to express their views. Having had the values of active citizenship instilled in their minds, fans felt empowered and further connected to Lady Gaga (Bennett 2014a).

Fan-initiated philanthropy is another form of fans' civic engagement. While celebrity charity in the Anglo-American world has been criticized for its capitalist ideology and depoliticizing tendency (Kapoor 2013), it has been suggested that fan-driven philanthropies in East Asia encourage more participatory and public-oriented youth cultures. For example, the fans of mainland Chinese pop music diva Li Yuchun (Chris Lee) have set up a charity fund and volunteer groups under her name, and have been actively engaged in domestic disaster relief and literacy campaigns, leading to endorsement from both the party-state and the celebrity herself (Jeffreys and Xu 2017:251–253). In South Korea, it has become common for K-pop fans to send rice sack mock-ups (known as dream rice) to show their support for their stars instead of sending wreaths of flowers (Jung 2012:2.2). For example, in 2010 fans of U-Know Yunho (a member of boy band TVXQ) sent 3.15 tonnes of rice to various welfare facilities under his name in order to celebrate his musical theatre debut.

Nonetheless, fan activism has its limitations and a number of criticisms have been voiced. Fans' participation in volunteering and philanthropy is often seen to be largely driven by the parasocial star–fan intimacy and the motivation to promote the public reputation of their stars rather than constituting an active or genuine form of civic engagement (Jeffreys and Xu 2017; Jung 2012). Moreover, the social networks and mobilization tactics that could have been used to influence existing power relations and lead to progressive social changes can work in the opposite direction. Queer fandom may not necessarily lead to fans' direct civic engagement with struggles for LGBT rights. For example, fans of a local queer singer-songwriter Denise Ho in quasi-democratic Hong Kong preferred heteronormative readings of her queerness (Li 2012; Li 2017). In more extreme cases, we must recognize that fans may mobilize themselves through nationalistic discourses in order to express blunt racist ideologies and engage in cyber vigilantism (Chen 2016; Jung 2012).

Transcultural and intersectional fandom

Since 2010, there have emerged two further interrelated strands of research that seek to unfold the operations of power and inequality in and beyond fandom. The first strand calls for a more transculturally oriented fan studies discipline. Researchers critique the work of English-language media fan studies scholars for dominating the debate and privileging assumptions in the field that tend to exoticize fan cultures originating in other parts of the world (Chin et al. 2018:300; Morimoto and Chin 2017). The shorter lag time between the domestic and international distribution of popular culture and greater accessibility facilitated by the Internet have further helped to contest the assumption that fan culture operates within a homological national/linguistic context (Chin and Morimoto 2013:104). Transculturally oriented fan studies aim to contextualize transnationally circulating media and to remain attuned to the nuanced sociohistorical, political and economic trajectories of consumption (Chin and Morimoto 2013:98). The term transcultural is employed to "allow for a transnational orientation, yet leaves open the possibility of other orientations that may inform, or even drive, cross-border fandom" (Chin and Morimoto 2013:93). In response, work that focuses on non-Western transcultural fandom has begun to emerge (Chen, L. 2017; Lavin et al. 2017). Chin and Morimoto (2013), for example, have explored the assimilation of media texts by fans in their close textual analysis of a Japanese fan text centring on Hong Kong star Leslie Cheung, which juxtaposes the star's persona, his film roles and fan artists' own Japanese cultural contexts. Fraser and Li (2017), on the other hand, have excavated the mediated cultural memory of Kowloon Walled City in Hong Kong, which was once known as "one of history's great anomalies", by tracing its fandom and global and regional flows of representation. Finally, studies are increasingly focusing attention on the fact that the West is by no means a homogeneous entity, as suggested by the language barrier experienced by Italian fans of American television shows, which discourages them from participating in English-language online fandom or communicating with producers (Benecchi 2015).

The second strand of scholarship calls for an intersectional approach that concerns race issues in particular. While fandom has always been a site for expressing intersectional identities of gender and sexual orientation, discussions of race remain under-theorized and have been "frequently treated as an add-on or as something that should be addressed somewhere later" (Wanzo 2015:1.6). Pande (2016) urges scholars to decolonize fandom studies by theorizing cyberspace fandom as a postcolonial space of contention and conflict. She problematizes the term fan of colour and challenges the default whiteness in fandom by examining fan campaigns that foreground diversity by non-white fans, such as those around the television series *Avatar: The Last Airbender* (2005–2008). On the other hand, social media platforms such as Twitter have offered non-white audiences a new counterpublic against which to define their own politics. For example, Chatman (2017) studies the online dynamics between black fans and anti-fans of ABC's primetime drama *Scandal* (2012–), which features a black woman in the lead. Warner (2018) explores how the release of preproduction news about *The Flash* movie by Warner Bros. led to the formation of 'The Iris West Defense Squad (IWDS)' on Twitter by a group of black American women. For the members of this squad, their collaborative efforts to protect the blackness of this television series character as well as the actress who portrays her serve a dual function: Iris West represents not only the object of their fannish love but also an identificatory model for their black female selves. De Kosnik (2018), on the other hand, delineates the concept of forced fandom, in which hegemonic cultural consumption is imposed on subordinate groups, with reference to American Filipinos' protests against television's stereotypical representations.

Transcultural and intersectional approaches to fandom aim to further unmask power relations and critically explore the infinite diversity of fan practices as citizen media. As Jenkins (1992:3) suggested, "there is nothing timeless and unchanging about this culture; fandom originates as a response to specific historical conditions". Those conditions emerge as shifting power relations among fans, between fans and media texts, and between fans and media producers. Through the lens of fandom, we are thus able to re-examine the intersectionality of power and reconceptualize the relations between individual citizens, communities, media, corporations and public(s) in the age of global media convergence.

See also: diversity; immaterial labour; self-mediation; user-generated content; wikis

Recommended reading

Click, M. A. and S. Scott (eds) (2018) *The Routledge Companion to Media Fandom*, New York: Routledge.

A comprehensive collection of essays that explore critical issues relevant to the study of media fandom. Key themes include methodology and pedagogy, ethics, technologies, identities, race, transcultural fandom, the media industry and the future of fandom studies.

Gray, J., C. Sandvoss and C. L. Harrington (2017) *Fandom: Identities and communities in a mediated world*, second edition, New York: New York University Press.

While the first edition, published in 2007, summarized and elaborated on the key themes that have shaped scholarship on fan phenomena, the second edition adds further to the ongoing discussion. It explores the materiality, spatiality and temporality of fandom, and includes essays on fan activism as well as on the hybrid and intersectional interactions between fan labour and fan producers.

Jenkins, H. (2006) *Fans, Bloggers, and Gamers: Exploring participatory culture*, New York: New York University Press.

Contains a collection of essays by Jenkins, one of the pioneers of the field of fan studies. Of particular note is his discussion of the notion of intervention analysis in which Jenkins explores the divided loyalties between fan communities and academia that scholars working on this topic are invariably forced to negotiate.

FILM STUDIES AND CITIZEN MEDIA

Jens Eder and Britta Hartmann

Film studies – also referred to as cinema studies or moving image studies – contributes to research on citizen media in two ways. Firstly, it provides conceptual and methodological tools to assist in the study of audiovisual media and moving images: their forms and histories, practices and contexts, institutions and organizations, genres and discourses, audiences and experiences, uses and effects, aesthetics and ethics. Secondly, film studies examines the whole range of audiovisual citizen media, including productions by non-professional filmmakers (ordinary citizens) seeking to effect social and/or political change. Examples of such works include radical films, counter-cinema, public access television programmes, activist videos available on VHS tapes and on the Internet, or 'artivist' video installations and screenings. Audiovisual citizen media may tell stories with a moral lesson; serve as visible depictions of injustice or even as admissible evidence in court; clarify critical issues and embrace a cause; promote campaigns and mobilize their audience; encourage protest and resistance by persuading and convincing their viewers; mock those in power and call for solidarity with the weak; and strengthen activists' sense of community and group identity.

In film studies, research on such topics is closely connected to interdisciplinary research on visual communication in the context of political protest, activism and social movements (McLagan and McKee 2012; Drew 2013; Doerr et al. 2014; Fahlenbrach 2017). In its various exchanges with other disciplines, film studies retains a focus on audiovisual media, including not only film and cinema, but also television, VHS, DVD, Blu-ray, online video, audiovisual installations, screenings in public places and any other means of presenting moving images. Film studies was one of the disciplinary origins of media studies and still has close ties to it, as well as to literary studies, history, rhetoric, linguistics and semiotics. In contrast to social-scientific disciplines like communication studies, it belongs to the humanities and focuses on connections between mediality, technology and aesthetics, on the one hand, and culture, ideology and power, on the other. Film studies makes use of a wide spectrum of theoretical approaches, and its various schools dedicate themselves to specific issues: e.g. film as art or as part of popular culture, empirical reconstructions of film experience, or feminist, postcolonial and queer film practices. The range of methods is similarly broad, encompassing case studies of paradigmatic films or audiences, close analyses of audiovisual forms, ethnographic research on audiovisual media practices, discourse analysis, as well as cultural and historical studies of filmmakers and filmmaking, stars, audiences, styles, genres and themes.

Film studies traditionally concentrated on cinema and television, as these relatively cost-intensive mass media require high standards of crafts(wo)manship and therefore privilege experts and trained professionals over ordinary citizens. Even in the field of non-professional production and DIY media, home movies or family films have received more attention than citizen media (Odin 1995; Zimmermann 1995; Schneider 2004). Since the second decade of the twenty-first century, however, this focus on Western mass media and non-political amateur films has begun to broaden and shift due to the ubiquitous spread of digital audiovisual media in the form of home cinema, public screenings and especially online video. Some authors call those new media practices post-cinematic (Denson and Leyda 2016; Hagener et al. 2016), a term that stresses their historical ties to cinema. As far as citizen media are concerned, however, digital technologies facilitate cheaper and simpler video production processes, global video distribution and reception as well as various forms of viewer interaction and participation.

These developments have contributed to a rapid and enormous increase in the amount and range of audiovisual forms created by citizens in their struggle for representation and social change. Film studies has expanded its research on audiovisual citizen media accordingly – but instead of adopting the term citizen media, it rather uses a variety of different labels. Askanius, for example, lists "[p]articipatory video, radical video, alternative video, community video, development video, guerrilla video, underground video, advocacy video, DIY video, subversive video, labor video journalism, [and] video for social change" (2014:453). Further terms focus on certain kinds of producers (e.g. workers' film, indigenous film, migrants' film, citizen witnessing), topics (amateur social problem film, protest film, black movement films, human rights film, sousveillance, video forensics), media (desktop documentaries, smartphone films, video activism) or differences from the mainstream (guerrilla film, guerrilla television, operative video). More general research surveys tend to include audiovisual citizen media under headings such as political film or film and politics (Tzioumakis and Molloy 2016).

In an attempt to find common threads that connect those different categories, we may distinguish between four strands of research in film studies that seem especially relevant to the field of citizen media: the political aesthetics of film, in the wide sense; the history of audiovisual citizen media, including its roots in socially engaged documentary and activist film; citizens' contemporary offline film practices, including the production, distribution and reception of radical films in alternative public spaces; and online video activism on the social web. The remainder of this entry explores these different strands of research – including relevant work from neighbouring disciplines – and directs attention to various kinds and examples of audiovisual citizen media.

The political aesthetics of citizen media

The first, mostly theoretical, strand of research examines the political aesthetics of audiovisual media. It deals with the question of how their specific mediality and their various forms are connected to filmmakers' practices and viewers' experiences in political contexts. Audiovisual media are characterized by specific mimetic, symbolic, affective and interactive potentials that influence their uses and effects (Grabe and Bucy 2009; Eder and Klonk 2017). Typically, moving images resemble the objects they represent, which makes understanding them comparatively fast and easy and endows them with the power to draw spontaneous attention, to cross language barriers and to provide large, untutored audiences with visible and audible evidence. Photographic images also bear physical traces of represented objects, which strengthens their indexical and evidential claims. In contrast to verbal texts, they directly address their audiences' immediate perceptions and bodily experiences and may therefore operate as particularly strong triggers of affect and emotions (Plantinga 2013). Their spontaneous affectivity impacts on the minds and memories of spectators and often induces them to act (Eder 2017). Moreover, audiovisual media allow complex information to be condensed into concise symbols and vivid metaphors (Coëgnarts and Kravanja 2012) which may be immediately grasped and easily remembered, while at the same time being open to controversial reinterpretations and remediations. Certain kinds of images (e.g. computer simulations, web interfaces, interactive documentaries or VR productions) may even be used as interactive tools for discovery, comprehension and communication.

Beyond those general potentials of audiovisual media, their political aesthetics partly depends on their various forms and genres: audiovisual citizen media may be fictional or non-fictional, live-action or animated. They may take the form of fictional or journalistic genres or of essay films, found footage films, experimental films or various types of documentaries (Nichols 2001). Their structure may be narrative (telling a story), rhetorical (developing an argument; Hesling 1989), descriptive, categorical or associative (Bordwell and Thompson 2010). Even broader is their variety of audiovisual styles and the specific quality of their images, sounds, editing patterns, among other features. This entry examines three key aspects of a political aesthetics: political representation, viewers' experiences and filmmakers' and users' practices.

In terms of political representation, film and television studies have long dealt with questions about the representation of individuals and social groups – concerning, for example, issues such as class, gender, sexuality and ethnicity (Waugh 1984; Rabinowitz 1994; Fiske 1987). Citizen media usually focus on represented human beings: they show social actors in political documentaries, celebrate global icons like Arundhati Roy (Ghosh 2011), follow 'microcelebrity activists' on social media (Tufekci 2013) or carry out 'character work' by telling moral stories about victims, villains and heroes in the struggles of social movements (Jasper et al. 2018). Film studies offers conceptual frames and analytical tools to better understand such represented beings (Smith 1995; Eder 2010).

Film studies is also concerned with the experiences and responses of the audience. This goes far beyond discussions in cultural studies about the different readings of films (Hall 1980), and only some selective examples can be provided here. For instance, there are comprehensive theories of the relations between emotions and ethics in film (Plantinga 2018), and many of their insights about topics like rhetoric, immersion, engagement or morality may be applied also to audiovisual citizen media. Phenomenological and other approaches (Sobchack 1992) stress that film experience is simultaneously aesthetic and political, and that it contributes to a specific distribution of the sensible (Rancière 2004), or a sensible politics (McLagan and McKee 2012): audiovisual media may disclose or hide certain political fields of experience to the viewers' senses. Documentary film researchers have examined how this bodily and affective experience of the political might translate into action, e.g. by way of a political mimesis (Gaines 1999) when viewers experience a bodily reaction to fighting and suffering on the screen. To explain why viewers' affective responses to political films and videos diverge so widely, some authors analyse films' affective strategies and their relations to viewers' diverging interests and intergroup conflicts (Eder 2017, 2018). Several other scholarly approaches have a narrower focus. For instance, Andén-Papadopoulos examines videos shot by camera witnesses and stresses how "the ritualized employment of the mobile camera as a personal witnessing device to provide a public record of embodied actions of political dissent" records not only these actions of dissent or the violence against protesters but also the witnesses' own dangerous act of participation, thereby fostering new forms of connection between witnesses, victims and viewers (Andén-Papadopoulos 2014b:120). The specific relations between citizen media and their users' emotions are an important field of investigation in contemporary film and media studies and have also been explored in several anthologies (e.g. Karatzogianni and Kuntsman 2012).

The practices of audiovisual citizen media have been researched by scholars who analyse films in their contexts – immediate contexts of media technologies and institutions or dispositifs (Kessler 2007), as well as wider, sociocultural and political ones. Such practices may pertain to film production, such as crowdsourcing and collaborative filmmaking (Bondebjerg 2016; Jenkins et al. 2016; McIntosh 2016); distribution and circulation, for instance, via festivals, microcinema screenings, shared tapes, video files, social media or free streaming sites on the Internet (Knight and Thomas 2011; Iordanova and Torchin 2012); reception, appropriation and remediation by active viewsers or produsers, for instance in the context of fan cultures (Jenkins 2006b); and citizens' attempts at 'greening the media' and avoiding environmental problems (Maxwell and Miller 2012; Cubitt 2016).

History of audiovisual citizen media

A second strand of research reconstructs the historical development of audiovisual citizen media. In addition to the *Encyclopedia of Social Movement Media* (Downing 2011), which includes several entries on film and video groups worldwide and may serve as a basic guide to this area of scholarship, there are various other works focusing on the use of video by political groups and movements in individual countries like the UK, Germany, Switzerland or the US (Dickinson 1999; Fountain 2008; Cain 2011; Hoffmann and Wottrich 2015; Nigg 2017; Drew 2013), as well as a host of more specific publications concentrating on certain episodes, media practices and actors.

Most crucially, not only the availability, but also the relative importance of audiovisual media in citizens' struggles for social change have increased dramatically. For a long time, audiovisual media content was much more expensive and difficult to produce and use than was the case with other citizen media, such as posters, leaflets, graffiti, street performances. Most means of audiovisual production and distribution – technological equipment, film studios and theatres, and television stations – were in possession of the power bloc (Hall 1980), i.e. big corporations, government organizations, wealthy individuals and functional elites. Therefore, producing and distributing audiovisual citizen media has always been an unequal David vs. Goliath struggle, with small alternative groups or grassroots initiatives struggling against dominant mass media, and political narrowcasting against the broadcasting of news and entertainment. Citizens were not only dependent on their own inventiveness, creativity and the availability of affordable, easy-to-use production equipment, such as cameras, sound recorders, editing systems. They also needed funding, professional expertise and opportunities to distribute their work. This meant they had either to confine themselves to amateur work in small niches or to cooperate with professional filmmakers, unions, parties, corporations or governmental institutions.

Therefore, the rise of audiovisual citizen media was connected to the emergence of new media technologies and organizations in the context of certain cultural and political developments. For instance, the introduction of analogue video, the proliferation of affordable digital cameras, and later the emergence of smartphones, the social web and video platforms marked turning points in the work of political groups and civic initiatives. New technological conditions and the resulting practical aesthetic possibilities were quickly explored and systematically used as a means of expression in political conflict (Aguayo 2006, 2019). Several phases can be distinguished: the development of audiovisual citizen media initially made substantial progress in the 1930s with the production of political or committed documentaries (Waugh 1984), workers' films and newsreel programmes, like those of the Workers Film and Photo League in the US, founded in 1930 and supported by the International Workers' Relief (Internationale Arbeiterhilfe). During the Great Depression, the League regularly coordinated the labour movement's documentary film productions (Alexander 1981). Weekly newsreel features documented the hunger marches of the unemployed, the gatherings of workers on strike and police brutality. As silent eyewitness films, they documented injustice, violence and hunger, calling for solidarity and resistance. Comparable working-class newsreels also became popular in Germany and other European countries. Professional filmmakers such as Joris Ivens demonstrated their support for the struggles of the working poor with films like *Borinage* (1934) (Waugh 2016).

The film movements of the early 1930s gave global impetus to various approaches to political film work. In this context, the evolution of political documentary films went hand in hand with ethical issues regarding representation in the broadest sense. More specifically, it also widened the range of ways in which films can voice the concerns of the socially oppressed, so that these are empowered to become active subjects, confidently standing up for their own political agendas. The interrelationship between voice, communicative power and representation was reconsidered in various ways, e.g. by collaboratively involving those affected in film production and thus making them co-producers (Waugh 2011; Nichols 2016), or by having experienced filmmakers guide them in shooting their own films.

In the 1960s, against the backdrop of new developments in film technology, these approaches gained new momentum in the context of civil rights; women's and student movements; and liberation movements in several countries of the so-called Third World. The interdependence of technology and political aesthetics was reflected in the American Direct Cinema movement (Beyerle and Brinckmann 1991; Saunders 2007) and its programmes of candid filmmaking for the sake of immediacy, authenticity and realness. Direct Cinema owes its development to the introduction of enhanced 16mm camera technology, as well as zoom lenses, light-sensitive film stock, directional microphones and pilot tone synchronization. Although the observational mode of Direct Cinema lead to films of a hitherto unknown intimacy, it was fiercely attacked because of filmmakers' belief in uncontrolled cinema and its ties to television. In France, the ethnographer and filmmaker Jean Rouch and the sociologist Edgar Morin used the same film technology for different ends: their cinéma vérité, developed while they were working on *Chronique d'un été* (1960), fed an interactive and reflexive mode. The subjects in front of the camera became collaborators in the filmmaking process and openly allowed the audience to look at their supposedly private lives and experience it as political.

In the USA, Emile de Antonio and other documentary filmmakers likewise contrasted the observational approach of Direct Cinema with an independent radical cinema that encompasses films like *In the Year of the Pig* (1968) on the causes of the Vietnam War, *Underground* (1976) on the Weather Underground Organization, and the political satire *Milhouse: A White Comedy* (1971) on Richard Nixon's politics (Beyerle and Brinckmann 1991; Waugh 2011). Around the same time, Latin American Third Cinema also called for both formal and political radicalism (Solanas and Getino 2000; Burton 1986; Martin 1997; Wayne 2016). Its representatives, such as Fernando Solanas and Octavio Getino with their influential Argentinian film *La Hora de los Hornos* (*The Hour of the Furnaces*, 1968), strove for close cooperation with the people, while considering film teams as tightly managed guerrilla units. And in the USA, the student movement gave birth to the leftist film collective Newsreel (Nichols 1980). The concerns of both radical documentaries and feature films range from the struggle for women's and gay rights to anti-colonial and black liberation movements. Film serves as a publicity tool in labour disputes; it is a vehicle for peace and anti-nuclear movements; it is used by environmental groups, by people fighting to preserve their neighbourhoods, and by youth groups demanding alternative youth centres (Pentley 2015; Zutavern 2015).

A major turning point in the history of activist filmmaking was 1967, when SONY launched in the USA market its so-called video rover or portapak video system, consisting of a camera and recorder connected by a cable. It took several years for the portapak system to reach the European market. The possibilities of this comparably cheap, convenient and easy-to-handle amateur video technology were enthusiastically welcomed and systematically explored by non-professional filmmakers for artistic and political purposes. Numerous video activist or guerrilla TV groups such as TVTV or videofreex emerged and sought to develop new forms of, and public spaces for, leftist or radical filmmaking practices as alternative voices against the politically controlled and/or capitalist media system (Boyle 1997; Fountain 2008; Hoffmann and Wottrich 2015; Nigg 2017).

In the case of West Germany, approximately seventy video initiatives, video cooperatives and alternative media centres were founded between 1969 and 1984 (Köhler 1980). Some were established at universities by students, others by filmmakers; they were formed in youth centres or sprang from the political work of local initiatives. These productions differed greatly both in terms of approach and aesthetic quality: as a form of bottom-up history, alternative reports were produced alongside conversations with contemporary witnesses at the time. Calls for solidarity could be found next to long-term documentaries about citizen movements. Video activists instructed amateurs, offered training to school teachers, gave courses at schools and youth centres, and produced technical manuals; they set up film distribution networks, compiled lists of video groups and published political works on film. While the first generation of video activists emerged in West Germany, the rigid political system of the German Democratic Republic was criticized by the East German Super-8 filmmakers of the 1970s, followed by the video pirates of the 1980s, who sometimes managed to send their footage to West German TV stations (Fritzsche and Löser 1996; Löser 2011).

Of course, these are only a few selected examples of historical research on audiovisual citizen media. Much more could be said about topics as different as, for instance, the influences of experimental directors like Peter Wollen or Jean-Luc Godard on British film cooperatives in the mid-1970s (Fountain 2008), camcorder AIDS activism since the mid-1980s (Juhasz and Gund 1995), and especially on the history of audiovisual citizen media in non-European countries (see several contributions in Downing 2011; Tzioumakis and Molloy 2016), including the development of 'indigenous counterpublics' (Ginsburg 2012; Wilson 2016).

What those different developments have in common, however, is a continuous struggle to distribute non-professional audiovisual citizen productions to wider publics. As will be elaborated in the next section, alternative cinemas, video libraries, screening circles and festivals have made significant contributions, and the introduction of public access television in the US and several European countries since the 1970s has opened up further spaces (Dowmunt 2002; Halleck 2002). Even in those cases, however, showing and watching audiovisual citizen media are still dependent on funding and control by the state, cultural organizations or commercial companies. This only changed when the World Wide Web made the online circulation of audiovisual content possible.

Citizens' current film practices in offline spaces

Research in the third strand studies how alternative films and television programmes are produced, distributed and watched in the contemporary offline world. Its topics range from feature-length documentaries by non-professionals (see several contributions in Juhasz and Lebow 2015) to self-organized activist film festivals (Iordanova and Torchin 2012; Tascón and Wils 2017). As "it is not the media that makes the revolution, but revolutionaries" (Chanan 2012:225), it seems of crucial importance to study how citizens and amateur or semi-professional filmmakers form public spaces in which they collectively produce, watch and discuss films and develop ideas on how to actively use them as a means of bringing about social change or encouraging political resistance.

In the initial video movement of the 1970s and 80s, these alternative or radical public spaces existed in youth centres, meeting rooms and local stores. Gradually, this concept was broadened to include film festivals, which – despite initial opposition from established filmmakers – included low-quality videos in their programmes. Even though the alternative scene nowadays mostly organizes itself online, it nonetheless continues to rely on offline spaces such as specialized festivals, exhibitions, political group work or street screenings – e.g. *A Wall Is a Screen* (n.d.). As a result, online and offline practices are closely intertwined (consider, for instance, the activities of Occupy Cinema), an observation that has more generally been stressed by sociologists and political scientists like Castells (2012).

Of major importance among those counter-cultural film practices are some that are considered leftist, neo-Marxist or radical (Mazierska and Kristensen 2015; Presence et al. 2020) – e.g. the work of associations like the Radical Film Network (RFN) (n.d.), founded in London in 2013 by a group of activists, academics, filmmakers and curators in the field of activist political film culture who shared a vital interest in supporting radical approaches to film and sustaining the work of radical filmmakers. According to its own website, RFN today connects 133 affiliated organizations in twenty-four countries, ranging from Argentina to the US, from the Amber Film & Photography Collective based in northern England, which started its radical film work forty-five years ago, to Zimmedia, originally founded in Zimbabwe and working for cinema and television as well as for various NGOs. RFN brings together not only producers of radical film and citizen media but also distributors and exhibitioners, including (online) archives, film festivals and individual cinemas throughout the world that publicly screen radical documentaries and feature films. The connection between those groups is primarily Internet-based but also leads to many collective offline activities. Among other things, RFN hosts an annual conference in different countries as well as the Workers Unite! Film Festival.

Other festivals focus thematically on human rights, such as the Human Rights Watch International Festival, the Amnesty International Film Festival (now Movies that Matter), One World – International Human Rights Documentary Film Festival and the Activist Film Festival, to name but a few (Iordanova and Torchin 2012; de Valck et al. 2016; Tascón and Wils 2017). Festival publics offer forums for exchange; they raise awareness for the violation of human rights in distant areas of the world, stoke anger, evoke affects and give birth to protest movements. They also serve as platforms to introduce and promote new projects that need organizational and financial support. From an academic perspective, film studies attempts to participate in such processes with recurring events like Visible Evidence, the world's largest conference on documentary film that regularly deals with forms of citizen media.

On the film production side, crowdsourcing of documentaries has become a vital trend that calls to mind Joris Ivens' idea of close collaboration between filmmakers and their protagonists or social actors. Theatrical films like the *One Day on Earth* trilogy (Kyle Ruddick 2010, 2011, 2012) or *Life in a Day* (Kevin MacDonald 2010), the latter compiled on the basis of 80,000 videos uploaded to YouTube and totalling over 4,500 hours, may be seen as forms of citizen media produced by volunteer filmmakers from (nearly) all over the planet. Although most of the films used in *Life in a Day* are essentially unpolitical and often narcissistic, their linkage stresses human connectedness and cosmopolitan citizenship (Gotto 2011; McIntosh 2016). In an overview article on the relationship between documentary and video activism, Daniel Marcus refers to "closer models of participation and interaction" in documentary films which connect political movements, subjects and their audiences (Marcus 2016:187).

Compared to the typical forms of video activism on the social web outlined below, the majority of films primarily produced for offline spaces follow a different logic: usually, they are longer, claim to have some artistic value and aim to fulfil aesthetic and ethical criteria of professional independent filmmaking. They deal with a wide range of topics such as human rights in general as well as feminist, queer/LGBTI+, environmental and animal rights, with anti-consumerist, anti-corporate and anti-war sentiments. Despite the generally accepted term citizen media, which refers to the utopia of a cosmopolitan civil society and its means of political participation, we face the paradoxical situation that many citizen media filmmakers are deprived of civil rights in the countries where they live and work – for example, because they belong to a suppressed ethnic minority without any political representation. The so-called 'accented cinema', i.e. films mostly produced by migrants subject to displacement, exile, (forced) migration and life in the diaspora, reflect those conditions and seek aesthetic strategies to express their feelings of isolation and alienation as a result of living and working in a foreign culture (Naficy 2001; Ezra 2006). Another example is films produced by indigenous peoples fighting for cultural recognition as well as political representation and self-government (Ginsburg 1999, 2012; Monani 2014; Wilson 2016).

The old question of how to win over larger publics for political agendas has in recent years led activists, advocacy groups and NGOs to produce and design their films more strategically and target previously defined audiences. In collaboration with story consultants and professional filmmakers, more and more films are being produced that make use of strong storytelling and particularly effective dramatic as well as stylistic devices. 'Strategic impact documentaries' draw upon affective rhetoric and persuasive techniques in order to achieve the greatest possible impact (Finneran 2014; Nash and Corner 2016). The new alliances of media professionals and amateurs, pursuing the goal of effectively spreading awareness of political issues, are likely to transform activist film practices in the coming years.

Contemporary video activism on the social web

The fourth line of research on audiovisual citizen media explores the emerging field of video activism on the social web. Despite the recency of this new mode of production, distribution and participation, some publications already provide first surveys of its history, forms, practices and publics (Askanius 2013, 2014; Presence 2015; Eder et al. 2020). This research differs markedly from the aforementioned work on offline film practices, as it deals with videos made primarily for online distribution and strongly draws upon interdisciplinary social media studies.

As described above, distributing audiovisual productions to wider audiences used to be a constant problem for citizen activists. This situation changed with the spread of wireless broadband access, social media platforms, smartphones with video cameras, editing software, streaming apps, etc. A complex system of web video production, distribution and reception emerged. Since the World Wide Web first allowed the circulation of videos online, their distribution evolved through a series of stages: citizen activists initially used static homepages to disseminate their message, then they presented their productions on collective websites such as Indymedia, followed by channels on video sharing platforms such as YouTube or Vimeo (Snickars and Vonderau 2009), and finally they set up profiles on social media like Facebook or microblogs like Twitter. Since around 2005, this has resulted not only in a much more visible flow of densely interlinked videos, but also in a greater dependency on algorithmic networks on commercial platforms.

Nevertheless, in the age of mass self-communication (Castells 2012), connective action (Bennett and Segerberg 2012) and hybrid media systems (Chadwick 2017), video activists have become "key players in a hegemonic struggle" (Killick 2016:77) and take part in the construction of affective publics (Papacharissi 2015a) or networks of outrage and hope (Castells 2012). YouTube is the world's largest database of audiovisual material, and billions of videos are watched each day on Facebook (Meeker 2016:78). While such platforms are dominated by music videos and other forms of entertainment, some activist videos also reach millions of users and have a demonstrably strong impact, for instance *Kony2012* (Jason Russell 2012), the controversial campaign video about a Ugandan warlord (Torchin 2016; Engelhardt and Jansz 2014), or the Chinese environmental webdoc *Under the Dome* (Chai Jing 2015). Many researchers have noted the crucial roles that videos played in mobilizing and documenting the uprisings in Iran in 2009 (Simons 2012) and in the MENA region around 2011 (Taher 2011; Snowdon 2014; Westmoreland 2016), as well as in the anti-capitalist protests by groups like Occupy or the Indignados (Thorson et al. 2013; Razsa 2014; Thorburn 2014; Robé 2016). Other authors have examined how videos by eyewitnesses of police violence became starting points for the BlackLivesMatter movement and other protests. NGOs like WITNESS specialize in educating and training citizens in the often dangerous practice of citizen camera witnessing (Andén-Papadopoulos 2014b; Gregory 2012, 2017). Today, there is an enormous variety of international video activist groups (see the links in Eder et al. 2015). Some groups document war and violence (Abounadarra Films, Syrian Archive), while others focus on repressive regimes (The Mosireen Collective, Sendika.org), anti-capitalist protest and leftist counter-journalism (e.g. Leftvision, Labournet, Global Uprisings, Occupycinema), migration (Crosspoint, Let's Stay), political art (!Mediengruppe Bitnik, Peng! Collective) or LGBT rights (see Tedjasukmana 2018). So far, the productions and specific practices of such groups have only rarely been examined (with notable exceptions such as Presence 2015).

The distribution of activist videos is networked, less restricted by gatekeepers, potentially global and characterized by potential for interaction, commentary, co-creation and participation (Carpentier 2012): individual videos may be collected in channels and playlists, sorted into categories, tagged, recommended, linked to other videos, searched, selected, aggregated, liked, commented on, re-edited, shared and spread across platforms. All these practices follow the rules and algorithms of corporations, which again are regulated by government laws; to some degree, they are "under the spatial, temporal and financial command of neoliberalism" (Killick 2016:76) and platform capitalism (Srnicek 2017). Citizens' profiles, channels and productions can be controlled and deleted, state actors can gain possession of citizens' personal data (Tufekci 2017a). And even when critical videos such as WikiLeaks' *Collateral Murder* about a war crime committed by the US army in Iraq attract an unusual amount of attention online, they may trigger widely divergent responses (Eder 2018) and be ignored by most mainstream media (Christensen 2017). Another dark side of today's video activism is the proliferation of hate videos, fake news, right-wing extremism and jihadism (Ekman 2014).

When film studies participates in the interdisciplinary discussions of these issues, it also considers the crucial roles of videos as audiovisual texts, their rhetoric, aesthetics and affectivity (theorized in the first strand of research mentioned above). Compared to film and television, the reception of online videos is characterized by usually smaller, often mobile screens, by the difficulty in restricting access or censoring content, and by a higher degree of viewer distraction. Video activism favours small media (Chanan 2012). Coover summarizes some typical online video characteristics, noting that "what works in streaming and in new media are short works; they are works accompanied by text; they are works from different people contributing to a common space; they are fragmented; they are multiply linked" (Coover 2009:244, quoted in Ivakhiv 2016:727). Ivakhiv adds that "[d]igital video eliminates the intensive productive labor involved in filmmaking in favor of a light and spontaneous *caméra-stylo*, a 'camera-pen' that can capture reality effortlessly anywhere. Yet digital video paradoxically also provides the possibility of total control of the image" (2016:727; emphasis added).

In the new media environment, various new audiovisual forms and genres are emerging, and several authors distinguish between different kinds of activist videos (Eder et al. 2020). For instance, Mateos and Gaona (2015) propose a typology that includes documents, rally calls, reactions, construction of identity/self-representation and didactic videos, while Askanius (2013) distinguishes between mobilization, witness and documentation videos, as well as archived radical videos and political mash-up videos. Such forms are connected to typical uses, for instance as evidence in courtrooms and international war tribunals, quasi-judicial settings and UN bodies, as a direct form of address to decision-makers, or as mobilization videos in community-mobilizing campaigns (Askanius 2014), but also as viral marketing, political entertainment, community building and community branding or for individual expression and communication (Reichert 2012). As all those kinds of activist videos compete for attention in their oversaturated media environments, their affective forms (Eder 2017) and their distribution strategies (e.g. tagging, seeding) become crucial.

Audiovisual citizen media will likely be a growing field in film studies. There is much to be done when it comes to describing, analysing and explaining its multitude of forms and genres, its rhetorical strategies, its impact and effects. In doing so, film studies will contribute to the urgently needed discussion of the complex political and ethical issues arising from the use of moving images as tools or weapons in the struggle for social change.

See also: activism; documentary filmmaking; social movement studies and citizen media; YouTube

Recommended reading

Aguayo, A. J. (2019) *Documentary Resistance: Social change and participatory media*, New York: Oxford University Press.

Explores how political ideas drive participatory action and proposes a new understanding of the networked capacity of documentary media to create public spheres and social change. The book is accompanied by a website hosting a sample of the more than sixty interviews with documentary filmmakers, critics, funders, activists and distributors on documentary resistance.

Askanius, T. (2014) 'Video for Change', in K. Gwinn Wilkins, R. Obregon and T. Tufte (eds) *The Handbook of Development Communication and Social Change*, Malden, MA: John Wiley, 453–470.

Provides a concise yet fairly comprehensive overview of video activism or video for change, with a focus on left-wing and social justice activism on YouTube. It connects understandings of video activism as alternative news, collective empowerment or documentation of injustices.

Eder, J., B. Hartmann and C. Tedjasukmana (2020) *Bewegungsbilder. Politische Videos in Sozialen Medien*, Berlin: Bertz + Fischer.

Offers a concise overview of the contexts, forms and functions of political videos on social media, with a focus on video activism and its contribution to the formation of political public spheres and the mobilization of protest movements. Analyses the diversity of activist web videos, their types and themes, their aesthetic and rhetorical strategies, their affective impact potentials, their historical models and production contexts.

FLASH MOBS

Christian DuComb

The first flash mob took place at the flagship Macy's department store in midtown Manhattan on 17 June 2003. An anonymous, widely forwarded email, which was later known to have been written by *Harper's Magazine* editor Bill Wasik, directed potential participants to gather at one of several unobtrusive bars near the store. "Ten minutes before the appointed time", writes Wasik, "slips of paper bearing the final destination were distributed at the bars". That final destination was Macy's rug department, "where, all at once, 200 people wandered over to the carpet in the back left corner and, as instructed, informed clerks that they all lived together in a Long Island City commune and were looking for a 'love rug'" (Wasik 2006:57). Wasik assembled this crowd of twenty- and thirty-something artists and professionals to perform a subtle critique of the conformism of hipster culture. Blogger Sean Savage dubbed the gathering a flash mob (Savage 2003); the name stuck, and Wasik organized six more flash mobs over the next three months, all while remaining semi-anonymous. A radio reporter had discovered only his first name at an aborted flash mob in May 2003.

The term flash mob has since been used to refer to a large crowd – organized through email, blog, mobile phone or social media – which assembles suddenly and inexplicably to perform a coordinated action, only to dissipate a few minutes later. As flash mobs spread to cities throughout the world in the summer of 2003, the phenomenon elicited a surfeit of commentary. *Time Magazine* wondered whether flash mobs were "an incipient form of social protest" (Shnayerson and Goldstein 2003), whereas the *New York Times* consigned flash mobbing to "the prank tradition of phone-booth stuffing, streaking, flagpole sitting and goldfish-swallowing" (Harmon 2003). Theatre scholar John Muse likewise situates flash mobs in the prank tradition, but he also argues that even the most playful flash mob performances "take advantage of the anxieties of a post-9/11 populace who have been primed by threat-level warnings to expect the unexpected" (Muse 2010:14). The whiff of danger that clings to the ostensibly harmless practice of flash mobbing may account for the persistence of the flash mob as a form of citizen media, which neither Wasik nor the commentariat anticipated. Wasik saw the flash mob as "a vacuous fad" and considered the arc of its development complete by 2005, when the Ford Motor Company and Sony Pictures Digital commercialized the concept by staging a series of *Fusion Flash Concerts* in cities throughout the United States (Wasik 2006:57–61). But despite the appropriation of the flash mob phenomenon by corporate interests, flash mobbing has endured as a form of sociality that brings together people who do not know one another in public performances – usually whimsical, but occasionally raucous or violent – which "spontaneously emerge and radically interrupt the flow of everyday life" (Gerecke and Levin 2018:8).

Flash mob genealogies

Wasik (2006) traces the origins of the flash mob to social psychologist Stanley Milgram's experiment on the drawing power of crowds, conducted on the streets of New York in 1968. For this experiment, Milgram gathered small groups of people, ranging in number from one to fifteen, to engage in observable and easily imitated public actions. For example, a group, or stimulus crowd, would stand on the pavement and look up at the window of a nearby building, and Milgram and his collaborators would measure the proportion of passers by who either looked up or stopped in response (Milgram et al. 1969:79–80). Unsurprisingly, a large percentage of pedestrians either looked up or stopped, for as Elias Canetti and others have observed, even small and unassuming crowds have a tendency to grow. "A few people may [be] standing together", writes Canetti; "nothing has been announced, nothing is expected. Suddenly everywhere is black with people and more come streaming from all sides" (1960/1984:16). Thus, the value of Milgram's experiment lies not in its predictable scientific results but rather in its performative interruption of the regular flow of people along the street, which Wasik compares to a "Fluxus-style happening" (2006:60).

Wasik is not the only flash mob exponent to locate the genre in the performance art tradition. Walker (2013:116, 127) argues that flash mobs arise out of "the historical avant-garde and [its] performance-inclined offspring", drawing specific connections between flash mobs and Dada, surrealism, the Situationist International and the happenings created by John Cage and Allan Kaprow. Muse extends this flash mob genealogy to include not only performance art and its antecedents but also various forms of surprise theatre, from Augusto Boal's street-side political interventions to television prank shows like *Candid Camera* (2010:11). Although the roots of the flash mob phenomenon may lie in social psychology, performance art and surprise theatre, most writers on the subject concur that recent technological developments helped the flash mob to emerge as a distinct performance genre, inspired most immediately by the use of mobile phones to coordinate massive anti-globalization protests during the World Trade Organization conference in Seattle in 1999 (Nicholson 2005).

Rheingold (2002:xii–xiii) coined the term smart mob to describe groups of people "who are able to act in concert even if they don't know each other … because they carry devices that possess both communication and computing capabilities". A year before the first flash mob and five years before the release of the iPhone, Rheingold anticipated that "these devices will help people coordinate actions with others around the world – and, perhaps more importantly, with people nearby. Groups of people using these tools will gain new forms of social power, new ways to organize their interactions and exchanges" (ibid.). As Rheingold predicted, the ubiquity of smart phones has galvanized the formation of smart mobs from the Arab uprisings to Hong Kong's Umbrella Revolution, but not all early instances of flash mobbing made use of mobile technology and social media to coordinate crowd actions. Wasik, for example, argues that technology played only a minor role in the flash mobs he organized in the summer of 2003. Since Wasik recruited flash mobbers through anonymous emails sent about a week in advance of each gathering, he believes that "one could have passed around flyers on the street … to roughly similar effect" (2006:58). However, he also acknowledges "the role that blogs played in the spread of flash mobs" as a "surprise" to him – a technological intervention that changed the mob's pattern of assembly from "an unbroken network of acquaintanceship" (ibid.:63), linked by a cross-section of email chains, into something more diffuse: a network of both acquaintances and strangers. In the years since Wasik's first flash mob in Macy's rug department, flash mobs have increasingly come to rely not only on email and blogs but also on smart phones and social media to coordinate activity among unaffiliated citizens. Mobile communication and computing technology form a kind of digital scaffolding which enables flash mobbers to gather, act and physically intervene at a specific time and place, with little or no in-person contact before the mob assembles.

Flash mobs as citizen media

Flash mobs have been staged on every continent including Antarctica, and in venues as diverse as train stations, museums and shopping malls. Some flash mobs, like *International Pillow Fight Day* – which took place in twenty-six cities around the world in 2008 and spread to seventy cities the following year – have become global phenomena. Despite eschewing the term flash mob, the New York-based comedy collective Improv Everywhere has created some of the most iconic examples of the genre with performances like *Frozen Grand Central* in 2006, in which 200 people froze at the exact same moment in the middle of one of the world's busiest train stations. Improv Everywhere's carefully produced video of *Frozen Grand Central* shows excerpts from the performance along with audience reactions, which ranged from frustration to bewilderment to delight. Almost invariably, passers by slowed down or stopped to watch the frozen crowd, restricting foot traffic through Grand Central Terminal much as Milgram's crowd experiment had restricted the flow of people along the street. When the mob began moving again after precisely five minutes of stillness, many of the onlookers applauded, tacitly acknowledging that they had been cast as the unwitting audience for a surprise performance (Improv Everywhere 2008).

The video of *Frozen Grand Central* includes surreptitiously recorded conversations among spectators, one of whom remarks, "I think it's some kind of protest, probably", to which his companion replies, "It's either that or an acting class" (Improv Everywhere 2008). Although it partakes of both of these traditions, *Frozen Grand Central*, like most flash mobs, had no specific political agenda (Muse 2010:9). Nonetheless, Brejzek argues that "interdisciplinary urban interventions" like the flash mob "have the capacity to open up public space to alternative uses and to refract the existing axis of viewing" (2010:110). According to Improv Everywhere, *Frozen Grand Central* was meant to encourage "commuters and tourists alike … to notice what was happening around them", changing how they see the city (Improv Everywhere 2008). Building on Brejzek's argument, it can be claimed that flash mobs today refract the axis of viewing not only by astonishing a live audience with an unexpected performance but also by shifting the site (and sight) of reception to a digital platform. Improv Everywhere's video of *Frozen Grand Central* exemplifies this phenomenon. It has racked up over 36 million views since it was first posted on YouTube in 2008, dwarfing the live audience for the event.

As Muse remarks, flash mobs "almost invariably film themselves … and later distribute the video online". This mode of distribution in turn creates "secondary mass audiences" through video-sharing sites such as YouTube and Vimeo, leading to "the diffusion of audience" (2010:9–11). Elsewhere, DuComb and Benmen (2014) have shown that the diffusion of audience through media that capture and reframe live events for secondary audiences is not new: it dates back at least to the American War of Independence, if not earlier. According to historian David Waldstreicher, in August 1776 readers in Philadelphia learned that in Savannah, more than 1,100 kilometres away, "'a greater number of people than ever …' came out to witness the symbolic funeral of [King] George III" (1997:33). Throughout the war years, American newspapers reported widely on local, patriotic performances in distant parts of the United States, often recounting these events in great detail. Waldstreicher contends that this transmediation of performance and print spread nationalism throughout the newly independent nation and created extra-local audiences for local uses of the street. These audiences "improvised upon events they read about [in the newspapers] and then publicized their own interventions in public life" (ibid.:11). The globalization of the flash mob, epitomized by gatherings like *International Pillow Fight Day*, reflects a similar pattern on a larger geographic scale: the transmediation of performance and online video generates secondary audiences, which then stage similar events of their own. Unlike the mobs that assemble for *International Pillow Fight Day*, the revolutionary crowds that Waldstreicher documents had concrete political goals and lacked the intentional irrelevance and digitally enabled anonymity of a typical flash mob. But through the old-fashioned vehicle of the newspaper they participated in a form of citizen media that presages the diffusion of audience enacted by flash mobs in the twenty-first century. This historical parallel "[unsettles] the presumption that flash mobs are a new species of performance, uniquely enabled by digital technology" (DuComb and Benmen 2014:35), but technology is nonetheless a shaping force in the coordination, execution and reception of flash mobs. Mobile phones were incorporated into flash mobbing as early as August 2003, when the participants in a Berlin flash mob shouted "yes, yes!" into their handsets before applauding and dispersing (Nicholson 2005). A later sub-genre of the flash mob known as mobile clubbing turned the smart phone or mp3 player into an indispensable element of the performance. In these flash mobs, which proliferated between 2006 and 2008, participants don headphones and dance to the music of their choice, creating a spectacle that looks from the outside like a silent rave or disco. At one mobile clubbing flash mob in October 2007, "more than 200 people followed the Twitter call to turn up and dance to their favourite sounds" in the Turbine Hall of the Tate Modern in London. Brejzek remarks that these dancers both performed for one another and "collectively produced a performance for the gallery visitors", suggesting the coexistence of multiple forms of spectatorship within the live event (2010:113–114).

A possible question here is whether by dancing to distinct, private playlists mobile clubbing flash mobbers disrupt the circuits of communication among performers – as well as between performers and audience – that are necessary for publics to cohere. Drawing on Hardt and Negri, Kaulingfreks and Warren (2010) characterize the mobile club not as a public but as a multitude, "composed of innumerable internal differences that can never be reduced to a unity or a single identity" (Hardt and Negri 2004:xiv). Because of its heterogeneity, Hardt and Negri argue that the multitude, unlike the masses, cannot be effectively controlled by state or corporate power. From this perspective, the potential of the flash mob as a form of citizen media "to effect aesthetic or socio-political change or express personal desires and aspirations" (Baker and Blaagaard 2016a:16) lies neither in the precise organization nor in the massive online audience of an event like *Frozen Grand Central*. Rather, the most radically generative flash mobs more closely resemble the emergent and improvisational character of mobile clubbing, where "being in common or being with others is demonstrated precisely where there is no communal identity" – nor even a common, scripted action – to unify the mob (Kaulingfreks and Warren 2010:218).

Branded entertainment, youth violence and the future of the flash mob

On 15 January 2009, the mobile telephone company T-Mobile staged a flash mob in the main concourse of London's Liverpool Street Station as part of its *Life's for Sharing* promotional campaign. Recorded by ten hidden cameras, the T-Mobile flash mob "began with the movement of a single disguised commuter, and [built] to include 350 dancers all performing in sync to a medley of classic and contemporary chart hits before suddenly stopping and dissolving into the assembled crowd" (Grainge 2011:166). Unlike the lacklustre *Fusion Flash Concerts* put on by Ford and Sony in 2005, T-Mobile's flash mob performance won professional and popular accolades. The *Life's for Sharing* television spot was named TV commercial of the year at the 2010 British Television Advertising Awards (Sweney 2010), and the longer YouTube version of the advertisement has been viewed over 41 million times (T-Mobile 2009). Lotz (2014:192) describes promotional campaigns like *Life's for Sharing* as "branded entertainment", a form of advertising "of such merit or interest that the audience actively seeks it out". But in the years since the viral craze for *Life's for Sharing*, most corporate attempts to stage, record and stream flash mobs as branded entertainment have failed. As Grant and Boon argue, "consumers have an aversion toward corporations and are less likely to share a video when they realize that it is made for commercial reasons" (2013:190). In other words, citizens are not naïve to the corporate appropriation of citizen media.

Less than six months after the T-Mobile flash mob in London, thousands of teenagers in Philadelphia descended on South Street, a central city shopping district, in a turbulent gathering that the local press quickly labelled a flash mob. The *Philadelphia Daily News* reported that these "rampaging teens" used "text-messaging and [social media] sites like Twitter, Facebook, MySpace, and the new OurSpace, which caters to African-Americans, to assemble the mob" (DiFilippo 2009:3). A few dozen outliers in this adolescent crowd turned violent, ransacking a convenience store, pulling drivers from their cars and, in the most serious incident of the evening, assaulting a cyclist and leaving him unconscious and bleeding in the street. In the first three months of 2010, four more violent flash mobs erupted in the city, including a second disturbance on South Street. Similar events in Brooklyn, Boston and South Orange, New Jersey mirrored this spate of violent crowd actions in Philadelphia. Wasik (2011) questions whether the term flash mob accurately describes such events, but DuComb and Benmen maintain that "the structural similarity of violent and non-violent flash mobbing runs deeper than the use of email, social media and mobile technology as organizing tools" (2014:34). On 29 March 2010, the gossip website *Gawker* reported that "the majority" of Philadelphia's teen flash mobs "were actually street performances organized by Team Nike", an amateur dance group with an extensive presence on MySpace (Chen 2010). A YouTube video shows Team Nike engaged in a spirited but peaceful dance party in the midst of the second South Street flash mob on 20 March 2010 (Team Nike 2010). Sadly, an unprovoked attack on a pedestrian marred this gathering. *Gawker*'s close reading of court testimony on the events of 20 March reveals Team Nike's planned performance as the kernel of the second South Street flash mob. This performance, publicized online with no calls for violence and no specified aesthetic or political intent, echoes the deliberate pointlessness of a Wasik flash mob, minus the precise coordination and hipster cache.

The practice of flash mobbing thus expanded in 2009–2010 to include branded entertainment and other corporate appropriations, as well as raucous street performances with the potential for violence. Nonetheless, flash mobbing has proved resilient as a medium for groups of unaffiliated citizens to playfully and peacefully express political grievances to a wide audience, both live and online. A cursory Google search for the term flash mob 2017 turns up videos of flash mobbers advocating for gay rights in St Petersburg, mocking Donald Trump's tax legislation in Texas, and singing at the Women's March in Washington, DC. The decentralized, anonymous planning and short duration of most flash mobs allows these performances to evade police surveillance and state censorship, making them a nimble vehicle for political protest. But more importantly, the very form of the flash mob affirms citizen agency through the enactment of what Kaulingfreks and Warren (2010:218), following Jean-Luc Nancy, call "being in common". Hipster frivolity, commercial appropriation and youth violence may bend the flash mob toward regressive ends, but its efficacy as a citizen medium remains intact.

See also: co-optation; mobile technologies; performance studies and citizen media

Recommended reading

DuComb, C. and J. Benmen (2014) 'Flash Mobs, Violence, and the Turbulent Crowd', *Performance Research* 19(5): 34–40.

Investigates the relationship between flash mobbing and violence in response to a spate of raucous youth flash mobs in Philadelphia and other American cities in the summer of 2009. Builds upon this investigation to argue that flash mobs are not uniquely enabled by digital technology, but instead have an historical precedent in the transmediation of print culture and street performance that began during the American War of Independence (1775–1783).

Gerecke, A. and L. Levin (eds) (2018) *Choreographies of Assembly*, special issue of Canadian Theatre Review 176: 5–99.

Brings together sixteen short articles on flash mobs and other pop-up performances – such as women's marches, #blacklivesmatter protests and the gatherings of the Occupy movement – all considered through a choreographic lens. Contributors address choreography as both a practice of performing scripted moves and as a metaphor for the structures of power and affect that shape contemporary forms of public assembly.

Kaulingfreks, R. and S. Warren (2010) 'SWARM: Flash mobs, mobile clubbing, and the city', *Culture and Organization* 16(3): 211–227.

Explores the phenomenon of mobile clubbing – in which flash mobbers wear headphones and dance to the music of their choice – as an example of citizen self-organization that allows participants to embody an alternative form of urban community. Argues that mobile clubbing can be politically generative insofar as it encourages a diversity of individual experiences within the shared context of a group action.

Muse, J. (2010) 'Flash Mobs and the Diffusion of Audience', *Theater* 40(3): 9–23.

Analyses several early and well-known flash mobs convened by magazine editor Bill Wasik and comedy collective Improv Everywhere in order to demonstrate that flash mob performances are directed toward both live and online audiences. Draws upon this analysis to update prevailing theories of spectatorship in theatre and performance studies by arguing that audiences are both spatially and temporally diffuse in the digital age.

GRAFFITI AND STREET ART

Myrto Tsilimpounidi and Konstantinos Avramidis

From the early inception of subway graffiti in 1970s New York, to the twenty-first-century hype surrounding street art, especially in gentrified urban areas, writings on the wall have superimposed layers of meaning on the urban fabric. Graffiti and street art appear as visual markers of shifting, complex counter-cultural discourses, of power struggles and marginality; they establish a new reality that insists upon being seen and heard. As an art form, then, graffiti and street art are largely connected to and inspired by existing social conditions in urban environments.

Graffiti and street art capture the ways humans inhabit and (in)form their spaces in the ever-growing metropolitan worlds we now live in; as such, they offer a valuable research lens through which to unpack some of the tensions and contradictions of urban life. Ephemeral by nature, graffiti and street art pieces are left to interact with other structures of the city, manifesting the multiplicity of urban realities: as sometimes vibrant and colourful, and at other times faded images in a perpetual process of urban renewal and social metamorphosis. To paraphrase Athenian street artist Bleeps, graffiti and street art can be viewed as a "visual diary on public display" (Tsilimpounidi 2017:143). Each city has its own visual diary; but even in the absence of such a diary – or, to put it more correctly, in the presence of an urban diary with its pages left blank – the study of graffiti and street art provides an entry point into investigating levels of surveillance, repression and securitization of public space in a given urban milieu. Thus, in what follows the focus is not on defining street art but rather on examining what it does. In particular, the entry explores the function of graffiti and street art in a city in a time of crisis, upheaval and rapid change. As such, it conceptualizes graffiti and street art as a kind of utterance and discusses how, in their being-in-the-world, public spaces are transformed into places (de Certeau 1980/1984).

Street artists refuse to accept the prevailing urban aesthetics and instead impose their own aesthetics on city walls. Urban creativity becomes a means of empowerment for the individual's identity; opposed to hegemonic urban design, it favours the creation of alternative, envisioned communities. The performance of tagging visual messages on the wall can be viewed as a form of resistance to the extent that it re-appropriates 'proper' acts of citizenship. Hence, street art performs a redefinition of the relation of the individual to (public/private) space, since by using public space as a surface for interaction and communication, artists create alternative spaces in the city (Avramidis 2014). These performative interventions in public space contribute to the creation of new visual vocabularies; they address viewers – city dwellers and passers-by – with powerful political statements, critiques and calls for action (Ferrell 1993, 1995).

Graffiti and street art practices are visually compelling; they garner media attention, and have achieved nearly mythological status among members of various subcultures, art historians and the public. Depending on one's point of view, they are synonymous with vandalism, marginalized youth and delinquency; or are seen as signs of free expression. What is important in the critical analysis of these phenomena is an awareness of the constant tensions generated by the fact that pieces of street art are often seen simultaneously as social problems, political acts and artistic practices that can be co-opted and depoliticized. These three tenets will be used to structure the entry in order to explain why these urban inscriptions have gained scholarly and artistic significance.

Graffiti and street art as a social problem

In the 1970s, New York City (NYC) was undoubtedly the centre of the contemporary boom in graffiti culture. The city was in decline and facing bankruptcy. This financial crisis – directly associated with the collapse of real estate markets – resulted in the growth of ghettos and accelerated processes of extreme social and spatial segregation. In this context, the dynamic inscriptions of marginalized urban youths – mainly belonging to racialized groups, living in poor neighbourhoods – were seen not as symptoms but as causes of the decay that NYC was experiencing. In other words, graffiti writing became a scapegoat for precisely those social conditions it was protesting against; it was constructed as an urban problem, and deemed responsible for the extensive degradation of the urban environment. Cresswell (1992) focuses on the reactions to graffiti in NYC in the early 1970s in order to examine wider issues of social formation and the meaning of place in urban centres. He analyses state discourses of disorder, while noting the displacement of graffiti from walls to galleries in Manhattan, elaborating the mutual relationship between place and ideology which has since become a central problematic in the scholarly study of graffiti and street art.

Austin (2001) builds on Creswell's work to trace how NYC graffiti was constructed by the media and the municipal authorities as an urban problem, one that created a moral panic that legitimated and paved the way for the infamous and extremely expensive "war on graffiti" (ibid.:5). He shows how the war on graffiti provided the necessary justification for the intense policing and militarization of what were presented as unsafe and unruly areas in NYC. The crucial point here is that the presence of graffiti and street art came to be constructed as justification for further social marginalization and state control of populations living in certain urban areas.

Such developments are not unique to New York in the 1970s; a similar logic has manifested itself in the urban exemplar of the post-2008 European crisis: Athens. After almost a decade of crisis, with severe consequences for the social and urban fabric – including rising homelessness, massive youth unemployment rates, brain drain, collapsing buildings and failure to rehabilitate public monuments – Amalia Zepou, an adviser to the municipality's programme for regulating graffiti and street art, claimed that "when a city collapses, and has been tagged everywhere, we have an obligation to stop it. Once graffiti becomes commissioned art, it is a signal of the beginning of the end of the financial or social crisis that the city has gone through" (2014, cited in Alderman 2014).

The timing of this statement should not be overlooked. It was made just a few days after the opening of the *No Respect* exhibition, the largest and most prestigious graffiti exhibition that Greece has ever seen. Zepou's statement reveals the deep aesthetic disturbance and revolutionary potential of uncommissioned graffiti and street art in urban spaces. When the removal and elimination of illicit graffiti and street art is used as a metaphor for the end of the economic, social and political crisis in the city, the neoliberal preoccupation with surfaces rather than substance becomes clear. Moreover, it is only when illegal street art is replaced with commissioned art that its aesthetic effects on the city can cease to challenge hegemonic narratives. This is not the only example of the insidious ways in which neoliberal ideologies have attempted to penetrate and hijack subversive social actions and subcultures. As with other subversive artistic expressions in the past, such as punk music, the neoliberal strategy used to suppress dissent involves capitalizing on its creativity and commodifying it.

Ferrell's (1993) seminal book *Crimes of Style: Urban graffiti and the politics of criminality* is the only text to provide access to the underground graffiti scene in Denver, Colorado. Based on four years of intensive fieldwork, it raises questions regarding the authority and positionality of the researcher in relation to a practice that is sometimes illegal and dangerous; for instance, during his fieldwork, the author was himself arrested on charges of graffiti vandalism. Most importantly, Ferrell's book offers an insider's perspective into this subculture and the ways in which graffiti writers circumnavigate state control and surveillance. It examines the interrelationship between the political message of graffiti and the writer's attempt to resist the commodification of their art. Pressure from official authorities, Ferrell suggests, transforms writing into an illicit pleasure of staging confrontations on city walls. Commodification, on the other hand, imposes order and control over messages: when graffiti and street art practices become commissioned art, they are automatically transformed into a colourful façade supporting the grand narratives of the city. In this sense, they are no longer disruptive, but instead constitute a sign of normativity. Ferrell even shows how commissioned street art can boost property values in a district, how it is marketed as hip and trendy, and thus how it often contributes to gentrification processes.

These findings pose interdisciplinary research questions concerning whether or not illicit graffiti and street art by definition constitute acts of vandalism, and what happens when exactly the same art piece (or act of vandalism) is transferred into a museum or onto a commissioned wall. Does the graffiti become high art, and if so, what gives meaning to an art piece: the socio-legal parameters of its context of production or the skills of the artists involved. Young (2014) carefully unpacks these questions and the conflicting and paradoxical responses to them, claiming that street art practices are a means of challenging conventional understandings of law, crime, culture and art. Young's approach is multi-layered, as it engages both with the artists' intentions of placing a piece on public display, and with the responses of passers-by to these pieces. Young suggests that street art in public spaces enriches the urban imagination and has the potential to prefigure a different kind of urban reality, which she calls "the public city" (ibid.:3).

Graffiti and street art as political praxis

Urban walls narrate many untold stories. That which remains invisible in the mass media or is not otherwise allowed to be expressed can be tagged on the wall. In other words, street art breaks the conspiracy of silence. It is a decentralized, democratic form with few barriers to access; the substantive control over messages lies with its producers. As Chaffee (1993:3) puts it, "[i]t is a barometer that registers the spectrum of thinking". Chaffee argues that mass communication should not be limited to professionals wielding technology or gadgets; there are "other significant processes and cultural settings involved in the flow of political information, that often, not exclusively, originate from below by grass-roots groups" (ibid.:3–4). Political street art offers a visual critique of the current sociopolitical status quo; or, to put it differently, it makes claims by tagging a space where those relegated to the margins can be represented. Iveson (2007) puts forward a similar argument as he identifies graffiti as a form of public address, a record of popular history through which the public is able to read social, cultural and political struggles in a given milieu. According to Iveson, it is through graffiti writings that we are able to identify the lack, the need and the struggle for social space, communication and representation that is experienced by marginal groups.

Political street art, as a term, encompasses art pieces that offer a conceptual engagement with social issues (Avramidis 2012; Tsilimpounidi 2012; Tsilimpounidi and Walsh 2010). What constitutes political versus apolitical street art cannot be determined solely in relation to aesthetic categories. We thus highlight those practices that have the capacity to cause disturbance, since what makes political street art 'political' is precisely its capacity to disturb the hegemonic conventions of cityscapes. It is this notion of disturbance that is perhaps the most forceful argument for the value(s) inherent to the practice of street art in times of riots, upheavals and crisis – not least because the visual disruption of urban spaces documents, archives and reflects the milieu of growing social uncertainties and spatial inequalities.

For example, walking along Panepistimiou Avenue, one of the most central and iconic streets of Athens and the street that became the epicentre of massive citizen mobilization during the anti-austerity protests in 2011, the passer-by can read some of the inscriptions on the Athenian fabric: 'Forever A Loan', 'Merry Crisis-Mas and a Happy New Fear', and 'Life, Not Survival'. These slogans refer to pressing social issues inspired by the political and economic climate of crisis and austerity in which they were produced. Street art can thus be seen to capture the need for self-expression in a changing environment; but, as a voice coming from the margins of social production, it also forces passers-by to witness something they would not otherwise ordinarily confront. Street artists do not need authorization from the dominant culture. On the contrary, what makes their art effective is the prior knowledge that it exists without permission. Since passers-by know it is illegal, they are aware there is no sanctioning of the message by official bodies, corporations or the state itself: the message comes directly from the margins.

Street art has been viewed as a barometer of freedom. Thus, it is said to have flowered on the Western side of the Berlin Wall, but was strictly suppressed in the East (Cresswell 1996; Iveson 2010:131). More recent examples that highlight the relationship between state oppression and street art can be found in Egypt, Tunisia and Libya since the beginning of the Arab uprisings (Mulholland 2011; Steavenson 2011; Zoo Project 2011). In particular, following the overthrow of the Mubarak regime, politically engaged street art proliferated in Egypt, transforming the fixed landscape of Egyptian cities into platforms for negotiation and dialogue. At the same time, however, scholars such as Abaza (2017) have argued that the overconsumption of revolutionary graffiti can strip it of its political potential, turning it into an exotic commodity to please the eyes of Western viewers.

Co-optation of graffiti and street art as artistic practice

To the extent that graffiti and street art practices have been co-opted by the neoliberal order, they are increasingly depoliticized and feature prominently in mainstream galleries and museums. Even though graffiti is by nature antithetical to orthodoxies, it risks becoming one when it is presented in a gallery, designated by an institution as art. This tension has been particularly evident in Athens since 2009.

The *No Respect* exhibition is a paradigmatic case of the institutionalization of graffiti. It was organized by the Onassis Cultural Centre in April 2014 and took place in the basement of its imposing headquarters in Athens. The exhibition showcased forty works executed on panels and vehicles that were brought inside and freely arranged to convey a street atmosphere. On each allocated space individual artists, or groups of artists, executed their pieces. The outcome was a series of pieces sitting in perfect order, the one next to the other, each bearing the signature style of its producer. The aim of *No Respect* was to attract to the Onassis Cultural Centre a young 'alternative' audience that is sympathetic to this kind of illicit urban imagery. In order to achieve this goal, the exhibition sought to draw a connection with the streets, capturing the dynamism of the graffiti phenomenon as it manifests on Athenian walls by replicating its aesthetics indoors (Karra 2014). In so doing, however, the exhibition undermined graffiti's critical potency, transforming it into an aesthetic object and, even, into an empty gesture. In *No Respect*, graffiti was detached from its actual political and historical context – that of the Athenian streets – and inserted into another context – that of the gallery – governed by different rules of viewing and ways of being: in other words, it was de-situated. The dissociation of graffiti from its original contexts meant that it became something different: visually arresting, but devoid of political meaning.

Curators who work for well-funded galleries and museums in various parts of the world are always looking for new trends and approaching acclaimed street artists. The bigger the subcultural capital of a street artist, the more profitable it is for the curators to institutionalize them. One example is *pixadores*, São Paolo's distinctive street writers. *Pixação* is a direct act of protest against inequality in Brazilian urban centres and constitutes one of the most skilful and dangerous forms of wall writing. Considerable tensions have emerged between the *pixadores* who 'sold out' and those still fighting on a daily basis to create their pieces on São Paolo's streets (Lamazares 2017). Another example of resistance against the co-optation and institutionalization of street art is the renowned street artist Blu. In March 2016, Blu covered all his artworks in the city of Bologna with grey paint; these works were created during twenty years of uncommissioned and persistent wall painting. Blu has become internationally well known for his political art pieces in Berlin, Los Angeles, Rome and São Paolo, but his earlier and most politicized pieces were still found in his hometown of Bologna. The erasure of the work was an act of protest against the exhibition *Street Art: Banksy & Co*, organized by Fondazione Carisbo and funded by Bologna's main bank foundation (Vimercati 2016). In this frenzy of capitalist co-optation and re-appropriation of graffiti and street art, Blu proposed erasure as an act of resistance. By erasing some of his most famous pieces, he signalled that art cannot be measured by the neoliberal standards of an artistic career, but rather by the social and political dynamics it instigates in grassroots spaces, meetings and gatherings.

Graffiti and street art constitute multi-layered and polyvalent performances produced by urban dwellers who want to communicate their messages and their own aesthetics in public spaces. Taking inspiration from Blu's act of erasure, we might also suggest that uncommissioned, politicized graffiti and street art creates a mirror through which citizens can recognize their own concerns. What makes it remarkable is the personal and at the same time deeply collective voice that emerges from each piece. These voices transmit a common message: their words act as living bridges between street artists and the urban populace in an attempt to disrupt and disturb the hegemonic monopoly on what is allowed to be visible in public space.

See also: anthropology and citizen media; co-optation; media

Recommended reading

Avramidis, K. and M. Tsilimpounidi (eds) (2017) *Graffiti and Street Art: Reading, writing and representing the city*, London: Routledge.

This edited volume focuses on the relationship between graffiti, street art and urban space. Rather than examining what the physical texts say, the essays in this collection place the emphasis on exploring what graffiti and street art do in different contexts. It thus opens up questions on the nature of public space and the right to the city, asking in particular who holds these rights and in relation to what city.

Iveson, K. (ed.) (2010) *Graffiti, Street Art and the City*. Special issue of *City: Analysis of Urban Trends, Culture, Theory, Policy, Action* 14(1–2): 25–134.

This special issue offers analyses related to graffiti in different urban settings. It focuses primarily on the tension between the production and eradication of this imagery in urban spaces and the lessons that derive from this struggle, with special emphasis on the concept of the right to the city. Issues related to contemporary processes of urbanization – such as control, commodification and surveillance – are raised in the articles.

Ross, J. I. (ed.) (2016) *Routledge Handbook of Graffiti and Street Art*, London: Routledge.

The *Routledge Handbook of Graffiti and Street Art* presents a polyvocal critical analysis of these urban phenomena. Its contributors highlight graffiti and street art simultaneously as a sign of social tensions, as a political act in public space and as an art form produced by highly skilled practitioners.

HACKING AND HACKTIVISM

Julia Rone

Hacking – usually defined as the process of gaining unauthorized access into a computer or network – is a practice that redetermines information technologies and infrastructures and repurposes them through playful exploration, craft and modification (Jordan 2016). In his classic work on early computer researchers at MIT, Levy (1984) defines the central tenets of hacker ethics as sharing, openness, decentralization, free access to computers and world improvement. Coleman (2013:17) suggests, however, that having taken into account all the varieties of hacking and hackers' diverse attitudes to work, money and networks, the core of common values can be reduced to only three – freedom, privacy and access – combined with an ambiguous relationship to legality. For Coleman (2013:13), "[f]iercely pragmatic and utilitarian" hackers are also "fiercely poetic and repeatedly affirm the artistic elements of their work". Hackers relate to technology with joy and passion in search of inventive solutions and clever hacks (Himanen et al. 2001:4–7). In this respect, they embody a Romantic vision of expressive individualism that produces not only software as a technical artefact but also particular social relations and institutions, such as the Free and Open Source Software (FOSS) community (Coleman 2013:14). As Torvalds notes,

> [a] "hacker" is a person who has gone past using the computer for survival ("I bring home the bread by programming") to the next two stages. He (or, in theory but all too seldom in practice, she) uses the computer for his social ties – e-mail and the Net are great ways to have a community. But to the hacker a computer is also entertainment.
>
> *Himanen et al. 2001:xvii*

That said, most of the time hacking involves mundane technical work and dealing with "constantly malfunctioning technology" (Coleman 2013:11).

An important characteristic of the hacking community, as suggested in Torvalds' definition above, is its strongly-gendered nature. Hackers have been predominantly males or female-performing males (Coleman 2014:175), often maintaining male-only stereotypes (Tanczer 2015) and occasionally blurring the lines between misogyny and non-gender-related trolling (Taylor 2003). Indeed, pervasive metaphors of the web as the Wild West and as an electronic frontier appeal to notions of transgression and frontier masculinity (Adam 2005). Nevertheless, media activists have increasingly challenged gendered and raced technologies with varying success and resonances with wider societal struggles (Dunbar-Hester 2010, 2017). Feminist thinking, moreover, has enriched notions of hacking by coupling them with the concept of "making" as both a method and a framework with which to "introduce new kinds of expertise, such as craft and care, into conversations of information technology" (SSL Nagbot 2016).

The multiple tensions within the hacking community are often expressed in binary distinctions between hackers and crackers (the latter being those engaged in unethical criminal hacking – Perin 2009), or through the use of finer categories such as black hat, white hat, grey hat, red hat and state-sponsored hackers, script kiddies (also known as skiddies), hacktivists, whistleblowers and others (Aukta 2018). The proliferation of such internal distinctions and sects has led some authors to abandon hacking as a monolithic analytical concept and to focus instead on "genealogies of hacking" (Jordan 2016:2–3) or "genres of hacking" (Coleman 2013:18), reflecting the historical development of different attitudes towards secrecy, computer security, criminality and political engagement. In addition, there have been calls to address the multiple origins of hacking and to acknowledge the contributions of groups such as the phone phreaks, who have tapped into the phone system in the US (Coleman 2014), and the hacker communities operating outside of the US such as the Chaos Computer Club and XS4ALL, based in Germany and the Netherlands respectively (Jordan 2016). Such sensitivity towards the multiple origins of this phenomenon is even more important considering the different ways in which hacking has been conceptualized in relation to politics: for example, while in the US hackers have tended to focus on the politics of technology and have attempted to provide an internal critique of liberalism (Coleman 2013), in countries such as Greece and Spain, hacking has been actively incorporated into the techno-imaginary of the leftist-anarchist scene (Treré et al. 2017). Regardless of its particularities within each national context, however, hacking has been marked by a series of shared trends that often run parallel to each other: these can be defined as commercialization, criminalization and politicization. The following sections explore each of these trends in turn.

Commercialization

Since the early days of this subculture, hacking has been persistently professionalized and integrated into the process of economic production. In the mid-1980s and in reaction against the increasing commercialization of system software, Stallman (1987) wrote the GNU Manifesto, proposing an alternative free operating system. Stallman's (2002) free software movement protected the core hacking values of sharing and free access by defining four freedoms, namely, the freedom to run a program, the freedom to study and change a program, the freedom to redistribute a program, and finally the freedom to redistribute modifications of the program. Free software was thus defined as free in the sense of free speech, not free beer (Free Software Foundation 2018); the emphasis is on an absence of use restrictions, not on its cost to the user. While this movement successfully pushed back against corporate encroachments upon cyberspace, in the late 1990s some of its members split in order to create the Open Source Initiative, avoiding the political and ideological connotations of the word *free* and accentuating instead, in a more business-friendly way, the superior characteristics of open modes of production (Raymond 1999).

While the free software movement has continued to exist and attract new supporters, the appropriation of hacker ethics by Silicon Valley businesses in order to further their own commercial goals has been widespread. This trend has culminated in ever more popular hackathons, where participants write code and build apps in intense events often promoted as recruiting opportunities but which are above all a promotion tool for particular technology brands. The promise of technological innovation and the joyfulness and creativity of hacking are intertwined at hackathons with a desire for self-promotion that often leads to self-exploitation (Zukin and Papadantonakis 2017). The creative disruption that hackathons aim to promote in fact contributes to the legitimation of a mode of labour that is short-term and highly insecure (ibid.). Additionally, in contexts such as India, for example, hackathons have been shown to rehearse a particular type of entrepreneurial citizenship that favours quick and unproblematic collaboration with socially similar actors (for instance, other members of the middle classes) instead of more long-term democratic engagement and attempts at coalition making between diverse actors (Irani 2015).

At the same time, hacking practices have become increasingly popularized and integrated in the culture of consumption. As users have developed ways of modifying non-Apple computers to make them compatible with Apple software, hacking-related cultural references and discourses have grown in terms of their visibility among new segments of the population, including not only software experts and computer geeks, but also amateurs, laypersons and non-experts (Magaudda 2012). Since the early 2010s, hacking has entered the language of a variety of fields, with experts offering life-hacks, growth-hacks and even happiness-hacks (Yagoda 2014). Thus, the original meaning of *hack* has clearly been expanded and the term is now used widely in management and lifestyle contexts in a move that represents a significant shift away from its earlier technology-related connotations.

Criminalization

Another important trend in the history of hacking has been the turn to criminality among certain hackers who have started employing their skills for illegal and unethical purposes. The criminalization of hacking reached its first peak in the 1990s, often dubbed the "golden age of cracking" (Jordan 2016:7). The now ubiquitous use of digital technologies in everyday life, as well as in more specialized fields such as robotics, nanotechnology and artificial intelligence, provides even more opportunities for hackers and the invention of new types of 'future crimes' (Goodman 2015). As a result, both hackers themselves and researchers have attempted to distinguish between ethical and unethical hacking practices: this has generally been achieved by separating hackers from crackers, or by contrasting white hat hackers, who specialize in testing methodologies to ensure the security of particular information systems, with black hat hackers, who break into computer systems for personal or financial gain, blackmail or simply to wreak havoc (Moore 2010).

Examples of unethical hacking include practices such as carding – stealing credit card details (Glenny 2011) – but also Distributed Denial of Service Attacks (DDoS), ransom demands and identity theft (Alexander 2013). A key set of tools facilitating such criminal activities are data encryption and anonymity protection software packages such as Tor, originally promoted as a means of providing "privacy for the weak and transparency for the powerful" (Assange et al. 2012), but often used by black hat hackers, along with drug dealers and child pornography distributers on the Dark Net (Bartlett 2014). Thus, the methods and tools of hacking cannot be categorically defined as either liberating and progressive or criminal. There is a fundamental ambiguity in the hacker's craft that makes the question of goals crucial in determining the morality of any instance of hacking.

Finally, the criminal dimensions of hacking should be situated in their broader societal contexts, as cybercrime is not randomly distributed around the world but emerges from particular localities and social groups (Lusthaus and Varese 2017). Rather than operating in some ethereal virtual realm, disentangled from reality, both white hat and black hat hackers operate from specific offline locations and in concrete national and class contexts that need to be taken into account in order to understand the turn to criminality. As discussed in the next section, these local contexts are also important when considering the more political dimensions of hacking.

Politicization

When hacking is combined with grassroots political protest and activism, it is typically referred to as hacktivism (Jordan and Taylor 2004). One of the most prominent hacktivist groups to have emerged online has been Anonymous, best known for its use of a Guy Fawkes mask as its symbol. This collective had its rather unconventional origins on the pages of the image board 4Chan, where anonymous posting, trolling, humorous deviance and doing things for the *lulz* (a corruption of the phrase *laughing out loud*) were the norm (Coleman 2014). It was only in 2008, after a couple of years of existence, that Anonymous became politicized through its organization of a mass action against the Church of Scientology. Following Project Chanology, as the mass action became known, Anonymous continued engaging in online trolling but also started embracing political causes (such as siding with protesters in Tunisia, for example), employing a variety of legal and mainly illegal techniques that included DDoS attacks, doxing (researching and broadcasting private and identifiable information) and providing technical assistance to on-the-ground activists (ibid.). The cyborg-activism of Anonymous has since exploited and reconfigured tensions between equality and hierarchy, reason and emotion, nihilism and idealism (Asenbaum 2017).

Similar to other collective names such as Ned Lud, the legendary leader of the Luddite movement in Britain in the late eighteenth and nineteenth century, the 'improper name' of Anonymous has become a terrain for multiple contestations over the use of its symbolic power (Deseriis 2015). Far from being a homogeneous collective composed only of white, libertarian Anglo-Saxon youths, Anonymous has spread across the globe, attracting enthusiasts from diverse countries, backgrounds and levels of technical expertise (Coleman 2014). While one of the basic features of Anonymous has been the collective's inclusivity and the claim that 'everyone can be Anonymous', in particular national contexts the group has also been associated with nationalistic attitudes and exclusionary discourse (Rone 2014).

Hacktivism has a long history that predates the appearance of Anonymous, however. The very term hacktivism was invented in the 1990s by a member of the hackers and do-it-yourself media group The Cult of the Dead Cow that formed the offshoots Ninja Strike Force and Hacktivismo, the latter seeking to apply to the Internet the Universal Declaration of Human Rights and the International Covenant on Civil and Political Rights (cDa Communications and Hacktivismo 2001; Shantz and Tomblin 2014). Closely related to the practices of these groups has been the notion of electronic civil disobedience invented by the Critical Art Ensemble to describe the performance of non-violent disruptive protest through technical means, including DDoS attacks and virtual sit-ins (Critical Art Ensemble 1994). Also in the 1990s, the Mexico-based Zapatista movement and alterglobalization activists started using technology creatively in order to achieve political impact, forging new and important blends of hacking and social movement mobilization (Jordan and Taylor 2004).

All in all, this was a period of unprecedented growth for independent media (Indymedia being a prominent example) and radical servers, understood as "anti-capitalist, anti-hierarchy, autonomous revolutionary collectives" that provided free or mutual aid services to radical and grassroots activists (Riseup 1999). If this first period of hacktivism can be described as cyber-autonomism, the 2011–2013 movement of the squares after the financial crisis was marked rather by cyber-populism, treating the Internet as a popular space, populated by ordinary citizens who feel comfortable using proprietary platforms such as Twitter and Facebook for citizen activism and protest (Gerbaudo 2017a, 2017b). Hacktivist collectives such as Anonymous engaging in DDoS attacks, defacing and digital disobedience entered in complex cooperative and occasionally conflictual relations with these mass protest movements using commercial platforms (Rone 2018). At the same time, new data activist projects and collectives appeared that focused not so much on hacking in the sense of unauthorized access to computers but on hacking politics: they sought to create new types of collaborative free and open source platforms for activism and whistleblower websites with the aim of opening up governments' actions for public scrutiny.

Another highly visible example of hacktivism has been the Cypherpunk movement (Levy 2002), which from the 1980s onwards sought to wield cryptography as a weapon of freedom, autonomy and privacy that would "fundamentally and inexorably reshape social, economic, and political power structures" (Narayanan 2013:76). The cryptographic quest to ensure privacy for citizens and transparency for governments culminated in the founding of WikiLeaks by cypherpunk Julian Assange (Assange et al. 2012). Since its launch in 2006, WikiLeaks has provided a secure way for whistleblowers such as Chelsea Manning to share sensitive government information and has worked in collaboration with established media such as *The Guardian*, *The New York Times* and *Der Spiegel* in order to make this information accessible to the general public (WikiLeaks 2015). The site published leaks dealing with issues such as government corruption in

Peru, civilian casualties in Afghanistan, torture in Guantanamo and the internal machinations of the US Democratic Party.

Needless to say, the disruptive activities of WikiLeaks attracted the attention of law enforcement authorities. Already back in 2010, the US government launched a criminal investigation against Assange and the Swedish government issued an international arrest warrant because of allegations he had engaged in sexual assault. Assange sought asylum and spent seven years in the Ecuadorian Embassy in London until, in 2019, his asylum status was withdrawn and his indictment unsealed by Donald Trump's Justice Department. Assange was accused among other things of hacking because he tried to help whistleblower Chelsea Manning to cover her tracks (Greenwald and Lee 2019). However, leading journalists have noted that this so-called hacking was in fact a common journalistic practice of protecting sources. Thus, the US government's indictment of Assange "poses grave threats to press freedom" (Greenwald and Lee 2019). Meanwhile, a large number of defections, insufficient funding and negative press coverage, especially after the leaks concerning the Democratic Party, have meant that the future of the entire WikiLeaks project is under serious threat (Lynch 2019).

Other hacktivist projects have been less controversial and have increasingly tried to combine the use of tech expertise with street protest, legal action and even art in order to address broader economic and political issues. The Spanish data activist collective X-Net, for example, engaged in advocacy for free culture and net neutrality but also took active part in the Indignados movement and established a secure anonymous mailbox for corruption-related leaks. After receiving a leak with the emails of one of Spain's top bankers, X-Net not only started a court case, financed by crowdfunding, but also staged the data-based theatre play 'Become a Banker', with which they toured the country (Rone 2017; X-Net 2018).

Finally, in light of revelations regarding both autocratic and democratic governments' programmes for mass surveillance (Bauman et al. 2014; Morozov 2011), many hacktivists have started developing ways to empower and protect protesters and secure their data. In a world in which state-sponsored hacking, espionage and surveillance are the practice rather than an exception (Woolley and Howard 2017; Zetter 2015), new projects such as Security Without Borders (Guarnieri 2017) and Security in a Box (Tactical Tech 2018) aim to offer secure technologies to citizen activists and journalists. As both governments and corporations increasingly store and analyse big data, new social practices also emerge that adopt a critical approach to data collection and exploitation. Drawing on the heritage of hacking and the Free Software Movement, data activists find technical fixes to resist the threats to civil and human rights caused by mass surveillance (reactive data activism), but also use the possibilities that big data offers for civic engagement, advocacy and campaigning (proactive data activism) (Milan and van der Velden 2016). Examples such as iOS 'jail-breaking', that is, the removal of software restrictions imposed by Apple on its operating systems, allow users to customize their devices, circumnavigate top-down modalities of information protection, improve privacy control and sometimes even gain additional insights into data flows that would otherwise remain opaque (Cooke 2018; Dimitrov and Chow 2013).

As commercialization, criminalization and politicization of hacking practices have unfolded since the 1990s, hacking has moved from the fringes to the mainstream of public attention. The control and use of data will be crucial in forthcoming battles for freedom of expression, recognition and empowerment. Thus, hacking and hacktivism, understood as part of broader social and political trends, are here to stay. Rather than remaining simply a subculture, hacking has become a vital skill for securing free citizen participation in politics, culture and society.

See also: anthropology and citizen media; big data; civil disobedience; co-optation; sousveillance; surveillance

Recommended reading

Coleman, E. G. (2013) *Coding Freedom: The ethics and aesthetics of hacking*, Princeton, NJ: Princeton University Press.

Provides a detailed discussion of the history, ethics, aesthetics and politics of hacking, with a special focus on the Free and Open Source Software community and the particular strains of liberalism that have informed its practice.

Jordan, T. (2016) 'A Genealogy of Hacking', *Convergence: The International Journal of Research into New Technologies* 23(5): 528–544.

Outlines four key historical phases in the development of hacking, beginning with the activities of phone phreaks and other do-it-yourself enthusiasts, and culminating in the rise of state-sponsored hacking, maker labs, hackathons and the wider cultural diffusion of hacking as a practice.

Lusthaus, J. and F. Varese (2017) 'Offline and Local: The hidden face of cybercrime', *Policing: A Journal of Policy and Practice*, pax042: 1–11.

Shows that cybercrime is not only an anonymous activity that exists in cyberspace but there are significant offline, human and contextual elements to take into account.

Stallman, R. (2002) *Free Software, Free Society: Selected essays of Richard Stallman*, Boston, MA: Free Software Society.

A collection of essays that define free software and trace the origins and philosophy of the Free Software Foundation. Stallman offers a timely critique of patents and copyright regulation, and defends the importance of four essential freedoms: the ability to run a programme as one wishes, to study how it works, to modify it and to redistribute it.

Zukin, S. and M. Papadantonakis (2017) 'Hackathons as Co-optation Ritual: Socializing workers and institutionalizing innovation in the "new" economy', in A. L. Kalleberg and S. P. Vallas (eds) Precarious Work, Bingley, UK: Emerald Publishing, 157–181.

Explores hackathons as a powerful strategy used to legitimize precarious labour. The authors argue that while participants benefit from the chance to network and learn new skills, corporate sponsors frequently organize such events as a means of enhancing their own reputation, outsourcing work and crowdsourcing innovation.

HIP-HOP

Christopher Vito

Hip-hop is a cultural movement whose origins can be traced to a tiny seven-mile circle in the South Bronx, where it grew out of the vocal and musical expression of culture by African American, Afro Caribbean and Latino men and women in the 1970s (Chang 2005). The formation of alternative local identities, which were represented through rapping (a variant of hip-hop that features rhyming speech delivered over the beat of music), DJing, breakdancing and graffiti, created a form of expression that reflected these men's and women's economic and social hardships as the throwaways of America's capitalism (Au 2005; Forman 2000; Morgan and Bennett 2011; Pough 2004; Stapleton 1998). Early hip-hoppers also offered a form of social critique against racism that supported the goals of the civil rights movement (Alridge 2005).

Since its inception, hip-hop has faced a concomitant struggle between maintaining its spontaneity and locality in New York and falling prey to commodification and co-optation (Pough 2004). On the one hand, hip-hop culture has been able to address numerous social issues, ranging from the prison industrial complex to political movements involving the Black Panthers and the Young Lords. On the other hand, the rise of new technologies, including cassettes, CD players and burners, as well as the Internet and social media, has allowed for a much easier flow of information and music production that has spurred a rise in commodification (Dennis 2006; Harrison 2006). Commodification, in turn, has helped hip-hop move from the margins to the mainstream and become incorporated into a music industry based on large conglomerates that homogenize music production, the distribution process and the means of consumption (Rose 2008).

Hip-hop artists have consequently utilized technological advancements and concordant shifts in the marketplace to create avenues for success despite competition from major players. Scholars such as Watkins (2005) have attempted to understand this complex, multifaceted, politically conflicting and much debated history of hip-hop, focusing especially on accounts of its resistance to major record labels and the current wave of independent artists adopting a DIY ethic that allows them a larger stake in music production, distribution and marketing.

A history of commodification and resistance

Hip-hop's history is characterized by cyclical patterns of commodification and resistance as the culture spread and evolved. Perkins (1996) argues that before 1979, hip-hop was a key component of the flourishing underground culture in the Bronx and upper Manhattan. Claimed by griots, ciphers and dance battles as their own, it became a breeding ground for discussion and debate on salient social issues. After

1979, however, its popularity began to draw the attention of the mainstream music industry, which set out to target it as a potential commodity (Oware 2018). This pattern was not new at the time nor specific to hip-hop. As George (2005) explains, independent black music – as evident in the case of both jazz and rock'n'roll – has traditionally represented an untapped space for growth for corporate labels. Similarly, Myer and Kleck (2007) and Johnson (2008) suggest that popular music has historically been driven by independent music. Indie labels are the key to finding what audiences want to hear, and are thought of as a means of feeling the pulse of the public. Large corporations are able to use sophisticated modes of production and distribution to take advantage of consumer tastes in order to make a profit, ultimately mimicking popular independent music until music sales decline, and then moving on to the next proven commodity in the indie scene to maintain record sales.

In the case of hip-hop, the 1990s became an important decade marked by a distinct shift in the culture. From 1979 to the mid-1990s, in what is known as the golden era, there was lyrical mastery, innovation in beat production, diversity in style and content, and a subsequent meteoric rise of hip-hop music in the mainstream media (Johnson 2008). The move from the underground into the mainstream reached a peak in the post-golden era, in the late 1990s, where the corporatization of hip-hop, or the full immersion of corporations in the creation of the music, undermined the diversity and cultural influences of the underground (Myer and Kleck 2007). By 1998, eighty-one million hip-hop records had been sold but 70 per cent of purchases were made by whites in mainstream culture (Rose 2008).

Myer and Kleck (2007) report that since hip-hop's corporatization there has been a decrease in one-hit wonders. As corporations invested more and more money in artists and albums, they moved to securing their investments by creating monotony and standardization in music through a tested model that maximized the potential to sell records and gain profits. The model now adopted by companies used indie labels and artists to test the market, and then with little risk these labels and artists could be bought out by major labels to boost record sales. For example, Bad Boy Records was bought out by Sony-BMG once they had obtained enough capital and parity to contend for record sales with larger corporations. In addition, major companies bought out radio stations and flooded the market with musicians they chose through the 'pay to play' system. The average cost to get a song played on the radio rose to USD 1,000 per song per station (Ball 2009). Thus, independent artists have historically had a much more difficult time getting spins on the radio, a situation that pushes them towards signing with major labels to obtain financial support.

Rose (2008) argues that corporatization has had detrimental effects on the hip-hop community, explaining that copyright ownership begins with the recording contract that transfers ownership of rights to the studio to sell, promote and benefit from copyrighted materials. This means that the power of ownership becomes more concentrated in the hands of the few, typically represented by elite white men. White males in the music industry's three major record labels are overrepresented in managerial and ownership positions (Myer and Kleck 2007); the infiltration and subsequent ownership of major hip-hop labels by whites date back to the 1970s, with white executives such as Malcolm McLaren, Rick Rubin and Tom Silverman being good examples (Chang 2005; George 2005). Furthermore, this very small segment of society occupies multiple positions of power in the music industry, a situation that appears more serious when we consider that three companies either own or distribute more than 85 per cent of the music circulating globally (Rose 2008).

Against this backdrop, and since hip-hop's rise in 1979, two dominant themes have emerged in the literature (Lena 2006; Lena and Peterson 2008; Lena 2013; Tickner 2008): conscious rap and gangster rap. Conscious rap was associated with the representation of the experiences of a marginalized, subordinated population and was geared towards building a sense of activism among its constituents. In particular, these musicians emphasized the local environment and expressed hostility towards corporate music production. Gangster rap portrayed the same representations of the ghetto but upheld values of consumerism and patriarchy. Its musicians also blended street credibility with commercial success in the form of a hustler protagonist. While these two themes are not mutually exclusive, gangster rap became mostly associated with mainstream hip-hop while conscious rap was predominantly produced underground. Yet the mainstream can encompass both gangster rap and conscious rap, and underground and independent hip-hop can express themes of gangster rap. In this respect, Harrison (2006) clarifies what is meant by underground hip-hop, stating that while it encompasses a wide variation of topics, it thrives in the hip-hop community closer to the end of the popularity spectrum, where audiences consist largely of friends, family and other associates. As underground artists gain more notoriety, their fan base extends beyond their circle and they gain access to new areas, such as record label formations and documentation of album sales. Artists continue to face a crossroads in deciding to either remain underground, signing with or creating an indie label, or signing with a large mainstream corporation.

During the post-golden era, the dominant model used by major corporations was gangster rap. Gangster rap emphasized the nihilistic, macho and violent side of ghetto life, despite the fact that its originating impulse was its disgust with the hypocrisy of mainstream culture. And indeed, Wells-Wilbon et al.'s (2010) analysis of the life of one of gangster rap's icons, American rapper Tupac Shakur, demonstrates this ambivalence as it highlights the importance of his legacy for youth popular culture, which was characterized by both radical critiques of mainstream culture and the reproduction of dominant cultural depictions of gangsters in urban neighbourhoods. Ultimately, this ambivalence means that hip-hop, especially gangster rap, mirrors the values, violence and hypocrisy of modern culture and represents some "ugly truths about everything society is and is not" (Taylor and Taylor 2007:213).

Unlike gangster rap, conscious rap has traditionally thrived in the underground and independent scene. In the post-golden era, underground and independent hip-hop act as a tool for legitimizing the authenticity of mainstream hip-hop when some artists sign with major labels; at the same time, many other musicians challenge it by distancing themselves from mainstream commercialization (Maher 2005; Rose 2008). Morgan and Bennett (2011) point out that commercial hip-hop still only represents a fraction of artistic production and performance. In the alternative underground space, there is more room for conscious rap to critically challenge the conventional norms of traditional, mainstream hip-hop and distinguish itself from mainstream culture. As Harrison (2006) points out, the underground tends to be united in embracing the progressive politics of subcultural inclusion and resisting co-optation by large corporations.

Ball (2009) discusses the importance of other material conditions salient to music creation and production. In order to understand the politics of hip-hop culture, he argues, scholars must examine how underground and independent hip-hop culture navigates a way out of corporatization by major companies. For example, resistance to major record sales might come from mixtapes that are distributed through various channels outside corporate outlets. This includes handing out music in person on the streets, circulating it via airtime at clubs and independent radio stations, or through online websites and social networks. Ball looks specifically at FreeMix Radio, which is a freely distributed monthly radio programme distributed on compact disc and acting as an alternative means of expression, free from the filters of mainstream media. Harkness (2012) similarly explores the Chicago underground hip-hop scene, focusing on how it removes itself from corporate infrastructure through self-production. Maher (2005) discusses the importance of self-production with indie rappers Dead Prez, who speak freely about their ghetto-centric and Afro-centric experiences: their experiences in urban low-income neighbourhoods and as African Americans.

Hip-hop ultimately negotiates a complex cultural and political landscape in its attempt to create a sense of identity (Ogbar 2007; Adams and Fuller 2006; Vito 2015). It is not monolithic and unitary, but rather diverse and highly dependent upon historical context and the ways in which the hip-hop community experiences and understands life at any point in time. This is the most salient perspective on the dichotomy between the mainstream and the independent. It is thus necessary to study mainstream as well as independent hip-hop culture and to acknowledge hip-hop as a complex and contested site of both resistance and domination (Dyson 2010; Hill 2009).

Hip-hop spreads across the globe

Scholars such as Harkness (2012) and Perry (2004) argue that hip-hop is a black form of culture and music. Here, race is treated as a fixed category that acts as the ultimate arbiter of authenticity; this presents a problem for scholars who also recognize the appropriation of hip-hop by other racial and ethnic groups in the United States underground (Kubrin 2005). Most notably, Filipino youths in the Bay area engage in DJing and dancing (Wang 2014). Similarly, this viewpoint is problematic in addressing the growing body of literature that focuses on cultural appropriation via the diaspora of hip-hop culture, whose members introduce it into new communities globally while simultaneously addressing pertinent local issues (Bennett 1999; Mitchell 2000).

Androutsopoulos and Scholz (2003) suggest that hip-hop's move from the margins to the mainstream means that it has become a site of cultural appropriation, extending the reach of rap music to new social and linguistic environments. Dennis' (2006) ethnographic research involving artists such as Choc Quib Town and Voodoo SoulJas shows how Afro-Colombians have appropriated hip-hop culture to rework traditional concepts of race and ethnicity. In particular, he demonstrates how they challenge the superiority of mestizos and bring questions of racism to the epicentre of culture. Omoniyi's (2006) work on Nigerian hip-hop song lyrics reveals a similar process in which Nigerian musicians produce hip-hop culture with significant variants that include phonological variation, codeswitching, cross-referencing, nicknaming, colloquialisms and reinterpretation. Lin (2006) shows that independent Hong Kong hip-hop artists challenge the capitalist practices of the pop culture music industry and produce music within niche spaces for both cultural survival and innovative cultural production. Other works have focused on European (Hesmondhalgh and Melville 2002; Mitchell 2000) and Australian hip-hop (Maxwell 1994).

Global hip-hop is best explained through the concept of glocalization – the simultaneous interaction between the global and local dynamics that takes a double-helix form (Smith 1997; Tickner 2008). This is important because local groups can appropriate hip-hop culture to address a wide range of issues, ranging from the individual and local to the global (Baker 2005; Bennett 1999). Global hip-hop is inherently translocal because it represents complex cultural dialogues between local innovations in diverse hip-hop forms, including transcultural interactions outside the United States and interactions between the United States and local spaces (Morgan and Bennett 2011).

Forman and Neal (2004) point out that there has been a persistent theme of authenticity in hip-hop culture. Within the United States, this is a complex issue as traditionally those who are considered authentic include men of colour while those considered fake tend to be white, female and/or belonging to the upper class. Yet bearing in mind the ongoing processes of cultural appropriation discussed here, Harkness (2012) shows that authenticity is highly malleable. For example, in the United States a poor white male may be perceived as more authentic than a Japanese middle-class male, but still less authentic than any black male artist. Outside the United States, the idea of cultural appropriation again brings into question the notion of authenticity (Hesmondhalgh and Melville 2002; Mitchell 1996). As Pennycook (2007) and Tickner (2008) point out, the relocation of hip-hop to other contexts contradicts the traditional theme of 'keeping it real' by presenting it in a form that is adapted to local contexts, languages, cultures and understandings of the real. Nonetheless, while glocalization has largely destabilized the idea of authenticity in relation to race, women of colour continue to be marginalized and to occupy a precarious position in hip-hop culture despite their increased representation (Dennis 2006).

The cycle of the indies

The emergence of self-owned record labels in the post-golden era has been explained as a response to the spread of bogus 360 degree contracts, management conflicts and poor economic relations between musicians and labels (Forman 2000), resulting in hip-hop artists becoming more familiar with the production and management side of the industry and acting as entrepreneurs in the rap game.

Large record companies have traditionally made money from selling physical copies of music in the form of compact discs. After hip-hop's boom in the late 1990s, the music industry took a sharp turn towards the digital age in the early 2000s (Karubian 2009), exponentially increasing the number of digital downloads and enabling direct distribution via online websites. First, online programs such as Napster and Kazaa created illegal P2P (person to person) music file sharing, which promoted easier access to digital content. Easy access to Torrent websites such as Torrentz and The Pirate Bay further supported illegal downloading of music. Second, Apple's iTunes provided access to legal digital downloads. Third, streaming music formats such as Apple Music and Spotify now allow users to pay a monthly fee to access a large library of music. Finally, direct distribution models have been used by artists to sell their music via online websites. Companies such as Tunecore (Byrd 2014) and CD Baby have eliminated the need for a record label by selling music online direct to the consumer for a small percentage of the profits. Many corporations thus faced loss of profits due to the decline in CD sales and the proliferation of online distribution (Ostrove 2014). By 2014, the RIAA was reporting that music sales had declined by approximately 65 per cent since their high point in 1999 (Resnikoff 2014). Nevertheless, companies soon adapted to changes in the industry by reasserting control over online sales, for instance by adopting 360 degree contracts. First appearing in an early form in 2002 with Robbie Williams' deal with EMI (Marshall 2013; Stahl and Meier 2012), 360 degree contracts allow a record label to claim a percentage of the income from all activities undertaken by an artist, rather than only from sales of records produced by the label. As Day (2011) indicates, these contracts have become standard practice in the music industry. By 2010, over half of Warner Music Group's acts had been signed to 360 deals (Ostrove 2014).

This dialectical struggle between major and independent labels is not new: the music industry as a whole is characterized by periods of innovation and diversity followed by periods of homogenization (Lopes 1992). For example, in the 1920s and 30s race-based indie labels were part of the blues music scene, and indie brands such as Motown and Stax emerged in the post-Second World War era (Forman 2000). Similarly, 1970s punk music in the United States, the United Kingdom and Australia pushed back against growing multinational corporations with a DIY (do-it-yourself) approach to the music industry. They emphasized the democratization of music by allowing more artists to own shares of the business. Strachan (2007) points out that this DIY ethic still informs UK indie labels that create a niche market by engaging in small-scale music production and critiques of global corporate media. In the early 1990s, during the golden era, hip-hop adopted a similar approach with the formation of smaller labels such as Ruthless Records, Def Jam and Tommy Boy (Forman 2000). These labels were able to create a niche for themselves by staying close to their fan base and the street, which allowed them to remain tuned to the latest sounds, trends, fashions and dances (Negus 2002).

Moving on to the second decade of this century, Tech N9ne has made headlines by choosing to stay independent rather than sign with a major record label. According to *Forbes* magazine, Tech N9ne made an estimated USD 7.5 million in 2012, which is more than the income earned by mainstream artists 50 Cent, Mac Miller and Rick Ross in the same year (Greenburg 2013). Tech N9ne attributes a large part of his success to building a strong fan base without the help of a major record label. His business model, which includes a deal with Isolation Network's independent distribution company Fontana, ultimately produces high profit margins with relatively low cost, as noted in his song 'Crybaby' (2008). Tech N9ne's success has been seen as an exception, as indie labels traditionally owned only 10–15 per cent of the market share (Day 2011). But this figure is steadily increasing, with independents now making up as much as 30 per cent of the market share (Moore 2013); examples include American rappers Macklemore & Ryan Lewis and Chance the Rapper, and UK artists Lowkey, Mic Righteous and English Frank. Ultimately, a historical analysis of hip-hop's struggle with patterns of commodification and resistance demonstrates the tentative nature of the culture and simultaneously its unlimited potential for instigating sociopolitical change.

See also: authenticity; co-optation; race & ethnicity studies and citizen media; subjectivity

Recommended reading

Chang, J. (2005) *Can't Stop Won't Stop: A history of the hip-hop generation*, New York: St. Martin's Press.

Based on original interviews with DJs, b-boys, rappers, graffiti writers, activists and gang members, with unforgettable portraits of many of hip-hop's forebears, founders and mavericks, this book chronicles the events, the ideas, the music and the art that marked the hip-hop generation's rise from the ashes of the 1960s into the new millennium. The epic story it narrates has never been told with this kind of breadth, insight and style.

Oware, M. (2018) *I Got Something to Say: Gender, race, and social consciousness in rap music*, Cham, Switzerland: Palgrave Macmillan.

Drawing on the lyrics of over 700 songs from contemporary rap artists and using innovative research techniques, this book reveals how emcees perpetuate and challenge gendered and racialized constructions of masculinity, femininity and sexuality. It demonstrates that although complicated and contradictory in many ways, rap remains a powerful medium for social commentary.

Vito, C. (2015) 'Who Said Hip-Hop Was Dead? The politics of hip-hop culture in Immortal Technique's lyrics', *International Journal of Cultural Studies* 18(4): 395–411.

Using hermeneutic methods, three central themes are identified in Immortal Technique's lyrics that illustrate how he expresses resistance to class domination: the idea that class conflict occurs in hip-hop and thus there needs to be a pull away from major corporations; the need for independent hip-hop to escape from false consciousness and resist hegemony; and how creating knowledge through independent hip-hop culture and language can offer a means of resisting class domination.

HYPERLOCAL MEDIA

Jerome Turner and David Harte

Since the early 2000s, hyperlocal media has emerged as something of a catch-all term to describe a series of commercial and community-led citizen media initiatives. The oft-cited definition by Damian Radcliffe (2012:9) that hyperlocal media is "[o]nline news or content services pertaining to a town, village, single postcode or other small, geographically defined community" may seem to offer a clear characterization of the phenomenon. However, that has not prevented a wide range of actors interested in prophesying the future of news from using it as an umbrella term for locally oriented information services that can be very different in operation and intent (Metzgar et al. 2011:774). Pitched as an answer to both the commercial woes of the local press as it entered the digital age, as well as a way to characterize citizen-led local media initiatives, hyperlocal media might be seen more as a useful buzzword than a coherent set of practices. Ali (2017:49) has argued that "local" has become an "empty signifier" in policy debates about the future of public service media; the term hyperlocal has arguably operated at a similar discursive level. Questions of terminology aside, however, Harte et al. are clear that there is a common desire among scholars, commentators and policy makers to identify the value of hyperlocal media in terms of developing new models of journalism enterprise, addressing the democratic deficit resulting from the decline of the local press, and mitigating the diminishing of social capital within communities (2018:10). The focus in this entry is therefore on the use of the term hyperlocal in relation to the issue of providing a small geographic area with news services that speak to these values.

Enterprise and motivation

The political economy of hyperlocal news has attracted considerable research interest, much of it consisting of an ongoing critical commentary on the viability of large-scale, commercial hyperlocal media operations which have either exploited the labour of community contributors or applied template-driven solutions in order to draw local audiences to online environments that feature advertisements. Research in the United Kingdom (Thurman et al. 2011; Baines 2012; Price 2010) has looked at attempts by newspaper groups to operationalize hyperlocal media, many of which have ended in failure, while in the United States scholars have examined the involvement of large Internet companies in hyperlocal media initiatives. As an example of the latter and in common with other critiques, St. John et al.'s (2014) analysis of the local news and information platform Patch in the US suggests that it lacked a "community sensibility" (ibid.:208), over-relying on official sources, limiting reader interaction and failing to offer community-driven perspectives.

By 2011 the multinational media corporation AOL had employed 800 journalists in 850 communities covering twenty-two states (Auletta 2011) as a result of its investment in Patch. However, by January 2014, after losing around USD 200 million, AOL had sold its majority shareholding. The citizen journalism website Backfence is an example of an earlier attempt to generate hyperlocal revenue in the US. Backfence was active from 2005 to 2007 (Kaye and Quinn 2010:45), but after only two years it effectively lost its entire investment of USD 3 million. One of the few relatively successful attempts at creating a bespoke hyperlocal media solution online is Nextdoor, which was closer to a social networking system than a news service (Masden et al. 2014).

This is not to say that bottom-up hyperlocal media run by citizens rather than commercial companies are any less likely to fold, but when they do it is often for different reasons. Van Kerkhoven and Bakker (2014) identified 350 hyperlocal news websites publishing in 199 municipalities in the Netherlands, while Harte (2013) found 432 active UK hyperlocal organizations. In both cases the sites identified were notable for being citizen led, but there was a sense of precarity about the ventures. They are thus seen as "both promising and vulnerable" (van Kerkhoven and Bakker 2014:307). Citizen-led hyperlocal media are usually run by individuals on a voluntary basis, but in some cases they attempt to generate income through advertising or other means. Those who have explored such avenues tend to acknowledge the difficulty of producing quality content once a financial model is put in place, concluding that they were more comfortable running their service as a hobby after all (Harte et al. 2016). Ultimately, the precarity of both financed hyperlocal organizations and those run as a hobby means that it is not uncommon for such organizations to cease operating altogether, despite the optimism of some scholars (Kurpius et al. 2010) and occasional evidence of hyperlocal organizations effectively generating revenues in inventive ways (Cook et al. 2016).

In terms of the motivations for citizen participation in hyperlocal media, Hess and Waller (2016:194) note that the networked or franchised hyperlocal operations set up by larger commercial media companies have been largely unsuccessful in making an adequate return on investment, continually failing to recognize that the appeal of hyperlocal media comes from participation in a cultural activity as much as it does from producing content. Other researchers have examined citizen-led ventures in order to highlight issues of representation, inclusion and motivation, and found that community news projects are driven by sheer enjoyment of the work and a commitment to the relevant community (Glaser 2010:585). Fröhlich et al. (2012) and Bruns (2010a) studied the large German community news network MyHeimat; Fröhlich et al. (2012:1059) note that creativity and fun are key motivations for citizens' participation in this initiative. Other studies have explored news projects that are typically set up by residents, who may draw on skills such as social media and web design but are otherwise not trained as journalists and might even balk at the suggestion that they should be understood in such terms (Firmstone and Coleman 2015:128; Harte et al. 2016). A wide range of digital delivery platforms and online publishing forms are used in these local news projects, in line with the increasingly open and participatory character of the Internet. Informed by an analysis of community websites based in London, Flouch and Harris (2010) offer a typology that usefully draws a distinction between two types of community news networks: "placeblogs" and "public social spaces" (ibid.:5–7). Citizens involved in producing placeblogs are motivated by a belief that highlighting issues of local concern is capable of driving local change (ibid.:5), and often publish across multiple blog posts based on investigations spanning several months. Flouch and Harris' second category, public social spaces, tended to be more conversational in character, with the focus of those involved being on building relationships through cursory updates of immediate local events on social media accounts. Domingo and Heinonen (2008:7–8) use the term "citizen blogs" to refer to a wide range of journalistic practices which are initiated by members of the public. In writing about local events, citizens contributing to such blogs are motivated to provide coverage of events and issues that are overlooked by the media, thereby contributing to creating a "micro-level public sphere" (ibid.:8).

Chronicling the everyday

Hyperlocal media is not regulated or constrained by the need to adhere to the "standardised and limited repertoire of news" characteristic of mainstream media due to its "conglomerated nature" (Atton and Hamilton 2008:79; Hartley 1982; Harcup and O'Neill 2016). This leaves room for hyperlocal editors and writers as well as the audience to create and shape the space together and in tune with their own tastes and desires (Turner 2017). Hyperlocal media may thus be thought of as an online "third place" (Oldenburg 2001:2; Oldenburg 1997; Soukup 2006) situated away from work or the home and used for engaging in discussions of civic issues with peers – discussions historically found in neighbourhood cafes, bars and barbers. Postill's (2011) study of hyperlocal media similarly identifies such third places in the collective behaviour of hyperlocal editors and their audiences, drawing on Bourdieu's (1979/2010) field theory to explain that all parties must observe best practice and participate in certain appropriate ways to maintain a viable and valuable neighbourhood resource. Set against current concerns that physical, neighbourhood community spaces are in decline or underused (Hickman 2013; Putnam 2000), hyperlocal media thus has the potential to fill an important vacuum in society (Hampton and Wellman 2003; Dahlgren 2009).

While hyperlocal media might be understood as practised within its own online spaces and according to its own set of rules, then, it is important to contextualize and situate it in relation to the wider neighbourhood as well as other media and forms of social communication. This approach echoes Dahlgren's (2005:148) definition of the public sphere as a "constellation of communicative spaces" rather than one distinct space and Castells' theorizing of the Internet as a "space of flows" (2000:407). While Castells focuses on communication practices rather than the practitioners themselves, however, Pink (2012) alerts us to the interweaving, entangled nature of an individual's media use across numerous sources, technologies and platforms. Individuals make use of multiple hyperlocal online sources as well as mainstream newspapers, radio and television and less formalized face-to-face communication with friends, family and colleagues. Hyperlocal media is therefore embedded in an intricate network of relationships, forms of activism and participation that cut across the categories of state and non-state, citizen and organization, and constitutes just one of many overlapping media and communication practices. In the context of studies of the everyday (Certeau 1980/1984), and the acceptance and incorporation of social media platforms in almost all areas of social life, studying the practices of hyperlocal media can offer insight into some aspects of a resident's experiences of their neighbourhood.

Although the spaces discussed here are social in nature, we cannot assume that audiences of hyperlocal media are driven to participate by the promise of accruing social capital as traditionally understood (Coleman 1988; Bourdieu 1983/2011; Hanifan 1916; Putnam 2000), i.e. by a belief that acting altruistically will lead to a change of status for the participant and an expectation that a future reciprocated act will follow. While some change in status might be achieved, audience participation is best described in terms of accruing capital in the form of local knowledge (Turner 2017), and while this form of capital is immediately recognized it does not endure. This currency of local knowledge is most clearly demonstrated in highly participatory hyperlocal spaces such as Facebook Pages, where audiences typically discuss local information without framing it as news (ibid.). Audiences moreover demonstrate agency in shaping such spaces as sites of affective experiences of connection and belonging rather than merely as functional, with a focus on information gathering and sharing. Photos of sunsets, thanking others for kind deeds and stories of reunited pets or lost wallets are just as common as event listings and information about school closure, demonstrating that hyperlocal media encapsulates all aspects of everyday neighbourhood life.

In terms of potency, it is often claimed in policy literature that hyperlocal media is as effective as the local press in holding local power to account. Although it is possible to cite specific examples of effective campaigning and investigations (Williams et al. 2015), the geographic patchiness and dissonance of hyperlocal media initiatives suggest that their contribution to the public sphere is variable at best. Yet there is value in examining the way in which everyday topics such as lost pets, traffic delays and charity events might sustain more participatory modes of communication. The banal and the everyday, it could be argued, is where hyperlocal media become active spaces of co-creation. Bruns' (2006) definition of the produser as someone who both consumes and contributes to the production of a given media output falls short of explaining the nature of such spaces, where the reality is far messier for both the individuals who are active across multiple media forms and within a given hyperlocal media platform. Some audience members will read content without visibly contributing, while others will be more active by posting comments, sharing posts and 'liking' stories. Ultimately, sustained scholarly engagement with hyperlocal media must be seen as part of an ongoing interest in the participatory promise and DIY (Do-It-Yourself) approaches of citizen-led journalism. Looking at the US provincial community press of the 1970s, Eisendrath (1979) noted how what he called the mom-and-pop press tended to be more personal than political in tone. Such initiatives were sustained by local residents investing their energies into covering the minutiae of community life. The result was "intensely local, rather than personal coverage" (ibid.:72). It is this appealing community image that is the focus of much of the discussion about hyperlocal media.

Future directions

Hyperlocal media are being pulled in multiple directions and called upon to serve not only community but also economic and policy interests. As far as scholarship is concerned, discussion is couched in prior taxonomies of participatory media (Bowman and Willis 2003; Bruns 2005; Domingo and Heinonen 2008; Flouch and Harris 2010; Gillmor 2004), and there is an expectation that hyperlocal media will fill the gap created by a receding mainstream local media, specifically local newspapers (Metzgar et al. 2011; Nielsen 2015). Hyperlocal media is further seen as providing spaces to extend civic and political engagement and activism (Hampton and Wellman 2003; Mesch and Levanon 2003; Metzgar et al. 2011; Väätäjä 2012); this is partly a legacy of long held concerns about the decline of neighbourhood community places (Putnam 2000) and spaces to develop a public sphere (Habermas 1962/1989). Research in the UK has found that those who run hyperlocal media initiatives in the country are highly motivated by the chance to fulfil this public sphere role (Harte et al. 2017). This is evident in the ways in which practitioners are often fiercely defensive of the reputational geography of their area or patch, walking a fine line between offering a heavy dose of good news to their audiences and effectively challenging power and the effects of austerity. The UK overview by Williams et al. (2015) shows that topics covered by hyperlocal publishers are generally in line with what one might expect from a local newspaper. There is, then, a potential for hyperlocal publishers to play a part in averting the democratic deficit, because readers get "a large amount of information about politics, particularly the politics of local government, which relates to the news' ability to foster informed citizenship" (Williams et al. 2015:689). Researching local rural communications in the UK, Baines found that hyperlocal media could be a kind of glue, providing content around which the community can come together (2012:152). However, this ideology of hyperlocal media being civic in nature, and with the intention of instilling a notion of reciprocal community in those participating, is questioned when audience responses are investigated more closely. Ethnographic studies of two urban UK hyperlocal media organizations (Turner 2017) noted that as much as editors sometimes made attempts to engage or mobilize residents in civic activism, these spaces are also significantly shaped by the audience's desire for a more banal, everyday flavour of citizen media that is for and *by* the people, in line with Baker and Blaagaard's distinction (2016a). To an extent, the UK-based research reflects that country's context of the political tensions of austerity and the shift away from paternalistic media policy-making. Such concerns tap into policy interests in the UK relating to the sustainability of existing models of public service broadcasting, with the UK media regulator Ofcom arguing that these sites have "the potential to support and broaden the range of local media content available to citizens and consumers at a time when traditional local media providers continue to find themselves under financial pressure" (Ofcom 2012:103). It is clear that policy makers, funders and researchers have largely shared similar concerns in relation to the need for communities to have mechanisms to hold authority to account.

Despite what is a growing volume of research on hyperlocal media, it remains a slightly elusive, imprecise site of study. However, there is much to be gained from continued investigation into how active citizens are seeking to sustain independent media operations in their neighbourhoods. Whether through the chronicling of everyday life or a more fervent public service approach, they are giving voice to communities who have often been abandoned by a mainstream media that has found profit at the local level difficult to generate. Finding solutions to sustaining hyperlocal citizen-led media is problematic but the prize of reinvigorated local media ecologies makes the issue worth pursuing.

See also: citizen journalism; community media; journalism studies and citizen media; public sphere

Recommended reading

Harte, D., R. Howells and A. Williams (2018) *Hyperlocal Journalism: The decline of local newspapers and the rise of online community news*, London: Routledge.

UK-wide research into the development of hyperlocal media, set against the decline of UK local mainstream press. Offers rich detail on working practices of hyperlocal journalists and raises critical questions about the sustainability of hyperlocal media.

Hess, K. and L. Waller (2016) 'Hip to be Hyper', *Digital Journalism* 4(2): 193–210.

An important contribution to debates about the value of hyperlocal media. Hess and Waller argue that the lens of subcultural theory can offer new insights into hyperlocal media as a marginalized practice that challenges the mainstream.

Metzgar, E. T., D. D. Kurpius and K. M. Rowley (2011) 'Defining Hyperlocal Media: Proposing a framework for discussion', *New Media & Society* 13(5): 772–787.

An attempt to define hyperlocal media in order to facilitate scholarly debate. Sets the parameters for further research from a public sphere perspective, arguing that hyperlocal media sets out to fill the gaps in existing journalism and is the latest in a long tradition of citizen-led civic journalism.

IMMATERIAL LABOUR

Dario Lolli

Immaterial labour is a theoretical concept developed in the early 1990s within Italian post-operaism – a political current of autonomist Marxism, or *operaismo* – to rethink "the nature of labour" in light of the new forms of flexible and automated production that emerged during post-Fordism (Lazzarato 1997:7). During the 1960s the factory had represented a crucial battleground for *operaismo* but, from the early 1980s, a new generation of politically engaged scholars had to confront the restructuring of labour along more flexible and decentralized lines. Accordingly, the context of struggle that Italian post-operaism was concerned with was no longer centred in the factory, but extended inside and outside of it by increased informatization, mobility and industrial fragmentation. Along with categories such as job insecurity (*precariato*) and self-entrepreneurship (*lavoro autonomo*), the concept of immaterial labour was collectively developed in this context as a tool to theorize these transformations and exert a political impact on them.

In the current debate, immaterial labour has come to define all the communicational, interpretive and affective skills that are required at multiple levels of networked and flexible production cycles. For instance, it includes the knowledge-intensive, creative or scientific work that lies at the heart of some of the most lucrative and strategic operations of contemporary capitalism, such as the development of brands, patents and intellectual properties. In spite of its emphasis on knowledge and communication, however, the consistent ascendancy of immaterial labour in several productive processes should not be considered in opposition to physical or industrial labour. Indeed, immaterial labour cannot transcend the industrial manufacturing of personal computers and other technological instruments that are required to carry out the networked and informational activities that it comprises. Immaterial production, therefore, does not displace material work, but rather activates and reorganizes globally extended value chains that encompass the physical exploitation of natural resources and multiple forms of labour power (Dyer-Witheford 2015).

What the hegemonic ascendancy of immaterial labour implies, however, is that capitalism does not only extract surplus value from material work, but also enhances the productive potential of a heterogeneous set of communicative, intellectual and affective abilities. These include not only problem solving and networking, but also the passionate attitude and interpersonal skills that the service sector demands. Because of their immaterial nature, these productive skills are employed beyond the temporal and spatial constraints of traditional factories or workplaces. As a result, every moment of life potentially "becomes raw material for capital accumulation" (Ross 2012:25). As Brophy and de Peuter (2008:179) put it,

> [i]mmaterial labor not only refers to the labor of the call center worker as she manages affect via a headset during a poorly paid work time at a cubicle; it also speaks of the unremunerated work of the person called when responding to, say, a consumer research survey.

Immaterial labour, therefore, encompasses not only the intellectual and decision-making aptitudes that are increasingly required by automated work in big factories and call centres, but also the diffuse and often unwaged activities of research, innovation and product qualification extended across society by networked systems of media and communication.

Cutting across debates on the information society, the creative industries and the knowledge economy, the concept of immaterial labour has been widely discussed internationally – not least because it draws attention to new forms of exploitation that emerge when productive activities are extended and unequally remunerated well beyond the bounded sphere of the traditional workplace. In this respect, the notion of immaterial labour has been particularly helpful in pointing at "both the multiplication of precarious, unstable, insecure forms of living and, simultaneously, new forms of political struggles and solidarity that reach beyond the traditional model of the political party or trade union" (Gill and Pratt 2008:3). In addition, immaterial labour has proved to be a central notion in the interdisciplinary field of citizen media as it presciently anticipated the emergence of an economic model based on the harnessing of social knowledge and cooperation that has become explicitly visible with the rise of convergent digital media and Web 2.0. As such, this concept raises the question of what counts as labour in a range of value-generating cultural practices that are not usually recognized as such – including interning for free in the hope of securing a rewarding career in the media industries, or simply blogging, uploading videos on YouTube and tagging friends on social media platforms.

This entry begins by explaining how immaterial labour was first conceptualized in the context of Italian political struggles. It then addresses criticisms of this concept and explores its most useful applications in the field of citizen media. These include debates on creativity and precarity in the media industries (Hesmondhalgh and Baker 2011; McRobbie 2016), free labour (Terranova 2004) and other modes of harnessing social cooperation that have emerged in the digital economy of Web 2.0 (Scholz 2012a; Srnicek 2017).

Genealogy of immaterial labour

The thesis of immaterial labour is most notably associated with the work of Italian post-operaist thinker Maurizio Lazzarato, author of an anthology on this concept (1997) partially available in the English language (2006, 2007). The genealogy of immaterial labour, however, should be considered against a larger and diffuse debate comprising several other voices and often discordant opinions (Wright 2007; Mezzadra 2009). These voices are not necessarily restricted to those of well-known post-operaist thinkers such as Virno (2004), Marazzi (2011) and Negri, co-author with Lazzarato of several pieces on immaterial labour (1993, 1994, 1997). In the 1970s, for example, Italian feminists played a crucial but often unacknowledged role in the theoretical development of this concept (Fortunati 2007). In dialogue with the militant enquiries on workers' conditions and political struggles that characterized operaist Marxism, Italian feminism contributed to reframe the political debate on labour by looking at the gendered work carried out in the domestic sphere. In contrast to the material male labour of the factories, they conceptualized immaterial labour as the domestic work of education, communication, care and, in particular, "the supply of love, affection and sex" (Fortunati 2007:146) – that had received very little attention from Marx but was regarded by feminists as an indispensable force of social and capitalist reproduction.

Drawing on these critical premises, in its later post-operaist articulations the category of immaterial labour was further extended to encompass all "the productive synergies" (Lazzarato 2006:139) activated by the shared, diffused and affective nature of post-Fordist labour. If the reforms introduced by capital in the name of flexible organization, or post-Fordism, were dissolving previous boundaries between labour time and leisure time, the factory and the house, what was at stake for post-operaist thinkers was nothing less than life itself. For Hardt and Negri, for example, immaterial labour always entails a biopolitical production that is founded on the very social capacity to communicate, collaborate, innovate and reproduce in common (Hardt and Negri 2001). Clearly, these communicational, interpretive and affective skills have always played a central role in the work carried out in the arts or the cultural industries – what Lazzarato names the "classic forms of 'immaterial' production" (2006:136), or the creative work of cinema, music, television, advertising, fashion and so on. Immaterial labour, however, was not primarily conceived as a theoretical tool to critique the culture industry (Adorno and Horkheimer 1944/1997). The thesis of immaterial labour argued that the combination of affective, intellectual, manual and entrepreneurial skills that were once typical of the creative sectors is now increasingly characterizing labour, to a larger or lesser extent, across all the spheres of post-Fordist production.

The deployment of these skills is considered productive even if labour providers are not paid or their activity is not categorized as labour by traditional standards. The productivity of immaterial labour assumes that the use value of contemporary commodities – in opposition to their standardized Fordist equivalents – resides primarily in the value of their informational, affective and cultural content. It is therefore possible, for example, to conceptualize branding as one of the most important developments in the management of the symbolic value of services, cultural products and mass-produced objects in today's global economy (Arvidsson 2006; Lury 2004). From this perspective, even consumption – once regarded as the final moment of the productive cycle – is a form of immaterial labour. Indeed, immaterial labour does not conclude with the destruction of the commodity, but extends into the production and transformation of the subjective and cultural sphere of the consumers themselves (Lazzarato 2006:137). Post-Fordist production therefore entails iterative processes of social qualification and singularization of commodities that cannot be supplied already made, as they are in a permanent state of development and flux, and necessarily involve both producers and consumers. As Lazzarato himself puts it (2006:144; emphasis added), "*[r]eception is thus, from this point of view, a creative act* and an integrative part of the product".

As Dyer-Witheford and de Peuter observe, "[i]mmaterial labor is less about the production of things and more about the production of subjectivity, or better, about the way the production of subjectivity and things are in contemporary capitalism deeply intertwined" (2009:4). Given that the production of active subjects of consumers/communicators in sites of immaterial labour coincides with the production of value, the view advanced by post-operaist authors represents a radical reconceptualization of the capital–labour relation and of the possible forms of political organization required to confront power under these changed circumstances. Labour activities are no longer confined to the factory or imposed as rules by the owners of capitals. Instead, they are increasingly incorporating a shared and common intellectuality – or "general intellect", as Italian autonomists would put it following the Marx of the Grundrisse (1993:706). With the emergence of precarious, networked and self-employed labour, therefore, leisure time and work time tend to overlap and the factory is *de facto* extended across society. In the social factory that emerges from these developments, the becoming hegemonic of immaterial labour is taken as both a source of domination and one of potential liberation. It is at once the exploitation of active, mobile and often precarious subjects and the radical potential of these subjects to socially and politically challenge capitalism by virtue of the autonomy of their diffuse and collaborative labour (Lazzarato and Negri 1997).

Following the debate on immaterial labour that emerged in the 1990s within Italian and French post-operaist circles (e.g. in journals such as *Luogo Comune* and *Futur Antérieur*), this concept found widespread use in the English language – not least because of the global success of Hardt and Negri's book *Empire* (2001) and the interest in autonomist Marxism that this volume generated. The debate on immaterial labour also paved the way for the seminal discussion on creative labour, precarity and post-operaist thought published in the special issues of journals such as *ephemera* (Dowling et al. 2007), *SubStance* (Mecchia and Henninger 2007) and *Theory, Culture and Society* (2007). In its new travelling articulations, immaterial labour has been widely adopted to identify and test the nature of precarious and casualized work in the cultural and creative industries – drawing on new empirical evidence and from perspectives that sometimes clash openly with post-operaist positions. It has also been used to account for the set of socioeconomic transformations that have emerged with the diffusion of interconnected digital media and their underlying economic model. The concept, however, has also been criticized for being too generic to effectively map the heterogeneous subjectivities of labourers working across extended and networked value chains. This criticism has encouraged some later amendments, especially those concerning the affective dimensions of immaterial labour (Hardt and Negri 2004).

Still, areas of debate and disagreement abound. Some scholars, for example, have argued that this concept needs to acknowledge and emphasize more strongly the enduring gendered and racial division of global labour under post-Fordism (McRobbie 2011; Dyer-Witheford 2015). Others, like anthropologist David Graeber (2008), have completely dismissed its usefulness, criticizing the epochal narrative of historical rupture deployed by Italian post-operaist thinkers as well as their arbitrary conflation of informational with domestic work. Political philosopher George Caffentzis (2005) has instead questioned a crucial theoretical axiom of Negri and Hardt's understanding of immaterial labour, the claim that the value of commodities would be now beyond measure, since it no longer strictly depends on "socially necessary labour time" (ibid.:99), but rather on a general and diffused social productivity that is constantly exploited by capital. According to Caffentzis, the difficulty in measuring value is not a new challenge introduced by post-Fordism but an enduring problem already faced by Marx in his days (ibid.:101). Solving this problem by getting rid of Marx's law of value, however, is not a viable option, as this would turn accumulation into a sort of mysticism unable to explain why "the average rate of profit [remains] positive" for the owners of capital (ibid.:104). This analysis calls for a more careful assessment of Marx's law of value and other systems of measurement and valuation (Adkins and Lury 2012); and for the realization that certain forms of value extraction from immaterial labour might not be radically new, but rather more intensive features of what capital has always exploited at different historical moments, whether in the nineteenth century or today.

Precariousness in the creative industries

The use of the adjective immaterial to identify major transformations in contemporary capitalism has also been an object of dispute and a source of misunderstanding. One of the most common has been the association of this concept with liberal discourses that have postulated the post-industrialization of society (Bell 1976) and the rise of an alleged creative class (Florida 2003). Although these perspectives have also focused on the dismantling of old industrial complexes in wealthy economies, they fundamentally differ from post-operaist theorizations in that they articulate a positive bias towards immaterial production without a critical analysis of the exploitation of precarious and self-employed labour that its emergence implied. In addition, while claiming that services and creative practices have increasingly expanded under post-Fordism, to distinguish and establish a hierarchy between intellectual activities and manual labour was far from the objective of its autonomist proponents (Hardt and Negri 2004:109). On the contrary, the debate on immaterial labour started precisely as a political reflection on the inability of the distinction between manual and intellectual labour and other dichotomies – producers vs. consumers, authors vs. audiences, waged vs. unwaged labourers – to come to terms with the "new nature of the productive activity" in the post-Fordist economy (Lazzarato 2006:133). From a post-operaist perspective, therefore, the point is not to give priority to one form of labour over another, but to recognize that "all forms of labor are today socially productive, [that] they produce in common, and share too a common potential to resist the domination of capital" (Hardt and Negri 2004:106–107). Indeed, as Nunes (2007:184) observes, immaterial labour

> has the common as both its ground – the general human capacities to affect and be affected, to communicate, to cooperate, to reproduce and innovate; and social relations themselves, symbolic

and affective codes etc. – and its result: it produces new being, i.e., new subjectivities, new enunciations, new forms of social life.

The objective of this theorization, therefore, is to open up a common ground for political struggles in an era characterized not so much by class identification as by increasingly flexible, mobile and precarious labour intensified under neoliberal reforms reducing subjects to active, responsible and competing entrepreneurs of themselves (Foucault 2008).

One area of research in which this notion has been widely applied is the study of the media and culture industries as eminent sites for the production of precarious immaterial labour and neoliberal subjectivities. Intervening in this area of research, Dyer-Witheford and Greg de Peuter (2005, 2009) have analysed the rise of the global video game industry as exemplary of these socioeconomic transformations across the globe. In the "scientific know-how, hi-tech proficiency, cultural creativity, human sociability, and cooperative interactivity" (Dyer-Witheford and de Peuter 2005) deployed in the production of video games, they find instances of immaterial labour, in which the promise of work as play is predicated on exhausting working shifts, fierce competition and intensive cycles of technological innovation, production and disposal. At the same time, however, the cooperative intelligence put to work by the gaming industry is simultaneously generating ambivalent tendencies that are able to destabilize and re-orient the exploitative forces of capitalist accumulation. This is the case, for example, of the anarcho-communist principles of hackers fighting against the monopoly over intellectual properties or the novel uses of video games for social activism and ideological contestation emerging from the bottom up (Dyer-Witheford and de Peuter 2009).

Other studies have further challenged celebratory views of creative work by stressing how exploitative conditions of labour have become the high cost to pay in exchange for the prestige and personal gratification that working in the creative sector is meant to bring (Gregg 2011; Hesmondhalgh and Baker 2011; McRobbie 2016; Ross 2012). Opening the study of the creative industries to a feminist perspective, for example, McRobbie (2016) has focused on the tensions and anxieties endured by young European middle-class women in their attempt to build a career in fashion and other creative industries. Connecting the concept of immaterial labour with recent neoliberal reforms, she argues that the precarious experiences of these women illustrate the internalization of an injunction to be creative as a paradoxical answer to the dismantling of previous systems of labour rights and security. She highlights how this affective and self-disciplining injunction to work harder and "discover one's own capabilities" (2016:15) does not translate easily into success, upward mobility and self-fulfilment for these young women; instead, it would appear to impact on their lives in the form of a biopolitical regime of permanent training, high stress levels and delayed or no reproduction.

In a book on precarity originally published in 2009 and translated into English much later, Lazzarato (2018) no longer regards the concept of immaterial labour as an effective tool to tackle political struggles pertaining to workers' rights and autonomy. His book analyses the ultimately failed attempt by French creative labourers (*intermittents du spectacle*) to mobilize against the reform of the benefit system for workers in the entertainment industry. By means of engaged participant observation, Lazzarato theorizes on the autonomous, horizontal and strategic formation of a struggling movement that stood resolute in rejecting prestige hierarchies between labourers. While the unions chose to articulate the workers' demands on account of the creative nature of their work, the movement tried to establish a united front with precarious labourers engaged in more technical and material jobs. For Lazzarato, this situation exposes the limits of the concept of immaterial labour to account for a diverse front of precarious workers without resorting to divisive class identifications based on the nature of their labour activity. Instead, he envisions the collaborative and strategic foundations of the French struggles as an antidote against the cultural and economic impoverishment of life engineered by neoliberalism, and a powerful example of horizontal forms of political organization capable of emerging from the shared intellectuality of cooperating labourers.

Free labour and the digital economy

The area in which the concept of immaterial labour has proved more fruitful and long-lasting is probably the study of digital media and their underlying economic system. As the notion of immaterial labour developed from an analysis of the role played by cybernetics and informatization in capturing and valorizing social cooperation outside the workplace, it has drawn attention to a set of sociotechnical transformations that have become more readily visible with the rise of convergent digital media and Web 2.0 – a time in which the social behaviour of web users is monetized by a plethora of digital platforms, algorithms and data mining technologies (Scholz 2012a).

In a seminal investigation into the digital economy inspired by post-operaist theories, Terranova (2004) has famously coined the term free labour to designate the range of often unwaged technical and cultural activities that have enabled the development and sustainability of the web – including web design, software modification, multimedia production, chats, blogs, newsletters and so on. Although monetized by the owners of network infrastructures and services, this free labour has not been "produced by capitalism in any direct, cause-and-effect fashion" (ibid.:79), nor does it necessarily amount to "exploited labour" (ibid.:91). Indeed, the emergence of the Internet has ambivalently relied on *both* social cooperation and capitalist principles; on a gift economy grounded in a subjective "desire for creative production … and the current capitalist emphasis on knowledge as the main source of added value" (ibid.:77). Terranova, however, warns against celebrating the democratizing potential of the Internet without simultaneously addressing how the web "effectively functions as a channel through which 'human intelligence' renews its capacity to produce" (ibid.:79), and how this capacity is captured and converted into value by its technological apparatus.

The celebratory stance underpins early discourses on media convergence that have hailed the emergence of online co-creative and participatory practices as an empowering effect of new digital technologies (Jenkins 2008). The growing ubiquity of interconnected instruments to edit and share media content has allowed online users to blur clearcut distinctions between media producers and consumers by effectively experimenting with "the same multimodal language that frames cultural products formerly made exclusively in studios" (van Dijck 2009:43). Yet, the more permissive and decentred environment that digital media enable has only exacerbated the reliance of previous systems of media production on audiences as crucial providers of cultural labour – for example, by making their life stories available to TV programmes and magazines, or by responding to contests and consumer surveys (Terranova 2004:95).

Attempts by media corporations to extract value from the collective productivity of active media consumption, which were already documented in the era of broadcast television and print media (Lazzarato 2006), have become more pervasive since the advent of digitization. In his analysis of multimedia franchises, for example, Johnson (2013) has highlighted how the cultural and affective creations of media fans online – usually known as user-generated content – have been not only encouraged by the media industries, but even fully integrated into their very models of serialized production, promotion and distribution. As Scholz notes, even "Wikipedia and other projects whose contributors are not driven by profit motives are not outside the dynamics of the digital economy" (Scholz 2012b:49). In this sense, the emergence of participatory forms of online co-creativity should not be mistaken for a radical challenge to capitalism, but rather seen as a manifestation of the new ways in which networked technologies capture immaterial labour.

The Internet, often considered by scholars as a model of gift economy (Jenkins et al. 2013:65), is therefore not immune to commodification and class inequalities. As McKenzie Wark observes, "the limit to making a gift of culture to everyone is that doing so adds value to the vector through which it is distributed, and that is not free" (2012:72). Wark names the new owners of informational infrastructures the "vectorialist class" (2019:45), techno-giants like Google that make profit not so much by controlling production as by monopolizing the technologies, logistics and supply chains enabling the global flow of information. Indeed, the interpellation of web users as simultaneously consumers and producers of information represents the very condition of existence for several digital platforms based on the Web 2.0 protocol (Gehl 2011). Compared to the nascent digital economy analysed by Terranova in the early 2000s, today's enclosed, subscription-based social networking platforms like Facebook, Instagram and Twitter are far better equipped to capitalize on the collective productivity of their users. By means of registering and storing not only user-generated content, but also metrics based on all the social connections, preferences and interactions of their users, these platforms effectively monetize these data by selling them to advertisers, security services and "operators like insurance companies, mortgage banks and employers" (Arvidsson 2016:6). Digital companies, in turn, utilize these data to perfect algorithms whereby they can fix the most favourable prices for their products, based on speculative predictive models rather than their actual demand on the market (Srnicek 2017:47).

Future directions

Contemporary forms of capitalist accumulation based on the extraction of data, therefore, reach far beyond the sphere of entertainment and social media. Indeed, the pervasiveness of enclosed digital platforms with the capacity to generate profit by controlling huge amounts of data has even led some commentators to rename the contemporary digital economy as platform capitalism (Srnicek 2017). Platforms are digital infrastructures capitalizing on the provision of logistics and services to a great number of user groups (Gillespie 2010), and are equally owned and maintained by tech giants such as Google and Apple, assets-free start-ups such as Uber and Airbnb, or old transnational corporations such as Siemens, Pfizer and Unilever (Srnicek 2017; Wark 2019). The emergence of platforms as both technological infrastructures and a popular discourse about their function and management is a global phenomenon that extends far beyond the Silicon Valley and the 'Western' world (Steinberg 2019). However, while platforms and their many innovative services – including crowdsourcing, crowdfunding and the so-called sharing economy (Sundararajan 2016) – are presented as an allegedly radical break with capitalist principles, they are often the very drivers of increasingly cheaper and casualized labour, of an emerging gig economy (Woodcock and Graham 2020) that allows companies to externalize their risks and fixed costs onto freelance temporary workers.

By offering a prism through which to examine the implications of these transformations, the debate on immaterial labour provides citizen media scholars with a critical insight into the digital economy, challenging the premise underpinning techno-utopian and liberal discourses that the emergence of participatory digital networks is a necessarily progressive and empowering development (Fisher 2010). It is in light of the enduring value of data – understood as socially produced information, affect and cooperation – that the notion of immaterial labour remains a crucial category to make sense of capitalist accumulation in the digital age.

See also: crowdsourcing and crowdfunding; fandom; mobile technologies; precarity; user-generated content

Recommended reading

McRobbie, A. (2016) *Be Creative: Making a living in the new culture industries*, Cambridge, UK and Malden, MA: Polity Press.

Extending feminist perspectives on immaterial labour, it investigates the dark side of the fashion industry in the context of neoliberal austerity. Through examples from the UK, Germany and Italy, McRobbie exposes the self-disciplining force that the contemporary discourse on creativity exerts on young women, while searching instances of alternative forms of creativity and social organization in their projects and subjective experiences.

Scholz, T. (ed.) (2012a) *Digital Labor: The internet as playground and factory*, New York: Routledge.

In dialogue with post-operaist theories, it investigates immaterial labour in the context of current digital transformations. Through contributions on hackers, free labour, fandom and algorithms, this collection provides a wide-ranging overview on how digital technologies capture value from social cooperation, information and data.

Virno, P. and M. Hardt (eds) (2006) *Radical Thought in Italy: A potential politics*, Minneapolis, MN: University of Minnesota Press.

Edited by two key post-operaist thinkers, this volume brings together some of the most important voices of the Italian political debate shortly outlined in this entry. The collection includes an early piece on immaterial labour by Lazzarato as well as contributions by fellow post-operaist thinkers Antonio Negri, Paolo Virno and Carlo Vercellone.

Wark, M. (2019) *Capital Is Dead: Is this something worse?* London and New York: Verso.

Updates and develops further the idea of a vectorialist class initially presented in their *Hacker Manifesto* (2004). By interrogating the vectorialist monopoly over data, patents and infrastructures, Wark reformulates several Marxists tenets for their application in the information age.

INDYMEDIA

Dorothy Kidd

The global Indymedia Center Network (IMC) represents the largest transnational experiment in democratic participatory media-making ever (Stringer 2013). Initially lauded for its technological innovations in building the world's first website for user-generated content, the global IMC is now better recognized for its radical prefigurative politics and horizontal organizational form. Operated independently from governments, corporations and the dominant media, this global news network shared its Internet servers, open source code, common site name, similar visual configuration and technical support team among its local sites. In contrast to the legacy news organization's vertical hierarchies of professional journalists, the governance and news production of Indymedia was distributed across self-organized autonomous local collectives and transnational project teams who operated with high levels of consensus decision-making.

The first Indymedia Center was founded in Seattle in late 1999 during protests against the World Trade Organization (WTO) Ministerial Conference. At the time, the dominant US news media maintained control of the news agenda; they had provided very little coverage of the neoliberal globalization policies of the WTO and other multilateral agencies and almost none about the growing resistance to these policies, inside and especially outside the US (Martin 2003). A coalition of media-makers, artists and computer techies thus designed Indymedia to break through this media silence. Their goal was to provide the protesters' view from the street, and as importantly, to convey the collective wisdom of citizen groups and social justice movements about the negative impact of neoliberal policies on a host of public concerns such as jobs, the environment, social programmes and democratic governance.

Working out of a downtown storefront, volunteer radio, video and print media teams assembled to report on the week of street protests and citizen-led educational forums, and especially on the repressive and violent actions of the police. At the same time, a tech team collaborated in cyberspace to design new open source software that allowed the media teams to upload their reports to the Internet, and, even more significantly, opened up the site to anyone with robust Internet access to download and upload their own media. Very quickly, over a million people across the planet accessed the website. Combining high levels of cooperation, media skills and innovative tech design, the Seattle Indymedia Center took the dominant news media by surprise, outflanking them in the number of stories and investigative reports produced and audiences reached and pressuring them to follow suit. Partly in response to the IMC coverage, the mainstream US news media began to report on the police violence, and to include spokespeople from national citizens' organizations in their coverage of what became known as the Battle of Seattle (Kidd 2003a).

Emboldened by its success, the global IMC rapidly expanded, growing to almost 200 local collectives and transnational media project teams working in some thirty languages on six continents. As the movements organizing against neoliberal capitalist policies coalesced into the global justice movement, the global IMC became its trademark media network, setting up tactical media centres and hack labs at a succession of counter-summits during meetings of state leaders (G7 and G8), multilateral organizations (WTO, IMF, World Bank) and United Nations forums such as the World Conference on Racism, and then moving to coverage of local and national events. IMC collectives provided "world-wide news of resistance at a click" (Stringer 2013:331), facilitating the networking of related struggles (Sullivan et al. 2011). Their postings affirmed the connection between local grassroots movements and regional and global contexts. At the same time, this new aggregator of stories of resistance represented another vision of global possibility, collectively challenging Margaret Thatcher's famous assertion that there was no alternative to capitalism.

The origins of the IMC

The IMC represented a new cycle of citizen and social movement media activism in which at least four overlapping precursor waves converged. First, the origins of the IMC can be traced to the US media democracy movement of the 1990s and early 2000s, whose nascent national coalitions – the Cultural Environment Movement and two national Media and Democracy Congresses – critiqued the commercialization and consolidation of global corporate media and called for citizen-based alternatives (Hackett and Carroll 2006:13). Second, alternative media activists brought with them their operational savvy from older national and transnational networks of community and pirate radio and television, independent newspapers, documentary video, graffiti and fanzine cultures; their radical media-making practices promoted first person reporting from communities and political perspectives systemically marginalized by the dominant media. Third, radical techies from the Free and Open Source Software movement infused the project with the copyleft ideas of collaborative hands-on experimentation, horizontal sharing of knowledge and resources and resistance to privatizing information (Sullivan et al. 2011). Finally, the global IMC drew on the anarcho-autonomist ideas of the emerging global justice movement. They built on what autonomist Marxist Harry Cleaver (1994, 1995) has dubbed the electronic fabric of struggle, the communications network formed in the 1990s by labour, environmental, feminist, indigenous and human rights groups mobilizing against trade liberalization, structural adjustment and other policies of the Washington Consensus.

One of the keystones of that network were the Zapatistas, who had first emerged in Chiapas, Mexico in January 1994 to challenge the North American Free Trade Agreement (NAFTA). The Zapatistas' courageous intervention and visionary strategy had inspired activists around the world, and many of the IMC founders acknowledged Zapatista Subcomandante Marco and his call for an intercontinental network of alternative communication against neoliberalism and for values of self-determination and direct democracy as a primary influence (Wolfson 2012; Kidd 2003a; Stringer 2013). Following the Zapatista vision of autonomy, the IMC focused their efforts on building a network of self-organized and governed local commons spaces, rather than attempting to reform existing state and corporate institutions (Kidd 2003b; Gerbaudo 2017a). Notably, and distinct from contemporary activist practice, the IMC were adamant about building a network that is autonomous from all systems of centralized power, and thus maintained their own Internet servers and designed their own software, which was tailored to the needs of their user communities, rather than depending on the digital corporations and emerging commercial controllers of the Internet.

The global IMC Network represented a quantum leap that surpassed all previous efforts of media activists in its scope and scale of media production and circulation, and in its singular focus on providing news and information of, from and for the global justice movement. Before the IMC, the media strategy of most citizen groups and movements was limited to brokering space within the dominant commercial or public service media (Kidd 2003b). After the founding of the IMC, activist media spaces became an accepted part of the repertoire of protests, providing the narratives and analyses of participating groups and countering law enforcement's repressive actions with practices of sousveillance, or citizen monitoring and video reporting on authorities (Robé 2016). More profoundly, most citizens and social movements now integrate media-making practices into their overall strategies and everyday organizational routines.

Key features of Indymedia

At its outset, Indymedia was lauded for its high-tech innovations and its ambitious vision of democratic web-based public space. Birthed in Seattle, the home of Microsoft and Amazon, Indymedia drew on a corps of high-tech designers and software engineers. At the time, there was no plug-and-play media sharing and Internet communications were limited to text-based email and Usenet groups. Indymedia's open publishing design was one of the first user-friendly, multimedia interfaces, built before Web 2.0 and commercial social media. The site enabled individuals and collectives, with the appropriate media-making and computer equipment, to report on stories in any medium from around the world, and allowed sites to produce and edit materials collaboratively across long distances. These two radical innovations afforded greater media power for citizens and social movements and disrupted the control of the news agenda by a small number of transnational corporations and state-run institutions. The constraints on citizen communications at the cusp of the millennium were not only imposed by the technical limits of the Internet, but also by social and communications inequalities which stretched across rich and poor countries and between rich and poor communities everywhere. Recognizing these deep divides of communications power, the Indymedia tech teams set about to provide both the equipment and the training for local IMC centres to operate. They recycled computers, loaded them with open source software and shipped them to sites around the world. At the same time, they set up hack labs to train volunteers in tech design and operation in sites such as the West Bank, Andean indigenous and *campesino* communities, camps of the landless movement in Brazil and squatted banks and community centres of the *piquetero* movement of unemployed workers in Argentina (Herndon 2003).

Indymedia's do-it-yourself (DIY) inventiveness was not limited to the digital domain. Local Indymedia centres experimented with the media instruments and communications practices most appropriate to the people and media ecology of their local context. For example, the Chiapas Mexico IMC only used the Internet to gather and circulate regional and international news, which they then repurposed for audiotape distribution and community radio. In Rio de Janeiro and São Paolo, local crews produced video documentaries and set up makeshift screens in the public squares and *favela* streets to foster discussion; they also distributed enlarged photocopied newssheets and posted them on walls all over the city (Kidd 2003b). In Australia, IMC volunteers created tactical media interfaces to connect with prison detainees; they designed the Indymedia Phone Patch for audio reports over the telephone, and set up micro-radio transmitters to broadcast from outside the prison walls (Lowenthal 2004/2006).

Key to Indymedia's repertoire was and is its experimentation with radical news content and participatory grassroots reporting. Indymedia volunteers shared a critique of the dominant media, especially the legacy news media routines that privileged the status quo and the perspective of official government or corporate spokespeople. Many of the volunteers were active in the global justice movement or other social movement protests and were particularly motivated by the need to challenge the dominant media's protest frame, which pictured protesters as violent, disruptive freaks rather than legitimate sources of critique of sociopolitical inequalities and issues (Xu 2013:2414). Summarized in the tag line 'Don't hate the media, become the media', the Indymedia approach focused on direct representation, encouraging activists to select sources and report on stories from populations and perspectives that were seldom given space in the dominant media.

The IMC Network highlighted the importance and connections between the common concerns of citizens and social justice movements on the global scale. Much of the Network, especially in the first few years, focused primarily on providing counter-publicity at the counter-summits and related demonstrations of the global justice movement. Nevertheless, many of the local, national and regional collectives, especially the Indymedia centres that have survived and continue to operate, worked with and reported on the protests and ongoing organizing of residual and emerging groups and movements in their home communities. For example, many US sites began by publicizing reports of ongoing protests at major national and international events; some now focus on work with local activists to publicize citizen critiques of local police and criminal justice operations (Robé 2016). In other cases, such as Champaign, Illinois, they encompass a wide range of media and arts activities. In Germany, activists from the anti-nuclear movements were involved in starting the early centres (Hintz 2003); in 2017, Linksuten was the main German independent media website offering a space for people to post anonymously during the counter-protests against the G20 summit in Hamburg. Indymedia Argentina was involved in the network from the outset. The early IMC there was closely connected with the *piquetero* movement of unemployed workers (Boido 2003); it continues to report on grassroots groups of labour and squatters, as well as struggles involving food, education and state repression (Giraud 2014:430) and against mining and resource extraction located primarily in indigenous communities. Established in about 2008, the Athens IMC has continued to play an important role in providing a counterpoint to the dominant Greek media in their circulation of news about campaigns against the European Union and austerity (Siapera and Theodosiadis 2017), police brutality and racist attacks against immigrants (Galis and Neumayer 2016).

Much of the more recent academic interest in Indymedia revolves around their collective practices and values, which prefigured a less hierarchical and more horizontal model of organization (Benski et al. 2013; Galis and Neumayer 2016; Gerbaudo 2017a, 2014; Lievrouw 2011; McDonald 2015; Pickard 2013; Stringer 2013; Sullivan et al. 2011; Wolfson 2012). All member centres signed a common set of principles of unity that recognized a commitment to "equality, decentralization and local autonomy", "non-hierarchical relationships" and the "development of a direct, participatory and democratic process that is transparent to its [Indymedia's] members" (Lievrouw 2011:138–139). Consensus decision-making was also a key principle, and was practised in reaching decisions about governance, pre-production planning and post-production editing on all levels, from the local to the transnational; decisions were reached via face-to-face discussions, discussions via email lists, Internet Relay Chats (IRC) and wikis. Finally, openness was not only a technological value, but was interpreted more broadly as openness to all to participate in the network and included outreach to local and especially marginalized social groups (Giraud 2014).

Overall, the network not only functioned as an international news site, but as "a web of mutual solidarity" and a space to share stories and circulate knowledge (Lowenthal 2004/2006). It described itself on its various websites as a

> network of collectively run media outlets for the creation of radical, accurate, and passionate tellings of the truth. We work out of a love and inspiration for people who continue to work for a better world, despite corporate media's distortions and unwillingness to cover the efforts to free humanity.
>
> *Stringer 2013:323*

The waning of the network

By 2008, as the global justice movement began to wane so too did the IMC. As of 2019, only a handful of local sites were still operating. Many arguments have been put forward to explain this reversal in fortunes. One explanation is that citizens and social justice movements switched to commercial platforms, blogs and social media once these sites became more user friendly (Lievrouw 2011; Pickard 2013). Another is that a volunteer site is unsustainable in the long run; many sites did not effectively recruit and train new volunteers or media teams in their local centres, contributing to high rates of burnout and turnover of members (Kidd 2003a). Volunteers also reported leaving due to the increase in repressive measures from state and corporate authorities as Indymedia centres became subject to surveillance, arrest of members, seizure of their servers, corporate law suits and other repressive measures (Stringer 2013; Giraud 2014). For example, in 2017 the German government raided and shut down Linksuten, the main German independent website, partly in response to mobilization against the G20 summit (Indymedia 2017b).

Political and cultural differences within the global network were also very difficult to overcome. The naïve beliefs in the transformative values of a horizontal, highly decentralized network structure governed through consensus-based decision-making did not allow for much proactive decision-making (Wolfson 2014:191). Reaching a decision in large translocal discussions was complicated and took a long time, which led to inaction and/or to the emergence of informal hierarchies. The most well publicized political conflict revealed a deep rift between the Champaign, Illinois IMC and allied US activists, and those belonging to the Argentina IMC. The US sites, which operate within a voluntary sector sustained by state and private foundation funding, wanted to accept funding from the Ford Foundation; the Argentinians, very mindful of Ford's history in the dirty wars of the Argentine Military Dictatorship in the late 1970s and early 1980s, vetoed the decision (Wolfson 2013).

Some commentators have highlighted internal problems among volunteers, exacerbated by longstanding residual power differences involving sexism, classism, racism and colonialism. Brooten and Hadl (2008), for example, documented the response of women in several centres to the dominance by professional-class men within collectives. Writing about the Vancouver IMC, Uzelman (2011) argues that the assumption that free access to information and the ability to post anonymously would produce radical democracy was challenged as women members of the collective began to leave the group, complaining of sexist practices, hate speech, spam and trolling online. Wolfson (2012) makes the point that the simplistic importation of Zapatista ideas to the very different contexts of urban rich countries led to static protocols and practices that prevented the network from expanding beyond the activist community. In addition, there was a reliance on technological fixes rather than investing sufficiently in building relationships and developing more innovative ways to work with existing movements, citizen groups and media organizations in their home communities.

Nevertheless, the legacy of the IMC continues in those sites that remain, their hundreds of alumni, and in the values and practices of radical media-making and democratic self-governance taken up by more recent social justice movements (Costanza-Chock 2012; Juris 2012; Pickard 2013; Stringer 2013; Giraud 2014; Wolfson 2014; McDonald 2015). The centres in Latin America, Oceania, Western Europe and the United States that continue to operate perhaps do so because they are more deeply embedded in their local and regional social and media movements (Giraud 2014). A legion of citizen reporters, videographers and techies trained by Indymedia now work in alternative, public service and commercial media, as teachers, advocates and policy researchers, and/or are still active in social and political justice movements.

Perhaps most importantly, Indymedia provides a powerful example of how to open up autonomous public media spaces, especially urgent at a time of increasing state repression around the world. Moreover, building secure independent media platforms that are not subject to corporate and/or state control is proving ever more necessary as the Internet becomes increasingly subject to the dominance of a small number of economically powerful transnational corporations, and ever more surveillance by nation-states. Finally, Indymedia provides a living example of news-making from, by and of grassroots, citizens and movements from every region of the world, representing a radical vision of planetary possibility.

See also: autonomous movements; citizen journalism; community media; documentary filmmaking; journalism studies and citizen media; prefiguration; World Social Forum

Recommended reading

Giraud, E. (2014) 'Has Radical Participatory Online Media Really "Failed"? Indymedia and its legacies', *Convergence* 20(4): 419–437.

Provides a contemporary assessment of Indymedia, especially reviewing its legacy regarding participation in media-making and democratic governance. Addressing more recent critiques, it reviews the contribution of Indymedia centres in a range of regions (Latin and North America, Africa, West Asia and Western Europe), outlining how they dealt with problems and ending with a discussion of the continued importance of Indymedia practices in contemporary social movements.

Stringer, V. (2013) 'This Is What Democracy Looked Like', in J. Juris and A. Khasnabish (eds) *Insurgent Encounters: Transnational activism, ethnography, and the political*, Durham, NC: Duke University Press, 318–341.

A valuable review of Indymedia from the perspective of an active participant (1999 to 2006) in the global video collective and US IMC sites in Texas. Emphasizes the highly participatory and collaborative production and distribution of videos and radio programmes by teams embedded within the global justice movement, arguing that the IMC contributed to changing how stories are produced and training thousands of activists in journalism and media-making.

Wolfson, T. (2014) *Digital Rebellion: The birth of the cyber left*, Urbana, IL: University of Illinois Press.

Charts the intellectual and technological history of the global Indymedia network, locating it within theoretical debates about participatory communications, democratic governance and social movement organization and politics. Outlining the lineage of the IMC from the Zapatista movement on through the Occupy Wall Street movement, it provides a valuable critique of the limitations of horizontalism, decentralization and the privileging of technological fixes for organizational problems to a powerful, inclusive and democratic media operation.

JOURNALISM STUDIES AND CITIZEN MEDIA

Karin Wahl-Jorgensen

In the field of journalism studies, examples of citizen media have been a major focal point for discussion, and have variously been viewed as an opportunity for audience engagement, an alternative to mainstream media or a threat to professional journalistic practice.

The main fault line within this body of scholarship can be found in the tension between understanding citizen media as a phenomenon which can be successfully integrated into the content and platforms of mainstream news organizations, as opposed to one which relies on bottom-up, organically emerging efforts by unaffiliated and non-professional citizens.

This entry begins by examining how mainstream news organizations have historically prioritized the creation of spaces and genres of audience participation. It shows that these initiatives, while based on commitments to democratic ideals, have been shaped by the logics of professional journalism, frequently constructing citizens as consumers and enabling them to make themselves heard only in reaction to the news organizations' priorities and agendas. The second half of the entry then explores how forms of citizen journalism have thrived outside the ambit of conventional media, due to the affordances of digital communications technologies. With the rise of blogging and social media platforms such as Twitter and Facebook, participatory opportunities have proliferated and diversified, and become unmoored from the practices, genres and epistemologies of professional journalism. Ultimately, scholarship in journalism studies highlights how citizen media are shaped by the power relations and political economy underpinning a dynamic and always-changing institution.

Citizen media and mainstream news organizations: historical contexts

The longstanding emphasis in journalism on facilitating citizen participation in the public sphere is not coincidental. As scholars have observed, such an emphasis is central to professional self-understandings and is tied to broader normative commitments to democratic ideals (Strömbäck 2005). James Carey (1987:5) famously described the public as the 'God term' of journalism, explaining that "[i]nsofar as journalism is grounded, it is grounded in the public. Insofar as journalism has a client, the client is the public. The press justifies itself in the name of the public". Carey (ibid.) therefore argues that the public is the totem and talisman of the professional journalist.

If the public is an object of ritual homage for journalism, it is also the case that the profession has always struggled with the challenge of how best to serve this all-important, yet elusive and contested category. The nature of journalists' relationship with the public – a troubled marriage if there ever was one – has been explored from numerous angles within journalism studies. For example, newspapers' letters to the editor sections have been conceptualized as important spaces for the expression and promotion of ordinary citizens' views and as arenas for broader public debate on matters of shared concern (Wahl-Jorgensen 2007). Letters to the editor have always been curated by news organizations and have therefore been fundamentally shaped by practices of selection and other professional routines, rather than emerging organically (Wahl-Jorgensen 2002b). Nonetheless, they have provided a means both for including personal stories that might otherwise remain unheard, and for forming communities of opinion and information sharing. As the journalism historian David Nord (2001) showed, letters to the editor in the *Federal Gazette* in late eighteenth-century Philadelphia played a vital role in promoting active citizenship during an epidemic of yellow fever. Community members wrote to the newspaper "passing along rumours, offering folk cures and remedies, speculating on the religious meaning of the disease, sharing their fears and their sorrows" (Nord 2001:200–201). Letters to the editor thus facilitated a form of citizen media which contributed to the emergence of imagined communities (Anderson 1991), whether local, regional, national or global.

Even if letters were, for a long time, the most prominent site for citizen contributions, other mainstream mass media, particularly radio and television, have for many years hosted forums and opportunities for public participation. In radio, participatory genres have been around as long as the medium itself, and have offered a space for the representation of the public, claiming to speak for the people (Loviglio 2002). For instance, the network radio programme *Vox Pop*, broadcast between 1932 and 1948, searched for the voice of the American people by interviewing individuals in the streets, with the stated intention of "posing questions of 'spectacular unimportance'" (ibid.:91). Broadcasters have, especially from the 1980s onwards, continued to make extensive use of vox pop interviews as a way of representing public opinion and including a variety of voices in the news (Lewis et al. 2005). Scholarship has shown, however, that vox pops do not reflect the balance of opinion on topics, with a large majority of news items based on such interviews presenting only one point of view (Beckers et al. 2018). While journalists endeavour to select vox pops representative of the population in demographic terms, and succeed in doing so on the basis of age and gender, they generally fail to represent "minority groups such as ethnic-cultural minorities and people with disabilities" (Beckers 2017). More fundamentally, news items that include vox pops have been shown to construct an inherently reactive and depoliticized public (Brookes et al. 2004; Lewis et al. 2005). Interviewees are invited to respond to political developments or policy proposals – rather than coming up with their own – and are usually included primarily on the basis of their personal experience and emotions, rather than their informed political opinions. This ultimately positions audience members as consumers rather than citizens.

Participatory genres emerging from within mainstream media organizations – commercial and public service-driven ones alike – have tended to promote audience engagement and voice on the basis of both normative and economic motivations (Lewis et al. 2005). With respect to normative motivations, the desire to include the voices of their publics has been driven by ideal understandings of journalism's role in society. These include the facilitation of popular participation and the representation of a diversity of views to counter the otherwise elite-focused content of news media. However, economic justifications are frequently more salient than normative ones in the minds of newsworkers (Wahl-Jorgensen 2002a). According to this logic, journalists often suggest that creating opportunities for audience participation is a way of ensuring loyalty among consumers.

While giving citizens a voice has thus been coupled with concerns about maintaining audience loyalty in the context of both public service and commercial media, community radio stations around the world have provided an important alternative model which is primarily focused on empowerment. Community radio has provided a way for otherwise disenfranchised groups to gain a foothold in the public sphere (Fairchild 2001:89; Riismandel 2002). As a 'third-tier' type of radio station, existing alongside public service and private broadcasters, community radio stations are "managed, run, controlled and owned by a community for the benefit of the community" (Nirmala 2015:41). Such stations, usually funded by governmental bodies or aid agencies, have played a particularly key role in mobilizing communities in developing countries (Manyozo 2010; Megwa 2007). Radio has long been the dominant medium across Africa because it is "pervasive, local, extensive, flexible, available, readily understood, personal, portable, speedy, and efficient" (Manyozo 2010:1). Though particularly widespread in Africa, community radio has also been important in India, where numerous studies (Nirmala 2015; Sharma and Kashyap 2015) have pointed to its role in empowering women by raising awareness, providing information and developing skills.

The forms of citizen media discussed so far, despite their diversity, are all driven by the logics of professional routines and shaped from the top down, rather than emerging from the independent efforts of citizens. The power relations underpinning the inclusion of citizen voices in this way have not been uncontroversial. If scholars have, as discussed here, documented the top-down nature of forms of citizen participation in mainstream news organizations, these concerns have also found purchase among journalists. Perhaps the most prominent attempts at addressing the power imbalances underpinning conventional journalistic practices can be found in the public journalism movement, which came out of the United States and reached its zenith in the 1990s. Arising out of concerns that the public was becoming increasingly disengaged with politics, the movement was premised on a commitment to enhance citizen participation in democratic politics (Rosen 2000:68). The public journalism movement sought out ways to report stories "from the perspective of ordinary citizens rather than articulating the viewpoints of senior political figures or local elites" (Franklin et al. 2005:214). To mention just a few examples, newspapers applied the movement's principles by inviting civic groups to audit their coverage, holding public forums to deliberate on key issues, and creating citizen panels to enhance election reporting (Rosen 1999).

Public journalism opened up a broader debate of lasting significance about the place of citizen participation in journalism and the structural limitations of existing forums. At the same time, it occasioned significant critical discussion (Davis 2000; Glasser 2000).

Scholars and practitioners alike expressed concerns that far from proposing a revolutionary challenge to the power structures of journalism, the movement represented a "market-driven gimmick to boost circulation" at a time of declining profits (Shepard 1994:30).

Citizen journalism: challenging professional authority from the bottom up

Given the tensions between the logics of mainstream news organizations and the normative ideal of promoting citizen participation, it is not surprising that those citizen media initiatives which have emerged from the bottom up, as a result of the organic self-organization of members of the public, have been the subject of significant scholarly interest. This has been particularly evident in attention to practices of what is frequently referred to as citizen journalism, or "journalism that is produced not by professionals but by those outside mainstream media organisations" (Atton 2009:265). Citizen journalists "typically have little or no training or professional qualifications; they write and report from their positions as citizens, as members of communities, as activists, as fans" (Atton 2008/2015, n.p.). The term citizen journalism was first used in the early 2000s and became a key focus of scholarship from the mid-2000s onwards (Allan and Thorsen 2009). Studies addressing this phenomenon were particularly drawn to exploring the ways in which the affordances of digital technologies – including not just the Internet but also smartphones and social media – have facilitated "the ordinary person's capacity to bear witness" (Allan 2009:18; see also Allan 2013). This, in turn, was seen to transform "what was once considered to be the exclusive domain of the professional", enabling *everyone* to be a journalist – even if only an "accidental" one (Allan 2009:18).

Despite the phenomenal growth in scholarship on citizen journalism during the first two decades of the twenty-first century, examples of citizens producing content outside mainstream media institutions can in fact be traced much further back in history. Different labels have been attached to such activities, each signifying their departure from conventional journalism: these include alternative journalism (Atton 2009), radical media (Downing 1984), participatory or amateur journalism (Lasica 2003), independent media, community media (Howley 2009) or citizens' media (Rodríguez 2001).

Scholars of these practices have historically taken an interest in media produced by progressive groups oriented towards social change. Examples include fanzines, radical leftist newspapers, political cartooning and the publications of social movements (Atton 2009). In the eyes of those studying these phenomena, such 'rebellious communication' is important not just because of its content, but also because of its mode of organization, production and distribution, which contrasts to the hierarchical structures of the mainstream media (Downing et al. 2001). For Rodríguez (2001), the significance of such media practices is in their pedagogical potential, providing a means for empowerment by enabling ordinary people to represent themselves and their communities. Indeed, even if citizen journalism faces many of the same problems as conventional media, the means of finding solutions to these problems are based on a "radically different interpretation of journalistic ideology", as Platon and Deuze (2003:336) found in their study of the Indymedia movement. The ideology of the Indymedia movement entailed, among other features, a non-hierarchical relationship with audiences and a strong commitment to transparency (ibid.). Scholarly approaches to alternative media, then, single out the ways in which such practices challenge prevailing power relations between journalists and citizens.

However, the potential for citizen media to emerge as a credible alternative to mainstream journalism, signalled by the rise of the citizen journalist, was dramatically transformed by the Internet, and subsequently social media. For example, blogging platforms were heralded as key tools for facilitating the participation of ordinary citizens: bloggers were seen to be able to share their opinions and insights in an egalitarian media ecology which challenged the authority of journalism (Domingo and Heinonen 2008). As the BBC's head of news at the time, Helen Boaden, said in 2008, with "blogs in particular – but also podcasts and videoblogs –the ability of the public to express opinion in public has exploded – especially in the USA – and they no longer need to be 'hosted by broadcasters'" (Boaden 2008, n.p.). Since then, blogs have become a stable and institutionalized part of the media landscape. This has meant, among other things, that they have been thoroughly integrated into the toolbox of both citizen media producers and professional journalists. As a result, scholarly assessments of the ability of blogs to transform forms of citizen participation have been somewhat tempered by evidence that the blogs which are most widely read are primarily those hosted by established media organizations and digital native media such as the *Huffington Post*, rather than those written by ordinary citizens (King 2015). Nonetheless, numerous fashion, sports and 'mummy' bloggers have made successful and lucrative careers (Hunter 2016; Pedroni 2015), while political bloggers have been seen to make a difference to global and national public debates, providing a distinctive space for public debate (Russell and Echchaibi 2009).

Contemporary practices of hyperlocal journalism often draw on the platforms, modes of presentation and epistemologies of blogging (Newman et al. 2012). Hyperlocal journalism, usually produced by non-professionals operating on a shoestring budget, seeks to fill the gap left by the decline in legacy media of coverage of issues relevant to local communities. David Kurpius and his colleagues, who have done extensive work on hyperlocals in the US context, define them as "geographically-based news organisations that operate largely in big metropolitan areas and cover a narrow range of location-specific topics. Such sites allow input from citizens through content contribution, blogs, and other feedback loops" (Kurpius et al. 2010:360). As such, hyperlocal journalism reflects the growth of networked non-professional journalistic practices within a complex media ecology. The ecology of the 'networked fourth estate' (Benkler 2011) encompasses legacy media and digital native players operating in complementary ways with and alongside practices of citizen journalism. This, in turn, reflects a monumental shift in power relations between journalism professionals and citizens.

Future directions

Since the mid-2000s, social media sites including Twitter, YouTube, Facebook and Instagram have enabled users to share their own content independently of the intervention of media organizations. This development has been seen to offer the promise of a "networked citizen-centred perspective providing opportunities to connect the private sphere of autonomous political identity to a multitude of chosen political spaces" (Loader and Mercea 2012:2). At the same time, as Papacharissi's research on social movements' use of Twitter showed, the affordances of the social media site have often paved the way for the emergence of "affective news streams" which are simultaneously humorous, cacophonous, collaborative and antagonistic (Papacharissi 2015a:113). The notion of affective news streams helpfully adds to a growing scholarly vocabulary which signals the distinctive forms of discourse that emerge when citizens co-create meaning and narratives in the public sphere. These, in turn, increasingly spill over into the epistemologies of professional journalism. Such narrative practices tend to be characterized by a challenge to the journalistic ideal of objectivity, and the introduction of situated, embodied and political registers (Blaagaard 2013a). More than anything, the scholarly consensus on citizen journalism suggests that it has the potential to reshape conventional understandings of journalism as detached, impartial and objective, by developing forms of storytelling that are emotional and personal and have the potential to cultivate empathy and hence generate a new moral imagination (Chouliaraki 2010a; Blaagaard 2013a, 2013b; Wahl-Jorgensen 2018). The proliferating opportunities, sites and circulation of citizen media, then, represent not merely a challenge to the forms of production, content and distribution of journalism, but also to its ways of knowing.

Scholarship in journalism studies has long been preoccupied with citizen media. Despite the clear affinity between practices of citizen media on the one hand, and journalism on the other, the relationship has been a challenging one, reflecting struggles over the power and authority to shape debate in the public sphere. Mainstream news organizations have always endeavoured to create spaces for citizen participation, often seeking to direct and carefully manage such spaces in ways that fit within existing agendas. In contrast to this top-down approach, scholars have long shown an interest in practices emerging outside mainstream news media, created by non-professional citizen journalists. Such citizen journalism has taken on particular force in the digital and social media era, working to contest both the practices and the epistemologies of conventional journalism.

See also: citizen journalism; community media; convergence; hyperlocal media; Indymedia; performance studies and citizen media; political science and citizen media; public sphere; wikis

Recommended reading

Allan, S. (2013) *Citizen Witnessing: Revisioning journalism in times of crisis*, Cambridge: Polity Press.

Taking the concept of witnessing as its starting point, Allan's book explores the intersections of ordinary individuals' reporting in crises and journalistic practices. In this way, Allan evaluates the literature on the topic of citizen journalism, conceptualizes citizen witnessing, and argues for a rethinking of journalism's public service in light of participatory culture.

Papacharissi, Z. (2015) *Affective Publics: Sentiment, technology, and politics*, Oxford: Oxford University Press.

Papacharissi argues that technology brings us into networks, but affects connect us to each other. The importance of affectivity to the creation and support of online publics is developed through the exploration of the cases of the Egyptian uprising of 2011–2013, the social movement of Occupy Wall Street on Twitter and in everyday expressions online.

MEDIA

Marwan M. Kraidy

In everyday language, media has come to refer to a familiar range of technologies and institutions that produce and circulate news and entertainment. Returning to the etymological origins of the term enables us to accommodate a broader range of meanings. The singular form *medium* has been used in the English language since the seventeenth century, and typically "refers to something that lies between two objects and links them" (Miller and Kraidy 2016:4); this sense derives from Latin, where *medium* originally meant 'middle layer'. Drawing on these roots, the term media is now used by researchers based across the humanities and social sciences to encompass any material or immaterial tool, object or environment that enables or facilitates the storage, exchange, dissemination and reception of information of any kind. This may include, for example, instruments like heliographs, the large mirrors that the ancient Greeks used to reflect sunrays in systematic rhythms to communicate over distance, in addition to the latest technological gadgets designed in Silicon Valley. It also encompasses the material and immaterial tools, objects and natural environments that individuals and groups have used at different times in history and in various locations around the world for purposes of expressing themselves in public, making social demands and engaging in political struggles.

The narrower, albeit more pervasive, use of media to designate the instruments of mass communication, such as newspapers, radio and television, was established in the middle of the twentieth century. Prior to this date, it was generally the term *the press* which was used to refer collectively to such phenomena. The academic study of the means of public communication mirrored these changes, as departments that were once focused on the press, mass communication or journalism changed their names to become departments of media studies or media and communication studies. The same shift occurred with titles of journals: for example, it is telling that in 1999, *Critical Studies in Mass Communication*, the flagship media journal of the United States-based National Communication Association, was renamed *Critical Studies in Media Communication*. The term social media has come to designate digital networked platforms such as Facebook, Twitter and YouTube, as if film and television were not profoundly social means of communication. The word *platform* was deliberately deployed by the public relations agents of companies like Facebook and YouTube, who argued that they were not media – that is, producers or distributors of content – but merely platforms that other media companies and ordinary people use (Gillespie 2010). Though the strategic goal of these for-profit companies in distinguishing platforms from media was to avoid government regulation, the distinction does not withstand critical scrutiny. As facilitators and amplifiers of information exchange between people, platforms are media; as corporations supported through advertising revenues, they are no different from other commercial media.

Other technologies have come only more gradually to be defined as media. In its early days, the telephone was not considered a media form, even though it did make possible the exchange of spoken words over long distances. This was largely because of its initially limited capacity and scope: the technology permitted only the exchange of voice signals between two individuals connected by a network of physical cables. The advent of satellite and cellular technology, on the other hand, in addition to developments in computerization and miniaturization, dramatically expanded the potential capacity and uses of the telephone. By the late twentieth century, the global growth of mobile telephony and the transformative rise of smart mobile devices turned the telephone into a media form (May and Hearn 2005). Arguably, in the early twenty-first century mobile phones are the single most pervasive type of media worldwide. This is due to the decreasing costs of mobile devices and cellular services, just as much as it is due to the exploding range of things that human beings can do with their phones: speaking, writing, taking photographs, banking and shopping, as well as a variety of political practices within both democracies and autocracies. It is through mobile devices, for example, that citizens mobilize voters, organize protests, call elected representatives, circulate jokes ridiculing rulers and undertake many other such activities.

With the critical-cultural turn that entered the field of communication and media studies in the 1980s, the ground was set for a broader understanding of media. This can be attributed to some extent to the legacy of the Canadian media scholar Marshall McLuhan, who described various objects as media, and of the American cultural anthropologist Clifford Geertz, which gave rise to a cottage industry of studies based on his notion of culture as text. This body of scholarship influenced in turn the subdiscipline of rhetorical theory and criticism, and encouraged researchers in this field to become more directly concerned with media, rather than their traditional interest in oratory. At the same time, the development of British Cultural Studies and its variants in the US, Canada and Australia focused attention on a wide spectrum of popular culture personalities, artefacts and dynamics – from fashion to popular music – which similarly could be discussed as media in the broader field of culture. For example, Fiske (1989) introduced the notion of denim clothing, and jeans in particular, as a type of media, that is, as instruments of human expression, while Hebdige (1979) explored the notion of style, especially in the domain of fashion, as a key medium of subcultural identity. Feminist scholars argued that the body itself can be productively analysed as a medium (Durham 2011), and the spatial turn gave rise to a consideration of public space – city squares, public parks, major avenues – as media (Amin 2008; Kittler 1996; Chikamori 2009; Khalaf 2013; McQuire 2006). Every disciplinary development and, more importantly, every emerging interdisciplinary field, has thus broadened our understanding of media. By the early twenty-first century, in addition to those working in the fields of media and communication studies, scholars from linguistics and literature, sociology and anthropology, political science and international relations, religious studies and even engineering, had begun to study media.

The extent of interdisciplinary interest in media was manifested most clearly during the 2010s, the contentious decade that saw the rise of popular protests worldwide, from the Arab uprisings to Occupy Wall Street, from Podemos in Spain to the Umbrella Movement in Hong Kong. Scholarship on these movements has shown that citizens who rose up against political autocracies and socioeconomic injustice harnessed a stunningly wide array of media, mixing materials, objects, platforms and repertoires: umbrellas in Hong Kong, cooking pots in Egypt, flowers in Lebanon, bread loaves in Tunisia, tents on Wall Street (Feigenbaum 2014; Kraidy 2016; Tsui 2015). Although we would not, at face value, think of these objects as media, citizens who deployed them, sometimes for exceedingly practical purposes – in Cairo demonstrators wore pots as helmets to protect their heads from police truncheons and bullets – imbued them with deep symbolism and used them to convey social aspirations and political messages.

Most citizen acts of communication deploy a configuration of various media in order to accomplish their objectives, stemming from the human body and connecting bodies to public space. Scholars have been aware of this for at least a half century; the Canadian literature professor and media theorist Marshall McLuhan (1964; McLuhan and Zingrone 1995) achieved fame in the 1960s partly by arguing that media were extensions of the human body. Since newspapers, radio, television and online platforms enjoy a hegemonic semantic status when we think of media, the remainder of this entry will focus on other, less taken-for-granted media, their interconnections, and the way they link up to traditional print and electronic media: the human body, public space and the natural environment writ large.

Bodies as media

Most communication processes begin with the human body. We move our tongue and open and close our mouth to speak, sing or shout political slogans. We raise our eyebrows to convey puzzlement, clench our facial muscles when we are angry, and raise our fists as we move our bodies in public space to convey political defiance. Even digital communication begins with the body: we apply our finger to a screen to start our device with fingerprint recognition, we use our fingers to type a text or our voice to dictate it. We switch music or television with a remote control. We type on keyboards. These are merely some of the basic communicative embodied practices that make our body a medium, even and perhaps particularly in the digital age (Kraidy 2013). Indeed, as the physical operator of other media and the perceptual and experiential processor of their output, the body is a meta-medium that connects various media in communicative configurations the emergence, operation and impact of which are highly dependent on context. As Durham (2011:55) explains, bodies not only operate media from keyboards to print matter, but bodies are also "represented and signified through the media", and the diversity of these bodies spawn complex "relationships, their rather singular joinings, with the media they engage". The best way to enable an epistemologically broad understanding of media, then, is to consider interactions between various media and the role of the human body in connecting them to one another.

The popular protests that exploded in the Arab world in 2011 provide excellent examples of how the human body interacts with other media, particularly platforms like Twitter, Facebook and YouTube. These cases show how the physical travails of bodies can be publicized by digital communication, or conversely, how bodies can launch themselves in discursive space through media. The 'We Are All Khaled Said' Facebook page is widely understood to have played an important role in starting the 25 January 2011 revolution in Egypt, given that it was here that pictures were posted of a smiling Khaled Said next to a picture of his face mangled by the police as he was being fatally beaten (Herrera 2014; Kraidy 2016; Poell et al. 2016). Words describing, and pictures and videos showing the aftermath of the Tunisian fruit vendor Mohamed Bouazizi's self-immolation are widely credited with having triggered the Tunisian revolution in December 2010. Although we do not have photos or videos of his act, a picture of his agonizing, bandaged body in the hospital, and later his portrait, became emblematic of the Jasmine revolution (Kraidy 2016; Zayani 2015). In contrast to Bouazizi's radical, death-causing act, we may think of Egyptian communication student Aliaa Al-Mahdy, who posted nude self-portraits with political messages on her blog in November 2011, and the resulting scandal in Egypt's revolutionary public sphere, as a more gradual activist act. If Mohammed Bouazizi's self-immolation demonstrates how offline revolutionary bodily acts may reverberate translocally across various media, Aliaa Al-Mahdy's act of online self-mediation proves that the reverse is also true: bodily speech acts can migrate from cyberspace to have real-life consequences by triggering heated debates, affecting political events, and revealing latent red lines in social norms and political ideologies in the supposedly non-ideological Arab uprisings (Kraidy 2010).

It is important to consider a broad spectrum of media because a narrow understanding of media as only print and electronic conveyors of text and images is a normative and exclusive space, predominantly Western, white and male. Aided by digital communication, the body is at the centre of an alternative conception of the public sphere which, while it applies with particular acuity to the Arab uprisings, can be extended to public discourse in other times and places. Whatever we may think of their practices, Bouazizi's and Al-Mahdy's provocations (Kraidy 2016), but also the activism of Femen (Reestorff 2014) and Pussy Riot (Bernstein 2013), exemplify a performative and contentious approach to the public sphere, one where the human body is at once a medium of expression and a symbolic battlefield. Though these activists achieved publicity by linking their body to social media, the same kinds of contentious and performative citizen engagement arise around entertainment television (Kraidy 2010; van Zoonen 2004). By accounting for the means of public communication outside of the historically normative, elitist, masculine, cerebral and deliberative norms of public discourse, as reflected for example in Habermas' theory of the public sphere, emphasizing the political potential of the human body yields a more inclusive model of public discourse. Connections between human bodies, technological platforms (Twitter, Facebook, YouTube) and media genres (entertainment, news, reality television) are central to understanding contemporary media.

Public space as media

In addition to the human body, public space can operate directly as a medium (Amin 2008; Khalaf 2013; McQuire 2006). City space forms collective urban culture (Amin 2008), facilitates and shapes human interactions with physical objects (Kittler 1996), and can be a canvas for human expression (Kraidy 2016; Chikamori 2009; Kittler 1996). In the age of drones, when our perception of the city gains the aerial angle, the city literally becomes a surface for citizen creativity: in October 2017 the Beirut-based rap and graffiti crew Ashekman completed a monumental piece of public art by painting the word *salam*, Arabic for 'peace', on the rooftops of eighty structures across more than one kilometre in the Lebanese city of Tripoli (Macguire 2017). Various forms of wall painting, from stencil graffiti to freestyle tagging to mural art, have long been one of the most prevalent forms of media in urban environments. Graffiti is understood as a practice of inscription, given that it connects material surfaces with human language and imagery (Chmielewska 2007). Graffiti is a medium because it exists

> at the point of convergence and tension, the point of juncture of the material and the immaterial – public *space* and the public *sphere* – a public *domain* appears, which is the territory of shared attention and the field of distribution of immediate and mediated visibilities.
>
> *Brighenti 2010:326*

Graffiti is an important form of media partly because the barriers to entry are extremely low, and so anyone with time, some paint and some graphic skills can theoretically inscribe their thoughts on a wall. This is significant in a context where, as the Argentinian-Mexican cultural critic García-Canclini put it, "the majority of the visual messages found in a city have been imposed" (1998:215). Against the state's monuments and capitalism's billboards, graffiti enables citizen expression. This became evident during the Arab uprisings of the early 2010s as cities like Cairo, Damascus, Manama, Tunis and even Beirut were covered with various kinds of graffiti (Abaza 2013a; Georgeon 2012; Khalaf 2013; Kraidy 2016). There is an important lesson here about how precarity and danger shape different kinds of graffiti, and by extension, media. During the uprisings, the level of danger clearly influenced the types of revolutionary graffiti displayed. Tunisia and Egypt presented larger-sized freestyle political graffiti, even murals, most famously on Mohammed Mahmoud Street in Cairo, but in the vastly more dangerous and deadly Syrian revolution, stencil graffiti dominated (Kraidy 2016).

Environments as media

Beyond the human body and public space, scholars from Gaston Bachelard (1938/1964) to John Durham Peters (2015) have argued that natural environments operate as media or shape natural objects as media. These approaches challenge anthropocentric understandings of media and communication, to include a variety of animals and natural phenomena (Parikha 2010; Peters 2015), in addition to articulations of the relationship between human and non-human actors developed by French theorists (Deleuze and Guattari 1980; Latour 2006). For insects, air is a medium, since it not only provides them with information about weather conditions, guiding where and how to fly, but air is also the medium itself through which insects move from one breeding ground to another, following the rhythm of the seasons. For dolphins, the ocean is a medium, since water enables them to exchange information with each other by transporting a specific language made of sounds that individual dolphins emit. The water of the ocean is also the physical matter through which dolphins move their bodies.

Water and air are not the only elemental media, as scholars call environments as media (Bachelard 1938/1964; Peters 2015). The scholarly examination of fire and flame as media has a long and illustrious history: research on elemental media draws on an old and rich history of thought in theology, philosophy and psychoanalysis. The sacred texts of the three monotheistic religions contain ample evidence of the use of fire as a medium, typically deployed to convey divine wrath and inflict punishment (Grant 2015; Labahn 2006; Rustomji 2009). The works of Bachelard (1938/1964, 1961) established a conceptual framework to understand fire as a medium: Bachelard defined the flame as a great operator of images and fire as a potent exciter of the human imagination. As is typically the case, the elements englobe other media to create 'aviation media' like drones (LaFlamme 2017) or 'atmospheric' and 'stratospheric media' (McCormack 2017). In other words, elements like air, water and fire are meta-media that shape the operations of all other media.

Media, then, encompass a broader spectrum that ranges from paper and electronic means of communication (from the print newspaper to digital networks), to the human body and its varied modalities, to public space and the natural environment writ large. Throughout history, subjects and citizens, peasants and aristocrats have used all these, typically in some kind of combination, to express themselves in the private and public spheres.

See also: activism; graffiti and street art; performance studies and citizen media

Recommended reading

Bachelard, G. (1938/1964) *The Psychoanalysis of Fire*, translated by A. Ross, Boston, MA: Beacon Press.

Offers a historically deep and geographically broad explanation of the communicative powers of fire, articulating the use of the flame in various cultures in connection with rituals of belonging, norms of social interdiction and the human imagination. Provides a framework for understanding the importance of natural elements as expressive media.

Kraidy, M. M. (2016) *The Naked Blogger of Cairo: Creative insurgency in the Arab world*, Cambridge, MA: Harvard University Press.

Brings together a variety of activists and citizen media practices, including digital memes, puppetry, naked activism and self-immolation, to emphasize the central role of the human body in citizen media, and identifies two broad domains of mediated activism, one gradual and another radical.

Zayani, M. (2015) *Networked Publics and Digital Contention: The politics of everyday life in Tunisia*, Oxford: Oxford University Press.

Uses the Tunisian uprising of 2010–2011 to identify, analyse and contextualize patterns of digital networking and contention in everyday life and popular culture. Provides a nuanced analysis of the interactions between citizens, media infrastructures and structures of political power.

MEDIA ECOLOGIES

Emiliano Treré

A number of theories conceive communication technologies as ecologies. The medium theory approach invites us to understand media as environments and ecologies, to look at them from a macro evolutionary perspective and explore the reasons for their extinction or survival, focusing on the coexistence and coevolution of multiple media. Adopting a more micro, less media-centric perspective, the information ecology approach sees an ecology as "a system of people, practices, values, and technologies in a particular local environment" and puts the emphasis "not on technology, but on human activities that are served by technology" (Nardi and O'Day 1999:49). Unlike the medium theory approach, which is traditionally more interested in media effects, the communicative ecology perspective – similarly to information ecology – underlines "meaning that can be derived from the socio-cultural framing and analysis of the local context" of communication (Hearn and Foth 2007). A communicative ecology is conceived as a milieu of agents who are connected in various ways by different exchanges of mediated and unmediated forms of communication, along three different layers: technological, covering the devices and connecting media that enable communication and interaction; social, which includes people and their social modes of organization; and discursive, which consists of the very content of communication (Tacchi et al. 2003). Inspired by the work of Felix Guattari, on the other hand, Fuller criticizes the use of the environment metaphor in the medium theory approach, because it suggests "a state of equilibrium" and conceptualizes media ecologies as rather static and stable. He proposes instead that media ecologies be defined in terms of "dynamic interrelation of processes and objects, beings and things, patterns and matter" (2005:2).

Since each approach stems from often conflicting traditions, these media ecological perspectives differ in various respects. For instance, the medium theory maintains a macro-level approach to the analysis of technology and society that often disregards social practices and appropriations, and tends to conceive technologies as the driving forces of societal change. It has hence been accused of technological determinism by diverse authors and traditions, including Williams (1974), the 'Social Shaping of Technology' approach (MacKenzie and Wajcman 1999) and theorists working within the mediatization paradigm (Hepp 2013). Despite these critiques, medium theory continues to offer many valuable insights. In particular, it presents a convincing critique of additive media change, deploys powerful metaphors of media as ecologies and environments, provides a welcome focus on the materiality of technological artefacts and insists on looking at media from a more holistic point of view. Information ecology is able to supplement medium theory by relocating concepts like coevolution and coexistence from the macro level to the micro/local dimension of analysis, placing attention on people's practices, needs and the values they attribute to technology. In a similar vein, communicative ecology brings more attention to the study of the complex interplay between the technological, the social and the discursive levels in situated social contexts. Finally, Fuller's approach to media ecology, which draws specifically on Guattari's work, makes three key contributions: it introduces a higher degree of dynamism and unpredictability within the media ecology; it reinstates the significance of the political nature of the ecology; and finally, similarly to medium theory, it demands that the materiality of communication technologies be seriously analysed. Although these approaches differ in various respects, their many strengths can be combined and their similarities foregrounded in order to highlight the fact that they all urge us to address media from a holistic perspective – to go beyond specific media instances, appreciating the complexity of media as empirical phenomena, and recognizing the importance of studying the interconnections between their materiality, the values attributed to them, the practices developed around them and the contents conveyed through them.

Attracted by the promises of a media ecology framework, scholars interested in citizen media, communication technologies and social movements have increasingly started to use the ecology metaphor in their studies. The remainder of this entry addresses the why and who of ecological adoption, charting the reasons of this interest in various research fields. It then discusses the how dimension, mapping the ways through which media ecology has been used with different degrees of intensity. The what dimension, finally, outlines the benefits of media ecology through various examples.

The why and the who of media ecologies

The metaphor of media ecology is increasingly adopted to make sense of the uses of communication technologies in protests and mobilizations, and more generally in various processes aimed at social change and political transformation. Because most of the authors who adopt the media ecology lens have a background in social movement studies, their understanding of citizen media is both similar to and different from that of citizen media scholars such as Rodríguez and Baker and Blaagaard.

Rodríguez calls for the study of media at the margins, focusing on processes of cross-pollination, adaptation, hybridization and replication that are often visible in grassroots media. She urges citizen media scholars to explore communication rhizospheres, meaning the micro-structures and processes that promote a range of media and communication practices within social movements. Grassroots communicators, she argues, operate in a communication rhizosphere where

> media use is not determined by whether technology is old or new, digital or not digital; what determines media use is a flux of historical information and communication needs, and how embedded community communicators employ available technologies to address these needs.
>
> *2017:58*

Baker and Blaagaard, on the other hand, define citizen media as encompassing

> the physical artefacts, digital content, practices, performative interventions and discursive formations of affective sociality produced by unaffiliated citizens as they act in public space(s) to effect aesthetic or socio-political change or express personal desires and aspirations, without the involvement of a third party or benefactor [as well as] the sets of values and agendas that influence and drive the practices and discourses through which individuals and collectivities position themselves within and in relation to society and participate in the creation of diverse publics.
>
> *2016a:16*

The similarities between Rodríguez' and Baker and Blaagaard's conceptualizations on the one hand, and those of social movement scholars using the ecological lens on the other are strong. They are both able to overcome the three fallacies that have plagued research on citizen media, social movement communication and digital activism (Treré 2018:43): the fallacy of spatial dualism, the fallacy of technological presentism, and the fallacy of alternativeness. The first fallacy refers to the tendency to treat the digital/virtual/cyber and the offline/physical/real differently (Lim 2015). By paying attention to face-to-face relationships, foregrounding the relevance of materiality and the continuous interplay between the digital and the physical spheres in contemporary protests, both approaches transcend this fallacy. The second fallacy refers to the tendency to overemphasize the importance of the latest technology or platform. By adopting a holistic approach to communication that is not restricted to the digital sphere and in particular to social media platforms, both approaches are able to overcome this fallacy. The last fallacy concerns the tendency to remain uncritical with respect to the corporate nature of digital platforms, overlooking the contradictions of social media power and their implications for activist practices. Both approaches consider the pervasive role that corporate social media platforms play in contemporary activism, emphasizing the hybridity and the impurity of the media ecologies that activists constantly navigate in their protest-related activities, along with the contradictions and the ambivalences that this navigation entails. Both reflect on the consequences of the different forms of appropriation of citizen media power by capitalism for the redefinition of citizen media themselves, and on the reconfiguration of the "relationship between the private and the public, the local and the global, mainstream and alternative media, corporations and citizens, the state and the individual" (Baker and Blaagaard 2016a:19).

However, there are also a number of differences between the approach of citizen media scholars and that adopted by social movement scholars. First, the most obvious is that while the former does not usually use the concept of media ecology, the latter has embraced it, though from different standpoints and with different degrees of sophistication. Second, social movement researchers who adopt the media ecology trope usually conceive citizen media only as part of the broader communication strategies of protest movements. Thus, their main focus of analysis is the whole array of social movement media practices, and only subsequently practices associated with citizen media. Consequently, not all the media used by social movements might be considered citizen media. Activists might rely on and appropriate mainstream media, sympathetic media such as progressive magazines and left-wing newspapers, and corporate digital media, including Facebook, Twitter and YouTube. In the social movement literature, this type of media would not usually be understood as citizen media – although, as the case study from Harlow (2016) below clearly demonstrates, corporate social media can be reconfigured as a form of citizen media in specific contexts. Third, while citizen media scholars tend to foreground unaffiliated citizens and collectivities, social movement studies scholars privilege the collective nature of activist media practices and include citizens associated with established organizations in their purview.

How media ecologies are adopted

A critical mapping of this emergent area of research reveals a very diverse scenario, with no clear boundaries. In particular, clear explanations of what constitutes media ecologies are lacking, and the elements that would constitute them are either not clearly specified or differ greatly from author to author. There are two major reasons for the absence of a shared definition of what constitutes a media ecology. The first is the diverse background and approaches of scholars who are adopting this conceptual lens. While social movement and citizen media scholars tend to be at the forefront of this emergent field, scholars from disciplines as varied as political science, media sociology, cultural studies, media studies, Internet studies and science and technology studies are also embracing this perspective. While this diversity of approaches is enriching and challenging, each brings to the debate a different interpretation and a set of academic jargon. The second reason has to do with the variable levels of engagement with earlier conceptual traditions that adopted and developed the concept of media ecology. Here, as has been noted (Treré 2019), the literature varies significantly and can be appraised along a continuum, with a rather generic adoption of the concept at one end, and a profound engagement with media ecologies at the other.

Some studies only evoke media ecology to highlight the increasingly complex interplay between traditional and digital media in contemporary protests. Reflecting on the 2010–2011 uprisings, for instance, Darmon (2013:1) states that portable devices like smartphones and social media platforms were combined with more traditional mass media channels, giving rise to "new media ecologies". Other work on the Arab uprisings refers to hybrid media ecologies which emerged through the combination of older and newer media technologies (Robertson 2013; Wilson and Dunn 2011). Thorson et al. (2013:421) suggest that the Occupy Wall Street movement's mobilizations in the US were characterized by "a loosely bound media ecology" in which digital material circulated across different social media platforms. In a similar vein, analysing media activism in the G20 protests in Toronto, Poell and Borra (2012:700) speak of an "activist social media ecology". In all these works, media ecologies are used generically to evoke the complexity inherent in the blending of old and new media and that characterizes contemporary activism. However, there is no attempt to provide more details on the composition of these ecologies, or to investigate the theoretical implications of adopting the metaphor.

Other studies develop a more structured understanding of media ecologies and explore some of their constitutive elements. For example, Srinivasan and Fish used the media ecology metaphor to describe the Kyrgyzstan uprisings of 2010, where a multiplicity of digital platforms were deployed in combination with low-tech media channels, stimulating the formation of community networks and grassroots coordination "through the re-mediation of messages via posters, megaphones, and word-of-mouth" (2011:3). Here, a close analysis of the media ecology reveals the ability of activists to effectively navigate these media ecologies to turn their local protest narratives into transnational discourses that inform multiple publics about political events. This was possible thanks to the role played by mainstream media networks such as CNN, Free Speech TV and Al Jazeera in rebroadcasting the content of activists' citizen media. Al Jazeera's role was particularly critical: it built alliances with social media activists and used digital media to share information via live Internet streams, and through a rebroadcasting agreement with Free Speech TV, a US-based non-profit satellite network broadcasting to 35 million American homes.

Similarly, in their analysis of Egypt's Tahrir Square protests, Tufekci and Wilson (2012) stress the need to move beyond a reductive focus on social media revolutions, and instead to consider the connectivity infrastructure of these events as a complex, intermeshed media ecology rather than in terms of specific devices or platforms. This ecology is constituted by three interrelated elements. First, satellite TV channels such as Al Jazeera and its key contribution in the formation of a new kind of public sphere in the Arab World. Second, the Internet, especially social media platforms such as Facebook and Twitter, with their capacity to alter the infrastructure of social connectivity. Finally, the spread and adoption of mobile devices, with their capabilities for dispersed communication and the increased possibilities they offer for engaging in practices of citizen journalism.

Another set of studies explicitly engage with earlier theorizations of media ecologies. For instance, Kahn and Kellner (2008) link their understanding of the role of citizen media to the media ecology tradition of the Toronto School, which theorized media as complex environments. Kahn and Kellner seek to expand the concept of media ecology as theorized by the Toronto School in order to accommodate newer technologies and to retheorize media ecologies from a critical and reconstructive standpoint: critical of corporate and mainstream uses of technology, and reconstructive in the sense of advocating for appropriations of technology that can advance social and political struggles (ibid.:23). Peeples and Mitchell's (2007) research on the protests against the 1999 WTO summit draws on the communicative ecology perspective (Tacchi et al. 2003). To investigate the role of the media in this context, they focus on three interconnected layers of the media ecology – the technological, the social and the discursive – to capture the organizational dynamics within activist networks, and the communication themes that transpired from the discussions that took place among activists. Other authors, such as Treré (2012) and Barassi (2015a), rely on the information ecology perspective developed by Nardi and O'Day (1999), connecting activists' practices, imaginaries and cultures with the material affordances of technologies. Their work shows how protesters choose their technologies, informed by their understanding of the risks of surveillance and commodification that some of these platforms intensify. These activists often use corporate social media in critical ways, simultaneously relying on a plethora of online and offline citizen media that – despite their more limited reach – protect their anonymity and digital rights without commodifying their protest activities.

Feigenbaum et al. (2013) borrow the language of media ecology to make sense of the multiple relations among social actors, things and environmental conditions in the context of protest camps. Guattari's reflections on the political value of media ecologies allow the authors to go beyond a mere environmental conception such as the one developed by the Toronto School, and to place the social and the political instead at the centre of ecological thinking. Feigenbaum et al. (2013) suggest that an ecological viewpoint can transform the ways in which activists think about their own positions and interactions within the media ecology, allowing us to "navigate the ways in which social movement ideologies are exchanged and carried into the reproduction of protest camps' infrastructures and practices" (ibid.:72). This aspect also allows us to trace these media ecologies as spaces where activists create, invent and experiment with citizen media.

What media ecology offers to citizen media studies

Exploring the full spectrum of practices around various media technologies allows the researcher to better appraise the effective participatory potential of each technology (Foust and Hoyt 2018; Mercea et al. 2016). This may reveal that some old technologies still play a fundamental role in contemporary activism, counteracting the uncritical celebration of the benefits of the latest technological platforms to appear on the scene. For instance, in his study of the student movement Anomalous Wave, which emerged in 2008 in Italy, Treré (2012) revealed the political relevance of mailing lists used by student collectives as a form of citizen media. This supposedly old digital platform played a key role in activists' protest communication, in contrast with the image portrayed by various Italian newspapers and academic accounts, which overemphasized the revolutionary role of Facebook and Twitter. A similar finding was reached by Bonini (2017), who demonstrates the significance of radios in the protests that took place in Turkey in 2013. Bonini studied the role played by *Açık Radyo*, the only independent and listener-supported radio station based in Istanbul, in the Gezi Park protests, concluding that radio has not lost its value as citizen media, but has only repositioned itself within the changing media ecology, blending itself with social media in order to continue amplifying radical political discourses and enabling activists to network. Similarly, in her comparative study of the media ecologies of various political organizations in Spain and the UK, Barassi (2013) emphasized the enduring political relevance of print magazines. In the digital era, these traditional forms of citizen media continue to operate and are continuously redefining their role in order to compete within a crowded media ecology where social and mobile media are given more prominence in relation to the spread of political messages.

As these examples illustrate, by embedding citizen media within a history of continuous adaptations, rejections and displacements, a media ecology approach allows us to appreciate how, why and under what circumstances their role changes. Hence, a media ecology approach is particularly suited to cross-temporal analyses that examine how the role of particular activist technologies has developed within specific social, cultural, economic and political contexts (Rinke and Röder 2011). For example, in her study of the media of anti-capitalist food activism in the UK, Giraud (2017) demonstrated how Indymedia, one of the most emblematic citizen media initiatives during the first half of the 2000s, has changed significantly due to shifts in activist media practices and in the broader media ecology. Giraud illustrates that in the context of food activism in the UK, Indymedia has not vanished but continues to fulfil an archival function, alongside other newer media that are used for coordinating more pressing political actions. Likewise, in her ethnographic case study of the Salvadoran group Activista and the launch of its Todos Somos Agua campaign, Harlow (2016) demonstrated that online social media, especially Facebook, were reconfigured as a form of activist citizen media in El Salvador. The activists she interviewed pointed out that they believed Facebook offered a space that allowed people with non-mainstream views to voice an opinion, making it possible for them to share news about mining, water contamination and other social issues that the public would otherwise never learn about. Furthermore, they saw Facebook as a reclaimed media territory for youths, who are normally excluded by mainstream media. Harlow's study shows that by adopting a media ecology approach we can better investigate how, despite enduring digital inequalities, social media can be appropriated in non-hegemonic and alternative ways.

Finally, a media ecology perspective can help us expand our conception of what traditionally constitutes citizen media, pushing us to pay more attention to what Feigenbaum (2014) calls the other media of activism. This includes, for instance, tents and tear gas in protest camps, as well as food-provision infrastructures in the case of the food activism movement described by Giraud (2017).

Future directions

While the media ecology perspective offers several advantages for the study of citizen media, there are some limitations to its application. It tends in some contexts to overemphasize the complexity of the citizen media used by activist collectives, which often rely solely on the production of video or use no more than a few media outlets (Rodríguez 2017). But its key weakness is that most studies approach media ecologies in a rather generic way, instead of providing more nuanced, detailed analyses. The result is that various studies still rely on an environmental and superficial conception of media ecologies, ultimately achieving no more than acknowledging and naming the diverse media technologies with which activists engage. In order to address the complexity of activist media, however, a more dynamic, practice-based understanding of ecologies is needed (Barassi 2015a; Mattoni 2017; Treré 2019), one that goes beyond the mere recognition of the multiplicity of today's protest media to investigate negotiations, choices, adoptions, rejections, displacements and appropriations of diverse media from a holistic and diachronic viewpoint. Such an approach would allow us to critically assess the shifting roles of citizen media in the digital age. It would also open up promising avenues of conceptual convergence with similar approaches developed in political communication, such as Chadwick's (2017) hybrid media system, which posits a move from a single mainstream media logic to a more complex ecosystem where competing media logics coexist and interact.

See also: convergence; disability media; documentary filmmaking; media practices; public sphere; social movement studies and citizen media

Recommended reading

Foust, C. R. and K. D. Hoyt (2018) 'Social Movement 2.0: Integrating and assessing scholarship on social media and movement', *Review of Communication* 18(1): 37–55.

Offers a critical review of the literature on the relationship between social movements and digital technologies from 2011 to 2017. Building upon media ecology scholarship, it critiques analyses of contemporary social movements that leverage technological determinism or isolationism to address the power of media in effecting social change and calls for greater complexity in scholarship at the intersections of movement and media.

Mercea, D., L. Iannelli and B. Loader (2016) 'Protest Communication Ecologies', *Information, Communication & Society* 19(3): 279–289.

This special issue recovers the ecological trope and accounts for its fall from grace in media and communication studies. Contributions underline its significance and illustrate how to reinsert it into the empirical terrain of interlocking media, communication practices and protest.

Treré, E. (2019) *Hybrid Media Activism: Ecologies, imaginaries, algorithms*, London: Routledge, 31–103.

Explores the key contributions of a media ecology approach for the study of the media/movement dynamic to overcome the communicative reductionism of the literature on social movement and communication. The five contributions are: media hybridity, media multiplicity, media evolution, media complexity and media criticality. A media ecology lens is then used to investigate the communicative complexity of various social movements empirically.

MEDIA EVENT

Evgenia Nim

A media event is a preplanned or spontaneous event which is represented by the media as socially significant and therefore attracts considerable public attention. Compared to other news, a media event possesses specific characteristics that are interpreted in various ways by different scholars. The theory of media events was first outlined by Daniel Dayan and Elihu Katz in their 1992 book *Media Events: The Live Broadcasting of History*. Since then it has been developed considerably as a result of numerous critical revisions. Among its potential extensions is the examination of media events within the context of citizen media, thus allowing an exploration of the concept of citizen media events.

The original definition of media events

The seminal work by Dayan and Katz (1992) places the concept of media events firmly in media studies. It interprets media events as televised public ceremonies of a historical nature, organized along the lines of three basic scripts: conquests, contests and coronations. Conquests refer to humankind's greatest achievements, such as successful peace negotiations or first pilots of space missions. Contests are remarkable events in the spheres of sports and politics, such as the Olympic Games and presidential debates. Coronations mostly take the form of parades, for example, weddings and funerals of royal family members. The drama of conquests revolves around the question 'will the hero succeed?'; of contests, 'who will win?'; of coronations, 'will the ritual succeed?'.

According to Dayan and Katz, media events have a number of features which distinguish them from all other events covered by mass media: the live broadcast; the interruption of everyday life and everyday broadcasting; the preplanned and scripted character of the event; the huge audience; the normative expectation that viewing is obligatory; the reverent, awe-filled character of the narration; the function of the event as integrative and conciliatory, bringing members of a society together (Katz and Liebes 2007:158). Unlike great news events, which focus on accidents and disruption, great ceremonial events are "the high holidays of mass communication" (Dayan and Katz 1992:1), celebrating order and its restoration. At the same time, media events do not offer a simple representation of socially significant events which would have taken place even without the presence of TV cameras. It is precisely the live transmission that contributes to the ceremony being perceived as highly important. The presence of the TV camera can further influence the way these events transpire, as well as their social effects. Thus, media events as understood by Dayan and Katz are different from pseudoevents as discussed by Boorstin (1961/1992), who focuses on events that are created exclusively for media coverage, such as press releases and news conferences.

Although ceremonial media events are initiated by society's elite or establishment, their success depends on all the actors involved: organizers, broadcasters and audiences. In European and North American societies, the media are not directly influenced by governmental organizations and can follow their own corporate standards and interests, which a media event has to adhere to. Broadcasters can choose which media events to cover and decide how to present them critically. Viewers also have veto power; they can apply alternative readings of a media event or ignore it completely. Thus, the hegemonic character of a public ceremony must be sanctioned by the general public, which imposes restrictions on the political elite's ability to manipulate audiences.

The three basic scripts of media events follow dramatic narratives based on the structure of the *rite de passage*, which includes three phases: separation, liminality and incorporation (Turner 1969/2008; van Gennep 1909/2010). Given the nature of live broadcasting, the audience as well as participants in the event interrupt their normal routine and enter a liminal space, a rite of passage, from which they emerge renewed. Media events are events that change the course of history. In this sense they possess a strong transformative potential, in certain cases realized in protests and rebellions such as the student riots in Czechoslovakia in 1989, which led to a peaceful dismantling of the socialist regime in that country. Analysing this Velvet Revolution, which received wide media coverage at the time, Dayan and Katz write that although television did not start the revolution, "it framed revolutionary actions as symbolic gestures that, together, had results in the real world" (1992:158). Such social dramas are not a typical form of ceremonial media events, which usually associate the elite with the sacred centre of the society and legitimize its dominance. However, all media events produce liminal moments during which the public plays out alternative scripts of the present and the future, opening up a utopian dimension.

Rethinking media events

Dayan and Katz' conceptualization of media events has been highly influential but has also been repeatedly contested and reinterpreted by other researchers, including revisions suggested by Dayan and Katz themselves (Sonnevend 2018). Criticism of the original approach mainly focuses on three issues. The first concerns the limitations of the functionalist paradigm and neo-Durkheimian reading of media events, which emphasizes the integrative role of mediatized ceremonies (Hepp and Couldry 2010). On the one hand, opponents of the ritual approach believe that modern societies are too complex and fragmented to posit the existence of a centre and of universally shared values which can be reproduced and confirmed by means of ritual media events. On the other hand, media events generate a "myth of the mediated centre" (ibid.:5), and in this sense function as "centering performances" (ibid.:12) that participate in establishing power relations and defining thematic priorities for specific agendas. The mechanisms for constructing this mediated centre, including the interactions between the hegemonic discourse and alternative discourses, must be critically examined. In addition, the Durkheimian sociology of national integration is scarcely applicable in the era of globalization, which is characterized by an intensive growth of translocal and transcultural connections; these connections blur the boundaries of nation-states and hence of national audiences. In these circumstances, media events are also globalized and become capable of centring not only certain societies or the whole world, but also various online communities.

The second issue concerns the narrow definition of the genre of media events and its three scripts. This critique points to the possibility of interpreting media events differently and to the need for broadening the meaning of the term (Cottle 2006; Hepp and Couldry 2010). In particular, it has been suggested that media events can include media scandals (Lull 1997), mediatized public crises (Alexander and Jacobs 1998), media spectacle (Kellner 2010) and popular media events (Hepp and Vogelgesang 2003). Some interpretations of media events are close to Boorstin's pseudoevents (1961) and Baudrillard's simulacrum (1981/1983). Thus, John Fiske's (1994:2) understanding of media events rejects a clear distinction between a real event and its mediated representation. At the same time, a media event is seen as a discursive event, not a discourse about an event. The meaning of the same social events can be contested by various media and audiences, which makes their integrative potential questionable. Looking beyond Dayan and Katz' assumption of media events as festive viewing makes it possible to distinguish between such genres as disruptive, traumatic or conflictual media events, including, first and foremost, terror, disaster and war (Katz and Liebes 2007). Finally, another type of event – namely, strikes and protests – also targets social and cultural change but does not exactly fit the integrative/disruptive opposition originally proposed by Dayan and Katz.

A final criticism concerns treating media events exclusively as television events, without taking into account other types of media and their interaction. Although the focus on television as a key transmitter of such theatrical media events in the early 1990s may have been justified, over the following quarter of a century media ecology, media technologies and practices of communication changed dramatically. In the era of deep mediatization (Couldry and Hepp 2017:53), professional journalists are not the only producers of content; digital media users are also able to produce and circulate multimedia content. Media events circulate across multiple media platforms, integrated into the digital environment through the Internet and mobile communications. In this context, it seems useful to consider media events as new media events (Neverson and Adeyanju 2017), transmedia events (Bacallao-Pino 2016), hybrid media events (Sumiala et al. 2016) or user-generated media events (Mitu 2016), reflecting their heterogeneity and the complex constellation of actors involved. Indeed, the development of artificial intelligence technologies and various processes of media robotization means that it is already possible to forecast the emergence of algorithmic or automated media events.

The prospects of integrative media events, with which the research of mediatized social dramas began, seem increasingly limited, although such ceremonies as the first Obama inauguration (2009), the wedding of Prince William and Kate Middleton (2011) and the Nelson Mandela memorial (2013) can still excite and move many people. As Dayan and Katz (1992:152) suggest, the rapid demise of common media events is largely connected with the emergence of new media, but chiefly with the increasing cynicism of the audience, for whom the sacred dimension of celebration, which requires suspending disbelief, has lost much of its appeal.

Overall, criticism of the initial approach to media events allowed a reconsideration of their ontology, broadening their typology and outlining new directions of research (Couldry et al. 2010; Fox 2016; Mitu and Poulakidakos 2016; Sonnevend 2016). However, such a broad understanding of media events leads to difficulties in identifying them, because the boundaries between media events and non-media events become harder to discern. These difficulties also arise in relation to transformative media events, a concept which is crucial for the subsequent definition and conceptualization of citizen media events.

Transformative media events

Dayan and Katz (1992) suggest that some media events possess a pronounced transformative function. Although any social event effects an alteration, whether qualitative or quantitative, of the initial given reality, in certain cases these changes, shifts and perturbations prove critical enough for the social order or particular elements of it to undergo a radical revision. Such turning-point events are not transformative in themselves, but they are defined as such by their participants and observers. Transformative media events gain their transformative power and significance precisely through the media; they are constructed by means of expressive images and narratives which circulate in a given media environment.

An attempt to identify transformative media events leads to numerous questions concerning both the definition of transformation and the definition of media events. The first question concerns where the boundary between transformative and non-transformative events lies. Dayan and Katz distinguish transformative media events from three other types of ceremonies: commemorations (national holidays and jubilees), responses (society's reactions to urgent situations) and restorations (such as royal weddings, associated with the era of monarchies). Unlike these ceremonies, transformative media events represent challenges connected with unresolved, longstanding social conflicts. Such events as riots and revolutions have an antistructural, subversive character. Beginning spontaneously, they pass through five stages: latency, signalling, modelling, framing and evaluating. The work of the media is particularly important at the three intermediate stages, when an alternative image of the desirable future is shaped. Transformative media events have "a causative power" (Dayan and Katz 1992:155) akin to that of a shaman: symbolic gestures help to define and shape a new reality. For Dayan and Katz, a transformative media event must necessarily be a success. However, it is not clear whether a protest that is suppressed or does not reach its original goals renders the media event in question non-transformative, or whether its success can be measured by the level of media response and audience involvement, as well as its less noticeable and more remote social consequences. The depth of social transformations and the degree of fulfilment of expectations remain somewhat unclear criteria for defining a transformative media event.

The second question concerns whether transformative media events should have a certain vector; in particular, whether they should be integrative and progressive. Dayan and Katz' approach suggests an affirmative answer in both cases, since they tend to refer to non-violent social changes that are supported by the majority, and which are directly or indirectly identified with the triumph of democratic values. For instance, Watergate, which resulted in US President Richard Nixon's resignation, publicly denounced the violation of American democratic principles and led to their reestablishment. The 1989 Velvet Revolution in Czechoslovakia was perceived not only as evidence of the collapse of socialism, but also as evidence of the country's return to the fold of the Western world. Similarly, the Egyptian revolution of 2011 was seen by many as a liberation movement fighting against tyranny. However, riots rarely take place without casualties; the society may be deeply divided; and the new regime's legitimacy is not always self-evident. A revolution can lead to a counter-revolution or a civil war, as in the case of Syria. And these are transformations, too. Transformative media events are often multi-vectored, because they are built around conflicting social interests, agendas and discourses. This also applies to events that take place in developed democracies, as in the case of the UK's 2016 decision to leave the European Union and Catalonia's attempts at separation from Spain. Opponents and proponents of the exit option in both cases have different definitions of what progress or regress means for their region, country or the international community. Vectors of social change are thus ultimately discursive in nature.

The third question is whether transformative media events are predominantly political phenomena, connected to a sudden replacement of elites and changes of social and political regime, or whether certain events in other segments of social and cultural life may also count as transformative media events. The scandal involving Hollywood producer Harvey Weinstein, which erupted on 5 October 2017 when *The New York Times* reported allegations of sexual harassment of film actresses by Weinstein, sparked an international campaign against sexual assault. The Internet flash mob which used the hashtag #MeToo to encourage victims of sexual harassment to share their experiences on Twitter and other social media attracted millions of people all over the world, including a number of celebrities. The so-called Weinstein Effect, which manifested itself in a series of high-profile public accusations of sexual misconduct, led to rapid dismissal of well-known cultural producers, journalists and politicians. Although sex scandals and protests against sexual assault frequently make the news, the Weinstein case can be considered a transformative media event because it led to a strong gesture against sexual abuse in general and of sexual victimization as an institutionalized mechanism of career growth, a gesture that had tangible consequences.

Finally, there is the question of the scale of transformative media events – whether only events happening at a global or national level may count as such, or whether those taking place in local communities, organizations and individual lives may also qualify. Transformations occur at all levels of social reality: micro, meso, macro and global. Mediatized performances often aim to attract translocal audiences in order to solve local or even personal problems. For example, an individual may overcome red tape or police abuse or a personal traumatic experience with the help of public campaigns, amateur videos, online petitions and discussions in social networks, raising the question of whether these may be considered transformative media events. An approach to media events that considers such instances transformative would imply that greatness, magnitude and historical significance are no longer considered indispensable prerequisites in defining the concept. Nevertheless, the status of any event is a discursive construction rather than an objective given, and even a personal story can become an important symbol or a trigger of more far-reaching social changes. From this perspective, almost any mediated utterance or action has the potential of becoming a media event. And the more deeply mediatized the social world becomes, the more likely it is to be perceived as a continuous flow of media events defined by various actors as big or small, significant or trivial. Nevertheless, the notion of (transformative) media events remains relevant today, despite the need for considerable modifications. On the whole, transformative media events can be regarded as mediatized manifestations of counter-power, which aim either at reestablishing a broken order or state, or redefining and renewing it radically.

Transformative media events and citizen media

Transformative media events initiated by citizens in public spaces to express their disagreement with certain social conditions and/or desire to change them may be understood as citizen media events. The concept was first introduced in passing by Baker and Blaagaard (2016a:10) in the course of defining citizen media. As an example of such an event, they discuss Ellie Cosgrave's dance against sexual harassment in a London Underground tube carriage on International Women's Day in 2013. Over a year after being sexually assaulted, Cosgrave returned to protest by means of dancing in the very same carriage. Her performance was filmed and uploaded on YouTube, and her story was recounted by *The Guardian*, which later launched its Project Guardian for victims of sexual offences on public transport to report similar incidents. Cosgrave's example encouraged other women to oppose harassment more actively, and the traumatic experience and public protest ritual of overcoming a painful experience evidently changed Cosgrave's own life. An engineer and researcher in urban innovation, she went on to produce urban design that can protect women from sexual violence in cities. In 2017, she appeared on the BBC's *100 Women Challenge* list as one of the most influential and inspirational women who have contributed to fighting harassment in public spaces.

Baker and Blaagaard (2016a) mention Ellie Cosgrave's dance performance as "a specific example of an actual citizen media event", thus clearly justifying the inclusion of individual public actions within the notion of citizen media. Conceptualizing citizen media as encompassing any artefacts, practices and discourses by means of which unaffiliated citizens acting in public spaces strive to effect sociopolitical or aesthetic change, as well as to express their personal desires, means that blogging, digital witnessing, graffiti, flash mobs, fansubbing, amateur filmmaking, hacking, online petitions and many other forms of civic engagement can be considered as citizen media. It is therefore necessary to distinguish between citizen media and citizen media events, and to explore the features of the latter.

A starting point would be to distinguish between content-oriented and action-oriented citizen media. Two practices characterize content-oriented citizen media. The first is citizen witnessing (Allan 2013) or sousveillance (Bakir 2010). This practice is based on video filming and other means of documenting various incidents by witnesses. The second practice consists of the production of specific counter-power content in blogs and memes, graffiti and mockumentary. Action-oriented citizen media, on the other hand, are above all mediatized actions in specific physical locations: protest actions, performances and interventions, such as hacker attacks, the creation of programmes to bypass blockages or game modding. The boundaries separating content-oriented and action-oriented citizen media are flexible: for example, an (action-oriented) participant in a political demonstration simultaneously becomes a (content-oriented) witness of the events he or she films on a mobile phone.

Various forms of citizen media are likely to generate different types of citizen media events. However, action-oriented citizen media, specifically mediatized public actions (protests, processions, performances, flash mobs), come closest to the classic definition of media event in the sense that they all have a ritualistic nature and are held in accordance with a preplanned script. In this case, citizen media, understood as forms of individual or collective expressions of agency, and the media events they produce are basically one and the same: a mediatized dance as a symbol of protest is simultaneously a typified practice and a unique event. More specifically, it is the dancing body that works as citizen media here, while dancing is only one among its many possible states. It is a defining feature of citizen media of this type that they are unstable and ephemeral, being no more than a temporary mode of a certain object or process. Just as individuals become citizens only when they are socially and politically active, human bodies and media platforms gain features of citizen media only by producing specific counter-discourses. Citizen media events emerge in the process of actualizing this transformative discourse (Nim 2016).

Content-oriented citizen media, especially digital witnessing, mostly involve registering external events initiated by third parties. For example, uploading mobile phone footage documenting an instance of discrimination in the workplace may result in an independent citizen media event if it succeeds in setting the agenda for one or more public debates and produces certain social effects. The same applies to any other content generated by citizen media. At the same time, many citizen media images and narratives, as well as various performances, can be part of larger scale transformative media events such as mass protests or civil wars, which are constructed at the intersection of numerous media platforms, discourses, actors and practices. Against this background, citizen media events should ultimately be analysed as part of the complex architecture of transmedia events circulating in a convergent media environment.

See also: media ecologies; mediatization

Recommended reading

Couldry, N., A. Hepp and F. Krotz (eds) (2010) *Media Events in a Global Age*, Abingdon, UK: Routledge.

A collection of in-depth critical essays offering a reassessment and extension of Dayan and Katz' original approach to media events, applying alternative theoretical optics and taking into consideration the rapid digitalization and globalization of media environment.

Dayan, D. and E. Katz (1992) *Media Events: The live broadcasting of history*, Cambridge, MA: Harvard University Press.

Excellent innovative text, which has become canonical and laid the foundation for the development of media events theory. The book conceptualizes "historical" media events as televised public rituals, which maintain social solidarity but at the same time have a transformative potential.

Mitu, B. and S. Poulakidakos (eds) (2016) *Media Events: A critical contemporary approach*, London: Palgrave Macmillan.

A collection of critical essays, where considerable attention is paid to the analysis of protest media events of the 2010s (Tahrir 2011, Euromaidan 2013). A number of contributions reflect upon the changing ontology of media events in the digital era, introducing new concepts such as transmedia events, user-generated media events, mobile media events.

Sonnevend, J. (ed.) (2018) 'Special Section: Media Events', *Media, Culture & Society* 40(1): 110–157.

The special section of *Media, Culture & Society* dedicated to the twenty-fifth anniversary of the publication of Daniel Dayan's and Elihu Katz' book *Media Events* includes ten short essays which reflect on the impact of the volume on media theory. The special section also contains a discussion by Dayan and Katz of the concept's endurance in the age of social media and audience scepticism.

MEDIA PRACTICES

Hilde C. Stephansen and Emiliano Treré

Once a marginal topic in academic research, citizen media is now a burgeoning interdisciplinary area of study that is attracting growing interest in a variety of fields, including journalism, social movement studies, drama and performance, translation studies and political communication. Much of this surge of interest has been prompted by the rise of mass self-communication (Castells 2009) in the form of blogs, social media and other digital technologies that enable unaffiliated individuals and groups to produce and disseminate their own media within global communication networks. For example, research on citizen journalism has examined how Web 2.0 technologies have enabled ordinary people to participate in news production (Allan 2013; Allan and Thorsen 2009; Thorsen and Allan 2014b; Wall 2012), social movement scholars have critically examined the use of social media platforms for protest mobilizations (Bennett and Segerberg 2013; Gerbaudo 2012; Juris 2012), and research on digital storytelling has emphasized the potential of digital technologies to give voice to marginalized groups (Lundby 2008b; Hartley and McWilliam 2009b).

Much commentary on citizen media has focused on the possibilities and limitations of digital technologies as a means for citizens to disseminate media content that makes visible hidden realities or challenges dominant discourses. However, there is also a growing interest in material and embodied aspects of citizen media and the social processes that surround their production and consumption (Stephansen and Treré 2019). As defined in this encyclopedia, the concept of citizen media encompasses not only digital contents and technologies but also physical artefacts, performative interventions and social practices and relationships. This entry introduces the concept of media practices as a tool for exploring the socially situated, embodied practices that relate to citizen media. Inspired by the so-called practice turn in the social sciences, an understanding of media as practice has been adopted by media researchers to develop a more socially grounded analysis of the media's significance in contemporary societies. The entry first situates the concept of media practices in a broader theoretical context by providing a brief introduction to practice theory and an overview of how the practice approach has been taken up within media and communication studies. It then offers a more detailed discussion of the relevance of the media practice approach for the study of citizen media, drawing on empirical examples, and ends by reflecting on some of the limitations of the media practice approach and highlighting some avenues for further research.

The practice turn in the social sciences

The turn of the century has witnessed the resurgence of practice theory as a challenger to prevalent ways of thinking about human life and sociality, as an attempt to transcend the "dualisms of structure and agency, determinism and voluntarism" (Shove et al. 2012:3). Stretching back as far as Wittgenstein and Heidegger, theories of practice have long roots in social theory and comprise a variety of approaches, including first generation practice theories such as Giddens' (1984) structuration theory and Bourdieu's (1972/1977, 1980/1990) field theory, as well as later second generation theories which have sought to systematize, elaborate and extend practice theory, notably Schatzki (1996, 2001), Schatzki et al. (2001) and Reckwitz (2002). Postill (2010) and Shove et al. (2012) offer helpful overviews.

There is no single, agreed-upon understanding of practices, but one widely used definition presents them as "embodied, materially mediated arrays of human activity centrally organized around shared practical understanding" (Schatzki 2001:11). Put slightly differently, practices are organized constellations of material activities performed by multiple agents (Schatzki 2012:14). Reckwitz (2002:253) defines practices as routinized bodily performances that also involve mental activities – interpretations, knowledge, emotions and motivations – and material and cultural objects. Practices can thus be situated along three key interrelated dimensions: meanings and representations; objects, technologies and material culture; and embodied competences, activities and 'doings' (Shove and Pantzar 2005; Shove et al. 2007; Magaudda 2011). Hence, a practice forms a block which depends on the existence and specific interconnectedness of these elements (Reckwitz 2002:249–250).

Practice-based approaches place materiality, process and knowledgeability at the centre of social analysis, allowing us to explore materiality as one of the many elements interacting with wider processes of emergence, transformation and decline of socially embedded practices. The aim of practice theory is not to provide general laws or explain causal or associative relationships between constructs, but rather to generate a set of discursive resources capable of producing accounts and analyses that enrich our understanding of social phenomena (Nicolini 2017).

The practice approach in media and communication studies

Prompted by this turn to practice in the social sciences, an understanding of media as practice has gained prominence in media and communication studies since the turn of the century. Though this interest in practice had various antecedents, not least within Latin American communication studies (Stephansen and Treré 2019), explicit engagement with practice theory was sparked by Couldry's (2004) article, which proposed an approach that understands media as practice rather than as texts or structures of production. For Couldry (2004), this implied a change of paradigm in media studies, as it shifts the focus from textual analysis or political economy to what people are "doing in relation to media across a wide range of situations and contexts" (Couldry 2012:37). Couldry defines media practices as the "open set of practices relating to, or oriented around, media" (2004:117), further distinguishing between "acts aimed specifically at media, acts performed through media, and acts whose preconditions are media" (2012:57). Practice theory thus offers a holistic approach to understanding the media's social significance, and allows us to grasp the distinctive types of social processes that are enacted through media-related practices. The focus on media practices allows us to see media at work in a number of contexts and situations, and to understand how media practices arrange, combine and more generally intersect with other social practices (Couldry 2004, 2012).

This turn to practice has antecedents within two broad strands of media research: audience studies and media anthropology. The question of what people do with media was the starting point for the Uses and Gratifications approach (Couldry 2012), and the cultural turn in the social sciences of the 1980s and 90s prompted the rise of qualitative research on active audiences. However, while Uses and Gratifications focused on "individual usage of bounded objects called media" (Cammaerts and Couldry 2016:327), practice theory differs in its social emphasis and focus on relations beyond the use of specific technologies (Cammaerts and Couldry 2016). The turn to practice theory can thus be understood as a response to a crisis in audience studies (Couldry 2012, citing Ang 1996), one that resulted from the growing ubiquity and embeddedness of media in everyday life. In a media-saturated world, practice theory, due to its openness, is seen to offer a more adequate framework for capturing the diversity of everyday practices that involve media (Couldry 2012).

A concern with practices has also been central in media anthropology, which takes people and their social relations – rather than texts or technology – as a starting point for analysing media as a social form (Ginsburg 1994:13). Within both fields, media anthropology and audience studies, scholarship on media audiences has moved away from a focus on direct engagement with texts toward "a consideration of multiple articulations with media in everyday life" (Bird 2010:85, 2003). However, although the notion of practices had been used widely by media anthropologists, it had rarely been defined or problematized, and there had been little explicit engagement with practice theory (Postill 2010). The edited collection *Theorising Media and Practice* (Bräuchler and Postill 2010) sought to remedy this by bringing media anthropology into explicit conversation with practice theory, as a response to Couldry's (2004) intervention. Contributors to this volume explored the value of practice theory for understanding different ways of engaging with media, from practices of newspaper readers (Peterson 2010) and news journalists (Rao 2010) in India, to uses of information and communication technologies by Norwegian (Helle-Valle 2010) and Danish (Christensen and Røpke 2010) families, amateur audiovisual production (Ardévol et al. 2010) and free software activism (Kelty 2010). The practice approach has also been adopted by digital ethnographers as a methodological framework for multi-sited, multi-platform studies of interrelated digital practices (Gómez Cruz and Ardévol 2013a, 2013b; Ardévol and Gómez Cruz 2013).

Understanding citizen media as practice

An understanding of media as practice has important implications for the study of citizen media. The media practice approach enables us to move beyond the tendency in citizen media studies to focus on content and to explore instead a much broader range of socially situated practices that relate to citizen media. More specifically, as Stephansen has argued, an understanding of citizen media as practice enables researchers to ask three broader questions: "What do people do, say and think in relation to citizen media?"; "What *kinds* of practices do people engage in that are oriented towards citizen media?"; and "What might the role of citizen media practices be in structuring other practices?" (2016:29–30; emphasis in original).

In drawing attention to social relationships and processes, the media practice approach resonates with a longer tradition of scholarship on alternative media that predates the current preoccupation with digital technologies. Although scholars in this field have not drawn explicitly on practice theory, social relations and organizational processes have been central to their conceptualizations of alternative media. Atton (2002) highlighted how alternative media producers engage in a range of practices aimed at transforming the social and economic relations involved in media production, distribution and consumption; these practices include non-hierarchical, collective forms of organization and anti-copyright publishing. Downing (2001) similarly emphasized the prefigurative character of radical media, showing how media activists attempt to practice what they preach by implementing radical democratic principles in their modes of organization. Rodríguez (2001) coined the term citizens' media to refer to media through which citizenship is performed or enacted. Countering the then-dominant framing of alternative and community media in terms of counter-information, Rodríguez showed how communication practices can empower communities and individuals, strengthen social bonds and thus act as a catalyst for social change.

In this sense, the media practice approach can be said to provide a new theoretical framework for addressing longstanding concerns in the literature on alternative, radical and citizens' media. The concept of media practices has proved particularly popular among scholars studying social movements and media. In this specific field of citizen media research, which has grown exponentially in response to the rise of Web 2.0 technologies and their adoption by protest movements, the media practice approach has been used to develop non-media-centric analyses of the role of media in movements for social and political change. Highlighting the wide range of media practices that activists engage in, this strand of literature has sought to challenge technological determinism and the tendency to focus on a single media platform (Treré 2012; Mattoni and Treré 2014; Treré 2019), in order to develop more nuanced analyses of the intersections between protest and media (Barassi 2015a; Kaun 2016a; Kubitschko 2015; Martínez Martínez 2017; McCurdy 2011).

One of the first and most comprehensive studies in the social movements literature to adopt the media practice approach was Mattoni's research on the media practices of the precarious workers' movement in Italy. Mattoni (2012:159) defined activist media practices as "both routinised and creative social practices that … include interactions with media objects (such as mobile phones, laptops, pieces of paper) and media subjects (such as journalists, public relations managers, other activists)" and "draw on how media objects and media subjects are perceived and how the media environment is understood and known". Blending social movement studies, media studies and the sociology of practice, Mattoni contrasts a media-centric approach that selects *a priori* the types of media that will be investigated (for instance, citizen or mainstream), with a media-practice approach whose strength lies instead in exploring how activists map, understand and then actively navigate the media environment with which they interact during their protest activities. Mattoni's approach is thus able to transcend the limitations of most social movement studies that either focus on how mainstream media cover protests and neglect the role of citizen media produced by protesters, or isolate the diverse media produced by activists from their multiple interactions with the press and mainstream media. Her focus on the practices of activists and social movement groups reveals instead that Italian citizen media like Indymedia or Global Project are part of a wider repertoire of communication, understood as the entire set of media practices that social movement actors may conceive as possible and use to reach social actors within and beyond the social movement milieu (Mattoni 2013a:50). In a given social and political context, activists' perception of the opportunities and constraints of the media environment informs their media choices: while at times citizen media can be better suited to communicating critical content to specific publics or organizing collective mobilizations, at other times it is more productive to negotiate with the mainstream media in order to reach a wider constituency.

Stephansen (2016) also analyses citizen media through the lens of media practice, focusing on media activism in the World Social Forum (WSF). Drawing on Couldry (2004, 2012) and Rodríguez (2001), she argues that the media practice framework brings into view a broad range of citizen media practices, beyond those directly related to the production and circulation of content, thus enabling researchers to explore the social fabric that such practices can help generate and the forms of agency they make possible. Stephansen describes the emergence within the WSF of what activists refer to as 'shared communication' – an approach to media activism that emphasizes sharing of content but also collaborative production processes and the exchange of knowledge and experience. She identifies four distinct types of media practices among media activists in the WSF (2016:33): organizational practices aimed at enabling collaborative production processes that stimulate an exchange of skills, experiences and ideas; capacity-building practices such as training grassroots activists to produce their own media; networking practices such as setting up dedicated spaces for alternative and citizen media at social forums that bring communicators from different parts of the world together; and movement-building practices that help develop a sense of shared purpose and collective identity – for example, seminars to share knowledge and debate strategies for media democratization. Stephansen shows how such practices create the preconditions for distinct forms of agency to emerge: they offer lived experience of 'another communication', build solidarity that can provide a source of strength for activists operating in difficult contexts, and generate a sense of individual and collective identity.

Scholars have also sought to combine media practices with other theoretical concepts and approaches, usually in order to complement perceived deficiencies in the media practices perspective. For instance, Mattoni and Treré (2014) argue that the media practice approach is particularly insightful for studying social movements at the micro-level, but less so for grasping meso- and macro-level processes, a point also made by Postill (2010). They therefore propose a conceptual framework that integrates media practices with two other concepts: mediation, defined as "a social process in which media supports the flow of discourses, meanings, and interpretations in societies" (Mattoni and Treré 2014:260, drawing on Couldry 2008a and Silverstone 2002), for studying the meso-level; and mediatization, "a concept used to analyse critically the interrelation between changes in media and communications on the one hand, and changes in culture and society on the other" (Couldry and Hepp 2013:197), for investigating macro-level dynamics.

Treré (2011, 2012, 2019) and Mattoni (2017) have also sought to combine media practice with media ecology approaches that pay attention to the complex, hybrid and multifaceted nature of the media systems within which social movement actors operate. Treré (2011, 2012, 2019) points out that these two conceptual lenses implicate – and reinforce – each other: on the one hand, an analytical approach anchored in practice theory puts us in a position to ask holistic questions regarding the whole spectrum of media used by activists; on the other, a media ecology perspective sheds light on the complex interrelations among multiple types of media (old and new, corporate and alternative, online and offline). Mattoni (2017:2) contends that, together, media practice and media ecology approaches are powerful because they recognize the wider range of technologies, actors and contents that activists interact with, historicize social movements' use of technologies, and emphasize activists' agency vis-à-vis media technologies.

Future directions

Practice theory has emerged as a productive framework for research on citizen media, allowing scholars to develop holistic analyses of activists' uses and understandings of media technologies, as well as the broader social contexts within which these are situated. However, although significant conceptual advances have been made, there is a need for deeper debate about the theoretical underpinnings of media practice. One key question concerns whether the practice approach can be considered a new paradigm (Couldry 2004), or whether it is more appropriately understood as a new conceptual or methodological lens, to be used pragmatically alongside other perspectives. A common criticism of practice theory is that while it is useful for studying the micro-level of social interaction, it cannot adequately account for large-scale political processes, and therefore cannot by itself provide an overarching theoretical framework (Postill 2010; Mattoni and Treré 2014). This criticism has been challenged, however (Hui et al. 2017).

Another important question relates to the relationship between practices and the extent to which media practices might anchor other practices by enacting new patterns of action that in turn prompt changes in other practices (Couldry 2004; Swidler 2001). In order to understand the role of citizen media in processes of social change, more focused empirical research is needed on the specificities of citizen media practices and their relationship to other social practices. This includes research on the audiences of citizen and social movement media – thus far a neglected area (Downing 2003). Scholars have long highlighted how such media blur the boundaries between audiences and producers, and, as already noted, the very concept of audiences is challenged by the ubiquity and embeddedness of media in everyday life. Practice theory seems particularly suited to grasping the complex array of practices situated at the intersections of production and use, but more empirical research is needed in this area.

An ongoing preoccupation in research on citizen media and media activism concerns the dynamic between media practices and the technological affordances of media platforms. Given the growing intensification of processes of datafication, this line of enquiry will only become more relevant and urgent. A key question here is how (and to what extent) the media practice approach can help theorize agency and social change in this context. One promising approach involves expanding the notion of media practices to include 'acting on media' – that is, to include practices explicitly concerned with politicizing media technologies and infrastructures (Kubitschko 2017). Such an approach could be further strengthened by articulating the media practice framework with scholarship on media democracy movements (Hackett and Carroll 2006; Milan 2013; Kidd et al. 2009) and data activism (Milan and van der Velden 2016). In an age increasingly dominated by corporate, algorithmically driven social media platforms, more work is also needed to integrate research on media practices with critical analyses of political economy.

See also: activism; anthropology and citizen media; media ecologies; networks and networked society; process vs. event; social movement studies and citizen media

Recommended reading

Mattoni, A. (2012) *Media Practices and Protest Politics: How precarious workers mobilise*, Farnham, UK: Ashgate.

Relying on the analysis of precarious workers mobilizations in Italy, Mattoni develops the concept of 'activist media practices' and compares different categories of media technologies, organizations and outlets, from the printed press to the Internet, and from mainstream to alternative media. Explaining how activists perceive and understand the media environment in which they are embedded, she discusses how social movement actors interact with a diverse range of media professionals and technologies and investigates how mainstream, radical left-wing and alternative media represent protests.

Stephansen, H. C. and E. Treré (eds) (2019) *Citizen Media and Practice: Connections, currents, challenges*, London: Routledge.

An edited collection that critically interrogates the relevance of the concept of media practices for the study of citizen and activist media, bringing together contributions from leading scholars in sociology, media and communication studies, social movement and critical data studies. A substantial introductory chapter by the editors traces the emergence of the media practice framework in Latin American and Anglophone scholarship, critically assesses how the media practice framework has been used in studies of activist and citizen media, and reflects on challenges and future directions for this interdisciplinary field.

Treré, E. (2019) *Hybrid Media Activism: Ecologies, imaginaries, algorithms*, London: Routledge.

Investigates the complexities, ambiguities and shortcomings of contemporary digital activism. It deconstructs the reductionism of the literature on social movements and communication, proposing a new conceptual vocabulary based on practices, ecologies, imaginaries and algorithms to account for the communicative complexity of protest movements. Treré develops a practice-based ecological understanding of social movements' communication, showing how the combined use of the media practice and the media ecology approach is able to restore the agency of social movement actors.

MEDIATIZATION

Andreas Hepp

Mediatization refers to the relationship between the transformation of media and communication on the one hand and culture and society on the other (Couldry and Hepp 2013:197). In terms of everyday experience, mediatization has quantitative as well as qualitative aspects. In regard to its quantitative aspects, mediatization refers to the increasing temporal, spatial and social spread of mediated communication. Temporally, media such as television were once only available at certain times but can now be accessed twenty-four hours a day. Spatially speaking, in the past media such as telephony was only accessible from one location and from a singular fixed device; communication (and other services) are now available through the use of multiple portable devices. Socially, our practices have become increasingly entangled with media. Some media scholars have argued that the quantitative spread of media permeates every social domain and that media have become so pervasive that we can now argue in favour of the "mediation of everything" (Livingstone 2009:1). In regard to media's qualitative aspects, mediatization refers to the specificity of certain media within sociocultural change. From this point of view the consequences of the saturation of everyday life by media and the extent to which this saturation relates to social and cultural change is of fundamental importance (Lundby 2014).

Crucially, mediatization research does not deal with the effects of individual media content, rather, it is more concerned with the ways in which the changing media environment is an element within the overall transformation of culture and society. This contrasts with the term mediation, which, however hard it tries, only really grasps a very general characteristic of communication; specifically, how communication mediates, or intervenes, between multiple actors (Silverstone 2005).

Taking this definition as its starting point, this entry begins by outlining a more detailed explanation of mediatization. It goes on to describe the current stage it is at as one of 'deep mediatization' and, finally, it discusses the importance of the concept for research into and the practice of citizen media.

Mediatization as a sensitizing concept

Mediatization, as defined above, can be understood as a "sensitizing concept" (Jensen 2013:206) that "gives the user a general sense of reference and guidance in approaching empirical instances" (Blumer 1954:7) and draws our attention to (current) changes in culture and society. On these terms, mediatization sensitizes us to the fundamental trends of change we experience in our media environment (Hepp and Hasebrink 2017): First, the differentiation in the vast amount of available digital media; second, the increasing connectivity made possible through these media that offer opportunities to individually and collectively connect across time and space; third, media's ever-increasing omnipresence which stimulates the possibility of being permanently connected from multiple locales; fourth, a rapid pace of innovation, that is, the emergence of new media and services over ever-shorter periods of time; and fifth, datafication, the representation of social life through computerized data arrayed across media devices and their underlying software and infrastructures.

However, sensitizing concepts that are used to place emphasis on certain phenomena have to be complemented by other analytical, "definitive concepts" (Blumer 1954:7) to ensure rigour and clarity in the research process. The main analytical principles of mediatization research derive from both the institutionalist and social-constructivist traditions (Couldry and Hepp 2013:195–198), however, there are differences between the ways in which these traditions conceptualize mediatization. Put simply, the institutionalist tradition finds its roots in mass communications research that understood media as an independent institution with its own set of rules. Mediatization in this tradition refers, therefore, to the ways in which different social fields might adapt to these institutionalized rules. The social-constructivist tradition, on the other hand, highlights the role various media play in the communicative construction of social reality and approaches the idea of mediatization as a means of analysing the ways in which media may operate across that process.

Research carried out in the institutionalist tradition focuses on the role that media – understood as a social institution – plays in influencing other cultural and societal domains, more often than not, positioned exteriorly to media; this is often referred to as a media logic. Originally coined by David Altheide and Robert Snow in 1979, the term media logic describes the influence that discrete mass-media formats exercise over other domains of society such as politics or religion.

More recently, media logic has been utilized more broadly and is often pluralized to take into consideration the existence of a variety of media logics depending on the particular kind of media under investigation (Strömbäck and Esser 2014; Thimm et al. 2018). Media logics understood in this way become "a metaphor and shorthand for the various modi operandi that characterize the workings of the media" (Hjarvard 2017:11). Typically, media logics are linked to more specific analytical concepts such as forms of interaction, organizational rules or media affordances, with an emphasis on how these forms, rules and affordances shape media use. Media's influence is not conceptualized as having a (more-or-less direct) effect on other institutions and domains; non-media actors bring with them their own logics which, in turn, potentially work against media logics resulting in inertia and resistance despite transformations in the media environment.

By contrast, the social-constructivist tradition emphasizes the role played by media in the construction of social and cultural reality and explores mediatization predominantly from the perspective of everyday actors (Knoblauch 2013; Krotz 2014). Researchers in this tradition question how our social practices are altered as they are entangled with media. Attempts to theorize media's influence here proceed by considering them as means of communication that mould our practices through processes of institutionalization and materialization (Couldry and Hepp 2017:32).

Generally speaking, institutionalization refers to the patterns through which actors mutually adjust their expectations of each other: what Berger and Luckmann call the reciprocal typification of habitualized actions (Berger and Luckmann 1966:72). Even the family is an institution that typifies particular forms and expected patterns of practice in terms of the types or the roles played by the actors involved (father, mother, partner, child, aunt and so forth). Using the idea of institutionalization when considering media does not only mean that we are thinking about media organizations. At stake are processes of institutionalization that are much more far-reaching in everyday life, such as mobile telephony, which institutionalizes a communicative triadic relationship between caller, called and bystander. Media institutionalize our communicative practices at a variety of levels: how we arrange ourselves in media use situations, our communicative forms, that is, our interactions through media, and the organization of technology and infrastructure.

This all operates in concert with processes of materialization which means that social practices become inscribed into the media technologies that we use and the infrastructures that accommodate them. Messenger software, for example, materializes a certain way of talking through its software-based user interface. These influences do not, however, travel unidirectionally but instead move in a more cyclical fashion. Within each social domain there exists an orientation toward everyday practice which may or may not be altered by media. Take the family, for example, or school; these social domains are constructed by an ongoing flow of social practices. These practices are enduring and, to some degree, inert. However, media can institutionalize and materialize practices and in this relationship they mould (shape, influence) these social domains. This is not to say that media have the capacity to transform them completely. As already mentioned, social domains are characterized by a sense of inertia as humans' orientations in everyday practice have the potential to resist and even alter media themselves.

Despite divergent origins and different conceptualizations of mediatization, however, researchers from both traditions have approached increasing rapprochement in their understanding and application of the term (Hepp et al. 2015). First, they both consider mediatization as a long-term "meta process" (Krotz 2014:131) of transformation that is accompanied by other transformational meta processes, such as individualization, globalization and commercialization. Second, both traditions share the assumption that mediatization does not operate in the same way across all social domains. Instead, the specific way in which mediatization unfolds differs significantly from one social domain to another (Lunt and Livingstone 2016:465). It is for this reason that empirical research on mediatization is always contextualized and aims to describe and critique specific forms of mediatization. Third, both traditions train their sights on the ways in which media as means of communication affect change or transform culture and society. Their principal interest is not, therefore, in the effects of media content and other more direct manifestations of media influence (Hjarvard 2017:1–3) but, rather, in the fundamental, transformational phenomena that occur within individual domains of society in parallel with the media and communication change. Finally, it is common practice to consider perception as one facet of these transformations: as long as people orient their practices to what influence they assume media exercise, media will have an influence, albeit indirect, on changing practices (Nölleke and Scheu 2018).

The challenges of deep mediatization

A more recent focus within mediatization research is media's increasingly digital character and the challenges it presents when we are faced with the necessity to rethink mediatization. While initial contributions to the subject have addressed the issue in fairly general terms (Finnemann 2014; Miller 2014), the discussion has recently intensified and has become more specific as digitalization rapidly evolves mediatization processes.

The reasons for this are multifaceted. Mediatization researchers have become increasingly aware that current change is less about the dominance of one (digital) medium but more about the differentiation of intricately connected digital media. Focus has shifted to the "polymedia" (Madianou 2014:323) or the "media-manifold" (Couldry and Hepp 2013:34) character of today's media environment. From this point of view, in order to understand how media mould each societal domain, it becomes necessary to consider digital media in terms of their intimacy with one another; that is, to reflect on the cross-media character of mediatization. Furthermore, being digital, media are not simply means of communication anymore; they are at one and the same time means of generating data while serving more general communicative functions. The data they generate is, in turn, harnessed to support various types of automated analysis which have become a fundamental part of the construction of the social world.

Digitalization has thus introduced a new stage of mediatization which has been referred to as "deep mediatization" (Couldry and Hepp 2013:7, Hepp 2020). This term describes an advanced stage of mediatization in which all elements of our social world are intricately related to media and their overarching infrastructures. Researching deep mediatization represents a challenge because it must also incorporate an analysis of algorithms and digital infrastructures. Through media's shift into the digital, mediatization research has developed various connections with more general research on the influence data has on society (Beer 2016; Gillespie et al. 2014). It has become necessary to investigate algorithms because deep mediatization invariably involves facets of the mediated construction of the social world that takes place through automated data analyses (Gillespie 2014). For example, the automated classification of users into certain interest groups as they shop online and presenting them with personal recommendations is only made possible through the use of algorithmic systems and the same can be said in reference to friend or follower recommendations on many online social platforms. More attention needs to be paid to the digital infrastructures that underpin contemporary media (Mosco 2017). Understanding mediatization as a concept that sensitizes us to media change means that we must rethink the relevance of specific research paths which, in turn, forces us to further integrate more detailed analytical concepts into the field.

Mediatization and citizen media

The question of mediatization has so far received scant attention within the literature on citizen media. However, for a while now there has been relatively intense discussion taking place on the subject of mediatization and social movements.

In their overview of media and the study of social movements, Mattoni and Treré (2014) argue that mediatization offers a wider frame within which we can contextualize existing research on individual and collective actors in social movements and their social practices. A mediatization approach renders possible the prospect of adopting a more long-term perspective on the transformation of social movements in a constantly shifting media environment, allowing us to reflect on what they call "epochs of contention … in which certain templates for collective action are available to protesters" (ibid.:257), and on how "social-movement families" transform in accordance with the process through which they carry out their social engagement (ibid.:258). As mediatization has advanced and expanded, especially in light of increased digitization, social movements in Italy, for example, have adapted their practices in terms of the new ways in which individual members participate, and how they are called upon and informed about actions: Mattoni and Treré go on to say that "[a]ctivists of the 2008 movement could … rely on a broader media ecology that … offered plenty of alternatives for multiple activities and possible media recombinations that students of the early 1990s could not even imagine" (ibid.:265).

In the field of social movement research, the discussion on deep mediatization is related to one particular important and necessary shift: while in traditional social movement research media were most commonly considered as a means of organizing and communicating a movement's objectives and interventions, media have now become an issue of social engagement in their own right. The hacker movement, for example, can be understood as a social movement that understands digital media and their infrastructures as an object of political engagement when, for example, the Chaos Computer Club hacks a voting machine to raise public awareness of the various security problems related to their adoption. This orientation toward media on the part of social movements has been referred to by Kannengießer and Kubitschko as "acting on media" (2017:1). Acting on media involves multiple civil society actors taking an active part in establishing and shaping media organizations, infrastructures, technologies and regulation that form parts of the fabric of everyday life. This process can be understood as a further manifestation of deep mediatization: as digital media and their infrastructures become more crucial in the construction of the social world, social movements also turn to making media a pertinent issue for their engagement in civil society.

Another issue relevant to citizen media scholars concerns the role that "pioneer communities" (Hepp 2016:918) play in the processes of deep mediatization. Consider the Maker movement, a community oriented toward practices of manufacturing and tinkering (Davies 2017:28–42). Through local makerspaces in which they provide a variety of technologies, including 3D printers and laser cutters, members of this movement have established a close relationship to civic tech and have contributed to the emergence of a potential new form of citizen media that is closely associated with DIY practices and self-directed manufacturing. However, pioneer communities do not necessarily represent social movements in the original sense of the word. Having a movement-like structure but relating to politics and industry much like think tanks do, they operate more akin to a hybrid whose specific nature lies in their orientation towards media-related change. In this sense, they also "act on media", but more on the level of everyday practice: they experiment with state-of-the-art media technologies to develop new forms of social practice.

Future directions

There is still much to learn about the latest changes in citizen media by adopting a mediatization approach. As we have seen, mediatization offers a broader frame through which we can contextualize contemporary social transformation. In today's deeply mediatized world, we can no longer understand citizen media without reflecting on their embeddedness within the media manifold. An understanding of how media have become a tool for data generation can be useful for citizen media actors when they are challenged, for example, by discourses on the smart city (Mosco 2019) and the role citizens might play in these technoscientific visions of the urban environment. Discussing these issues and framing them through the lens of mediatization provides an orientation that travels in the direction of more fundamental shifts that create the analytical space to form a critical analysis.

Reflecting on deep mediatization and pioneer communities is a useful approach when studying citizen media as the latter are often understood as pioneering in their methods of media appropriation and use. The main challenge here is the extent to which it is possible to identify pioneering practices in the citizen media realm and how any subsequent findings could be contextualized. To help address this conundrum we can learn from research on pioneer communities that has identified how the social construction of forerunners in a specific realm of media-related change is, first and foremost, one that is performed by the respective collectivities themselves (Hepp 2020:42–52). This means that it is vital that researchers remain critical in relation to such self-related constructions. At the same time, these kinds of constructions are, as the Maker movement demonstrates, partly dependent on wider media coverage, an issue which can also be considered when investigating citizen media and the role they play in specific locales.

The debate around acting on media in general, as outlined in this entry, also poses a range of challenges for citizen media researchers. In the context of deep mediatization, questions raised by emerging (digital) media and the infrastructures on which they depend are of critical importance. Establishing new forms of citizen media may involve moments of acting on media, which means becoming engaged with the discussion on a city or a region's (digital) media infrastructures, public support or resistance to these infrastructures, and the questions raised by new forms of regulation to manage them in the public interest.

See also: media ecologies; media event; social movement studies and citizen media

Recommended reading

Hepp, A. (2020) *Deep Mediatization*, London: Routledge.

A general introduction into mediatization research with a special focus on digital media and their infrastructures. The book addresses the making of deep mediatization, its main trends, how to theorize media-related transformations, the re-figuration of society, the individual in times of deep mediatization and how far deep mediatization can support the "good life".

Hjarvard, S. (2017) 'Mediatization', in C. A. Hoffner and L.V. Zoonen (eds) *International Encyclopedia of Media Effects*, New York: Wiley, 1221–1240.

A short overview of mediatization research in an institutionalist perspective, addressing questions of 'media logics', the difference between mediatization and mediation and institutional interdependency and struggles.

Lundby, K. (ed.) (2014) *Mediatization of Communication*, Berlin and New York: de Gruyter.

A handbook on mediatization research which addresses – besides conceptual questions – the mediatization of various social domains such as education, families, financial markets, law, religion, etc.

MIGRATION STUDIES AND CITIZEN MEDIA

Moira Inghilleri

Citizen media has been conceptualized through a number of academic disciplines, including media and communication studies, global economics, sociology, urban and cultural geography, as well as anthropology and the performance arts. Migration studies, on the other hand, has been late in providing a fuller context for considering citizen media events related to migration though some scholars in this field have considered so-called ethnic/minority media in the context of macro geopolitical issues and social policy. Features of citizen media are in evidence, for example, in the body of research that engages with migrants' use of web-based and other technologies for maintaining transnational ties (Parreñas 2005; Thompson 2009; Madianou 2012) and constructing diasporic identities (Cunningham and Nguyen 2003; Alinejad 2011; Harney 2013; Halilovich 2013), although they are mostly alluded to rather than explored. The same can be said for the coverage of migrations concerned with the perilous departures (Tazzioli 2016) and the not so welcoming arrivals (Vandevoordt 2016) often associated with them. Cultural and media studies have produced more significant migration-studies-relevant research in which citizen media broadly defined have been the object of research and analysis, despite persistent gaps.

Citizen media and migration

In some instances, citizen media produced by migrants about migration bear the hallmarks of what Naficy (2001) termed 'accented' media in that they involve performances of identity typical of displaced exiles living in a diaspora that will never be home. Sun (2002) offers a similar analysis of the formation of a global diasporic Chinese 'transnational mediasphere' within which meanings of 'Chineseness' are constructed and contested. In other contexts, citizen media practices serve as a means for migrants to reclaim or recodify their present experiences without nostalgia or a strong sense of displacement from their homeland (Inghilleri 2017). Certain media are specifically targeted at established residents to educate them about the reasons newly arrived groups of migrants seek sanctuary within their communities (Jahangard and Duffy 2016). With regard to citizen media produced by and for more well-established migrant communities, notions like ethnicity, diaspora and exile are understood not as "some existential state of 'being' but in more fluid terms of 'gazing'/back toward the old home, in toward the new home, and around about the world of new diasporas" (Sreberny 2005:445–446; Deuze 2006a). Whether viewed from a distance or in close proximity, all these creative expressions produce an impression; they seek to elicit some kind of response or reaction. And like all signifiers, there can be gaps between their intention and their interpretation. Although an explicit aim of citizen media is to interrupt prior assumptions, opening up audiences to new or revised interpretations of particular social and cultural phenomena, in some cases the indexical or symbolic functions of the message can easily be obscured.

A distinguishing feature of the social and interactional aspects of citizen media is that they are situated outside commercial and certain not-for-profit third party institutions in order to diminish the potential for co-optation or appropriation by corporate gatekeepers seeking new areas for capital investment. Its producers aim to ensure that their creative processes and products remain outside the power and influence of the mass media which, as Delhaye notes in relation to the field of high art, subjects artists and institutions to "complex procedures of selection, valuation and canonisation by many different gatekeepers [who] have the power to make newcomers visible or invisible, to include or to exclude them, to put them centre-stage or to relegate them to the margins" (2008:1303). The determination and evaluation of whether or not a particular endeavour meets the criteria for citizen media can be complicated nevertheless. Significant overlapping issues to consider include: what counts as commercial and non-commercial support or uptake; how or whether the media produced is designed explicitly to link producer and audience across space and time – i.e. across differences in age, cultural identities, social histories and so on; the impact of the participation of migrant artists and writers on their claim for greater recognition and inclusion where this is the goal; and how the relationship between the quality of the collaborative political process and the quality of the art produced is understood.

Citizen media produced by or with migrants, its messages and aims, are specific to the category to which a migrant or group of migrants belong, where unique spatial and temporal considerations are also involved. In popular usage, the terms migrant and immigrant are sometimes used interchangeably to refer to individuals and families, without much thought given to the many distinct categories of migration that exist. These include permanent residents who once were migrants, long- or short-term undocumented residents, temporary guest workers, refugees and asylum seekers, and labour migrants (Inghilleri 2017; Agier 2013/2016). Whatever the motivation to leave, departures are rarely the beginning of these migrants' trajectories. Forced migrations are one kind of urgent pressure that results in leaving one's homeland, while another impetus, initiated by a pioneering individual or by earlier generations, originates in a belief in the ability to forge a new life elsewhere. Across these different categories of migrants there is some degree of overlap in the factors that determine the migration. For many individuals, the fear and uncertainty brought on by chronic unemployment and poverty compel them to seek better opportunities elsewhere. For refugees and asylum seekers, these conditions often exist as a consequence of war, political turmoil or specific practices of discrimination targeted against them. There are significant differences between refugees and asylum seekers, however. Refugee applications are generally approved on humanitarian grounds, based on preestablished resettlement quotas determined by individual nation states or regional agreements. Depending on the circumstances and conditions in the country from which they are fleeing, some refugees are provided with permanent resettlement while others are granted only temporary protection. Asylum seekers must apply for protection from within a country, either immediately upon arrival or after they have entered and resided in the country, legally or illegally, for a period of time. There is no pre-authorized recognition of their asylum status: they must demonstrate their fear of persecution based on an individual claim and provide ample supporting documentary evidence. Despite the circumstances surrounding their departure, asylum seekers are increasingly being viewed and treated the same as any undocumented migrant, that is, as a category of uncontrolled immigration. Upon making a claim, many are held, sometimes for months or even years, in makeshift detention centres under dehumanizing conditions while awaiting a decision on their case (Luiselli 2017; Koser 2007).

Some migrant groups who have been permanently relocated become legal residents upon arrival in a destination country, others are indefinitely detained in refugee camps or other holding facilities for prolonged periods of time during which their future is uncertain, and labour migrants with overseas foreign worker status, whether they have short- or long-term contracts, are intentionally kept isolated from the local population (Longhi 2013). This includes domestic workers, despite the fact that they live in the homes of the families they serve. The opportunity for these different groups to represent themselves will depend on a number of interrelated dimensions of the migrant experience, including: access to networks of support; ability to communicate in the language of the host country, coupled with access to translation or interpreting services; the degree of common ground, for example in relation to class and cultural background, among migrants from the same country or between migrants from different regions; and their overall visibility in the wider social or political context in which they function (Inghilleri 2017). Such factors can have significant influence on the extent to which a citizen media project can be initiated and can sustain support.

Following Baker and Blaagaard (2016a:15), the citizen media discussed in this entry include "material and immaterial artefacts and devices that may be endowed with expressive power by human agents and used to communicate information as well as emotions, values, narratives". Each of the projects described represents a unique opportunity to give voice to a distinct experience of migration through poetry, the human body, the written and spoken word, and material objects that become endowed with symbolic content through their context of use.

The quotidian nature of citizen media

The notion of citizens' media (rather than citizen media) was first introduced by Rodríguez as a collectivity "*enacting* its citizenship by actively intervening and transforming the established mediascape … contesting social codes, legitimized identities, and institutionalized social relations and … empowering the community involved, to the point where these transformations and changes are possible" (2001:20; emphasis in original). Through case studies of citizens' media, Rodríguez analysed both the processes and products of collaboratively produced media in print, radio, video and television by socially and economically marginalized members in four different societies. An important finding of this groundbreaking research was that throughout the process of media production it was the everyday micro-political dimension of their lives that was foregrounded in the local discourses of the participants, not the dissident macro-politics that framed their narratives more broadly. Rodríguez likened their localized expression of experience to that of

> living organisms that evolve and develop uniquely in permanent interaction with their complex environments/contexts: at some point they strengthen their struggle against one target, but later they can abandon that target and take on a new one, which, in turn can be abandoned to take on a third one.

This observation led her to conclude that the value of citizens' media lay in its freedom and ability to articulate the quotidian and not as a "one-dimensional static platform" from which to aim at unified goals (ibid.:158).

Rodríguez' early work drew on political theorist Chantal Mouffe's proposed concept of citizenship as a "form of identification, a type of political identity: something to be constructed, not empirically given" (Mouffe 1992:232). This proposal is part of a broader attempt to reconcile competing theories of justice between communitarian and universalist criteria for what defines who is a moral subject, who has ethical standing, and who merits ethical consideration within a given polity. With regard to migration, such questions merge with the issue of territorial boundaries and the effectiveness of national laws in organizing and controlling the increasing movement of people in a globalized world. The decision over who is granted temporary or permanent rights of residence remains with individual nations, which have increasingly tightened control over immigration through stricter border controls, mass deportation and the construction of detention facilities designed to cordon off migrants from the rest of the population under the guise of protecting national security and stability. Under these conditions, citizen media can become an important space for the exploration of political identity construction within migrant communities and in the shared public spaces where communities of all kinds interact. Public articulations of a migrant presence can throw a spotlight on the complex recurring processes of assimilation, transnationalism and the enduring ambivalence of the migrant experience (Iannelli and Musarò 2017).

The case of overseas foreign workers in Singapore is illustrative. An estimate of foreign-born workers in Singapore, not including permanent residents, provided by Rubdy and McKay (2013) is 36 per cent out of a population of 5.2 million. Of these, 15 per cent are highly skilled and 85 per cent are semi- or unskilled (ibid.:158). Most male migrants come from Bangladesh, Nepal and other poor South Asian countries to work in construction, shipyard work and manufacturing. The majority of female migrants are domestic workers from the Philippines and Indonesia. In 2014, Shivaji Das, an established travel writer and management consultant originally from India, established a poetry competition for migrants (Migrant Worker Poetry Competition 2019) while he was volunteering with a local nonprofit Singaporean organization, Transient Workers Count Too (TWC2). The competition has evolved into a popular platform for other migrant workers to share their stories through poetry written and performed in their own languages. According to Das, the most encouraging aspect of the increase in popularity of the competition is "the way these migrant workers themselves have volunteered in organizing the event itself while working in cross-cultural groups" (The Online Citizen 2017). The finalists compete at the National Gallery in Singapore, delivering powerful oral performances. The top three receive a cash prize, with the winner receiving around USD 350. The organizing committee comprises a group of volunteers, among them A. K. M. Mohsin, Editor-in-Chief of *Bangla Kantha* (Voice of Bengal), the largest Bengali newspaper in Singapore, which is distributed free of charge among manual worker and domestic labour migrants. A major goal of the paper is to educate Singaporean society about how migrants' work benefits their national economy and raises awareness of the diverse talents and skills the migrants bring to their country. Mohsin is also founder of Dibashram (*diba* meaning 'day' in Indian Bengali and *ashram* meaning 'hermitage' in Indian Sanskrit), described on its website as "a space that draws people together – both the migrant and resident populations of Singapore – and spreads a message of mutual understanding and respect" (Dibashram: A Space for the Migrant Worker in Singapore, n.d.). Although much of this collective activity is oriented toward expressing their personal desires and aspirations (Beretta 2016), as opposed to direct political activism, this practice of citizen media gives expression to aspects of those lives that would otherwise remain hidden.

Citizen media, materiality and migration

Materiality as a concept is used across many different disciplines to focus attention on the impact of material or physical factors in relation to a particular phenomenon. It has long been recognized that the form in which people transmit ideas is consequential in and of itself. Different means of communication have the power to shape human association and action, and the message of any new medium is the change of scale, pace or pattern it introduces into human affairs (McLuhan 1964). Within official institutions involved in migration control, the procedures that individuals are obliged to fulfil, involving the filling of forms and an oral interview, generally produce truncated narratives of migrants' experiences that, particularly in the case of refugees and asylum seekers, prevent them from adequately conveying their reasons for fleeing their place of origin. The citizen media accounts of displacement, detention and the conditions of labour migrants described here offer examples of alternative discursive forms (and translational norms) through which narrative content can be framed. These different modes of practice illuminate more effectively the problematic categories of 'migrant', persistent biases toward certain groups, and the inhumane policies associated with reception in the context of twenty-first-century mass migration.

In his memoir *The Translator* (2009), Daoud Hari, a Zaghawa tribesman from Western Sudan, provided a first-hand account of the brutal genocide in Darfur. In 2003, he was among the hundreds of thousands of villagers attacked and driven from their homes by Sudanese government-backed militia groups. With his village burned to the ground, and his family decimated and dispersed, Hari escaped, eventually finding safety across the border in Chad. Despite official denials at the time, the evidence was overwhelming that the Sudanese government had been training, arming and paying the Janjaweed militias to kill non-Arabs in Darfur and clear them off the oil-rich land. For others who had also been eyewitnesses to these atrocities, words were not the preferred mode of expressing their memories. In 2007, during a fact-finding mission to eastern Chad led by the human rights organization Waging Peace, Darfuri women in a refugee camp told a researcher that their children had witnessed horrendous events when their villages were attacked. On their urging, the researcher spoke to the children, many of whom had been forced from their homes three or four years earlier. With the help of interpreters who spoke Arabic and the languages of Darfur, she asked the children to write down their memories of events. One of them asked if they could draw instead. They drew pictures showing their villages full of tanks and armed men on horseback, houses on fire and helicopters circling the skies. Villagers are shown under attack, women are led off in chains, and civilians are shot at, and try to defend themselves with spears and arrows. Helicopters bear the markings of military aircraft and the men in camouflage are labelled by the children as Janjaweed militia. These drawings were submitted to the International Criminal Court in the Hague by Waging Peace and accepted as contextual evidence of the crimes committed in Darfur. They were used in the trials as illustration of the atrocities and helped confirm that the Darfur population had been attacked by the Janjaweed militias (International State Crime Initiative 2010).

Another example of innovative media, in this case made public via social media, involves one man's unique protest against Australia's practice of indefinite refugee detention on nearby offshore islands. In 2014, the Baptist pastor Tri Nguyen, an Australian citizen who had arrived as a refugee from the Vietnam war in 1982, embarked on a project he called 'The Gift of Refuge' in response to the growing number of refugees and asylum seekers who were showing up at his church, many of them Muslims seeking a temporary spiritual home. Their presence heightened his awareness of the government's harsh policies and negative treatment of migrants, and he took it upon himself to go to Canberra, the capital city, on a walking pilgrimage. Nguyen pulled behind him a small replica of a boat he and his father had built years before to commemorate their escape from Vietnam. For this journey, he added wheels and painted the words 'Thank You' on the side of the boat for the welcoming treatment he had received decades earlier (Nguyen 2014). His aim was to gift the boat to Parliament while calling attention to and critiquing its current treatment of migrants, which involves forced return and/or lengthy detention on offshore islands. Through social media and local contacts, he invited anyone who wanted to join him or to contribute their time, money, accommodation, vehicles or publicity in support of his trek, which covered 670 kilometres/416 miles (The Gift of Refuge, n.d.).

Both the Darfuri children's drawings and Nguyen's wooden boat are these citizens' attempts to represent unutterable trauma through media that make the emotional content of their experiences translatable and accessible to a wider audience. Similarly, in his book *No Friend but the Mountains* (2018), the Iranian Kurdish writer, journalist, filmmaker and political refugee Behrouz Boochani experiments with different forms of communication (and a different medium of communication, the iPhone) to describe his experience as a detainee on Manus Island, one of Australia's offshore detention centres. Although his medium is the written word, his juxtaposition of the mundane objects that surround the prisoners and the way power and control were exercised in the prison add a unique visual materiality to the text. Specific objects become a mode of translation for communicating the inner thoughts and feelings of the detainees, as when he describes the role of a generator in maintaining order and control within the prison (Boochani 2018:176–177; emphasis in original):

> As midnight draws closer the prisoners retire to their foam mattresses, return to sleep after a day of commotion. Suddenly, the generator quits again. Another hammer clobbers the head of the prison. Everyone's hopes and dreams manoeuvre in tandem with the unbearable intensity of heat … seeping alongside each other into the tapestry of nightmares. Startled out of sleep, the prisoners wake. Sweating, heads bursting as though in a furnace. Not to forget the mosquitos. Now, without any fans operating, the mosquitos venture more ruthlessly into the rooms. Within minutes the prisoners escape the rooms and drop directly into the black of night, unleashing a monsoon of profanities. The swearing echoes through the abyss of darkness that restrains the prison.
>
> …
>
> The fear of the officers results in getting the generator up and running a lot quicker this time. When the lights go out, the prison transforms into a dangerous beast. At any moment an uncontrollable situation can emerge. The generator has a face with the following features:
>
> *A device resembling someone of old age/*
> *Constituted by an intricate system of deteriorating wires/*
> *Poles and pipes of rusting metal/*
> *Probably within a dingy space/*
> *Somewhere worse than the prison/*
> *Covered by an old cloth/*
> *Under the protection of a rag/*
> *A rag that is withering away/*
> *The generator is withering away.*
>
> I want to believe the generator is a living being, with a soul, an organism that takes pleasure in throwing the prison into disarray whenever it feels like it.

Each of these media projects foreground the politics of the everyday, a consequence of the effects of different migration experiences, lived as a cluster of culturally, socially and linguistically managed encounters and entanglements with others. With his unique writing style, Boochani paints a powerful picture of how totalizing institutions function (Goffman 1961). From inside their refugee camp, the children of Darfur testify to the atrocities they witnessed in a medium that gives power to their young voices in an international court. The established links between the Poetry Competition, *Bangla Kantha* and the Dibashram community centre, all mediascapes of a kind, became crucial in providing these migrants with a voice. And as Tri Nguyen's pilgrimage demonstrates, even one individual with the imaginative use of his or her body can alter and challenge prior perceptions in public space (Fusco 2001; Culp 2013). Whether or not his gift was accepted or acknowledged by the Australian state, the thousands who came out to support his protest became his primary audience.

These projects are direct expressions of an acutely felt micro-politics of marginalization and exclusion. This is what marks them as citizen media, in contrast to the types of partnerships imagined and created between artists and local migrant communities that tend to be initiated by the artists and that create what Iannelli and Marelli (2017:71) describe as "a flexible, temporary 'we', whose borders are continuously redrawn in response to issues and conflicts that arise in different places and at different times". The processes and products of migrant-directed citizen media are not aimed at such fleeting moments of solidarity. Their objective is to establish a newly constituted discursive space created by migrants in which alternative modes of perception regarding notions of identity, belonging and participatory and human rights can be contemplated and constructed.

See also: citizenship; philosophy and citizen media; precarity; race & ethnicity studies and citizen media; witnessing/testimony

Recommended reading

Agier, M. (2013/2016) *Borderlands: Towards and anthropology of the cosmopolitan condition*, translated by M. Fernbach, Cambridge: Polity Press.

An insightful anthropological perspective on the topic of what the author calls border situations, in reference to the global context of migration. Agier discusses the border and border walls as socially constituted spaces – sites of conflict as well as encounter. The book also examines the (re)appearance of essentialism as a political strategy and the implications of how the concept of identity is utilized by disparate groups in the current politics of migration.

Boochani, B. (2018) *No Friend but the Mountains*, translated by O. Tofighian, Sydney: Picador.

This award-winning book offers a searing portrayal of life inside one of Australia's island prisons. Written in real time and through recall, Boochani provides an intelligent and inspired interpretation of his own and others' departure and detention. His ability to situate himself both inside and outside of these experiences, a synergy that is beautifully captured in the translation by Omid Tofighian, has produced a profound account of his and other detainees' experiences of the Manus Island camp.

Iannelli, L. and P. Musarò (eds) (2017) *Performance Citizenship: Public art, urban design, and political participation*, Milan: Mimesis International.

The essays in this book represent a number of disciplinary approaches to different forms of citizen participation, originating in the field of contemporary public art and urban design. Although the authors work in diverse regions across the globe, their common objective is to improve the democratic character of contemporary societies and to enhance citizens' participation in decision-making processes, particularly those individuals or groups who are economically, politically, socially and culturally marginalized.

MOBILE TECHNOLOGIES

Michael S. Daubs

This entry provides a brief history of mobile technologies, ranging from rudimentary mobile telephone systems first developed in the late nineteenth century to modern Internet-enabled smartphones that have given access to a range of tools and specialized mobile applications since the early 2000s. It then discusses some of the social, political and cultural developments related to the emergence of mobile technologies, including the blurring of the formerly distinct divisions between presence and absence, public and private, work and leisure. The final sections present two case studies examining the use and value of mobile technologies for the production and dissemination of citizen media. The first case study explores Instagram influencers, that is, Instagram users with significantly large numbers of followers who use their position of prominence within the social media platform to promote commercial products. While some influencers partner with companies and produce sponsored posts, others use their online popularity to launch their own businesses, demonstrating how Instagram can act as a form of citizen media that is "actively intervening and transforming the established mediascape ... empowering [those] involved" (Baker and Blaagaard 2016a:9). The second case study examines how protestors in the Occupy Wall Street movement created and engaged with citizen media as part of their protests in New York in 2011. The analysis shows the extent to which mobile technologies played an integral role in enabling activists to orchestrate the mix of online and offline actions that were staged during their occupation of Liberty Square.

A brief history of mobile technologies

The origins of mobile communication technologies can be traced back primarily to Italian inventor Guglielmo Marconi's unveiling of his wireless telegraphy device in England in 1896 (Goggin 2006:24). This technology was used by both the British and the Boer armies during the South African War (1899–1902) and during the First World War (Austin 2017:17; Huurdeman 2003:286). Military and civil uses dominated the development and expansion of rudimentary portable telephony systems throughout the early twentieth century. During the Second World War, for example, American forces used portable two-way radio telephones called handie-talkies or walkie-talkies (ibid.). Similar systems would power mobile radio-based telephones in cars and trucks during the postwar years.

Cellular mobile telephony was not developed until the 1970s, when Motorola engineer Martin Cooper combined advancements in miniaturization and automation to develop the DynaTAC, the world's first modern mobile phone. Due to its 1.1 kilogram weight and rectangular shape, this was nicknamed 'the

Brick' (Deuze and The Janissary Collective 2012:297). Nevertheless, the DynaTAC successfully launched the first generation of analogue mobile phones; these enabled direct voice communication between users but offered few other features. The next big shift in mobile technologies involved digitalization, which increased the multifunctional capabilities of mobile technologies at the end of the twentieth century. Canadian mobile technology firm Research in Motion (RIM), for example, developed a series of Blackberry handheld devices which were popular with business executives and politicians because they combined mobile telephony with mobile email access (Manzerolle 2013:147).

Starting with the introduction of Apple's first iPhone in 2007, however, mobile technology makers shifted focus to a mass consumer user base. The iPhone, and other smartphones such as those running Google's Android mobile operating system, expanded upon the multifunctionality offered by Blackberry devices by incorporating WiFi and mobile broadband connection capabilities, sizeable hard drives, more sophisticated digital cameras and advanced processors capable of running increasingly sophisticated mobile applications. Smartphones have thus evolved into Internet-enabled computers that provide access to a range of tools and services including instant messaging, social media, image capture and editing, mobile content streaming and video games.

Thanks to their portability, users have increasingly integrated these devices into their daily lives, relying on them for communication and the organization of many everyday activities (Deuze and The Janissary Collective 2012:300). Once a mere status symbol or business tool, the mobile phone has developed into a "citizenship commodity" – a must-have device for modern-day citizens (Fortunati 2002:526). The combination of instant communication, information access, and content creation and sharing has become integral to citizen media activities in the digital age (Goggin and Clark 2009). Yet, as outlined below, these capabilities have at the same time blurred the lines between previously distinct sociocultural concepts including presence and absence, public and private, work and leisure.

Sociocultural impacts of mobile technology

One of the sociocultural changes visible since the introduction of the first mobile phones is a complication of the notions of presence and absence. Mobile technologies not only allow people to be less tied to a particular physical location, but also enable constant and immediate communication with distant others. The possibility of maintaining constant communication can sometimes lead to burdensome expectations that people are always available, reachable or contactable (Deuze and The Janissary Collective 2012:304), which can affect their engagement with their physical environment and those around them. The ring of a mobile phone can pull someone's attention away from a physical space and instead shift it to their communication network (Fortunati 2002:518–519). At the same time, the possibility of instant communication with others allows people to connect with absent friends and family when they find the physical environment around them to be unimportant or intimidating (ibid.:515). Mobile technologies can therefore act as a lifeline that keeps people tethered to a community of friends and family (Deuze and The Janissary Collective 2012:304). In short, the connectivity afforded by mobile devices means that individuals apparently present in a given place may only be half-present, with their attention focused elsewhere.

The ability of mobile technologies to shift attention between present and distant people or spaces has additionally contributed to a destabilization of the dividing lines between public and private. Mobile technologies offer the possibility of

> choosing between the public space of streets, stations, means of transport and the private space of interpersonal relationships, between chance socialness which may develop with those who happen to be passing by, and chosen socialness (e.g. with the friends you decide to call on your mobile phone).
>
> *Fortunati 2002:515*

Smartphones, and the array of tools they offer, also allow users to access information on the go, leading to cultures and societies that assume the ubiquity of information (Goggin 2011:149). Public spaces can thus be emptied of significance as attention is pulled elsewhere, but mobile technologies can also fill spaces with meaning, since personal connections, data, relationships and meanings can be mapped on to a physical space via mobile technologies.

Finally, access to both interpersonal networks and information via mobile devices can blur distinctions between work time and leisure time. Early discussions concerning mobile technologies and wireless data networks focused upon their importance to mobile workforces that could be available "anywhere and anytime" (Manzerolle 2013:215). Mobile providers framed the ability to conduct work at any time and from any location as flexibility for employees. That framing, coupled with additional productivity engendered by apps and tools available on modern smartphones, naturalized the erosion of leisure time. Moreover, leisure activities involving mobile technologies have increasingly been reframed as value-generating labour. For example, social networks such as Facebook, often accessed via mobile apps on smartphones, track and record data generated by their users in exchange for free access to the network. Some of these data, including basic demographic and relationship information, are willingly volunteered by the user, while other data, such as location (particularly when accessed on GPS-enabled smartphones), political leanings, interests, hobbies and preferred cultural products, are inferred from their interactions and activities, that is, from the new forms of labour performed on these social network platforms. These data are frequently commodified and used to allow for targeted advertising and tailored content that generate profit for the host companies (Fuchs 2012b:146).

It is in this evolving context that we must consider the use of smartphones for the production and dissemination of citizen media. These activities entail labour, even if that work is often enjoyable for the content creators and typically constitutes a hobby rather than an obligation. The content that is generated is largely made to share with friends and family (van Dijck 2009) but, like contributions to social networks, can also generate value for the platforms that host the content. At the same time, this free labour can provide its users with benefits beyond monetary remuneration, including community recognition, exposure, improved skills and participation in public and/or social debates (Hesmondhalgh 2010:278). Such practices highlight how mobile technologies can create media "through which individuals *become* citizens" (Stephansen 2016:28; italics in original). The increasing ubiquity and utility of mobile technologies, as well as the social, political and cultural dynamics they can produce, are visible in two case studies: the use of mobile technologies by Instagram influencers to promote commercial products and the role of mobile technologies in large-scale protest actions such as Occupy Wall Street (OWS).

Instagram influencers

The social networking site Instagram allows users to 'self-brand' by sharing images via an individualized profile page (Hearn 2017). Images posted to the site can have accompanying captions in which users can include hashtags – keywords which help categorize posts and make them show up more easily in searches – and tag other users using the @ symbol followed by a unique username. Although Instagram is accessible via the web, the social network is most often accessed via mobile apps such as those made for popular Apple/iOS and Google/Android mobile operating systems (Marwick 2015:142).

The preference for the mobile app is tied to the multifunction capabilities of modern smartphones; using built in cameras and Internet connectivity via WiFi connections or mobile broadband, users can capture and upload images instantly from any location where data services are available. Users regularly take and upload rehearsed and posed digital self-portraits (selfies), which enhance Instagram's immediacy in order to help users form stronger relationships with their followers (Marwick 2015:139). Instagram thus enables users to turn what might otherwise be candid, private moments into public displays for wider consumption in an attempt to foster a connection with other users. This merging of public and private generates a sense of presence or immediacy between people who might be geographically distant.

Popular Instagram users, sometimes referred to as microcelebrities (Marwick 2015) or influencers (Duffy 2017), engage in and showcase activities they enjoy for perceived personal benefits. A few successful influencers are able to benefit monetarily by gaining brand sponsors (Duffy 2017:6). Alternatively, they may be successful in translating popularity into support for personal business ventures and partnerships, as was the case with beauty influencer Huda Kattan's partnership with Sephora (Backaler 2018:127). Instagram influencers thus exemplify the merging of work (value-generating labour) and leisure (enjoyable activities) that can result from the use of mobile technologies. For many influencers, microcelebrity is a potential path to self-employment and financial independence. For the vast majority of Instagram influencers, however, the work put into producing photos to gather followers can be viewed as a form of aspirational labour, that is, under- or unpaid labour performed in the present with the goal of attaining a career in the future (Duffy 2017:4). Aspirational labour is a way to improve and market skills; the number of followers and likes represent a form of social capital that can provide future financial or career benefits. Engaging in aspirational labour can therefore be an empowering experience.

Influencers similarly demonstrate how mobile technologies can blur distinctions between public and private; while influencers often self-brand by adopting a public persona informed by commercial marketing techniques and "conventional status hierarchies of luxury, celebrity, and popularity" (Marwick 2015:139), their popularity is also dependent upon the public perception that they are "just like us" (Duffy 2017:3). Self-branding therefore involves "trying to bridge the inherent contradictions between one's real-life behaviour and the more restrained behaviour required by workplaces" (Marwick 2013:203). In order to build a public identity, influencers rely on selfies and other candid (or posed to appear candid) photos which capitalize on a personal, familiar aesthetic. While many scholars focus on the economic aspects of microcelebrity (Duffy 2017; Hearn and Schoenhoff 2016, Khamis et al. 2017), influencers' self-branding activities can also be empowering, acting as a form of self-realization or self-writing (Marwick 2013:202). For many users, Instagram and other social media platforms are a "modern technological tool for self-revelation, confession, self-management and self-improvement" (Sauter 2014:830). In this way, Instagram influencers' self-writing and self-branding activities subvert Instagram's intended function to act as "a personal nostalgic archive" of memories and spontaneously captured moments (Abidin 2016:7); they allow the user to carefully craft an image and engage in subversive frivolity, "the under-visibilized and under-estimated generative power of an object or practice arising from its (populist) discursive framing as marginal, inconsequential, and unproductive" (Abidin 2016:1). In other words, while selfies and self-writing Instagram influencers are sometimes dismissed as frivolous or vain, they can also be viewed as citizen media that allow the influencer to "look inward, to the self … and express his or her personal desires and aspirations" (Baker and Blaagaard 2016a:11).

Mobile technologies in protests: Occupy Wall Street

While Instagram influencers may generate citizen media that is reflexive and empowering in nature, the media practices of activists in the Occupy Wall Street movement (OWS), which emerged in 2011, demonstrate how mobile technologies allow collectives to "position themselves within and in relation to society and participate in the creation of diverse publics" (Baker and Blaagaard 2016a:16). OWS arose in response to the world economic crisis, brought about by the abuses of multinational banks, and a growing awareness of income disparities between the world's richest and poorest citizens. The official start of OWS is attributable to a 13 July 2011 blog entry simply titled '#OCCUPYWALLSTREET' posted on the website of the Canadian anti-consumerist magazine *Adbusters* (Adbusters 2011). That post, and the activist sentiment it engendered, resulted in a weeks-long takeover of Zucotti Park (renamed by protestors as Liberty Square) in Lower Manhattan. *AdBuster*'s inclusion of the hashtag symbol in the title of the post signals the importance of online communication to this activist movement. In the weeks between July, when the post was made, and 17 September 2011, the date selected for the start of the occupation, protestors on Twitter used hashtags such as #occupywallstreet and #ows to mark tweets related to the protest that were meant to raise awareness, coordinate actions, share information and build solidarity.

Mobile phones in particular played a key role in the protests (Costanza-Chock 2012). During the occupation of Liberty Square, protestors used several mobile applications developed for OWS. One app for Apple devices simply called Occupy compiled news, videos and photos from a variety of online sources (Faust III 2011). Protestors would use this app to keep up to date on the occupation and related protests, while those who were sympathetic to the cause but unable to participate used the app to follow developments and, via the videos and photos, participate vicariously. Another app, Shouty, turned smartphones into streaming media servers (Kingkade 2011). Since loudspeakers in Liberty Square were banned, protestors used this app so that people at the back of a crowd – and those on the other side of the globe who sympathized with OWS – could listen in on protest speeches. In short, those unable to participate directly in the protest could be virtually present through their smartphones, thus engaging in acts of citizenship through media.

Activists and protestors also used smartphone apps developed for existing web services such as YouTube, Twitter and Facebook. This software allowed participants to rapidly disseminate information, images and videos of the Occupy protests with their support networks of friends, family and other activists. In this sense, mobile media acted as a site of convergent journalism that, as in previous protests such as the Egyptian uprising in 2011, acted both as "a platform of political advocacy" and as an "interface between online communication and an offline activism of mass demonstrations on the ground" (Chouliaraki 2013a:275). Activists additionally used mobile technologies to create content for websites such as OccupyTogether.org, which provided background information on OWS as well as a directory of assemblies, related organizations and protests. Protestors "uploaded video and commentary to a variety of social media sites almost instantly, so that it was almost as if the incident were being streamed" (Maslin Nir 2011). At times, protests were literally streamed online in real-time thanks to a channel on LiveStream.com called Global Revolution (2011). These forms of citizen media allowed protestors to share their own narratives of the protest in order to counter or correct what they felt were misleading or incorrect characterizations in mainstream media coverage. At the same time, these examples demonstrate the overlap between personal, private networks and public spaces of protest, as well as the dissolving barrier between presence and absence enabled by mobile technologies.

Finally, many protestors used social media sites like Facebook and Tumblr to share stories about their personal economic hardship in a series of videos, photos and posts that usually ended with the phrase 'I am the 99%'. These personal perspectives, featured on various OWS-related social media channels and protestor-run websites such as Occupy.com, allowed protestors to turn personal experiences and events into issues of public debate. Slogans such as 'We are the 99%' emphasized the diversity of opinion within the movement while simultaneously asserting its inclusiveness, a collectivity not based on a single "substantial identity – race, ethnicity, religion, nationality" but one that asserts collectivity as "the 'we' of a divided people, the people divided between expropriators and expropriated" (Dean 2011). Mobile technologies were integral to the emergence of that collective.

Future directions

Although these two case studies exhibit significant differences, the media activities of Instagram influencers and OWS activists help to illuminate how mobile technologies contribute to a citizen media framework that "encompasses the physical artefacts, digital content, practices, performative interventions and discursive formations of affective sociality produced by unaffiliated citizens as they act in public space(s) to effect aesthetic or socio-political change or express personal desires and aspirations" (Baker and Blaagaard 2016a:16). However, each case also offers intriguing avenues for future consideration. For example, while influencers can indeed use social media platforms such as Instagram for self-branding to explore their identity, the influencer model also naturalizes the commoditization of the self and reinforces a neoliberal model of constant productivity, which problematizes claims of independence and freedom often espoused by influencers. An ongoing question regarding mobile technologies in protest movements is whether the kind of connectivity that such devices allow truly enables the strong kind of solidarity required to enact change or whether their use simply facilitates the production of weak ties and a different form of self-branding through which people give the appearance of being politically active without expending any meaningful effort, a phenomenon sometimes referred to as slacktivism (Christensen, H. S. 2011; Kristofferson et al. 2014). As these questions suggest, critical discussions of the proliferating use of mobile technologies in citizen media contexts will need to engage with both the positive and negative affordances that these devices offer.

See also: activism; flash mobs; immaterial labour; photography

Recommended reading

Costanza-Chock, S. (2012) 'Mic check! Media cultures and the Occupy movement', *Social Movement Studies* 11(3–4): 375–385.

Using a mix of qualitative and quantitative analysis, this article offers an in-depth investigation of the use of media tools by protestors and activists during the Occupy movement. This includes a useful examination of how media inform the organization of the protest community and vice versa, as well as the specific practices and tools activists develop to create and disseminate content across media platforms.

Duffy, B. E. (2017) *(Not) Getting Paid to Do What You Love: Gender, social media, and aspirational work*, New Haven, CT: Yale University Press.

An expansive study that frames social media influencers as engaging in 'aspirational labour' or unpaid creative work. Through numerous interviews and fieldwork, Duffy examines the gendered aspects of this aspirational labour and the often exploitative relationships between social media influencers and corporate brands. A key text for those who want to better understand the motivations of and issues facing influencers and microcelebrities.

Goggin, G. and J. Clark (2009) 'Mobile Phones and Community Development: A contact zone between media and citizenship', *Development in Practice* 19(4–5): 585–597.

Provides a useful overview of how mobile technologies – and mobile phones in particular – have been used for community development and activism, resulting in new forms of organization, democracy and governance that nonetheless incorporate traditional understandings of community and citizenship. A good primer for highlighting the relevance of mobile technologies to studies of citizen media.

NETWORKS AND NETWORKED SOCIETY

Dorismilda Flores-Márquez

As a noun, verb or a theoretical concept, the term network refers to connections among nodes. Today, it is usually associated with digital culture, because the networked nature of the Internet makes evident the nodes and connections among them. More broadly, however, it is used to explain different processes of the natural and social world, as Castells argues, "[n]etworks constitute the fundamental pattern of life, of all kinds of life" (2004:4). We can thus speak of neural networks, electric networks, transport networks, social networks, computer networks, social-digital networks and many more, given that "[t]here is a path between any two neurons in our brain, between any two companies in the world, between two chemicals in our body. Nothing is excluded from this highly interconnected web of life" (Barabási 2002:18). Each network has particular features, but the logic of nodes and connections is common to all networks.

This entry provides an overview of different approaches to the notion of networks by highlighting some basic elements of social and social-digital networks, as well as of the networked society, at the micro and the macro levels. Four dimensions are proposed for the analysis of the networked society: actors in the networked society – that is, citizens and their participation through/with/in media; practices – especially media practices – as the means of participation that enable citizens to take part in public space; space and time as inseparable dimensions that frame our experience of citizenship and participation; and power, in the sense of the tension between dominant structures and the empowerment of citizens.

Approaches to the notion of networks

Although much scholarly work concerns itself with networks, what is meant by the term varies among different perspectives.

The micro-level approach considers social networks in everyday life and interpersonal interactions. Rheingold (1993, 2008, 2012) – a pioneering American scholar of virtual communities – provided a vocabulary with which to discuss the Internet and its social implications as far back as the early 1990s. He did so by reflecting on his own experience as a member of virtual communities, which he defined as groups of people connected through bulletin boards and networks. Rheingold argued that computer networks could change the lives of people on three main levels: the personal level of individuals as human beings, the social level of person-to-person interaction, and the political level of citizens. However, he emphasized the difference between networks and communities in terms of quality, continuity and degree of commitment, stressing that the community is stronger than the network. Papacharissi (2011b), on the other hand, has contributed to developing a micro-level approach to social networks by analysing them in terms of identity and community. Drawing on Erving Goffman's work on the presentation of self in everyday life (Goffman 1959/1990), she introduced the notion of the networked self to highlight the incorporation of digital technologies in everyday life and the consequent shift in sociability. Both perspectives, Rheingold's and Papacharissi's, share a focus on individuals who are part of communities, approaching them as nodes that are part of a network enabled by digital technologies. The main transformations discussed correspond to the self and its everyday practices in Goffman's terms, but in Rheingold's and Papacharissi's models these would not occur without the interaction with technologies.

Other scholars approach networks from a macro-level perspective to understand the intersection between economic, political and cultural transformations on the one hand, and information and communication technologies on the other. According to Castells (2004, 2009, 2010) and van Dijck (2006/2012), the networked society emerged as a new form of society in the information age, in which the social structure is made up of networks powered by technologies that become "the nervous system of our society" (van Dijck 2006/2012:253). Both authors emphasize the network as a global system that characterizes our time, but they differ in terms of what they treat as the unit of analysis. While Castells (2004) considers the network itself as the unit of analysis, van Dijck (2006/2012) focuses on individuals, groups and organizations that are linked by media networks. Drawing on Castells, Cardoso (2006) proposes another micro-level approach that is focused on communication, emphasizing that the network society implies networked communication. He argues that the media system is structured around two main networks: television and the Internet. Television involves a logic of low interactivity that contrasts but coexists with the high interactivity of the Internet. Cardoso thus attempts to account for the connection between the interpersonal and mass communication dimensions and the coexistence of features of industrial and information societies. Understanding media as part of the network society implies taking into account the global level, in which media systems are interconnected among countries, companies and other structures.

The micro and macro levels both conceptualize the network as a way of understanding social relations and dynamics. The micro-level approach risks being blind to power structures, however, while the macro-level approach can blur the potential of small actors to challenge these power structures.

From networks to the networked society

A network is defined as two or more nodes and the connections among them, which produce an open and flexible structure (Castells 2004; Rheingold 2012). A society, on the other hand, is "an integrated system of social structures and functions" (Ritzer 2011:2), where the social refers to the people who form these social structures and live together within different kinds of organizations. Although the term social network is often associated with the Internet, it is by no means a new concept. Social scientists have used this metaphor for more than a century, studying different types of community as social networks (Wellman 1999). Rainie and Wellman (2012:21) define the social network as "a set of relations among network members – be they people, organizations, or nations". According to them, the social network connects individuals; these individuals, in turn, may be members of multiple networks.

However, the Internet has always been conceived as a network, at least in technical terms. Its typical definition refers to computers and their connections through protocols (Barabási 2002; Castells 2010). The Internet acquires the shape of a social network when digital platforms enable users to connect, and some scholars therefore emphasize the social dimension of digital networks. Rheingold (1993) points out that the technical foundation of the Internet is irrelevant to people in everyday life, but that technology allows people to come together and enhances the spirit of cooperation (Juris 2004; Rheingold 1993). Other authors use the terms social-technical or social-digital networks to highlight the intertwining of the digital and social dimensions of networks as a constitutive part of our social life (Trejo Delarbre 2015). Digital technologies contribute to extending and amplifying the reach of social networks that were traditionally based only on speech as a medium of interaction (Rheingold 2012; Scolari 2009; van Dijck 2006/2012).

The networked society is structured around networks powered by information and communication technologies (Castells 2004, 2010), resulting in a structural change within the networked information. This change has implications for the liberal markets and for political freedom (Barabási 2002; Benkler 2006; Castells 2010; van Dijck 2006/2012; Rheingold 1993, 2012), and reflects the paradoxical dual origins of the Internet: governmental control – ARPANET, the precursor network of Internet, was a project of the United States Department of Defense – and the libertarian initiatives of tech-savvy groups who sought to construct decentralized networks, as outlined in 'A Contract for the Web' (World Wide Web Foundation 2019). The four dimensions discussed below allow us to approach this tension and the connections between the micro- and the macro-level approaches to social networks by highlighting the empowerment of citizen actors and their potential for effecting transformation within structural conditions that do not always support them.

Four dimensions of the networked society

The first dimension concerns actors, and the need to identify who acts in the networked society and who participates in it. It is possible to argue that everyone is involved, because the networked logic affects even those individuals and groups who are not included in the network. All of us, willingly or unwillingly, are part of the networked society. At the same time, not all of us have the appropriate conditions, interests and capabilities to participate in it. In representative democracies, the conventional ways of participation revolve around political elections. Some citizens, such as activist groups, prefer more direct ways of participation, such as protesting, demonstrating or intervening in other ways to effect change. Taking part in movements constitutes a willingness to become an actor (Pleyers 2010).

Focusing on networks, Rheingold (1993) points to a tension between the potential of citizens to break the monopolies of the broadcasting logic through digital technologies, and the ease with which the commodification of networks can limit freedoms. The key issue here is participation. Participatory culture refers to the possibility of people taking part in the production of cultural materials through making posts and comments, tagging, blogging and similar activities. Even though this revolutionary information technology is introduced by the state and tech-savvy groups, common people have found ways to appropriate digital technologies to communicate through them, in order to transform and create their own networks. Participation thus refers to the involvement of non-professionals in the production of meaning. Although digital media has participatory potential, media itself does not determine the level of participation; instead, it is influenced by the uses to which it is put and the skills of those who use it (Carpentier and De Cleen 2008). Citizens, organizations, fictional characters, trolls and bots coexist in the networked society, restricting the potential and opportunities for participation.

The second dimension focuses on the practices in which actors engage. A practice is "a routinized type of behavior" (Reckwitz 2002:249) which connects different elements, from bodily activities to emotions and knowledge. Media practices are characterized by regularity, are social, relate to human needs, and provide a basis for thinking normatively about our relations to media (Couldry 2012). Rather than a strategy of communication, users of digital media have internalized the presence of technologies in everyday life to the point of taking them for granted. Users often participate in online public expression because that is where they find their audience and community. They engage in citizen media practices because these practices enable them to participate and to develop their own voice (Baker and Blaagaard 2016a; Stephansen 2016), which in turn "contribute[s] to the emergence of a civic culture that supports critical, emancipatory processes of knowledge production" (Stephansen 2016:12). In the global age, most activists participate on the Internet as a form of struggle to achieve recognition of their worldviews (Flores-Márquez 2019; Juris

2004; Milan 2013). The paradox is that while digital media is increasingly easier to use, the media practices of most users tend to revolve around consumption (Dragomir and Thompson 2014).

The third dimension is space and time, given that actors perform practices in specific spatial-temporal contexts. On the macro level, we need to consider that the diffusion of digital technologies was initiated in specific countries: the Internet as a technical network was created in the United States. However, the invention of the World Wide Web, which made the Internet more accessible, took place in Geneva, as a result of the work of Tim Berners-Lee, Robert Cailliau and their team (Castells 2010). Digital technologies thus originated in the United States and Europe. From there the Internet spread to the rest of the world and allowed the formation of connections among localities in a global space; Castells asserts that "[t]he key spatial feature of the network society is the networked connection between the local and the global" (2010:xxxv). At the same time, this logic highlights inequities among regions and conditions of human development, as well as in relation to the access and use of digital technologies. Hence, as already stated, although everyone is affected by this global logic, not everyone is included in the networked society. Global networks imply a set of connections mainly among metropolitan regions, with specific nodes for specific issues, such as Silicon Valley as a major node of technological innovation (Castells 2010). As a consequence, the networked society has become more unequal than mass society, defined by van Dijck (2006/2012:20) as "a social formation with an infrastructure of *groups, organizations and communities* ('masses') shaping its prime mode of organization at all levels (individual, group/organizational and societal)" (emphasis in original).

Three communication spaces may be identified as key elements of the networked society: the Internet, the media and the streets. The net refers to online public expression practices framed by technologies as the backbone of the networked society. Media refers to mainstream media practices, mainly consumption practices, framed by the growing trend of the concentration of media ownership in the hands of a small number of players. The streets refer to the appropriation of urban public space by citizens, which has been a feature of global mobilizations as well as of local initiatives of everyday activism for a number of decades. This account overlaps with Cardoso's (2006) argument about the Internet and television as the nodes that structure the networked society. Mainstream media is still relevant, and television continues to be the most popular media worldwide, while Internet access and usage are unequally distributed across different regions of the globe (Dragomir and Thompson 2014). But this model accounts for co-presence as a key element that enables small actors to form their own networks in order to take part in wider and stronger networks. Furthermore, we are witnessing the overlapping of different kinds of networks. The technological level is perhaps the most evident, since a single entity – whether institution, collective or individual – will often have a website, a Facebook page, Twitter and Instagram accounts and a YouTube channel, among other platforms. Overlap exists at the social level too, since a single citizen is generally a member of different groups, and each group interacts with other groups in the same locale or globally, across different locales (Papacharissi 2011b; van Dijck 2006/2012).

Time also matters, since the networked society is a specific social form of our era. The impact of the network logic is now evident in economics and politics – for instance in international agreements and flows – as well as in different kinds of cultural exchanges. Even the social movements and media projects of the current era are taking the shape of a network. At the same time, as Cardoso (2006) affirms, there is a coexistence of features among networked societies, industrial societies and even predominantly agrarian societies, suggesting the persistence of different conceptions and levels of development that are associated with social inequities and with different worldviews.

The fourth dimension is power. It is commonplace to refer to the Internet as a network of networks, with an emphasis on its structure and an assumption of no central control being exercised (Castells 2004; Rainie and Wellman 2012; Rheingold 1993). Several authors adopt optimistic perspectives on the reconfiguration of power in the information age, arguing that the network paradigm is based on the power of many-to-many (Castells 2010; Rheingold 1993). Castells (2004:12), for instance, argues that "the capacity for any communicating subject to act on the communication network gives people and organizations the possibility of reconfiguring the network according to their needs, desires, and projects". But despite the pervasiveness of networks, power is not equally distributed, and it does not operate in a democratic or fair manner. Some nodes acquire more connections and power, while other nodes become dependent on the stronger ones. Citizen media practices intervene in this dynamic by empowering citizens to create and distribute their own content and hence develop alternative perspectives (Stephansen 2016). Research on bloggers, urban groups and activists highlights the determination of individuals to use digital media to express themselves on specific issues (Flores-Márquez 2010, 2013a, 2013b, 2019). Most actors are critical of mainstream media and develop their own content, learning by doing, as a form of activist/citizen media practice (Cardon and Granjon 2010; Lievrouw 2011; Stephansen 2016). They do so in specific urban and digital spaces that connect different contexts and levels, including the net, streets and media at the local level, as well as interact with activists committed to similar causes across the world. These practices also occur in a specific time characterized by immediacy and real-time media logic. Ultimately, the character of the networked society is reinforced by expanding the possibilities of participation and interaction among actors (Cardoso 2006; Castells 2009; Rheingold 2012; van Dijck 2006/2012).

Future directions

The concepts of networks and the networked society are helpful for explaining some of the ways in which we live together in the current era, which is characterized by the pervasiveness of digital media. Although networks have been present throughout human history, this specific historical moment makes the network logic more evident and requires scholars to accord it more critical attention.

A number of issues are particularly important to pursue and call for further research. First, more attention needs to be paid to the paradoxes of the networked society, which includes and excludes at the same time and has sharpened the inequities among regions and actors in terms of access, use, levels of knowledge, skills, potential for connection and political conditions. Second, there is the tension between the libertarian and authoritarian aspects of the networked society, as the same network that enables the empowerment of citizens through digital media practices also facilitates permanent surveillance, control and censorship through digital networks as well as the dissemination of fake news. As early as 1993, Rheingold highlighted the risks associated with specific interests controlling the Internet in the future. Now that future is our present and his warning rings true in these post-truth times.

See also: media practices; space and place

Recommended reading

Cardoso, G. (2006) *The Media in the Network Society: Browsing, news, filters and citizenship*, Lisbon: Centre for Research and Studies in Sociology.

Approaches the communication dimension of the network society by analysing the organization of the media system. The book positions media – and especially television and the Internet – as a key part of the network society and raises various issues relating to questions of citizenship.

Castells, M. (2010) *The Information Age*, Volume 1, *The Rise of the Network Society*, second edition, Malden, MA: Wiley-Blackwell.

Originally published in 1996, the first of three volumes on the information age presents a sociological analysis of the shift from the industrial society to an informational society structured around networks, focusing on economic, social and cultural dimensions of this shift.

Dijck, J. van (2006/2012) *The Network Society: Social aspects of new media*, third edition, London: Sage.

Originally published in Dutch in 1991 as *De Netwerkmaatschappij*, this book analyses the link between new media and society, including technological, economic, political, legal, social, cultural and psychological aspects, and offers policy recommendations relating to the network society.

PARKOUR

Michael Atkinson and Kristina Smith

Parkour continues to grow rapidly as a physical culture around the globe. Once an activity practised by marginalized suburban youth, it has become a widespread fixture of modern metropolitan life. There are parkour street crews, parkour gym classes, parkour courses in physical education and even parkour-inspired fashion shops (Gilchrist and Wheaton 2011). While the physical culture has seeped into many modern institutional spaces and has become ideologically diluted in its representation as simply an alternative form of youth exercise, many urban participants remain loyal to the original philosophy guiding parkour practice. More specifically, as Atkinson (2009, 2013a) has noted, parkour carries with it a geo-spatial-corporeal-political ethos for many practitioners (known as traceurs). The physical culture of parkour is not merely about daring gymnastic performances such as building-to-building leaps, vaults, climbs and rolls; it is concerned with the deliberate use and contestation of urban space as a place wherein youth may articulate their own political interests, philosophies, experiences and values outside the frameworks of mainstream organized sport and leisure (Kidder 2012, 2017). As such, parkour remains a physical culture through which members may express their social beliefs, political ideologies and lived realities. Important at this time, given the manifold uses of the practice and in the age of engaged citizenship through digital media, is the fact that many practitioners have become meaning-makers through the use of documentary, YouTube style videos and vlogs. This entry will also discuss the burgeoning importance of doing 'digital parkour' as a central technique for representing the practice at this pivotal historical juncture.

The historical underpinnings of parkour

The historical roots of parkour as a physical cultural practice date back to the early twentieth century. Parkour is a particular offshoot of a style of training called Hébertism, which emerged in the 1910s and 20s through the athletic philosophies of French naval officer George Hébert. Hébert believed that dangerous outdoor regimes of training could help prepare his students for the moral requirements of everyday life (courage, confidence, truth, calmness, other-orientation and oneness). Practitioners were simply told to run through the woods, over bushes, through streams, climb up and down trees and traverse fields. This Natural Method demanded that one possesses sufficient energy, willpower, courage, coolness and *fermeté* (strength) to conquer any physical or mental obstacle. To this end, Hébert firmly believed that a major part of life should be lived, in contemporary terms, entirely off grid (Atkinson 2013a, 2013b). Hébert became the earliest proponent of what the French military would later call the *parcours* (obstacle course) method of training. Among the French soldiers exposed to the Natural Method and *parcours* was Raymond Belle. Belle taught his son David the principles of the Natural Method. After moving to the Parisian suburb Lisses, David Belle explored the rigours and benefits of the Natural Method with his friend Sébastien Foucan. By the age of fifteen, Belle and Foucan had developed their own suburban style of the Natural Method they termed parkour. The moniker parkour clearly derives from Hébert's use of the term *parcours*, and the French military term *parcours du combattant*. The nine founding members of the original Lisses crew were David Belle, Sébastien Foucan, Châu Belle Dinh, William Belle, Yann Hnautra, Laurent Piemontesi, Guylain N'Guba Boyeke, Malik Diouf and Charles Perriére. The group became known as the Yamakasi.

While often ignored in most academic accounts of parkour, especially in analyses of the group's genesis, the group has been highly active in producing their own media representations (Angel 2011, 2016). Rather than expressing prototypical subcultural resistance or ambivalence towards popular media, the Yamakasi were generational pioneers in embracing and utilizing media to highlight their practices and ideologies. Their popular cultural media debut might be located in one particular mediated exhibition that they staged in 1997, when David Belle's brother Jean-François invited the group to perform for a firefighter demonstration in Paris. Here, they took to the stage to articulate the physical practice as *l'art du déplacement* (the art of movement). The Yamakasi were not in collective agreement regarding the inherent politics and social risks associated with showcasing their culture in such a highly performative, watered down and spectacular fashion. While keen to promote the practice and its core values of hard work, dedication and character building to other Parisian youth, the real threat of drawing unwanted attention from young thrill-seekers and others could not be ignored. Jean-François Belle fuelled controversy among members of the group by sending still pictures of the event and video snippets of the performance to a popular French TV programme, and the popularity of parkour grew almost immediately. In today's media vernacular, the event and its representation went viral. The footage of the performance appeared in television programmes, newspaper features and on websites, and the Yamakasi became minor celebrities in France, the UK, the Netherlands and a host of other European countries. This formed a watershed media moment which set in motion the group's active involvement in both soliciting media coverage of their practice and becoming citizen producers in their own right. The event, however, would create a foundational crack in the group's collective consciousness that would define parkour for generations to come.

The Yamakasi caught the attention of French film director Luc Besson, who later solicited the group's counsel and involvement for a parkour-inspired action sequence in the 1998 movie *Taxi 2*. Besson became so enamoured with the group he documented their physical culture in *Yamakasi* (Zeitoun and Seri 2001) and its sequel *Les Fils du Vent* (Seri 2004). Besson also showcased parkour-inspired action in the film *District 13* (Morel 2004), this time featuring traceurs David Belle and Cyril Raffaelli; and yet again in the film's sequel *District 13: Ultimatum* (Alessandrin 2009) (remade in English as *Brick Mansions* – Delamarre 2014). While the physical culture had already been featured in a spate of media stories, accounts and even film, parkour historians often cite the 2003 Channel 4 documentary *Jump London* (Christie 2003) as the flashpoint in the group's global mass mediation and the trigger for a significant groundswell in membership. Audiences were enthralled by the new physical culture and the dazzling displays of urban acrobatics offered by participants.

Belle and Foucan quickly gathered recruits and followers in Paris and then across Europe. By the end of the 2000s, media in France, the UK and the Netherlands had documented the emerging lifestyle movement. Media reports predictably undermined the group's core philosophy and consistently framed the practice as a vacuous and style-oriented urban youth counter-culture. Specifically, the media cautioned against the inherent physical risks associated with the flamboyant movements, and questioned whether parkour might be a gang, while paying little attention to its core tenets (Atkinson 2009; Gilchrist and Wheaton 2011). Belle referred to parkour's mediatization during the late 1990s as part of a generational "prostitution [and destruction] of the art" (Christie 2003). He made these comments even before witnessing the lifestyle's mass mediation and commercialization in the movies *Breaking and Entering*

(Minghella 2006), *Casino Royale* (Campbell 2006), *The Bourne Ultimatum* (Greengrass 2007), *Live Free or Die Hard* (Wiseman 2007), *Prince of Persia: The Sands of Time* (Newell 2010), *FreeRunner* (Silverstein 2011), *Run* (Bartesaghi 2013), and *Tracers* (Benmayor 2015). The group were similarly showcased in a two-season MTV programme entitled 'Ultimate Parkour Challenge' and in nearly thirty video games either based on the practice directly or featuring parkour-based movements. As a result of media attention, widespread youth interest, the commercialization of parkour images and identities, and movement away from the spiritual towards the spectacular aspects of parkour throughout the early 2000s, Foucan and Belle disagreed vehemently over the vision and purpose of parkour.

The ideological split between Belle and Foucan and the eventual break up of their original Lisses crew into separate parkour factions is rather stereotypical for a youth subcultural movement, especially in the wake of the lifestyle's popular cultural ascendance as a commodity. Parkour is a unique youth counterculture which has long engaged in practices of self-promotion through media production, but as such, the group has been historically cleaved by different members' perceptions of the importance, value and place of media for signifying parkour. Since the millennium, we can distinguish three relatively distinct traceur contingents that have emerged with their own collective relationship with media consumption and self-representations.

First, there remains the self-labelled tribe of Natural Method traditionalists who believe that training in woodland is the only true method of parkour practice. Geographically, these individuals live in close proximity to densely wooded areas, or possess the requisite economic and social capital to travel to wilderness regions on a semi-regular basis. They are, however, a small and esoteric wing of the larger parkour movement, and they share a very weak interest in producing or consuming parkour-related media (Atkinson 2009). They may consult parkour websites, YouTube videos featuring other Natural Method enthusiasts, or record their own exploits while practising in the woods, but only for the purposes of teaching and documenting their lifestyle for likeminded others. For them, parkour is meant to be experienced and lived offline, and these individuals view the ongoing commercialization of the physical culture as a perversion of its essence (Atkinson 2012, 2013a).

Second, (sub)urban traceurs also view themselves as parkour traditionalists but firmly eschew the idea that one needs to train in the wild in order to reap the social and psychological benefits of parkour. Instead, they see their own neighbourhoods and cityscapes as important physical environments to explore, experience, understand and deconstruct through athletic movement (Atkinson 2009). They are moderately involved in consuming and producing parkour media, and tend to exist and practice in small crews and attend regional jams. Third, and following the lead of Foucan, an entire generation of competitive, media-savvy freerunners has emerged (Atkinson 2012). These individuals are indeed globally oriented, thrill and sports competition omnivores who may understand (and even partially embrace) elements of parkour's spiritual and moral essence, but who are nevertheless predominantly interested in the spectacular physical tricks, mediation and commodification of the practice. Competitive freerunners have become the largest demographic in the parkour global network. They tend to live in urban environments on the fringes of downtown city cores; they are predominantly males, they share lower-middle or upper-working-class backgrounds and they value alternative sports cultures as important spaces in which they can connect with others (Atkinson 2009).

There is a small but growing body of scholarship which examines all of the parkour enthusiast varieties outlined above and which describes how the practice has emerged from an esoteric, hidden and specialized physical culture into the popular, athletic, artistic, socially challenging, contested and globally mediated practice of today (Ameel and Tani 2011; Clegg and Butryn 2012; Edwards 2009; Geyh 2006; Gilchrist and Wheaton 2011; Guss 2011; Kidder 2012, 2017; Mould 2009; Ortuzar 2009; Thibault 2013). Within the research literature, save for landmark studies by Gilchrist and Wheaton (2011, 2013) and Thorpe and Ahmed (2015), few have paid critical attention to how traceurs proactively employ media to venture beyond simplistic mass mediations of the group in cinema and video games, or to denigrate outsider claims and moral panic surrounding the group as urban vigilantes and malcontents. In what follows we briefly highlight how traceurs digitally relate to others and attempt to give meaning to the physical culture on their own terms.

Documenting parkour

Notwithstanding the differences between the various parkour factions and their relationships to media as outlined above, the impact of mobile, user-friendly, high-quality and shared digital media technologies and platforms has changed the face of the physical culture. First, regionally, nationally and globally attuned websites have helped create a sense of community among parkour enthusiasts and have been the primary vehicles for the dissemination of the physical culture. As a quintessential late-1990s subcultural movement, the first and second generations of 'Parkour kids' grew up online and, like many of their peers, shared an appetite for self-mediation (Atkinson 2009). The development of community organizing websites such as UrbanFreeflow.com and Parkour.com quickly became both ideological beacons for the movement and anchors for learning, teaching and preaching the benefits of the parkour movement. Further still, the global flow of parkour to such places as the United States, Mexico, India, Thailand, Turkey, Russia, Brazil and others has been heavily shaped by the development of not-for-profit organizations, including the International Parkour Federation (IPF). Representing nearly fifty Parkour National Governing Bodies in countries around the world, the IPF's mission is to create an international coalition devoted to establishing parkour as a recognized sport. The IPF should be viewed not only as an official body promoting parkour as a sport but also as a major media hub for spreading information about the global proliferation of the practice. The Internet has been instrumental to the production and circulation of parkour manuals, learning guides and information about parkour traditions on an international level. Through individual and collective efforts, traceurs actively create their own global library of subcultural meaning in real time; or, in Guss' (2011:73) terms, they create a complicated, nuanced and truly diverse online image of "the [parkour] multitude".

In the early days of digital parkour, enthusiasts used the web primarily as a space in which to create a virtual representation of their communities for others to see, but also as a site for advertising when particular jams (parkour sessions) would be held, and even as a tool for recruiting new potential members. Research has consistently shown that traceurs feel their crew websites also help to normalize the subculture for parents, teachers, police and others who might not know anything about the group but nevertheless have fears or concerns about them. Creators and curators of the sites have been able to post photos, make announcements, provide links to their own or others' videos, as a way of expressing their own thoughts and constructions of the practice, and as a means of removing cultural veils and myths about the group (Gilchrist and Wheaton 2013; Guss 2011). Simple representations of parkour along these lines have paved the way for the foundation of parkour communities in every corner of the globe. The Wall Runners of Gaza group (or Gaza Parkour), for example, formed in the mid-2000s in the Khan Yunis refugee camp when Mohammed Aljakhabir and Ahmad Matar watched a YouTube video featuring traceurs (Thorpe and Ahmed 2015). For a generation of young Palestinians who had grown up in an environment of near-constant conflict, underemployment and disenfranchisement in Gaza, parkour became a method of self-expression, an escape and a way of life. In a context of local social ridicule and misunderstanding, the original members of Gaza Parkour utilized social media both to dampen stigma associated with the practice among their fellow Palestinians and to promote the physical culture as a method of personal salvation and social development (Thorpe and Ahmed 2015).

Second, the advent of affordable smartphone technology and the ubiquitous spread of social media has provided varied practitioners with an ability to promote parkour for themselves and to defend it against complaints from within that the practice has become excessively commercialized (Atkinson 2012, 2013a; Kidder 2017). As with any subcultural formation, when the practice is taken up in a diluted form (as a form of entertainment and sport) or *en masse* as a commodity (in movies, video games, YouTube videos), the aims of the practice may become muddied. The self-proclaimed lifestyle members of the parkour community take to media space to lament and tactically resist parkour's incorporation into popular culture. In particular, lifestyle members use practices of self-mediation to problematize the ongoing mainstream appropriation and dilution of parkour. For example, in their vlogs and blogs they speak out against the growing popularity of Parkour for Everyone classes in local fitness centres, the promotion of the practice by local governments as a means of combatting obesity, the use of parkour in commercial advertising campaigns and the transformation of parkour into a hierarchical global sport (Atkinson 2013a, 2013b). Traceurs also use carefully shot mini documentaries on their mobile phones (often shared as Instagram posts) to capture the real street face of parkour to enrich public understandings of the meanings of parkour before its commodification. Indeed, past and ongoing research in the Canadian context illustrates how this category of local practitioners attempt almost nostalgically and perhaps romantically to challenge participants into focusing on the moral virtues and character-building potential fostered through the physical culture, rather than on its spectacular physical techniques that act to dilute the very essence of parkour (Atkinson 2009, 2012, 2013a, 2013b).

More traditionalist parkour enthusiasts interviewed and researched by Atkinson (2013a) caution against the ongoing fascination with self-mediating parkour *ad nauseam*. For some traceurs based in Canada, recording a training or jam session, filming new tricks and/or recording video diaries explaining one's thoughts about the practice not only distracts one from the actual practice of parkour in the here-and-now, but can also become self-indulgent and distract from the community-building ethos of parkour (Atkinson 2013a, 2013b). Traceurs may try to 'one up' each other, innovate with increasingly more dangerous tricks, or view themselves as the next parkour celebrity. Self-reflexive traceurs additionally note that YouTube videos of seasoned veterans performing stunts make the practice look easy to those who fail to appreciate the depth and breadth of skill, training and experience required. Video representation of any physical culture, they argue, distorts time and space, allowing one to forget that any move may take weeks, months or even years to perfect. As a result, young people might attempt to replicate the tricks and injure themselves needlessly. Parkour in this sense is transformed from a lifestyle with a deep history and philosophy into something reduced to risky jumps across buildings, leaps into space and running through the streets.

See also: mediatization; self-mediation

Recommended reading

Angel, J. (2011) *Ciné Parkour*, London: CreateSpace Independent Publishing Platform.

Places important emphasis on theoretical understandings of the practice of parkour. Provides a robust critical analysis of the activism and creativity involved in the subcultural practice, and the role of social media in its ongoing development.

Edwards, D. (2009) *The Parkour and Freerunning Handbook*, New York: Dey Street Books.

Fundamental and comprehensive guide that illustrates the philosophy and cultural development of parkour, as well as providing a how-to manual teaching the techniques of freerunning. This is key reading for novice traceurs interested in building their skills and confidence within the sport.

Kidder, J. (2017) *Parkour and the City: Risk, masculinity and meaning in a postmodern sport*, New Brunswick, NJ: Rutgers University Press.

Based on years of participant observation, the author delves into parkour as a social phenomenon and explores its origins, identity politics and the legitimization of the practice as a sport. This is an important text for anyone seeking to explore the interactions between danger, masculinity and urban space, as well as the role of the media in shaping subcultural practices.

Thibault, V. (2013) *Parkour and the Art du Déplacement: Strength, dignity, community*, Montreal: Baraka Books.

Explains the parkour philosophy while examining the sociocultural aspects of the sport, as well as the spiritual side to the practice. This book addresses the many motivations of freerunners and is especially helpful for anyone new to the subculture.

PERFORMANCE STUDIES AND CITIZEN MEDIA

Maria Chatzichristodoulou

Performance studies is an interdisciplinary field that draws on theatre studies and performance art, sociology, anthropology, media studies and other areas of scholarship. It studies acts of performance, while also using the concept of performance as a lens to study the world. A performance studies approach can be applied to most disciplines since, according to Schechner (2006), a pioneer of performance studies, any human action either *is* performance or can be studied *as* performance. This perspective on performance has influenced the humanities and social sciences since the 1990s; indeed, performance is increasingly used as a heuristic method to understand and study human behaviour. In view of this comprehensive notion of performance that Schechner and other performance studies scholars have put forward, "the narrower foci of traditional theatre studies ... become important specific strands in the nexus of cultural metacommentaries" (Zarrilli 1986:372). The discipline of theatre studies is thus understood as just one part of the new paradigm of performance (Shepherd and Wallis 2004:151–160).

The related concept of performativity has also been influential; first proposed by philosopher J. L. Austin (1962) to explain the power of language to make things happen or effect change through speech acts, it was consequently used by Butler (1990, 1993) in her analysis of the social construction of gender identities. The concept has been widely applied since then, in a paradigmatic shift known as the performative turn (Dirksmeier and Helbrecht 2008). Thus, for example, journalism has been studied as performative by Broersma (2010), among others, who argues that features such as form and style give journalism a performative power that allows it to legitimize and impose specific representations of the world. Similarly, citizenship has been discussed as performative by Iannelli and Muscarò (2017) in the context of contemporary public art and urban design, as has media, in its many manifestations, as in Foellmer et al.'s (2018) *Media Practices, Social Movements, and Performativity*, which examines some of the uses of digital media in movements such as Occupy and the Arab uprisings as performative acts.

Citizen media is defined *by Oxford Living Dictionaries* (2018) as "(a collective name for) blogs, podcasts, and other forms of media produced by members of the public or published outside traditional media channels, especially on the Internet", while the online *Free Dictionary* (2003–2018) redirects searches for citizen media to its entry on citizen journalism, thus collapsing the two terms into a single practice. Baker and Blaagaard, however, have argued for a much wider and more inclusive definition that encompasses "physical artefacts, digital content, practices, performative interventions and discursive formations of affective sociality produced by unaffiliated citizens" (2016a:16). Their invitation to conceptualize citizen media as, among other things, "performative interventions" suggests that, by default, citizen media and performance often overlap. Understanding citizen media as performative and using performance as a lens through which to study it grounds it in specific bodies – with specific gender, size, shape, colour, age, sexuality, ability – and thus highlights its political dimension. It also connects it with notions of embodiment, space, place and temporality; with issues of presentation and representation; of agency, power and social dynamics; and enables an understanding of performance and performativity as constitutive acts, that is, acts that can make things happen and effect change.

Baker and Blaagaard's expanded definition of citizen media also suggests that acts of citizen media may set out "to effect aesthetic or socio-political change or express personal desires and aspirations" (2016a:16), and may include a diverse range of practices that are physical and digital, community-led and individual, informative and affective, thus creating new opportunities for connections with and across disciplinary confines. This widening risks rendering the definition meaningless in relation to performance: following on from this it could be argued that every performance or performative act constitutes citizen media because it expresses personal desires and aspirations – given that performance, as a medium of expression, is bound to express personal desires. It might therefore be more meaningful to focus, in relation to citizen media, on performances which aspire to "effect … socio-political change", regardless of how this aspiration is pursued or, indeed, of its outcomes.

Rehearsing for the revolution

Forms of theatrical or performance interventions that strive to effect social change traditionally either have participatory elements built into their dramaturgy or are entirely dependent on audience participation. Applied theatre forms, such as the Theatre of the Oppressed developed by Brazilian director Augusto Boal, aspired to operate as a "rehearsal for the revolution" (1974/1979:122). Theatre of the Oppressed is not a form of theatre to be watched by audiences but one to be performed by them, as 'spectactors' (Boal 1974/1979), in a process of acting out different scenarios through which they can address real-world problems or concerns. At its core, it challenges theatre as an art form to be produced and performed by professionals, as this inevitably reinforces representations of the world as it is; instead, it offers a theatre performed by the people and for the people in an attempt to try out different ways of transforming their world. In this context, citizens can rehearse behavioural change for themselves and find ways to actively tackle issues that oppress them (ibid.). The fact that the theatrical process is entirely disconnected from the professionals and passed on to the people as a medium that can help them enact social change echoes the definition of citizen media adopted in this volume in a form that is embodied and enacted.

The Theatre of the Oppressed has functioned as a model for a range of community theatre practices, such as Chris Johnston's Citizen Theatre technique, a form of improvisatory, workshop-based theatre that aims to confront participants with their own behaviours in order to effect change (Johnston 2005, 2017). Johnston used this method to work with local communities, prisoners and those on probation. Other techniques developed by Boal himself as part of the Theatre of the Oppressed, such as Newspaper Theatre, have direct connections to both journalism and citizen media. Newspaper Theatre consists of techniques for "transforming daily news items, or any other non-dramatic material, into theatrical performances" (Boal 1974/1979:121). It was a technique used by non-professionals, which aimed to "demystify the media, and educate people to question the notion of objectivity" (Boland and Cameron 2005). Newspaper Theatre stemmed from the practice of Living Newspaper, a very similar form of theatre that originated in Russia during the Bolshevik Revolution and which aimed to present factual information on current events to a popular audience. Living Newspapers were also known to encourage activist action, and were open to the use of multimedia in order to achieve their aim of engaging the people. The best known company developing this type of theatre was the Federal Theatre Project (FTP), which operated in the US in the 1930s, during the Great Depression. Sponsored by the Works Progress Administration, part of the New Deal created by President Franklin D. Roosevelt to combat the Great Depression, FTP was developed not as a cultural activity but as a relief measure, to employ artists and cultural workers (Perry 1950; Vacha 1986).

The Theatre of the Oppressed originates from a long tradition of political theatre. It can be traced back to Epic Theatre as created by Erwin Piscator and Bertolt Brecht and further developed by Brecht himself. It also draws on the Marxist practice of Agitprop: in the 1920s and 30s, after the Russian Revolution, the USSR's Department of Agitation and Propaganda employed theatre groups to stage performances with the aim of communicating news and current affairs in an accessible manner to a largely illiterate population (Brown 2013). At the same time, in Germany, Erwin Piscator used theatre as a means of communicating political messages, and was experimenting with incorporating primary source material such as verbatim interviews and documentary film footage into his spectacles. Influenced by Dada, he argued in favour of a more overtly political form of theatre aligned to the struggles of the proletariat. He founded the Proletarian Theatre in 1920 (Bryant-Bertail 2010), and in 1925 he wrote *In Spite of Everything! (Trotz Alledem!)*, a piece deriving entirely from political documents, which is often cited as the first documentary drama (Mason 1977). An innovator in form as much as content, Piscator used media and journalism, including film projections and newsreels, to develop a form of total theatre that we might today describe as immersive, in order to communicate mass events. He inserted verbatim political documents, news reports or direct quotations from public figures in several of his productions (Innes 2010; Styan 1981:128–138). Piscator thus arguably used theatre as a means of performing citizen media, in order to give voice to a range of citizens whose issues and concerns were not represented through the mainstream media of the time.

The use of verbatim material in theatre has since spread exponentially as a practice. Paget was the originator of the term Verbatim Theatre in the 1980s (Paget 1987), and among the first to research the practice, examining the work of practitioners such as John Cheeseman, Rony Robinson and David Thacker, who operated in Britain in the 1950s and 60s, and the seminal work of Joan Littlewood (Gibson 2011:3). Today, Verbatim Theatre is an almost ubiquitous technique, and plays developed by established artists such as Moisés Kaufman, Alecky Blythe, Anna Deavere Smith, Paul Brown, David Hare and Max Stafford-Clark incorporate words of real people as spoken by them into the drama. As Gibson points out (2011:1), Verbatim Theatre aspires to tell the stories of others, "especially those marginalised or oppressed by virtue of their race, class, gender, religion, sexual orientation, (dis)abilities and so on". Providing "an arena for the marginalised" (ibid.:3) or giving a "voice to the voiceless" (Hare 2005:112), such works can be conceptualized as citizen media, the latter also often discussed as a means of "bringing diverse voices into pluralist politics" (Pettit et al. 2009:443). However, acts of citizen media empower citizens to take direct control of the means of distributing information and affect in order to tell their own stories – or voice their own voices – and effect some form of change. Indeed, citizen media can achieve more than just broadcasting a variety of voices: Pettit et al. argue that it can "contribute to processes of social and cultural construction, redefining norms and power relations that exclude people" (2009:443). Verbatim Theatre, on the other hand, is about people's real stories being told by a host of professional others, including the playwright, actors and director of the play. Unlike the Theatre of the Oppressed, which consciously and explicitly sought to remove the aesthetic, narrative and performative processes from the professional artists' and producers' hands in order to create theatre by and for the people, Verbatim Theatre is made by and for the people through the mediation of professional theatre makers. Although the aim is to stage and broadcast people's voices, the agents of this staging are not the people whose stories are being told. This dislocation of agency raises a number of ethical issues concerning the practice of documentary and Verbatim Theatre, including questions around accuracy, (mis)representation, (mis)appropriation and (dis)empowerment. Several scholars have argued that Verbatim Theatre is a "problematic performance methodology" (Jeffers 2006:2) due to its claims to authenticity – claims questioned by Stuart Fisher, who argues that the truth of a traumatic event is not "transparent, knowable or even communicable" (2011:112). Alcoff (1991) and Hazou (2009) further suggest that many Verbatim Theatre practitioners speak *for* rather than *with* others, while Beck (2016) raises questions around the ethics of selecting, editing and aestheticizing testimonies, particularly in relation to traumatic events.

All the forms of theatre discussed so far, such as Boal's Theatre of the Oppressed and Johnston's Citizen Theatre, focus on facilitating interventions around issues and situations that concern specific communities.

Although the theatrical action often takes place in a public or community space, it does not constitute an intervention in its own right; instead, it aims to function as a rehearsal for an intervention that is waiting to happen, with the constitutive power of theatre making this intervention possible through enacting its potential. In that sense, Theatre of the Oppressed and the tradition of practices it has inspired have become "communication spaces where citizens can learn to manipulate their own languages, codes, signs and symbols, empowering them to name the world in their own terms" (Rodríguez 2011:24). Other forms, which are in the same historical lineage of political theatre but still rely on professional storytellers mediating real people's narratives, such as Verbatim Theatre, aim to raise awareness and visibility around issues which are being sidelined or suppressed, thus giving voice to the voiceless. These can be understood as forms of citizen media in the more narrow definition of the term, as early articulations of embodied and enacted citizen journalism.

Gesturing the revolution

Some of the most powerful performance interventions that take place in the public sphere and aspire to generate social change aim to raise awareness around issues not through communicating factual information but through a gesture that can generate affect. These are often, as Shalson puts it, "public expressions of dissent against prevailing systems" which "demand change" (2017:8). The use of theatricality and performance in protest situations has a long tradition. According to Garfinkel (1967:11), all "expressions and other practical actions" are "contingent ongoing accomplishments of organized artful practices of everyday life". Saward argues that the same applies to the "structures and actions of political life" (2015:217) and proposes the term performative politics to describe the phenomenon within which political grammars are enacted and performed (ibid.). Performative politics, Saward suggests, is "rehearsed or repeated citational action" that is designed to draw attention to "both the theatrical and productive (or constitutive) elements of performance in and of politics" (ibid.). These actions do not necessarily follow the same lineage of political performance as the practices discussed above. Their lineage is not to be found in theatrical traditions of political theatre but in visual art and performance practices such as the Events and Happenings of the 1960s and 70s (Kirby 1965), which "combined elements of painting, poetry, music, dance, and theatre and staged them as a live action" (Wainwright 2008/2011), and the Live Art and Performance Art traditions. In keeping with these traditions, protest actions use bodies rather than text as their primary means of communication.

Following Agamben's (1996/2000) characterization of gesture as pure means, which rejects the separation of action into means and ends, Hughes and Parry (2016:80) use the term protest gesture "to focus on the relationship between bodies and politics", arguing that protest actions or protest gestures are characterized by the imaginative and disruptive-of-social-norms use of bodies in public space. Speaking of "gestural grammars of protest" (ibid.:79), they suggest that our times are characterized by "an extraordinary proliferation of protest movements" (ibid.), and that these movements "draw on theatre and performance to enhance their potency" (ibid.:80). Among the examples they offer are the occupation of Seattle in 1999, the series of uprisings in the Arab World in 2010–2012, and the 2014 US demonstrations following the shooting of Michael Brown, a young black man, by a white police officer in Ferguson. The latter demonstrations are characteristic of how gesture is used in protest to generate affect: following the shooting, protesters from St Louis to Times Square in New York City marched with their arms in the air chanting 'hands up, don't shoot' – Michael Brown's last words before he was shot down by the police officer (ibid.:82). This theatrical gesture, which repeatedly embodied on a mass scale Brown's last actions and words, served as a powerful and emotive way of raising awareness about the incident, using theatre's ghosting technique to haunt America's streets. Ghosting, a term coined by Carlson (2003:2), suggests that every theatrical performance incorporates elements of a previous one, a *déjà vu*, as the "repository of cultural memory". Hughes and Parry "focus on moments of protest that contest the framing of some bodies as outside of the normative political order and its modes of citizenship" (2016:79) in order to theorize such protest gestures as citizen media. This is in tune with Rodríguez' (2011:24) definition of citizen media as "communication spaces" that empower citizens to "name the world in their own terms", but in the case of the theatrical protest gesture or performance action as protest, the process of empowerment takes place through bodies: bodies enact the world in their own terms, using the constitutive power of performance to embody practical alternatives to oppression.

The wave of demonstrations that spread across Turkey in May 2013 in response to urban development plans for Istanbul's Taksim Gezi park provide further examples of theatrical protest gestures. The violent eviction of a peaceful sit-in at Gezi sparked protests across Turkey around issues of freedom of the press, freedom of expression and the right to peaceful assembly. Three-and-a-half million people are thought to have taken part in the demonstrations across Turkey, in which eleven people were killed and more than 8,000 injured (Özdemir 2017). During the weekend of 15 and 16 June in particular, Turkish police brutally attacked demonstrators, including medics and staff who were treating the wounded in Taksim Square (Seymour 2013). On Monday 17 June, performance artist Erdem Gündüz captured international attention when he responded to the brutality of the police by merely standing still. Gündüz simply stood on his own, facing the Ataturk Cultural Centre, from 6 pm until 2 am local time. As people began to notice what he was doing, some started to join him in Taksim Square, but also across the country, in cities like Ankara and Izmir (ibid.). As with the hands up protests of Ferguson, Gündüz' simple gesture of standing still over a prolonged period of time went viral on social media, with the handle #duranadam (the standing man) trending on Twitter. Suddenly, a lone, quiet gesture of defiance became larger than life, spreading through social media and manifesting itself in public space through bodies standing still. The Standing Man gesture enacted an alternative response to the violence inflicted upon protesting bodies by the state apparatus. As a "gesture of exception" (Hughes and Parry 2016:82), it excluded the bodies that stood still from the political grammar of protest as understood by the government, causing confusion and succeeding not only in raising international awareness of the demonstrations, but also in moving international publics through affect.

A final example of performative gestures of protest also emerged in the context of protesting urban developments in Skopje, capital of the Republic of North Macedonia. Raspeani Skopjani, a self-organized and inclusive choir, responded to a government project planned with a lack of transparency, which aimed to transform the city centre of Skopje "into an area of concentrated development" (Marchevska 2017a), by singing. Members of the choir would democratically choose songs for each situation, "then sing *in situ* in a guerrilla action, which is recorded and then posted on YouTube" (Jakimovska, in Marchevska 2017b). Although their singing interventions were neither advertised nor framed as performance events, and the only people who attended were those who happened to be at the particular place and time a guerrilla intervention took place, the performances reached a large audience through social media. Choir members documented their interventions, edited them and posted them on YouTube, where they maintain an archive under the YouTube channel *Plostad Sloboda* (City Square Freedom). As Marchevska (2017a:397) explains, "Raspeani Skopjani stage a form of tactical media protest animated by an implicit agenda to reclaim public space (both physical and virtual) from state control". Their protest gestures are not confrontational or literal, and they cannot be easily read – and dealt with – by the regime as forms of disobedience. Raspeani Skopjani's interventions can thus be understood as political gestures of domesticity in Hughes and Parry's (2016) terms, despite the fact that they take place in public space. The singers appear in various spaces in the city of Skopje, spaces which are rendered familiar through public debates concerning their redevelopment; they treat Skopje as their home, and indeed, the home of all its citizens, reminding publics that this is their city and that they should therefore exercise agency in relation to how it develops. The imaginative use of bodies in space has the power, in those mundane acts of singing, to speak out and speak up in a language and a tune that authorities find difficult to decode and respond to, its performance a gesture that baffles, confuses and unsettles what Boal calls the oppressors. In this instance, guerrilla performance, public intervention and citizen media collide to raise awareness about an issue of public interest, disrupt daily business-as-usual, confront accepted practices and demand social as well as aesthetic change. Through their protest gestures, both the Standing Man and Raspeani Skopjani succeeded in bringing international attention to a specific localized context that would not, otherwise, have received much attention through international mainstream media.

Citizen media as performance activism

In 2012, Rabih Mroué presented *The Pixelated Revolution*, a piece about the Syrian revolution and ensuing civil war, at Documenta 13 in Kassel, Germany. It consisted of a lecture performed in front of a collage of images and videos taken from the Internet, which had been posted by civilians in an attempt to document the state's acts of violence towards them. While analysing the stylistic conventions of documenting violence within the context of a revolution, Mroué presented, among other documents of war, a short video of a sniper shot taken from the rooftop of a building. The audience see the sniper looking up, clocking the film-maker, aiming towards him, and shooting. The video then goes blurry as the camera phone appears to drop to the ground. Shouting is heard, and the video ends abruptly. This is the record of an activist documenting what is likely his own violent death. As a performance of citizen media, it is a protest gesture – a defiant gesture of documenting and broadcasting the atrocities of war even as the violence is directed towards the documenter, with what we assume to be fatal consequences. But it is also simultaneously a rehearsal for a revolution, in the context of Mroué's performance-lecture itself, which reframed and recirculated these documents among international audiences, drily breaking down their aesthetic conventions as a new activist grammar. One possible message of this particular activist – or heroic – act of citizen-media-turned-performance is that if citizens take it upon themselves to document acts of violence and make them publicly accessible, they may be able to bring the perpetrators to justice. If citizens can collectively hold them accountable for the atrocities they perpetrate, perhaps those acts of violence will no longer occur. In this sense, both Mroué and the activist documenting his own shooting embody, in different ways, the world as we want it to be, and as the performance of citizen media can contribute to making it – a space that is open, transparent and free of intimidation.

See also: activism; citizenship; flash mobs; media

Recommended reading

Boal, A. (1974/1979) *Theatre of the Oppressed*, translated by A. Charles, M.-O. Leal McBride and E. Fryer, second edition, London: Pluto Press.

A hugely influential book, written at a time when Brazil was under a brutal military dictatorship, this is the philosophical foundation of activist Brazilian director Augusto Boal's Theatre of the Oppressed. Boal argues that all theatre is political, and that theatre can be used as a weapon in people's fight for liberation. Tracing theatre's history through ancient Greece and beyond, he argues that theatre was originally by and for the people, but was consequently used by the aristocracy to establish divisions between the people through the separation of roles between the protagonist(s) and the chorus – the one/few and the many. Boal proposes a system for a theatre that destroys the barrier between protagonist(s) and chorus, and advocates a poetics of the oppressed which aims at the "conquest of the means of theatrical production" (p. xxiv).

Hughes, J. and S. Parry (2016) 'Theatricality and Gesture as Citizen Media: Composure on a precipice', in M. Baker and B. B. Blaagaard (eds) (2016) *Citizen Media and Public Spaces: Diverse expressions of citizenship and dissent*, London and New York: Routledge, 79–95.

A book chapter that examines contemporary theatrical forms of activism, looking at the role of artists in protest events across the globe, from the Occupy movement to the uprisings across the Arab world. Exploring the concept of the theatrical protest gesture, Hughes and Parry propose three gestural grammars of protest: gestures of exception, gestures of domesticity and the ecological gesture.

Johnston, C. (2017) *Disobedient Theatre: Alternative ways to inspire, animate and play*, London and New York: Bloomsbury.

This book, the last by influential theatre practitioner Chris Johnston (written a year before his death), asks what principles and tools make theatre effective and powerful when it argues back to society, challenging and upsetting societal norms. It draws on both theory and practice and is informed by conversations with practitioners from across the UK; it also offers possibilities of a 'disobedient culture' through games and exercises for creative practitioners.

Paget, D. (1987) '"Verbatim Theatre": Oral history and documentary techniques', *New Theatre Quarterly* 3(12): 317–336.

This article introduced the term Verbatim Theatre to describe documentary theatre that makes use of "taped actuality recording" (p. 317) as its primary source material. Paget traces the development of Verbatim Theatre as a form, explores its methodology and working practices, and examines the scope of its influence within the context of professional theatre.

PHILOSOPHY AND CITIZEN MEDIA

Omid Tofighian

Many instances of what is defined as citizen media in this volume involve transhistorical and cross-cultural intellectual work and constitute philosophical practices that have traditionally motivated and cultivated what Medina (2012) refers to as epistemic friction within philosophical debates. Such practices provide evidence of a longstanding, anti-hegemonic and decolonial philosophical tradition (Quijano 2000; Moreton-Robinson 2003; Dotson 2015, 2016; Smith 1999; Denzin and Lincoln 2014; Grosfoguel 2011; Cooper 2008, 2012; Blackwell 2011; Buck-Morss 2009). Citizen media has thus always been a legitimate and formative interlocutor within philosophy, but new conversations are needed to address its role in philosophical debates, as well as its exclusion, undermining or erasure throughout divergent philosophical spaces, discourses and activities. Recognition of and deeper engagement with the philosophical potency of citizen media is an emancipatory act and a source for epistemic insurgency (Medina 2017).

This entry examines the interdependent relationship between community advocacy, resistance, media and philosophical ways of knowing. It makes two major interventions and draws on three examples of citizen media: the political and intellectual work of the Black Panther Party; the strategic resistance and cultural dynamics associated with the Haitian revolution; and the prison writing and cultural production of Behrouz Boochani. The first intervention challenges the limitations of professional philosophy by elucidating the philosophical work done through citizen media. The second intervention illustrates how the social, cultural and political spaces and communicative practices of citizen media constitute sites for philosophical discovery.

The three case studies offer rich examples of citizen media as defined by Baker and Blaagaard (2016a). They function as models for future work on the interdependence between citizen media and continental philosophy, as well as for exploring connections between citizen media and analytic philosophy. The entry demonstrates how strategies, objectives and visions advanced through the three examples motivate, respectively, critical rethinking of methodology, tradition and the canon.

Philosophical methods: the Black Panthers and silencing strategies

Heiner (2008) argues that Foucault's transition from his archaeological work of the 1960s – including *The Order of Things* (1966) and *The Archaeology of Knowledge* (1969) – to his genealogical period, represented by *Discipline and Punish* (1975) and *History of Sexuality* (1976) and the *1976 Lectures*, was influenced by his visit to the US and encounter with race and class struggles there. More significantly, Heiner explains how Foucault's meetings with members of the Black Panther Party (BPP) and his exposure to their writings and activism impacted many of his central ideas and approaches to power, incarceration, social and discursive formations, and knowledge. The BPP, particularly the philosophies and activism of Angela Davis and George Jackson, were key to shaping the development and trajectory of Foucault's philosophical methodology. There are two interrelated sources of influence to investigate here: Foucault's interaction with BPP writings and ideas, and his involvement in activism. Both of these correlative encounters illuminate the critical role of citizen media in philosophical approaches and discovery.

Insight into the philosophical dimensions of citizen media practices is no doubt one of the features that attracted Foucault and others in Europe to inquire into and support the BPP. In addition to the scholarly work of its members and their political activism, the organization established officially as the Black Panther Party for Self Defence conducted sixty-five successful community programmes between 1966 and 1982. These included Sickle Cell Anemia Research Foundation, The Black Panther Newspaper, Free Food Program, Free Breakfast for Children Program, Campaign for Community Control of Police and numerous art, culture, youth and wellbeing initiatives. Their various practices of citizen media and commitment to community were multidimensional: the BPP created sites where activism, philosophy, culture and heritage, self-determination and community pride came together. As a philosophical movement, the BPP was inspired by Marxism and the works of Frantz Fanon, Malcolm X, Marcus Garvey, Nat Turner and Martin Robison Delaney. Although it cannot be divorced from their community empowerment and protection initiatives, it is the BPP's critical insight into the US prison system and their Black liberationist counter-history that are relevant to understanding their influence on Foucault's genealogical method.

The interweaving of citizen media and philosophical theorizing was not unfamiliar to Foucault before his acquaintance with the BPP and trip to the US. In particular, he was already a member of the Groupe d'Information sur les Prisons (GIP) and actively protested prison reform, provided support for prisoners and pressed for more transparency regarding how prisons were operated. The GIP had published translations of writings by and interviews with members of the BPP. Building a relationship with and researching the works of the BPP through his role in the GIP was a logical extension of his political activism and thinking at that particular time (Heiner 2008:332–335). The GIP had also published a series of pamphlets critical of the prison system after a number of their activists visited the US. The knowledge transferred to Europe as a result of these visits was studied intensely by Foucault. Indeed, the development of the genealogical method is evident in the introduction to the GIP's first pamphlet – which was primarily written by Foucault – on the group's investigations into the Parisian prison system (ibid.:320). As his approach transitioned from archaeology to genealogy, Foucault expounded the notion of the will to truth, introducing the idea of a regulative system designed to constrain and exclude (Foucault 1972:215–237). The will to truth is mutable and its conditions are determined by historical and institutional factors. For Foucault, it regulates by conspicuously determining and adjudicating, within the confines of accepted social formations, which statements can be considered truth-bearing. The will to truth is a principle that governs the production of knowledge, how it is rendered, its transmission and distribution, and who has access to it.

It was through the works of Angela Davis and George Jackson – possibly also Huey Newton, Bobby Seale, Eldridge Cleaver and others – that Foucault realized the profound interlocking relationship between persecution of political activists, the plight of poor people, resistance by racialized groups and the prison-industrial complex (Heiner 2008:319). And it is through the theorizing and community advocacy of Black Panther members – methods and activities that highlight the perpetual colonial dimensions of US history and emphasize the relationship between race, class and the prison system – that Foucault initiated and reified his genealogical method. Foucault himself acknowledged his uncritical reading of class struggle prior to his genealogical period, before becoming more familiar with the situation in the US and the BPP (Foucault 1989:73), admitting his naiveté regarding the gap between affluence and poverty in society and improvements in the general standard of living in Europe. Heiner (2008:320–321) points to key moments and comments by Foucault that reflect the revelation he experienced after engaging with the BPP and their critique of the prison system. It is true, as Heiner explains, that Foucault was involved in activism before his visit to the US, but those experiences did not help develop his intellectual work in the same radical way. Foucault the philosopher of struggle, including his critique of the prison system, emerged after he engaged with the thought and actions of the BPP (ibid.:317).

Angela Davis' (1971) comparison between contemporary liberation struggles and the struggle led by Nat Turner, leader of the most important slave rebellion in the US, who was sentenced to death in 1831, illuminates the epistemic and political resources Foucault drew on. She argues that Turner's fight for freedom and the philosophy driving it were criminalized in the same way that movements dedicated to emancipation are quashed today. Writing from prison, Davis describes strategies used by the state to identify struggles against oppression and purposely subvert them. For BPP philosophers such as Davis, the prison system is an organ of coloniality that constitutes part of a legacy reaching back to chattel slavery (Heiner 2008:319). A close reading of Foucault's work during the genealogical period and the writings of the BPP reveals clear examples of inspiration, influence and interactions, and Heiner is able to show that the genealogical method is deeply indebted to the intellectual work and political strategies and struggles of activists such as Davis (ibid.:332–344). And yet, Foucault's writings throughout the 1970s neglect to mention the BPP and, ultimately, contribute to marginalizing and excluding their philosophical legacy. The epistemic injustice Heiner highlights in the work of Foucault continues to this day. The scholarship on Foucault's middle period (1970–1976), including the genealogical approach he is credited for, continues to "bar access to the actual historic-political insensitivities and creations that in fact motivated the genealogical project" (ibid.:314). Heiner describes the undermining and erasure of the BPP by both Foucault and scholars specializing in his thought as a form of testimonial injustice (Fricker 2007). Dotson (2012) refers to the same pattern as contributory injustice, and Pohlhaus (2012) as wilful hermeneutical ignorance. The kind of harm inflicted on the BPP here can ultimately be attributed to a combination of (conscious) personal failings and structural factors. Both can be understood as disciplinary and methodological issues that distort and hamper a philosophical appreciation of citizen media.

Epistemic injustice committed against the BPP aside, studying the genealogy of Foucault's genealogical method reveals the philosophical potency of citizen media. By recognizing citizen media as a necessary and central philosophical voice, it is possible to establish the sociocultural and intellectual conditions that can support a fuller methodological repertoire in the discipline.

History of philosophy: the Haitian revolution and transcultural philosophical networks

In the course of highlighting a number of significant claims related to agency and self-determination and critiquing the presuppositions of the Enlightenment worldview, Jean-Marie (2013:242) asks: "What can a philosophical interpretation hope to contribute to the historical interpretation of the Haitian revolution?" His insights can be expanded by asking an additional question: what can a transcultural reading of the Haitian revolution as a paramount historical movement contribute to reinterpretations of the history of philosophy?

Focusing on the Eurocentric and patriarchal assumptions and goals of the European Enlightenment, Jean-Marie argues that for an Enlightenment thinker such as Kant, the rational human being presupposed in his intellectual project, particularly as presented through the essay 'What Is Enlightenment?' (Kant 1784/1996:11–22), did not accommodate enslaved people. Essentially, for thinkers such as Kant, humanity is racialized as white, male and free from bondage (Jean-Marie 2013:246). The Enlightenment paradigm thus relies on a racial distinction; how one is racialized determines their humanity. Jean-Marie draws on Trouillot's critique of the racial thinking conditioning the European Enlightenment. Trouillot (2013) explains how the paradigm espoused by Kant and others meant that a phenomenon such as the Haitian revolution was seen as unfathomable, unthinkable even by the slaves and leaders involved; the latter – military leaders such as Toussaint L'Ouverture and Jean-Jacque Dessalines – were free Black men when the revolt erupted. Of particular relevance to a discussion of citizen media is Jean-Marie's argument that Voodoo cosmology, its related transcultural practices and associated social networks provided the structural and organizational basis for revolution (2013:249–252). This leads him to disagree with Trouillot's view that the revolution was unthinkable to the Haitians involved (ibid.:248). Instead, Jean-Marie examines how the religious and linguistic conditions, together with the shared ambitions for emancipation, established the foundations for political organizing and state building. These factors also helped form the notion of universal humanity as part of the philosophical schema of the Haitian slaves and leaders, in contrast to the exclusionary preconceptions of the European Enlightenment.

In the lead up to the battle against the European slavers, a Voodoo ritual was conducted in a maroon society (a community of fugitive slaves) in Bois-Caiman, led by Dutty Boukman (Buck-Morss 2000:fn. 38), both a maroon leader and religious figurehead. The practice of this particular religious rite provided a significant impetus and philosophical foundation for the Haitian revolution (Nesbitt 2008; Buck-Morss 2009). Haitian Voodoo cosmology is a secular interpretation and sociopolitical expression of African-based syncretic religious rites. The maroon communities reformulated and reconfigured elements from diverse African, as well as Taino and Roman Catholic, belief systems into a new institution that functioned to unite disparate cultures and languages among enslaved Blacks. This sociopolitical-religious order formed the basis for resistance against the *Code Noir* – a draconian degree passed by the French king in 1685 which, among other things, outlawed all religions other than Roman Catholicism – and against the colonial enterprise, and it also gave birth to the very notion of an independent Black nation and Haitian citizenship (Jean-Marie 2013:251–252).

The Haitian revolution challenged the knowledge system that undergirded much of modern European thinking. Ethnically distinct Black peoples forcibly transported to European colonies engaged in sophisticated reasoning, negotiation and collective planning in order to achieve social cohesion and create political principles necessary for a successful decolonial movement. A shift in consciousness had occurred and a particular notion of Black subjectivity created; mapping strategies for revolution involved the development of social agency, a sense of community pride that carried emotive force, expressed clear sociopolitical ambitions, and created an organizational structure with purpose and direction. This would not have been possible without the role of Haitian Creole, a language with an inherent place-based logic and semantics (Jean-Marie 2013:253–255). As an embodied epistemology of resistance, the creation of the republic of Haiti through armed rebellion was therefore "a radical Enlightenment … both a major critique of the European Enlightenment and the completion of its ideals" (ibid.:255).

Jean-Marie's conclusion, and his critique of Trouillot, may be extended to suggest that not only did the enslaved Black and revolutionary leaders have the capacity to think the revolution, but only they – the slaves and leaders – could have thought the Haitian revolution given the context, setting and dynamics at play in the uprising. What occurred in Haiti is inseparable from the philosophical and political thinking and narrative of the slaves and leaders, supported by innumerable and interweaving features and factors, including the particular kind of decolonial revolution that took place and its development over years (Buck-Morss 2000 833, fn. 38), its outcome as the first Black independent nation, and the emergence of a Black subjectivity with multidimensional resonance and transnational consequences (Jean-Marie 2013:242). The conclusion that only the slaves and leaders could have thought the revolution is derived from a profound appreciation of citizen media practices, a series of their unique revelatory encounters and exceptional determinants (Buck-Morss 2000:835–836). In the case of the Haitian revolution, being, knowing and doing are interdependent with the landscape; a lived experience of history and visceral familiarity with coloniality; an emergent sense of religious-social-political agency; the establishment of a secret society of maroons hidden in the mountains; the remapping of the mountains as imaginal dwellings constituted by freedom; and the formation of an emancipatory social imaginary exclusive to the African diaspora.

When analysed against the backdrop of the competing colonial forces ravaging what we know today as the Global South – a global system configured by indigenous genocide and the enslavement of African peoples – the exceptional nature of the Haitian revolution stands out further (Jean-Marie 2013:255; Nesbitt 2008). The idea that no human being should ever be subject to forced labour was articulated in theory and manifested in praxis, as evident in the series of Haitian constitutions promulgated by the leaders of the revolution in the early nineteenth century. The Haitian uprising disrupts constructions of Western history as "coherent narratives of human freedom" and self-contained national accounts conditioned by disciplinary isolation (Buck-Morss 2000:822). The history of philosophy is equally vulnerable to the same critique, and has been guilty of marginalizing counter evidence and valorizing select narratives and figures. Philosophers such as Rousseau and Hegel were cognizant of the conditions in the colonies and the struggles for freedom that were taking place, and yet failed to discuss the phenomena in their philosophical works on topics related to freedom and humanity (Buck-Morss 2000:831). Indeed, there is evidence that Hegel was deeply influenced by reports of the Haitian uprisings and victory, primarily through the journal *Minerva*; Buck-Morss (2000, 2009) argues convincingly that his slave–master dialectic is the result of his exposure to and reflection on the events in Haiti. And both the French Revolution and Western abolition movements were radicalized after the gains made by Blacks in the colony of Saint-Domingue (Buck-Morss 2000:833–834).

The positionality of the Haitians provided them with a form of epistemic privilege that played a pivotal role in the mobilization and success of the anti-colonial uprising. The unique nature and context of the revolution impacted the Western imagination and subsequent responses in ways that emphasize the value of the transcultural thinking that informed its formation and outcomes. The Haitian revolution's role as a philosophical event, an inauguration of a tradition, and a sociopolitical process provides a model for considering the significance of citizen media when reinterpreting and refining the history of philosophy.

The philosophy canon: Behrouz Boochani and introducing new knowledge into philosophy

Behrouz Boochani is a Kurdish-Iranian writer, journalist, scholar and human rights defender who was held in the Australian-run refugee detention centre on Manus Island, Papua New Guinea (officially called Manus Regional Processing Centre) from 2013 until 2019. He had fled Iran in 2013 due to political and ethnic persecution and sought refuge in Australia via a boat journey from Indonesia. After being temporarily detained for approximately one month while undergoing processing on Christmas Island, Australia, he was exiled by force to Manus Island. Boochani studied political science, political geography and geopolitics in Iran, and is an award-winning journalist with publications in Farsi and English. He also writes (and performs through song) in Kurdish, researches many topics related to his ethnic group, and is an advocate for Kurdish language and culture. In 2018 he was made non-resident Visiting Scholar at the Sydney Asia Pacific Migration Centre (SAPMiC), University of Sydney; in 2019, Adjunct Associate Professor of Social Sciences, University of New South Wales and Visiting Professor at Birkbeck, University of London; in 2020, Honorary (Principal Fellow), Social and Political Sciences, University of Melbourne. Boochani's articles have been published in *The Guardian* (Boochani 2017a), *The Saturday Paper* (Boochani 2017b), *New Matilda* (Boochani 2016a) and *Huffington Post* (Boochani 2016b), and his full-length documentary *Chauka, Please Tell Us the Time*, co-directed with Arash Kamali Sarvestani, had its world premiere at the Sydney International Film Festival and its international premiere at the BFI London Film Festival in 2017. The footage was smuggled out of the detention centre and sent to the Netherlands for co-direction and production; this was made possible largely through WhatsApp voice messaging. His first book, *No Friend but the Mountains: Writing from Manus Prison* (Boochani 2018), was published by Picador and won the prestigious Victorian Prize for Literature in Australia in the same year, followed by numerous other awards; it mixes genres such as folklore, journalism, autobiography, philosophy, poetry, psychology and political commentary.

Boochani's cultural productions, scholarship and unrelenting activism while in prison represent a remarkable example of citizen media that has potential to enhance the philosophy canon. This kind of reform can foster epistemic pride among those who have experienced or are experiencing various forms of displacement, exile and incarceration. His body of work cannot simply be reduced to refugee memoirs or other forms of refugee narrative; it has the potential to contribute to genres such as clandestine philosophical literature, prison narratives, philosophical fiction, dissident writing, transnational literature, decolonial writing and political commentary pertaining to the regions he inhabited – Iran, Kurdistan, Manus Island. Reflection on the philosophical and literary qualities of Boochani's texts and cultural productions creates new spaces and frameworks for interpreting refugeehood as an intellectual standpoint. Issues pertaining to displacement and exile are relevant to philosophy and the writing, art and activism produced through these spaces and experiences impacts philosophical discourse and the nature of inquiry. Understanding the interdependent relationship between philosophy and citizen media can reveal new philosophical approaches toward refugee standpoints, in addition to enabling us to think through appropriate and nuanced philosophical concepts pertaining to borders, bordering practices and forced migration.

This philosophical project is a political act, an acknowledgement of responsibility toward the growing number of displaced peoples. It is also a testimony to philosophers from other eras who suffered similar kinds of uprooting, exile and loss of community or homeland. In the context of the contemporary world, this undertaking is an application of the moral imagination, one that validates and foregrounds philosophical perspectives that have been marginalized, or philosophical ambitions that have been forestalled due to a combination of precarious location, transience, restricted movement, statelessness and non-citizen status.

A collaborative and equitable philosophical programme dedicated to rethinking the canon and sensitive to discussions about forced migration is crucial. In many disciplines and much of scholarship, the dominant narratives used to frame experiences of displaced and exiled peoples are largely constructed for them, rather than by them, which limits interpretations of accounts of forced movement; it also demands – particularly in a contemporary context – limited and damaging performances of refugeehood. Certain recurring tropes pervade government policy, advocacy, scholarly discourse and cultural representations; these tropes reduce multifarious and fluid identities, and unique lived experiences, to problematic socially and politically constructed categories (Tofighian 2018b). Narratives, arguments, images and testimonies are arranged, mediated and widely circulated in discursive texts and aesthetic works by institutions, academics, the media, official state discourse, NGOs and community development actors using norms and tropes of authenticity and identity familiar to host nations, funding bodies and analysts with citizen privilege. Expanding the philosophy canon to include commentary, scholarship and literature written from refugee standpoints, as in the case of Boochani's output, can enrich points of reference; allow reevaluation of categories and descriptions; introduce novel theoretical frameworks; expand and transform research constellations; generate projects characterized by shared philosophical activity (Tofighian 2018a); and create epistemic pride (Barnes 2016).

Serious consideration of citizen media produced by refugee scholars and creative writers begins philosophically first by reclaiming narratives of displacement, and second by deconstructing the often racialized nature of victim–saviour configurations. Radically rethinking the canon in this way opens spaces for incorporating socially, culturally and philosophically richer intellectual tools and subject matter. Engaging with citizen media created by the diverse standpoints of displaced and exiled peoples further represents an imperative moral position, irrespective of any desired theoretical outcomes, a mode of engagement that does not succumb to the normatively laden deficit/surplus dichotomy in contrasting the sociocultural positions of refugees with that of others.

Boochani's mobilizing activities inside the Manus detention centre, his online activism, journalism, the film *Chauka, Please Tell Us the Time* (shot clandestinely on a smartphone) and the book *No Friend but the Mountains* (written via text messaging) document an important phase in the history of migration to Australia. They also reflect an extraordinary critique that draws on a unique interweaving of experiences, insights, previous research training and a form of interaction with the natural environment only possible for someone in Boochani's predicament. The embodied knowledge he conveys gives visceral access to details of the systematic torture and strategies of punishment adopted by Australia's racialized border regime. The works produced under these circumstances act as valuable philosophical interlocutors and help to create new ecologies of knowledge with a commitment to place; however, for their potential to be realized, professional philosophy needs to practise greater openness, beginning with the make-up of the canon.

Citizen media practices have the potential to contribute new knowledge, and may therefore be interpreted as exclusive philosophies, as representing excluded epistemologies, and as unacknowledged knowledge systems. Further, countering and decentring hegemonic and exclusory traditions, institutional structures and practices in philosophy in the ways proposed here can result in liberating philosophical discoveries for marginalized and stigmatized groups. Taking up such challenges in professional philosophy today, transforming its key tenets and commitments by foregrounding expressions of resistance by marginalized, racialized and colonized peoples, generates important new insights into issues of freedom and wellbeing. Such an approach to philosophy also draws attention to the interdependent relationship between epistemic standpoints, physical and virtual borders, and the regulative power schemata integral to nation-states. Whether it contributes to methodology, enriches the philosophical tradition or enhances the canon, citizen media functions as a legitimate and empowering discourse in this context.

See also: migration studies and citizen media; race & ethnicity studies and citizen media; witnessing/testimony

Recommended reading

Boochani, B. (2018) *No Friend but the Mountains: Writing from Manus Prison*, translated by O. Tofighian, with translator's note and essay, Sydney: Picador.

Boochani wrote this novel in Farsi using WhatsApp text messaging while incarcerated in the Australian-run immigration detention centre on Manus Island, Papua New Guinea. His text messages were translated and edited through a complex process that involved consultation, collaboration and sharing. The book narrates and critiques neocolonial strategies which maintain and reinforce border violence; it communicates the lived experience of systematic torture in Australia's notorious offshore detention centres and critically analyses the technologies of oppression, domination and submission. The majority of the book details the atrocious conditions and demoralizing encounters in prison by telling stories about various human beings locked up there or by unpacking the 'Kyriarchal System' that drives and sustains Australia's border regime. Interrelated themes include a critique of coloniality, the application of a style described as 'horrific surrealism', and the notion of anti-genre.

Buck-Morss, S. (2000) 'Hegel and Haiti', *Critical Inquiry* 26(4): 821–865.

This article introduces some of Buck-Morss' most significant critiques of scholarly marginalization and exclusion regarding the role of the Haitian revolution in the development of Hegel's philosophy. It argues that the profound impact of the Haitian revolution on Hegel's thinking about master/slave associations has been largely ignored within scholarly narratives and academic disciplines engaging with the European Enlightenment and the notion of freedom espoused by Enlightenment philosophers.

Davis, A.Y. (1971) 'Political Prisoners, Prisons, and Black Liberation'. Available online: www.historyisaweapon.com/defcon1/davispoprprblli.html.

Written in Marin Country Jail, Angela Davis' essay critically analyses the US prison-industrial complex in connection with racial oppression. She draws attention to pervasive contradictory perceptions of resistance by comparing slave revolts with the American War of Independence and points out that revolutionaries such as Nat Turner were defined as criminals while revolutionaries fighting against the British were praised for their struggle for liberation. This early example of prison writing reflects an essential feature of Davis' overall thought and work – she avoids individualizing her plight and foregrounds the historical, ideological and structural issues affecting the black community as a collective.

PHOTOGRAPHY

Karen Cross

The increased visibility of citizen media in the twenty-first century is often linked to the rise of mobile phones and the capacity to share images online. This was reflected in early scholarship on mobile camera phone technology published in the mid-2000s (Daisuke and Ito 2003; Van House et al. 2005). Rather than focus on technology alone, this entry broadens the discussion to consider the intimate relationship between photography and the citizen from a more historical perspective. It shows how contemporary photographic practices, ranging from the non-professional to the professional, and from the journalistic to the artistic, sustain connections with the past, and especially with analogue traditions of photography that pre-date the digital. Finally, it demonstrates why an understanding of the history of photography might be important when thinking about citizen media today, and foregrounds the rootedness of the medium in more global shared experiences that are ultimately political in nature.

The critical history of photography

Photography entered the social and cultural imaginary even before the invention of the first cameras, reflecting a long-held desire to capture permanent traces of people and things through a process of representation that is independent of the artist's hand. Various theorists writing during the first half of the twentieth century began to reflect on this trajectory and on the social and political importance of photography to society as a whole. Among these, one of the best known is Walter Benjamin, who wrote his seminal essays 'A Short History of Photography' and 'The Work of Art in the Age of Mechanical Reproduction' in 1931 and 1936, respectively; the English translations appeared in 1972 and 1969. In these essays, Benjamin explores how photography ripped at the aura of more traditional art forms, such as drawing and painting, and argues that it could be considered an inherently social and political medium. Describing it as "unconscious optics" (1972:7), he also draws a parallel between photography and psycho-analysis, thus foregrounding the therapeutic possibilities of the photograph as a medium which brings into view visible evidence that may otherwise remain hidden.

Following Benjamin, other scholars sought to complicate the idea of photography as an inherently truthful technology and began to explore in further depth histories of the medium that had not previously made their way into the canon (Sekula 1986; Tagg 1988). These scholars showed how photography became an important tool of social and political representation, not just an art form, a tool which could be used by groups and organizations to further their own political goals. They additionally sought to draw attention to the way in which the medium operates as part of a wider play of knowledge and power; during the second half of the twentieth century, Benjamin's socialist perspective was thus combined with Foucauldian thinking (Sekula 1986). This new critical development focused attention on the way that photography was implicated in the growth of panoptic state and commercial concerns. With this, it also became apparent, however, that the functional and practical aspects of photography were beginning to be sidelined. The way in which photography could provide "a form of *practical* knowledge, an inscription of, and an intervention in, a socially divided world" (Roberts 1998:4; emphasis in original) became radically undermined, and this posed a particular challenge to documentary photographers. Photographic media were further challenged by the advent of digital forms of image production and reproduction in the 1990s, which severed the connection between the photograph and its referent. It became increasingly apparent that computers could produce images entirely independently of the camera, and this led to claims that we were living in a "post-photographic era" (Mitchell 1994:224), and that software was rapidly taking over (Manovich 2013).

In spite of these developments, the conceptual connection between photography and the notion of 'the real' has never been entirely broken, and the social nature of photography has been repeatedly asserted. By coining such concepts as "the networked image" (Rubinstein and Sluis 2008:9), theorists have attempted to grasp the new symbolic transactions that arise in the turn to digital technologies. Online communications signal how the photographic image functions as a kind of connective tissue through which social experience and presence are performed. The digital vernaculars that often accompany the latest forms of visual communication, such as hashtags and geolocation tools, allow for new forms of social relational experience to be performed across geographic and temporal distances and divides which could not always be bridged previously. The online circulation of images in social revolutionary contexts, such as during the Arab uprising in 2011–2013, provides an example of how photography has been crucial in establishing a sense of shared experience of global political change. Beyond social movements, the reporting of events such as the London Bombings in July 2005 and Hurricane Katrina in August of the same year also saw news organizations beginning to provide a wider platform for images created by members of the public.

All of these developments have significantly shifted the role of mainstream media organizations and their relationship to the audience. What appears to have been set into motion is a new way of imagining the citizen and their role within the public domain. Far from simply levelling the playing field between the amateur and professional, however, what has become apparent through these changes is that we are witnessing the emergence of a new relationship between the media and the audience. Professionals, and those aspiring to become professionals, have taken on a more curatorial role in relation to images, and have entered into a newly emotionalized dynamics of care and recuperation with otherwise marginalized and lost experiences. Moreover, beyond simply documenting what they see, professional documentary photographers can also be seen to have shifted their practices to align with amateur mediations of citizen experience and the aesthetics of social media that are steadily proliferating through everyday life.

More explicit visualizations of the interiority and subjecthood of the photographing 'I' have become apparent within photographic practices, but, again, these are not necessarily an effect of new technologies. Rather, they have served to foreground the growing interest in the contemporary avant-garde with its particular focus on popular culture. Critical photographic projects, recalling the work of Bill Owens and Stephen Shore, for instance, cast their world in a much more vernacular light and seek to connect more deeply to citizen concerns. Photographers such as Gilles Peres and those making a more explicit departure from documentary reality, including Martha Rosler and Allan Sekula, who make use of 'ready-made' and 'found' images, have also sought to unravel the entrenched spectacularizing displays of public reporting. The work of these artists involves what Ritchin refers to as a process of "bending the frame" (2013:1) to perform more overtly "useful" (ibid.:8) kinds of representation that could be considered social interventions rather than just reflections of the ordinary and everyday.

Photographic citizen media

Contemporary photographers following this trend have begun experimenting with strategies of dissemination that often seek out more co-participatory models of production that connect with marginal positions. In different ways, photographers have attempted to engage the viewer through often long-term investments in community building, typically involving the use of social media. They cover a wide range of subject matter and thematic bases, including humanitarian issues and the environment. James Balog's Extreme Ice Survey (EIS), begun in 2007, presents one such example, in which the intention is to provide what the homepage of the project website refers to as a "visual voice" for the planet's changing ecosystem. The time-lapse pictures, which can be viewed on the website, animate the shocking reduction of ice and provide undeniable evidence of human impact on the environment.

This principle of bringing to light that which may otherwise go unchallenged is also evident in the work of the French artist JR, whose project Inside Out (begun in 2011) draws attention to humanitarian issues. Through his macro-scale creations that begin life as photo-booth creations, he seeks to foreground the stories of groups and people who have suffered injustices, often as a consequence of government failures. Various photographic projects are staged as interventions in public spaces which represent a wide range of distinct, yet often connected, causes. Portraits line the walls of the Mexican-American border, pictures of mothers who have lost children due to violence appear in Caracas, Venezuela, as do images of those who were the victims of the *Charlie Hebdo* terrorist attack in Paris, France. Images appear as graffiti on walls and other infrastructure, thus locating the discussion at street level; but JR also makes use of social media, including Instagram, to connect causes together on a more international and global scale.

Alternatively, artists have sought to expose the inner workings of digital media and the way our social communications are heavily monitored by the surveillance technologies that form the ubiquitous architecture of governmental organizational and defence systems. These systems form the focus of the work of artists such as Trevor Paglan, who uses a method of sousveillance (Vierkant 2009), meaning under sight, or surveillance from below. In 2016, Paglan was awarded the Deutsche Börse Photography Foundation Prize for images documenting American reconnaissance satellites in Earth's orbit, as well as his pictures of the offshore network cables documented in the Snowden files. It is no surprise then that given its capacity to document the infrastructure of state surveillance theorists and artists today continue to defend photography and its value for promoting citizen interests.

It is not just through the simple use of the camera to document the material signs of power and control directly that photography is shown to be an important representational tool. Photography is conceptually important within the domain of digital media today, as evident not only through the proliferation of photographic images across the web, but also via the ongoing attachment to the photographic materialities and the aesthetics of popular analogue frames that now feature within social media. The demonstrable presence of the amateur lens allows for certain kinds of departure from history to be made, but attachments to the past also complicate contemporary experiences of citizenship. Well known examples where photographers have adopted the amateur aesthetic in the field of reporting include images of soldiers taken in Afghanistan by reporters Erin Trieb (2009), using a Holga toy camera, and Damon Winter (2011), whose award winning pictures made with Hipstamatic caused something of a controversy within the photojournalistic community.

Having apparently polluted the purity of photojournalism, Winter found himself needing to defend his use of the camera phone, which, he argued, allowed for unfettered access to the daily lives of troops. While no doubt the camera phone allowed the journalist to embed himself into the everyday backdrop of life at war, what his use of retro lenses and editing features highlighted was how photojournalism has always involved a series of subjective and stylistic choices. Photography never offers a pure and unadulterated view. Furthermore, the use of retro-filters and familial photographic lenses can be understood as evidence of the persistence of nostalgia that is often performed in relation to personalized and domestic forms of photography, such as the family album. Yet, the particular choices made by the professional are those which enable such kinds of image to function as a form of memory and witnessing, rather than as straightforward documentary evidence.

Perhaps a more useful concept to explain what is at play in such new media practices is that of hypermediation (Bolter and Grusin 1999), understood as a process via which media represent themselves through digital creativity. By replicating the analogue past, familiar forms of social representation are secured and citizen forms of identification, premised on orientations to the ordinary and everyday, can be authenticated. Underlying these trends, one could also argue, is an attempt to come to terms with the losses of the past and the rise of a therapeutic culture of recuperation and repair. This is in part related to the controlling mechanisms that see the professional internalizing analogue aesthetics within their own practice, or the reification of amateur tools, but it is also partly related to a desire to manifest newer, more widely representative views of society and the power plays that it involves.

Exploring the conceptual limits of photography as memory, Simon Norfolk's work *Chronotopia*, initially published in 2002, functions as a reminder of the long-running battle over how to represent war adequately in spaces that do not support historical forms of consciousness. This is an especially pressing issue within certain contexts such as Afghanistan, where the continuing spectral presences of military intervention impose themselves irreparably on the landscape. By capturing the overlay of different battles on the landscape through the use of large format analogue cameras, the photographer attempts to challenge the general lack of attention to the continuing influence of Western power within this region. Yet, such critical work is heavily aligned with artistic and intellectual traditions and necessarily finds its audience within the academy, its archives and the museum. The argument for striving for better understanding of the unknown and unnamed citizen photographer persists. So too does the question of the wider role of photography in the context of citizen media.

There are other ways in which the experiences of citizens come to light through the domain of professionalized and artistic practice, and other ways in which the intimacy provided by the analogue frame becomes useful. Also nominated for the 2016 Deutsche Börse prize was the photo book *In the Shadow of the Pyramids* by photographer Laura El-Tantawy, whose images depicting the Egyptian Revolution seek to marry personal experience with the wider events of the street protests. The combination of documentary shots with those found in her personal family album remind us of the human experience of revolution, as well as the ongoing struggle and search for identity in times of crisis and change. The images, which represent fleeting moments of history as it is unfolding, also work against the iconicity of modern-day reporting, and the way that even citizen images can stand to represent events from a singular standpoint. They remind us of the often impressionistic and messy aspects of experience, which cannot be packed neatly into documentary forms, and which exceed the limited professionalized visions of citizen viewpoints and experience.

Future directions

Through the discussion of various examples, this entry has shown that, in response to the widespread emergence of mobile media and the plethora of images that now circulate online, we need ways of accounting for the processes of mediation and governing systems that are involved in sustaining our investments in photography. But beyond the new critical directions that are apparent in the practices of professionals and artists, it is also important to return to the question of what it is that photography offers, not just in terms of images but also in terms of the process itself. This question prompts a consideration of more than just the works of particular photographers and photographic projects. The temptation is to cling to old ways of thinking, to describe shifts and changes in terms that are recognizable within the scholarly frame of artistic master narratives. However, it is important to consider how photography is fast becoming foregrounded once again as a medium of 'the people', just as it was following its invention at the beginning of the nineteenth century. Artists and professionals are deeply aware of this today.

Perhaps it is not more filters of understanding that we need, or ideologically aligned perspectives, but rather different ways of sensing what it is that continually draws us to photography as a means of documenting our lives. But then again, it is not just the creation of content and archives that are important. Photography is not simply a tool of representation. More than ever, it is the very experience via which we gain our sense of self in the world, in ways that enable social forms of relating to one another and that permit the creation of politicized forms of presence. These developments shift the focus of attention away from the question of technological artefacts, instead placing emphasis on the ontological aspects of photography as a process, condition or even a state of being – that which Ariella Azoulay frames in terms of "the citizenry of photography" (2007/2008:118).

There is, we might conclude, no citizenship without photography, and this is what makes photography a potent agent of change. This is not necessarily to argue for or against a technologically determinist perspective but, perhaps like Benjamin and others following in a similar critical vein, to accept that photography invites an ethical confrontation between the self and other, photographer and subject, along with the eventual viewer(s) of images. Uses of the camera may be subject to many different perversions, but photography does not allow us to evade the presence of others. This may sometimes be uncomfortable, but it is in that discomfort that we are afforded the opportunity to understand the kinds of projections we make. The function of photography is not simply to represent other people's stories. It is also to bring about an encounter with the way that we and other people see the world.

See also: amateur; mobile technologies; selfies

Recommended reading

Azoulay, A. (2007/2008) *The Civil Contract of Photography*, translated by R. Mazali and R. Danieli, New York: Zone Books.

Offers an ontology of photography based on a critical discussion of its ethical status and the power relations that make photographic meaning. The work brings a particular focus on the powerless and on situations where citizenship is limited or made impossible.

Benjamin, W. (1931/1972) 'A Short History of Photography', *Screen* 13(1): 5–26.

Outlines Benjamin's critical position on photography and reflects on the wider social implications of the medium.

Ritchin, F. (2013) *Bending the Frame: Photojournalism, documentary and the citizen*, New York: Aperture.

Provides a wide-ranging and detailed discussion of contemporary photographers, with a special attention to political work and to opportunities for critically reframing photojournalistic practice in the digital age.

POLITICAL SCIENCE AND CITIZEN MEDIA

Mette Marie Roslyng

Political science has so far paid little attention to citizen media as a phenomenon that concerns how citizens engage in political activities through their use of different forms of media produced for and by them. And yet, engaging with the emerging body of work on citizen media can arguably offer political science insights into both citizen participation and the role of media in democratic politics.

First, the liberal democratic imaginary relies on a social contract between citizens and their elected representatives as power holders (Held 2006). However, views of the precise role played by the citizen within this contractual commitment varies greatly across the field. While there is no doubt that some form of citizen involvement is crucial and necessary in all understandings of democracy, the level, intensity and form of such engagements diverge according to the model of democracy and the democratic practices studied and conceptualized (ibid.). Second, the role played by the media in setting the conditions for critical and engaged public spheres in which more or less active citizens may play a part in politics is acknowledged in political science. The work of Jürgen Habermas (1992/1996), most notably, has been seminal in articulating a particular understanding of the role played by (news) media in the deliberative democratic model. Later work at the intersection between political science and media studies has also been occupied with the role of (mostly) mainstream news media, but also in some instances with alternative and critical media, in relation to citizen participation (Dahlgren 2006; Carpentier 2011; Dahlberg and Phelan 2011; Curran et al. 2012).

An understanding of citizen media as involving unaffiliated citizens actively intervening in public space through diverse forms of media practices (Baker and Blaagaard 2016a:16) can provide new perspectives on addressing some of these issues. This widens the concept of citizen participation, which would otherwise be conceptualized narrowly as tied to the liberal democratic social contract, at the same time as allowing a broader understanding of media that moves beyond traditional journalism and beyond the traditional view of media as a tool and a space reserved mostly for political and cultural elites.

The liberal democratic imaginary

Political science can be read as a discipline linked to the emergence and development of a liberal democratic imaginary that revolves around several crucial debates. Along one axis run discussions about direct and representative forms of democracy; along others run questions of conflict or consensus and republicanism versus communitarianism (Mackenzie 2009). The different variations or models of liberal democracy, which developed as theory and practice within the field of political science, have a shared focus on politics founded on a social contract between a sovereign people, the citizens, and a governing, acting state (Rousseau 1762/1992). A focus on liberal rights naturally follows from this contractual foundation of democratic discourse. For democratic pluralists such as Dahl (1961), citizens wishing to claim their political rights have ample opportunity to influence the political system, either through elections or through more active forms of engagement in interest groups or as members of political parties. These forms of political agency rely on the notion of representativity and make it possible to define the role of the citizen in public affairs and in democratic decision-making. However, elitist, later followed by corporatist, neo-corporatist and neo-Marxist democratic critiques, have allocated much less space to citizen participation in Western democracies, where power is concentrated in closed professional networks. Here, asymmetric access to power is understood to exclude citizens from institutionalized political processes (Held 2006).

A similar development may be traced in media and communication studies, where an emphasis on media effects and news media's agenda-setting power can be tied to a pluralist and mostly institutional conception of politics (Dearing and Rodgers 1996; Iyengar 2010). Although citizens within these later approaches can be seen as active and critical media users, they most often remain fixed in their role as media consumers and will therefore usually take no active part in creating and distributing news. The propaganda model (Herman and Chomsky 1988; Herman 2000) offers an alternative approach that acknowledges and analyses the structural and ideological workings of mainstream media. Within this model, news and journalistic practices rely heavily and uncritically on input from elite interests in society and are embedded in the capitalist market system (Herman 2000:102). Media, more often than not, become a mouthpiece for elites and power holders and therefore play an important role in maintaining the popular support that sustains elitist networks of power. Some studies of media framing also include critical analyses of the hegemonic potential inherent in journalistic and editorial framing of news and events in ways that are biased, for instance, against New Left activist groups (Gitlin 2003a). The hegemonic workings of the media explain how elites succeed in maintaining general support as well as why and how critical social movements tend to be excluded in mainstream media. The dominant debates in studies of citizen participation in relation to media power and political communication thus seem to implicitly (and sometimes explicitly) mirror the theoretical disputes between pluralists and their critics, which have played a determining and fundamental role in the development of political science as a modern discipline.

The liberal mode of democracy has been advanced in terms of understanding the deliberative potential of a strong and dynamic public sphere granting access to a variety of viewpoints, interests and rationalities, with the potential of forming a public opinion as an ideally consensual expression of the *demos* or a people (Habermas 1962/1989, 1992/1996). The general will of the *demos* is most often expressed as a national will tied to the boundaries of the nation-state (Fraser 1990), with news media playing a crucial role in the formation of a national public sphere. However, the role of alternative or citizen media, for instance in relation to otherwise excluded subcultures, subversive artistic expressions and the myriad of emerging digital and non-digital networks (Roslyng and Blaagaard 2018), remains less clear in the liberal mode of democracy. These forms of media practices, it can be argued, however, can contribute to the development of alternative publics or counterpublics that may challenge general public opinion and political elites (Fraser 1990; Warner 2002).

Despite its importance for political science and related disciplines, the deliberative model of democracy has been criticized precisely for its inability to account for conflictual and antagonistic processes in society. Mouffe (2000a) turns the deliberative democratic argument for inclusion on its head when she points out that the attempt to neutralize and overcome conflict in consensual politics is itself founded on conflicts which supress some interests. Consensus is always the expression of hegemony, in which some rights are privileged while others are excluded (ibid.:49). For Mouffe, the boundaries of democratic citizenship are under constant renegotiation. Democratic citizenship, despite being based on the principles of equal rights, therefore contains a moment of closure, a moment of exclusion of those who are not citizens.

Following Mouffe and other authors critical of deliberative democracy, Brown (2015) demonstrates how the concept is permeated by a neoliberal imaginary that favours a governmental, economist rationality over the democratic potential for engagement and participation. This imaginary "construes subjects as relentlessly economic actors" (ibid.:32) rather than as citizens. As a result, the liberal democratic premise of a social contract between citizens and power holders is revealed as a myth. Not only is the liberal social imaginary an expression of a moment of discursive closure – "a partial fixation" of meaning (Laclau and Mouffe 1985:112) – in which citizenship is in practice exclusionary and the universality of equal rights is put into question, casting democracy in terms of economic governance also serves to close the space for political expression even further. This economization of politics and democracy undermines political subjectivity and stands in stark contrast to citizens' attempts to participate through various media in productive, active and creative ways.

Radical democracy and citizen media

In contrast to the deliberative goal of eliminating power struggles and disagreement through processual and rationalist means, Mouffe argues for the need to develop a radical or agonistic model of democracy in which power imbalance and conflict are recognized as constitutive of democratic relations, and therefore also of citizenship and citizen participation through and with media. In agonistic pluralism, antagonisms are accepted as a necessary part of political relations but may be recast in democratic terms in such a way that antagonism is transferred into agonism. Within agonistic relations, the protagonists are seen as adversaries rather than enemies and will not seek to eliminate or destroy one another. Despite disagreements, the adversaries share a common (democratic) space within which they can operate (Mouffe 2000a:100ff). Mouffe has since moved on to see the diverse political struggles as part of a new, anti-hegemonic "left populism" which makes use of a reference to the people, or indeed the citizen (Mouffe 2018:9). A liberal democratic imaginary can therefore be replaced with a radical democratic imaginary (Smith 1998; Brown 2015) in which citizen participation can be found somewhere in between antagonistic and agonistic spaces.

Citizen media, as developed and defined by Baker and Blaagaard (2016a) and Blaagaard and Allan (2016), among others, shares several assumptions with the radical, agonistic democratic imaginary. Citizens' performative and creative use of media is cast as profoundly political rather than as an expression of politics and policy. By political here is meant the way in which traces of power are made visible and antagonism brought to the foreground, whereas politics is understood as the institutionalization and maintenance of order and coexistence (Mouffe 2000a:21, 101). Citizen media does not operate from within institutional politics but will often seek to redefine the conditions for talking and thinking about something as political and for regaining a political space, often in innovative and creative ways, from which to speak outside or on the fringes of the system of professional politics and journalism.

Examples of mediated citizen engagements through creative, artistic and political activist practices in agonistic public spheres can be found at the intersection of political science and media studies. These examples allow for a broadened conceptualization of media that goes beyond the professional and closed journalistic forms familiar in traditional media studies and can demonstrate agonistic and antagonistic practices at work. For Mouffe (2007:4), for instance, "critical art is art that foments dissensus, that makes visible what the dominant consensus tends to obscure and obliterate". The Yes Men, her example of how critical and activist art intervenes and becomes political, is also a prime example of how unaffiliated citizens engage in activist and performative ways with digital and analogue media as well as with on-site protest happenings. The Yes Men engage in culture-jamming practices, and are known for staging a number of satirical, anti-capitalist events through and with digital and other media. As a particular form of counter-hegemonic activist practice, culture jamming works through the appropriation, or indeed reverse appropriation, of corporate or governmental rationalities and methods in order to subvert them (Cammaerts 2007b;

Baker and Blaagaard 2016a). The Yes Men most famously received attention for creating a spoof website for the World Trade Organization (WTO) with 'corrected' content. Represented by Jacques Servin and Igor Vamos posing as Andy Bichlbaum and Mike Bonanno, the Yes Men then received several invitations to talk at seminars and conferences, where they impersonated WTO spokespeople. Delivered with stone-faced coolness in front of a disengaged audience, the duo – with Andy Bichlbaum impersonating WTO spokesman Hank Hardy Unruh – presented an alternative history of textile, replete with blatantly racist and cynical statements, including the claim that, if left to market forces, slavery would eventually have evolved into the current exchange of labour with Third World countries. The grand finale included a demonstration of a gold suit that allows managers to monitor workers worldwide with the use of sensors implanted into bodies.

The case of the Yes Men shows how creative and satirical use of media can contribute to constructing new, activist identities that challenge the status quo. Impersonating professional, corporate or governmental official stakeholders, the activist performers mock the deliberative space available within international organizations and in mainstream media for citizens to speak rationally, yet on deeply unethical terms. During the first part of their presentation, in which they use governmental and corporate rhetoric to present arguments that seem to endorse slavery and modern forms of exploitation, the audience does not seem in the least perturbed. Interestingly, this activist intervention started out as a digital mediation, in the form of a hoax website, became an on-site happening at a conference, which was then recorded, posted, distributed and shared on YouTube and other sites. The culture-jamming practices being performed closely mirror the way corporate and international institutions use interconnected professional media platforms – both digital, as in the case of official websites, and non-digital, including conferences and network meetings.

In another 2004 prank, Jacques Servin – alias Andy Bichlbaum – impersonates Jude Finisterra, a spokesperson from Dow Chemicals. Dow had a few years earlier taken over the company responsible for the tragic and very extensive gas leak disaster in Bhopal, India in 1984. Performing live on BBC World, the impersonated Jude Finisterra takes full responsibility for the leak on behalf of Dow and announces that the company will compensate all victims. Here again, the activists expose a corporate rationality as cynical, and the event can in many ways be seen as an interruptive revelation of economization at work in governmental practices. Economist rationality is also shown to run parallel with the increasing commercialization of mainstream media, art and popular culture. This development plays an important role in obscuring the workings of asymmetrical power in democratic politics. Citizen media, on the other hand, pursues the opposite aim. As unaffiliated citizens, the Yes Men set out to reveal the networks of corporate and government actors to be deeply political and vested in cynical and dehumanizing practices. The role played by many, if not most, forms of citizen media practices may therefore come close to the role that Mouffe attributes to critical art: to illuminate the political as the "ever present possibility of antagonism" (2007:2).

Fenton (2011:186) draws on the Peoples' Global Action (PGA) network in her analysis of how the political mobilization of citizen groups changed with the advent of the Internet. Although she does not conceptualize the PGA and its use of digital media specifically as citizen media, the PGA does fit the definition given its emphasis on direct action, the distance it places between the movement and institutional state politics, and its disassociation from mainstream media. These networked forms of political action, vested in the logics of new social movements, also tend to bypass traditional systems of political representation. The PGA has its roots in the Zapatista struggle against the North American Free Trade Agreement's (NAFTA) engagement in a poor Mexican region in the 1990s, but this struggle was from the outset discursively set in global terms, as focused on anti-capitalism, anti-imperialism and anti-feudalism. Covering a range of topics, from environmentalism and migration to social and economic justice, the PGA is an example of citizen political action that is locally situated but calls for global explanations and solutions through the use of digital media. The networked organization of the PGA structured as a complex interplay between digital presence and direct physical action and bypassing traditional institutional politics as well as traditional media is made possible by the way the Internet has changed the conditions for citizenship.

Similarly, Reclaim the Streets (RTS) developed during the 1990s and 2000s as an attempt to allow local citizens to reclaim urban spaces for community use, employing a mixture of mediated communication and local happenings involving the communities in the area. Both RTS's and PGA's non-hierarchical and anti-bureaucratic format resonate with other types of citizen media in which citizens form loose collectives focused on particular causes and use media creatively to act as citizens. As evident in these examples, citizen media initiatives can also be more focused on the act of protest than on the presentation of a coherent and predefined ideological project. They are often momentary and media-oriented rather than directed at building a permanent organization or project, and are characterized by a weak or loose political structure that favours non-representation and a radical openness in terms of the topics deemed to require action (Fenton 2011, 2012b).

The commitment to autonomy and non-representation and the global perspective of the PGA, the RTS and the Yes Men as ensembles of citizens are at odds with the liberal idea of a unifying *demos* within a nation-state. They reveal the presence of public spheres that do not only go beyond liberal pluralism but also reject the processes of deliberative rationalities within which well reasoned arguments are put to scrutiny in a public sphere. These activist practices express a radically antagonistic form of political protest because they reject the conditions of liberal pluralism as a form of governance that supports the capitalist and imperialist structures the movements are fighting against. Moreover, the economic regime of which NAFTA and the WTO form a part is subject to the economization identified by Brown (2015) as a governmental regime which closes the space for the political rather than allowing for a plural and inclusive debate. From the perspective of the citizen movements examined here, the international governance regime represents one hegemonic version of global economic and social policy which is maintained by national politics, corporate behaviour and mainstream media. This regime can therefore only be opposed via independent media formats and platforms that allow citizen groups to present the world with their own voice and on their own terms. The practices of these groups therefore provide good examples of how citizens can bypass traditional media and exploit the autonomy offered by a digital media landscape, thus creating the opportunity to develop alternative or counter media as forms of expression as well as a way of elaborating political positions (Downing 2001). According to Marchart (2011:65), we may here distinguish between alternative media, which are expressions of alternative networks for debate and organization situated completely outside mainstream news media, and counter media, which engage with mass media and the general public in the presentation of counter-hegemonic demands and critique. Both forms of media practice are central to establishing and maintaining processes of identification within protest groups.

Another example of citizen media practice, but from a nationalist perspective, can be found in De Cleen's (2015) study of citizens protesting, during and through the organization of a traditional Flemish Songfest, against authoritarian and racist expressions of Flemish nationalism in order to rearticulate a new tolerant and democratic nationalist discourse. While this struggle to define Flemish nationalism took place on the fringes of institutionalized (party) politics, ordinary citizens, artists and other actors outside parliamentary politics participated actively in defining the terms under which the struggle is played out and mediated – as audiences, performers and organizers. At first sight this may appear to indicate an attempt to find an expression of a *demos*, a unified people brought together by the nationalistic political overall aim of the event. However, as De Cleen points out, this political attempt to rearticulate nationalism contains elements of radical antagonism or agonism. Flemish nationalism cannot be both tolerant/democratic and racist/exclusionary at the same time. Since the political struggle over how to define the Flemish nationalist and separatist project takes place on a communal platform, a tradition-bound concert, between adversaries who presumably share a reliance on the democratic rules of the game, the protests are expressed as agonistic democratic political identity. The struggle to define and delimit the *demos* within a Flemish public sphere thus ultimately reveals the particularity of any nationalistic practice. It also shows that the expressions of a *demos*, of a people, take the form of processes of identification which involve a negotiation about the boundaries of who can belong to 'the people'.

Future directions

Although these examples were not originally conceptualized as instances of citizen media, they all seem to cover, or to come very close to, citizen media defined in active, performative and participatory terms. The examples are connected in terms of the way citizens engage in political action with and through different forms of media, and they go beyond an understanding of citizen media as witnessing or recording incidents (Peters 2011), although it has to be said that both recording and witnessing form an important part of the activist political performance. This ties in with political science's concern with public participation as collective and/or individual political agency. While political science as a discipline has engaged mostly with traditional forms of news media and their role in democracies, the political role of the arts and other creative expressions realized through a variety of media by and for citizens has received some attention (Mouffe 2007). Drawing on recent literature on citizen media may contribute to broadening the field's understanding of citizen participation and the democratic potential that can be found in citizens' creative use of digital and other forms of media in a social world seen as both networked and mediatized.

Citizen media as political intervention dovetails most closely with the radical agonistic or antagonistic model of democracy. This allows for an understanding of citizen media as alternative voices that develop in independent public spaces outside institutionalized politics and mainstream, professional media. It also reveals an extended concept of media that goes beyond the role played by mainstream, institutional and professional forms of news media as the main channel available for citizens to express their grievances and concerns. Citizen media is therefore a form of engagement with media that recognizes the political as an antagonistic or agonistic terrain. This understanding calls for a rearticulation not only of the contractual relation between citizens and their political representatives, but also of the very notion of political representation.

Moreover, Brown's critique of the liberal democratic imaginary and its reliance on a governmental and economist rationality which delimit rather than expand the potential for citizen engagement suggests that citizen media may take on the role of revealing this very rationality as political – the Yes Men example discussed earlier seems to support this view. Citizen media initiatives may therefore trigger a renegotiation and contestation of what may constitute a legitimate *demos*, and of the very nature of the social contract between sovereign citizens and the state. For this reason, it can be argued that political science's very sparse engagement with new forms of citizen media thus far has consequences for the field's ability to adequately explore the role of citizens and the democratic potential of taking citizen participation seriously. Addressing this democratic potential means not just engaging citizens by inviting them to contribute within established institutional politics and media but allowing people a space within which they can organize their own positions and protests. Citizen media as practice and phenomenon may thus also involve moving between antagonistic and agonistic spaces to engage politically in democratic relations.

See also: citizenship; journalism studies and citizen media; postcolonial studies and citizen media; prefiguration; public sphere

Recommended reading

Curran, J., N. Fenton and D. Freedman (eds) (2012) *Misunderstanding the Internet*, London and New York: Routledge.

A collected volume that develops an approach to Internet policy and digital media that is explicitly political and therefore of great relevance to any work on citizen media that is concerned with the politics of protest and democratic participation. The authors explore the intersection between digital media and other forms of participation, contributing particularly to the field of digital media politics by drawing on political economy and notions of power to offer a critical understanding of recent media developments.

Dahlberg L. and S. Phelan (eds) (2011) *Discourse Theory and Critical Media Politics*, Basingstoke, UK: Palgrave Macmillan.

A collected volume that offers a promising attempt to bring together new directions in critical media theory, discourse theory and radical politics, which is essential to expanding our understanding of how citizen media can contribute to new democratic directions in areas of political science that engage with digital, social, activist and participatory media. Offering theoretical contributions to the field of critical media politics, the authors also draw on several useful and illuminating case studies that can be seen as examples of citizen media practice.

Mouffe, C. (2018) *For a Left Populism*, London: Verso.

Elaborates a theoretical thesis that can work to support and substantiate citizen media studies. If citizen media contributes to developing radical democratic practices, it can be argued that these may be conceptualized as a form of left-wing populism that brings together numerous kinds of struggle in opposition to right-wing nationalist populism as well as neoliberal hegemonic policies. Examples of citizen struggles discussed include Syriza in Greece and Podemos in Spain.

POPULAR CULTURE AND CITIZEN MEDIA

Randa Aboubakr

Popular culture is one of the most difficult terms to define in the field of cultural studies (Storey 2012:1), for two reasons. Popular culture cannot be understood without first establishing what is 'unpopular' culture (Frith 1997a:415), and it often implies its otherness in relation to categories such as mass culture and folk culture (Storey 2012:1). Whereas constructing popular culture in opposition to unpopular culture represents a daunting theoretical challenge (Frith 1997a:415), thinking about popular culture in terms of otherness and overlap with conceptualizations of folk and mass cultures can significantly facilitate engagement with the concept.

Mass culture is primarily understood as involving the production and dissemination of cultural commodities imbued with some sort of prefabricated meaning in industrial societies by means of mechanical reproduction (Frith 1997a:415; Adorno and Horkheimer 1944/1979). What is popular about mass culture therefore concerns the mass consumption and appeal of such commodities (Frith 1997a; Storey 2012). Folk culture, on the other hand, is generally understood as the cultural practices and artefacts of heritage that date back to preindustrial times, mainly disseminated through historical records and orality (Frith 1997b:199–200; Storey 2003:2–4). The popular in folk culture thus relates to its wide distribution, especially at the level of consciousness, among masses of people. These two conceptualizations of mass culture and folk culture are seen to divert attention from people's actual lived experiences, either by locating them in the past in the case of folk culture (Storey 2003:4), or in the cultural commodity itself in the case of mass culture (Frith 1997a; Adorno and Horkheimer 1944/1979; Storey 2012). However, both conceptualizations of lived experience are useful for understanding popular culture as the property of ordinary people who engage with culture through their daily personal and social experiences, rather than as a set of exclusive practices consecrated in (bourgeois) institutions (Storey 2003:15; Williams 1965:63).

The term popular culture might therefore be best approached by scrutinizing its two constitutive notions: culture and the people. Since the late nineteenth century, the concept of culture has been variously conceptualized in cultural thought. The rise of the working classes as a demographic force in urban areas in Western Europe threatened the order of the latter part of the nineteenth century with the vanishing of a common culture. In order to defend that common culture against the emerging forces of industrialism, there emerged an understanding of culture as a reified sphere comprising the 'best' of what refined intellect had produced in terms of "knowledge and truth" (Arnold 1960–77:519). At the time, culture was a particular way of life, the property of a chosen elite who set the standards to be followed by the rest of the people (Leavis and Thompson 1933:5), and were in charge of both producing culture and disseminating it to other sectors of society.

On the other hand, with the emergence of the working classes in urban centres, popular culture became the label used to describe the cultural practices of that rising social group (Arnold 1960). These practices were strongly tied to the material conditions constraining the lives of the working classes, and hence were markedly different from the cultural practices of the enlightened (Hoggart 1957; Thompson 1963). Whereas Arnold (1960) and Leavis and Thompson (1933) sought to emphasize the threat that working-class culture represented for the continuity of a common culture, Hoggart (1957) defends working-class culture, in the face of such accusations and the overtaking by mass, i.e. mechanically reproduced culture, of the field of cultural production and representation. Thompson (1963) also places the category of class at the centre of his conceptualization of the popular, and extends his understanding of culture to encompass practices produced by the working classes within the course of their daily lives to facilitate their resistance against industrialism. However, in most early conceptualizations of popular culture as the culture of the working classes, those practices highlight escape and passive distraction (Storey 2012:50).

Subsequent, particularly post-Marxist conceptualizations of the term culture opened up the scope of the concept by significantly minimizing the role played by social class in determining its referent. They also introduced a democratizing view of culture as driven by ordinary, everyday practices of a larger community, thus reflecting changes in relations of domination in society (Williams 1991:415; 1989:4). This politicization of the term, which can be seen to have taken momentum in the mid-1980s, paved the way for the reconfiguration of popular culture as symptomatic of the political situation and disorder in a society (Storey 2012:21). In this conceptualization, popular culture encompasses both the practices and the products of the cultural industry as people read and make sense of them in their everyday lives (Bulliet 1998; Fiske 1989; Storey 2012; Barker 2000). Particular emphasis is therefore placed on creative consumption, in response to earlier attacks on mass culture by Adorno and Horkheimer (1944/1979) and the Frankfurt School on account of its alleged uniformity, lack of originality and its passivity and compliance (Adorno 1941). Ultimately, creative consumption becomes the resistive use of otherwise hegemonic cultural commodities such as television or a pair of jeans (Fiske 1989:26, 29).

Popular culture and citizen media

Under the post-Marxist/Gramscian turn in cultural studies, the departure from a uniform cultural expression is lauded as key to understanding popular culture as a site of struggle over power, especially in industrial societies (Williams 1991; Hall 2006). Cultural production reflects and transmits the hegemony of a dominant group, either in the form of cultural commodities or cultural practices which seek to challenge domination by creating an interventionist repertoire of practices derived from ordinary people's everyday activities. Whether working from within the schema of consumerism to subvert it, as evident in the creative use of cultural products, or from outside it altogether, as illustrated by the emergence of new forms of cultural expression such as graffiti and street performances, this understanding of popular culture intersects at several points with Baker and Blaagaard's (2016a) understanding of citizen media. The discourses of popular culture and citizen media thus understood belong to the realm of the marginalized and disenfranchised (Hall et al. 1978:11; Baker and Blaagaard 2016a), and are therefore inherently oppositional or interventionist. Consequently, analysing marginalized and oppositional cultural artefacts yields insights about dominated and dominant groups (Williams 1991:415).

Against this backdrop, culture becomes a socially created category, and the boundaries between high and low culture are no longer helpful categories for the study of cultural production. Aided by the postmodern suspicion of established canons and master narratives, the study of popular culture is more concerned with describing the phenomena it investigates than with evaluating or relegating them to specific categories of worth (Barker 2000:41). Insofar as the fields of popular cultural production and citizen media are not dominated by an orthodoxy, they are noticeably tolerant to questions of distinction and are evolutionary in nature (Bourdieu 1977/1993:83–84) – as is also the case with other marginalized fields of cultural production. Debates about the question of value, therefore, continue to be informed by the interests of the dominant ideology (Williams 2009).

For both citizen media and popular culture studies, in the absence of aesthetic criteria with which to approach and evaluate popular cultural production, the worth and importance of a product or a practice are largely ideological. Also driven by ideology is the analysis of such production with a view to uncovering social and political values embedded in the work; understanding the relations of power holding in a given society at a particular moment in time; and assessing how this power is negotiated materially and symbolically in and through cultural products (Eagleton 1984).

Actors, artefacts and practices of resistance

Focusing on the confrontation between people's everyday practices and the power bloc invites an examination of the place of resistance in popular cultural production from two main perspectives: the creative/resistive potential of consumption (Bulliet 1998:41; Fiske 1989:25), and the forms of resistance originating outside consumerist practices (Williams 1991; Storey 2012).

North American cultural studies have conceptualized popular cultural production primarily in terms of consumerist capitalist goods – whether in the form of material goods or as symbolic goods, such as those produced by the entertainment industry. The meaning of resistance in this scholarly strand is to be found in the very interface between the industrial product and people's everyday use of it (Fiske 1989:28). By contrast, British cultural studies – and, in particular, the work of scholars affiliated to the Birmingham Centre for Contemporary Cultural Studies – has generally adopted an understanding of popular culture as a set of resistive or interventionist practices carried out by subordinate groups. These practices are not necessarily tied to the process of capitalist production that the goods originated from; rather, the analytical focus is on people's everyday-life practices with which they respond to authority (Williams 1991; Storey 2012). This form of cultural materialism is based on a reworked understanding of the relationship between base and superstructure in Marxist thought, according to which cultural products and practices are not seen as the output of the economic sphere alone; instead, they are regarded as the product of multiple layers of determination forming "homologous structures" (Williams 1991:408). The base can be seen as a process involving a range of cultural practices that constitute people's social existence, rather than as relationships of production and economic activities. The superstructure, on the other hand, becomes a sphere where cultural and ideological structures correspond to, rather than merely reflect or reproduce, the social structures and activities contained in the base (Williams 1991; Althusser 1965/1969, 1971). This latter understanding of cultural production as resistance intersects with Baker and Blaagaard's (2016a) understanding of the sphere of production of citizen media, where the stress on creative consumption and the reworking of the products of the culture industry is not paramount. The emphasis that post-structuralist poetics places on the absence of an author who controls the meaning of a text explains why cultural studies has prioritized the concept of practice in cultural production, vis-à-vis that of a finished product endowed with a 'meaning'. Arguments in favour of the dispersal of meaning in cultural artefacts and practices (Barthes 1957/1973) are also behind the concept of creative consumption in resistant cultural practices. From this perspective, the meaning of a cultural product is not determined *a priori* by the cultural industry; rather, it is dispersed and varied: there are as many meanings as there are consumers.

Discussing resistance and ideological affiliation in popular culture therefore entails investigating the message and meaning. As popular cultural practices are part of people's everyday lives, meaning is constantly being (re)created. For instance, the particular meanings that an individual or a social group derive from a TV commercial or a piece of graffiti depend far more on the recipient's own reconstruction of meaning than on a given set of meanings handed down by the TV show's producers or the graffiti artist (Barker 2000:4; Fiske 1989:32). This act of creative consumption also highlights the locality and social circumscription of popular culture (Bulliet 1998:41; Fiske 1989:25), and renders the range of potential meanings it is expected to produce dependent on a set of shared experiences and knowledge. The resulting polysemy is crucial in the process of negotiation with the power bloc that popular cultural production entails.

Discussions of meaning are linked to the idea of intentionality and ideological commitment in both popular culture and citizen media studies. In that respect, citizen media seem to be strongly focused on an ideological message emanating from a particular context of contestation (Baker and Blaagaard 2016a:8, 10). It could therefore be assumed that polysemy and the dispersal of meaning in popular culture make popular cultural production more open to various interpretations because it is not necessarily focused on a particular message or emanating from a particular incident. As popular culture encompasses varied forms of everyday practices that are not necessarily aimed at conveying an ideological message (de Certeau 1980/1984), resistance does not have to be achieved necessarily through a wilfully interventionist practice, but can also be managed through less 'productive' measures such as dodging ideological engagement or going out of control altogether (Fiske 1989:69).

Taking intentionality into account throws more light on the varying conceptualizations of the message in popular culture and citizen media. The latter seems to flag up the message more strongly than the domain of popular culture does. Although more contemporary understandings of citizen media seek to free the term from its classical link with the (mainstream) media and journalism in general, and locate it in a wider repertoire of social practices, citizen media remains strongly tied to the sphere of news – understood as the dissemination of an activist ideological message (Baker and Blaagaard 2016a; Hughes and Parry 2016). Thus, whereas some of the practices that fall under the two concepts – such as street performances, graffiti, blogging – are the same, the message-driven nature of citizen media practices signals them as primarily aiming at conveying a message, rather than as practices generated more spontaneously within the course of people's everyday lives which do not necessarily have a message attached to them. Consequently, the 'political' dimension is more strongly highlighted in conceptualizations of citizen media (Baker and Blaagaard 2016a:10) than it is in popular culture.

The reference to citizenship in citizen media and to people in popular culture is another example of convergence between both disciplinary domains. But while the use of citizen in the former is meant to indicate practices contributing to the creation of citizenhood (Baker and Blaagaard 2016a:14), the people is not a category that evolves through practice. Rather, it is a more neutral designation of a population or group of people pursuing their daily activities, whether intentionally engaged in acts of protest or not (Ross 1989:3; Williams 1965:63). Despite the differences in the conceptualization of people in popular culture and citizens in citizen media, the two notions share qualities of fluidity and collectivity. Less coherent than a social movement in the theoretical sense of the term (Tilly 2004), producers of popular culture and citizen media are often structured as loose networks of unaffiliated citizens who produce a repertoire of collective action (Baker and Blaagaard 2016a). They resort to such channels of participation and mediatization because they do not have access to or are not represented in more institutional and corporate channels, including sites of sponsored cultural production (ibid.) and/or political organizations. Like citizen media, popular culture is an open arena for the participation of ordinary (non-professional) citizens, and accommodates an unlimited range of participatory and representational practice.

Most authoritative conceptualizations of popular culture seem to stress its collective nature (Williams 1991; Thompson 1963; Hoggart 1957; Storey 2012; Fiske 1989), in line with earlier conceptualizations of culture – whether culture of the elite, or working-class culture – as a construct determined by class. Although culture studies largely abandoned the category of class in its conceptualization of popular culture subsequently, the emphasis on producers of popular culture as a group sharing a common background persists. This point would not appear to be as strongly emphasized in conceptualizations of citizen media, where the act of mediatization seems to be equally powerful, regardless of whether the media content under scrutiny has been produced by an individual or a group (Blaagaard and Allan 2016).

Both popular culture and citizen media studies stress the loose collectivity of producers as well as the absence of demarcation between producer and consumer (user). Indeed, the producers of popular cultural artefacts and practices are also the users of such artefacts and practices in material space and, in particular, in the digital sphere. The term user-created content, describing digital content produced outside of professional practices, signals this conflation between producer and consumer (Gaudeul and Peroni 2010). This, in turn, re-shapes the link between cultural production and the culture industry. The (re)creation of content in the digital realm, which is a product of technological industry, does not reflect the sameness and monopoly criticized by Adorno and Horkheimer (1944/1979), nor the element of creative consumption foregrounded by Fiske (1989). The stress rather falls more heavily on issues of reach and the emergence of new narratives.

Spatial interventions: between the local and the global

Popular cultural practices and citizen media are space-based and seek to interact with and in space. They seek to provide disenfranchised sections of the community with an expansion of the range of representation. Graffiti, street art, parkour and hip-hop are among the many forms of popular cultural production that stress this reciprocity with space (Chaffee 1993; Abaza 2011; Sanders 2012), and are among the oldest practices in the history of popular cultural production.

Bakhtin drew attention to the practice of carnivalesque performances carried out by ordinary people during the season of carnival in the streets of the medieval city to invert social hierarchies (Bakhtin 1965/1984). In its opposition to medieval official feasts, carnival worked as an alternative reality articulating people's opposition to authority (Morris 1994:195–198). A contemporary vision of carnival can be seen in the immediate aftermath of revolutions and uprisings, when people literally take to the streets in a material act of reclaiming space and a symbolic act of (re)asserting authority and access to representation. The practice of occupying spaces at times of political upheaval is a symbolic act of smoothing striated and reified spaces (Deleuze and Guattari 1980/1987), and hence of enhancing the heterogeneous character of those spaces (ibid.:488–492).

In this kind of spatial intervention, the body becomes central. In early conceptualizations of popular culture, the body was the site of contestation between the people – more specifically, the working classes – and the authorities (Fiske 1989:81), so representation and resistance revolved around it. From this perspective, the degradation of the body in carnival practices stood as a sign of equality and defiance (Fiske 1989:83; Morris 1994:195–196, 204). This understanding of the body also has special spatial manifestations. From a spatial perspective, the figure of the carnival clown featuring in medieval popular festivities (Bakhtin 1965/1984) seeks to reclaim both space and representation. In the context of the modern capitalist city, the role of the carnival clown is comparable to the figures of the trickster and the guileful ruse (de Certeau 1980/1984:98–99; Hyde 1998), who raid spaces and invert their intended use, and with it the hegemonic role of the modern state apparatus in (re)organizing space. In a more contemporary moment, the occupation of streets through silent encroachment (Bayat 2010), particularly in neoliberal cities, is a manifestation of emerging nodes of agency and activism.

Although early conceptualizations of popular culture stress the locally circumscribed nature of the concept (Bulliet 1998; Fiske 1989:25), in contemporary digital culture it is difficult to draw boundaries between the local and the global, with transnational and transethnic cultural spaces emerging as new sites of shared meaning production. This is evident, for instance, in attempts to view raï, reggae, hip-hop and rap as part of the global beat sound (Shoup 2013:46) or to consider TV formulas such as *Who Wants to Be a Millionaire* as part of a global network of mutual (re)definition and (re)interpretations (Condry 2001). Indeed, the turn of the twenty-first century has witnessed the proliferation of popular cultural production that can more easily be disseminated and connected in the digital realm (Hamilton and Shepherd 2016; Iskandar 2014; Allagui 2014; Berenger 2013; van de Donk et al. 2004). It is also evident that citizen media rely heavily on digital production or on the dissemination of reports about material cultural production through the digital medium (Nim 2016; Pérez-González 2016). The stress on practices shared globally also makes it more possible to view the issue of class in larger terms, and to consider general issues of marginalization – such as race, gender, age, sexuality and political disenfranchisement (Hall et al. 1978:11) – as representing the ideological struggle between dominant and subordinate groups. In this sense, popular culture, like citizen media, encompasses varying forms of contentious cultural production motivated by that very ideological struggle.

The growing politicization of ordinary people in the contemporary world – as illustrated, for instance, in the various interconnected Occupy movements the world witnessed in 2011 and 2012 – raises important questions pertaining to the future impact of popular cultural production and citizen media on local and global activism, and the extent to which stronger forms of ideological intervention will strengthen and enhance the connectivity within movements for change.

See also: performance studies and citizen media

Recommended reading

Barker, C. (2000) *Cultural Studies: Theory and practice*, London: Sage.

A comprehensive guide to cultural studies tracing the evolution of the discipline, and covering key theories and concepts. Later editions engage with relevant issues such as the relationship between culture and evolution (second edition) and digital culture (fifth edition, co-authored with Emma A. Jane).

Donk, W. van de, B. Loader, P. Nixon and D. Rucht (eds) (2004) *Cyberprotest: New media, citizens, and social movements*, London and New York: Routledge.

In-depth exploration of the relationship between information and communication technologies and social movements. Introducing a more inclusive understanding of social movements in the information age, the book traces both the historical evolution of the use of mass media by protest movements, and the varied strategies they employ and adapt. Introduces case studies of digital activism from places as far apart as East Timor, Portugal and Australia.

Storey, J. (2012) *Cultural Theory and Popular Culture: An introduction*, sixth edition, Harlow, UK: Pearson Education.

A compelling exploration of the place of popular culture in cultural studies, drawing on various disciplines such as cultural studies, media and communication studies, sociology and anthropology, philosophy, psychoanalysis and literary theory. The book outlines diverse conceptualizations of popular culture, surveys examples of popular cultural practices, and provides extensive description and commentary on them.

POSTCOLONIAL STUDIES AND CITIZEN MEDIA

Bolette B. Blaagaard

As a critical framework which concerns itself with the deconstruction of Western-centric discourses and systems of power, postcolonial theory both questions the concept of the citizen as it relates to the nation-state and explores the potential of media as a tool for diversifying public participation in culture and politics. Consequently, fruitful intersections between the fields of postcolonial studies and citizen media are available to scholars and activists who wish to develop an understanding of how, and with what implications and limitations, mediated, marginal voices may affect social change. Beginning with an introduction to the work of Gayatri Spivak (1988, 2005) on subaltern voices, the entry presents a discussion of the ways in which postcolonial theory can help challenge media studies' dependency on the idea of the unity of the nation and on the concept of the citizen. The second half of the entry then seeks to highlight a number of points of potential dialogue between the fields of citizen media and postcolonial studies, exploring the interactions between these two areas of study on multiple political, creative and critical levels.

Postcolonial media studies

The fundamental focus of postcolonial theory is the critical deconstruction of modernity and of Western hegemonic knowledge production (Said 1978; Bhabha 1994; Bhambra 2014). Western knowledge is produced by claiming an invisible but universal standpoint which constructs the Western subject as the norm and everyone else as other – thus sustaining the position of power held by the Western subject and culture. This critique of Western epistemology is laid out by one of the major voices in postcolonial theory, Spivak (1988), in her article 'Can the Subaltern Speak?', a critical reading of the poststructuralist work of philosophers Gilles Deleuze and Michel Foucault. Here, she argues that Deleuze and Foucault are guilty of re-inaugurating a transparent European subject despite claiming to dismantle this subject. Taking as her starting point a published conversation between the two philosophers, Spivak (ibid.:272) critiques their thinking in which, she states, they "systematically ignore the question of ideology and their own implication in intellectual and economic history", thereby rendering invisible their complicity in upholding Western philosophical hegemony. If the subject of power is transparent, it follows that Deleuze and Foucault leave it up to the subaltern to speak on equal terms. But, Spivak argues, this is in fact impossible for subaltern subjects: there is no voice outside Western discourse, and ignoring this fact is a failing incompatible with postcolonial thinking and critique. How the challenge of this postcolonial condition is overcome or made available to creative forces is the topic of postcolonial analyses and practices. Spivak (2005:478) suggests that we may try to listen differently and thereby make "unrecognisable resistance recognisable". Spivak's postcolonial project is, then, to practice an aspiration and desire for unlearning privilege in order to attempt to place oneself in the position of the other – however daunting and difficult the task.

The postcolonial critique Spivak levels at poststructuralist thinkers may likewise be levelled at scholars of media studies and media institutions. While media studies has existed in various forms since the beginning of the twentieth century, when the Chicago School began to focus on the relationship between democracy and communication (Dewey 1927), the discipline is often mainly associated with media effect studies (Katz and Lazarsfeld 1995; Potter 2011), audience studies (Katz and Liebes 1989) and mediatization theory (Couldry and Hepp 2013). Studies of media from these traditional perspectives are characterized by empirical and pragmatic approaches, and therefore when introduced issues of difference generally take the form of questions interrogating media representations of difference – gendered, racial or sexual – and/or how these representations are decoded by audiences (Hall 1980, 1997a). It should also be noted that the development of media studies is closely linked with the emergence of the nation-state and modernity at large. Media scholars have explored the ways in which media help create the imagined communities which are fundamental both to the progress of political and democratic institutions, and to the cultural understanding of national belonging (Anderson 1991; Habermas 1962/1989, 2008/2009). Thus, media institutions have been seen to play a significant role in supporting and sustaining the invisible but universal position of Western epistemology. Media studies' pragmatic approach to the study of national cultures and communities depends upon an understanding of such structures as first and foremost politically neutral; postcolonial theory, on the other hand, deconstructs the concepts of objectivity and universalism, and insists that all communication is political (Grossberg 2002:367–368).

For postcolonial scholars, postcolonial studies can be viewed as both a continuation of and a challenge to media studies (Grossberg 2002). Firstly, postcolonial studies can be seen as a continuation of media studies because media studies implicitly works with the understanding that all communicative expression is contextual; all communication is situated and is inseparable from the relations in which it is located (ibid.). Since no context can escape the historical, political and economic fact of colonialism, postcolonial theorists argue that their field of study offers media studies a pertinent opportunity for timely reflection. Once colonialism is seen as part and parcel of modernity, it becomes necessary for media studies to acknowledge the flawed but self-sustaining relationship between the idea of the nation-state and media, and to reflect on its implications. Secondly, while the idea of the nation is mounted on an idea of otherness and difference supporting a 'them' vs. 'us' dichotomy, postcolonial theory challenges media communication by offering an affirmative definition of difference and otherness. Rather than continuing the binary opposition of either-or difference – for instance, either them or us – and of both-and difference – both them and us – postcolonial theory offers instead 'x and y and …', that is, a more positive and nuanced system of identification (ibid.). In the words of Stuart Hall (1995:247), "[postcolonial theory's] theoretical value therefore lies precisely in its refusal of this 'here' and 'there', 'then' and 'now', 'home' and 'abroad' perspective". The focus on difference as multiplicity encouraged by postcolonial theory allows for other perspectives and voices to emerge in the narratives of history, communities and identities. However, here postcolonial studies' continuation of and challenge to media studies are in tension, because these narratives should be conceived of only and always in relation to and as a critique of the way in which knowledge is produced through particular positions of power constructed by Western regimes and which media institutions arguably reproduce (Hall 1995:251; Foucault 1976/1990). It is far from an easy task to change a structure of binary oppositional pairs into multiple voices.

Postcolonial theory is poststructuralist and deconstructivist at its core rather than rational and successively logical (Hall 1995:255). This consequently positions the field in opposition to the dominantly pragmatic approaches of media studies (Cere 2011; Grossberg 2002; Hall 1995). While postcolonial theory rejects the institutional perspectives of media studies, it also offers rich alternatives. These alternatives, however, typically call on activist approaches to media, a non-binary understanding of national belonging, and a different kind of listening. For instance, it follows from the deconstruction of difference and otherness that the concept of the nation is in question too. Once the concept of the nation has been revealed to rely on an imagined, political but arbitrary sense of community (Kumar 2014; Anderson 1991), the concept of the citizen can be operationalized outside the designated borders of a nation. From this perspective, the citizen is unaffiliated and emerges through her political acts for social change. Importantly, however, this is only a conceptual construction, while actual citizens or unaffiliated individuals remain bound by the constitutional laws dividing nations and peoples. Taking a cue from postcolonial theory and cultural theories of representation, the fact that contextual power is never transparent or inconsequential allows media producers and scholars alike to explore such tensions and to engage with other narratives, voices and perspectives. Navigating by postcolonial theoretical perspectives thus encourages media scholars to move beyond their traditional disciplinary focus on media institutions and to engage more substantially with citizen media projects and practices.

If postcolonial critique aims to bring to light alternative, subaltern sites of knowledge production, these are not necessarily presented in traditional academic or journalistic forms and therefore they typically require a different kind of listening in order to be comprehended. Arguably, the political critiques offered by postcolonial studies are not only compatible with, but are illustrative of the performative, creative and political practices of knowledge and cultural production in which unaffiliated citizens engage. In the following sections, therefore, the entry highlights points of synergy between postcolonial studies and citizen media as defined by Baker and Blaagaard (2016a). Specifically, it is argued that postcolonial and decolonial practices proposed by various scholars for delinking knowledge production from Western hegemony and thereby challenging Western epistemic traditions are helpful for understanding practices of citizen media. In order to illustrate this point, the next section first introduces the concept of vernacular expressions, discussed by Paul Gilroy (1993a) as embodied practices of remembering colonial culture and continuing postcolonial struggles through art, before turning to the practice of counter-reading literature and historical accounts, developed by Edward Said (1978) and others as a means of reimagining and reconstructing hegemonic narratives. Lastly, the practice of border-thinking is examined from a citizen media perspective. The concept of border-thinking has been developed by Walter Mignolo (2013) as a means of engaging with the nomadic and migrant positioning adopted by many individuals and activist groups in relation to knowledge production: border-thinking aims to allow for epistemic disobedience, that is, a rejection of the structural subordination of vernacular or subaltern knowledge productions and formats. In different ways, these three selected practices are in dialogue with Spivak's seminal question of the subaltern's ability to speak within commercialized and modern contexts.

Practices of vernacular expressions and commercialization

Due to the dichotomous nature of Western modernity, subaltern voices are often heard only in the context of stories of history and tradition. Whereas Western culture is understood to be forward looking and entrepreneurial, subaltern cultures are perceived as backwards and belonging to the past. As a postcolonial scholar, Gilroy (1993a:198) redefines tradition as "the living memory of the changing same". This "non-traditional tradition" does not stand in opposition to modernity but rather is produced through modernity and is in fact "an irreducibly modern, ex-centric, unstable, asymmetrical, cultural ensemble" (ibid.). Gilroy argues that tradition needs to be understood in this manner in order to rescue the concept from the "obsession with origins" (Lott 1994:57), in which inspirational interconnections between different times and places are neglected in favour of stories of purity and "unblemished exceptionalism" (ibid.). In this theoretical framework, then, subaltern expressions are no longer viewed as existing outside of modernity, but as a key constitutive part of the evolving ensemble.

In order to illustrate this conceptualization of tradition, Gilroy uses the example of African American music as a mnemonic device that reiterates cultural and traditional knowledge in song and genre, but always continually within the contemporary political context. Gilroy argues that tradition as the changing same is practised in the music and in the culture that surrounds the diaspora (Gilroy 1993a:198). While Gilroy focuses on African American music and culture in his early work, the concept of the changing same that is expressed through a wide range of vernacular cultural expressions is additionally brought into continental European and British contexts (Gilroy 1993b, 2004). Both in terms of rhythm and of lyrics, music is viewed as a form of vernacular expression which calls on a counter-culture – or counterpublic – of active listeners, who are engulfed in the memory-making of the present as the music is drawing in (and on) historicity and social memory. Subaltern publics or counterpublics are thus framed as important sites for the circulation of knowledge produced in resistance to hegemonic epistemologies (Fraser 1990; Warner 2002). Popular culture such as music is neither just play nor immediately political, but it has political consequences (Gilroy, in Lott 1994:81). While the artists Gilroy discusses as vernacular expressions care less about being heard by Western media than about enacting a political community and culture in and of itself, the potential critique of national unity, universalism and objectivity within the practice is only recognized as resistance if listening occurs. That is, they can only be seen as expressions of resistance against the dominant power-knowledge discourse if they are understood in relation to this same discourse that structures the idea of modernity.

Cultural products – such as music production – imply engagement with both the local and the global in specific ways in a practice which is based on a participatory culture (Ponzanesi 2014:2–3). The Frankfurt School of Critical Theory generally portrayed the cultural industry as homogenizing and oppressing culture and politics through the commercialization of aesthetics and of art. However, taking her cue from Gilroy, Ponzanesi insists on a more nuanced view, which recognizes the intersections of power relations between different actors in the postcolonial cultural industry field. Her theoretical framework is an "optic of analysis, through which the trajectory of the postcolonial artefacts is detected, accounted for and elaborated in its multiple functions" (ibid.:47), and which allows her to analyse cultural products as practices that co-shape the parameters for the appreciation of postcolonial art and that operate within a complex field of interactions. Yet, the aim of postcolonial studies must be political in scope while simultaneously linking the political and "the definitions of aesthetics with the ideology of aesthetics and with hegemony" (ibid.:45) through which the theory questions the Western logics and canon "with a sincere demand for a multicultural curriculum" (ibid.:45). On the one hand, postcolonial creative expressions such as literature, film, music, art, games and fashion may strike back against the Western commercial logic of consumption of the other by critiquing, mimicking and mocking dominant and oppressive culture as well as insisting on other forms of communication besides textuality (Gilroy 1993a:77–78, 1993b). On the other hand, these practices may easily fall prey to co-optation and the cannibalistic consumer logic according to which minority cultures are displayed for commercial purposes. Ponzanesi believes that it is in a space in-between these binary positions of analysis that postcolonial critique is found. Parallel to Ponzanesi's argument, Gilroy argues that postcolonial expressions are imbibed with a doubleness: both "inside and outside the conventions, assumptions, and aesthetic rules which distinguish and periodise modernity" (1993a:73). The in-betweenness or doubleness of modern cultural and political production is pivotal to understanding what may be tentatively termed postcolonial citizen media. Through critical postcolonial thinking, both Ponzanesi's and Gilroy's analyses bring to the fore the point that citizen media is always situated within a dominant culture – within as well as against which resistance is expressed. In order to identify the trajectories of aesthetic, political products, their production as well as consumption, analyses of citizen media following a postcolonial critique must draw a cartography of the functions and meaning-making of such expressions.

Counter-reading and border-thinking

Part of the mnemonic agency of music relies on counter-readings of history and historical knowledge. Counter-reading history is a conscious rejection of modernity's teleological and successive narrative by way of acknowledging that knowledge is always situational, interpreted and an expression of interest (Said

1997). Postcolonial critique of Western hegemonic knowledge claims was theorized by Edward Said (1978) through the concept of orientalism in his ground-breaking book of the same name. Orientalism is based on the asymmetrical power relation between the colonizer and the colonized in which the former regards the latter from a position of dominance. Writing history and cultural narratives from this position allows the colonizer to imagine the culture and knowledge of the colonized as absolute. In contrast, Said (1997:163–167) argues, counter-reading requires the reader of other cultures to understand, firstly, the situational quality of the narrative. The narrative is never absolute or universal, rather the counter-reader must align herself with the culture which she seeks to understand by studying the narrative from within the culture which is foreign to her. In other words, she must 'delink' from the epistemes and paradigms of Western thinking (Mignolo 2013). Secondly, knowledge of the social world must be understood as an interpretation and, as such, as always particular to the individual reader. Counter-reading requires the reader to overcome the distance between the situation of the narrative and that of her own through awareness of her own determining situation and context. Finally, Said asserts, interpretation is never without interest – be it economic, political or cultural. We are all products of the histories we are told and taught, and an awareness of their constructed and affiliated nature is necessary in order to perform a postcolonial counter-reading. In short, Said urges scholars, laypeople and media to understand the world from the perspective of colonial history and its political and economic implications. While Said may be criticized for upholding a binary view of them vs. us, a view in which the other(ed) culture is easily identifiable and delineated, his critique of Western truth claims is important. Without critical awareness of the Western power-knowledge discourse and its impact on the world, our understanding and praise of vernacular expressions may come across as a naïve generalization. Keeping Said's critique in mind, practices of counter-reading potentially provide grounds for the analysis and exploration of vernacular expressions that reject the assumed purity of the other(ed) and subaltern.

Taking this critique one step further, when readers have come to grips with the colonial condition, they delink (Mignolo 2013). Delinking is the process of insisting on the validity of othered knowledge produced from another position than the dominant one and of rejecting the idea of others' constructed inferiority. Mignolo is clear about the discursive construction of the other and uses the realization of this construction to argue for the legitimacy and necessity of delinking. Delinking in turn leads to border-thinking. Border-thinking is independent thinking, writes Mignolo (ibid.:136). The borders in question are not geographical, but the borders between epistemes and ontologies. They are borders between the colonial perspective and modernity, and border-thinking rejects modernity's epistemic heritage in favour of the senses and bodies of the decolonial, thus committing epistemic disobedience (ibid.:136–137). Mignolo suggests that border-thinking may be practised in academia in opposition to the neoliberal university and institutionalization of knowledge. Equally, border-thinking as well as counter-reading and expressions of vernacular culture could be imagined as key instances of the political acts through which individuals and collectives become citizens by "transforming their sense of self and their environment" (Baker and Blaagaard 2016a:13). Indeed, like the unaffiliated citizens conceptualized by Baker and Blaagaard, postcolonial theoretical concepts take their point of departure in oppositional and countering strategies against institutional, economic and political power brokers.

To illustrate this point, it is useful to refer to Blaagaard (2016a, 2016b), who uses postcolonial critique and counter-reading to understand journalism as vernacular culture in a colonial setting by counter-reading the historical context within which it was produced. Taking her starting point in a newspaper published between 1915 and 1925 by an African Caribbean editor in the last years of Danish colonial possession of the Caribbean islands of St. Thomas, St. John and St. Croix, Blaagaard argues that this newspaper constitutes an intriguing example of vernacular expression. While journalism is a professional practice which has grown in tandem with modernity and which rests on assumptions of objectivity and universality, making the practice work for social reforms and labourers' rights in the colonial setting and from the perspective and embodiment of the labourers is arguably a counter-reading, if not an act of border-thinking. It would be a mistake to conclude that the other of colonialism spoke through the newspaper only in the voice of professional, modern journalism. Rather Blaagaard asserts that the space created within the pages of the newspaper developed into a postcolonial as well as a cosmopolitan space: not only was it situated in a politically precarious geopolitical space, but it also sought to evoke cosmopolitan bridges to other othered voices, for instance, to the civil rights activists of early twentieth-century New York.

Digital postcolonial critique

The methodology of counter-reading and border-thinking is applicable to historical as well as contemporary readings of citizen media. Dutch media scholar Koen Leurs (2015, 2016) identifies the postcolonial digital humanities as a methodology adapted from postcolonial studies and which can be applied especially productively to migrant youth's use of digital media platforms. Postcolonial digital humanities investigate ways in which "postcolonial studies has evolved through different phases of internet culture" (Risam and Koh, n.d.) and ensures a "digital commitment to social justice" (Leurs 2016:253) by embedding postcolonial politics within the digital processes. For instance, Risam and Koh founded and maintain the website Postcolonial Digital Humanities, which operates and supports projects such as The Rewriting Wikipedia Project. This project argues that Wikipedia is biased due to its large majority of white, male editors. In response, Risam and Koh (n.d.) work towards "interven[ing] in what postcolonial studies critics have termed colonial paradigms of knowledge production and imperialist hierarchies of information". Similarly, Leurs (2015) uses gender and postcolonial theories to highlight implicit sociocultural biases present within new media platforms and the ways in which these affect migrant youth's uses of them for performing their hybrid identifications. While Leurs uses postcolonial studies to aid his own research by introducing an ethics of shared knowledge production with his informants, he moreover identifies tactics employed by Moroccan-Dutch youth (2015) as well as urban youth in London (2016). The first tactic Leurs discovers is the "act of racial passing" (2015:254), which he links to Homi Bhabha's (1994) notion of mimicry. Mimicry is a performance of passing as belonging to the dominant, white culture but with a difference that gently mocks the imperial master culture while also leaving it unscathed. Another tactic used by online youth is one of appropriating a space in which the minority becomes the majority. Leurs investigates media platforms that are directed at Moroccan-Dutch youth specifically and which allow young people of ethnic minority background to be in a safe space of their own making. Leurs thinks of these spaces as "subaltern counter-publics" (Fraser 1990; Leurs 2015:255). Finally, Leurs finds that the media platforms make "hybridized belongings across Internet applications" (Leurs 2015:256) available to the youth. Hybridized belonging is a form of everyday conviviality (Gilroy 2004) that is foundational for the multicultural society, but which operates "away from state-organized initiatives" (Leurs 2015:259).

While the tactics employed by Dutch migrant and London youth and identified by Leurs are not explicitly political or activist in nature, his methodological considerations may be seen as allowing political acts to develop. The interstitial spaces which Leurs theorizes "reveal not only the formation of new digital consumers, but also the emergence of new political subjectivities as young urbanites learn to live with difference" (2016:268). Both Leurs and his informants develop strategies of incorporating critiques of modern logics through their uses of collective media platforms, present the reader with alternative, subaltern knowledge, and illustrate tactics of everyday conviviality and resistance to the identity labels posited upon the subaltern by the majority. Postcolonial theory thus informs both research and everyday practices of media use and consumption.

Can the subaltern speak by means of citizen media?

This entry has argued for an implicit link between postcolonial studies and citizen media practices based primarily in a postcolonial critique of modernity and media studies. However, criticism also goes in the other direction: media scholars have chastised postcolonial studies for not focusing on economic issues associated with media production and for entertaining an elitist focus on literature while downplaying the importance of more popular media (Shome 2016). Both of these critiques are valid. Only very slowly are scholars opening their eyes to the importance of media – popular, citizen and otherwise – to a postcolonial project. This entry has presented a few examples of how scholars have started to work conceptually with the interstices of postcolonial theory and citizen media. From the discussions above it is clear that there are affinities between citizen media and postcolonial critique. Both exhibit an ability to deconstruct the transparency of Western power, question the concept of the citizen as it relates to national belongings by speaking from and to in-between spaces and across national borders, and both argue for a multiplicity of voices as political tools for social change. Postcolonial critique, then, may serve as a conceptual framework through which counter-cultural music, journalism, digital interactions and other citizen media productions may be seen to provide in-between spaces that double up as creative forces demanding political subjectivity and recognition. However, despite the velocity and volume with which ideas and voices spread, citizen media alone cannot be seen as a remedy for asymmetrical power relations and postcolonial struggle. If citizen media is to "mak[e] unrecognisable resistance recognisable" (Spivak 2005:478), a new way of listening remains key.

See also: citizenship; political science and citizen media; race & ethnicity studies and citizen media; subjectivity

Recommended reading

Anderson, B. (1991) *Imagined Communities*, London and New York: Verso Books.

Imagined Communities explores the historical and political role of the printing press in colonial societies and how it worked to establish a sense of national belonging despite the colonial context. The book is a milestone in thinking the politics of colonial life and media together.

Gilroy, P. (1993a) *The Black Atlantic*, London: Verso Books.
Gilroy, P. (1993b) *Small Acts: Thoughts on the politics of black cultures*, London and New York: Serpent's Tail.

These two books conceptualize and operationalize the idea of vernacular cultures and explore how they produce a living and changeable memory of past and present struggles. Whereas the former title is mainly theoretical, the second is a collection of essays that illustrate the potential of citizen media.

Said, E. (1997) *Covering Islam: How the media and the experts determine how we see the rest of the world*, London and New York: Vintage.

In *Covering Islam*, Said brings his Foucauldian critique of orientalism to bear on the news industry and its coverage of Muslims in the West. Whereas *Orientalism* (1978) focused on representations in classical literature, *Covering Islam* investigates the processes by which the otherness of the non-Western subject continues to have a discursive hold on the public – now through media discourses.

PRECARITY

Jacob Breslow

In one of its senses, the term precarity describes a shift in advanced capitalism wherein workers in increasingly diffuse sectors of the economy are engaged in informal, irregular and casual labour. This shift has come about through the increasing reach of globalization, transnational outsourcing, new developments in communication technologies and global changes to social and economic governance. There are a number of schools of thought which have worked to make sense of and contextualize this shift, variously describing this increasing precarity as an effect of post-Fordism, post-industrialization, the new economy, new capitalism, information society, network society and/or risk society (Beck 2000; Beck and Ritter 1992; Beck et al. 2000; Bell 1973; Boltanski and Chiapello 2005; Brown 2003; Castells 1996; Sennett 2006). Most of these contrasting approaches tend to agree, however, that precaritization has taken on a new lease of life in the wake of the 2008 financial crisis. What is seen to be different about this post-crisis moment is that, in the wake of the crash, more and more industries which traditionally offered well-paid and secure work are showing increasing levels of precarity.

Capitalism, it should be noted, has always relied on undervaluing labour, on precarious conditions, casualization, inequality and the marking of particular communities and populations as disposable (Wright 2006). As Lorey (2012:165) has argued, "precarity in capitalism is nothing new. The normalization of neoliberal precarity has a long history in industrial capitalism, where insecurity in working and living has, for a lot of people, been the norm, and the welfare state is the exception". This understanding of precarity as having always been a central facet of capitalism and neoliberalism need look no further for evidence than the unpaid domestic labour that has always and continues to be done primarily by women within the household (Fraser 1997; Glenn 1992; Himmelweit 2007; Morini 2007). Precarity, as such, has always been gendered. Furthermore, by including migrants into the historical and contemporary analytical framing of precarity, we might suggest that in order to understand precarity as a new phenomenon one would need to disavow those who have always been marginal to, or excluded from, the production of wealth and security. Another example of a category of labourers who have always been defined by precarity is that of cultural and creative workers (Gill and Pratt 2008). In this sense, the professional producers of citizen media – artists, media workers of both traditional and new varieties, cultural labourers, activists and creative intellectuals – might be understood as "iconic representatives" of the "brave new world of work" (Gill and Pratt 2008:2–3), and as the vanguard of a new political subject called the precariat. For this entry's unpacking of precarity, attention will thus be paid to communities for whom inclusion within capitalism has always been precarious: sex workers, migrants, racialized and indigenous communities, industrial labourers and those facing violence at the hands of the state.

An alternative theorization of precarity expands the realm of interrogation and analysis, moving away from a focus on the precarious conditions of labour, and towards an understanding of embodiment and relationality which foregrounds the interdependence, frailty, mortality and sociality of humankind (Berlant 2011; Butler 2004; Butler and Athanasiou 2013; Ettlinger 2007; McCormack and Salmenniemi 2016). Here, as Butler writes,

> [w]e can make the broad existential claim, namely, that everyone is precarious, and this follows from our social existence as bodily beings who depend upon one another for shelter and sustenance and who, therefore, are at risk of statelessness, homelessness, and destitution under unjust and unequal political conditions.
>
> *2012:170*

While these two understandings of precarity (precaritization and precariousness, respectively) generally have different bodies of scholarship attached to them, they also have overlaps which are important to foreground. Indeed, as indicated in the quote above from Butler, the politics of responding to precariousness demands that we attend to the material realities of precaritization: homelessness, destitution, statelessness and inequality.

Precarity and citizen media

The relationship between precarity and citizen media is similarly complex and multidimensional. On the one hand, developments in communication technologies (one of the factors that has arguably led to the apparent increase in precaritization) have also led to a proliferation of citizen media related to precarity. Alongside the boom in mobile apps, technology industries and Internet companies such as Uber, a taxi company which is notorious for disbanding traditional employment contracts and operating solely through the hiring of subcontractors, is the simultaneous proliferation of (some) individuals' access to apps which facilitate global connection, communication and media production. Sharing content globally through social media platforms such as Twitter, Facebook and YouTube has both allowed for a democratization of content creation and opened up new channels through which the plights of precarious communities might be made known to others. As discussed below, this increase in the digital production and viral circulation of images and narratives relating to precarity has led, at times, to the globalization of "affective sociality produced by unaffiliated citizens as they act in public space(s) to effect aesthetic or socio-political change or express personal desires and aspirations, without the involvement of a third party or benefactor" (Baker and Blaagaard 2016a:16).

That said, there are many under-acknowledged dimensions to this democratization of content creation and increasing access to social media. These include for example the ramifications for those precarious workers who now find themselves employed by social media companies or by the outsourcing companies that these industries rely on. As Roberts (2016b) notes, the rise of digital technology companies has been accompanied by a massive increase in e-waste, both material and immaterial, and the practices involved in disposing of this digital trash are deeply concerning. The disposal of material e-waste – such as broken or obsolete PCs, smartphones and tablets – is often managed by shipping it "to developing countries where crude and inefficient techniques are often used to extract materials and components … pos[ing] dangers to poorly protected workers and the local natural environment" (Baldé et al. 2015:4). Similarly, the disposal of immaterial e-waste – for example, user-generated content that has been flagged for removal from social media platforms for violating content agreements and community standards – has developed into a form of (often precarious) labour called commercial content moderation. This labour is increasingly being outsourced to subcontracted workers based across the Global South who are required to view and erase thousands of violent, pornographic, traumatizing and banal images and videos each day (Breslow 2018).

In an age where online digital citizen media is often both a means of documenting and resisting violence, and simultaneously a product legally owned by multinational technology companies, many social media platforms are struggling to find a balance between upholding community standards by removing footage of graphic violence, and staying competitive by being on the forefront of independent and participatory journalism. Responding, for example, to a video of the death of a young woman named Neda Agha-Soltan, a 26 year-old who had been shot dead while protesting against the presidential victory of Mahmoud Ahmadinejad in Iran in 2009, Buni and Chemaly (2016) describe the dilemma posed by YouTube's content moderation team: "[the video] was fueling important conversations about free speech and human rights on a global scale and was quickly turning into a viral symbol of the movement … And the clip was already available elsewhere, driving massive traffic to competing platforms". Here, footage of a young woman's death at a protest becomes evidence of precarity, a catalyst for political solidarity with those who are experiencing precaritization and precariousness, and a product which must be managed by precarious labourers, many of whom are located in the Global South (Malkowski 2017).

The remainder of this entry will introduce three case studies exploring the relation between citizen media precaritization, precariousness and the precariat. These examples include and extend beyond the digital. They involve protests, manifestos, performances, artwork, documentaries, the creation of online spaces and the coming together of bodies in public space. Global accounts of precarity politics are discussed at length in Lee and Kofman (2012) and Munck (2013).

European Mayday Parade

Beginning in Milan in 2001 and spreading to eighteen European cities by 2005, the European Mayday Parade protest campaign is one of the most visible and well known examples of a collective movement that has mobilized against increasing European precaritization. There were many developments which triggered the Mayday Parade, including a series of legislative, discursive and sociopolitical shifts towards flexible labour and economic inequality. In Italy, where the movement began, for example, legislative changes in the decade leading up to the parade "progressively expanded the categories of short-term contracts that employers could use to hire workers and extended the number and range of labor market sectors in which they could be used" (Mattoni and Vogiatzoglou 2014). In response, activist collectives such as (ex) Deposito Bulk and the Chainworkers Crew organized the Mayday Parade and shifted the discourse in Italy from flexibilization to precaritization. Citizen media were central to the successes of this movement, and they took multiple forms. As Mattoni and Doerr write, "from the very beginning, activists within the EMP (European Mayday Parade) considered posters, postcards and other visual creations as central to the construction and diffusion of the parade" (2007:131). Activists utilized the aesthetics and "cultural grammar of popular religiosity" (Tari and Vani 2005) to create new subversive imagery and symbols of precaritization. *San Precario*, an icon of the protest, was invented as the protector saint of precarious workers. "In the aesthetic style of traditional religious processions for Catholic saints", Mattoni and Doerr write, "activists brought a *San Precario* statue into the procession" wherein he performed a miracle: "the costs of products had been reduced immediately" (2007:131). *San Precario* was represented through various forms of media: statues at protests, stickers, saint cards, viral images, protest signs and leaflets. He is, as Tari and Vani (2005) write, "a floating signifier. Rather than being, the saint becomes, constructing lines of flight according to need, personal inclination and group affiliation". Another example of citizen media created for the European Mayday Parade was the production of the Euro Mayday Net-Parade. This online parade, developed collectively, became what Mattoni and Doerr have described as a "composite visual representation of precarious workers, with each and every protestor able to create his or her virtual alter ego, the so-called avatar, in their preferred style, dress and languages of expression" (2007:132).

Sex workers

Sex work has routinely been criminalized, and sex workers face multiple forms of precarity on a daily basis. As sex work activists have argued, this precarity is even evident in legislative contexts, such as under the so-called Swedish Model, wherein some forms of sex work have allegedly been legalized. Policies which criminalize the purchasing of sex work, rather than the sale of sex, might seem to decriminalize the sex worker, but in practice they make the actual selling of sex legally impossible (Amnesty International 2016; Kulik 2003; Lyon, W. 2014; SCOT-PEP 2015). In response to the criminalizing policies which target sex workers and which make sex work a precarious profession, sex worker activists have produced citizen media as a strategy for resistance. One such example is the Sex Worker's Opera, a mixed-genre production written and performed by sex workers and their friends (Bush 2017; Filar 2016). The production, which is compiled from over sixty stories collected from sex workers around the world, utilizes performance, storytelling and humour to shift the narratives around sex work. The Opera's website also contains a growing digital archive of curated stories written by sex workers across the globe. The Sex Worker's Opera and its archive are themselves forms of citizen media: they bring together the lived experiences of affiliated and unaffiliated individuals who are working against their conditions of precarity to reclaim both online and offline spaces.

The publication and dissemination of the Declaration on the Rights of Sex Workers in Europe (ICRSE 2005a) also worked to challenge the forms of precarity that are forced onto sex workers by regressive, criminalizing and stigmatizing policies and discourses. This declaration was put together by 200 delegates from twenty-eight European countries who gathered in Brussels for the European Conference on Sex Work, Human Rights, Labour and Migration. Upon completing the declaration and the Sex Workers in Europe Manifesto (ICRSE 2005b), the declaration was presented in the European Parliament alongside a series of recommendations on sex work and public policy. The declaration specifically challenges the conditions of precarity faced by sex workers: discriminatory practices and legislation pertaining to health care, border and migration policies, social security, housing, employment, education and criminal justice that target sex workers. In creating this document and presenting it to the European Parliament, "sex workers in Brussels strategically enacted themselves as active citizens in order to make their political voices heard in a context where they are marginalized from the rights associated with free movement across the EU" (Andrijasevic et al. 2012:503). Andrijasevic et al. (2012:511; emphasis in original) argue that this attempt at mobilizing sex workers – by bringing a collective of people together in public and producing these two texts – "simultaneously *disrupts* active citizenship by questioning the distinctions [of active and passive citizenship] and leveling the hierarchies embedded in the institution of EU citizenship and by challenging the limitations of citizenship in its territorially or culturally bounded form".

The massacre at the Marikana platinum mine

Unlike protests organized in Europe against the rise in precaritization, resistance movements in the Global South have frequently been confronted with deadly state violence. On 10 August 2012, miners in Marikana, South Africa went on strike as part of efforts to secure safer working conditions, housing and better pay. Describing their conditions, the International Labour Organization reported that despite significant steps having been taken in other industries since the end of Apartheid, miners in South Africa continued to be subject to "a variety of safety hazards: falling rocks, exposure to dust, intensive noises, fumes and high temperatures … Many miners also suffer from diseases such as silicosis and tuberculosis" (ILO 2012). Between the early 1990s and the mid-2000s, South Africa formally adopted a neoliberal macro-economic policy that brought about economic instability and currency crashes in 1996, 1998, 2001, 2006, 2008 and 2011. These crashes increased the number of South Africans in debt by 2.2 million between 2007 and 2009, creating circumstances in which workers, like those at the Marikana mine, "were subject to garnishee orders which reduced their take-home pay to a fraction of their ongoing living expenses" (Bond and Mottiar 2013:287; Bateman 2012). Six days into the strike, South African security forces opened fire on a group of strikers, killing thirty-four and wounding at least seventy-eight miners in a matter of minutes. As Grahm (2016:836) writes, the aftermath of the massacre can only be understood as a "truly Kafkaesque turn of events": the National Prosecuting Authority utilized an Apartheid-era law to charge 270 of the survivors with the murder of their fellow workers who had been killed by the police.

Following the massacre, conflicting narratives about the violence were promulgated by the popular press, 97 per cent of which, Duncan (2013) reports, failed to include accounts from the workers themselves. Recent independent documentaries such as *Night Is Coming: A Threnody for the Victims of Marikana* (Kaganof 2014), *Miners Shot Down* (Desai 2014) and *Mama Marikana* (also known as *Strike a Rock* – Saragas 2017), have sought to tell the story of these striking miners and their families. Given our interest here in precarity and citizen media, *Mama Marikana* is of particular note for the way the film sutures together footage of the miners' protest with a performance by the women of Marikana. The film opens with three scenes being intercut with one another: footage of a group of women singing and performing the event of the massacre, both in private and in public, are spliced together with footage of the massacre itself. "What becomes clear", Lucy Grahm (2016:845) writes, is that the mediated splicing of these scenes deployed by Aliki Saragas, the director, "enables a response that is different from watching direct footage of the massacre".

Here, the combination of community performance and documentary filmmaking highlights performative enactments of mourning, recognition and a refusal to be marked as disposable. There are thus multiple stagings and scales of citizen media at work in this example: a local community's performance, an independent film being distributed at film festivals across the world and marketed online, complex multi-layered filmic representations of a protest against precaritization and debt which turned deadly, and bodies coming together in song, gesture and dance to tell a narrative about struggle and perseverance.

As all three examples demonstrate, citizen media is of central importance in contemporary social justice movements that are responding to and challenging longstanding and exceptional conditions of precaritization and precariousness. Under conditions of precarity, the production of citizen media fosters connections between previously unaffiliated citizens, creating the potential for change and refusal. Citizen media can thus advance a critique of material and symbolic precaritization, rendering the normalization of precarity and precariousness insidious and violent. At the same time, incorporating an analysis of precarity into our understanding of citizen media has additionally shown the concepts of citizen and media to be ambivalent. Centring the processes of precaritization and precariousness into our considerations of both of these concepts demands attending to the subjects who are relegated to the margins of these structures. It means holding onto the unstable yet necessary work that these two concepts do as they are put to use by individuals and communities demanding their right to flourish.

See also: immaterial labour

Recommended reading

Butler, J. and A. Athanasiou (2013) *Dispossession: The performative in the political*, Cambridge: Polity Press.

Writing during the Egyptian Revolution and the Greek financial crisis, Butler and Athanasiou are in dialogue about the forms of dispossession that accompany, and structure, migration, colonialism, homelessness and statelessness. This is a key text for thinking through the ways in which performativity, activism and collective movements can resist state violences globally.

Gill, R. and A. Pratt (2008) 'In the Social Factory? Immaterial labour, precariousness and cultural work', *Theory, Culture & Society* 25: 1–30.

An introduction to a special section of *Theory, Culture & Society*, in which Gill and Pratt argue that the precarity of cultural workers is symbolic of, and symptomatic of, recent transformations in advanced capitalism.

Lee, C. K. and Y. Kofman (2012) 'The Politics of Precarity: Views beyond the United States', *Work and Occupations* 39(4): 388–408.

This sweeping text documents various forms of precarity globally, providing case studies from the Middle East, South Africa, China, Southeast Asia, Europe and the United States. Doing so, the authors argue that any policy interventions in the Global North must account for the ways in which precarity is a global phenomenon, one that disproportionately impacts the Global South.

PREFIGURATION

Marianne Maeckelbergh

The term prefiguration refers to a political practice in which social movements attempt to bring about a desired goal or vision for society by enacting and embodying that vision themselves in their own daily organizing. The principles and values of the desired future society are what guide the movement's own organizational logic. Many political ideals can be prefigured, but the concept of prefiguration has surged in popularity since the turn of the century, mainly due to the rise of social movements, such as the alterglobalization and Occupy movements, that prefigure a form of horizontal politics. Horizontality is both a value and a practice: it refers to the active creation of non-hierarchical relations through fluid, open and egalitarian decision-making processes that are perpetually open to transformation based on critiques from within. Horizontal politics prefigures structures of democracy that are based on decentralized networks instead of nation-states.

Prefiguration is "inherently experimental and experiential" (van de Sande 2015:189) since movements create and try out new political and economic forms *in situ*. These new forms are "both actual and imagined, emergent but unguaranteed" (Bonilla 2010:135) and must be viewed in relation to the context in which they emerge. Contemporary movements' use of prefigurative politics is built upon a political trajectory that draws on centuries of anarchist praxis and on the anti-systemic impulse of the 1960s global wave of social movement organizing. This entry discusses prefiguration in relation to this abbreviated history and through the lens of three interconnected dimensions of the concept: first, conflation of means and ends; second, the relation to radically participatory and egalitarian horizontal politics; and finally, the relation to the state.

Merging means and ends

A key historical source of contemporary prefigurative politics is anarchist theory and practice. Indeed, prefigurative politics is commonly listed among the defining features of anarchist praxis (May 1994; Graeber 2002; Franks 2003; Bowen and Purkis 2004), and although the concept of prefiguration has been used by anarchists to denote a variety of ideas, it remains "a core concept in contemporary anarchist thinking" (Kinna 2016:198). Despite references to prefiguration in anarchist thought and practice dating back centuries, what is meant by the term has changed over the years. Although it was once perceived as perfectly consistent with anarchist ideals to emphasize the political relation between means and ends while also seeing society as a pyramid and maintaining complex systems of power hierarchies (Marshall, P. 1992:629), today prefiguration refers specifically to the creation of internal movement organizational forms that reflect the ideal and goal of creating a less hierarchical society. In practice this means the elaboration of complex systems to ensure egalitarian decision-making.

To understand the significance of a prefigurative strategy of social change, we must place it in relation to historically dominant theories of social change. Traditional left theories of social change rest on the idea of a linear process that involves a future moment of transition when the revolutionary goal will finally be achieved. For the old left, "actions [were] judged by whether they assist[ed] or hinder[ed] the revolutionary goal" (Franks 2003:20), invoking what Franks calls a logic of "Leninist consequentialism" (ibid.:16) that privileges the ends of revolutionary action over the means. In this way, the desired future is perpetually deferred to an unspecified moment in the future after a "period of readjustment" (Yates 2015a:2). Prefiguration, on the other hand, is based on the idea that the ends and means are equally important, and the intention (over time, and temporarily) is to make ends and means indistinguishable by enacting the desired future in the present moment. The theory of social change that is invoked by prefigurative politics assumes that the means – the process of social movement organizing itself – is a key site of political innovation: it is the space where the movement's ideals and goals are formed and where the future society is built. Many recent movements have therefore expended a great deal of energy on the creation of elaborate procedures for egalitarian, non-hierarchical, inclusive and open forms of decision-making within the movement itself.

The contemporary emphasis on internal movement democracy came to hold a central place in prefigurative politics through a gradual transition over time, but a key turning point was the movements of the long 1960s (1956–mid-1970s). These movements were characterized by a shift in social movement praxis away from linear, programmatic, unitary theories of social change as embodied in the modernist paradigms of both Communist theory and development theory, towards more experimental, diverse and open practices of social change. The new left emerged as a key political force that defined itself specifically in contrast to the old left and its hierarchical, programmatic theory of social change (Horn 2007). The new left emphasized notions of participatory democracy and the pursuit of multiple movement goals, was sceptical of authority, was anti-systemic, and rejected the idea of a centralized, unitary, predetermined revolutionary programme (Horn 2007; Maeckelbergh 2011; Miller 1994; Polletta 2002; Wallerstein 1989).

Horizontal prefiguration, specifically, has its predecessor in the 1960s movements' attempts to reinvent democracy by elaborating theories of participatory democracy. The emphasis on prefiguration, however, emerged more out of the critiques of how this participatory democracy was being embodied than it did out of the articulation of an ideal. The critiques offered by women and minorities within the movements of the 1960s pointed out how hypocritical the movements were to demand participatory democracy and equality for all, while also, for example, expecting women to cook, clean, make tea and take minutes of meetings (Evans 1980; Mansbridge 1983:21; Gitlin 1987/1993:362–376; Miller 1994:204–206). These critiques led to an elaboration of movements' understanding of what equality meant and how to ensure it, as well as movements' understanding of the relationship between their goals and their everyday practice. It is mainly from these, mostly feminist critiques that the idea emerged that the means used to organize within the movement needed to reflect the ideals desired outside the movement if those ideals were to fulfil their political potential.

Due to these critiques from within the movements themselves, participatory democracy transitioned from a theory of how democracy should work into elaborate egalitarian decision-making processes (Graeber 2009; Horn 2007; Polletta 2002). These participatory processes were improved upon in the following decades, by the feminist movements, anti-nuclear movements, peace movements, DIY (do-it-yourself), and autonomous movements of the 1980s that were characterized by anti-authoritarianism, independence from political parties, decentralized organizational forms and an emphasis on direct action (Katsiaficas 2006:3; Grauwacke 2004). The lifestyle politics of DIY movements invoked prefiguration as a mode of social change in one's everyday life beyond confrontational protest – a form that persists today (McKay 1998; Power 2016; Yates 2015b) despite being underrepresented in the literature. Everyday prefiguration takes many forms but includes alternative educational projects (Amsler 2015; Trott 2016), alternative food provision networks (Cucco and Fonte 2015), experiments in work place autonomy (Kokkinidis 2015) and various citizen media initiatives (Kulick 2014).

The autonomous movements, the anti-nuclear movement, the feminist and peace movements were particularly important because they developed procedures for consensus-based decision-making, including the spokescouncil structure for decision-making, which is a large assembly where (individuals from) smaller groups come together to coordinate action. In the 1990s, these decision-making practices merged with the democratic impulse inspired by the *encuentros* of the Zapatistas in Mexico and similar assembly-based projects throughout Latin America (Dinerstein 2015). The Zapatistas' poetic notions of "walking we ask questions" (quoted in Holloway 2004) and "the question everywhere is the same, but the answers may be different" (Marcos, quoted in Flores and Tanaka 2001) perfectly capture the open-ended form of horizontal politics that continued into the alterglobalization and Occupy movements. This horizontality envisions that the goals of movement organizing will come from the process of organizing itself; that these goals will be different from one location to the next; and that they will remain open to intervention and, therefore, transform over time. With the global rise of the alterglobalization movement in the early 2000s, these consensus-based structures and procedures for participatory democracy came to dominate social movement organizing practice in Europe and North America – and elsewhere, to varying degrees.

Prefiguration in the alterglobalization and Occupy movements

The alterglobalization movement came into prominence in November 1999, when tens of thousands of protestors descended on Seattle and successfully blockaded the summit meeting of the World Trade Organization, bringing this global power to a halt for a few days and derailing its economic agenda significantly in the long run. In the years that followed, people around the world would mobilize to protest nearly every summit meeting of the world's leaders. Among the targets were the Group of Eight (G8), the World Bank/International Monetary Fund (WB/IMF) and the North Atlantic Treaty Organization (NATO). In addition to the anti-summit mobilizations, the movement would organize large-scale gatherings which they called social forums to exchange ideas, experiences and strategize common action. The alterglobalization movement employed elaborate decision-making procedures for explicitly horizontal prefiguration (Graeber 2002; Juris 2008; Maeckelbergh 2009).

Although this movement stole headlines with its spectacular confrontations with state power, these events were often viewed by movement actors as secondary. "For me", one of the participants in the London South East Assembly Meeting (28 May 2005) stated at a public meeting attended by the author, "it's not so important what happens at the summit, but more what we create, how we organize. The ecovillage will become a model of how we want to organize our world ourselves, a new way of working together". At each of these summit protest events, the activists would create elaborate 'villages' to house those coming from around the world to protest. Rather than being viewed as a practical matter, this housing was perceived as a chance to create, in the current moment, the type of future society they desired. These villages would include self-organized kitchens, ecologically sustainable compost toilets, medical centres, spaces for children and an elaborate media infrastructure. The alterglobalization movement created Indymedia – a platform for user-generated news content. The first Indymedia centre (IMC) was set up during the WTO protests in 1999 at a time when the Internet was still in its infancy and social media did not yet exist. Indymedia centres coordinated the dissemination of news through multiple platforms: the web, public access TV, community-based radio and a daily print publication, and rapidly became a popular model for numerous media projects that help citizens "become the media" and embody "a model for more democratic communications through the development of alternative media structures, practices and content" (Kulick 2014:366). In the process of organizing a protest mobilization, therefore, the alterglobalization movement did much more than protest – they developed enduring structures of democratic participation, from meeting procedures to media platforms. Boéri and Hodkinson (2004), two activist scholars active within the Social Forum organization process, describe the importance of prefigurative politics for Babels, the network of volunteer interpreters and translators that made the Social Forum possible: "the real story of Babels lies in its embodiment of the innovatory but difficult process of 'pre-figurative politics'", they point out; "By attempting to put into practice the principles of solidarity, pluralism, equality, and horizontality, Babels is creating not only alternative systems and practices to free-market capitalist society, but also the social counter-power needed to defend and embed them permanently".

Prefiguring horizontality in the alterglobalization movement was always a conflictive process – many actors rejected a prefigurative approach to social change, and those who advocated it often found themselves in a position of repeatedly pointing out the movement's own shortcomings. Nevertheless, when the occupation and assembly-based movements – from the Arab uprisings to Occupy – swept across the world in 2011 (and after), many scholars noted the importance of prefiguration to some aspects of these new mobilizations (Baker 2016; Graeber 2013; Juris 2012; Lin and Liu 2016; Maeckelbergh 2012a, 2012b; Polletta and Hoban 2016; Razsa and Kurnik 2012; van de Sande 2015). For example, Occupy Wall Street (2011) drafted principles of solidarity, of which the first four were: engaging in direct and transparent participatory democracy; exercising personal and collective responsibility; recognizing individuals' inherent privilege and the influence it has on all interactions; and empowering one another against all forms of oppression.

Just as in the movements of the 1960s and the alterglobalization movement before them, however, the importance of prefiguration to the Occupy movements presents itself mainly in the form of critique that the movement is reproducing power hierarchies and discrimination. In Occupy Oakland, a group emerged that referred to itself as Decolonize Oakland: it called upon the Occupy movement to recognize that "any movement that doesn't confront the continuing force of colonization, patriarchy, hetero-normativity, and white supremacy replicates these oppressions" (Decolonize Oakland 2012). This is an example of an important principle of horizontality, namely that structural inequality must be actively confronted in order to avoid reproducing it. At Occupy Wall Street in New York City, many people referred to the centralized structure of the General Assembly, not as a liberating, horizontal and participatory space, but rather as bureaucratizing the movement. Consequently, a critique emerged of the fetishization of the General Assembly and of the prioritization of horizontal form over content. Many movement actors tried to reintroduce networked decentralization into their political structures by setting up meetings of neighbourhood assemblies or spokescouncil meetings consisting only of members from the different working groups, but these new structures never fully resolved the problems. In December 2011, Occupy Portland circulated a proposal to "make changes that will facilitate more space for open discussion, connection and affinity building, instead of process heavy, decision making meetings that sometimes stifle new ideas that need a more open space to foster". This was an attempt to diversify the political process and alleviate the feeling many people had that the General Assembly stifled one's ability to act. These critiques did not result in a full rejection of horizontal politics, but they did challenge the idea that merely creating a horizontal decision-making process – structures without politics – would be enough to bring about a radically non-hierarchical political practice.

Prefiguration vs. state-driven social change

Horizontal decision-making is directed not at taking power, but at changing the way power operates. Part of the process of changing the way power operates involves changing the way problems and solutions are envisioned. In horizontal decision-making, the ideal future remains open to input from participants, through collective dialogue. A transformation of this magnitude to the way people commonly do politics requires a transformation of existing institutions and therefore cannot be brought about by the politicians currently operating within these institutions, nor can it be created by taking control of the existing institutions. Horizontal prefiguration draws on anarchist ideas and practices in its emphasis on the creation of a political practice that unfolds outside of, and in some cases explicitly against, the state. For some, the ultimate aim of horizontal prefiguration is to create forms of political, social and economic organization that – theoretically, in the long run – would come to replace the state. As Sitrin, an activist scholar active in both the alterglobalization movement and Occupy Wall Street, argues, prefigurative movements are "movements that are creating the future in their present social relationships ... social change isn't deferred to a later date by demanding reforms from the state, or by taking state power and eventually instituting these reforms" (Sitrin 2006:4).

Prefiguration's emphasis on social relationships as an important source of power – as opposed to emphasizing the state as the key political actor – has led to much critique being levelled against social movements that employ a prefigurative strategy. These movements are often dismissed as not being 'for' anything, as having no strategy at all, as having no clear demands, as being ineffective. However, these critiques ignore two realities of movement organizing. Firstly, that the goal of horizontal democracy, of a society without hierarchy, necessarily cannot be brought about through the hierarchy of a state apparatus unless that state is prepared to eliminate itself (quite unlikely), and secondly, that employing a prefigurative strategy does not preclude a movement from having multiple types of goals, many of which involve demands being made of the state. A prefigurative process can even be used to formulate such demands. The Occupy movement had many clearly articulated demands which existed in harmony with the aim of transforming all of society, but movement actors often intentionally chose not to emphasize the demands as a way of challenging the premise of such critiques. As one participant active in the movement argues in the film *Occupy Wall Street: One Year Later* (Jourdan and Maeckelbergh 2012),

> [f]or people that think that economic, political and social justice is reducible to a set of demands, they're just wrong. This is a structural issue that is happening globally and it is a crisis of capitalism. We want to ask better questions and be in dialogue about what the answers are going to be but we're in no rush.

The exact articulation of prefigurative politics in relation to the state apparatus varies from one movement to the next, largely depending on each movement's ultimate and immediate aims. Swatuk and Vale state that in the case of the #Feesmustfall student movement in Southern Africa, despite a number of activities "demonstrating features of prefigurative politics" (2016:332), the students often had "direct engagement with the established centres of power for specific ends" (ibid.:343), and that "prefiguring the 'new' – a better life for all – requires hanging on to the old, that is, a government ... continually packaged as the only means to overcome race and class based inequalities across the country" (ibid.:337). In the case of the urban feminist organization City for All Women Initiative/Initiative: une ville pour toutes les femmes (CAWI/IVTF) in Ottawa, Siltanen et al. argue that prefiguration can be understood as "a resource for feminist organizing that facilitates meaningful political engagement with the local state by providing experiences of alternative ways of being together and doing politics, and by helping to mitigate the risks of co-optation" (2015:263). Cooper offers a unique addition to the literature on prefiguration-state relations in her study of 1980s British municipal radicalism, suggesting that this movement represents an attempt to "prefigure the state" by "fashioning a new state conception" (Cooper, D. 2017:340) through a "radical vision of pluralism" at the municipal level (ibid.:341), and thus raising the question of whether the state can be re-imagined in a more plural form and then prefigured from within itself.

Future directions: Confrontational prefiguration?

With increased scholarly and activist interest in the practice of prefigurative politics, many have begun to ponder the question of what the relationship might be between prefiguring an alternative society and the act of protest itself. Prefiguration is often described as belonging to the process used to plan a protest, but is not theorized as part and parcel of the act of confrontation that protest often involves. In this narrative, prefiguration is relegated to spaces that are carved out as safe, even if taken by force initially. Today, some of the most interesting acts of prefiguration are deeply interconnected with the act of confrontation itself, especially confrontations that are embedded in people's everyday lives. This calls for new analyses that explore the "crossfertilization of prefigurative and contentious politics" more deeply (Tadros 2015:1345) and that engage the embodied dimension of prefiguration to explore "the linkages between bodily and structural violence" (Lin et al. 2016:309), linkages that are necessarily "deeply raced, gendered and classed" (ibid.:315). The Black Lives Matter (BLM) movement in the United States, for example, emerging from the quotidian life-threatening violence of police shootings, prefigures the empowerment of black people in part by creating organizing structures that promote black leadership. Tadros (2015) examines the prefigurative aspects of the anti-sexual harassment movements in Egypt between 2011 and 2013 when women faced violent sexual assaults, including rape, in the form of mob attacks during street mobilizations (Maeckelbergh and Jourdan 2013). Tadros argues that those who created patrols to intervene when such mobs formed were not only fulfilling a crucial instrumental role, but were also prefigurative of

> a social reality in which women were free to exercise their right to voice their political opinions without fear of assault. They did not act in order to call upon the government to change its security practices, rather they self-organized to create safe spaces for themselves.
>
> *2015:1355*

Similarly, the Kurdish movement for democratic autonomy has grown to iconic status among movement actors internationally in part because of its insistence on creating a decentralized political system that actively promotes the empowerment of women and the inclusion of multiple religions and ethnicities. These new political structures are being forged in the midst of full blown wars. The result has been an articulation of the prefiguration of democratic autonomy with elaborate structures and ideologies of self-defence, including women's self-defence.

Under the Donald Trump administration in the US, the threat of violent attacks from the alt right, far right, neo-Nazi and fascist groups has led many of those previously involved in Occupy and similar movements to reconsider the importance of community self-defence as a political imperative. Confronting fascists in the street might not seem prefigurative at first sight (Franks 2014); and the same can be said of the anti-sexual harassment patrols in Egypt. But in all these struggles for collective self-defence we see a prefigurative desire first, to take matters into one's own hands rather than relying on the state, and second, to reproduce in the social relations of the physical conflict the movement's egalitarian ideals. Each of these movements articulates a relationship between prefigurative politics and a "politics of survival" (Lin et al. 2016:314) that was not present in the alterglobalization movement, and not present in the same way in most Occupy-era movements. The particular configuration that each of these movements invokes of the relationship between confrontational self-defence and prefiguration is a fruitful area for future research that could potentially transform many of the assumptions we hold about what prefigurative politics is and how it can bring about social change.

See also: activism; autonomous movements; direct action; diversity; political science and citizen media; process vs. event; social movement studies and citizen media; solidarity; World Social Forum

Recommended reading

Graeber, D. (2002) 'The New Anarchists', *New Left Review* 13: 61–73.

Outlines the importance of prefigurative politics for the globalization movements of the 1990s and early 2000s, and offers examples of the democratic practices and structures that were being prefigured within the globalization movement, and the transformative effect of prefigurative politics on those involved.

Maeckelbergh, M. (2009) *The Will of the Many: How the alterglobalization movement is changing the face of democracy*, London and New York: Pluto Press.

A full-length ethnography of how prefigurative politics was practised and contested within the alterglobalization movement in the early 2000s. It explores the elaborate procedures of democracy within movement organizing processes to show that the horizontal politics being developed by the movement can provide important insights for our understanding of democracy in general, especially in a globally interconnected world.

Polletta, F. (2002) *Freedom Is an Endless Meeting*, Chicago, IL: University of Chicago Press.

Explores movement practices of participatory democracy to show that social movements have been using bottom-up decision-making practices to organize effectively for decades. Through a detailed historical analysis, it describes the complexity of participatory democracy within social movements throughout the twentieth century, showing the internal contradictions that such processes bring forth. The history of participatory democracy provided is essential for an understanding of how prefigurative politics is practised in social movements today.

Yates, L. (2015) 'Rethinking Prefiguration: Alternatives, micropolitics and goals in social movements', *Social Movement Studies* 14(1): 1–21.

Surveys the literature on prefiguration up to 2014 to argue that prefigurative politics combines five processes: collective experimentation; the imagining, production and circulation of political meanings; the creating of new and future-oriented social norms; their consolidation in movement infrastructure; and the diffusion and contamination of ideas, messages and goals to wider networks and constituencies. Particularly interesting for its literature review, but also draws on original research within autonomous social centres in Barcelona.

PROCESS VS. EVENT

Carlie D. Trott

Despite their interdependence, the concepts of process and event are often distinguished from one another in social movement studies. Though both consist of various spatiotemporally located actions, as ontologically distinct categories (Steward 1997:94–97) process and event offer dual lenses through which to view and interpret the field of citizen media. Where events are conceptualized as discrete, temporally bounded, externally directed products of collective struggle, processes encompass their continuous, ever-shifting activities and underlying democratic potentialities. Processes and events form the complex building blocks of movement activity and serve as analytical tools which frame approaches to, and with which to make sense of, the lived experiences and actions of "unaffiliated individuals and collectives as they reclaim public and digital spaces in the pursuit of non-institutionalized agendas" (Baker and Blaagaard 2016b:1). Events, as unitary entities, may consist of particular projects or products in citizen media, whereas processes, as fluid arrangements, encompass a diverse range of ongoing activities and practices that surround – and make possible – the production and circulation of content (Maeckelbergh 2009; Stephansen 2016). Processes and events are thus structurally complementary (Mayo 1961) in the citizen media landscape, as together they form an integrated whole.

Though the process–event distinction is useful in tracing the history and contours of citizen media scholarship, the boundaries between process and event can be blurry. Due to their mutually constitutive nature, not only are these concepts bound up with one another in theorizing and research, but distinguishing process from event is rarely a straightforward exercise even in practical contexts, where events are often only understood as such in retrospect. Parsing process from event is especially thorny in citizen media initiatives that practise prefigurative politics, where the methods employed by political actors reflect the goals of the movement – that is, where activists aim for means-ends consistency (Kulick 2014; Yates 2015a).

Process and event as concepts and themes in citizen media scholarship

The process–event distinction has been theorized and examined across a range of disciplines, most notably in philosophy, linguistics and the cognitive sciences (Gill 1993; Steward 1997). In their most basic form, events are conceptualized as discrete entities of occurrence, or "things that happen", according to the definition in Merriam-Webster (2018). Being a somewhat amorphous category on their own, events are often conceptualized relationally in scholarship. That is, rather than being defined in and of themselves, events are identified in direct relationship or opposition to other, more familiar metaphysical categories (Casati and Varzi 2015). For example, events take on shape and meaning as they are ontologically contrasted with other units such as objects, facts, properties or times. Events are also defined spatiotemporally, by not being processes. To qualify as such, events, in their most fundamental philosophical sense, need not be spectacular. For example, an event, defined as something that happens, can be as ordinary as a sunrise or a smile. However, in the study of social movements in general, and citizen media in particular, events of interest follow a more restricted application of the term. Specifically, an event of scholarly significance is usually a "noteworthy happening" (Merriam-Webster 2018).

Events of scholarly significance in citizen media tend to be discussed along two dimensions. First, events may refer to specific actions by citizen media actors, such as carrying out campaigns, protests or performances (Baker and Blaagaard 2016b). As units of analysis in scholarship, these types of event can offer a useful means by which to segment and dissect extended arcs of struggle. For example, in order to document and examine divergent political cultures within the alterglobalization movement, De Angelis (2005) analysed one event, namely the European Social Forum (ESF) held in London in October 2004. Alternatively, events in citizen media may refer to specific incidents of regional, national or global significance – embedded within broader sociocultural, political, economic and/or environmental realities – which provide context and fodder for citizen media activities. Along this second dimension, events may offer a focused arena in which to identify and explore diverse examples of citizen media in action. Examinations of the citizen-led circulation of firsthand information about global events via digital technologies – cell phone photography, video, social media – offer numerous examples, including Cross' (2016) examination of these technologies during the 2005 London bombings, Blaagaard and Allan's (2016) analysis of an Israeli air strike on Gaza in 2014, and Chouliaraki's (2016) discussion of the use of digital media throughout the post-Arab Spring conflicts in Libya and Syria. Events are often thought to be the unidirectional outcomes of processes, but as these case studies indicate, events can be novel occurrences that serve as incubators for creativity, learning and sharing, leading to insights or changes in process.

Relative to events as noteworthy happenings which transpire from time to time in social movements, processes refer to "continuous activit[ies]" or "something which is *going on through time*" (Steward 1997:95; original emphasis). The event–process distinction is implied in a functional definition of process as "a series of actions or operations conducing to an end" (Merriam-Webster 2018), whereby processes – as procedural means or intervening modes – are conceptualized in relation to consequential products or outcomes. Like events, processes are said to occur or take place, though unlike events, their spatial and temporal boundaries are less clear. For example, it seems appropriate to ask where and when a demonstration took place, but not where or when collective agency occurred (Stapleton and Froese 2015). On the other hand, processes can be "continuous, ongoing, constant, incessant, perpetual, unremitting, sporadic, intermittent, irregular, [or] steady" (Steward 1997:98), while events are less straightforwardly so. This is because events take place in temporal instants, or over demarcated intervals, whereas processes take place or unfold continuously over time. Processes and events alike are spatiotemporally located, proceeding through various stages at different times though in distinct ways. As their less visible support structures, processes are thought to underlie and animate events.

In citizen media, processes consist of ongoing activities by political actors that envelop, guide and constrain decisions, dialogue and action in movement contexts. Although processes (as continuous activities) may theoretically refer to any number of contextual goings-on within citizen media spaces, processes of scholarly significance typically fall into a few broad, intersecting and simultaneously occurring categories. A first category covers media-centric processes, or those involving the ongoing nuts and bolts activities that amount to the production and dissemination of self-produced media content, for example via blogging, hacking and performing. A second category, surrounding and making possible the production and dissemination of content, is citizen-centric processes, or those taking place within and between individuals and groups involved in citizen media. These relatively less visible processes concern the internal dynamics within movement contexts – communication, decision-making, relationship-building – and comprise intra- and interpersonal processes of development, negotiation and adaptation. Finally, a third category, which considers the range of possibilities associated with citizen media as a political endeavour, consists of change-centric processes. These are rooted in the transformative potential of citizen media to alter social, cultural or political realities. For example, the diverse activities that citizen media actors engage in are often considered to instantiate an array of democratic potentialities grounded in visions of a more just society. This broad set of processes filter through all aspects of citizen media, and variously overlap in both meaning and time.

A closely-related concept to process in citizen media is that of practice. At each process level – media-, citizen- and change-centric – there is a diversity of associated practices undertaken by unaffiliated individuals and groups as they generate and circulate media content. For example, practices may include core activities culminating in the creation and distribution of content, while others encompass "practices oriented *towards* citizen media – from training programmes for reporters to participatory production processes to neighbourhood screenings" (Stephansen 2016:28; original emphasis). Like processes, practices refer to "continuous" and/or "repeated" activities (Merriam-Webster 2018), though unlike processes, a more inclusive term, practices consist exclusively of modes of doing. As such, practices represent a key dimension of citizen media process, and consist of socially situated and often goal-directed actions.

Explorations of process in citizen media are typically embedded within event-based studies, with events serving as narrative anchors or contexts within which to examine individual and collective processes. At the same time, case studies focused on specific spaces for the self-produced or collective production of oppositional media may foreground process over product as a means to emphasize the significance of the former to the latter (Juris et al. 2013). For example, Kulick (2014) explored the strategic dilemmas faced by young people as they managed an independent media outlet, finding that internal tensions such as conflicting perspectives and external limitations relating, for instance, to resources simultaneously exerted pressures that shaped the collective process and its creative outputs. Other examples include the examination of communication practices in the World Social Forum (Stephansen 2016), learning processes around consensus-based decision-making during the Occupy movement (Szolucha 2013b), and internal contrasts and debates around the practice of democracy, inclusivity and participation in the ESF process (De Angelis 2005).

Beyond serving as a means to classify and describe the various dimensions of citizen media action, the process–event distinction also provides a framework for understanding the history and contours of scholarship on social movements. In citizen media, as in other forms of resistance, disproportionate scholarly attention has been paid to the tangible, visible, outwardly facing products of social movement struggle, rather than to the more abstract, less visible, internal dynamics of action that facilitate or constrain the former (Goodwin and Jasper 2004; Haiven and Khasnabish 2014). Most often, this translates to analyses of specific events rather than ongoing processes, despite the practical inseparability of the two.

This preoccupation with publicly facing dimensions of citizen media appears to be less rooted in scholars' outsider status relative to social movements, as many scholars are active in the social movements they study (De Angelis 2005; Stryker et al. 2000), and may have more to do with the power of language in shaping the ways in which humans understand and interpret the world. The process–event distinction is rooted in "dedicated linguistic devices (e.g., verb tenses) [that] are tuned to events and event structures, as opposed to entities and structures of other sorts" (Casati and Varzi 2015:1; Steward 1997). Scholars are, after all, human, and our "ability to think about the temporal and causal aspects of the world *seems to require* parsing those aspects in terms of events and their descriptions" (Casati and Varzi 2015:1; emphasis added). With this explanation in mind, it seems almost intuitive that events hold greater appeal as the primary framing device in social movement scholarship. As individual happenings, events are conceived of as concrete, temporally and spatially located entities organized into part–whole hierarchies. As such, they can be counted, mapped, measured, contrasted, labelled and variously interpreted and described. For events, but not necessarily processes, there is a clear 'how many?', 'when?', and 'how long?' (Steward 1997). These characteristics make events more attractive as objects of study because they cohere as phenomena with beginnings and endings, successes and failures, structures and locations, compared to the relatively amorphous processes that run through or permeate all of these same aspects. In increasingly quantitatively oriented academic spaces, processes may be overlooked as "activit[ies] of which there can *be more or less*, but not *one or two*" (Steward 1997:96; original emphasis).

Since the turn of the century, however, the everyday, temporally extended social practices and tensions within social movements have garnered increasing attention (Yates 2015a). Citizen media events are increasingly understood as physical and digital extensions – or public expressions – of ongoing, deliberative processes, rather than as spontaneous eruptions of creative resistance. As a result, citizen media processes such as networking and collaboration are accredited with simultaneously shaping the tangible outputs – i.e. media content – created and disseminated by unaffiliated individuals as well as the democratic potential of various arrangements (Juris et al. 2013; Stephansen 2016).

Attending to process in citizen media has opened up new possibilities for, and in many cases has necessitated, the use of qualitatively different frameworks for understanding and assessing citizen media activities. Evaluating movement success or failure offers one example: beyond the clear bounds of noteworthy happenings such as campaigns, demonstrations and performances, what qualifies as success is less straightforward because processes such as creativity, cooperation and discourse are ongoing activities within movement contexts, often without clear beginnings or endings. As such, judging process-based achievement or impact requires different perspectives, and a new language tied to appropriate markers. In describing the broad spectrum of processual victories that should be recognized and celebrated as such, De Angelis emphasizes everyday wins in processes of communication, relationship-building and cultivating spaces of empowerment, in addition to achieving larger movement goals such as "major concessions of the state" (2005:203–204).

With greater attention to processes, events within citizen media are increasingly recognized as points along a continuum: not only as products of preceding activities, but as sites of process-oriented (re)generation. Moreover, understanding events as rooted in diverse citizen media processes more realistically identifies noteworthy happenings as punctuation marks dotting much lengthier narratives of ongoing struggle, which are in turn rich in historical, technological, structural and social storylines. As articulated by Stephansen, adopting a practice-based approach to citizen media scholarship "highlights the *social* foundations of publics, by bringing into view the broad range of socially situated practices that form around – and are necessary for – the production and circulation of media content" (2016:37; original emphasis). In brief, by attending to process as well as product, such approaches render a more comprehensive view of the citizen media landscape as made up of diverse and intersecting processes in which events are firmly rooted.

Process vs. event and prefiguration

Beyond their practical inseparability, the boundaries between process and event are further blurred in citizen media initiatives guided by the prefigurative ethos, whereby the medium itself embodies the emancipatory message of movement actors (Trott 2016). Simply stated, in prefiguration process is "a goal in and of itself" (Maeckelbergh 2009:83). Those engaged in prefigurative politics offer a vision of what social change looks like "within the live practice of the movement" (Breines 1981:6; Futrell and Simi 2004). From this perspective, it could be argued that citizen media is inherently a prefigurative endeavour.

In prefigurative spaces, the process–event distinction fails to meaningfully cohere insofar as the process itself instantiates the social change aspirations of those engaged in social movements (Kulick 2014). In prefigurative settings, it could be said, the process *is* the event, thus raising the question of whether processes are causal agents of which events are effects or vice versa. The very praxis of prefiguration rests on a conundrum faced by philosophers – specifically that process and event may be considered "two ways of looking at the same thing" (Steward 1997:91). As Steward argues, "[l]ooking back over any given period of time … it is not instantly obvious that happenings of two entirely distinct varieties were occurring throughout", but rather "[a]ll we *really* have, it might be said, are events succeeding other events in time – and different ways of looking at those events" (ibid.:95). As such, events and processes can be distinguished in retrospect, but not in lived moments. From a prefigurative standpoint, the world consists of various processes which inevitably shape specific occurrences, from daily interactions to global events, throughout history. In any scenario, the ends are inherent within the means. To offer the prefigurative counterpoint to Steward's critique, all we really have are processes.

To complicate matters further, studies of prefiguration tend to emphasize the complex, ever-shifting nature of social processes within movements, whereby prefigurative processes are never fixed but instead are subject to modifications in response to specific encounters, individual learning and a given groups' internal dynamics. As De Angelis explains in his attempt to document the process-oriented clash between two groups participating in the ESF (the horizontals and the verticals), horizontality and verticality are not "states of being" but "*modes of doing*" (2005:195; original emphasis). Thus, the social processes distinguishing the two groups were not "a static set of procedural rules" to be followed uniformly by adherents, but rather they were mutable, performative modes that "develop and emerge from the interacting agents themselves" (ibid.). In prefiguration, this perpetual cycle of process-oriented modification and adaptation has been referred to as "permanent provisionality" (Fielding and Moss 2011:155), in recognition of the fact that "the struggle for social change is not a horizon *event*, but rather an ongoing *process*" (Trott 2016:270; emphasis added).

Future directions

The field of citizen media is made up of a multitude of events and processes that are intertwined and mutually reinforcing. As such, distinguishing process from event in citizen media scholarship and practice is often not as straightforward as their respective definitions might imply. Given that processes permeate all aspects of events – as they precede, mediate and follow discrete actions – they are not independent components. Further, there is no inherent relationship of directionality between the two: processes may give rise to, obstruct or give meaning to a particular event, just as they may result from, react to or cease at the end of an event. In theory as in citizen media practice, events derive their very existence from continuous, complex creative processes. As mutually constitutive entities, neither exists without the other. More importantly, neither exists beyond interpretation.

Despite their interdependence as constructs with boundaries that are at times ill-defined, the process–event distinction can be instructive. For scholars of citizen media, distinguishing process from event can be useful in classifying and describing various movement activities. Moreover, as analytical tools, the categories of process and event are useful in segmenting and examining themes in citizen media scholarship. For example, despite increased attention to process in citizen media studies, some processes have received greater attention than others. Citizen media scholars have historically focused on the production and circulation of content, and the democratic possibilities inherent in the use of alternative platforms to challenge dominant discourses, over the socially situated practices that surround citizen media (Stephansen 2016). To comprehend, replicate and build upon influential events that have taken place within the diverse field of citizen media, one must look to the full range of processes underlying them, and indeed creating the conditions for the possibility of their existence.

See also: activism; autonomous movements; media practices; prefiguration; social movement studies and citizen media

Recommended reading

Angelis, M. de (2005) 'PR Like Process! Strategy from the bottom-up', *Ephemera* 5(2): 193–204.

Offers an examination of ongoing processes – the internal dynamics and relations between movement actors – within the boundaries of an event, the 2004 European Social Forum. In contrasting the political cultures of so-called horizontals and verticals, the author draws connections between social processes taking place within the alterglobalization movement, the language used to ascribe meaning to movement processes, and the need for alternative frameworks to better align processes with movement goals.

Stephansen, H. C. (2016) 'Understanding Citizen Media as Practice: Agents, processes, publics', in M. Baker and B. B. Blaagaard (eds) *Citizen Media and Public Spaces: Diverse expressions of citizenship and dissent*, London: Routledge, 25–41.

Shifts focus from the content of citizen media to the many and varied practices that develop around its production and dissemination. In the language of process vs. event, the focus on socially situated practices makes visible the often hidden, and less often studied, processes underlying citizen media events as well as the democratic potentialities – for example, for agency, formation of publics – that arise in their wake.

Steward, H. (1997) *The Ontology of Mind: Events, processes, and states*, Oxford: Oxford University Press.

Provides an introduction to the theoretical distinctions between mental events, processes and states from the branch of philosophy known as the philosophy of mind. In questioning and clarifying the event–process distinction, language is asserted as central to our retrospective discrimination between these two categories of happenings.

PUBLIC SPHERE

Petros Iosifidis

The public sphere, as articulated in particular by German political theorist, philosopher and sociologist Jürgen Habermas (1962/1989), is a space for rational and universalist politics that is distinct from both the state and the economy. It is a scene of activity in which people are addressed as citizens, as rational political beings, and not merely as consumers. In other words, the public sphere provides a space in which public communication may be conducted. Public communication comprises "those processes of information and cultural exchange between media institutions, products and publics which are socially shared, widely available and communal in character" (Ferguson 1986:ix).

The concept of the public sphere constitutes a central analytical tool for making sense of the relationship between the media and democracy, particularly in terms of civic engagement. In his *Structural Transformation of the Public Sphere*, originally written in German in 1962, Habermas explained that in the late eighteenth century a new political class, the bourgeoisie, came to the fore, especially in Britain; this political class formed a public body which, in sharp contrast with the old authorities, notably the state and the church, provided the conditions for the development and dissemination of reason-based, public opinion. The creation of a network of institutions by the bourgeoisie within civil society, especially the launch of a number of newspapers, provided the means through which private thoughts could become public. The printed press, libraries, coffeehouses and universities became places for public debate and criticism of government policy. The new public sphere was in principle open to all and was protected from the power of both the church and the state.

However, in his historical analysis of the evolution of the public sphere, Habermas argued that the space for rational and universalist politics created by the capitalist market was damaged by both the extension of the state and the growth of monopoly capitalism. The formation of large private institutions – in particular, advertising agencies and public relations firms – and the deals they made with each other and with the state while excluding the public, led to the replacement of rational public discourse by power politics. As Habermas argues, these large organizations "strive for political compromise with the state and with each other, excluding the public sphere whenever possible" (1974:54). He thus concluded that "the idea of the public sphere, preserved in the social welfare state mass democracy, an idea which calls for a rationalization of power through the medium of public discussion among private individuals, threatens to disintegrate with the structural transformation of the public sphere itself" (ibid.:55). The role of the media has been central to the replacement of what Habermas terms the "ideal speech situation" (Habermas 1983/1990:86) by conditions of distorted communication. The ideal speech situation that enables rational public opinion takes place under conditions where all citizens have access to information and there is freedom of expression and publication of opinions about matters of general interest, which implies freedom from economic and political control (Habermas 1962/1989:27, 1964/1989:136). In Habermas' view, whereas the development of the independent press at the beginning of the nineteenth century had opened up the possibility of rational public debate and public decision-making on political and judicial matters, it later came to function increasingly as a tool for managing and manipulating public opinion. Indeed, Habermas and his many followers since have argued that the public sphere has now principally become a platform for advertising and public relations (Webster 2006; Schlesinger 2009). Around the world, control of the news media is used to reinforce the power of autocratic regimes and to deter criticism of the government by independent journalists. This may be achieved through official government censorship, state ownership of the main radio and television channels, legal restrictions on freedom of expression and publication (such as stringent libel laws and restrictive official secrets acts), limited competition through oligopolies in commercial ownership, and the use of outright violence and intimidation against journalists and broadcasters (Sussman 2001). Referring to the Internet and social networks era, Papacharissi (2010) argues that new, online and personalized technologies enabled citizens to expand the scope of their social activity, thereby rearranging the boundaries between public and private spaces. As discussed below, her work provides a fresh angle with which to examine the circulation of information among the public in the digital age and calls for a blurring and redefinition of the boundaries between the public and private space.

Assessment and critique

Despite these developments, Habermas' theory merits consideration because he carefully conceptualized the nature of the public sphere, viewing it as an achievement of the new bourgeois (or capitalist) class in Europe, and an outcome of this class's successful struggle against feudalism and church or state oppression. At the same time, however, Habermas' thesis has been questioned on historical grounds. Many argued that he idealized the early period of history to which he referred, and particularly questioned the notion that the eighteenth-century press was politically independent (Mortensen 1977; Hohendahl 1979; Curran 1991a, 1991b). Koss (1981, 1984), in his analysis of the British political press, contended that political control by proprietary interests was in fact exercised over a large part of the press from as early as the eighteenth century. Koss' analysis showed that the early British press was not independent to the extent that Habermas described, and therefore, one could argue, did not contribute to freedom of expression to the degree suggested by Habermas.

Habermas has also been criticized for his argument that the public sphere enabled rational debate. Curran (1991a:35) suggests that "the newspapers celebrated by Habermas were engines of propaganda for the bourgeoisie rather than the embodiment of disinterested rationality". Dutton (2007:13) has similarly found the concept of the public sphere productive, but "too closely tied to a romantic view of the past". In today's global multicultural society, criticisms of Habermas' ideal public sphere could additionally include its universalizing angle and apparent neglect of difference, its emphasis on the national rather than global space, and its normative understanding of a unified national sphere signified and constituted by the media. According to Fraser (2007), there is little consensus on what constitutes a common good, nor can any values truly be described as universal: rather, cultural values are increasingly shaped by exchanges between various local, national and transnational actors, including the state, corporations, civil society actors, citizens and consumers, mainstream and marginalized groups.

Although the historical account in Habermas' theory may be questionable, he nevertheless pioneered a novel line of inquiry in pointing out that the public sphere – a conceptual rather than physical space – and democracy – expressed through engagement in rational discussion – are closely connected. Habermas' thinking thus provides a valuable set of theoretical resources with which to advance important issues relating to democratic society in the contemporary era. His work also offers a starting point for understanding the media's role in public communication, highlighting its influential position in shaping people's understanding of social and political issues, practices and identities. Garnham (1986), for example, although critical of Habermas' historical assumptions, has adopted his central thesis and, by connecting the notion of the public sphere to that of public service, has used it to construct arguments in favour of public service broadcasting.

According to Habermas' ideals, the media should facilitate the process of rational argumentation by providing a context for public discourse which is essential for the formation of free and reason-based public opinion. The media should help encourage debates over political ideas, contribute to the circulation of information among the public, and thus help maintain the strength and vitality of democracy. The free circulation of information among the public is important both for expressing the common interest and for enabling citizens to take part in debates relating to that common interest. However, the arguments made in favour of a vibrant and open public sphere are not just political arguments. Habermas' theory focuses exclusively on the implications for politics, but there are countless ways in which we might consider the importance of the public sphere for society. There are arguments about cultural heritage, environmental preservation, public health and universal education, to mention but a few. Over the course of time, these desirable objectives have been interpreted and characterized as public goods, that is, goods or services whose consumption by an individual does not reduce the overall availability of the good or service for the rest of the citizenry. One version of the public interest argument, for example, has found its fulfilment in the provision of universal education in most Western European countries since the nineteenth century (Smith 1989). Offering universal education is now considered a public good, that is, good not only for the individual concerned, but for the whole society. Another version of the public interest argument has valued the right to authentic cultural expression and the right to participate in defining the historical development of a given culture (White 1994). Provision for the arts, in particular, has often rested in the hands of the state, both because cultural heritage was regarded as a service that needs to be preserved for future generations and because the state could help ensure that all social classes are able to gain access. Additionally, public service media can contribute to the public interest and enhanced civic engagement in at least three broad areas: information – particularly factuality and accuracy of news and public representations; cultural representation, in the sense of creating a pluralistic social and cultural community; and universality, assuming public service media are available to all at the point of reception at low cost (Iosifidis 2014). The notion of public interest has thus been broadened to include important public services at zero or low cost for the interests concerned.

The structural transformation of the public sphere

The idea of the public sphere has been the focus of renewed interest as a result of the advent of the Internet and other networked digital technologies which can provide new communication spaces in which public debate can be conducted. The online forums and social spaces of the web differ substantially from the platforms for public debate constructed by traditional broadcast media in a number of ways. First, they attract many more people than traditional media (Iosifidis and Wheeler 2016). In 2019, 45 per cent of the world's population (or 3.5 billion people) were social media users (We Are Social 2019). These numbers are out of reach for traditional media such as radio and television stations. But it is not only numbers or scale that matter, for social networks allow much greater interactivity as well as the possibility of many-to-many communication on a global scale, rather than one-to-many as is the case with broadcast media. Moreover, networked media is not constrained by national borders to the same extent as traditional media. The emergence of the Internet and social media has thus led to the globalization of the public sphere and public opinion. The space for public discourse has expanded and the formation of public opinion increasingly takes place in a transnational context that crosses national boundaries. Whereas the traditional media in the form of the newspaper press and public television has been an integral part of the creation of a national public sphere, there is now a widespread assumption that new spheres of communication networks can provide the basis for shared concerns, common tastes and political and cultural debates at a global level.

Most significantly, however, some scholars have sought to explore how the Internet and online digital media shapes, and in turn how it is shaped by contemporary forms of democracy, and how the new media ecology alters the process of civic engagement. In her book *A Private Sphere*, Papacharissi (2010) discusses the way in which new technologies are embedded in individuals' routines and how the new media ecology alters the process of civic engagement. She suggests (ibid.:165) that "democracy is more than a political system of government since it combines personal trajectories of success and failure in everyday life through a shared system of decision-making". People discuss politics alongside other things, and these practices help them connect politics to essential parts of their daily lives. In this sense, democracy is viewed as resolving the individual's relationship to the public and the private. At the same time, Papacharissi suggests that online technologies reshape contemporary democracy by blurring and redefining the borders of public and private. Online technologies afford people both public and private spaces, rather than merely a public sphere. Indeed, Papacharissi suggests, the spaces presented by online technologies are hybrid spaces, simultaneously public and private: thus, "[n]ew technologies create a new civic vernacular for individuals, allowing an actualization of civic identity in tropes distinct from the deliberate model of the public sphere" (2010:130). Papacharissi contends that this new civic everyday language operates in the private sphere. She adds that in postmodern democracy, civic identity can materialize outside the deliberative model of the public sphere. As the border between the public and private space has been blurred, she concludes that "the private sphere describes and explains the mechanisms for civic connection in contemporary democracies" (ibid.:167).

According to Dutton (2007), while the rise of traditional news media enabled the development of the Fourth Estate (for example, the investigative journalism conducted by the *Washington Post*, *Time* and *The New York Times* to publicize the Watergate scandal in the US in the 1970s), the growing use of the Internet and related digital technologies can also be seen to have promoted a new source of accountability in government, politics and other sectors. Dutton explains how this emerging 'Fifth Estate' is being established and why this could challenge the influence of other more established bases of institutional authority and help support the vitality of liberal democratic societies. Indeed, as the new media disrupt the industrial model of information, citizens now have the power to oversee the actions of their elective representatives, thereby enabling a more direct form of democracy to emerge (Iosifidis and Wheeler 2016). In the same vein, Dahlgren (2005:160) argues that the Internet may expand the public sphere by "allowing engaged citizens to play a role in the development of new democratic politics".

A democratic social system can be defined as a system in which the supreme power is vested in the people. The origins of democracy can be traced back around 2,500 years to Athens, Greece, where important political decisions were made in person by (male, property-owning) citizens voting in public assemblies. Today, most democracies around the world are categorized as representative systems because the people usually choose from a selection of candidates the individual or party that they wish to represent them in parliament. The Internet seems to challenge such hierarchical structures as it provides a powerful means for direct citizen involvement in public life and politics. It appears to offer the possibility of new forms of post-electoral democracy.

In particular, the availability of information via social media such as Facebook and Twitter, and the rise of user-generated content such as personal blogs have enhanced citizens' ability to communicate and self-organize. The emerging citizen movements around the world thus serve as a check and balance on the prerogatives of government. Millions of citizens have taken to the streets of Sao Paulo, Tel Aviv, Manila, Madrid and Bangkok demanding good governance and an end to corruption. Demonstrators temporarily swept away autocratic governments in many Arab countries such as Tunisia, Egypt and Libya, although most of them were later replaced by even more repressive regimes. Citizens in Southern Europe have called for an end of austerity measures that lead to economic exploitation and hopeless poverty. Indians have demanded protection from rape. In China tens of millions of bloggers have become a virtual citizens lobby pushing for environmental change, blocking huge new dams and petrochemical plants. Hoffman (2013) acknowledges that there are certainly risks that these newly empowered citizens could become pawns for populist demagogues, but it could also be argued that this is far more likely to happen when the media are controlled by a few than when there are multiple and independent sources of information.

In today's network society, power is multidimensional and is organized around digital, interactive and self-expanding networks whose participants have very diverse interests and values. In direct contrast to power relations that are embedded in the institutions of society, especially those of the state, social movements exercise counter-power by constructing themselves initially through a process of autonomous communication, free from the control of those holding institutional power. As Castells (2012:9) contends, "because mass media are largely controlled by governments and media corporations, in the network society communicative autonomy is primarily constructed in the Internet networks and in the platforms of wireless communication". These social networks carve out a new public space for deliberation, distinct from the constitutionally designated space which is occupied by the dominant political and economic elites. But it remains debatable whether these new media truly enhance democracy and contribute to political participation.

Indeed, many attempts to extol the democratizing and empowering potential of the Internet and social media have been dubbed naïve and idealistic (Nieminen 2009:40). Not surprisingly, the attempt to ground Habermas' ideal speech situation in the web has been met with scepticism. Coleman (1999) suggests that much online discussion is characterized as bad-tempered, perhaps as a result of the decline in public debate in physical spaces such as open meetings and street corners, where people first learned to argue effectively. Wilhelm (1999) also refers to the dangers of poor dialogue and a skewed distribution of contributors in cyberspace. Moreover, as Boeder (2005) argues, it is often the case that major decisions and actions concerning transnational matters continue to occur without intense public engagement.

See also: anthropology and citizen media; civil society; community media; hyperlocal media; media ecologies; political science and citizen media; social media; YouTube

Recommended reading

Dahlgren, P. (2005) 'The Internet, Public Spheres, and Political Communication: Dispersion and deliberation', *Political Communication* 22(2): 147–162.

Deals with key issues and difficulties facing democracy, such as the destabilization of political communication systems in the Internet era, from the perspective of the public sphere.

Habermas, J. (1962/1989] *The Structural Transformation of the Public Sphere: An inquiry into a category of bourgeois society*, translated by T. Burger and F. Lawrence, Cambridge: Polity Press.

Originally published in German under the title *Strukturwandel der Öffentlichkeit. Untersuchungen zu einer Kategorie der bürgerlichen Gesellschaft*, this is a landmark contribution to modern understanding of democracy and the role of the media in creating a rational space for public dialogue, distinct from both political and vested economic interests.

Habermas, J. (1964/1974) 'The Public Sphere: An encyclopaedia article', translated by S. Lennox and F. Lennox, *New German Critique* 3: 49–55.

This later work by Habermas acknowledges that the ideal of the public sphere, a space for rational political debate among an informed citizenry, has been damaged by the expanded role of the state and the appearance of large private institutions, such as advertising agencies and public relations firms.

Papacharissi, Z. (2010) *A Private Sphere: Democracy in a digital age*, Cambridge: Polity Press.

Explains the way in which new online, personalized media technologies have expanded the scope of citizens' social activity, thereby altering the process of civic engagement and creating a private space for discussion.

PUBLICS (AND NETWORKED PUBLICS)

Elaine Yuan

The origins of the concept of the public and publicness can be traced back to the European Enlightenment (Habermas 1962/1989). In this context, the concept was initially developed in opposition against the social and political structures of traditional authoritarian power in pre-modern European society. The public in this sense engenders two different understandings. First, it implies that public policy must stem from reasoned citizen action expressed through public debate, which embody autonomous power, otherwise those in authority do not have the moral legitimacy to rule. The second and contrasting implication is that not every person must express her or his reasoning in the public arena; s/he may merely shape that reasoning in communication with her or his fellow citizens, and reach some conclusion regarding the judgement of public opinion as a whole. This second, more pragmatic approach underlay the development of the interactionist 'co-orientation' model of Chicago School sociology (Splichal 2006:699). Similarly, the notion of public space too may be understood differently within different schools of political philosophy. A public space may be either one of "heterogeneous coexistence" or one of "inclusive solidarity" (Weintraub 1997:25–26). It may be "a space of symbolic display, of the complex blending of practical motives with interaction ritual and personal ties, of physical proximity coexisting with social distance", or alternatively "a space of discourse oriented to achieving rational consensus by communicative means to address common concerns" (Weintraub 1997:25–26).

Habermas' early theorization of publicness embraces the Enlightenment ideal of reason (Splichal 2006). Based on a historical account of the emerging bourgeois class in the eighteenth and nineteenth centuries in West Europe, Habermas developed the notion of the public sphere, which denotes a new social arena of discursive interaction (Habermas 1962/1989). This public sphere is where private individuals form the public by participating in public debates. Such a public served as the counterpart of modern public authority, which differed from the traditional feudalist power (Habermas 1962/1989). Habermas depicted his liberal bourgeois public as a single collective actor representing all members of society. This public, however, ultimately proved both exclusionary and transient. The public sphere in contemporary society is understood as comprising multiple publics engaged in complex relations, rather than being one entity in any singular, essential form (Calhoun 1998; Fraser 1990).

The concept of public(s) as a social category has also developed in line with the research tradition of the Chicago School of sociology. Within this body of scholarship, the public is often contrasted with two other modern collective entities: the crowd and the mass (Butsch 2011). During a time when concern about social upheavals marked the era of mass society, the term *crowd* emerged to describe mentally incapable and emotionally charged groups of anonymous individuals, unconscious of their actions and thus liable to spontaneous behavioural imitation – including that of violent behaviour (Le Bon 1895/1960:10). Masses, on the other hand, were made of anonymous, highly heterogeneous and isolated individuals. Although united by a common focus, masses are generally dispersed geographically and physically, and thus incapable of taking effective collective action. Critics suggest that modern conditions are more favourable to the production of masses than of public(s), in part due to widespread state control over mass media, which essentially eliminates any autonomous sphere of communication between individuals (Butsch 2011). A public has the critical ability to "think and reason", while a crowd can merely "feel and empathize" (Butsch 2011). The public can also be distinguished from the masses by its tendency toward communication that takes the form of disagreement and discussion around a certain issue. Therefore, the public differs from the crowd and the mass by its critical discussion mode. Consistent across comparisons between public, crowd and mass is an interest in finding out the role modern communication plays in shaping these social entities (Butsch 2011). Direct contact between individuals leads to the formation of a crowd, while masses (or mass audiences) often emerge through the one-way communication channels of mass media. A public cannot form without also being an audience, because media create the popularity and visibility that partly define the public concept. However, although they do engage in modest (and often ambivalent) levels of critical interpretation and identity formation, audiences fall short of the collective and direct action taken by a public (Livingstone 2005:31).

Publics and networked communication

Modern media have played important roles in the formation of publics. A communicative model for theorizing the public underscores the role of media in facilitating discursive actions and emphasizes that a public is formed through discussion among its constituents (Price 1992). Within this framework, publics are enacted by mediated publicity, that is, by a moral imperative that all actions which affect human rights must be made public (Splichal 2006). More importantly, media provide the practical means for publics to self-create and self-organize through discourses (Warner 2002).

While mediated processes of communication and association are crucial to defining publicness, the notion of the public sphere has inspired conceptualizations of network and mobile media since their advent. The key questions in the growing field of new media studies focus on how new forms of network and mobile communication affect public formation and the implications that these tools might carry for democracy. The Internet and related information and communication technologies offer new opportunities for participatory and deliberative democracy by enabling new channels and forms of citizen participation in public discourse. The application of the concept of the public sphere in new media studies, however, invites some potential caveats. First, the expansion of the Internet as a communication medium poses a formidable challenge to the primacy of traditional political and media institutions in organizing public discourse. No longer stable and homogenized, the symbolic space for public discourse has become fragmented and heterogeneous, comprising "spatially extensive networks of multiple publics and communities" (Calhoun 1992:37). Second, public discourse is not equally open to all. Media access and symbolic resources are often effectively concentrated in the hands of dominant social actors (Fraser 1990). In this light, it is important to examine the contested interactions between different players and identify the mechanisms that render some of them subordinate to others (ibid.). That is, we must examine the broader political meaning of the various symbolic struggles within the discursive space mediated by network media (Crossley and Roberts 2004). Indeed, network communication entails not only technological but also cultural changes. It calls for a new conceptualization of publicness whose qualities of openness and connectivity have been radically transformed.

The social public and Internet sociality

Conventionally, public is often defined in opposition to private. The dichotomous division is maintained through a series of symbolic distinctions between objects, people, practices and institutions. Managing such boundaries requires various interpretive strategies and traditions. The development of new forms of communication and association brings new strategies, infrastructures, practices and norms with which to define and maintain the boundaries that regulate emerging spaces and that do not fit neatly into the existing public/private categorization (boyd and Marwick 2011).

Two types of Internet-mediated sociality have lain at the centre of much scholarly interest: communities and networks. Community association emphasizes a shared identity, common goals and reciprocal moral obligations among members. The Internet is capable of facilitating solidarity by broadening and strengthening traditional neighbourhood communities based on categorical identities and common cultural traits. In addition, it can serve as a communal space for communication and interaction. While traditional communities typically form out of shared geographical locations and embodied practices, close-knit online communities can be based entirely on common interests and mutual liking (Rheingold 1993; Postill 2008). Internet sociality also takes the form of networks. The social logic of network sociality derives from voluntary, short-term, instrumental encounters among a large number of relatively loose social connections (Postill 2008; Wellman et al. 2003). While community associations were prevalent in the past, network sociality has become the main form of association in modern Western civil society (Taylor 1990).

Changes to mediated communication in contemporary society, however, cannot be reduced to a shift within a simplified dichotomy separating networks from communities (Couldry and Hepp 2017; Postill 2008). The Internet serves crucial bridging functions between diverse and distinct modes of communication that vary widely in scope, size and relational context. Internet sociality provides mechanisms for switching in and out of the established modalities of sociocultural interaction. It is capable of facilitating not only the maintenance of relatively bounded and stable complexes of institutions such as lifeworld communities, states, economies and civil societies, but also the emergence of new social interactions and movements (Bennett and Segerberg 2012; Emirbayer and Sheller 1999). The interspatial connectivity of the Internet is conducive to large-scale conversations that grow from communication organized around small, topically connected communities. In this way, personal issues may be elevated into public affairs through rapid network communication. Benkler (2006), for instance, has demonstrated how attention to social issues in the networked environment spreads from engaged groups of people via websites with heavy internal links, and moves through external links to reach other clusters (ibid.:259). Still more often, Internet sociality develops in a rhizomatic fashion that reflects non-linear, heterogeneous, decentralized and fluid network connections among diverse social actors. Online opinions and comments from myriad sources, stimulated by quickly spreading memes, hashtags and buzzwords, feed off each other to generate further momentum for participation (Meng 2011). Such moments can be affective as well as behavioural in nature, engendering structures of feelings and collective consciousness among participants (Papacharissi 2015a).

In fact, symbolic practices enabled by diverse forms of Internet sociality have great implications for contemporary social life at all levels. "What the Internet offers", Feenberg and Bakardjieva (2004:39) suggest,

> is a flexible communicative space that can be constructed and bent in an infinite number of ways by sufficiently motivated groups of people. The implications and significance of what these groups build depends on the shape that they give to the space they create, and the relationships produced within it.

Embodying the multi-layered complexity of contemporary communication, Internet sociality is fluid, porous and interstitial. In the lifeworld of personal and informal associations, the Internet has expanded the spatiotemporal conditions of everyday interaction and communication (Zhao 2006). Intimate personal relations can now be extended to mediated reach across large geographical space. Various types of personal relations including families, friendships and romantic partners can be maintained through diverse patterns of synchronous and asynchronous connection enabled by such network platforms as email, blogs, social networking sites and microblogging sites (Tong and Walther 2011). Consequently, the Internet has become a fertile ground for lifeworld discourses.

The structural change of mediated sociality calls for a revision of existing analytical categories. Corporeal co-presence is no longer the benchmark for evaluating all forms of human contact in the Internet era (Zhao 2006). Rather than having to work through separate realms of online or offline, virtual or real, or synchronous or asynchronous, social actors traverse the world as collaborative spaces across material and digital realms situated in intertwining social milieus "while simultaneously interacting with the cultural inscriptions written into the experience" (Farman 2012:45). Actions and relations in the virtual space are real so long as they bear actual consequences and effects upon social agents. Moreover, Internet sociality also resists rigid classification that previously separated interpersonal from mass communication, consumers of symbolic content from producers, or private communication from public discourse (Papacharissi 2010). Offline personal connections may be maintained as online networks. Private and personal relations may facilitate public and civic participation on the ground.

The structural transformation of mediated publics

In light of the transformation of mediated communication, media scholars have theorized the new possibilities for imagining collectivity and representing publics (Couldry and Hepp 2017). "Private spheres" (Papacharissi 2010:161), "networked publics" (Benkler 2006:11), "calculative publics" (Gillespie 2014:188) and situational "issue publics" (Marres 2007), for instance, are formed through the particular communicative architectures of the various digital media in which they emerge. Indeed, mediated communication in the network era has important implications for understanding publicness. The discursive universe mediated by network and mobile media is necessarily a socially organized space, "with characteristic lines of division, relationships of force, and other constitutive features" (Calhoun 1998:24). What defines this space is not consensus among interlocutors, but rather a process in which differences are articulated (Calhoun 1998). Such publics – bonded by disagreement and discussion – can take any form of discursive expression.

A public forms through the efforts of individuals who share an interest in a given issue and who represent themselves in wider society by debating among themselves and with other society members (Warner 2002). As Splichal (2006) has noted, this new definition is consistent with the idea promoted by Habermas that the public sphere is a kind of communication network that circulates public opinion. According to Habermas, the public sphere cannot be conceived of as an institution, an organization or a system. Rather it is a "network for communicating information and points of view" in which "the streams of communication are, in the process, filtered and synthesized in such a way that they coalesce into bundles of topically specified public opinions" (Habermas 1992/1996:360).

From this perspective, the public cannot be understood as a monolith. It is not the product of a single foundational act, but a relational entity, shaped by shifting conjunctures and alliances, as well as attempts to forge links between separate publics and different types of publicity. During such a process, "varied actors participate, speaking through different cultural understandings, never altogether agreeing on just what the public is, yet producing it continuously if incompletely through their very discourse" (Calhoun 1998:24). The public, therefore, is constituted by a bundle of temporary, conflictual and plural discourses in a process through which social identities become meaningful and have effects (Laclau and Mouffe 2001:966). Always in flux and forever reacting to new situations, the public constantly breeds new forms of communication and modalities of conflict while solidarities and enmities are realigned as new arenas of interaction emerge (Emirbayer and Sheller 1999:164).

Clearly, the transformations of publicness cannot be attributed solely to the media (Couldry and Hepp 2017). Such broad social changes as individualization (Beck and Beck-Gernsheim 2001), globalization (Giddens 1990;Tomlinson 1999) and neoliberalism (Harvey 2007) also have significant implications for the public(s). The development of a new social structure of public space and the formation of publics are mutually constitutive. Network media – especially the Internet – become an arena where symbolic practices of agreement, dispute and negotiation take place through diverse modes of communication. The publics, at the same time, take shape "at the moment of open communication and popular participation through which alternative directions for social life are collectively reflected upon and adjudicated" (Emirbayer and Sheller 1999:155). Emergent patterns of alliance or rivalry consolidate on public networks of viral memes, agenda-setting and media events. The amorphous concept of the public takes on cumulative weight during frequent invocations of the concept in all its different manifestations in symbolic space. This further raises public consciousness, so the notion of the public good rises to gain force as a legitimizing principle for the discursive space.

At the same time, various publics and their attendant discourses sustain the discursive space as a polycentric and variegated universe. The diverse goals and interests of plural publics enable the multiplication of interlocking networks (Bourdieu 1982/1991; Ferguson 1998). These emergent and fluid locales of symbolic interaction are often interstitial spaces intersecting with existing domains of relations and understandings, especially within the political and economic fields (Emirbayer and Sheller 1999:156). Three prominent interstices include interfaces with the state (political publics), the economy (economic publics) and society itself (civil publics). Political publics express themselves online through symbolic protests, petitions, publications and other actions. Economic publics occupy the space of communication for "collective bargaining and grievance procedures; workers' associations and forms of cooperation; moral or social economies; and structures of influence over corporate governance, consumer rights" (Emirbayer and Sheller 1999:156).

Civil publics encompass the communicative matrices through which members of the whole range of associations that make up the fabric of social life, from families, to professional associations, learned societies and religious organizations engage in conversation with each other and reflect on the life of their institutions. These communicative spaces often interpenetrate each other. As events unfold, publics emerging in collective action aimed at political inclusion or economic benefit are often associated with public reflection, questioning established roles and conventions within the social domain itself (ibid.:160–162). Within this line of argument, a wide array of approaches may be identified. Some see these publics in normative terms, others take a more descriptive stance, some adopt a historical perspective while others prefer a contemporary focus (ibid.:160). The Internet as a symbolic space is where different publics of diverse orientations toward political, economic or social institutions relate to each other through networks, communities and events using diverse modes of communication.

Future directions

In summary, both the practical and analytical boundaries that delineate the public in opposition to the private have been challenged by new developments in network communications technologies. The lines between the public realm of political community and citizenship, vis-à-vis the economic and political institutions, and the private realm of the individual and the family are blurred in a sphere of fluid and diverse forms of mediated sociability enabled by network tools. Such mediated sociality makes new forms of self-organization possible.

It is important to consider media both as technological infrastructures and as processes of sense-making (Couldry and Hepp 2017). Social networks not only afford individuals the possibility of connection but also provide architectures through which they organize and interact. The boundaries and constituents of a given public are structurally defined by access, visibility and participation enabled by network media. In this light, "networked publics are simultaneously (1) the space constructed through networked technologies and (2) the imagined community that emerges as a result of the intersection of people, technology, and practice" (boyd and Marwick 2011:7).

Contemporary publics, therefore, need to be understood in terms of "the power relations, the networks of communication, the topography of the issues, as well as the structure of influence" (Calhoun 1992:38). Habermas' singular and universalistic category of the public is reshaped and fashioned through the involvement of multiple but unequal discursive participants. Publics, from this perspective, take shape in "open-ended flows of communication" that bridge the distance between socially different interlocutors and allow them to formulate collective positions and forge practical alliances in pursuit of shared interests (Emirbayer and Sheller 1999:156–157). Emerging as arenas of public communication, forums of critical discourse and fields of collective decision-making, the publics created through these processes vary in scope, power, internal structure and interaction dynamic (Emirbayer and Sheller 1999).

See also: public sphere

Recommended reading

Calhoun, C. J. (ed.) (1992) *Habermas and the Public Sphere*, Cambridge, MA: The MIT Press.

In this edited volume, scholars from a wide range of disciplines ranging from critical theory, feminism, cultural studies, to democratic politics respond to Habermas' most directly relevant work on the relationship between civil society and public life, *The Structural Transformation of the Public Sphere.*

Habermas, J. (1962/1989) *The Structural Transformation of the Public Sphere: An inquiry into a category of bourgeois society*, translated by T. Burger, Cambridge, MA: The MIT Press.

A historical account of the transformation of the public sphere in early modern Europe.

Weintraub, J. and K. Kumar (eds) (1997) *Public and Private in Thought and Practice*, Chicago, IL: University of Chicago Press.

This edited volume explains the significance of the public/private distinction in an increasingly wide range of debates from social and political theory to historical sociology and cultural studies. The essays highlight crucial processes that have shaped the culture and institutions of modern societies by commenting on controversies surrounding such issues as abortion rights, identity politics and the requirements of democratization.

RACE & ETHNICITY STUDIES AND CITIZEN MEDIA

Gavan Titley

The idea and practice of race in social and political relations have historically been confronted through a mosaic of political movements and popular mobilizations for freedom, equality and justice. These have required building collective intellect and shared understandings, and convincing wider publics through significant communicative labour. In contexts where contemporary social formations remain marked by the often stark inequalities and cultural exclusions produced by forms of racism, these public cultures are marked by debates and conflicts between "those who acknowledge the shaping presence of race and coloniality as against those who deny it" (Shohat and Stam 2012:xix). As racism "never stands still" (Sivanandan 1990:64), anti-racist movements and actors have had to work hard to name racism, to make it knowable and to build shared, mobilizing understandings in order to confront it. These histories of undoing race and confronting racism are brimming with the kinds of communicative practices, political desires and public orientations that, having commanded limited scholarly attention within the social sciences, are now associated with the concept of citizen media. This lack of interest transcends the somewhat predictable absence of the organizing concept of citizen media from the research field of race and ethnicity studies and foregrounds the absence of sustained engagement with the grassroots media work and autonomous communicative practices of racialized populations and their networks of action and agitation.

This entry begins by underlining the general indifference of race and ethnicity studies to media, before moving on to examine a body of research where media and communication studies and race and ethnicity are integrated. Focusing on examples of historically focused analysis and contemporary media practices, the entry examines what is gained and what is lost by approaching this research through the lens of citizen media. The concluding section turns briefly to recent research on digital media, #BlackLivesMatter and anti-racist protest in the United States as an emergent area where the analytical preoccupations of citizen media studies most keenly overlap with race and ethnicity scholarship.

Media in race and ethnicity studies

This absence of a sustained focus on anti-racist media practices seems to cut against the grain of the contested yet durably critical orientation of race and ethnicity studies. While marked, since the 1990s, by an expansive diversity of conceptual and theoretical perspectives, recent intellectual surveys emphasize that the history of race and ethnicity studies has been significantly influenced by the civil rights and Black Power movements in the US (Hill Collins and Solomos 2009); the struggle against state and far-right racism in the postcolonial/industrial labour migration societies of postwar Western Europe (Murji and

Solomos 2015:2–5); and the political rupture shaped by anti-colonial movements and the "radicalization of the disciplines" that anti-colonial thought entailed (Shohat and Stam 2012:74–81). This critical orientation has fostered a commitment to recovering marginalized practices and perspectives as forms of practical knowledge about race. A key critical task of historical scholarship on race has thus involved recalling the histories of radical thought and action forged by those racialized within colonial modernity, in part as an attempt to undo forms of Eurocentrism that extend through the modern fixation on race to the postwar, late twentieth-century narrative that credits 'Western' progressiveness with the putative overcoming of racism (Hesse 2004). In contemporary contexts marked by the conjuncture of official commitments to anti-racism and racism's sociopolitical mutability, everyday practice has emerged as a key site for the study of the sociology of racisms, focusing on the experience of racism and how it is challenged in everyday life (Essed 1991).

Against this backdrop, it is difficult to identify the reasons why autonomous media and communication practices have failed to register consistently as a research-worthy aspect of these histories and experiential reflections. It could be argued that this failure is an inflection of a more general limitation of the social sciences, namely the tendency to regard media as sources, structures or agents, but rarely to integrate analysis of media processes or practices. Even recent and rich surveys of race and ethnicity studies (e.g. Murji and Solomos 2015) fail to include contributions that treat media and communications as a relevant, never mind central, focus or dimension of analysis.

Race and ethnicity in media and communication studies

There is also a pronounced inattention in media and communication studies to these media practices, though this does not stem from a lack of attention to questions of race and ethnicity. The starting point here is arguably the field's sustained engagement with representational practices and what Stuart Hall termed the 'spectacle of the Other' (1997b). In a review of 'research on racism, ethnicity and the media', Downing and Husband (2005) pointed to the dominance of textual research in relation to representation as a consequence of the relative ease of accessing media content for research as well as the discipline's own priorities: deciphering communication as a symbolic process and analysing meaning as a determining dimension of cultural and political significance (2005:26). Further, the process of symbolic production in the era of mass media has been predominantly influenced by few-to-many power relations embedded within societies structured in dominance (Hall 1996). Accordingly, while different media studies approaches emphasize the ideological, discursive or articulatory significance of media representations, these analyses have often been viewed, not always fancifully, as forms of intervention that work towards equipping citizens with forms of political literacy that denaturalize racializing frameworks and repertoires. In this context, the most insightful contributions to the study of media representations on race explore how we can "theorise the contradictions and account for the multiple implications of representations" without "losing sight of the fundamental question of power" or overlooking "the fundamental complexity and potential ambivalence and openness of representations" (Orgad 2012:35).

This relation between power and complexity has played an important role in steering research on the production of media representations towards examining how more accurate, progressive or nuanced frameworks and images may be generated by existing media institutions. The question of whom, however, is mainly considered as internal to the institution. The work of media, as Silverstone has argued, involves 'boundary work', and "the media, in their centripetal phase, articulated the boundaries of national and linguistic cultures" (2007:19). Consequently, a significant tradition of institutional research has explored how public service media mediate the community of postcolonial and migration nations, focusing on how shifting regimes of multiculturalism, cultural diversity or creative diversity have created ambivalent modes of formulaic and thematic inclusion in programme rationales and production (Malik 2013; Horsti et al. 2013).

Similarly, research on journalism and news production has long revolved around the question of newsroom diversity, while eschewing any causal or essentialist relation between an increased presence of black and ethnic minority media workers and the diversification of content and perspectives – particularly in relation to practices governed by the professional ideology and routinized practices of objectivity (Allan 2010:190–194). It is much rarer to encounter studies such as the one conducted by the 1980s academic/activist collaboration *Campaign Against Racism in the Media* (CARM) following a glut of anti-immigrant reporting in the British press. The CARM study emphasized the need for collaborative media work within and beyond media institutions to bring "pressure to bear from those on the receiving end of racist bias on those responsible for producing it" (Sheridan 1982:2). The multifaceted approach advocated by CARM involved building trade union support, organizing meetings between journalists and racialized communities, and producing campaigning material on media racism. Central to CARM was the Black Media Workers' Association (BMWA), which combated discrimination in mainstream media institutions while committing to building black media – understood both as independent entities and as one of the access programming formats offered by regional and local media. A stated aim of the campaign was to give voice to the concerns of the black community through various initiatives, including video training for campaign groups. Such initiatives have clear affinities with citizen media practices, although in a context of institutionalized racial discrimination, they are more focused on tackling underrepresentation in institutions and professional production as a necessary if insufficient condition for changes in the regime of symbolic representation. Nevertheless, it is important to draw attention to histories such as that of CARM and BMWA as future, historically attentive approaches to citizen media practices could examine the ways in which institutionally focused initiatives also created the space, need and capacities for more diffuse and autonomous media practices within the networks and fora of racialized communities.

Histories of public-building practices

Histories of autonomous and oppositional anti-racist and ethnic community media, scarce as they may be in the field, provide the most productive space for thinking about citizen media *avant la lettre*. As González and Torres (2012) argue, the media of minority communities in the United States has historically been shaped by a double exclusion. The sensationalist, stereotypical and racializing coverage of the mainstream commercial press, combined with the 'racial blind spot' of the radical and labour press shaped the development of a "separate, segregated wing of America's opposition press, more commonly referred to today as the 'minority' or 'ethnic' media". These media were actively marginalized and poorly archived, with most now "relegated to the footnotes of official journalism histories" (2012:11–12).

These histories are characterized by the cooperative, affective, value-driven and public-building practices that are now related to discussions of citizen media, but also by the forms of professionalization, massification and commercialism which are anathema to it. If, as Baker and Blaagaard (2016b) argue, the concept of citizen media does not rely on a community-centred theorization of communicative interventions, care should be taken to ensure that the study of histories of explicitly oppositional, community-building media do not, in turn, become an exercise in validating contemporary definitions of citizen media. Instead, the emphasis should be on whether and how the notion of citizen media helps to draw critical attention to diffuse and unconsolidated aspects of community/dissident/alternative media practices.

For example, the consolidation of a nineteenth-century new press "forged in direct opposition to racism and colonial conquest" (González and Torres 2012:65) was possible thanks to the mobilization and initiatives of free blacks, non-European migrants and Latinas. This web of responses ultimately fed into a scaled-up, commercial/movement-based mass circulation press. In this context, the concept of citizen media has a heuristic value. It draws attention to the expressive politics and communicative practices of initiatives that seek to challenge media exclusion and misrepresentation, and that develop as part of the non-linear processes through which alternative media or the dissident press take shape.

Squires (2002) provides a useful theoretical framework to study the interplay between citizen media and community/alternative media along these lines. Focusing also on the modern history of race in the United States, Squires argues for retaining the concept of a 'Black Public Sphere' as a political marker of the continued impact of race and sets out to resist the easy appropriation or marginalization of black-led political struggles. However, in Squires' view, the tendency to fix resistant and mobilizing practices within collectives as 'counterpublics' fails to reflect both the intersectional heterogeneity of marginalized groups, and the diversity of public-making practices. A black public, in her view, is not a straightforward expression of group identity. Instead, it is "an emergent collective composed of people who (a) engage in common discourses and negotiations of what it means to be Black, and (b) pursue particularly defined Black interests" (2002:454). This definition thus allows for the emergence of "heterogenous Black publics" and "for people who do not identify as Black, but are concerned with similar issues, to be involved in a coalition with Black people" (ibid.).

Squires' emphasis on relational and coalitional practices allows her to identify overlapping patterns of discursive and political response shaped by political context, internal dynamics and access to resources and institutions. A public can 'enclave' itself, focusing on internal debate and collective intellect, while avoiding repressive and derogatory treatment in the wider public sphere. Alternatively, it can respond as a counterpublic, engaging in wider debate and movement-building, or act as a satellite, deliberately building shared interests and capacities on its own terms (to the extent possible) while strategically engaging in periodic wider public engagement (2002:457–463). As an example of an enclave public, she turns to the same era of newspaper development as that explored by González and Torres (2012), but the emphasis is less on the successful establishment of titles than the hidden communication practices and networks of those in racialized hereditary chattel slavery and the networks of reading rooms, debate societies and meeting places maintained by free blacks to develop "Black protest and ideologies of self-determination" (Squires 2002:459).

Diasporic, transnational and migrant media studies

While migration studies has had an uneasy relationship with the field of race and ethnicity (Lentin 2014), communications research has long recognized that the experience and exigencies of migration, and the sociopolitical, economic and linguistic marginalization that accompanies racialized migration, stimulate often innovative forms of media production and practice (Karim 2003). Given that since the 1990s the availability of satellite and other transnational diasporic media has been taken to indicate a shift in the boundary work of the media, and audience engagement has often been politically coded as a form of self-segregation, both reception and ethnographic studies of migrant audiences, and studies of migrant media production have tended to focus on the agency involved in everyday media use and meaning-making. As Bailey et al. (2007:2) argue, "diasporic media are not set points of difference". The availability and presence of expansive transnational media within diasporic media cultures may well offer "new possibilities for expression and representation" (ibid.), but this does not mean that self-expression and representation necessarily follow any predictable community-oriented lines.

This latter point has taken on even greater significance as digital media use has been integrated to the field of study. In particular, the intersection of the significant movement of asylum-seekers across European borders in 2015 and 2016 with digital media proliferation has resulted in a renewal of the field, though studies of media framing of refugees vastly outnumber studies of how people seeking refuge have drawn on connective media platforms to coordinate routes, maintain social relations and act on their situations. This multiplicity of communicative practices in a context of hardship, oppression and politicized public attention has stimulated an attention to the ways in which digital practices are multivalent, acting across the public and private, navigating everyday life and encompassing the kinds of actions now associated with citizen media. In this approach, the danger of a banal retreat to social media empowerment narratives is countered by a sensitivity to how digital media provide contingent spaces of communicative action under particular circumstances.

Leurs (2017), for example, has argued that the digital archives accrued by recently arrived young refugees in the Netherlands – consisting of selfies, videos, message threads, social media updates – are "historical documentations of individual and collective experiences, feelings, traumas and aspirations" that, through cross-platform digital storytelling practices instantiate "articulations of social injustices and appeals to human rights" (2017:675–676). Godin and Doná (2016) have examined how young Congolese refugees in London have developed multimedia projects shaped by the desire to challenge the essentializing, depoliticized category of 'refugee voices', which also necessitated carving out their own representational space beyond the established Congolese diasporic media and activist networks.

These particular studies unpack how digital media – in a context where the possession of a smartphone by somebody seeking refuge has been coded as proof of their bogus status – provide possibilities for media practice shaped by the absence of other forms of media access. Nevertheless, Baker and Blaagaard's (2016a:8) critique of the "conflation of citizen media with social and digital media" – that they regard as "insensitive to the realities of the uneven global distribution of power and resources" (ibid.) – has a definite import in this area of research. In an ethnographic study of the immigrant workers' rights movement in Los Angeles, Costanza-Chock (2014) argues that taking social media as the analytical starting point risks blunting an assessment of what is truly innovative in digital communication practices within movements, while also extracting them from the mesh of practices that activists with unequal access to communications co-create in a complex media ecology. What she terms 'trans-media organizing' essentially involves social media coordination and strategic attempts to use connective platforms to access the news cycle and receive mainstream coverage. However, to limit the analysis of social movement media practices to digital media would be to extract them from their relations with a wider repertoire of media practices, including "graffiti, flyers, and posters; newspapers and broadsheets; community screenings and public projections; pirate radio stations and street theatre" as well as "other forms of media-making [that] abound within vibrant social movements" (Costanza-Chock 2014:9).

While this attention to scale, multiplicity and performative intervention suggests a productive affinity between the idea of transmedia organizing and citizen media, the political context of organizing, and the status of these organizers, suggests a rupture. The media actors in these last examples – from those enslaved in the racial capitalism of the plantation to those remaindered as excess humanity in the punitive warehousing of asylum systems – are clearly not citizens. It is one thing, as theorists of citizen media have argued, to articulate an idea of citizenship as a practice of political and public becoming beyond the political status conferred within the system of nation-state sovereignty. It is another to recognize that it is this very system of stratified rights that actively produces their non-citizen status; the European asylum system, as De Genova argues, "is a regime for the production of migrant illegality" (2018). That the citizen in citizen media holds open a space of practice and a sense of participation beyond a formalist understanding may not be enough to anchor this category in relation to processes where race is renewed in and through the distribution of citizenship.

Emergent practices

Discussions in race and ethnicity studies tend to default to the North American experience. Nevertheless, it is worth concluding by underling the rich body of work that is now emerging in relation to the communicative life of the protest networks that have organized around #BlackLivesMatter. In the first study to chart the contours and life of the #BlackLivesMatter movement, K.-Y. Taylor (2016:283) presents it as one of those "periodic ruptures in the US narrative of its triumph over racism as a defining feature of its society" – and hence a mobilization that is acutely aware of the need to contest the carefully curated postracialism of US public culture.

While a focus on movement communications should not be taken as a mediacentric account of political action, the intensive interest of media researchers in this milieu stems from its popular association with digital mobilization (Freelon et al. 2016b). The hashtag #BlackLivesMatter stands in for, and creates a narrative of, linear political action commencing in the aftermath of the killing of Trayvon Martin in 2012 and intensifying after the police murder of Michael Brown in Ferguson in 2014. However, the hashtag itself did not come to signify a movement until after the Ferguson protests (ibid.). As Kuo (2018) has argued, this period saw the proliferation of racial justice activist hashtags as sites of communicative action not necessarily emerging from specific mobilizations. According to Kuo, "their primary value may be in elevating and circulating discourse, but these hashtags can help establish grounds for participation, build individual and collective identity, and organize for collective action" (2018:496).

Florini's (2015) study of the podcast *The Week in Blackness* following the acquittal of George Zimmerman for Martin's murder is another important contribution to the study of this movement. Like Kuo, Florini shows awareness of the tensions between the utility of connective media and how the "architectures of digital media networks can reproduce the emphasis on the individual that is at the core of neoliberal racial ideologies, creating the potential for dominant racial logics to map easily onto digital networks" (2015:441). Consequently, a key theme in researching this collective action has been examining the ways in which networked activists have been able to exploit the dynamics of the hybrid media system (Chadwick 2013) to challenge not only media framings, but mainstream journalistic practices. The campaign for justice for Trayvon Martin has generated several studies, the most comprehensive of which is Graeff et al. (2014). It proposes the notion of network framing to capture how participatory media users benefit from the media dynamics of intensive media events, where the demand for coverage of a 'breaking news' event creates possibilities of mainstream entry and amplification for networked activism.

The extraordinary police repression meted out to protestors in Ferguson, Missouri in August 2014 was intensively mediated, and the impact of the proximity of professional journalists with a diverse milieu of activists and protestors has been a focus of media research. For Barnard (2018), a significant result of Twitter's temporal role as a hybrid journo-activist space during the protests was that networked publics succeeded not only in drawing attention to police brutality, but in documenting and amplifying, and thus linking together individual cases of violence against people of colour into an "undeniable pattern explained only by structural racism" (2018:2256).

A number of studies focus on distinct digital media practices which emerged in this period, including Jackson and Foucault Welles' (2016) study of everyday citizens as network initiators of dissent in a concerted media event. Among these, it is Mirzoeff's (2017) study of #BlackLivesMatter that proves more relevant to Baker and Blaagaard's (2016a) conceptualization of citizen media and, in particular, to their insistence on performance and the body as communicative resources. For Mirzoeff, the public occupation of space by racialized people and their allies creates the condition for a space of appearance "where you and I can appear to each other and create a politics" (2017:32). This fact of co-presence, and its affective value, provides the basis for media work, as "by our consent it is possible to mediate that dialogic space into materially shareable and distributable forms" (ibid.). Thus secured, the space of appearance is instantiated through three streams of visibility: witnessing in person and through machine-generated imagery; protest as forms of mobilization that render injustice visible; and mediatization which ensures a "co-presence between physical and digital spaces" (ibid.:90). Physically, mediatically and politically, the space of appearance refutes the police injunction that there is 'nothing to see here'.

In this recent literature on #BlackLivesMatter, the attentiveness to the layering and scale of communicative practices, shaped in and shaping of a political moment suggests a critical affinity with Baker and Blaagaard's (2016a) conceptualization of citizen media. However, given the irreducibly ambivalent resonance of citizenship in anti-racist struggles, it remains unlikely that the concept will gain significant resonance in the field of race and ethnicity studies. This is not necessarily a problem for either field, however, as interaction between the kinds of critical orientations discussed herein is more important than explicitly shared rubrics.

See also: diversity; community media; hip-hop; migration studies and citizen media; philosophy and citizen media

Recommended reading

Nikunen, K. (2018) *Media Solidarities: Emotions, power and justice in the digital age*, London: Sage.

Engages with the contingencies of solidarities engendered through mediated relations. In paying attention to media practices and cultures of production, it provides an important set of orientations for thinking about citizen media practices.

Saha, A. (2017) *Race and the Cultural Industries*, Cambridge: Polity Press.

This volume – not centrally concerned with debates about citizen media – focuses on how media systems, media structures and the political economy of media production constrain how race is mediated. It provides a very useful orientation to thinking about what autonomous media interventions may or may not contribute in this context.

Titley, G. (2019) *Racism and Media*, London: Sage.

Analyses how the deep integration of digital media alters the mediated politics of racism. By consistently paying attention to anti-racist media practices, it examines the conditions under which oppositional media practices may take shape and flourish.

REMEDIATION

Owen Gallagher

At the turn of the twenty-first century, Bolter and Grusin (1999:45) developed and published their theory of remediation, understood as the incorporation or representation of one medium in another medium. Their analyses of various media forms revealed four primary types of remediation. Faithful and transparent remediation involves, for example, producing a movie based on a novel. Improvement is a form of remediation where enhanced technology is used to improve the fidelity of the content, as in Blu-ray vs. DVD. Refashioning takes place when an older medium, like silent black and white cinema, becomes outdated and is aggressively replaced by Technicolor talkies. Finally, absorbing and repurposing types of mediation can be illustrated by virtual reality (VR) and massively multiplayer online games (MMOs), whereby blocks of code and game assets are frequently modified and recombined to create new interactive experiences. The concept of remediation built on the seminal work of Marshall McLuhan (1964/1994:23–24), who argued that the content of any medium is always another medium with its own grammar that codifies reality in its own particular way. For Bolter and Grusin (1999:2–19), immediacy and hypermediacy represent the "double logic of remediation". Some new media such as VR, attempt to be immersive and interfaceless, thus striving towards transparent immediacy – i.e. the sense that one is not experiencing mediation even when one actually is. By contrast, websites and mobile apps, which cannot be experienced without an interface, represent hypermediacy: they constantly remind users that interfaces exist between them and the content they are consuming. Both immediacy and hypermediacy gravitate towards normalized 'real' experiences whereby the user is either unaware of the interface or expects interfaces in all media interactions and no longer notices them. For Bolter and Grusin (1999:71–72), authenticity of experience is a key factor in the development of new media. Omar (2017:250–266) argues that new media remediate old media by borrowing form and content, creating potential disputes between what is considered authentic. Established media tend to have an authenticity that new media seek to imitate or refashion, so they often survive even when they have been improved or replaced by newer, more efficient forms.

Deuze (2006b:66) defines remediation as the "remix of old and new media" and describes how engaging in remediation involves the adoption, modification, manipulation and reformation of the consensual ways in which we understand reality. For Deuze (ibid.:66), the three principal components of digital culture are remediation, bricolage and participation, with bricolage representing the "highly personalized assembly, disassembly and re-assembly of mediated reality". Deuze argues that contemporary digital culture is built upon these identified components; however, as Deuze states, "none of this is new" but instead may be regarded as a "supercharged version of that which came before" (ibid.:72). While it is clear that digital culture builds upon the past, it is not accurate to state that there is nothing new under the sun nor to minimize the impact of new technologies on society and culture. Despite its scholarly focus on contemporary technologies and culture, remediation tends to maintain a strong relationship with tradition, often relying on old media as its foundation; however, as a concept its very existence relies upon the development of new media technologies in order to compare, contrast, incorporate and represent older forms in new ways. Deuze's analysis reifies digital culture, while simultaneously arguing for a processual remaking of it, through examples of increased hyper-individualization in politics (changing notions of civic engagement), economics (changing notions of consumerism) and journalism (changing notions of news reporting), leading to the emergence of new types of citizenship, participation and activism (Deuze 2006b:68–70).

Remediation also has another definition, which means to remedy something, as in remedial education. According to the *Cambridge English Dictionary* (2018), to remediate is to "correct something that is wrong or damaged or to improve a bad situation" and remediation means "the process of improving or correcting a situation". In the context of remediation and citizen media, the meanings blend and feed into one another. Citizen media, particularly video-based content such as critical remix videos (CRVs), seek to right wrongs and injustices, correct misinformation, combat fake news and promote positive ideal traits such as equality and liberty. With citizen media, the traditional gatekeepers of media are bypassed, circumvented or short-circuited as ordinary citizens find ways to carry out small acts of tactical resistance in their daily lives, using the tools and techniques of digital activism now available to the majority population through Internet-enabled devices.

Citizen media and remediated forms

Citizen media refers to a variety of content produced, published or performed by unaffiliated citizens with the aim of affecting sociopolitical change (Baker and Blaagaard 2016a:16). This can include many types of media, including material artworks, performative interventions and remediated digital content, such as online videos, memes, mashups and remixes (Navas et al. 2018:3–11). Many examples of citizen media – for example, when an engaged citizen uploads footage of witnessed police brutality recorded on their phone – do not use remediated content in their composition, at least not initially. The purpose of such material is often to expose the wrongdoings of those in power or their representatives and to bring change to a situation, perhaps through political pressure, or by having the offenders punished, incarcerated or relieved of duty.

As reported by *The New York Times* (Benzaquen et al. 2018), increasing quantities of raw video footage depicting excessive US police brutality have been uploaded online in recent years. Similar videos have emerged from many other countries, such as the police beatings of Rohingya civilians in Myanmar (O'Sullivan 2017:1), the brutal beating with clubs of street sellers in Kenya's Uasin Gishu county by local law enforcers (Lubanga and Ominde 2018:1), and the violent beating of a motorcyclist by New Delhi traffic police for simply not wearing a helmet (Times Now Bureau 2018:1). Such videos are potential source material and once they have been uploaded, inevitably end up being remediated. These videos have gone viral through remediation – a process that begins, for example, when someone shares one of the videos on their Facebook page, another embeds it in their blog, perhaps a local news network picks up the story and then Aljazeera or BBC International play the video as part of a news report. This may lead to a still frame of the footage appearing in a newspaper or magazine to illustrate an article about police brutality, thus exposing the original video to millions of new viewers. A remixer may subsequently download the video and use a clip as part of a longer edited composition, such as a supercut of similar atrocities, or they may sample a few seconds and loop it to create an animated GIF or a meme clip. The remixed video is reposted, shared and recirculated via social media and the remediation cycle begins again, ad infinitum.

Table 1 Textual manifestations of remediation: a typology in flux

No.	*Type*	*Description*
1.	Remix	a new mix of a preexisting video, song, image or text that alters, appropriates and recombines sampled material into new arrangements
2.	Mashup	a type of remix that combines two or more disparate sources into one
3.	Meme	an image, video, song or text, typically humorous in nature, that is copied and spread rapidly online, often with slight variations
4.	GIF	a type of meme that combines a number of GIF image files into a short animated image sequence, often to communicate a humorous reaction
5.	Fansub	a new version of a foreign film or TV show that has been translated by fans into a different language using subtitles and republished online
6.	Scanlation	a new version of a foreign comic or graphic novel that has been translated by fans into a different language and republished online
7.	Supercut	a type of video mashup featuring numerous clips of the same phenomenon edited together to reveal connections and patterns
8.	Vid	a fannish music video composed of re-edited clips from popular TV shows prominent in fan culture, often revealing hidden subtexts in the source material, created by vidders engaged in the act of 'vidding'
9.	Recut	a type of remix video that re-edits only from the same source material to produce a different cut, often dramatically altering the original meaning
10.	CRV	a digital video composed of previously published media elements, which have been appropriated, repurposed and reconfigured in the creation of a new work that communicates different messages and meanings than the source material, often highly critical of targets attempting to expose hidden information about the objects of criticism

There are many types of intertextual remediated material in current circulation, and as is the nature of such content, new strains and variations are constantly evolving and emerging from those that already exist. In Table 1, I propose a working typology of the most common textual manifestations of remediation in contemporary culture.

Remediation and authenticity

We are still living in a time when raw footage videos, like the police brutality examples mentioned earlier in this entry, are considered a strong form of admissible evidence. As Paul D. Butler, a law professor at Georgetown University and a former prosecutor put it, "[t]hese videos are smoking-gun evidence, both literally because they are very graphic, which generates outrage, and figuratively, because people believe their own eyes" (Benzaquen et al. 2018:1). However, times are changing and every day, the truth-telling quality of raw video footage is being eroded. Chouliaraki (2015:11) identifies this relatively new struggle of authenticity within the sphere of citizen media, specifically with regard to digital witnessing – mediatized spectacles from war and conflict zones, which invite publics to witness suffering and death as moral events that require a response. However, audiences can no longer trust the source of such videos, due to the proliferation of recording devices and platforms whereby many actors, each with competing interests, vie for the attention and reactions of audiences through the distribution of potentially shocking footage. Generally, people still tend to believe what they see, but advances in technology have made it increasingly possible to produce fake footage that appears to be real.

The film and animation industries have been working to perfect the art of 'faking' reality for more than a century (Welles 1973) and recent developments in cameras, computers, post-production visual effects and compositing software means that filmmakers now have more control than ever over how their footage appears to viewers. A growing number of fake news and documentary-style TV shows, from *Documentary Now!* (Hader and Armisen 2015) to *The Fake News with Ted Nelms* (Helms 2017), highlight a growing cultural concern within the public consciousness regarding the perceived reality of what we are seeing on our screens in so-called 'real' documentary films and news reports. Yet far more disturbingly, artificial intelligence assisted deep learning technologies, such as *FakeApp* face-swapping software (FakeApp, n.d.), are increasingly being used to produce deepfakes by remediating existing footage and recombining different faces, bodies and voices, in often seamless and believable ways, to make it appear as though people have done things they did not do in places they have never been (Gillespie, N. 2018:1; BBC News 2017:1). The logical conclusion of these trends is that the potential impact of citizen media videos may be eroded as it becomes easier for guilty parties to deflect targeted exposure by declaring smoking gun videos to be fake. This could cast doubt on the authenticity of such footage in the minds of those it seeks to convince, persuade, outrage or rally to action and ultimately allow more people to get away with crimes that have actually been committed. In theory it may be possible to reverse-engineer and potentially identify fakes using the same technologies that created them, but as artificial intelligence-assisted compositing software continues to increase in sophistication and verisimilitude, it will become progressively more difficult to distinguish between authentic and manufactured footage.

Remediated remix

Within the realm of remix, one of the most potent types of remediated citizen media is the critical remix video (CRV). These videos are inherently reliant on the reuse and manipulation of extant source material in their formal composition. CRVs may target many kinds of injustice, for example, videos that are critical of governments and political figures, the mainstream media and multinational corporations that engage in unethical or immoral behaviour causing harm or suffering to others, as perceived by the remixer (Gallagher 2018:2). Askanius (2013:5–8) offers a typology of activist online video genres, including the mobilization video, the witness video, the documentation video, the archived radical video and the political mashup video – a term that is also used to refer to CRVs. Askanius suggests that revisiting the analogue precursors of digital video may help us to contextualize and better understand new forms of video activism. She offers a whistle-stop history of radical, subversive, progressive and alternative filmmaking practices throughout the twentieth century, from the American underground cinema of the 1950s and 60s, to the protest cinema of the 60s and 70s, which documented the political activism of civil rights, anti-colonialism and anti-war protests, as well as women and gay liberation movements. Through the revolutionary aesthetics of the 1970s to the guerrilla television movement of the 1980s and 90s, Askanius traces these anti-systemic, leftist filmmaking practices to the present day, identifying a common thread of anti-capitalist sentiment and targeted criticism of neoliberalism as the root of all societal ills. For Askanius, contemporary political movements, while focusing on various specific targets, are united by "how they contest the neoliberal mechanism by which all aspects of political and social organization are increasingly based around the primacy of unregulated markets and economic growth" (Askanius 2013:4–5).

While Askanius' premise here is largely valid, it is perhaps too broad a claim that all political protest is essentially targeting the same concern, as there are many examples that focus on issues far removed from the neoliberal agenda – including those that hail from other parts of the political spectrum, which are equally entitled to be considered forms of political protest. CRVs demonstrate how shifts in power over the control of online video production and distribution have given rise to myriad new forms that combine aesthetic and discursive qualities in a variety of creative ways. Such videos enable activists to voice dissent on a public stage and participate in public protests against a variety of sociopolitical injustices. CRVs are inherently remediated forms, the latest in a long line of video and film activism modes stretching back to the earliest filmmaking technologies, while possessing their own unique aesthetic qualities for the digital age of sampling and remixing.

Critical remix as civic activism

Critical remix videos have a role to play as symbols of citizen empowerment, functioning as tools of political advocacy, forms of political protest and modes of political commentary (Edwards and Tryon 2009:1). Such videos can reach large audiences without having to go through the traditional gatekeepers of mainstream media – they can be picked up and shared extremely quickly via social media and find their way into popular discourse, influencing and contributing to public discussion, without ever having to be broadcast by traditional media channels. Thus, they can potentially operate beyond the control of corporate media and traditional party politics. CRVs are themselves frequently remediated by other politically engaged citizens who may feel the need to "remix the remix" (Amerika 2011:1) and produce alternative versions, combining multiple different sources and minor adjustments, in a similar way to how image/text memes develop, spread and proliferate online. Through the creation and redistribution of such videos, citizens are attempting to address social, political and ethical issues by adding their own distinctive voices to various issues and debates.

A clear example of this occurs in the masterfully executed *There Will Be Oil* (Valdez 2010), which remixes a wide variety of movie and TV sources with actual political broadcasts and news footage depicting oil fields, refineries and war in the Middle East. This CRV presents a searing critique of US foreign policy, specifically the invasion of Iraq for oil, which led to massive death tolls and ongoing destabilization within the region. The remix is a clever riff on Anderson's *There Will Be Blood* (2007), whereby the persuasive voice of Daniel Plainview (Daniel Day-Lewis) is appropriated and remediated in the service of manipulating indigenous people to get what he wants – their oil – at any cost. There are subtle nods to the nepotism of the Bush administration, echoing Plainview's words about his own son and the importance of "looking after family". By uploading this remix video, which has been viewed thousands of times, reuploaded, remixed and remastered by fans around the world, Valdez offers his own unique perspective and adds yet another critical voice to debates about the perceived legitimacy of US military intervention in foreign countries.

Another excellent example of this kind of global citizen engagement occurs in *BBC Reports on Royal Family Celebrations* (Smith 2015), which remixes footage of Queen Elizabeth II's public birthday celebrations in London with audio commentary from a very different kind of parade held in honour of North Korean leader Kim Jong-Un, marking the seventieth anniversary of the Workers' Party of Korea. The cognitive dissonance produced as a result of hearing commentary from one source mapped onto visuals from another is amusing at first, until it becomes apparent that this CRV is intended as a serious critique of the BBC, the British monarchy and North Korea's tyrannical leader. In this remediated context, Queen Elizabeth is referred to as a "cult leader", British soldiers as mindless, choreographed automatons and the British public as cowering subordinates (comments which were originally directed at the North Korean congregation). The video is a clever critique of both North Korea – a secretive dictatorship accused of significant human rights violations by the UN, responsible for dramatically increasing the potential threat of nuclear war in recent years; and the UK – a so-called democratic country with a monarchical head of state and church, suffering from severe social and economic problems and a highly uncertain future for its citizens. Yet both countries host extravagant annual displays of national identity where thousands of citizens throng the streets to show adoration for their respective heads of state. According to its producer, "while satirical in nature, the video was also meant to pose further questions about the BBC's role as a public service broadcaster and its remit and responsibility to impartiality" (Smith 2015:1). This is a recurring theme among CRVs, which tend to target those in power who are seen to have abused their positions, most frequently political figures, governments, large corporate entities, media organizations or influential celebrities. A common trait of such CRVs is the presence of humour or entertainment value, often intended to make serious subject matter more palatable for wider audiences.

This is highly evident in the remix *Quichael Collins* (Reilly 2016), with its clever remediation of source material, its playful representation of fan culture, and the seemingly cosmic confluence of events that enabled its creation. At first look, the video appears to be a well-known scene from *Star Wars: Episode I – The Phantom Menace* (Lucas 1999) where Qui-Gon Jinn (Liam Neeson) is killed by the evil Darth Maul; however, as with the previous example, the audio track has been replaced with dialogue from a very different source – in this case, the Irish film *Michael Collins* (Jordan 1996), also starring Neeson in the title role. As we witness the spectacle of Qui-Gon's execution, his protégé Ben Kenobi (Ewan McGregor) rushes to his aid, at which point the audio track is hilariously replaced with the dialogue from the final scene of *Michael Collins* where Collins' compatriot, Joe O'Reilly shouts "Ah Jesus, Mick! Ah Mick!" in a thick Cork Irish accent (as Neeson's character meets yet another untimely death).

The video was uploaded to YouTube in 2016, the centennial year of the 1916 Easter Rising (the rebellion and subsequent revolution that led directly to Ireland regaining her independence from England), which is the core focus of the remix. The first hint we receive that this is no ordinary *Star Wars* scene occurs in the iconic scrolling prologue text, which has been rewritten to refer to the (British) Empire and how its most troublesome colony has always been the one closest to it – Ireland. In 1916, it continues, a rebellion broke out led by 'mastermind', Michael Collins, whose "life and death defined the period in its triumph, terror and tragedy" (Reilly 2016:1). The symmetry and remediation of sampling two films starring Liam Neeson (who dies in both scenes), combined with the 100-year anniversary of the Easter Rising and the filming of new *Star Wars* scenes in Ireland – for *Star Wars: The Force Awakens* (Abrams 2015) and *Star Wars: The Last Jedi* (Johnson 2017) – not far from where Michael Collins was actually shot and killed in County Cork, create a unifying sense of completion in this piece. The video is critical of a number of varied targets, including the British Empire, Irish leader Éamon de Valera, Liam Neeson and the films *Michael Collins* and *Star Wars: Episode I – The Phantom Menace*; however, it is clear that its primary goal is to make people laugh at a serious topic, as its creator is a stand-up comedian, as well as a true fan of the *Star Wars* franchise and the Irish film industry.

This sense of humour and playful irreverence evident in so many CRVs can be a powerful political and social leveller. It can be used to highlight serious issues in an entertaining way, or to critique, expose or discredit illegitimate political leaders. The now infamous *Zenga Zenga Song* (Alooshe 2011), one of the most widely circulated CRVs ever produced, did exactly this. The video was produced at the height of the Arab Spring, when Libyan leader Muammar Gaddafi was in power and attempting to quell rebellions that had spread from neighbouring Tunisia and Egypt, but instead found himself facing a full-scale civil war. Gaddafi broadcast a speech claiming that there was no resistance in Libya and that he had the full support of the Libyan people – a complete falsehood and an attempt to use rhetoric to reassert his ideology and authority within the country. Alooshe's remix video appeared on YouTube the following day, remixing the Pitbull/T-Pain song *Hey Baby* (2010), and using auto-tune to match Gaddafi 's speech to the music track's rhythm and pitch. The resulting music video went viral, undermining Gaddafi 's authority, and became an anthem for the Libyan resistance, who would shout "Zenga! Zenga!" in the streets (Häkkinen and Leppänen 2014:1). The *Zenga Zenga Song* was subsequently remediated by another remixer using footage from a speech by Syrian leader, Bashar Al-Asaad, who is widely regarded as being complicit in the operation that led to Gaddafi 's eventual assassination.

For Alooshe, the experience of seeing how his remix video was received by the public and spread around the globe taught him that beyond its entertainment value, a remix video can have real impact on the world, changing the way people view world events and political leaders, and should not be treated lightly. The *Zenga Zenga Song* has been viewed over five million times on YouTube alone, and this is merely a fraction of the true number of times the video has been played, shared and viewed since its publication (Alloco 2018:90–95). CRVs like these are a form of citizen media that can have a real impact on what happens in people's lives and as such, remixers have a certain responsibility to consider when creating such work.

Future directions

Remediation is an important concept in discussions of citizen media and one that has become entwined in our everyday interactions with all media due to the widespread adoption of digital tools, online distribution networks and the proliferation of digital content. Citizens are increasingly seizing opportunities to participate more actively in civic life. This may involve the remediation of sampled clips from mainstream news sources by an individual citizen combined with personal commentary, or at the other end of the spectrum, the remediation of citizen media content by large networks for use in mass media broadcasts. Increasingly, edited clips from YouTube videos created or witnessed by individual citizens are being repurposed and redistributed by mainstream broadcasters to enhance news stories.

Many citizen media voices are expressed outside of mainstream media, however, through alternative, counter-hegemonic or community-driven initiatives, often established through a shared dissatisfaction or collective distrust of mainstream information sources. Such marginal groups face difficult challenges that large, well-funded media corporations do not have to contend with, but the potential remains for contributors of citizen media to facilitate participatory forms of communication aimed at transformative social change. The production of remediated citizen media content can be highly empowering for individuals, and can lead to transformative personal experiences, developing one's identity into a socially conscious activist or politically engaged citizen with a public voice. It also enables individuals to empower others, particularly those in positions of social disadvantage, by raising awareness of suffering, injustice or inequality and potentially inspiring people to do something to change their lives and the lives of others for the better.

In the third decade of the twenty-first century, we find ourselves living in a world where a reality TV star has been elected president of the USA; nations are leaving the European Union; climate change disasters are on the increase; civil wars rage around the world; poverty and unemployment are rising; regressive copyright proposals are working their way through the legal system; net neutrality and online privacy are threatened by corporate interests – and many people are reacting by watching, sharing and publishing more remediated content than ever before. There are many possibilities for the future of remix. CRVs are a potentially powerful tool for political change and can have a real impact on the world, as they can influence people, raise awareness, educate, alter opinions, and expose wrongdoings among those we elect to represent us and make decisions that affect society and our daily lives. By downloading some footage, editing it into a remix video, and then publishing and sharing it online, we are able to express our unique points of view as global citizens on whatever issues are most important to us and help make a positive difference in the world, however big or small.

See also: authenticity; convergence; self-mediation; video games

Recommended reading

Gallagher, O. (2018) *Reclaiming Critical Remix Video: The role of sampling in transformative works*, New York: Routledge.

A detailed analysis of critical remix video – i.e. user-created content composed of previously published media elements, which has been appropriated, repurposed and reconfigured in the creation of new works that communicate different meanings than the source material. These new messages are often highly critical of someone or something and attempt to expose hidden information about the object of criticism. CRVs have the potential to act as tools of resistance against perceived injustices, raising awareness of social issues, influencing public perceptions and offering visions of alternative world views.

Lessig, L. (2008) *Remix: Making art and commerce thrive in the hybrid economy*, New York: Penguin Press.

In this seminal book on the topic of remix, Lessig spotlights the harmful culture war waged against those who create and consume sample-based online creative content. America's copyright laws have ceased to perform their original, beneficial role: protecting artists' creations while allowing them to build on previous creative works. The current system, as it has evolved, criminalizes those very actions. Remix is an urgent, eloquent plea to end a war that harms every intrepid, creative user of new technologies. It also offers an inspiring vision of the postwar world where enormous opportunities await those who view art as a resource to be shared openly rather than a commodity to be hoarded.

Navas, E., O. Gallagher and x. burrough (eds) (2015) *The Routledge Companion to Remix Studies*, New York: Routledge.

A comprehensive volume comprising contemporary texts by key authors and artists active in the emerging field of remix studies. As an organic international movement, remix culture originated in the popular music culture of the 1970s, and has since grown into a rich cultural activity encompassing numerous forms of media. The act of recombining preexisting material brings up pressing questions of authenticity, reception, authorship, copyright and the technopolitics of media activism. This book approaches remix studies from various angles, including sections on history, aesthetics, ethics, politics and practice, and presents theoretical chapters alongside case studies of remix projects.

Navas, E., O. Gallagher and x. burrough (eds) (2018) *Keywords in Remix Studies*, New York: Routledge.

A collection of contemporary texts consisting of twenty-four chapters authored by researchers who share interests in remix studies and remix culture throughout the arts and humanities. The essays reflect on the critical, historical and theoretical lineage of remix to the technological production that makes contemporary forms of communication and creativity possible. Remix enjoys international attention as it continues to become a paradigm of reference across many disciplines, due in part to its interdisciplinary nature as an unexpectedly fragmented approach and method useful in various fields to expand specific research interests. The focus on a specific keyword for each essay enables contributors to expose culture and society's inconclusive relation with the creative process, and questions assumptions about authorship, plagiarism and originality.

SELFIES

Mette Mortensen

While photographic self-representations date back to the infancy of this medium in the first half of the nineteenth century, the concept of the selfie has only become widespread following the launch of the iPhone 4 with its front facing camera in 2010. Since then selfie culture has exploded (Senft 2015). To think selfies in terms of citizen media may seem counterintuitive in light of how this genre has often been associated with a culture of self-centredness as well as the spontaneous, transient and trivial (Murray 2015). However, selfies are not only used to track, record and perform one's own instant history. They also become a personal point of departure for making political statements and inscribing oneself into larger social contexts. Even if selfies necessarily focus on the self, they still raise questions concerning the relationship between individual and collective identities, subjective stories and shared concerns (Jerslev and Mortensen 2016). Selfies can in this sense be thought of as acts of witnessing of, by and through the self, which become embedded in and reconfigure "structures of public visibility – of who we see, how, and why" (Chouliaraki 2017:78).

This entry is concerned with selfies produced, disseminated and mobilized as communicative acts of civic engagement and resistance. Proceeding in three parts, it first offers a definition of selfies as expressions of citizen media and provides an overview of existing research on selfies of relevance to research on citizen media. The following sections present two main groups of citizen media selfies taken by, respectively, civilians in areas of armed conflict and citizens engaged in forms of connective action.

Selfies as citizen media

This entry is premised on a definition of citizen media selfies as photographs or videos produced with mobile media and disseminated via social media platforms, in which citizens performatively construct and represent themselves as a political act or in relation to a political context. This relatively narrow definition excludes many everyday media practices and artefacts, but still encompasses a diverse range of actors and circumstances: from victims and eyewitnesses sharing their experience of and critical stance towards terror attacks and other violent events (Nashmi 2017); to refugees recording their flight, life situation and community (Chouliaraki 2017); to activists documenting the cause for which they fight or depicting themselves taking part in protests (Wei 2015; Neumayer and Struthers 2018). Across these different contexts, selfies reconfigure the mediated visibility of these contested or conflictual situations from the bottom up.

This definition also links selfies to the ways in which the photographic portrait has historically been tied to regimes of visibility, to disciplining of the body and to negotiations between empowerment and disempowerment (Sekula 1986; Tagg 1988; Mortensen et al. 2003). Since the second half of the nineteenth century, the photographic medium has been used for social sorting (Lyon 2003), that is, as a means of identifying and classifying individuals and groups according to their gender, race, social status, etc. As the most important emblem of identity, representations of the face linger in between the personal and the political, the social and the cultural. The systematic, photographic recording of citizens' faces by the authorities thus supports the view that representations of the face are at once abstractions (coordinate systems for identification) and affective entities (coupled to the person's individuality and life story). While the use of photography for identification is connected to what might be termed facial politics as well as to state surveillance and control, selfies and other self-representations may at first glance seem to subvert this power dynamic by enabling the individual to stage and perform his or her own image. However, a more critical perspective on selfies highlights complex power relations that go beyond simple oppositions between disempowerment and empowerment, involuntary mediated visibility and affirmative mediated visibility (Thompson 2000, 2005; Butler, J. 2009). Selfies too are employed for purposes of citizen control through "the rise of biometric governance, and in particular the use of facial recognition in surveillance and policing of individuals and communities" (Kuntsman 2017b:15). When selfies circulate within networked environments, they represent individuals both as visual images and as data points (Rettberg 2017; Mortensen and Lomborg 2019).

In the literature on selfies, two main research strands of interest to scholars of citizen media have emerged. The first revolves around the question of which individuals and groups turn to selfies as a mode of self-expression as part of struggles for recognition and justice. Accordingly, this approach to studying selfies has been preoccupied with issues pertaining to race, gender and body politics (Senft 2015; Williams and Marquez 2015). Within this area of inquiry, selfie practices have been examined in connection with, for instance, political participation, citizenship and activism (Kuntsman 2017a), participatory journalism (Koliska and Roberts 2015), post feminism (Murray 2015) and queer storytelling (Vivienne and Burgess 2013). The second strand of research addresses the modes of representation by focusing on selfies as a visual genre which enables the performative enactment of the self (Hess 2015; Jerslev and Mortensen 2016). Selfies are inevitably self-referential and performative as they depict the individual taking the photograph in the very act of taking the photograph. In selfies, the individual is at one and the same time the object and subject of the image (Olszanowski 2014). They communicate the intimate act of the individual looking at him- or herself while also functioning as a vehicle for the "strategic management of the self" when they are shared on social media (Jerslev and Mortensen 2016:254).

Even if these two research strands have been presented as incompatible (Cruz and Thornham 2015), they are both vital to understanding selfies within a citizen media framework. Taken together, they enable analysis of how the self in selfies is visually performed and situated in diverse contexts and power relations (Frosh 2015; Chouliaraki 2017). Combining these two perspectives is important for understanding citizen media selfies as they negotiate between the individual and the collective, between representations of the self and embodied political communication. This duality between context and genre will be exemplified in the following sections, which present the two main, broadly defined groups producing and disseminating citizen media selfies: citizens in areas of armed conflict and citizens involved with or taking a stand on political issues.

The civilian perspective

Since the early 2010s, selfie photographs and videos have been created, distributed and mobilized by civilians experiencing scenes of war and terror. This trend exemplifies how digital media have transformed the mediated visibility of armed conflicts and complicated the relationship between media and politics.

Top-down processes of communication through mass media by political elites have increasingly been challenged and contested by bottom-up processes of citizens contributing to the flow of information via mobile and social media (Andén-Papadopoulos and Pantti 2011; Allan 2013; Mortensen 2015b; Blaagaard et al. 2017). These representations often transmit exclusive firsthand knowledge, but do so from the partial, subjective and decontextualized perspective of actors on the ground, who are themselves affected by or are active stakeholders in the conflict. These selfies emphasize the embodied presence of the image maker, often with the imminent pretext of humanitarian suffering, physical violence or material damage.

The war in Syria and the so-called refugee crisis that ensued has provided many examples of selfies showing civilian perspectives on war (Alabed 2016). A prominent case is the young Syrian girl Bana Alabed, who tweeted about the escalating humanitarian crisis befalling her family and other civilians in the besieged city of Aleppo. From the autumn of 2016, when she was seven years old, she started using Twitter with the help of her mother to make appeals for support to an international public with images and compelling tweets like the one from 26 November 2016: "[t]onight we have no house, it's bombed & I got in rubble. I saw death and I almost died. Bana#Aleppo". Alabed's selfies, as well as photos of her taken by others, were picked up by the mainstream news media, which closely followed the heartrending story of the family's flight from Aleppo and subsequent asylum in Turkey. At the same time, her images became a symbolic battleground for high politics: Syrian president Bashar al-Assad dismissed them as propaganda, while Turkish president Recep Tayyip Erdoğan warmly greeted Bana Alabed and her family on their arrival in Ankara (Taylor, A. 2016). Her image tweets gave exclusive insight into the Syrian war from a child's point of view, but also raise the question of why she gained so much attention in comparison with the many other civilians who similarly transmitted selfie eyewitness testimonies from Aleppo during this period. One partial explanation could be that this case aligned with the news media's preference for human interest stories and familiar frames, which manifested itself in their repeated claims that the young Syrian girl had turned into "our era's Anne Frank" (Madani 2017).

Other examples of civilians disseminating selfies from conflict zones include images created in the aftermath of terror attacks, for example those in Barcelona, London and Manchester in 2017. Situating citizens at the crime scene, these selfies function as acts of witnessing and as memorials to the loss of life inflicted by the violent events. They also work as a way to reclaim the destroyed space and condemn the terrorist actions. When circulated under hashtags such as #wearenotafraid, this intention becomes particularly evident, because they emphasize civilian resistance to the violence and destruction enforced by terrorists – and the continued threat hereof. Meanwhile, other selfies by civilians from conflict zones appear to be profoundly out of place, to borrow Nunes' (2017) expression. They put on display performances that seem to be inappropriate for the context, for instance by documenting violence without a clear purpose or communicative intent. In the most extreme case, this involves posing next to dead bodies. Other selfies may come across as ambiguous because they are taken out of context and become difficult to read, not least for distant media users. An example of this surfaced in 2014 during the Russian invasion of the Crimean Peninsula when Ukrainian citizens posed for selfies with Russian soldiers. In the photos, Ukrainian people were smiling and making a V-sign with their fingers while standing next to heavily armed, uniformed Russian soldiers. Finding this footage puzzling in its lack of a well-defined message, the news media pondered whether these images were to be regarded as a "tone-deaf faux pas or a genius PR move" (Shim 2014).

Selfie activism

Individuals engaged in activism and other political struggles constitute the second main group creating citizen media selfies. Even if these selfies take their point of departure in the representation of individual identity and history, they have the potential to speak for larger collectives and open a space of appearance, to use Arendt's (1958/1998) expression, for groups which otherwise lack mediated visibility – or lack it on their own terms (Mirzoeff 2017). Regarded as citizen media, selfies are always self-expressions, but they are never only self-expressions.

The online dissemination of selfies forms an open-ended archive, in which the sheer volume of images accumulating under certain hashtags becomes a statement in and of itself concerning, for instance, shared experiences across national borderlines or the extent of resistance (Murray 2015). When circulated and mobilized on social media, selfies generate and are themselves generated by spontaneous and fragmented publics. They are driven by more or less ephemeral emotional and political engagement, which has been discussed in terms of both connective action (Bennett and Segerberg 2013) and affective publics (Papacharissi 2015a). It is, however, important to bear in mind that when these selfies attract attention beyond the immediate circle of the person's social media friends and followers, they do so according to what van Dijck and Poell (2013) term social media logic. When and how certain selfies begin to trend or go viral across platforms is not only determined by the urgency of their political testimony (Chouliaraki 2017). It is, to some degree, also influenced by the corporate and commercial interests governing algorithmic filtering (Mortensen et al. 2018).

Selfies have been disseminated and mobilized in response to a variety of political issues, including questions of race, gender and sexuality. Black activists have for instance posted selfies under the hashtag #TheBlackout in order to generate more and more diverse images of black communities on social media (The Blackout 2017), while LGBTQ rights advocates have used snapshots of same-gender kissing in response to law prohibiting homosexuality in countries such as Russia and Chechnya (Gani 2015). Perhaps the most prominent example of this kind of citizen media selfies emerged in the autumn of 2017 when stories of sexual abuse and harassment were circulated under the hashtag #MeToo. What started as an isolated case – accusations against film producer Harvey Weinstein for systematic sexual aggression – evolved into an international movement. Some of these social media posts included selfies, which tied the experience of sexual transgression or violence to specific bodies. For instance, using this hashtag, Azerbaijani photojournalist Rena Effendi (2017) posted a picture of her bloodied face after a taxi driver had attacked her, because she wanted other victims of sexual violence "to feel less ashamed".

Selfies have additionally documented acts of civil disobedience, for example, when posted by women in Saudi Arabia driving cars when this was still forbidden (Senft 2015). However, what Kuntsman (2017b) refers to as selfie visibility may not always be desirable for individuals taking an active part in civic movements, if they want to stay under the radar of the authorities or general public for one reason or another. Selfie visibility can also be controversial in the context of citizen media when selfies appear to attempt to make no connection with the collective. This was the case with the so-called Riot Hipster, who prompted a strongly negative reaction when a photo was circulated of this male protester caught in the seemingly self-absorbed act of taking a selfie in front of a fire at a protest against the G20 summit in Hamburg 2017 (Neumayer and Struthers 2018). The photograph was disseminated alongside laconically critical remarks such as "[t]hat feeling when you're overthrowing capitalism but just can't resist taking a selfie on your iPhone 7" (cited in Neumayer and Struthers 2018:86). The example of the Riot Hipster controversy directs our attention towards the tensions between self-branding and political engagement, as well as those between entertainment and affective resonance, inherent in selfies deployed in connection with activism. They are frequently not merely "the embodiment of self-awareness, or the expression of a political standpoint, but also a self-branding tactic" (Wei 2015:7).

Selfies ultimately represent both social practices and networked cultural artefacts (Mogoş 2016). On the one hand, citizens produce and distribute selfies to situate themselves in diverse political contexts. Turning the camera towards oneself has become a ritual, a way to make sense of current events, take a stand on political issues, and join online publics of political engagement. On the other hand, selfies are also representations of the self, networked material artefacts, which are produced and mobilized in large numbers across social media platforms. Citizen media selfies in these online environments contribute to patterns of mediated visibility, meaning-making and political opinion formation according to logics of online activism and citizenship as well as to logics of marketization and algorithmic visibility.

Civilians situate themselves in areas of armed conflict through selfies, which tell their story, appeal for humanitarian relief and reclaim the space hit by airstrikes or terrorist atrocities. They on occasion enter into mainstream news coverage, in which case they are used to relay the bodily experience of being present as these violent events unfold. Selfies in other cases may seem to be out of place or to have been created for no apparent reason other than the mundane habit of taking and sharing selfies. In the framework of civic engagement and activism, selfies, as representations of the face, are inevitably politicized. They re-evoke the slogan from the 1960s and 70s that 'the personal is political' by sharing individual bodily experiences that contribute to collective and connective struggles for visibility, equality and justice.

See also: photography; self-mediation; witnessing/testimony

Recommended reading

Chouliaraki, L. (2017) 'Symbolic Bordering: The self-representation of migrants and refugees in digital news', *Popular Communication: The International Journal of Media and Culture* 15(2): 78–94.

This article analyses migrant and refugee selfies published in Western mainstream news media as instances of symbolic bordering, a practice that appropriates, marginalizes or displaces these digital testimonies of the self.

Frosh, P. (2015) 'The Gestural Image: The selfie, photography theory, and kinesthetic sociability', *International Journal of Communication* 9:1607–1628.

Offers a conceptual underpinning of selfies by critically applying key terms from photography theory such as indexicality, composition and reflection.

Kuntsman, A. (ed.) (2017) *Selfie Citizenship*, Cham, Switzerland: Palgrave Macmillan.

Explores the notion of selfie citizenship, that is, the political use of selfies in different geopolitical settings. This edited volume brings diverse perspectives on selfie citizenship as practices of activism, mobilization and witnessing.

SELF-MEDIATION

Katie Warfield

Discussing self-mediation involves entering into a discussion of concepts like subjectivity, identity and selfhoods (self) on the one hand, and products and processes of representation (mediation) on the other. Although they may be regarded as part of a recent phenomenon, Rettberg (2016) situates contemporary forms of self-mediation in relation to others that have existed for generations, like self-portraiture and diaries. In an age dominated by mass media (radio, TV, film and music), self-mediation was less a matter of selves representing selves, as Stuart Hall (1980) classically argued, and more a function of hegemonic forces resorting to major media institutions to present others in ways that served the interests of the majority. The rise of digital and networked modes of self-mediation, including citizen media led by and catering for "unaffiliated individuals and collectives" (Baker and Blaagaard 2016a:1), has greatly facilitated the dissemination of ideas, opinions, feelings and ways of being that challenge institutionalized and systemic norms.

The increasingly participatory, layered, mobile, complex and wearable nature of contemporary media has blurred the boundaries between classic categories like producers, texts and audiences, and contributed to the emergence of produsers (Bruns and Jacobs 2007) and prosumers (Toffler 1970). Against the backdrop of such developments, we may be experiencing the end of audiences (Livingstone and Das 2009) or perhaps, we are living a media life (Deuze 2012). Passive audiences are gradually evolving into active networked publics (Papacharissi 2015a; Thumim 2012) and counterpublics (Fraser 1990). In keeping with these changes, online texts are being regularly remixed and hacked, thus making it sometimes difficult to determine and locate the authentic or original text – whose importance is, in some cases, inferior to that of the novel redistributed products of networked creativity such as memes (Davison 2012).

Situating mediation theories

In *Self-representation and Digital Culture* (2012), Thumim maps four theoretical approaches that communication studies has developed to theorize mediation. The first approach, emerging from non-human technology and materiality traditions, is interested in the role of technology in meaning making. Whereas communication studies has classically separated interpersonal from mediated forms of mass communication, these categories are becoming entangled in the context of digital media – which allow personal tweets to achieve global reach. The second category of mediation theory attends to the sociocultural and discursive flows and forces at play in mediation processes, focusing on "the broader cultural context within which media meanings are made, remade and circulated" (ibid.:52). For instance, this tradition is interested in how social media hashtags challenge dominant normative representations. A case in point is #uglyselfies, which is used by everyday young women as an accompaniment to selfies that show a less polished self. These hashtagged selfies create a provocative contrast to how the female body is usually portrayed in mainstream visual culture. The third category of mediation theory examines close readings of representations in media attending to the "processes that come between those represented in a particular media text and their audiences" (ibid.:54). Examples of these processes could include analyses of identity management undertaken by microcelebrities (Abidin 2016) – a person famous within a niche group of users on a social media platform – and musicians (Baym 2015) who negotiate their self-presentation amid the desires of their fans and audiences. Finally the fourth category of mediation looks to the decentring of accountability, as dispersed intermediaries like citizen journalists and bloggers are becoming increasingly involved in shaping representations in the context of many networked mediation processes. In sum, Thumim's mapping proposes that mediation theory is interested in the material and discursive forces at play in mediation, as well as the changing nature of and dynamics within processes of mediation. These inquiries map in parallel ways onto contemporary examples of self-mediation in citizen media, with one exception: when mediation involves the body, the self and negotiations of subjectivity, affective and emotional forces emerge as an important third force alongside the material and discursive ones.

Self-mediation in citizen media

As discussed, the term mediation points simultaneously to shifts in the material and discursive products, modes and dynamics of media production in contemporary society. The products of self-mediation in citizen media include, for example, blogs, microblogs, selfies and self-imaging practices, representations of the quantified self via tracking devices like a fitbit, representations of the self on social media sites, representations of microcelebrities and their relationships with fans, and politically motivated mediations such as hashtag activism and networked solidarity movements. Importantly the modes of self-representation may be photographic, written, video, art-based (fan fiction), aural, iconic, indexical or symbolic (emojis), and most often include a networked assemblage of many or all of these modes stretched across multiples platforms and networked locations online and offline. These modes are socially constructed as material-discursive assemblages whose material form and affordances are sometimes threaded with racist, sexist, heteronormative, ablest and other forms of discursive inequalities (Wittkower 2016). The dynamics are also layered and enfold the profit motives of major social media platforms of major media stakeholders like Google and Facebook. Self-mediation involves these manifold fluid and changing networked dynamics.

An important difference between classic mediation theories and networked self-mediation is the role that affect, alongside material and discursive forces, plays in the latter. Since one of the primary devices of self-mediation is the cellphone, those publics involved in self-mediation are more intimately tied through their bodies to their devices and audiences in space and time. Their connections to others become importantly affective (Hillis et al. 2015) whereby individuals connect to others through affective means and political protests and movements progress along affective capacities (Massumi 2015). Building on Thumim's account of the material and discursive forces at play in mediation processes, self-mediation in networked society cannot be defined simply as the production and online distribution of an *a priori* offline self; instead, it can be more productively understood as the becoming of permeable online/offline networked subjectivities, amid and through the manifold intimate material, discursive and affective forces at play in networked society.

The next section examines the material, discursive and affective dynamics in a series of case studies of self-representations in citizen media which include: blogging, early social network sites, fandom, news and digital stories, the quantified self, and the spatial self via locative media.

Self-representations in citizen media

Early blogging challenged at once the classic and hegemonic relationship between producer, audiences and texts, as well as the affective relationships with publicly shared texts. Henry Jenkins describes the relationship between selfhood and media that obtains in the context of blogging as "more private and personal than traditional journalism, more public than diaries" (Jenkins 2006b:179). Blogging flourished in the early 2000s and started with a small handful of coders who kept and curated online sites that "pre-surfed" (Blood 2017:2) the web by reading widely and pulling out articles and commentaries from edges and cracks. Jill Walker Rettberg (2016) describes the changing dynamics of audience relations through the metaphors they use to define blogs.

> Where *homepage* remains a fixed noun, the word *blog* has rapidly become a verb as well ... A weblogger filters a mass of information, choosing the items that interest her or that are relevant to her chosen topic, commenting upon them, demonstrating connections between them and analysing them.
>
> *Rettberg 2016:252; emphasis in original*

The producer is therefore a curator-researcher of sorts, in a chain of other researcher-curators, who post, circulate and recirculate timely information. Importantly, blogging challenged the understanding of agency in previous conceptions of mediated subjectivity. Blood argues that whereas mass media's unidirectional flow and corporate motives typically position audiences as passive, blogging trades audiences for publics where affective and counter-discursive forces like a blogger's "sarcasm and fearless commentary reminds us to question the vested interests of our sources of information and the expertise of individual reporters as they file news stories about subjects they may not fully understand" (Blood 2017:2). Blogging as a form of self-mediation in citizen media permitted everyday people to become new intermediaries in the flagging, highlighting and popularizing of marginal discourses.

Microblogging, as illustrated by the sort of conversation that happens on Twitter, is a form of self-mediation that evolved through the establishment of platform vernaculars. Individual hashtags as a form of microblogging often piggy-backed on popular discourses creating a material attachment to a larger affective political dialogue and exchange. Jackson and Foucault Welles (2015) narrate how the New York Police Department's attempt at a positive public relations campaign via the hashtag #mynypd was hijacked by oppositional readings and counterpublics to present and distribute ironic, affective and intense images of the racial violence the police force brings against everyday people of colour in the city. Similar to blogs, microblogging reproduces conversational, curatorial as well as power dynamics. However, material differences in platform design changes the temporality of mediation in microblogging to create shorter bursts of commentary and response. Thus, individual voices possessing sufficient emotive and affective intensity can shift popular discursive dialogues in creative and oppositional ways.

Not to paint the emergence of blogging and microblogging as strictly a matter of technology and critical mass, Theresa Senft (2008) points out the range of content on blogs and in particular the affective and voyeuristic qualities of early blogging, which can be seen in the appeal of early livecamming. Livecamming involves the livestreaming of one's personal verbally and visually recorded thoughts, reflections, confessions and stories. The appeal of being afforded an intimate view of another person's diaries is facilitated by the balance between distance – accomplished via the material affordances of being able to remain anonymous – and remaining in affective proximity, i.e. intimate and close to taboo topics typically beyond the appropriate boundaries of offline everyday conversation.

In terms of early social network sites, danah boyd and Nicole Ellison (2007:2) famously traced a brief history of social network sites and defined them as

> web-based services that allow individuals to (1) construct a public or semi-public profile within a bounded system, (2) articulate a list of other users with whom they share a connection, and (3) view and traverse their list of connections and those made by others within the system.

In using such web-based services, audiences become networked publics (boyd and Ellison 2007) and socially mediated selves. Networked publics are not simply sterile entanglements of people seeking to establish connections based on discursive common interests (like news and politics); indeed, they are also affective publics (Papacharissi 2015a) sharing and connecting on emotional and affective planes. Early work on social media shows these sites as being spaces of networked community building by marginalized groups like geeks, freaks and queers (boyd 2017b:2).

The issue of visibility and invisibility is also important when discussing self-mediation on early social media networks. What users choose to disclose or hide on most social media platforms is dual-layered: on the one hand platforms determine how much user data will be made public or remain private based on details laid out in user agreements and default settings. These settings tend to make content as public as possible. On the other hand, users also determine community practices of disclosure and privacy, which often run contrary to platform definitions. Writing about self-imaging practices of women attracted to women users on hook up apps like Tinder, Stef Duguay (2016) argues that participants elastically stretch and contract their degree of visibility and disclosure. Duguay calls this practice "identity modulation" (Duguay 2016:86), which is a result of platform design, community practices and app affordances. Identity modulation, understood as a form of self-mediation, involves the affective sense of safety and comfort felt by someone displaying themselves online, alongside strategic manipulation of material platform affordances to make one's queerness more or less visible. In her writings about women sharing topless photos on the subreddit Girlsgonewild, Emily van der Nagel (2017b), who studies practices of pseudonymity, argues that users have very refined practices of disclosure and visibility in online spaces which are negotiated with and against platform affordance (materiality), community practices (discourses) and users' emotional sense of safety (affect).

The study of online fandom, located at the intersection of fandom studies and Internet studies, is broadly interested in the changing nature of, and the relationship between, celebrities/popular texts and online publics. Henry Jenkins' *Textual Poachers* (1992) borrows from the work of Michel de Certeau to suggest that the online production of fan texts involves tactics and strategies that allow produsers to negotiate their relationship with texts, audience, platform and selfhood. Although these fan productions seem to be more communicative than about self-mediation, they often reveal aspects of the digital subjectivities of fans. For instance Bennett (2013) discusses the important ways in which fan fiction can operate as a mode of self-reflection. Bennett suggests that "the fan object is deeply interwoven with our sense of self and who we would like to be" (Bennett 2013:154), thus implying that fans often project an idealized version of the self onto the object of fandom. Stein and Busse (2017) broaden the mechanics of fandom production to include non-human forces when they write that the creative forces behind fan fiction are the result of the original text, intertext and community relationships, as well as interface or technological constraints. This combination of forces creates new texts, which challenge classic ideas about authenticity, ownership and authorship (Stein and Busse 2017).

Self-mediation is also a political act with the potential to either empower citizens and publics or constrain their democratic participation. Chouliaraki (2010b) states that discourses on empowerment tend to emerge from self-mediation located outside established institutional structures of power, whereas discourses on constraint tend to emerge from within such systems (Chouliaraki 2010b). Thumim has also examined the process by which self-mediation variously constrains citizen freedom, but she argues that these degrees of freedom and constraint are deeply entangled with the modality (i.e. the materiality) and genre of mediation (mass media, online media, public media) (Thumim 2012). Self-mediation can provide counter-discourses via more diverse depictions of everyday lives, histories and the social situations of marginalized people (Yefimova et al. 2015). Self-mediation can also present the authentic faces of celebrities and politicians, whereby authenticity can be at once a challenge to the typical visual depictions of people in positions of power and evoke more intimate emotional responses between celebrities and audiences (Baishya 2015). Further, compilations gathered by news networks of images taken with handheld digital material phones by everyday citizens can act as grassroots witnesses to provide multiple perspectives to the reporting of a news event (Roberts and Koliska 2015; Brantner and Lobinger 2015). Kuntsman and Stein (2015) have written extensively on the political uses of selfies and particularly on selfie militarism, which marked the Israeli military's first efforts to employ social media as public relations tools in the aftermath of 2008–2009 Israeli incursion into the Gaza Strip. Through manifold forms of collective networked self-mediation like hashtag activism (Williams 2015), hashtag hijacking (Jackson and Foucault Welles 2015) and networked solidarity (Papacharissi 2015a), networked counterpublics have emerged as collective forms of self-mediated resistance towards dominant systems of power.

The quantified self is a quantified form of self-mediation that plays an important role in examining the interplay between selfhood and mediation. Research on the quantified self examines technologies, tools, apps and systems that capture data about individuals' actions, habits and practices. These include trackers, such as fitbits, apps to measure quality of sleep or sex, as well as the repercussions of such technologies on one's relationship to one's sense of self and one's body or the becoming of the body via quantified tools of mediation. Van Dijck (2017) defines dataism as an ideology associated with the emergence of networked societies that seeks to quantify behaviours by drawing on various types of data about social phenomena. Markham and Baym (2009) argue that, under dataism, data is reified over that which it represents. The material, discursive and affective complexity of the phenomena behind the data matters less than the rational measurable/quantifiable numbers. The mediated self studied by proponents of dataism becomes the data double (Haggerty and Ericson 2000). Erasing affective motivation, the data double body is seen as an assemblage of distinct empirical quantified flows that can be mined for various imperatives, but mainly commercial as is often done with data circulating on social media platforms and collected via such tracking apps.

With the increasing pervasiveness of mobile technologies concurrent with location-based media and apps, self-mediation must also consider mediated selfhood as a spatial practice. Jason Farman (2012) suggests a co-constitutive conception of space and, following Henri Lefebvre, argues that spaces and bodies "produce one another and this production must be theorized with cultural and physiological specificity. We don't all have the same bodies, nor do we have the same experience of embodiment/embodied space" (ibid.:18). Thus location-based technology theory, and the study of mobile apps, is generally interested in an approach to technologies influenced by phenomenology: how technologies are used by radically unique individuals in different offline places, locations, at different times of the day, which themselves are also punctured by radically unique affective and discursive flows. Apart from location-based technologies and mobile apps (Sutko and de Souza e Silva 2011; Schwartz and Halegoua 2015), there are also theorists interested in concepts of emplacement or the experience of identity at the intersections of offline and online spaces (Farman 2012; Hjorth and Pink 2012; Hjorth and Hendry 2015). Researchers interested in concepts of emplacement focus on the varying interactions among bodies, offline spaces, technologies, often from a phenomenologically informed position. Wargo (2015) looks at the elastic nature of selfhood that comes about in different offline spaces via self-images and photos of the everyday taken and shared on Snapchat. Both groups are interested in self-mediation where a core component of both selfhood and media is a consideration of material and ephemeral spaces: online and offline, material and discursive, and affective, socioculturally situated and individually felt.

Future directions

This entry began by arguing that as personal mediation devices become smaller, more mobile and thus more integrated into our everyday, the material, discursive and affective forces will become more tightly entangled and folded under the skin of the user over time. This will continue to be the case with the advent and mass marketing of even more novel personal mediation devices like AI, immersive 3D video and livestreaming geo-location apps.

See also: fandom; parkour; remediation; selfies; space and place

Recommended reading

Baym, N. (2015) *Personal Connections in the Digital Age*, second edition, Cambridge: Polity Press.

Baym discusses the impact of digital technologies on our personal lives. She speaks with nuance about the macro forces that shape digital self-presentation and digital social relations and relationships.

Rettberg, J. (2016) *Seeing Ourselves through Technology*, Basingstoke, UK: Palgrave Macmillan.

This text provides a detailed history of the historical mode of self-mediation, like diaries, and compares them to contemporary manifestations of self-mediation like selfies, blogs and wearable devices.

Thumim, N. (2012) *Self-Representation and Digital Culture*, Basingstoke, UK: Palgrave Macmillan.

Provides a companion text to Rettberg, but offers more wide-ranging and non-digital examples of contemporary storytelling via media like reality TV, digital storytelling and oral histories in museums.

SOCIAL MEDIA

Neil Sadler

Contemporary web platforms often described as social media are both numerous and diverse, rendering them notoriously difficult to define (Obar and Wildman 2015). For example, Facebook, Twitter, YouTube, Instagram, Snapchat, LinkedIn and Flickr are all popular in Western Europe, the US as well as many other parts of the world; VK and OK.ru are widely used in Russia; WeChat and Weibo are dominant in China. There are also many other more narrowly focused social media sites oriented to specific constituencies such as academics (academia.edu), gay men (Grindr), seekers of undiscovered musicians (Soundcloud) and right-wing political activists (Gab). These sites vary greatly in terms of their technical affordances and user bases, the content types they foreground (video on YouTube and Vimeo, images on Instagram and Flickr and text on Twitter) and the types of relationship between users they promote (for instance, following on Twitter compared to reciprocal relationships on Facebook). At the time of writing, however, the technical affordances of social media have increasingly converged – text, images and video content are now all fully integrated on Instagram, Twitter and Facebook, for example, and Facebook has enabled Twitter-style monodirectional following. Nonetheless, certain features are prototypical of social media, allowing diverse sites to be viewed as at least a roughly coherent group. First, they all emphasize, and rely on, user-generated content. Second, in almost all cases, they are oriented to the generation of profit. Third, they enable two-way communication between geographically dispersed individuals. Fourth, they enable citizens to transmit messages to large audiences. This entry discusses in more detail the implications of these features as they pertain to citizen media.

User-generated content

Almost all social media emphasize communication between users rather than between site developers and users. Social media sites were not the first websites to do this and it would be wrong to characterize the pre-social media Internet as unsocial. During the 1990s, a wide range of formats emerged based on user–user communication, including fora, chatrooms and Internet relay chat, and it is from these formats that contemporary social media are derived. Yet Internet-based communication in the Web 1.0 era remained largely one-way. Site developers, be they governments, institutions, collectives or individuals, produced content and published it online to make it available for consumption by others. The technological and cultural barriers to creating websites meant that content creation remained the concern of a relatively small proportion of Internet users, themselves a relatively small proportion of broader populations. As such, this approach largely remained within the dominant pre-Internet paradigm of active authors and relatively passive readers.

The development of Web 2.0 in the early 2000s, based partly on the availability of greater bandwidth and partly on developing ideas regarding the uses to which the Internet could and should be put and by whom, represented a major shift in the sociality of the web. The success of pioneering social networking sites, such as Friends Reunited (2000), MySpace (2003) and Facebook (2004), transformed web-based communication from the domain principally of a limited group of technology enthusiasts into the social and cultural mainstream. With sites such as these, developers provided a platform only and relied entirely on users for the provision of content; Friends Reunited did not collect information on its members' education histories but provided a means for the site's users to upload this information. It was the content uploaded by users which was of interest to other users, rather than the platforms themselves. This model has continued unchanged to the present day – every post on Twitter, photo on Instagram and video on YouTube has been created by a user rather than by staff working for these sites. Consequently, the uses to which different social media platforms are put are defined almost entirely by users whose activities are influenced by the technical affordances of individual sites, but who are otherwise largely unconstrained. It is this flexibility that allows for sites such as Twitter to be used for purposes as diverse as maintaining discursive communities around particular interests (Dayter 2015); presenting the self as a personal brand (Page 2012); virally marketing products (Araujo et al. 2017); engaging in mainstream political debates (Conover et al. 2011); and conducting social activism (Ince et al. 2017).

Profit-orientation and digital labour

Although the uses to which social media are put by their users are extremely diverse, the service-providers in almost all cases have a single objective: the generation of profit. Social networking sites generate revenue in two main ways: advertising and the sale of user data. The former requires maximizing the amount of time users spend on their sites as this increases user exposure to advertisements, allowing higher prices to be charged to advertisers. The latter also requires maximizing the depth of user engagement – users employing sites in more ways results in the generation of more fine-grained data which allows for the creation of more detailed user profiles. More nuanced user profiles then allow for more tightly focused advertising. For example, a Facebook user who has liked the pages of their favourite businesses, checked in to their favourite restaurants, posted political opinions and uploaded their date of birth can be targeted precisely with advertisements, whether for products, political parties (a trend that began during the US election campaign of 2008) or services. Social media developers therefore ultimately care little about the purposes for which different users employ their sites, provided they are employing their sites and that their engagement is as deep as possible as this is the basis of their economic model.

Capturing users' attention in order to facilitate advertising is not, of course, new. This has been the central crutch of most newspapers for over a century and for commercial television for many decades, famously theorized as the audience commodity (Smythe 1977). The major innovation of social networking sites was in delegating the production of that content to users themselves – users become the producers of the product for which they pay, through their exposure to advertising, to access. This phenomenon, common to all social media sites, has been critically examined through the notion of prosumption (Bruns 2005, 2010c) as the once sharp distinction between producers and consumers becomes increasingly blurred. Although some have hailed the rise of prosumption as signalling a positive democratic shift (Jenkins 2006a), prosumption has also been heavily critiqued by Marxist scholars through the concept of digital labour (Arvidsson and Colleoni 2012; Fuchs 2015; Fuchs and Sandoval 2014; Scholz 2012a). Proponents of this critique argue that the encouragement and harnessing of prosumption on social media is intimately linked to the drive inherent to modern capitalism towards ever increased economic productivity and that it is a means for making leisure time generate economic value as play and labour are blurred into 'playbor' (Scholz 2012a). Although users may enjoy using social networking sites and be able to employ them for activist ends, they also generate surplus value for the sites themselves, ultimately collected as profit.

As many, if not most, users are not aware of this, many Marxist critics suggest that social media are intrinsically exploitative, regardless of the uses to which they are put.

There is disagreement, however, regarding the nature of the value thus created. Fuchs (2014a, 2014b, 2015) has argued repeatedly that the value produced is principally a function of the time users spend on social media and thus the volume of adverts to which they are exposed. Arvidsson and Colleoni (2012), on the other hand, argue that time by itself means little, and that value is principally generated through affective engagement with advertising material, rather than through simple exposure. Their argument is grounded in the contemporary economic value of intangible assets such as a brand and prevalent advertising models which, although they vary from site to site, in many cases are priced according to the number of users who actually click on adverts, rather than a raw number of impressions. This distinction is extremely important for citizen media use of social media as, from Fuchs' perspective, citizen media use of such sites which increases traffic and attracts user attention to them (within the so-called attention economy) is inevitably generating revenue for the corporations which own these sites. From Arvidsson and Colleoni's perspective, on the other hand, the value produced by attracting users to a site depends greatly on the activity of those users and the extent to which they meaningfully engage with the advertisements to which they are exposed. This suggests it is possible, at least in principle, for activists to minimize the extent to which their digital labour is exploited, if the users they attract do not engage with advertising material in the manner desired by site developers.

Facilitating dialogue, filter bubbles and the public sphere

There is no doubt that social media greatly facilitate communication between geographically dispersed individuals. In principle, they also facilitate dialogue between ideologically divergent groups by creating a space populated by people representing a wide range of ideological standpoints – mainstream social media sites such as Facebook, Twitter and YouTube are used by liberals and conservatives, doves and hawks, the religious and the secular. This has led to significant discussion of social media with reference to Habermas' notion of the public sphere (Jorba and Bimber 2012; Downey and Fenton 2003; Howard 2010; Papacharissi 2010; Tufekci and Wilson 2012). The public sphere, as Habermas (1962/1989:27) understands it, is a space where "private people come together as a public" to engage in a "debate over the general rules governing relations in the basically privatized but publicly relevant sphere of commodity exchange and social labor". In such spaces, which may be physical or virtual, citizens engage in free discussion and debate in order to address major issues of common concern to that society. In their ideal form, according to Habermas (ibid.) only rarely or never genuinely realized, such spaces are characterized by a lack of concern for status or social hierarchy and by a diversity of viewpoints, on the basis that the common concerns at stake prevent the participants from fracturing into cliques. Habermas' vision of an ideal public sphere is therefore closely aligned with early utopian imaginaries of the Internet, similarly characterized by a lack of hierarchy and as a site for constructive debate about matters of common concern.

Reality, however, is frequently somewhat removed from this ideal. Although social networking sites may be used by groups holding very different beliefs regarding matters of common concern, contact between such groups, and particularly meaningful contact based on respect for alternative viewpoints, is often limited. This is elegantly shown by Lotan (2014) in visualizations regarding interactions between Twitter users in the context of the 2014 Israeli invasion of Gaza which show significant activity by both supporters and opponents of Israel's actions yet remarkably little interaction between the two groups. Furthermore, offline hierarchies limiting the range of potential communicative interactions are often replicated on social media (Fuchs 2014c). In practice, social media sites, as experienced by users, have a strong tendency to devolve into filter bubbles (Pariser 2011) where, although a potentially vast range of perspectives are in principle available, users are in fact exposed to a relatively narrow range of viewpoints which often accord with attitudes they already hold.

There are two main reasons for this. The first, and more contentious, is the suggestion that a general human tendency towards homophily – that is, the desire to associate primarily with people who share similar traits and viewpoints (McPherson et al. 2001) – is replicated in social media. The available empirical research on this issue is, however, somewhat equivocal, with many studies finding a high degree of homophily on social media (Aiello et al. 2012; Krivitsky et al. 2009; Sunstein 2018; Wu et al. 2011), and many others finding that, on the contrary, social media users are exposed to a wide range of viewpoints (Conover et al. 2011; Fletcher and Nielsen 2017; Gruzd et al. 2011; Hermida et al. 2012; Yardi and boyd 2010). The second reason is the ever-increasing role of algorithms in shaping the content seen by users, a phenomenon which has led to the development of critical algorithm studies as a sub-field in its own right (Gillespie and Seaver 2016; Jürgenmeyer and Krenn 2016; Seaver 2017). For any given user, the advertisements for products and political and social causes seen on Facebook, the hashtags and keywords marked as trending on Twitter, and the videos recommended on YouTube are selected by opaque algorithms which choose relevant content on the basis of the user profiles almost all social media sites generate for their users on the basis of their behaviour. That this leads to users in most cases being presented with content with which they already agree is a consequence of the fact that commercial social networking sites are ultimately concerned with maximizing revenue through the mechanisms described in the previous section, rather than with serving a public need and facilitating the creation of a genuine public sphere.

Democratized mass communication

Where previously it was difficult, in practical terms, for citizen groups to transmit messages to large audiences, social media sites provide a ready means for doing this at minimal financial and organizational cost. This has undeniably empowered actors in new ways. For example, the Egyptian media collective Mosireen, which documented the Egyptian revolution of 2011 through film, was at one point the most watched non-profit YouTube channel in the world and its most popular videos have been watched hundreds of thousands of times. It is impossible to imagine a similar citizen group being able to distribute video content to such large audiences in the pre-social media age. Social media may not necessarily make it easy to reach and convince groups who hold opposing views but they do make it far easier to communicate with large numbers of like-minded people than was previously possible. Social media are thus a powerful tool for activists, regardless of the challenges of breaking in or out of users' filter bubbles or the difficulties of creating a genuine public sphere.

This idea underpinned cyber-optimist scholarship and media coverage during the late 2000s and early 2010s which suggested that social media would enable a new wave of counter-hegemonic citizen activism, as exemplified during the so-called movements of the squares in Egypt, Turkey and the USA. Proponents of this idea, notably Shirky (2008, 2011), argued that the availability of easy mass communication at minimal direct cost solved several problems which had plagued social movements up to that point. Haunss (2015) summarizes the key advantages of social media as proposed by the cyber-optimists: first, they reduce transaction costs, making it easier for geographically dispersed individuals to communicate and coordinate with one another. Second, they solve the 'rational choice' problem of collective action, enabling large, but weakly connected, groups to form easily without the need for strong collective identities (Bennett and Segerberg 2012). Third, they reflect the logic of the network society (Castells 1996, 2012) and align with broad shifts towards individualization and from collective to connective action. Finally, they enable new forms of organization which do not require strong centralized leadership.

On the other hand, techno-pessimists, a movement closely associated with Evgeny Morozov, argue that these factors are less significant than they first appear. For a start, while social media clearly reduce transaction costs, the larger question is whether transaction costs were previously a major barrier to effective campaigning (Haunss 2015). Second, while weak-tie groups may form more easily on social media sites than in offline campaigning, the resulting formations also dissipate more readily. Morozov (2011:220) goes as far as to suggest that the possibilities for online activism that social media offer may distract activists from more effective, if more dangerous, activism on the ground. Third, conceiving of citizen movements as networks is not new and did not require the emergence of social media. Moreover, in exacerbating social trends towards greater individualization, such sites may further weaken the power of citizens to work collectively to achieve change. Finally, debate continues as to the extent to which social-media-based leaderless organization is possible and effective, with significant research indicating that many horizontal movements retain some degree of hierarchy and continue to rely on the organizational expertise of committed activists (Abdelrahman 2014; Hardt and Negri 2011; Prentoulis and Thomassen 2013; Stavrides 2012).

Following this early dichotomy, debates on the links between social media and citizen movements have become more nuanced. Social media are now seen as one tool among many available to citizens (Brym et al. 2014; Gunning and Baron 2014). There is now something of a consensus that social media are neither a necessary nor sufficient condition for democratization but are nonetheless an important tool for contemporary citizen movements (Brym et al. 2014; Gunning and Baron 2014). They may not provide a technological quick fix for society's ills, but they do offer communicative possibilities to activist and other citizen groups that were previously unavailable and that have come to play an important role in enabling the formation and maintenance of transnational citizen groups, whether oriented towards political reform or simply based around shared interests and beliefs (Brym et al. 2014; Gunning and Baron 2014; Kim et al. 2017; Milošević-Đorđević and Žeželj 2017; Murthy 2018b; Papacharissi 2015a).

See also: activism; authenticity; civil society; content moderation and volunteer participation; Facebook; mobile technologies; public sphere; sousveillance; temporality; Twitter and hashtags; user-generated content; Weibo; YouTube

Recommended reading

Fuchs, C. (2017) *Social Media: A critical introduction*, second edition, London: Sage.

Based primarily in Marxist and Frankfurt School critical theory, Fuchs provides an ideal entry point to the now extensive body of literature on social media conducted from this perspective and effectively highlights important questions of power, dominance, ideology and the unequal relationships between citizens and capitalist states.

Haunss, S. (2015) 'Promise and Practice in Studies of Social Media and Movements', in L. Dencik and O. Leistert (eds) *Critical Perspectives on Social Media and Protest: Between control and emancipation*, London and Lanham, MD: Rowman & Littlefield, 13–34.

Part of an excellent collected volume, the introductory chapter by Haunss provides a clear and detailed overview of the development of key debates in scholarship of social media and protest from the 1990s to the time of publication.

Papacharissi, Z. (2015) *Affective Publics: Sentiment, technology, and politics*, Oxford and New York: Oxford University Press.

Presents a sophisticated account of the central role of affect in shaping social media communication, particularly with regard to the formation of issue-based publics. Papacharissi also provides an excellent example of how qualitative and large-scale quantitative analysis can be effectively integrated into social media research.

SOCIAL MOVEMENT STUDIES AND CITIZEN MEDIA

Tina Askanius

Social movement studies is an interdisciplinary subject area located at the intersections of political science, anthropology and sociology; it draws on a number of intellectual and methodological traditions to explain processes of social mobilization, collective action and collective identity formation. Social movements may be understood as "(a) mostly informal networks of interaction, based on (b) shared beliefs and solidarity, mobilized around (c) contentious themes through (d) the frequent use of various forms of protest" (della Porta and Mattoni 2015:1496). Importantly, media, mediation and a variety of communication processes are at the heart of each of these dimensions of social movements, and bringing attention to the intersections of media and movements is vital to understanding processes of collective action, political identity formation and cycles of contestation. As a conceptual and theoretical horizon, citizen media has often been deployed in scholarly analysis to understand this media–movement nexus and raise questions about how civil society actors organize collectively and engage in activism in, through and about the media (Mattoni 2013b).

Social movement studies has traditionally paid only "tangential attention to media dynamics", and a "persistent divorce" between media studies on the one hand and theory and research by sociologists, political scientists and historians on the other has created a certain fragmentation of knowledge on the topic (Downing 2008:41). This situation began to change at the turn of the century, and a growing body of work is now seeking to bridge these historic gaps. The notion of social movement media practices has proved a particularly fruitful unifying concept around which scholars are seeking to cross-pollinate perspectives from movement scholarship and media and communication studies. This entry focuses on social movement scholarship that approaches social movement media as a form of citizen media. Much of this work draws on the notion of media practices in contentious politics and proposes a turn to practice theory as a means of avoiding media-centrism, one-medium/platform reductionism and binary distinctions between digital and non-digital as well as old and new forms of citizen media (Mattoni 2017; Mattoni and Treré 2014; Kaun 2016a). Further, while ensuring that the broader historical trajectory of this relationship is not ignored, the entry pays particular attention to the adaptation and appropriation of digital media and later social media into social movement media repertoires and how this has influenced both strategic and expressive activities of social movement actors and organizations.

Finally, providing an account of citizen media from the perceptive of social movement studies implies that privilege is given to the role of citizen media in movement building, collective action and strategic communication at the expense of the appropriation of such media and media practices in looser, more individualized and ephemeral forms of civic engagement and activism of unaffiliated citizens that have also become widespread since the 2010s.

Bridging social movement studies and media and communication studies

Traditional approaches to social movements – works rooted in resource mobilization theories, political process approaches and new social movement theories – have paid little attention to communication processes and only sketchily broached issues around media, mediation and communication, while failing to systematically address their role in mobilization and movement building, with few exceptions such as Cammaerts et al. (2013), Downing (2001), van de Donk et al. (2004) and Earl and Rohlinger (2010). Kavada (2016) argues that communication is indeed the spectre haunting collective action theory. The spectre, however, lurks in the background of key works in social movement literature: it underlies the conception of collective identity as a process of interaction in the work of Melucci (1996), just as it is implicitly present in the work of Tilly (2005), who considers contentious politics as "an ongoing conversation between claim-makers and their targets' (Kavada 2016:9).

The absence of a cross-disciplinary framework with which to understand citizen media more fully can partly be explained by the very nature of both media and communication studies and social movement studies, as the two are rooted in different yet equally fragmented fields – including sociology, anthropology, political science and psychology – that seldom speak to one another (Roggeband and Klandermans 2007). An additional reason for the two fields' fragmentation and blind spots in bridging knowledge on these matters may have to do with the fact that collaborations between their respective scholars are only rarely established in any formalized or sustained way. There are, however, signs that this situation is beginning to change as new technologies giving rise to new forms of mediated protest and collective actions are forcing a rapprochement between sociology and communication (Earl and Garrett 2017).

The emergence of digital media has thus brought media and communication into sharper relief in social movement studies (Kavada 2016), as evident in the notion of social movement media practices around which scholars are seeking to cross-pollinate concepts and perspectives from both fields. The work of Mattoni (2012, 2017) and Treré and Mattoni (2016) has been particularly important in making the connections between the two disciplines visible, as has the growing body of work on media and mediation in the Occupy Wall Street movement, the anti-austerity protest movements across Southern Europe and other protest movements across the MENA (Middle East and North Africa) and BRIC (Brazil, Russia, India and China) regions. Important steps towards bridging the two fields have also been taken in the work of Kavada, who argues for the need to take communication seriously in social movement theory by proposing an understanding of collective action as "emerging in conversations and solidified in texts" (2016:9). Another explicit attempt to create cross-fertilization between the two fields involves deploying the concept of mediation opportunity structures, as proposed by Cammaerts (2012) and further developed by Uldam (2013) in her work on online mediation opportunity structures embedded in the anti-capitalist strands of the environmental movement. Work on mediation opportunity structures combines "theories of mediation with social movement theories that assess the opportunities and structural constraints for social movements as well as the logics they attribute to their protest actions" (Cammaerts 2012:122).

Others, like Earl and Garrett (2017), have sought to systematically integrate perspectives from political communication into social movement studies to explain how the rise of new forms of digital protest renders traditional assumptions about the media and movements increasingly irrelevant or outdated. Shifting the focus from strategic communication and media tactics to the audiences of social movement communication and how messages are received, they argue, can enrich our understanding of citizen media in the digital era and expand social movement scholarship more generally (ibid.:480). Flesher Fominaya and Gillan (2017) further propose approaching the issue from the perspective of a technology–media–movements complex (TMMC) to inquire into the use of media in social and political movements and allow for a better integration of the different intellectual traditions that are currently focused on technology, media and movements as separate strands.

Common to these perspectives is an explicit focus on how media, mediation and communication are involved in the very construction of collective struggle and identity formation, and hence cannot be written off as secondary or merely ancillary dimensions of action repertoires. Instead, these approaches place communication processes and the role of citizen media in these processes at the very heart of how we might understand collective action and protest movements today.

Citizen media and/in social movements

Early studies on media–movement dynamics are often divided into two strands of related inquiry, focusing either on mainstream/news media framing of social movements (Gitlin 1980; Halloran et al. 1970; Snow 2004), together with the effects or resonance of these frames (Snow and Benford 1988), or on the production and circulation of alternative, citizen media to create counter-frames and produce autonomous spaces and public spheres. In the first strand of literature, scholars have focused on the asymmetric relationship and power dynamics between mass media and social movements. Mass media have traditionally been crucial to social movements, not least because they "carry movement ideas to a broad audience and give activists leverage in institutional and political processes" (Rohlinger and Vaccaro 2013:736). Other studies have examined the reasons for and consequences of the tendency to frame social movements and protests negatively with reference to the practices, conventions, frameworks and business models inherent to journalism and the media system. The second strand of scholarship highlights instead the different creative, potentially subversive ways in which social movement actors create new or re-appropriate existing media for activist purposes, partly as a response to being misrepresented or ignored in mainstream media and partly based on the impetus to operate outside or in the margins of state-run or commercial media systems (Downing 2001; Rodríguez 2001). Beyond examining available tools to counter-frame and respond to mainstream media, scholars belonging to this strand have examined citizen media as a critical dimension of intra-movement communication and an important part of collective identity construction and symbolic meaning-making processes.

With the emergence of the Internet, attention to the role of media in social movement literature increased dramatically. When the Zapatista movement emerged in Mexico in 1994, it received considerable attention among both social movement and media scholars and is often highlighted as one of the first examples of a collective actor to engage in what were then new forms of digital and online-based citizen media practices. Explicitly drawing on the networks of communication offered by the Internet in their struggle for indigenous autonomy, the Zapatistas in turn inspired the alterglobalization movement, primarily in the US and Europe, which to a large extent built its communicative infrastructure around the online site Indymedia to protest neoliberal globalization in the late 1990s and into the first years of the millennium (Pickard 2006).

Yet another upsurge in academic attention to citizen media and social movements occurred with the 2009–2011 spread of protests worldwide, from the popular uprisings in the Middle East and the so-called movements of the crisis protesting austerity measures and education cuts in Europe (della Porta and Mattoni 2015) to the Tea Party Movement and Occupy Wall Street in the US. Often understood with reference to specific media technologies and labelled as the Facebook and Twitter revolutions of the so-called Arab Spring and beyond, some of the interest in questions around social mobilization and the media technologies which facilitated and to some degree orchestrated the protests were, however, saturated by media-centrism and some degree of Internet optimism. This gave rise not only to unfruitful and simplified perspectives that implied a direct causal relation between media and social change, but also disregarded the broader geopolitical contexts and historical trajectory into which these events were inscribed; Kaun and Uldam (2017) offer a critical discussion of this post-2010 hype. In the wake of these facile interpretations, and partly as a response, social movement media scholars have offered sobering empirical accounts of actual media practices – rather than technological affordances – and the possibilities and limitations offered by technological developments, along with in-depth ethnographies of the everyday workings of media technologies as experienced by social movement actors (Barassi 2015a). An extensive body of critical research has thus emerged on the role of social media in anti-austerity protests (Figueiras and do Espírito Santo 2015; Treré et al. 2017), including the Indignados protests/15M encampments (Postill 2014; Gerbaudo 2012), anti-capitalist movements (Barassi 2015a), the climate change movement (Askanius and Uldam 2011; Bennett and Segerberg 2011), the immigrant rights movement (Costanza-Chock 2014), feminist movements (Fotopoulou 2017) and the Occupy movement (Costanza-Chock 2012; Juris 2012; Kavada 2015; Kaun 2016b).

Gerbaudo (2017) argues that the emergence of social media marks a fundamental shift in social movement media practices, from cyber-autonomism to cyber-populism, and that the two periods of digital activism correspond not only to two phases of technological development of the Internet (the so-called Web 1.0 and Web 2.0), but also to two different protest waves, first the alterglobalization movement, and later the movement of the squares that began in 2011. Whereas the first wave was "informed by the 70s and 80s counter-culture, DIY culture, and the tradition of alternative media, from pirate radios to fanzines", the second wave adopted what he considers a technopolitical attitude which sees the Internet as "a space of mass mobilisation in which atomized individuals can be fused together in an inclusive and syncretic subjectivity" (ibid.:479). Social media are, however, not necessarily best understood as citizen media in the sense proposed by Baker and Blaagaard (2016a), or even as compatible with social movement activism. Some scholars suggest we should see social media as replacing or subsuming rather than embodying new forms of citizen media, as the latter are traditionally built on ideals of democracy and participation that are very different from those underpinning the business models and algorithmic design of platforms such as Facebook, YouTube or Twitter. Several studies have sought to draw attention to the dangers associated with allowing these corporate platforms to take over the spaces and tools that make up the motley array of independent and non-commercial citizen media and the problematic aspects of conflating citizen media with social media, conceptually and in practice. Activists are increasingly being subjected to surveillance and censorship or are inhibited in other ways and deterred from engaging online. In her studies on British Petroleum's surveillance of activists in the climate change movement, for example, Uldam (2018) shows how activists face increasing challenges in protecting their online security and privacy and argues that while social media may afford an unprecedented level of visibility for social movement actors, this visibility comes with the risk of being monitored not only by governments but increasingly also by corporations looking to manage and eliminate reputational risks. Others stress that the commercial logics of profit-driven social media are not only not designed for activism but are also potentially detrimental to the activist cause (Askanius and Gustafsson 2010; Barassi 2015a; Youmans and York 2012).

Finally, the widespread use of social media by social movement actors has raised questions relating to issues of historicity, time and democratic practice. These emerge from divergent temporalities between the slow political processes of movement building and organizing and the accelerated pace and cultures of immediacy of these technologies. In her work on three historical movements of the dispossessed in the US, Kaun (2016a) discusses the contradictions between the long-term organizing for social change in social movements and the time regimes of the media used for these purposes. She traces the increasing social acceleration related to media technologies employed by activists in relation to their political work and identifies a desynchronization between what she calls social media time on the one hand and political time on the other. This desynchronization essentially disrupts activists' ability to connect past and present struggles in meaningful ways and to build viable political projects. Similarly, Milan's (2015) work on cloud-protesting and collective identity demonstrates how social media are primarily built for social movements operating in the present moment, because these media work through algorithms that compute popularity and interactions and produce regimes of "real-timeness" (ibid.:890).

Future directions

Since the turn of the century, scholars have been calling for research on media–movements dynamics that avoids media-centrism/techno-determinism, goes beyond binary conceptualizations of offline/online and new/old media, ensures holism in approach and design, and pays attention to history by contextualizing studies of contemporary forms of collective action and the role of citizen media in them.

Avoiding media-centrism in the study of citizen media is among the most widely debated themes in the literature. Adopting practice-based approaches has been put forward as one way of pursuing a non-media-centric perspective. Couldry's (2004) much-cited call for media research to enter a new paradigm by adopting such an approach urged scholars to focus on what people are actually doing with media instead of assuming that technological affordances necessarily enable or empower citizens in specific ways. In social movement scholarship, numerous studies have since sought to marry media studies and practice theory, proposing concepts such as activist media practices (Mattoni 2012), social movement media practices (Costanza-Chock 2014) and citizen media practices (Stephansen 2016; Stephansen and Treré 2019) as overarching frameworks with which to approach the movement–media relationship analytically and conceptually. The reorientation towards practices offers a way of embracing all the processes involved in citizen media (production, content/text, reception, appropriation, recirculation/mediation, archival practices, memory work) in ways that challenge reductive accounts of citizen media as restricted to one process or practice. On a more fundamental level, a practice-based approach offers a means of avoiding communicative reductionism and of bringing a non-media-centric framework into the analysis of citizen media in social movement politics.

By increasingly approaching the study of social movements and citizen media from the perspective of practice theory, current scholarship is also concurrently challenging some of the tenacious binary oppositions that have long dominated the field. Such binaries include thinking about citizen media in terms of offline/online, real/virtual or analogue/digital. An early generation of studies which took an interest in online media and collective action had a tendency to draw up artificial boundaries between offline and online, seeing these as two separate spheres and sites of analysis. In moving beyond dichotomous and binary distinctions that gloss over the various ways in which the online/offline spheres tend to overlap and interact, scholars are increasingly encouraged to engage with flexible and organic research that spans both spaces (Treré and Mattoni 2016). Overcoming the online/offline dichotomy allows us to follow the changeable, criss-crossing patterns of activists' simultaneous engagement with a range of different platforms in both digital and physical contexts (Postill and Pink 2012).

Contemporary scholarship has also been critiqued for paying disproportionate attention to specific platforms such as Twitter and Facebook and specific types of citizen media instead of exploring their interdependence across different repertoires of communication within a broader media ecology. Research in the field to some extent suffers from the predominance of single case study approaches that prioritize the analysis of isolated platforms such as Facebook (Mercea 2013), Twitter (Penney and Dadas 2014), YouTube (Askanius 2014) or mailing lists (Kavada 2009), while failing to address the interrelated workings of these interfaces within a broader media ecology. Such studies run the risk of fragmenting the complex realities of the interplay between media and movements and of overlooking connections between multiple technologies, actors and their practices. Making practices rather than technologies the starting point for studying citizen media is one way of avoiding the one-medium/platform bias, and the concept of media ecologies offers a conceptual framework with which to ensure holism and understand the interdependence of different media. Media ecologies and neighbouring concepts such as protest communication ecologies (Mercea et al. 2016) and information ecologies (Treré 2012) thus increasingly form part of the vocabulary and toolbox of social movement studies.

Finally, future research on citizen media and social movements should avoid focusing too heavily on current outbursts of protest and the latest technological development as this risks losing sight of how the social forces and practices that generate movements vary across time and place. As Mattoni and Treré (2014:255) point out,

> although the disciplinary roots of social movement studies are well anchored in the work of historians, when it comes to research on media and social movements, literature appears diachronically fragmented due to the absence of a historical perspective on the development of the media/movement relation.

This predominance of presentism (Postill 2014) in contemporary scholarship – or what Melucci (1994) labelled the myopia of the present – "makes it difficult to spot underlying commonalities in the nature of communications, technological adoptions, agencies and power" (Flesher Fominaya and Gillan 2017:386). The lack of a historical perspective is partly due to the bias towards techno- and novelty-fetishism discussed above, but it is further exacerbated by the tendency to uphold the dichotomy of old versus new media. The focus on the very latest technologies, with little or no consideration given to the relevance prior media forms may continue to have in social movements, or to how new and old technologies converge and/or overlap in different contexts, results in reductive accounts that disregard how new forms of citizen media recruited for activist purposes may be reminiscent of historical forms, and how we might learn from these continuities. Being attentive to the broader historical trajectory of how media have shaped collective action requires investigating issues related to movement memory and legacies (Fernández-Savater et al. 2016; Zamponi 2018) and the role of digital media in activist archival practices (Askanius 2018; Kaun 2016b).

Ultimately, future scholarship on social movements and citizen media must strive for an understanding of the changing relationship between civic engagement and digital technologies; this understanding needs to be solidly anchored in research on the role of the analogue technologies and practices that came before these technologies, and with due attention to the historical trajectory of citizen media in earlier waves of contestation.

See also: activism; autonomous movements; diversity; film studies and citizen media; media ecologies; media practices; mediatization; prefiguration; social media; temporality

Recommended reading

Kaun, A. (2016) *Crisis and Critique: A history of media participation in times of crisis*, London: Zed Books.

Offering a valuable historical perspective on citizen media and social movements studies, Kaun explores changes in citizen media practices within the context of three major historic economic crises and related social movement struggles: the unemployed workers' movement during the Great Depression, the rent strike movement of the early 1970s, and the Occupy Wall Street protests following the Great Recession of 2008. This diachronic study provides an in-depth analysis of the cultural, economic and social consequences of media technologies and the changes and continuities in how they have shaped popular resistance to capitalism over time.

Mattoni, A. (2012) *Media Practices and Protest Politics: How precarious workers mobilise*, Farnham, UK: Ashgate.

As a social movement scholar focusing on processes of mediation and activist media practices, Mattoni's work offers a theoretical dialogue between social movement studies and media and communication studies. Although it examines the precarious workers' movement in Italy specifically, the book offers critical and broadly applicable perspectives on citizen media practices and social movement mobilizing and organizing more generally.

Stephansen, H. C. and E. Treré (eds) (2019) *Citizen Media and Practice: Currents, connections critiques*, London: Routledge.

An extensive collection of practice-oriented research on citizen and activist media. The volume continues and extends a growing tradition of scholarship within the media-as-practice paradigm, which took off around the mid-2000s in media studies, social movement studies and beyond.

SOLIDARITY

Alex Khasnabish

In the sprawling terrain of social justice struggles, both contemporary and historical, few terms occupy such pride of place or feature as ubiquitously across a diversity of movements and discourses as solidarity. According to the *Oxford English Dictionary*, "solidarity" means "unity or agreement of feeling or action, especially among individuals with a common interest; mutual support within a group". This notion of unity forged across difference in pursuit of common interests is at the heart of contemporary uses of the concept, particularly within social justice spaces and struggles. Despite its prominence, a cursory survey of its usage testifies to the fundamental ambiguity in the meaning and significance of solidarity, however. This ambiguity exists at both political and analytical levels. At the political level, solidarity is ambiguous in the sense that it does not refer to any specific type of political ideology or practice. As Kip notes,

> [i]n addition to the confusion that prevails with respect to the term's political commitments and practical implications for radical forces, invocations of 'solidarity' can also be found among conservative trade unions, within Catholic social teaching, and in nationalist and racist politics.
>
> *2016:391*

At the analytical level, invocations of solidarity are ambiguous because of a lack of clarity regarding the terms for its realization and whether it is an object to be achieved or a relationship that is enacted and contextual.

The roots of solidarity

The origins of the term solidarity lie in Roman law, and specifically in the concept of *obligato in solidum*, a reference to shared responsibility of debtors to repay a financial debt (Kip 2016; Roediger 2017). The significance of this concept lies in the fact that it does not rely on existing bonds of kinship to delineate shared responsibilities, rather, "it established shared liability among people who may have been strangers with heterogeneous interests" (Kip 2016:394). Solidarity would not acquire an explicitly political connotation until the nineteenth century, and even then it was at least as frequently mobilized by conservative interests as radical or progressive ones. While the French Revolution and revolutionary upsurge of the mid-nineteenth century brought the notion of *solidarité* to prominence and put it into direct competition with the kin- and feudal-based concept of *fraternité*, this development often glossed over the "extent to which existing patterns of racial divisions and uneven development produced solidarities compromised by their creation within industrial capitalism and imperial expansion" (Roediger 2017:169). Indeed, the inability of European radicals and revolutionaries to express and establish solidarity with revolutionary Haiti – the first successful slave rebellion leading to the founding of the first free Black republic in the Western hemisphere – speaks to the force of structured relations of exploitation and oppression and the limits they pose to the forging of solidarity across difference (Linebaugh and Rediker 2000). But as radical histories from below demonstrate, even as solidarity fails to manifest in one case, it flourishes unexpectedly in another. Exploring the everyday solidarities built between common folk – sailors, soldiers, slaves, domestic workers, sex workers and more – radical histories of the making of the modern white supremacist, heteropatriarchal, capitalist world show us the daily possibilities available for the enactment of radical and transformative relations of mutual aid (Federici 2004; Graeber 2012; Linebaugh 2008; Linebaugh and Rediker 2000; Rediker 2004, 2007). They also testify to the lengths to which dominant interests will go to snuff out such alternatives.

Featherstone defines solidarity as "a relation forged through political struggle which seeks to challenge forms of oppression" (2012:5). Featherstone's explorations of solidarity highlight several key characteristics: first, it is a transformative relationship for those involved in forging it, not a thing to be achieved; second, it is grassroots in nature and often constructed from the margins, not something imposed from above; third, it is border- and boundary-crossing, a political activity that cannot be reduced to the container of the nation-state; and fourth, it is subject to uneven power relations as a consequence of the differential placement of those constructing it (2012:5–6). Perhaps the most important insight offered by Featherstone in his critical study of solidarity is its mutually transformative potential. When solidarities are successfully enacted they do not leave the identities of those who bring them into being untouched or unchanged and they give birth to new possibilities and new ways of being together. This is critical because it means that solidarity is much more than a tool for resistance; it is in fact a key element in building other forms of life in common. Equally important, however, is the realization that individuals and groups do not enter into relation with one another on equal footing. While these differences in power, resources and privilege need not necessarily be abusive or destructive, they complicate the construction of solidarity and demand that we pay attention to the way power relations infect and inflect political work.

Crucially, Milstein (2015a) argues that empathy is the root of solidarity. One of the most often used tools in the state's arsenal of social control in defence of dominant interests is the sowing of distrust, competition and suspicion among movements. This is particularly common in struggles for social justice made up of more radical and more liberal groups since the orientation of their constituents to dominant institutions, social norms and relations of power and privilege are dramatically different. For example, liberal activists and organizers have frequently publicly denounced radicals and even collaborated with law enforcement to identify and criminalize those engaged in a variety of more militant, direct action-oriented tactics such as corporate property destruction, lockdowns and blockades, and black bloc actions (Graeber 2009; Milstein 2015b). Against the impulse to judge and police the social change tactics and strategies engaged in by others, something deeply shaped by our location in relation to larger structures of power and privilege, Milstein advocates for a revolutionary empathetic sensibility:

> [B]y taking the time to ask 'questions whose answers need to be listened to', we begin to truly see why people protesting alongside us choose a particular tactic on a particular night in a particular place. We see a widening 'horizon of context', complexity, and humanity.
>
> *2015a:153*

Beyond mere understanding, such revolutionary empathy at the root of real solidarity allows for the building of the foundations of robust and resilient movements capable of self-organization and self-determination.

Uneven terrains

While the term solidarity is frequently used by and associated with the labour movement, even a cursory exploration of this terrain reveals important lines of fracture and flight from any single conception of it. For example, contrast the kind of solidarity envisioned and practised by the Industrial Workers of the World (IWW) with that embodied by conventional trade unions. Founded in 1905, the IWW was committed to organizing all workers regardless of gender, nationality or ethnic or racialized identity and doing so on a basis that emphasized direct action on the job as a way to build the movement toward revolution capable of ending capitalism (Cole et al. 2017:4). The famous Preamble to the IWW's constitution and general bylaws lays out a vision of radical solidarity based on the fundamental recognition that "[t]he working class and the employing class have nothing in common" (Industrial Workers of the World 1905:3). While the employing class controls all material wealth, the working class regularly faces hunger and deprivation and so "[b]etween these two classes a struggle must go on until the workers of the world organise as a class, take possession of the means of production, abolish the wage system, and live in harmony with the earth" (ibid.). The Preamble concludes by affirming that "[i]t is the historic mission of the working class to do away with capitalism", reorganize production industrially for a post-capitalist world, and form "the structure of the new society within the shell of the old" (ibid.). While conventional craft unionism embodied by unions like the American Federation of Labor accepted the fragmentation of workers by workplace, frequently excluded workers from organizing on the basis of gender, nationality and racialized identity, and effectively served as an important element in the management of worker dissent on behalf of ruling-class interests, the IWW's approach to organizing for revolutionary social change was much more grassroots, expansive and radical. Predictably, the IWW was also subject to brutal repression by the state and capitalists as a result (Buhle and Schulman 2005; Cole et al. 2017). Nevertheless, the IWW has persisted and remains an example of revolutionary solidarity and social movement unionism that understands that people's power is built collectively and at the grassroots (Lynd and Grubačić 2008).

In a similar vein, Prashad (2008) explores the non-aligned movement (NAM) and the political project of the Third World beyond the notions of failed development it tends to evoke today. In his insightful history of the non-aligned movement, he explores the possibilities and pitfalls of an attempt on the part of decolonizing nations in the mid-twentieth century to forge an alternative political, economic and social path to the US/Soviet dichotomy. Beyond the hegemonies of the US and Soviet spheres of influence, Prashad's careful and inspiring history forwards a conception of the Third World representing not failure and dependency but a viable alternative developmental project based on a decolonial solidarity. The failure of the project to coalesce was due, in large part, to the powerful forces brought to bear against it by First and Second World interests, but also to the internal competition and lack of trust among constituent nations themselves (ibid.). In this case, the failure to realize alternatives to a white supremacist, heteropatriarchal, settler-colonial and imperialist world order is a result of the failure of solidarity itself.

Solidarity as powerful, fraught and unfinished practice

Waves of social movement activity on a global scale provide abundant examples through which to explore solidarity in theory and practice, but one of the most interesting and illuminating is the encounter between the Zapatista movement and the global justice movement. On 1 January 1994 an army of indigenous insurgents rose up in the far southeast of Mexico, in the state of Chiapas, and declared *¡Ya basta!* – enough! – to five centuries of oppression, exploitation, racism and genocide (Collier 1999; Khasnabish 2010; Muñoz Ramírez 2008; Womack 1999). Naming themselves the Zapatista Army of National Liberation after Emiliano Zapata, one of the most radical and revered heroes of the Mexican Revolution, this indigenous insurgency was the product of a decade of clandestine organizing in some of the poorest and most oppressed communities in Chiapas. But its roots stretch much farther than that to decades of radical organizing and action throughout Mexico and centuries of resistance to colonialism and imperialism (Womack 1999).

The Zapatista uprising was timed to coincide with the day the North American Free Trade Agreement was set to come into force, but what began as a seemingly familiar Latin American guerrilla insurgency quickly transformed into something bigger and more powerful. With no preexisting links, Mexican civil society responded to the uprising with incredible displays of solidarity, demanding the government call a unilateral ceasefire and enter into negotiations with the insurgents. While the Zapatista Army would remain an armed force, this encounter between the Zapatistas and diverse elements of Mexican civil society would mark the birth of a broad-based, grassroots, radical and non-violent Zapatista movement calling for justice, democracy and liberty for all Mexicans (Khasnabish 2010; Muñoz Ramírez 2008). Perhaps even more unexpectedly, the Zapatista struggle would also resonate far beyond the borders of Mexico to inspire the work of activists and organizers around the world (Conant 2010; Khasnabish 2008). Travelling routes from conventional print and television media to activist solidarity delegations and the Internet – only an emerging terrain of action and communication for activists at the time – the Zapatista struggle would spark a new era of social justice organizing and action in the shadow of the collapse of the Soviet Union and the apparent global ascendance of neoliberal capitalism. Indeed, the global justice movement that coalesced in the late 1990s would not have taken the shape it did without the vital spark provided by the Zapatistas (Khasnabish 2008; Midnight Notes 2001; Notes from Nowhere 2003). At the same time, the international attention generated in the context of this encounter between the Zapatistas and global civil society afforded a modicum of security to Zapatista base communities in Chiapas that were regularly threatened with military and paramilitary violence. Yet this encounter and the incredible possibilities it generated did not come without a downside. As Zapatista spokesperson Subcomandante Marcos (Marcos 2004) would write, too frequently the solidarity efforts directed at the Zapatistas took the shape of pity, charity, paternalism and appropriation rather than an authentic and egalitarian process of building real links of mutual aid and common struggle among diverse movements. As Roediger (2017) argues, this is precisely why it is so necessary to make solidarity uneasy since it is not merely a question of its presence or absence but of how it is constructed, by and for whom, on what terms and with what consequences.

A renewed commitment to grassroots, direct action and radically democratic action for social change was one of the hallmarks of the cycle of struggles commonly associated with the global justice movement and its afterlives in movements like Occupy. In Dixon's engaged research on what he labels the "anti-authoritarian current" of North American radical politics (2014:3), he explores how, why and with what consequences organizers and activists are putting "another politics" into practice (ibid.:5). Drawing from his own experience as an organizer and activist within the anti-authoritarian current as well as dozens of interviews conducted with others all across Canada and the United States engaged in this political current, Dixon's work illuminates how this diverse political current is itself a product of anti-racist feminism, prison abolitionism and a reconfigured anarchism. In a sense, solidarity as a constant, unfinished practice is central to the anti-authoritarian current because of its emphasis on a politics of prefiguration. Dixon argues there are four central features of the anti-authoritarian current: struggling against all forms of oppression, exploitation and domination; developing new social relations and forms of social organization through struggle; connecting improvements in people's daily lives to long-term transformative visions; and organizing that is bottom-up and grassroots (ibid.:6–7). These features clearly resonate with the complex practice and ideas surrounding solidarity outlined above. In this sense, solidarity is the name for a generative, radical and always unfolding relationship that is less about unity across difference than it is about a process of collective liberation. In a similar vein, Walia's work on collective efforts to confront, dismantle and build alternatives to border imperialism also emphasizes the need to "[undo] … the borders between one another" (2013:11). A radical decolonial framework is central to this vision, particularly because it places the everyday, relational and lived ways these power dynamics and the paths beyond them are worked out at the centre. Walia argues that decolonization movements struggle not only against "settler colonialism, border imperialism, capitalism, and oppression" but to advance "other ways of laboring, thinking, loving, stewarding, and living" (ibid.:12). Recognizing that "we are but one part of the land and its creation" nurtures an expansive and prefigurative radical political practice that "encourages us to constitute our kinship and movement networks based on shared affinities as well as responsible solidarities" (ibid.). This decolonial solidary practice shares powerful affinities with what Shotwell (2016:12) calls "aspirational solidarity" grounded in "collective conceptions of worlds that do not yet exist" and that might guide our action toward others, both human and non-human, with whom we share this planet. Solidarity is not only processual and transformative in this conception, it is also orienting and prefigurative. It recalls and expands upon the IWW's maxims of "an injury to one is an injury to all" as well as the need to "build a new world in the shell of the old" even as it highlights the fundamentally impure, unequal nature of the relations and positions from which we begin to work toward such possible worlds (Industrial Workers of the World 1905:3). It also resonates with Roediger's exhortation that we embrace "solidarity while simultaneously being uneasy about the assumptions it sometimes evokes" (2017:159). Roediger encourages us not to reconcile or dismiss this unease but to make use of it and question

> if solidarity is always a good thing, to recall what and whom solidarity leaves out, and how it is premised on those leavings out, to consider how solidarity works across differences in kinds and degrees of oppression, and to ask if the presence of solidarity is the logic of things or if for long periods it may be a treasured exception.
>
> *ibid.*

The Zapatistas use the phrase "walking we ask questions" to describe their approach to the pace and process of social change because collective liberation is a journey taken together and one without a fixed or final destination (Marcos 2002). Similarly, solidarity is a fraught, uneasy and never-finished practice of making and sustaining relationships that hold the potential not only to challenge powerful actors and institutions but to remake us and the worlds we inhabit.

See also: activism; prefiguration; social movement studies and citizen media

Recommended reading

Featherstone, D. (2012) *Solidarity: Hidden histories and geographies of internationalism*, London: Zed Books.

Presents a variety of case studies that highlight the relational, uneven and international nature of solidarity. This work offers a geographically broad and historically deep conception of solidarity as a relationship always in the making and never free from prevailing forms of power and injustice. It also does valuable work in rejecting nation-state centred conceptions of solidarity, looking instead to its international manifestations to guide its analysis.

Kip, M. (2016) 'Solidarity', in K. Fritsch, C. O'Connor and A. K. Thompson (eds) *Keywords for Radicals*, Oakland, CA: AK Press, 391–398.

A wide-ranging survey of solidarity focusing on its use in radical political theory and practice. An excellent review of the concept that offers considerable breadth but little depth, set in the specific context of solidarity's relationship to modern and primarily Anglo-American radical political usage.

Milstein, C. (ed.) (2015) *Taking Sides: Revolutionary solidarity and the poverty of liberalism*, Oakland, CA: AK Press.

An edited volume focusing on the politics and practice of militant solidarity across a variety of contemporary radical struggles for social justice and social change. Denouncing the liberal notion of allyship and the non-profit industrial complex to which it is connected, the chapters highlight the necessity of becoming accomplices in revolutionary struggle.

Roediger, D. R. (2017) 'Making Solidarity Uneasy: Cautions on a keyword from Black Lives Matter to the past', in *Class, Race, and Marxism*, New York and London: Verso, 157–188.

Presents a nuanced and radical exploration of solidarity as concept and practice in a richly historicized context. Roediger draws out the contradictions and complexity of solidarity across a variety of social movement struggles, urging the reader to attend to the unevenness and silences in its production as much as its promises and potential. This work is a powerful illustration of solidarity as an unfinished, uneasy relationship and one that is shaped by prevailing relations of power.

SOUSVEILLANCE

Paul Reilly

The ubiquity of smart phones, combining digital camera technology with high-speed connectivity to the Internet via broadband cellular networks, has provided unprecedented opportunities for recording and disseminating eyewitness perspectives that focus public attention on the conduct of authority figures. This entry explores how social media platforms can be used to create and share acts of sousveillance, broadly defined as a form of inverse surveillance that empowers citizens through their use of technology to "access and collect data about their surveillance" (Mann et al. 2003:333). The two primary forms of sousveillance, hierarchical and personal, are critically evaluated in this entry with reference to a number of prominent examples. These include the #BlackLivesMatter campaign which has sought to focus attention on violent police attacks on African Americans since 2014 (Freelon et al. 2016a), as well as the use of YouTube videos filmed by eyewitnesses to highlight alleged police brutality during the so-called Battle of Stokes Croft that occurred in Bristol, UK in April 2011. The entry considers how audience responses to acts of police brutality shared on platforms such as Facebook and Twitter are influenced by news media coverage of these incidents. Previous research has indicated that the sharing of sousveillance footage online may raise as many questions about the behaviour of the alleged victims as it does the police (Reilly 2015). The entry concludes by considering whether sousveillant practices facilitated via social media constitute a shift in informational power from elites to marginalized groups and individuals.

Sousveillance as response to surveillance society

The term sousveillance was coined by electrical engineer Steve Mann as part of a critique of pervasive organizational surveillance practices which, he argued, threaten the autonomy of individuals (Mann 2004; Stanley and Steinhardt 2003). In response to our contemporary surveillance society, in which the few watch the many through cameras mounted on or in buildings, Mann (2013:1) encourages the use of body-worn cameras for the purposes of enabling the many to watch the few. The rationale for this 'undersight' is that data generated by the surveillance of private citizens by entities in positions of power lacks integrity and provides evidence that, although widely used in court cases, is all too often "less than the full truth" (Mann 2017:3). The practice of sousveillance thus shares much in common with citizen journalism, the process whereby citizens play an active role in the "process of collecting, reporting, analyzing and disseminating news and information" (Bowman and Willis 2003:9). Indeed, citizens themselves are more likely to use the latter term to describe their use of technology to 'watch the watchers' due to its prominence in the lexicon of journalism and the frequency with which it is invoked by professionals to describe such activity.

Mann identifies two key forms of sousveillance: personal sousveillance refers to the use of cameras at "eye-level for human-centred recording of personal experiences", while hierarchical sousveillance is a more purposive, political activity that focuses on documenting the actions of authority figures such as the police (Mann 2004:1). The former practice has generally involved documenting personal experiences in order to bring communities together, "without necessarily involving a political agenda" (Bakir 2010:21). The latter shares the injunction to care that is an integral component of media witnessing, the term used to capture the ways in which digital media technologies transform people's capacity to bear witness to events and encourage others to engage with these perspectives (Allan 2013).

Irrespective of the motivations underlying such acts, the recording of personal experiences in public spaces can contribute towards the emergence of a situation of equiveillance, defined as an "equilibrium (balance) between surveillance and sousveillance" (Mann 2004:627). However, there are two caveats in relation to this sur/sousveillance distinction that should be noted. First, sousveillance was not always considered a countervailing force to the oversight of the architecture-centred surveillance of authority figures. It was not initially conceived as a mechanism with which to document incidents of police brutality for example, even if the term has become largely synonymous with such activities in the social media era. Indeed, Mann (2013:4) has suggested that the feedback loops created by citizen sousveillance might be equally capable of capturing evidence of police officers "doing acts of good", as was demonstrated by footage of New York Police Department (NYPD) officer Larry DePrimo purchasing a pair of shoes and giving them to a homeless man in November 2012. Second, we must recognize that the rise of sousveillance does not necessarily entail the decline of surveillance and in fact, in a society in which mutual monitoring by citizens is becoming more and more widespread, surveillance too may increase. The complexity of these developments is illustrated by Mann's (2013) Veillance Plane, an eight-point compass showing how the amount of surveillance and sousveillance in a physical space could rise or fall in response to changes such as an increase in the number people recording footage on smartphones. The eight directions of the compass point to different practices, from the use of surveillance cameras by business establishments which simultaneously prohibit customers from having or using their own recording technologies (dubbed McVeillance following incidents involving McDonalds restaurants), to the use of technology to prevent surveillance of citizen activity (anti-surveillance) (ibid.). Both sousveillance and surveillance are conceptualized as orthogonal vectors in this model, which suggests that increases in one might not be at the expense of the other. For example, Mann (2013:6) envisages a scenario in which the oversight provided by the installation of three extra surveillance cameras within a bar would, at least temporarily, be complemented by the undersight of six customers, who were recording their experiences using wearable cameras, leading to a situation of near-total mutual veillance. Therefore, reductive analyses that frame sousveillance as a panacea to surveillance should be replaced by more contextualized approaches that recognize their coexistence within contemporary societies.

The efficacy of sousveillance in focusing public attention upon the actions of authority figures may ultimately depend upon the size of the network through which it is distributed, although smaller communities of sousveillance officers can help facilitate dialogue about the power asymmetries it exposes (Mann and Ferenbok 2013). This was certainly the case in the two most prominent examples of sousveillance in the pre-social media era, the Rodney King assault and the Abu Ghraib abuse scandal. The former revolved around video footage of four Los Angeles police officers assaulting Rodney King on 4 March 1991. Eyewitness George Holliday covertly captured the assault on the African American taxi driver using his Sony Handycam and shared the video with local television station KTLA, which was later used as evidence in the trial of the LAPD officers charged with attacking King (Mann et al. 2003). The footage raised broader questions about police brutality towards African Americans and was repeatedly shown by US networks during the trial and subsequent acquittal of the four defendants, which resulted in five nights of rioting in Los Angeles that left fifty people dead and 2,000 injured. Holliday later sued five US news networks for copyright infringement in a landmark case which was rejected by Judge Irving Hill on the grounds that the only way the news media could "tell the complete story" of the King assault was through the broadcast of the video (Reis 1995:285). A similar theme emerged in relation to the leaking of photographs depicting the abuse of Iraqi prisoners by US military personnel in Abu Ghraib prison in 2003. Bakir (2010) explores how both personal and hierarchical sousveillance 'impulses' played a role in the production and dissemination of these torture photographs. The former was evident in the trophy shots, taken on digital cameras as mementos of the prison guards' time in Iraq, that were brought to the attention of the US Army's Criminal Investigation Command (CID) when they received two CDs containing the photographs from a whistleblower in January 2004 (Beier 2007). Three months passed before the torture photographs were leaked and published by CBS News and other mainstream media outlets in the United States. Their publication in news media were said to have fuelled the Iraqi insurgency in Iraq by providing irrefutable evidence of the hitherto covert US policy of torture in the war-torn country, despite the subsequent efforts of the US government to frame Abu Ghraib as an isolated incident (Bakir 2010). Like the Holliday video, the sousveillant potential of the Abu Ghraib images appeared to be determined by viewer perceptions of the events and issues depicted therein.

Social media and the intensification of sousveillance

The advent of social media and Internet-enabled smart phones has been linked to an "intensification of sousveillance and the rise of sousveillance cultures" since around 2010 (Bakir 2010:23). These participatory media form part of a sousveillant assemblage that is comparable, to a certain extent, to the surveillant assemblage developed by governments and businesses in the 2000s, which has deployed increasingly sophisticated technological systems to monitor and collect data on citizens (Bakir 2010; Mann 2017). Whereas George Holliday had to rely on his cumbersome Sony Handycam to record the Rodney King assault and local television networks to broadcast this footage, Internet-enabled smart phones empower eyewitnesses to record and share this footage almost instantly with a potential global audience via social media such as Facebook and Twitter. In doing so, citizens inadvertently provide yet more personal data for social media companies that engage in the mass surveillance of their users for the purposes of targeted advertising (Vaidhyanathan 2018a). While many social justice campaigners are cognisant of these privacy concerns, they have continued to leverage the connective affordances of social media to focus attention on hierarchical sousveillance on numerous occasions since 2010. Probably the most prominent examples of social media sousveillance have been deployed by Black Lives Matter, the campaign set up by activists Alicia Garza, Patrisse Cullors and Opal Tomet to highlight police killings of unarmed African American citizens in July 2013 (Bonilla and Rosa 2015). Emerging first as a hashtag and then transitioning into a much larger social justice movement that combined online activism with more traditional modes of protest such as public demonstrations, the group shared distressing footage showing several of these controversial killings (Fischer and Mohrman 2016). Most notably, one video showed Staten Island resident Eric Garner being held in a chokehold by several NYPD officers, despite him being heard repeating the phrase "I can't breathe" eleven times (*The Guardian* 2014). The subsequent coroner's report confirmed that the cause of death was the compressions on his neck and chest from NYPD Officer Daniel Pantaleo, who faced no indictment from the Staten Island grand jury for his actions. The eponymous hashtag was used to focus attention on this video, and to highlight the NYPD's disregard for Black lives (Freelon et al. 2016a). Twitter debates surrounding such footage facilitated "large-scale informal learning" about the tensions between the police and Black communities, particularly among conservatives who acknowledged for the first time that these killings were unjust (Freelon et al. 2016b:79). In this sense, the integration of hierarchical sousveillance into a social justice campaign appears to have enabled non-elites to shape public discourse around race and law enforcement in the United States.

Audiences often use heuristics and their own political views to decide whether any given eye-level perspective constitutes an instance of hierarchical sousveillance. This has implications for eyewitnesses who use social media to share sousveillance; viewers may not necessarily agree that this footage is prima facie evidence of misconduct by authority figures such as the police. We might consider, by way of example, how YouTubers responded to footage of alleged police brutality during the so-called Battle of Stokes Croft, a series of violent clashes occurring between police and members of the public in the Stokes Croft district of Bristol, UK on 21 April 2011 in the aftermath of a controversial police raid on a local squat (known as Telepathic Heights). Both local and national media were quick to frame the violence as a manifestation of the No Tesco in Stokes Croft campaign, which opposed the opening of a new Tesco supermarket in the area on the grounds that it threatened the future of a number of local independent traders and would destroy the "unique character" of the neighbourhood (People's Republic of Stokes Croft 2010). Its activists turned to social media to refute these allegations and to highlight the brutal police dispersal of a peaceful protest against the squat eviction that they blamed for the violence (Hall 2011). A key component of this strategy was the sharing of hierarchical sousveillance on YouTube, which appeared to provide support for the claims by local residents that the police tactics were heavy-handed. For example, one video showed a man being forcefully pushed out of the way by a police officer without any warning, while another captured footage of the helicopter that had been the subject of many complaints from local residents. Analysis of comments posted under these videos provided little evidence to suggest that this footage had successfully focused attention on police brutality and countered the media framing of the incident as an anti-Tesco riot (Reilly 2014). Some commentators agreed that the police tactics appeared to be 'unfathomable' and some claimed they had been brutalized during the events captured on camera (ibid.). However, the majority did not perceive these videos as evidence of police brutality; instead, officers were often criticized for not adopting more aggressive crowd control measures to disperse onlookers. Moreover, the antisocial behaviour of the crowd was generally subject to more criticism than the actions of the riot police, with many commentators conflating the violence with the anti-Tesco campaign. Indeed, the results of the study suggest that there was little rational debate about who was responsible for the violence and the views of many commentators appear to have been influenced by mainstream news media coverage of the riot (Reilly 2015).

Sousveillance as a shift in informational power

Although by no means unique to the social media age, acts of sousveillance have become increasingly prevalent in the era of Facebook, Twitter and YouTube. Citizens are now able to record and share their personal experiences online using a combination of Internet-enabled smart phones and social media platforms, with the resultant visual evidence providing alternative perspectives which contrast with the one-sided surveillance of public spaces by elites. That is not to say that the growth in participatory veillance envisaged by Mann will necessarily nullify this oversight. Rather, equiveillance is likely to remain elusive while state and non-state entities continue to invest in more and more sophisticated technological approaches to the mass surveillance of citizens. A more pertinent question is whether the use of social media for sousveillance constitutes a shift in informational power from elites to marginalized, peripheral actors. Certainly, the Black Lives Matter movement has illustrated how the integration of hierarchical sousveillance into advocacy campaigns can raise questions about the conduct of elites, while also facilitating informal learning about broader issues of social injustice. There are now unprecedented opportunities for citizens to use such footage to intervene in a political information cycle that hybridizes older and newer media logics (Chadwick 2013). In this sense, the use of social media by citizens as a novel means with which to engage in sousveillance does appear to have the power to disrupt established hierarchies within journalism and politics. Nevertheless, it is perhaps too early to tell whether sousveillance will significantly alter power relations in contemporary societies. The use of commercial social media to create and share sousveillance contributes to the surveillant assemblage by providing social media companies with personal data that can be used to monitor the actions and behaviour of citizens. The black box algorithms operated by platforms such as Facebook are also likely to mirror the editorial functions of professional news media insofar as they may limit the visibility and discoverability of such content. It is therefore perhaps ironic that eyewitnesses who share their experiences online may still need to capture the attention of mainstream media in order to amplify hierarchical sousveillance in such a way as to influence public attitudes towards authority figures such as the police. However, focusing attention upon their alleged misconduct does not in and of itself change attitudes towards elite actors. Even in those cases where visual evidence is widely shared online, there is still a strong likelihood that mainstream media depictions of the events captured on camera will continue to shape viewer perceptions of the alleged police brutality or elite wrongdoing it highlights. Furthermore, the incorporation of these digital acts into professional media coverage of such events may not necessarily provoke the emotional reaction in the audience sought by the witness. The formulaic mediation of suffering by professional journalists in news packages that incorporate such user-generated content may "numb rather than mobilise moral sensibilities", as has often been the case with coverage of natural disasters (Chouliaraki 2013b:35). Therefore, the potential of sousveillance to allow for public scrutiny of the actions of authority figures, both good and bad, depends on how well it is integrated into broader social justice campaigns, the salience of its narrative and the nature of the ways in which it is circulated via both traditional and social media.

See also: hacking and hacktivism; social media; surveillance

Recommended reading

Bakir, V. (2010) *Sousveillance, Media and Strategic Political Communication*, London: Continuum.

Brings a strategic political communication focus to the study of sousveillance in the era of web-based participatory media. Uses case studies from Iraq, the US and UK to examine the rise of the sousveillant assemblage, and explores how personal sousveillance carried out using Web 2.0 applications might inadvertently provide more data for the surveillant state.

Fischer, M. and K. Mohrman (2016) 'Black Deaths Matter? Sousveillance and the invisibility of black life', *Ada: A Journal of Gender, New Media, and Technology* 10. Available online: https://adanewmedia.org/2016/10/issue10-fischer-mohrman/.

Provides important insight into the integration of sousveillance technologies into social justice movements and campaigns. This study also highlights how the political economy constraints of online platforms may limit the distribution of images depicting police brutality.

Mann, S. (2013) 'Veillance and Reciprocal Transparency: Surveillance versus sousveillance, AR glass, lifeglogging, and wearable computing', *2013 IEEE International Symposium on Technology and Society (ISTAS): Social implications of wearable computing and augmediated reality in everyday life*, Toronto: IEEE, 1–12.

A seminal article that explains the Veillance Plane and contextualizes contemporary accounts of sousveillance with reference to related concepts of anti-surveillance, counter-veillance and McVeillance.

SPACE AND PLACE

Matilda Tudor

The concept of place did not assume prominence within modern philosophy until the twentieth century, when Heidegger (1954/1993) began to explore the *dasein* (being there) of human existence. Heidegger was interested in exploring how we exist in the world by making it meaningful and place-like. During the 1970s, his ideas were further elaborated by human geographers, who wanted to move beyond conceptualizations of place as material structure and towards an understanding of the meaningfulness of place. From their perspective, a distinction could be drawn between the "abstract realm of space and an experienced and felt world of place" (Cresswell 2009:172). It was then possible for space to acquire meaningfulness as place through senses of orientation and at-home-ness. These ideas were to dominate the field of human geography for several decades.

The temporal dimension of Heidegger's concept of dwelling as a way of making place was explored in some detail by Tuan (1977), who stressed that dwelling implies a pause. While space enables movement, acquiring a sense of place is only possible when we stop (ibid.:6). This distinction had major implications for postmodern geography, as mobility came to be considered an inherent threat to acquiring a sense of place. Thus, during the late twentieth century scholars began to theorize the postmodern capitalist society as placeless, arguing that increased mobility of information and people made any real attachment to place impossible (Augé 1995; O'Brien 1992; Ohmae 1995; Relph 1976). With the publication of Meyrowitz' *No Sense of Place* (1985), these ideas were extended to the field of media and communication technologies. Inspired on the one hand by Goffman's (1959/1990) theories of the social geography of everyday living, and on the other by McLuhan's (1964/1994) theory of the medium itself as the message, irrespective of its content, Meyrowitz aimed to capture some of the ways in which the rise of electronic media altered the relationship between space and place in everyday life. Recognizing the increased fluidity of information, content and positionality, he argued that electronic media have taken us towards the end of place-bound, segmented identities.

Media and the mobilities of place

On the whole, then, this branch of humanist geographic scholarship posed mobility as an inherent threat to the meaningfulness of place; Cresswell refers to this strand of thought as "sedentarist metaphysics" (2009:176). The rise of electronic media and the accompanying time–space compression were consequently read by many as leading to an increasing sense of placelessness. Others, however, challenged the association of place with pause. Seamon (1979), for example, drew on the work of another phenomenologist, Maurice Merleau-Ponty, to argue an opposing point of view. While intentionality in Heidegger's *dasein* is about 'being there', and thus the subject's relationship to her world, Merleau-Ponty (1962/1997) points out that this relationship is essentially embodied. Through intentionality, Merleau-Ponty argues, the body-subject is constantly transcending herself through an openness to the outside. It is this open transcendence between the body-subject and her world, which is purely habitual and pre-cognitive, that creates orientation. Staying within the experiential paradigm, Seamon thus used Merleau-Ponty's ideas to argue that if anything, space becomes meaningful and turns into place through habitual actions and thus through mobilities. Rather than being static scenes, places are entwined with such collective mobilities – what Seamon calls place-ballets – of people and of things, and these mobilities are fundamental to our sense of place.

Moores (2012, 2017) puts Seamon's place-ballet into dialogue with contemporary media society, arguing that rather than presenting an obstacle to place, media is part of the everyday habitual body-scheme orienting the body-subject in space and time. The small and large scale mobilities that constitute senses of place are increasingly entwined with mobile devices and screens carried as extensions of our bodies. Thus the habitual actions of scrolling through social media on the metro, or reading the same newspaper every morning while having coffee at a local café, are all part of what creates the meaningfulness of those places. It might even be that suddenly losing one's access to WiFi, or not getting hold of that same newspaper one is accustomed to reading every morning, turns the metro or the café into strange spaces, leading the subject to feel uncomfortable and out of place.

While Seamon challenged the sedentarist metaphysics of humanist geographers, others critiqued them for being insensitive to relations of power. Whereas the Heideggerian approach seems to regard place as more or less a natural, unmediated phenomenon, the 1980s Marxist and feminist scholars aimed to shed light on the social processes involved in the construction of place, emphasizing their conflictual aspects (Cresswell 2009:173). Space and place, they stressed, are socially constructed and relational, and thus embody inequalities and power struggles. Dominant attitudes and social orders such as sexism and capitalism bestow on places their normative dimensions by prescribing certain practices and identities and rendering others out of place. In line with arguments put forward by critical phenomenologists such as Fanon (1986) and Young (1980) during the same period, this meant that the capacity of bodies to experience space as meaningful and home-like was understood as relative to the social orientation of space and different for different kinds of bodies due to variation in dimensions such as race, class and gender. Hence, some bodies may enjoy mobility while others are confined to stasis within the same physical place. Further, as pointed out by Marxist human geographer Lefebvre (1974/1991), there will always be a dialectic of space oscillating between the reproduction of the normative and the micro-alterations introduced by those who choose to transgress or resist.

Once media is no longer understood to be at odds with senses of place, but rather as an agent in the mundane production of space, it becomes possible to think of the potential for citizen media to subvert the limitations of place, as the examples discussed later in this entry demonstrate. But first, the epistemological relationship between media and space/place are explored further to establish whether media and communication technologies merely constitute tools that assist human beings to renegotiate and extend themselves into space or whether they can also produce spaces and places in their own right.

Media as constitutive of space and place

With the arrival of the Internet, the sedentarist metaphysics of humanist geographers gained new ground, albeit in a more optimistic guise. Being online much of the time was in itself thought of as the epitome of a placeless universe – a cyberspace – where disembodied surfers could escape the restraints of physical, real territoriality (Batty 1993; Benedikt 1991; Rheingold 1993). Thus, the online was largely regarded as disconnected from the offline, as a space where one could go in order to be someone else, or to experiment with multiple and flexible identities (Turkle 1995). This view was, however, increasingly questioned as studies of everyday digital media use evolved and empirical work found that what people did online was by and large interconnected with their offline lives (Bakardjieva 2005; Miller and Slater 2000; Sundén 2003). On the one hand, the ways in which people brought the Internet into their lifeworlds were moulded by their offline circumstances (Bakardjieva 2005), and on the other hand, the Internet itself was not free from the constraints associated with materiality (Sundén 2003). New developments in the Web 2.0 environment further intensified the relationship between the online sphere and bodies in time and space. While early Internet resources such as IRC-channels and digital bulletin boards had been largely based on anonymous or pseudonymous interaction, and thus partly supported transgressive ways of being online, this was replaced by what we now know as 'the real name Internet' (Hogan 2012), with personal data at the centre of its economy. At this point many online forums went from community grounded to commercial enterprises, and pseudonyms and nicknames were exchanged for profiles attached to identifiable and coherent selves (van der Nagel 2017a).

It could be argued that rather than undermining our sense of place, the real name Internet serves to anchor local identities and experiences of place through social networks that often feed into offline relationships, or geolocation software that constantly reminds users of where they are and what and who is within geographical reach. As ever more mundane practices entwine with digital media, from dating to television watching to online banking, news reading and baby monitoring, it is increasingly less common to conceptualize the Internet as a separate space. Within the post-digital paradigm, people rather live what Deuze (2012) has termed medialives, where it is increasingly difficult to single out media as anything other than just life as we know it. Thus, it has become more common to argue that digital media should no longer be regarded as "a space", since "[w]hat happens via new technology is completely interwoven with what happens face-to-face and via other media – the telephone, the television, films, music, radio, print" (Baym 2010:153). From the perspective of citizen media, this suggests that media and communication technologies are conceptualized primarily as an infrastructure through which individuals and groups may come together, coordinate and disseminate information in real time across vast distances.

Moores (2004, 2012, 2017), however, argues the opposite by drawing on media phenomenologist Scannell's (1996, 2014) notion that there is a spatiotemporal doubleness to media practices. Media in this view "offers locations or spaces of sort to occupy, such as television channels and email inboxes", while media users simultaneously occupy an inescapable material environment (Moores 2012:x). Taking this as a starting point, Moores argues that instead of collapsing spaces such as the online and the offline into one space or ascribing primacy to the one over the other, the media user is best conceived of as occupying several spaces at once. Rather than thinking that a person who, for example, is watching a romantic movie is either furthermost in territorial space, because that is where her mortal body is situated, or conversely furthermost in the narrated space of the movie, because that is where she is emotionally absorbed, she should be seen, Moores argues, as co-present in several sites, each intimately interconnected and experienced through the living body. Moores stresses that this double – or perhaps multiple – positionality must be taken into consideration in explaining contemporary social experience; this is particularly important if we wish to understand the positionality and experiences of individuals and groups reaching out to one another through media technologies for support, empowerment and resistance. In particular, it may allow us to better capture the thickening of space as it becomes complexly layered by code into alternative spheres of action, and to capture the dialectics between the myriad sites occupied by the same individuals.

Brighenti's (2016) concept of interstices helps to clarify how such an understanding may support productive perspectives on citizen media's role in constituting space and place. Defined as "in-between spaces" (ibid.:xvi), the concept goes beyond the centre/periphery and core/margin dichotomies. It nevertheless engages with the issue of power and minoritarian struggles: the in-betweenness of interstices implies that they are surrounded by more institutionalized, recognizable or legally powerful spaces. Further, interstices are not to be regarded as gaps but as active and event-like phenomena produced by a multiplicity of actors. Mitchell (2016) has drawn on the concept, for example, in relation to inhabitants of tent cities, while also pointing out the importance of not assuming interstices to be overladen with positive or progressive overtones since these spaces might also be inherently repressive.

Being commonly consigned to invisibility within mainstream space, particularly outside the Western metropoles, queer populations use various types of new technology to make space for themselves within otherwise homophobic and intolerant environments, thus creating interstices in Brighenti's terms (Craig and McInroy 2014; DeHaan et al. 2013; Gray 2009; Hillier and Harrison 2007; Light 2016; Tudor 2018). In this context, the common use of so-called hook-up (or dating) apps may serve to produce cracks within the urban settings for queer bodies to locate one another and form a common, otherwise unseen geography. Using the geolocation function of any mobile device, the applications map out queer others across the city, allowing the user to establish contact, negotiate encounters and exchange pictures while on the move, in what is commonly felt to be a comparably safe and discrete space (Albury and Byron 2016; Batiste 2013). As such, the space constructed by the users of such applications can be read as a separate but connected surface, with intricate relationships to the surrounding, more institutionalized normative order; an interstice simultaneously exposing and hiding queer desires.

Digital dynamics of space for political action

As long as place is understood in opposition to mobility, media and communication technologies are likely to be regarded as a threat to senses of place. However, with a conceptualization of space and place which takes as its starting point the mobilities of bodies and things, media can be regarded as intimately entwined with the habitual actions that constitute place. Given that space and place are also now understood as socially produced through processes of power and as embedded in the constitutive dialectic of normativity and transgression, media must ultimately be regarded as tools for renegotiating place. Media further serve to constitute spaces and places in their own right, thus multiply positioning the media user in several spaces at once.

Massey and Snyder's (2012) analysis of the 2011 Occupy Wall Street movement provides an illustrative example of the many space/place dimensions potentially involved in citizen media initiatives. Highlighting activists' simultaneous engagements in online and offline spaces, they argue that participants acted in a "hypercity built of granite and asphalt, algorithms and information, appropriating its platforms and creating new structures within it". Using Twitter and Facebook as well as specifically tailored forums, activists were able to mobilize people in offline assemblies on the ground. Their online activity further served to anchor the particular presence of the actual occupation beyond its immediate locale by livestreaming meetings, posting event calendars and using discussion forums. Thousands of people who had no previous affiliation with social change movements came to engage with the occupation. At the same time, this dispersed set of agents also served to form their own online geographies which renegotiated common conceptions of a counterpublic, as well as the lines and patterns of the global map as the #OccupyMap charted Occupy activities across the globe through geolocative tagging and hashtags, thus exposing global bonds of radicalism.

Massey and Snyder's analysis is in tune with developments in the field of communication geography, which has emerged among media scholars and human geographers primarily from Scandinavia and the UK since the early 2000s (Adams et al. 2017; Falkheimer and Jansson 2006; Couldry and McCarthy 2004; Pink and Leder Mackley 2012, 2013, 2016). Communication geographers argue that rather than dismissing the spatiality of media in the post-digital age, we need a more thorough and grounded understanding of the particularity of place in media use within the materiality of everyday life. They address questions relating, for example, to how media interacts with "the physical conditions of acting and thinking" evident in "the actual character and physical limitations of the room" (Bengtsson 2006:190), and how it relates to the social and normative confines of possible and acceptable action. In one attempt to draw together thinkers from human geography as well as media studies, the multi-author volume *Communications/Media/Geographies* defines communication geography as uniting "phenomenological, micro-oriented perspectives with structural issues relating to, for example, global communication geometries and relations of social space" (Adams et al. 2017:7), in order to "enhance our understanding of how representational spaces and practices are complexly articulated to other material geographies and realities across a variety of scales", thus necessarily bringing in questions of power (ibid.:7–8).

Articulating media and communication technologies' capacity to constitute and renegotiate space and place also exposes tensions in digital culture that are easily glossed over and rarely addressed within the post-digital paradigm. As in the example of the Occupy movement, the spatiality of citizen media is too often understood as coherent with traditional representational models based on the politics of visibility, emphasizing the symbiotic relationship between online and offline activity. However, studies of citizen media should also be attentive to the micro-politics of everyday living for those without capacity and/or intention to overthrow systems or claim recognition, as in the case of many queer populations across the globe who attempt to form alternative spheres or interstices for existence, intimacy and (sometimes) community. Tudor (2018) argues that the online subversive activity of non-organized queer men in Russia, for example, might be more dependent on the promise of invisibility than visibility. The men interviewed in Tudor's study wished to conduct their lives along multiple contradictory investments, often striving to keep their online queer activity strictly separate from other spheres in life. The way in which they move across the online/offline boundary might thus be understood as discontinuous, with many experiencing it as an important dividing line that allows them to compartmentalize their lives while resisting societal pressures on enacting a straight masculinity. Even if that divide is never airtight, but is continuously negotiated and challenged, these queer men are able to adopt various strategies to prevent their online queer spaces from bleeding into other spheres in uncontrolled ways. The nature and ease of online/offline transcendence in the context of citizen media should therefore also be discussed in terms of normativity.

See also: networks and networked society; self-mediation

Recommended reading

Brighenti, A. M. (ed.) (2016) *Urban Interstices: The aesthetics and the politics of the in-between*, New York: Routledge.

A range of international scholars engaged in researching issues of spatial justice provide empirical case studies that draw on Brighenti's concept of interstices. Offering a series of intriguing examples from urban areas as diverse as Gaza, Quebec and India, the book draws a complex picture of such in-between spaces as potential sites for both resistance and governance, and as always intimately entwined with larger social and political processes.

Couldry, N. and A. McCarthy (2004) *MediaSpace: Place, scale and culture in a media age*, New York: Routledge.

A foundational collection of texts for communication geographers, with an interdisciplinary reach across media studies, anthropology and geography. While the empirical case studies may date over time, the theoretical scope of the first part of the anthology, including the introduction by the editors, remains essential reading for anyone interested in the intersection between media theory and spatial theory.

Moores, S. (2017) *Digital Orientations: Non-media-centric media studies and non-representational theories of practice*, New York: Peter Lang.

Moores has written extensively on the spatial dimensions of everyday media use, starting from matters of orientation and habitation. This book is a selection of his work since the early 2000s, largely focusing on his synthetization of human geography and phenomenology as a foundation for more non-media-centric media studies.

SUBJECTIVITY

Aoileann Ní Mhurchú

Subjectivity is not simply a noun indicating presence, although it does refer to *the subject* and thus to the presence of personhood. It also arguably refers to the qualities and conditions of something being a subject, through the use of the suffix '-ivity'. This is because we can link *the subject* to phenomena through which it comes into being – such as to perspectives, feelings, beliefs and desires. We have thus seen the term subjectivity usefully engaged to emphasize increasingly the active process of be(com)ing rather than a static idea of simple presence (e.g. Edkins et al. 1999; Isin 2002; Malik 2016). This entry will firstly examine the importance of several key interventions in decentring the Cartesian subject of Enlightenment thought in order to understand subjectivity as an active process. It explores how these ideas have been developed in particular in postcolonial/decolonial thought through the concepts of embodiment, identification and hybridity. It will then move on to focus on creative practices linked to vernacular language and music – increasingly recognized as interventions into static and more fixed understandings about identity, belonging and community. Finally, it considers how we can understand subjects from this perspective as made by the acts they engage in rather than preexisting those acts.

Decentring the Cartesian subject: embodiment, identification and hybridity

The Cartesian subject was believed to consist of an inner core from the moment of birth, which was understood to develop as the individual grew. This philosophical belief formed the basis for Enlightenment epistemology and the dominant understanding of what it is to be a subject as unified with a fixed presence. This Cartesian subject has been decentred in two stages (Hall 1992). The first stage involved the introduction of the sociological subject. Unlike the Enlightenment subject whose inner core functioned as the source of the subject's identity, the sociological subject developed an interactive conception of identity and self. However, the sociological subject retained "an inner core or essence that is the real me", albeit with a recognition that this could be modified through "dialogue with the cultural worlds outside and the identities which they offer" (ibid.:285).

The second and more radical stage of this decentring process was a move from the sociological subject to the postmodern subject, which is a subject without fixed, essential or permanent identity at its core and root. Stuart Hall emphasizes the importance of Marxist, feminist, psychoanalytic and poststructural contributions to this second phase of decentring. In particular, these contributions include the questioning of individual agency as separate from social structures led by Karl Marx, who stressed the significance of modes of production and exploitation of labour; the casting of doubt on the idea of consciousness as characteristically rational and accessible by Freud, who showed consciousness to be a particular aspect of the mind and not its most general feature; feminism's challenge to the dominance of a supposedly disembodied, sexless and genderblind form of subjectivity as a position of universal neutrality; and finally the emphasis on the intersection of knowledge and power and thus on language as a cultural practice rather than a neutral tool, developed in poststructuralism. These moves towards an understanding of subjectivity as an active process of becoming (rather than fixed presence) – which highlight the importance of social structures, the body, the unconscious and language – have been further developed through the ideas of embodiment, identification and hybridity in postcolonial/decolonial thought.

Postcolonial/decolonial thought points out that the cultural web of relations instilled under European colonialism is "grounded in the suppression of sensing and the body" (Mignolo 2011, n.p.). In contrast, it underscores the importance of thinking about the way in which people can be "'written' all over ... carved and tattooed with the sharp needles of experience" (Anzaldúa 1990:xv). What this type of thinking challenges is the colonial association of knowledge with the visual (what can be seen) and thus with presence, which ignores other sensory experiences and their role in developing knowledge. Postcolonial/decolonial thought has repeatedly pointed to a dominant understanding of a visibly transparent (knowable) world – of people, bodies and identity – produced under colonialism (Blunt and Rose 1994). This continued visual mapping manifests itself today in how divisions are drawn: between the good life and the ghetto, between the citizen and the migrant, and between the citizen and terrorist (Agathagnelou and Ling 2004; Sajed 2010). Postcolonial/decolonial work not only draws attention to this visual mapping but also explores resistances to these divisions and the possibility of new spaces of political engagement by emphasizing that the subject needs to be situated in terms of social structures: such as race, gender, ethnicity and class. It argues that different ways of being a subject come to be possible only through the construction of meaning in terms of the body and its senses – for example, through ideas of skin colour, hair texture, gestures, ways of speaking, modes of dressing. Such work therefore points to the way in which being a subject does not simply exist to be *seen* (as an essence) but is based on contestation and difference in meaning and variants of what is possible across alternative cultural contexts (hooks 1989).

Most notably, Stuart Hall has pointed out that a growing awareness of, and possibility for movement of people, ideas, experiences, cultures and commodities around the globe challenges the ability of a singular (and thus visible) national position or geographic origin to give us a definition of who we are. Emphasizing "identity as an endless ever unfinished conversation", he has pithily observed: "When I ask anyone where they are from I expect nowadays to be told an extremely long story" (Hall quoted in Akomfrah 2013). For Hall, therefore, we need to understand how subjects are structured around points of temporary identification. "Identities", Stuart Hall argues (1990:225), "are subject to the continuous 'play' of history, culture and power". He explains:

> I use the term identity ... precisely to try to identify that meeting point where the processes that constitute and continuously reform the subject have to act and speak in the social and cultural worlds ... They are unstable points of identification or suture ... Not an essence but a positioning.
>
> *ibid.*

This emphasis placed by Hall on needing to think about the constant *negotiation* of identity, developed insights in psychoanalysis – in particular the work of Jacques Lacan. Freud had demonstrated the primacy of the unconscious over the conscious and thus presented a theory of the subject based around absence and lack (of control of thought and action). Taking this further, Lacan (1997) developed a theory of the subject based around the idea that the self is always retrospectively produced through its surrounding social or symbolic order in an attempt to compensate for this lack. Hall builds on this by highlighting key aspects of the social order in culture and history so that we can think through this constitution of the subject.

Significantly, through his work and the importance he places on the notion that subjects are sutured around points of temporary identification, the idea of the individual standing outside of the social order – as a conscious, clearly defined gendered, ethnic, racial, linguistic presence – is demonstrated to be an illusion despite the fact that we 'see' individuals all around us. What Hall points to is the need to see identity as an aspect of subjectivity – something which contributes to the active process of becoming a subject (insofar as it provides an anchor point) but which is always and only temporary. Subjectivity here can be understood to include many forms of identity (unending positionings) – including potentially messy and contradictory ones – but subjectivity is not simply reducible to any of these meeting points given the ways in which they are negotiated around dynamic processes such as feelings, desires, beliefs and cultural perspectives – both unconscious and conscious ones.

Homi Bhabha's concept of hybridity helps us to understand much of what is at stake here given its emphasis on the link between "negotiation of meaning and representation" (1990:211) and the question of 'newness'. Bhabha argues that all cultural articulation is based upon acts of signification – i.e. on producing icons and symbols, myths and metaphors through which we live culture. In the case of language, the imprecision of this signifying practice creates a gap between the symbols and their meaning, between the signifier and the signified. As this gap produces different interpretations – for example, of what it means to be an American and how we should understand its icons of freedom, democracy and inclusivity (Weber 2011) – no culture is full unto itself. Culture is incomplete in itself not only because there are other cultures which can contradict its authority, but because of the gaps in its own symbol-forming and meaning-making activities and abilities. Because there is no culture which is full unto itself, the original basis of all cultures is open to translation. It is this process of translation "through displacement or liminality [which] opens up the possibility of articulating different, even incommensurable cultural practices and priorities" (Bhabha 1990:211). The potentiality for articulating newness for Bhabha therefore is not about a complete break from what came before, but is always mixed up with what has come before (and therefore familiarity) – it invokes what Freud referred to as the uncanny. It is this emphasis on mixture and liminality or in-betweenness that Bhabha has developed in his concept of hybridity. Bhabha continued his ideas of hybridity through the concepts of ambivalence and mimicry (Bhabha 2004:85–92). What Bhabha's theory of hybridity does is to undermine any secure foundations in personhood by pointing to the prevalence of in-betweenness and liminality.

Creating the subject in vernacular language and music

The concept of citizen media emphasizes performative interventions and actions in public space(s) linked to aesthetics and expressions of personal desires and aspirations (Baker and Blaagaard 2016a). What is particularly interesting here is the way this opens up the possibility for many creative forms which citizen media might take: for example, in the use of digital networks (Fotopoulou 2017), film (Tascón and Wils 2017), street art (Morayef 2016) or in music and language.

The relationship between music and intervention has been widely documented, in particular in relation to slave songs, postwar youth mod culture and rock music as well as hip-hop (Keyes 2002; Sullivan 2001). Hip-hop, in particular, has become synonymous with youth subculture and resistance, originating as it did in the US among disadvantaged youth of colour as a medium for bringing attention to the social and political realities in which they were forced to live. Elsewhere mixing languages has also for a long time been understood as something that can be used to signify difference by reworking and (often) parodying a dominant language (Britton 1999). Significantly, in recent years, a number of studies have focused upon developments around the dual engagement in hip-hop and vernacular language use by immigrant youth in diversely populated urban areas. Exposure to multiple cultures and linguistic influences in such spaces and increased global awareness have been shown to be developing novel practices of language mixture and hip-hop due to the specificities of the local-global influences in each given context (Nortier and Dorleijn 2013;

Nortier and Svendsen 2015). What is pointed to is the increasing *blending* of local, indigenous and migrant languages to form the base language for hip-hop (Schmitt 2005; Williams, Q. 2017). This reinforces the potential of this genre for "reframing the basis of citizenship" within such youth practices and putting in place "a complex set of political affiliations and social *boundary crossings*" (Maira 2005:69; emphasis added; Alim et al. 2009; Ní Mhurchú 2016). Therefore, hip-hop is not only seen to convey a message (and act as a medium) to express dissatisfaction with society and point to wrongs perpetuated against youths, but is also seen as a tool to "establish solidarity with fellow community members and to strengthen their position in society" through cross-cutting linguistic and musical influences (Cutler and Royneland 2015:140).

Verlan: Speaking in inversions through hip-hop

One of the more well known vernacular languages is Verlan – referred to in the lyrics of the following excerpt of TIS' rap song 'J'ai besoin' (I Need) below:

> [Translation] I need to make my parents proud
> even if I talk in Verlan
> and if I do talk in inversions
> I need [to do so] to construct my own paradise
> without passing through hell, I can't remake the world
> therefore, I need to create.
>
> *TIS 2007: 2,39 – 4,46 mins*

The origins of Verlan are in the growth of France's *banlieus*, which are peripheral housing projects and home to France's poorest immigrants, heavily populated by North African and African Arabs, their children and great/grandchildren. Verlan draws inspiration from the ancient French wordplay tradition called *Argot*, which involves inverting syllables of words or, as the lyrics above put it, speaking in inversions. Verlan – which is itself an inversion of the syllables in the French word for 'inversion' (*l'envers*) – uses standard French as a base language, but draws on vocabulary from other languages, including Arabic, Creole and Romani. By inverting French words and including vocabulary drawn from elsewhere, Verlan can be said to generate creative modifications of standard French *using* other languages (Nortier and Dorleijn 2013).

As is also the case with Rinkeby-Svenska in Sweden or Kiezdeutsch/Straßendeutsch in Germany, among other examples, this type of vernacular language has received some attention. A popular (or populist) view is that what we are simply seeing is "a random accumulation of errors" by migrant heritage youth or a form of youthspeech, which people grow out of (Wiese 2014:215; Radhhani 2016). In contrast, critical sociolinguists have controversially argued that these vernacular languages (officially called multiethnolects) are a form of creative expression of the self (Nortier and Dorleijn 2013; Wiese 2013). Labelled a *contact language* these are understood as developed in situations where the existing "repertoires of languages available to the people in contact did not provide a sufficiently effective tool for communication" (Bakker and Matras 2013:1). When asked about why she uses Verlan, one woman explained it thus: "we are looking for ourselves, and we can't find ourselves" (Doran 2004:93).

Language has played an important role in nation-building historically, but in recent decades it has become particularly influential in securing homogeneity in the form of standardized national languages – for example through the introduction of language tests for citizenship purposes (Blackledge and Creese 2010). Although some countries have several national languages there is still limited choice here – for example, the Republic of Ireland has English and Gaelic as their national languages; Belgium has French and Flemish. This limited choice ignores the reality of much wider linguistic diversity in most countries (Alexander et al. 2007; Pavlenko and Blackledge 2004). For example, despite the existence of a range of regional languages, standard French has historically in France served as "the major vehicle for the transmission of the national culture and … revolutionary ideology" (Safran 1992:793). This role was achieved through the establishment of the Académie Française in 1635 which was set up specifically with the aim of protecting (and thus prioritizing) standard French on the basis that this provided a basis for a solid French identity, and later through the institutionalization of standard French as the sole language of instruction in 1891. Today standard French continues to be seen as the transmitter of national unity in the face of growing diversity (Villard and Yan-Sayegh 2013). Any attempts to undermine the centrality of standard French are still viewed as threatening French identity by the Académie Française and the French Senate – which vetoed attempts to recognize regional languages in the French Constitution as recently as 2008 (ibid.:244) in response to efforts by regions like Brittany, the Basque country and Corsica at reviving old languages – a pattern reflected across Europe, amid a broader assertion of local identity. The Senate did make a concession at the time to recognize for the first time regional languages *in* the Constitution (under article 75-1); however as recent as February 2018 French Prime Minister Macron has continued to reject any idea that the senate will move towards recognition of regional languages on an equal footing to standard French. This "monoglot standardization" (Blackledge 2004:72) acts to reinforce the idea of nationality and French identity as singular because differences of language are presumed to be subsumable into a common French nationality rather than posing alternatives for how people identify as French.

Yet, we can argue that what is articulated in vernacular music and language is a refusal of the idea that being a national subject can be fit into one space *rather than* an Other space. Drissel (2009:121) points out that "French hip-hoppers are involved in a complex process of reconfiguring and synthesizing relevant idioms and vernaculars found not only in global hip-hop and their 'native' culture but also in their 'host' country of France". Indeed, French rap is a synthesis of influences from around the world including French, African American, North African and African Caribbean expressive culture; it draws on preexisting musical genres from French traditional music as well as rock, jazz and gospel. As hip-hop artist Stromae (whose name is a Verlan inversion of Maestro) notes in his song 'Bastard' ('Bâtard') (Stromae 2013a) you are told "either you are one or the other … Are you Flemish or Wallonian? … Are you white or brown, eh?" (Stromae 2013b: 08,02 – 11,45 mins). Having grown up in a French Belgian suburb with a Flemish Belgian mother and a Rwandan father, Stromae's upbringing does not fit into this national dualism. His refrain in this song – which is on an album entitled 'Square Root' (*Racine carreé*) combining Carribbean and African musical influences with 1990s-inspired dance beats – is to note that he refuses this dualism : "I am, always have been, and will remain neither one nor the other" (ibid.). Use of language mix and hip-hop can thus be understood both to reflect and to influence the *construction* of hybridized selves as subjects, which undermine spaces of national linguistic and ethnic homogeneity. What is being challenged here is both national linguistic homogeneity and the idea of singular ethnic identification. The emphasis on mixture and inversion (and thus the question of constant process) undermines the idea of a fixed space in which the subject can be grounded where a national language is linked to an originary or a limited number of overarching cultures and ethnicities.

Looking at such practices we move away from the idea of preexisting fixed cultures simply coming together, and towards ideas of process in the emphasis on movement, change and disruption; in particular it allows us to reflect on young people's agency and to think about how this (re)forms ideas of political identity and belonging given that they use and reference multiple languages and ethnicities rather than only one or several narrow linguistic and ethnic options. They reformulate dynamic understandings of subjecthood and move beyond static limited ideas of identity *via* their use of multiple languages and ethnic influences. These practices challenge the idea of pre-given ethnic, gendered, sexualized, indigenous or migrant groups simply expressing themselves through music and language. Instead, they show how subjects are *produced within* vernacular music and language; the way in which subjects are made by the acts they engage in rather than preexisting those acts.

This entry has contemplated creativity linked to the formation of subjects and subjectivity as expressed in vernacular music and language. It has done so to consider what fluid forms of performative intervention and aesthetic actions are at play here as forms of citizen media and why an understanding of subjectivity as process is therefore useful.

See also: immaterial labour; postcolonial studies and citizen media

Recommended reading

Ní Mhurchú, A. (2016) 'Unfamiliar Acts of Citizenship: Enacting citizenship in vernacular music and language', *Citizenship Studies* 20(2): 156–172.

Considers the importance of 'unfamiliarity' in how practices of vernacular music and unconventional use of language challenge conventional sites of citizenship; it also explores the need to expand existing understandings of 'resistance' in order to recognize the significance of such artistic practices.

Shapiro, M. (2013) *Studies in Transdisciplinary Methods*, London: Routledge.

Explores the importance of seeing a wide variety of artistic sources – from fine art to cinema and from literature to the blues – as integral to social science research and the methods it employs; in particular to understand questions surrounding who we are. It both shows how these artistic sources can be used in writing social science and as sources to understand the world better. A key text for understanding the philosophical as well as methodological potential of the aesthetic turn within the social sciences.

Williams, Q. (2017) *Remix Multilingualism: Hip-hop ethnography and performing marginalised voice*, London: Bloomsbury.

A compelling ethnographic study of how new forms of voice are enabled while others are silenced through innovative practices of language mixing by hip-hop practitioners, focusing in particular on multilingual participants in the local hip-hop culture of Cape Town in South Africa.

SURVEILLANCE

Arne Hintz

In October 2016, the *Bristol Cable* uncovered the widespread acquisition and use of 'IMSI catchers' by several British police forces (Aviram 2016). A sophisticated surveillance device, IMSI catchers can intercept mobile phone calls and locate users, and have been used to, for example, identify participants in public protests. British police had previously refused to acknowledge their use of such devices but the investigation by the *Cable* of purchase data and other records provided a detailed understanding of which police forces had acquired this technology and which companies had provided it. Larger media organizations, such as *The Guardian*, picked up the story and gave it nationwide prominence (Pegg and Evans 2016). The original investigation, however, was conducted by a local media cooperative and thus outside the realm of experienced journalism and without the resource base of a large publishing house. *The Cable* is a nonprofit, member-run and democratically organized group of part-time and volunteer journalists.

Small, alternative, nonprofit and citizen media have reported on surveillance, but they have also been subject to states monitoring their activities and to pervasive data collection by the companies that operate the platforms that are used by many citizen journalists. As journalists, they are dealing with sensitive information and are vulnerable to the increasing surveillance efforts by state institutions and businesses. Yet, as citizens, they often lack the legal protection, institutional backing and material resources to protect themselves. This entry will discuss how citizen media engage with and negotiate the surveillance of all our communications and much of our everyday activities. It will start with an introduction to the theme of surveillance, highlighting both state surveillance and data collection through the commercial infrastructure of social media that many citizen journalists use. It will then review recent research on how surveillance is reported by both mainstream and non-mainstream media and how it affects both citizens and journalists, discussing prominent concepts such as the chilling effect and digital resignation. Finally, it will explore how citizen media may offer avenues to respond to surveillance, for example through technological self-protection and what has been discussed as sousveillance and data activism. The entry will contextualize citizen media practices in the broader implications of surveillance for state–citizen relations and citizenship. While citizen media have long been a key component of active citizenship and its notion of empowerment, the power shifts implied by surveillance require us to rethink the role of citizen media, particularly in digital environments. While surveillance has been conducted across history and technological cycles, this entry will focus on contemporary practices of data collection and digital monitoring, and situate citizen media in the context of current digital media infrastructures.

Surveillance and data collection

The practice and condition of surveillance denotes the monitoring and tracking of people's communication, their movements and their activities. This can happen openly, for example through a surveillance camera, or more subtly through automated tracking of online exchanges. Traditional concepts of surveillance have been heavily influenced by the model of the Panopticon – an architectural structure that would enable comprehensive surveillance within a physical space, such as a prison (Bentham 1791). As a representation of disciplinary power, the Panopticon could not only increase the number of people who can be controlled but regulate the behaviour of social actors through the organization of space (Foucault 1975/1977). It expresses a form of social control that combines pervasive monitoring of people by an institution with the need for self-discipline and self-control by the surveilled. However, with the emergence and spread of electronic communication, the concept of the assemblage has gained prominence in capturing the complex nature of data flows in networks (Haggerty and Ericson 2000). With the increasing monitoring of digital interactions and people's data traces, scholars have observed the emergence of a surveillance society in which "all manner of everyday activities are recorded, checked, traced and monitored" (Lyon 2007:454). Surveillance has advanced from a specific action against defined targets to a social condition in which information is gathered about everything, everyone and all the time (Braman 2006). As data extracting infrastructures and surveillance technologies increasingly penetrate everyday life, surveillance becomes normalized and internalized (Murakami Wood 2009). It is at the core of contemporary governance mechanisms that identify, classify, assess and sort populations, both for commercial and government purposes (Gandy 1993; Lyon 2015).

State agencies continue to be among the main initiators and beneficiaries of surveillance, but they are increasingly complemented by the private sector, particularly commercial intermediaries and data brokers. Social media platforms and online services, such as Facebook and Google, operate on the basis of a business model of collecting and analysing user data. Detailed knowledge about user locations, activities, brand preferences and political orientations, as well as those of their friends and networks, is the foundation of their market value, and so they are designed to maximize (corporate) surveillance (Trottier and Lyon 2012). Users are tracked as they move across the web and subjected to multiple forms of identification. An industry of data brokers, profilers, optimizers and scoring agencies has emerged which analyse and monetize the vast range of user data that is collected by platforms, trackers and sensors (Christl 2017). The big data generated through business actors and commercial processes is thus at the heart of contemporary surveillance trends (Lyon, D. 2014). Yet, much of this data is generated by the users of digital communication networks who share personal information on social media, let their phones track their geographic location and health data, and check on each other through status updates and real-time interactions. Collecting and sharing data has become a social norm embedded in the contemporary culture of connectivity (van Dijck 2013) and desirable digital technologies lure us into participating in data extraction as a form of seductive surveillance (Troullinou 2016). While the hierarchical monitoring in our surveillance society continues, it is embedded in a broader surveillance culture (Lyon 2017) that values visibility and exposure and in which people are not just subject to, but also initiate, surveillance activities. The broader notion of veillance (Bakir 2015) has thus been employed to encompass not just traditional power relations between the watcher and the watched, but also mutual monitoring among subjects, and bottom-up investigations by the disempowered of the powerful.

While some surveillance practices are relatively transparent, for example the existence of CCTV cameras in public spaces, surveillance is more often conducted in secret and without subjects knowing exactly when, where and how they are monitored. We typically do not know who watches the CCTV recordings and views data traces, and by what means and to what extent our data is analysed. The revelations by whistleblower Edward Snowden in 2013 were therefore a historic moment as they demonstrated in great detail how intelligence agencies such as the US National Security Agency (NSA) and the British Government Communications Headquarters (GCHQ) operate and proved the unprecedented extent to which our every move in online environments is tracked, monitored, analysed and stored. According to Snowden and the media organizations that published his leaks, state agencies use a wide variety of means to collect and analyse information. They harvest data from the Internet's backbone cables and intercept communication that passes through key nodes and pipes, for example between the UK and North America (through programmes such as Tempora and Upstream). They collect data from the servers of large Internet companies such as Google, Facebook and Apple – sometimes with their consent, sometimes without (through programmes such as Prism, Muscular and Squeaky Dolphin) – and they collect millions of text messages and users' geolocations each day (through programmes such as Dishfire and Co-traveller). Analytical tools such as Xkeyscore allow them to sift through these vast amounts of data, search for specific types of communication and online activity, and profile people and networks. Complementing the interception of data flows, other programmes have been developed to break the encryption that protects our data exchanges with, for example, social media platforms and online banking, to weaken the security of software products, and to hack into telecommunications services (Greenwald 2014; Fidler 2015). By demonstrating the extent and exact practices of contemporary data-based surveillance, the Snowden leaks provided a rare spotlight on otherwise secretive processes and lifted the abstract and often speculative topic of surveillance out of obscurity.

Media coverage of surveillance

The Snowden leaks demonstrated the importance of journalism in exposing secretive practices and helping people understand the nature of their digital interactions. Traditional media institutions such as *The Guardian* and *The Washington Post* were instrumental in bringing the revelations to the attention of a wider public, sometimes with considerable risk to their own operations (as demonstrated by a raid conducted by intelligence personnel of *Guardian* offices and by the detention of *Guardian* journalists, shortly after the revelations began – Greenwald 2014). The media thereby raised citizen awareness of surveillance and its consequences (Wright and Kreissl 2013) and exposed threats to journalism and other professions in times of mass surveillance (Russell and Waisbord 2017). The Snowden leaks became a global media event and thereby elevated a previously abstract and obscure issue to a prominent public debate (Kunelius et al. 2017).

However, a closer look at the media coverage reveals a more nuanced picture. To start with, the analysis and interpretation of surveillance practices has been shaped by social, historical and political contexts, and affected by national and geopolitical interests. Media reporting in the US, for example, has differed significantly from that in Germany or in China (Kunelius et al. 2017). More significantly, the mediated public debate on surveillance has not necessarily led to a better understanding of the challenges of digital life, nor to a broader recognition of the risks of surveillance for citizens. As research on media coverage in the UK has demonstrated (Wahl-Jorgensen et al. 2017; Hintz et al. 2018), the leading role of *The Guardian* in exposing surveillance practices was overshadowed by support for state surveillance programmes in the coverage of most other newspapers and broadcasters. Monitoring and data collection have mostly been discussed in the context of national security threats, rather than threats to civil rights, and journalism's traditional reliance on official sources has underpinned the dominance of perspectives that have regarded surveillance as necessary and acceptable. Political elites have often been identified as the main targets of surveillance, and implications for international relations and business practices have been more widely discussed, while paying less attention to the consequences for the population as a whole. Concerns over civil rights and everyday digital citizenship have therefore been pushed to the margins of public debate and rendered invisible, whereas surveillance practices have often been discursively justified and thus normalized (Wahl-Jorgensen et al. 2017). A number of international studies on media coverage in the aftermath of the Snowden leaks, while considering specific national circumstances, confirmed this dominant trend

(Kunelius et al. 2017; Mols and Janssen 2017). In part, it demonstrates that journalists have struggled to communicate the complexities of digital life to their audiences, but in part, it also points to structural limitations. Traditional journalism has a propensity towards stories which can be personalized, simplified and polarized, and it typically avoids technical detail. It is subject to pressure from government and state institutions which, in the case of the Snowden leaks, exerted significant influence to downplay the implications of the revelations. It also relies heavily on official sources which, in this case, led to a strong presence of the views of security services and government spokespersons in the media coverage.

Non-mainstream media are less inhibited by these practices, norms and pressures, and accordingly their reporting on surveillance has differed substantially from predominant mainstream coverage. Wahl-Jorgensen et al. (2017) analysed the output by blogs and digital native news organizations and found that these media have provided a space for more critical arguments, questioning the assumptions underlying mass surveillance and exposing flaws in state surveillance programmes. Emerging web-based media organizations, such as *Vice* and *Huffington Post*, tech blogs such as *Gizmodo* and *Mashable*, and the wider range of blogs addressing social and political issues have raised significant concerns over a lack of transparency surrounding intelligence agencies as well as violations of privacy. They have typically foregrounded citizen concerns, and they have identified members of the public, rather than political elites, as the main targets of surveillance. Non-mainstream media have articulated a broader range of opinions and have thus provided greater scrutiny of political actors, expanding what the public can legitimately contest and dispute (Russell and Waisbord 2017; Hintz et al. 2018).

Impacts on (citizen) journalists

While the media coverage of surveillance has differed significantly according to the type of media, journalistic practices have been affected by surveillance across both mainstream and non-mainstream approaches. Yet here too, differences between professional and citizen journalism are recognizable. For everyone working in the field of journalism, surveillance represents a challenge to core practices and procedures. In a context in which it becomes increasingly difficult to secure the confidentiality of conversations with sources – whether online or face-to-face – any investigative work is facing fundamental risks. An increasing number of journalists – particularly those who have covered surveillance – are aware of the implications of widespread surveillance for their own communications practices. As James Ball – formerly the special projects editor for *The Guardian* and a WikiLeaks employee – put it, if you are a journalist "and do not know how to communicate securely, you are negligent" (Ball 2015). According to the Pew Center's report (Pew 2015), around two thirds of US investigative journalists believe the government has collected information about their activities, and nearly half have changed the ways they store sensitive information. This heightened awareness among journalists is, in many cases, in stark contrast with their reporting which has typically normalized surveillance. In research interviews, some journalists have been disparaging of arguments supporting widespread monitoring that were so prominent in the newspaper coverage and critical towards justifications for surveillance that apply arguments of national security. Facing the risks of increased monitoring of their activities, many (including journalists working for tabloids and conservative newspapers) have been concerned about the abuse of counterterrorism legislation for pervasive surveillance (Hintz et al. 2018).

If mainstream journalists are affected, citizen journalists are even more vulnerable. Without the support and resources of larger media organizations, they are often left without protection and with few means to defend themselves against surveillance measures. They typically do not enjoy the privileges of professional journalists, such as access rights to restricted institutions or areas, and the right to protect a source and collect certain types of information. Their legal position remains precarious as they may not be eligible for membership in journalist unions and thus to hold a press card. Organizations such as Article 19 and international institutions such as the UN Human Rights Committee have therefore argued that the same protections and privileges should apply to professional and citizen journalists (Article 19 2013). The risk of surveillance intersects with vulnerabilities to speech restrictions and concerns about physical safety. Legislative reform in many countries has tightened rules against the incitement to violence, crime and terrorism but has, at the same time, been vague and open to subjective interpretations. This has led, for example in the UK, to a steep rise in prosecutions against bloggers and social media users for comments posted online (Greenwald 2015). Activists and nonprofit organizations that provide infrastructure to citizen journalists have been subject to repression and incriminated through the use of antiterrorism legislation, and such infrastructure has, at times, been confiscated by police (Hintz and Milan 2009).

The widespread use of commercial platforms has enhanced the vulnerability of citizen journalists to thorough monitoring by both state and commercial entities. With data collection being the core business model of social media, activists and journalists using such platforms to publish their content are subject to the detailed analysis of both their own activities and their networks. States have used such data for surveillance purposes, and dissidents in many countries have been identified, monitored and sometimes arrested on the basis of their social media use. In many countries, authorities have used social media to collect the user data of potential protesters and regime critics, and have infected the computers of opposition supporters in order, for example, to capture webcam activity and thus spy on people (Villeneuve 2012). As the Snowden revelations demonstrated, such practices are also and increasingly used in Western democracies (Greenwald 2014). The mere possibility of surveillance has affected the online activities of citizens and the media content that they produce, as research has shown. Fears that engaging with controversial issues may cause them harm have led journalists and other writers to avoid or treat cautiously contentious issues, such as protest reporting and critical investigations into government policies (PEN 2013).

Chilling effect, confusion and resignation

Surveillance may thus deter people from engaging in (legal and legitimate) online activities because they fear exposure to authorities, punishment or criminal sanction (Penney 2016). This has been described as the chilling effect of surveillance. In the aftermath of the Snowden revelations, for examples, studies have detected a decline in privacy-sensitive search terms on Google (Marthews and Tucker 2015), a decline in page views of Wikipedia articles relating to terrorism (Penney 2016), and a "spiral of silence" in surveillance debates on social media (Hampton et al. 2014:1). Moreover, this chilling effect might not just affect actual activities but people's underlying assumptions and desires. As Greenwald (2014:177–178) notes, "the individual trains him- or herself to think only in line with what is expected and demanded". The pervasive monitoring of people's movements, actions and communication therefore undermines critical debates and dissident voices, as the United Nations Special Rapporteur on Freedom of Expression and Opinion, among many other scholars and institutions, has repeatedly emphasized (UN General Assembly 2013). As noted above, activists, advocacy groups and social movements have been particularly implicated in surveillance measures, and some have responded by ensuring that they remain sufficiently mainstream on the spectrum of political activism in order to evade the spotlight of state attention (Dencik and Cable 2017). This is a worrying outcome from a democratic perspective as democratic engagement and civic action are crucial factors for social change and democratic reform. It also provides significant challenges for citizen media which have often been important vehicles for the expression of dissent and discussions on social change.

The chilling effect is not necessarily based on an informed understanding of the practices and consequences of surveillance, and thus on a conscious choice by citizens. Rather, many studies have demonstrated widespread confusion and uncertainty among people about how data is collected, by whom and for what purpose (Dencik and Cable 2017). While most people have some awareness that their activities may be monitored and that data about them is generated and analysed, the exact practices and consequences of the condition of surveillance remain obscure. This reflects the secrecy around surveillance measures, the lack of informed media debate (as described above), and – on a more general level – the liquidity of modern surveillance assemblages (Bauman and Lyon 2012) in which a wider range of actors and practices intersect. As a consequence, the heightened visibility in contemporary datafied society is neither properly understood nor adequately explained.

Uncertainty over how data travels and how organizations collect, share and use data does not lead to widespread disengagement from online activities and digital infrastructures, but rather to a disillusioned perception of inevitability in pervasive data collection and surveillance. People wish to use digital infrastructures but see little possibility to challenge, circumvent or mediate the data collected on them. This response should not be misunderstood as either apathy or consent, but rather as an active pragmatism that is based on a general resignation to the status quo. As several studies have confirmed, many people do have prominent concerns and worries about surveillance and they express unease with the implications of using data-generating infrastructures. Yet the necessity of using these technologies to interact with contemporary social environments, coupled with a lack of control over their workings, lead to what has been called surveillance realism (Dencik and Cable 2017) and digital resignation (Draper and Turow 2019). People thereby resign to ubiquitous data collection and the inevitable reality of a surveillance society in order to participate in a social order that is increasingly based on data extraction and data sharing. This resignation is shaped, not least, by persistent secrecy around surveillance practices, obfuscation in privacy agreements between users and platforms, and by making services inaccessible if personal data is not shared.

The disempowerment of digital citizens poses significant challenges for the concept of citizen media. Playing a key role in earlier celebrations of liberation technology (Diamond 2010) and empowering "the people formerly known as the audience" (Rosen 2006), citizen media have been an important feature of the notion of digital citizenship which has typically focused on the enabling and democratizing aspects of digital infrastructure (Mossberger et al. 2007; McCosker et al. 2016). In a context of pervasive surveillance, however, the earlier enthusiasm that has surrounded citizen media meets the challenges of the chilling effect and digital resignation. Moreover, their role in empowering citizens to raise their voice and contribute their perspectives to public debate is complemented by the consequence of exposing citizens to, and thereby empowering, the state and commercial surveillance actors. This constitutes a new environment for citizen media and requires a rethinking of their roles and practices.

Responses to surveillance

While established understandings of citizen media are complicated through the emerging realities of data collection, citizen media continue to serve as vehicles to address, respond to and resist pervasive surveillance. To start with, they offer important avenues for critiquing and challenging surveillance regimes. As the example of the *Bristol Cable* at the beginning of the entry demonstrates, the vast range of digital, non-profit and grassroots media provide important means for critical investigations into surveillance and for expanding the public debate. As channels for citizen-based reporting and analysis, they allow those who are otherwise subject to state and corporate surveillance to gaze back at power 'from below', engage in their own monitoring and data collection, and thereby watch the watchers. This practice has been called sousveillance (Mann et al. 2003) and refers to the use of a wider range of technologies – from wearables to cameras on smartphones – which allow people to capture, process, store and transmit information and thereby monitor elites and document abuses of power. The use of data leaks by whistleblower platforms such as WikiLeaks and Xnet can be regarded as another, indirect form of sousveillance if they expose wrongdoing by power holders and reveal information about government or corporations that was previously unknown to the public. Similarly, blogs and alternative media that document malpractice and confront authorities can be important means for such bottom-up veillance.

Citizen-based and self-organized technologies have also been a key part of counterveillance measures to detect and block surveillance. The development of secure and anonymous communication infrastructures has become a prominent feature of digital engagement, with "numerous digital rights and internet freedom initiatives seizing the moment to propose new communication methods for activists (and everyday citizens) that are strengthened through encryption" (Aouragh et al. 2015:213). These have included privacy-enhancing tools such as the TOR browser, the GPG email encryption system and the encrypted phone and text messaging software Signal. Technical solutions to surveillance have been complemented by self-organized communications infrastructures as alternatives to corporate services such as Google and Facebook. Groups such as Riseup.net, Autistici and Sindomino have offered mailing lists, blog platforms and collaborative online workspaces that protect user privacy and are hosted on the groups' own secure servers. Indymedia, arguably the first social media platform, was run by activists in the same manner, and so were more recent non-commercial and privacy-enhancing social networks such as Crabgrass and Lorea. Such technological alternatives that reinforce autonomous and civil society-based media infrastructure have been a key part of both citizen media and anti-surveillance activism (Hintz and Milan 2009), even though their adoption has remained marginal in comparison to large corporate providers whose vast resources and ease of use have enabled a far more widespread uptake (Askanius and Uldam 2011; Terranova and Donovan 2013).

Exploring citizen media practices in a datafied age, data activism has applied alternative forms of data use and thereby subverted the dominant purposes and power relations of datafication. Data mapping in crisis situations, strategic hashtag uses in social movement mobilizations, and a wide range of activist initiatives of data gathering, analysis and visualization have served as data practices that draw from the ideas and strategies of sousveillance, activist technologies and citizen media. Yet whereas counterveillance tactics have responded to surveillance in a reactive manner in which citizens resist the threats to civil rights from pervasive monitoring by corporate actors and state agencies, proactive data activism appropriates data to foster social change. Its action repertoire includes "examining, manipulating, leveraging, and exploiting data, alongside with resisting and meddling their creation and use" (Milan 2017:153).

Using data as a means for social and political change, data activism seeks to reclaim citizen agency that, as noted above, has been challenged in the context of datafication and surveillance. It aims to "bring democratic agency back into the analysis of how big data affect contemporary society" (Milan 2017:153). By focusing on people's control over their data, the transparency of data analysis and popular comprehension of data output, it addresses concerns that are at the core of contemporary debates on surveillance. As such, it is both a strategy for action and a means for critically interrogating the politics of data. However, it does not change the underlying challenge of surveillance that is now inherent in every digital act. Data activism can, in this sense, add bottom-up sousveillance and dissident data uses to existing top-down surveillance, but it exposes activists to the same pressures of datafication that the wider digital citizenry faces.

Citizen media in a surveillance society

As means for citizen empowerment, citizen media have been prominent examples for the opportunities that new technologies can provide. From radio waves to the Internet, media activists and citizen journalists have applied communication technologies for their own ends and have thereby contributed to the shaping and development of each medium. Digital infrastructures have vastly expanded the opportunities for mediated citizen engagement and social participation, and a vast range of citizen media initiatives have applied digital channels of communication to express opinions, investigate power holders and create new spaces of activity. However, while digital environments have expanded the power of citizen media, they have also exposed participants to new forms of surveillance. They have offered important tools for citizen initiatives and, at the same time, have allowed states and corporate actors to collect more information on citizens than ever before. Digital citizenship is thus both an active and empowering, and a monitored and controlled form of citizenship. Citizen media are at the centre of this tension.

As this entry has shown, all journalism is vulnerable to surveillance, but citizen media are affected in particular ways. They enjoy fewer protections and resources than mainstream media organizations; the infrastructure they use is often precarious, particularly as the dominant commercial platforms are based on the business model of data collection and analysis; as dissident and oppositional voices, many of them face repression; and as 'citizen' media they are subject to the consequences that surveillance poses for all regular Internet users and citizens. In particular, the chilling effect of surveillance and the dynamics of digital resignation and surveillance realism go to the heart of citizen media as spaces for expression and reflection. However, citizen media are also prime avenues for addressing the challenges of surveillance. As we have seen, they offer more space for critical investigations into surveillance than classic journalist institutions. They are important means for the sousveillance of elites and for watching the watchers, and they provide privacy-enhancing media infrastructures that operate differently from the data mines of commercial social media. And finally, as forums for grassroots technological exploration, they offer significant means for counteracting surveillance measures and protecting citizens from pervasive data collection. Perhaps most significantly, surveillance places the question of citizen agency on the agenda as it serves to monitor, sort, assess and thus control citizens and their activities. In this context, citizen media offer means for reasserting agency and exploring its future trajectory.

See also: big data; hacking and hacktivism; sousveillance

Recommended reading

Forde, S. (2011) *Challenging the News: The journalism of alternative and community media*, Basingstoke, UK: Palgrave Macmillan.

A key contribution to the growing body of literature on alternative and community media, this book focuses on alternative news practices and thus helps us understand the different approaches adopted by citizen media producers towards issues such as surveillance.

Hintz, A., L. Dencik and K. Wahl-Jorgensen (2018) *Digital Citizenship in a Datafied Society*, Cambridge: Polity Press.

Explores the consequences of modern forms of surveillance for the concept and diverse practices of digital citizenship, including citizen media and the broader field of people's online engagement. It argues for a more complex understanding of digital citizenship that considers the transformation of state–citizen relations due to the permanent monitoring of citizen activities and the data-based categorizing, rating and scoring of citizens.

Lyon, D. (2015) *Surveillance after Snowden*, Cambridge: Polity Press.

Lyon, D. (2018) *The Culture of Surveillance: Watching as a way of life*, Cambridge: Polity Press.

Both of these books by one of the most eminent scholars in surveillance studies discuss our knowledge of surveillance in light of the political, social, cultural and technological developments of the twenty-first century – particularly, the Snowden revelations and pervasive data collection on social media platforms. These two titles thus offer valuable insight into the practices, infrastructures and implications of 'data-veillance'.

TEMPORALITY

Anne Kaun

This entry discusses the temporality of citizen media from two perspectives: firstly, in terms of the inherent temporalities of citizen media technologies (Virilio 1977/1986, 1995; Castells 2000; Frabetti 2015), and secondly, in terms of the temporalities of civic engagement, with a focus on social and protest movements (Keightley 2013; Postill 2013; Barassi 2015a). It examines some of the ways in which the temporalities of citizen media technologies intersect with and diverge from the temporalities of civic engagement and hence highlights the relevance of temporalizing aspects of citizen media for social change. Following and at the same time extending the definition of citizen media proposed by Baker and Blaagaard (2016a), media technologies are here understood as infrastructures that allow unaffiliated citizens to act in the public arena in order to further political change.

Focusing on the ways in which technology shapes many aspects of social life, as outlined by Wajcman (2015), allows us to consider the relationship between temporal regimes of media technologies and their consequences for civic agency. While acknowledging the crucial role of technology in shaping our perception of time, this focus makes it possible to question technological determinism and digital exceptionalism, both of which overemphasize the role of media technologies in political engagement (Marwick 2013). A dialectical approach to the temporality of media technologies is thus more productive; it involves understanding technology as a cultural form, as proposed by Raymond Williams (1974) – that is, as an expression of larger social and political structures rather than being independent of them. Hence it is important to consider the cultural, political, economic and social configuration within which certain media technologies, including citizen media, emerge. Time – that is, objectively measured units of time – and temporality, meaning subjectively experienced and relational aspects of time, are both closely intertwined with different media that help us keep track of time, preserve past experiences and plan the future (Keightley 2012). Lefebvre (1974/1991) provides an extended discussion of temporalizing practices that are enabled by media as a crucial part of modernity, where lived time disappears and is replaced by instruments that measure time; Rosa (2013) offers a useful summary of this argument. According to Lefebvre, time has become a resource to be consumed, deployed and exhausted. In that sense, our temporal experiences are part of and drive the general process of rationalization in late modernity.

Citizen media as infrastructures

The starting point for a discussion of the temporality of citizen media as infrastructure is that time as an abstract category has to be mediated in order to be experienced (Frabetti 2015). Time-mediating technologies such as calendars, clocks and diaries link the individual, lived experience of time to a shared sense of time, as well as to natural temporal cycles such as seasons (Lash and Urry 1994; Peters 2015). The subjective experience of time, or temporality, in turn emerges in relation to these mediating objects. An in-depth understanding of media technologies, and of citizen media more broadly, thus requires an examination of the consequences of their specific temporalities. This, in turn, requires us to extend the definition of citizen media suggested by Baker and Blaagaard (2016a) by considering citizen media as infrastructures that allow and support civic engagement. The term infrastructure is commonly understood as meaning "that which runs 'underneath' actual structures – railroad tracks, city plumbing and sewage, electricity, roads and highways, cable wires that connect to the broadcast grid and bring pictures to our TVs"; that is, "that upon which something else rides, or works, a platform of sorts" (Leigh Star and Bowker 2002:151). Edwards (2003) argues that infrastructures constitute the foundation of modern social worlds by linking macro, meso and micro scales of temporal, spatial and social organization. Infrastructures are thus "interrelated social, organizational, and technical components or systems" (Bowker and Leigh Star 1999:99). Theorists of infrastructures have suggested that these systems are often taken for granted and perceived as ready at hand; they remain largely invisible until we experience a breakdown, rupture, glitch or failure. Furthermore, infrastructures are based on extensive work that only rarely becomes visible to us, and yet Larkin (2018) has also pointed out that infrastructures have their own aesthetics, thus questioning the widespread assumption of their invisibility. Infrastructures fulfil different functions and therefore acquire different levels of visibility. Nevertheless, they are all fundamentally relational in nature: their meaning emerges in the practice and activities of people connected to technical structures.

If the meaning of infrastructures emerges at the interface of practices and technical structures, so does the meaning of temporality. In that sense, the mediation of temporality involves social and political aspects of our lives and consequently needs to be considered in the context of civic engagement. Furthermore, citizen media serve as infrastructures that coordinate and synchronize people and objects in time and space in order specifically to effect social change. At the same time, citizen media are a constitutive part of the temporal systems that are constructed in different cultures in different ways. Kaun and Uldam (2017), for example, argue that discussions of digital activism often follow the logic of digital universalism – that is, the idea that digital media are used in similar ways in different countries and cultures. However, research on digital activism around the world shows the importance of the social, political and economic context for shaping the ways in which digital media are employed for activism (ibid.). It is hence fruitful to adopt Raymond Williams' (1974) understanding of technology as cultural form, given that it theorizes technologies as expressions of larger social and political structures rather than as independent of them. Just as activism emerges within and in response to a variety of societal contexts, moreover, so do media technologies, including digital media. Hence, the character and form of media technologies that operate at any moment in time are shaped by specific social, political and economic needs and practices, and they in turn shape the very possibilities for self-expression, political participation and activism in each context (Winner 1980; Williams and Edge 1996; Joerges 1999; Wajcman 2015). Citizen media must therefore be understood as deeply situated within this dynamic, and both shaped by and shaping forms of civic engagement.

With corporate social media increasingly being referred to as citizen media infrastructures (Kaun and Uldam 2019), it is crucial to consider some of the problematic aspects of these platforms, including their temporality. One such problematic aspect that concerns the temporality of corporate social media is that they rely on a business model which valorizes user data. Driven by their main goal of increasing the amount of data gathered about and from users, the infrastructure of these corporate platforms follows a temporal logic that is based on producing ever more data through user engagement. Consequently, the platform is built around principles designed to increase user contributions, which in practice means it establishes a requirement for users to constantly update and engage in the platform, a requirement that highlights exchanges over content (Dean 2005) and is experienced as digital immediacy (Kaun 2015). In and of itself this is not problematic. It becomes problematic in the context of civic engagement and political participation that involve several layers of temporality and correspond with a multiplicity of temporal experiences, including those involving longer periods of latency and extensive deliberations (Kaun 2016a). If the platforms flatten users' temporal experience, we can speak of an increasing desynchronization of civic engagement and the civic media infrastructure that social media have become. Facebook and similar platforms which are increasingly being treated as citizen media infrastructures have temporal properties that run counter to certain forms of civic engagement and decision making, including participatory decision making that requires longer stretches of time (Polletta 2002). Hence, social media as one form of citizen media infrastructure lead to contradicting effects. On the one hand, they help collectives reach out and mobilize large numbers of supporters within a short period of time and in a cost-efficient manner (González-Bailón et al. 2011); on the other hand, they pose challenges for the process of political participation, reflection and elaboration (Kaun 2016a).

Temporalities of social movements

Tarrow (1994) offers a widely discussed conceptualization of protest cycles, or cycles of contention, that follow different intensities of mobilization and organization over time. These cycles and varying intensities of social movements are closely linked to the political opportunity structure, which makes it possible to effect change through particular means rather than others. Alongside periods of intense activism, there are also phases of latency, but these periods of slow development are as crucial to social movements as fast-paced periods of intense mobilization and rapid change. Tarrow's conceptualization of protest cycles assumes a linear temporal development of social movements. In contrast, ethnographic and phenomenological accounts often emphasize the multi-layered character of temporalities of civic engagement (Keightley 2013). Postill (2013), for instance, developed a theory of multiple temporal layers of protest which accompany the very diverse range of media technologies used in the organizational processes of activist initiatives. Postill, whose work is summarized by Kaun (2016a), relies on Sewell's (2005) conceptualization of historical sequences as temporally multi-layered social processes. Consequently, he develops events, trends and routines as important categories that constitute a heterogeneity of protest time. While events refer to concentrated sequences of action that transform social structures, trends are less concentrated – they are rather directional changes in social relations. Finally, routines are more or less stable schemata that reproduce social structures. Postill shows in his analysis of the Indignados movement in Spain that protests are made up of countless timelines that run parallel to each other. Barassi (2015b) similarly focuses on the multiplicity of protest time, but unlike Postill, she concludes that even if protests are temporally multi-layered, activists always have to relate to a hegemonic perception of time. In information-based, communication-oriented capitalism this hegemonic perception of time is connected with speed, presentness and immediacy.

The temporality of media practices that activists have to navigate is central not only to the possibility of protest, but also to the possibility of any critique or reflection being undertaken by the protesters. Any act of protest marks an accelerated temporality that is preceded by tedious preparations and organization on the part of activists. Protest in the form of marches or the setting up of protest camps, for example, further establishes a new kind of temporality for those directly involved and those affected in their daily routines and flows (Kaun 2015, 2016a). Hence protesters always act within a protest time that is entangled with a more general regime of temporality, or as Barassi (2015b) calls it, hegemonic perception of time. Finally, Mattoni and Treré (2014) conceptualize the role of time in the context of social movements around micro, meso and macro scales. On the micro level, they identify punctuated events that potentially transform the movement. Meso-level temporality relates to cycles, tides and waves of mobilization that go beyond single events; these can be considered as concluded sequences. Macro-level temporality concerns epochs of mobilization that are connected to a certain set or templates for mobilization that are shared, for example, across a generation of social movement actors.

Beyond the discussion of temporality and cycles in civic engagement, protest and political participation, there have been attempts to invigorate time as a political strategy. Sharma (2013), for example, calls for temporal insurgency: arguing that political practices are sensitive to and disruptive of temporal norms, she suggests that temporality should be considered as a means of invoking politics. Thus, with reference to the hegemony of certain temporal cycles such as nine to five as well as the slowness and boredom that pervade the life of service workers, both of which maintain the invisible infrastructure of high-speed capitalism, she develops the notion of precarious temporality as a strategy of political insurgency that can disrupt such cycles (Sharma, 2014). Similarly, Pink (2008) has investigated the practices of the Slow Cities movement, which values traditional ways of living in the city and resists the fast-paced life style of capitalism. While Sharma argues for dismantling the temporal order by adopting temporalizing practices that resist dominant patterns such as the nine-to-five work routine, the movement that Pink studied is focused on sustainable city development by locally oriented city organizations as an expression of resistance against fast capitalism. Temporality here carries symbolic meaning: slowness is a label for sustainability and long-term thinking rather than an individual temporal practice of, for example, moving slowly through the city space.

One topic related to the question of social movement temporalities that has gained growing attention, but that is still nascent in the field of social movement studies, concerns memories and the archiving practices of activists. Zamponi (2015) argues that collective memories can acquire crucial political relevance not only because they provide symbolic material for current mobilizations and identification, but also because they challenge dominant narratives of societal conflicts; Doerr (2014) makes a similar point. Remembering the past of previous social movements serves to preserve valuable knowledge about strategies and tactics, but beyond this tangible knowledge it can also become part of a political project that addresses questions of visibility and representation (Polletta 2006). Engaging with issues of memory and self-archiving practice thus adds an additional layer to questions of temporality in the nexus of media and social movements.

Memory work is strongly related to questions of how we preserve materials and narratives. Both the form in which the archive is held and the archiving practices of social movements are central here. Community archives emerged in the context of the 1950s and 60s civil rights movement, as an attempt to write and preserve histories that are close to the movements themselves. They are self-organized spaces for storing and working with ephemera, posters, books, badges, fanzines and other material forms that are crucial to protecting the voices of the movements beyond mainstream history writing. With the emergence of digital technologies that appear to allow anyone to preserve everything at very low costs, archiving has attracted renewed attention from activists and researchers. Flinn (2008), for example, speaks of an emerging archival movement, evident in a growing number of community archives and increased scholarly engagement. He defines community archives or community histories as "the grassroots activities of documenting, recording and exploring community heritage in which community participation, control and ownership of the project is essential" (ibid.:153). The definition of community archives is hence twofold. On the one hand, they involve collecting and preserving objects and narratives, and on the other they are about managing the process of preservation itself in ways that empower the community. Kaun (2016b), on the other hand, traces changes in temporalities of social movements by examining evolving notions of the archive in the context of protest movements over time, including official state archives, community and digital archives. She argues that although digital media have made archiving easier, the temporalities involved have political implications; other scholars have also pointed to the impact of digital ephemerality in this respect (Donovan 2013; Terranova and Donovan 2013). In community as well as official institutional archives, archiving work moves from the tedious process of preserving physical materials to the equally tedious work of indexing and ordering information while the data are constantly being updated. Kaun concludes that the digital archive, which establishes a temporal regime of immediation and permanent processing, is characterized by a loss of temporal zones that are important within social movements. The flattening of temporality and standardization of immediacy encompass a fundamental change in the politics of the archive, with consequences for activists. Preserving memories of alternative political imaginations requires us to reflect critically on the technological infrastructures that are used to generate narratives about such imaginations and to maintain the historical records of the movements.

Future directions

A number of contemporary technological developments pose new challenges to citizen media, specifically in relation to time and temporality. These developments all relate to datafication, that is, to procedures of quantification and the transformation of a growing number of objects into data, and the automation of processes of judgement, evaluation and decision making (Mayer-Schönberger and Cukier 2013; van Dijck 2014). Datafication is relevant in the context of citizen media not only because of the increased possibilities of surveillance, predictive policing and various methods developed to prevent protest (Elmer 2015; Uldam 2016), but also, and in particular, because of the changing epistemologies and ontologies of temporality.

The large-scale collection of data and the datafication of all aspects of our lives are linked to future-oriented temporalities of anticipation which attempt to "govern and secure on the basis of possible or potential futures that threaten some form of disruption to an existing social-spatial order" (Anderson 2010:782). These data-based forms of future-oriented governance are founded on previous behaviours and models of projected futures that foreclose the possibility of imagining the future as open-ended and constantly negotiated. Anderson argues that this temporality is informed by the logics of preemption, precaution and preparedness, which offer "a coherent way in which intervention in the here and now on the basis of the future is legitimized, guided and enacted" (ibid.:788). While precautionary action begins before an identified but uncertain threat has materialized, and before it "reaches a point of irreversibility" (ibid.:789), the form of intervention adopted in preemption is "incitatory … and justified on the basis of indeterminate potentiality" (ibid.:790). The logic of preparedness, finally, "prepares for the aftermath of events" brought about by the logic of preemption (ibid.:791). In terms of data, preemptive acts, in combination with predictive calculation, set out to realize a specific future created through paradigmatic techniques that include trend analysis, modelling and data mining (ibid.:787). Preemption and other anticipatory logics seem to constitute all future-oriented temporalities of datafication. However, as Cheney-Lippold (2017) argues, prediction, preemption and anticipation are inherently conservative and past oriented, as they are always based on historical data, models and assumptions about the social world. Consequently, rather than being future oriented in the way that social movements that work for a better world are, they reproduce and reinforce assessments and decisions made in the past and contribute to an inherently conservative vision of the future. Many anticipatory projects, in other words, are not geared to projecting a visionary future that needs to be actively constructed but instead reinforce and perpetuate established models drawn from the past.

Discussing the data temporalities of prediction and preemption, and focusing on predictive policing, Andrejevic concludes that we are witnessing a "shift in emphasis from past to future [that] displaces narratives of causation with the goal of predictive intervention"; consequently, he argues, "preemption does not rely on forms of disciplinary surveillance and its attendant forms of subjectification, but rather on post-panoptic logics of surveillance that substitute total monitoring for selective surveillance and prediction for explanation" (2017:880). Andrejevic further argues that preemption constitutes a foreclosure of politics and is decidedly post-political. In the context of citizen media, datafication confronts us with the new challenge of how to resist the obsession with numbers and how to provide vivid and valuable alternatives that can challenge emerging modes and temporalities which reinforce historical divides and inequalities. Future research should explore forms of resisting these emerging temporalities from within as well as attempts to develop utopian visions outside of the hegemonic temporal regime of digital immediacy and data-based anticipatory governance.

See also: archiving; social media; social movement studies and citizen media

Recommended reading

Barassi, V. (2015) *Activism on the Web: Everyday struggles against digital capitalism*, London: Routledge.

Presents Barassi's ethnographic work among political groups in the UK, Spain and Italy and highlights the dilemmas and challenges faced by contemporary social movements in the context of digital media. The book offers an extensive discussion and investigation of the hegemonic temporality of digital immediacy and acceleration with and through digital media that political groups now have to contend with.

Kaun, A. (2016) *Crisis and Critique: A brief history of media participation*, London: Zed Books.

Presents a short diachronic comparison of how media practices within social movements have changed over time, from mechanical speed during the 1930s to the temporality of flow in the 1970s, and finally the digital immediacy that characterizes contemporary societies and hence social movements.

Polletta, F. (2002) *Freedom Is an Endless Meeting: Democracy in American social movements*, Chicago, IL: University of Chicago Press.

A classic study of deliberative, participatory democracy. Although not explicitly focused on questions of temporalities, it highlights those aspects of participatory democracy that take time, but that are crucial for the political project of establishing political alternatives, including extensive meetings and discussions of goals and how to reach them.

Sharma, S. (2014) *In the Meantime: Temporality and cultural politics*, Durham, NC: Duke University Press.

Provides an in-depth study of the lived experience of marginalized and precarious temporalities that make fast capitalism possible, engaging with the experiences of taxi drivers, corporate yoga instructors as well as frequent-flyer business travellers and slow-food devotees. Sharma shows how different temporalities of waiting, being on hold and speeding up all depend on each other to allow for the hegemony of fast capitalism to emerge.

TWITTER AND HASHTAGS

Neil Sadler

This entry offers an overview of the key characteristics of the social networking site Twitter and of hashtags, as they are found both on Twitter and other sites. The first section discusses three main issues: Twitter as a resource for research, the uses to which Twitter has been put by citizen activists, and the extent to which Twitter causes a reconfiguration of the relationships between citizens, corporations and states. The second examines the main functions of hashtags in the context of citizen media.

Technical affordances of Twitter

Twitter users' communication practices, as they pertain to citizen media, are shaped by three key factors relating to the site's technical affordances: first, individual messages are tightly constrained in length – to 140 characters between 2006 and 2017, and to 280 characters from 2017 onwards. Second, relationships between users on the site need not be reciprocal, that is, it is possible for user A to follow user B without user B necessarily also following user A. Third, the site's technical affordances have continually changed since its launch in 2006.

The first feature means that communication on Twitter must employ highly economical language. This has led to the development of a distinctive communicative style, characterized by strongly elliptical writing and a high degree of fragmentation. The character limit makes it impossible for users to tell whole narratives or offer detailed descriptions in individual tweets. Consequently, they must rely heavily on implicature, depending on readers to creatively supply missing information (Sadler 2017). This has been a major factor in the growth of highly affective modes of communication, as compelling narratives embodying common understandings are traded for solidarity based on affective alignment, understood in the sense of shared states of physical intensity (Massumi 2002; Deleuze and Guattari 1980/1988) which can be more readily communicated in the limited space available on Twitter. The second factor means that communication on Twitter can take on something of a broadcasting quality (Meikle 2016:76), as the mandatory two-way relationships seen on earlier sites such as MySpace, and in earlier versions of Facebook, give way in many cases to largely one-way relationships where high-profile users are "unable to follow all their followers in return, even if they wanted to" (ibid.).

The third means that Twitter use has changed significantly since its launch as new features have been introduced and existing features changed. Key affordances such as @replies, mentions and retweets were not present at the site's launch, but were added later (Twitter 2018). Other major changes include: in 2010 adding the ability to embed video content directly within tweets; in 2013 launching the 'blue line' which causes tweets involved in the same conversation to be displayed together, breaking with the site's until that point consistent commitment to always presenting tweets in the chronological sequence of their posting; and in 2015 introducing the quote tweet feature, allowing users to retweet another user's tweet while also adding a comment of their own. Equally significant is Twitter's extensive use of algorithms to select the order in which tweets appear in searches and which hashtags and keywords are selected as trending for individual users, based on their popularity but also according to user characteristics such as physical location and whom they follow. These algorithms are both frequently changed and secret – no details are publicly released as to the criteria for favouring certain tweets over others at any one time. Twitter is consequently something of a moving target for both researchers and users, as both must frequently adapt to, at times major, changes in the way the site operates.

Twitter as a resource for research

While Twitter creates challenges for researchers, it is also a remarkable resource. The data it makes available, coupled with the development of new methodologies, has allowed for the generation of new insights. Twitter's principal advantages as a data source are its size, its computer readability, its rich metadata and the variety of uses to which it is put by its users. It is Twitter's size that has captured the imaginations of researchers more than any other aspect. As of late 2019, approximately 500 million tweets are posted daily (Internet Live Stats 2019) by over 330 million active users (Statista 2019). Yet, this would mean comparatively little for researchers were it not for the second advantage – all of this data is amenable to immediate computerized analysis without the need for lengthy processing as, for example, with much traditional corpus linguistics or archival research. Researchers have consequently been able to analyse very large volumes of data, often in near real-time. For example, Meraz and Papacharissi (2013) studied 1.5 million tweets in their analysis of interactions between users in news distribution; LaLone et al. (2017) analysed 10 million tweets to examine Twitter's potential for providing to-the-minute updates in disaster situations; and Kwak et al. (2010) analysed 106 million tweets in their exploration of Twitter as both a social networking site and a news source.

The value of Twitter data is further enhanced by the richness of the metadata embedded in all tweets. All tweets encode a variety of, usually invisible, data in addition to their visible content (see Baca 2016 for a detailed introduction to metadata). This information includes the author's biography at the time the tweet was posted, their geographical location, the number of people the user follows and who follow the user, the date on which their account was created and their selected language. Although most of this information is available from user profiles, the fact that it is also embedded within all tweets, in a readily computer-readable format, greatly facilitates quantitative analysis. Grieve et al. (2017; Huang et al. 2016), for example, have used the content of tweets combined with their geographical metadata to examine regional linguistic variation in the UK and US, analysing volumes of data that would have been practically impossible to acquire in the past.

The fourth major advantage Twitter offers researchers is the richness of its data as a result of the diversity of contexts in which the site is used. That users employ Twitter for such diverse purposes allows research drawing on Twitter as a data source to be similarly diverse. For example, previous studies have used Twitter data to analyse discussions on wine in the context of product marketing (Wilson and Quinton 2012); storytelling in the context of amateur ballet (Dayter 2015); and fan engagement with TV shows (Harrington et al. 2013), in addition to the various citizen media uses discussed elsewhere in this entry. In many cases, the interactions analysed in such work would have been largely inaccessible to researchers in the past, due to their being conducted orally and never recorded or viewed as private and not made publicly available.

Citizen media uses of Twitter

With regard to citizen media, Twitter has been put to three main purposes: as a megaphone (Brym et al. 2014) to communicate factual information and overtly subjective perspectives; to influence mainstream news agendas; and to facilitate the creation and maintenance of loose networks of affiliation. Social networking sites, and Twitter in particular, have undeniably democratized communication. Although early utopian visions of the Internet facilitating the free and uninhibited exchange of perspectives and information have long since lost their lustre, Twitter has enabled citizen groups to communicate with audiences on a scale previously open almost exclusively to institutionally backed actors through the mediation of news organizations. Twitter, in principle at least, affords anyone with an Internet connection (and in some contexts a proxy service to circumvent government access restrictions) the ability to share text, images and video to audiences around the world, breaking "the bottlenecks of print" (Levinson 2001:128) which gave traditional media organizations their dominant gatekeeping role. This possibility first rose to mainstream prominence during the so-called movements of the squares in the early 2010s, when Twitter was widely used by activists in countries including Iran, Spain, Greece, Romania, Tunisia and Egypt to directly broadcast their own perspectives on events in their local contexts to both geographically local and distant audiences without the mediation of traditional media institutions (Nunns and Idle 2011; Willis 2014; Gunning and Baron 2014).

Yet, Twitter's growth has not obviated citizen media groups' need for the traditional media. As media institutions' gatekeeping role has shrunk, a new gatewatching role has emerged for these organizations, involving the direction of users' attention, rather than directly policing access to information (Bruns 2005). Reflecting this changing role, many news agencies have integrated Twitter into their digital beats, as they search for what they perceive to be newsworthy content (Hermida 2010, 2013; Broersma and Graham 2012, 2013; Knight 2012). Traditional media institutions are by no means the only important gatewatchers on Twitter, and there are many examples of unaffiliated citizens being 'crowdsourced to prominence' (Meraz and Papacharissi 2013) gaining large followings and influence as gatewatchers by the actions of other ordinary Twitter users. The Egyptian Twitter user Zeinobia, for example, is an unaffiliated citizen who had a modest online following prior to the 2011 uprising in Egypt but who quickly gained a Twitter following of tens of thousands after the commencement of mass protests, without institutional support, due to her reporting on the demonstrations. Nonetheless, large organizations frequently still enjoy access to far larger audiences than either individuals or citizen groups (Ali and Fahmy 2013). Consequently, citizen use of Twitter often remains oriented towards capturing the attention of traditional news organizations and influencing their agendas by creating content for them to find, and in many cases it is only through the intervention of traditional media organizations that citizen groups are able to gain access to large audiences. For example, the Gaza-based parkour team PK Gaza have been regularly posting content in English on Twitter and other social networking sites since 2011, yet it was only following a flurry of attention from mainstream Anglophone media institutions including *The Telegraph*, *The Financial Times*, *The Guardian*, *CBS* and *Newsweek* between late 2014 and 2016 that they were able to attain (brief) widespread attention in the UK and US.

The third, and most controversial, role of Twitter within citizen media concerns the organization and maintenance of activist movements. Throughout the 2000s, a broad divide was maintained between so-called cyber-optimists, who broadly supported the idea that social networking sites would lead to a new wave of citizen activism, and cyber-pessimists, who argued the opposite. Optimists, most prominently Shirky (2011), proposed that the mass communication between citizens enabled by social networking sites such as Twitter made new forms of mass citizen organization possible; these were characterized by large but loose conglomerations of individuals connected by weak ties without the need for overarching institutional control or collective identities (Segerberg and Bennett 2011; Bennett and Segerberg 2012). Pessimists, on the other hand, were quick to point out that these conglomerations were as quick to collapse as they were to form (Gladwell 2010) and that the low barriers to entry with social media-based activism encouraged slacktivism (Morozov 2011) – superficial and fleeting engagement with political causes. More recent scholarship examining social media use in citizen movements in various countries has recognized that social networking sites have come to play a significant role in many contemporary social movements but that strong ties and networks of deep commitment and solidarity remain vital. For example, Gunning and Baron (2014) and Gerbaudo (2012) emphasize that experienced activists played a central role in coordinating the protests of the Egyptian revolution in 2011, using, but not relying on, social media. Theocaris et al. (2014) and Boulianne (2015), on the other hand, point to the complex relationship between social media use and engagement in civic and political life more broadly.

Although Twitter has allowed citizen activists to reach audiences of a previously impossible size, in other respects it has led to the replication of traditional hierarchies, particularly regarding the relationships between citizens and institutions. Although it has been extensively used by citizen media groups, Twitter is not a citizen media initiative. It remains a profit-oriented commercial venture. Accordingly, as the power of traditional media organizations has shrunk, the power of corporations such as Twitter has grown. The site is used by citizen and progressive forces as well as by reactionary and state-sponsored currents. As in the offline world, the power relations between these groups are asymmetrical and, as Fuchs (2017:232) notes, "those who have a lot of reputation, fame, money or power tend to have many more followers than everyday people. Their tweets also tend to be much more often re-tweeted than common people's tweets".

To date, Twitter has followed an approach to free speech grounded in the ideology of Silicon Valley technolibertarianism, characterized by a strong belief in the intrinsic value of the free interchange of information and data with minimal governance. At certain junctures, Twitter adopted measures intended to facilitate citizen media activity, against the wishes of state actors. For example, text2tweet and speak2tweet services were introduced to allow users to tweet by sending SMS messages and leaving voice mails during Internet shutdowns in Egypt in 2011 and Syria in 2012. Similarly, in 2013, automated Arabic–English translation was introduced for key Egyptian Twitter users tweeting in Arabic to enable non-Arabic speakers to follow their tweets (BBC 2013). Yet Twitter's developers have also extended similar support to more problematic groups. For example, high-profile American neo-Nazis using the site had their profiles formally verified, offering an implicit endorsement of their right to speak on the site, a policy which was changed only reluctantly in late 2017 following significant public opposition (White 2017). Twitter, as with other social networking sites, has been generally reluctant to accept responsibility for content posted on the site and has been slow to meaningfully engage in tackling the issue of the widespread use of the site by state-sponsored trolls (Bulut and Yörük 2017; Tanchak 2016).

The site's activities are also clearly driven by commercial imperatives. Since 2010, the company's business model has focused on maximizing the amount of time users spend on the site so that they can be targeted by advertisers. It is thus grounded in the logic of the attention economy, treating user attention as a commodity to be bought and sold (Tufekci 2017a). The content which attracts user attention is itself user-generated, engaging users in frequently invisible digital labour which has formed the basis of extensive critiques of the site (Fuchs 2014a; Fuchs and Sandoval 2014; Scholz 2012a). This tension is especially problematic for anti-capitalist citizen media movements, who, by creating content which attracts the attention of users, are generating revenue for neoliberal corporations. Market imperatives also create pressures for the developers to submit to state pressure in order to retain access to customers. Since 2012, the site has removed tweets at the request of governments, although not all requests are complied with and censored tweets remain accessible outside the jurisdiction of the government requesting censorship.

Hashtags

Hashtags are words or short phrases preceded by the # sign which, at their most basic, function as part of folksonomies – user designed systems for categorizing and organizing information (Potts et al. 2011). They were first used on Twitter in 2007 but built on earlier categorization methods used in Internet Relay Chat (Salazar 2017). Having since spread to other social networking sites, including Facebook and Instagram, hashtags were initially introduced by users themselves and only later formally integrated into the sites on which they are now used. Their categorizing function is frequently important within the tweets of even a single user: a user may post on a wide variety of topics and the inclusion of hashtags allows them to quickly signal the broad topic of a given post. They play a significant role in what Zappavigna (2012, 2015) terms searchable talk, that is, new modes of communication characterized by a need to make content easily searchable and retrievable which are grounded in the logic of the database (Manovich 2001), rather than by the back and forth exchanges of traditional oral communication or linear presentation of most traditional writing. Such systems are necessary on social networking sites due to the huge volumes of information involved, which make finding relevant information challenging, and the difficulty of providing contextual information in environments where brevity is emphasized either by user-norms, as on Facebook and Instagram, or by technical affordances, as with Twitter.

Beyond categorization, hashtags serve a number of additional functions. First, hashtags are central to connective storytelling practices on Twitter, whereby large numbers of users collectively produce narratives by tweeting about major events and tagging their tweets with a common hashtag. Papacharissi (2015a), for example, discusses how this happened in the context of the Occupy Wall Street protests of 2011 with users employing the hashtag #OWS to produce connectively, if not necessarily collaboratively, a common account of the protests, encoding a particular set of meanings. This also highlights the role hashtags have come to play in providing key nodes of connection in processes of collective sense-making on social networking sites. Users may additionally employ hashtags in order to contest what they see as the prevailing narrative of events, as a mechanism for inserting their contrary views into conversations (Conover et al. 2011; Yardi and boyd 2010). This technique has been employed by citizens as a means of striking back at institutions; in 2014, for example, the New York Police Department (NYPD) launched a Twitter campaign encouraging citizens of New York City to post pictures of themselves with NYPD officers with the tag #myNYPD to celebrate their positive experiences with police, only for users to hijack the hashtag with posts highlighting incidents of police brutality carried out by NYPD officers (Jackson and Foucault Welles 2015).

Second, hashtags also commonly function as empty signifiers, signs which can be filled with different meanings by different audiences (Gerbaudo 2014; Jeffares 2014:21–23; Trillò 2018; Papacharissi 2015b). This allows them to serve as rallying points for large numbers of people without requiring large-scale ideological or narrative alignment. For example, the hashtag #jan25, which rose to prominence during the Egyptian revolution of 2011, signalled broad opposition to the ruling regime and approval of mass protests calling for change without requiring agreement as to the details of what should follow. This flexibility facilitated its use by groups with widely differing understandings of the 25 January revolution including those who simply wished to see the end of then president Mubarak's rule, but not necessarily a wider overhaul of Egyptian society; radical anti-capitalists who wished to see a fundamental shift away from the neoliberal economics of the Mubarak era; Islamists who wished to see a greater role for religion in the day-to-day running of the country and so on. The situation was similar with the Indignados movement in Spain beginning in 2011 – the hashtag #15M was widely used to signal opposition to the situation of precarity and unemployment suffered by millions of young people in Spain yet said little about the desired solution. One reading of this situation is that the use of empty signifiers is driven by the wider shift to highly affective modes of communication (Papacharissi 2015a; Papacharissi and De Fatima Oliveira 2012; Meraz and Papacharissi 2013), with hashtags often communicating little ideational information but facilitating widespread affective alignment, enabling the production and, at least temporary, channelling of the intensity required for collective action (Gould 2010; Massumi 1995). Papacharissi (2015a) suggests that affect, often focused around hashtags, is the key factor which holds together otherwise diverse publics on the site, reducing the need for shared meanings and playing a central role in Twitter's capacity to be the focal point of collective action.

Third, and as Farman (2012:1) notes, hashtags are "not just slogans. They signal solidarity, and reclaim a sense of belonging and commonality. In other words they communicate and create identity". Using hashtags allows users to quickly and clearly signal their political, social or cultural affiliations. This can occur positively, by using hashtags associated with particular social or political identities, or negatively, to disassociate the tweeter from the identity with which that hashtag is associated and thus reaffirm an alternative identity. For example, Ince et al. (2017) show how the hashtag #BlackLivesMatter was, and is, used by both supporters and opponents of the movement for both positive and negative affirmation of their collective identities. Alternatively, counter-hashtags may emerge in response to prevalent hashtags, again facilitating a negative assertion of identity, for example in the case of #alllivesmatter developed in response to #blacklivesmatter (Carney 2016).

In many cases, these functions overlap and individual hashtags fulfil several or all of them simultaneously. For example, the hashtag #gezipark, widely used by Turkish activists protesting against the government's urban development plans for Istanbul in 2013, made it easy for users of social networking sites to find content relating to the protests, whether to follow events on the ground, see how those events were represented by activists, or acquire concrete information to facilitate their own involvement. It facilitated the creation of a collective narrative of the protests, enabling the gradual accretion of a loosely coherent and many-authored account of the protest movement. It powerfully communicated the affective state of the protestors, serving as shorthand for the anger and frustration felt by those involved. Nonetheless, it remained as something of an empty signifier, channelling a shared affective state in response to the actions of the ruling regime but without demanding that those who used it share precisely the same understanding of the government's actions, nor of their desired outcomes. Closely related to the previous two functions, it enabled users to clearly and quickly take on and reaffirm a collective identity based in part on the collective narrative built up around the hashtag and in part through opposition to the perceived neoliberal and authoritarian identity of the Turkish regime.

See also: content moderation and volunteer participation; social media

Recommended reading

Fuchs, C. (2017) *Social Media: A critical introduction*, second edition, London, Thousand Oaks, CA, New Delhi and Singapore: Sage.

Based primarily in Marxist and Frankfurt School critical theory, Fuchs provides an ideal entry point to the now extensive body of literature on Twitter and other social media conducted from this perspective highlighting questions of power, dominance, ideology and the unequal relationships between citizens and capitalist states. Although only one chapter focuses specifically on Twitter, most of the issues explored in the other chapters are also relevant.

Gerbaudo, P. (2012) *Tweets and the Streets: Social media and contemporary activism*, London: Pluto Press.

A leading scholar of social media and activism, Gerbaudo offers a detailed and richly theorized account of the interplay between social media, including a strong emphasis on Twitter, and offline citizen activism in the context of the movement of the squares during the early 2010s through a series of case studies.

Meraz, S. and Z. Papacharissi (2013) 'Networked Gatekeeping and Networked Framing on #egypt', *The International Journal of Press/Politics* 18(2): 138–166.

Provides a compelling account of the interplay between unaffiliated citizens and elites in the production of news about and framing of the 2011 Egyptian revolution. It also provides an excellent example of the ways in which large-scale quantitative analysis can be effectively integrated with close qualitative reading in Twitter research.

Murthy, D. (2018) *Twitter: Social communication in the Twitter age*, second edition, Cambridge, UK and Malden, MA: Polity Press.

Provides a wide-ranging account of the uses to which Twitter has been put in citizen media contexts including journalism, activism and the mediation of disasters, as well as describing the platform's evolution since its launch in 2006. An ideal starting point for anyone new to the study of Twitter.

USER-GENERATED CONTENT

Melissa Wall

From citizen-made videos of natural disasters to fan fiction queering commercially produced stories, user-generated content has changed the ways information is produced and circulated around the world. Optimists suggest that this emerging phenomenon provides a means of generating a more open, collaborative space for the creation and dissemination of media content (Bruns 2008; Jenkins 2012b). According to this view, user-generated content is often depicted as transformative, forcing traditional institutions – whether they serve public or commercial interests – to adapt. These changes mean that individuals creating their own media artefacts can present the world from the viewpoint of underrepresented communities, foster their own individual creativity or even express resistance to the status quo (Shirky 2008). The incorporation of citizen voices into journalism, for example, is viewed as a disruption that has dramatically affected the news industry, enabling new forms of journalism that are potentially more democratic (Hanusch 2017; Matheson and Allan 2009; Robinson 2009; Thorsen and Allan 2014a; Wall 2019). Likewise, this optimistic perspective often draws attention to the ways in which corporate media content is appropriated and remixed by amateurs in order to critique and sometimes subvert existing power structures (Busse 2017; Coppa 2017).

A less enthusiastic view of the changes wrought by the rise of user-generated content suggests a different reality. Critics believe that these forms of citizen media production all too often fail to live up to claims of cultural resistance. Instead, Internet users appear to have been enlisted as a source of free labour for the commercial entertainment media industry (Bird 2011). In fields such as journalism, critics suggest that user-generated content encourages de-professionalization, that is, the replacement of highly trained, skilled and experienced journalists by non-specialist amateurs (Nicey 2016). Even more pessimistic views suggest that content is frequently labelled as user-generated when in fact it amounts to state propaganda. Deployed by authoritarian governments, these forms of media production may be contributing to a growing lack of trust in traditional institutions and possibly even democracy itself (Morozov 2012; Woolley and Howard 2017).

This entry begins with a summary of the competing definitions for user-generated content, finding that the term holds an uneasy position within scholarship on citizen media. It then explores several key arenas within which such content is created (fan fiction, journalism and social movements). The third section outlines the variety of ways in which professional institutions have responded to this new media phenomenon, with strategies ranging from co-optation to repression. The entry concludes with an expression of hope that the term will ultimately fade from usage, having been replaced by more liberating language.

Definitions

What exactly user-generated content consists of is a matter of some debate, and the term itself, although widely used, brings with it some limiting assumptions. Alacovska (2017) argues that examples of user-generated content can be traced back in history prior to the advent of digitization, yet it is generally understood that user-generated content is first and foremost a digital product made possible by the affordances of convergent media technologies. Hence, this category has been defined by some observers in quite broad terms as "content that is voluntarily developed by an individual or a consortium and distributed through an online platform" (McKenzie et al. 2012). More discerning critics vehemently argue that user-generated content is essentially a "technicist and politically evacuated" notion which corresponds more closely with corporate media's commoditizing interests than with the aim of authentic citizen participation (Cottle 2014:xi).

Some definitions identify user-generated content as a "modular and atomistic form of content" that can easily be redistributed or appropriated (Nielsen 2016:89). This could include practices such as merely commenting on another user's social media post or providing a product rating or review. It could even comprise crowdsourced contributions that can provide valuable insight in the aggregate such as when users contribute data points on a map. Still others argue that the term can only describe certain types of media content (Wardle and Dubberly 2014). Even here there is disagreement. Some explanations insist that user-generated content consists primarily of visual content such as photos and videos whether captured by the user themselves or created by mashing up others' work (Ferrari 2017), while others include a wider variety of modes such as text messages or even citizen-generated verification reports (Norris 2017; Pain 2017). Still other definitions focus on who is producing the content, asserting that user-generated content is best defined as material produced by non-professionals who do not work for commercial entities (Kim and Lowrey 2015).

Whichever argument we find most compelling, the term user-generated content positions the person creating the content as a user. Such terminology appears to suggest that we view such individuals merely as operators of someone else's media platform or techniques. While they may be actively generating or creating material, they are not the creator of the enabling device or method; they are only making use of it. This might appear to be purely a matter of semantics. Nevertheless, the term may be said to domesticate or even devalue the contributions of individual citizens. Indeed, user-generated content often works as a neutralizing label that tamps down the expectations we hold in relation to non-professional content (Pantti and Andén-Papadopoulos 2011).

Dissatisfied with the term, some critics have offered alternatives. The Organization for Economic Co-operation and Development suggests replacing it with user-created content, which they define as "content made publicly available over the Internet, which reflects a certain amount of creative efforts and which is created outside of professional routines and practice" (Vickery and Wunsch-Vincent 2007:4). Meanwhile, Jenkins et al. (2013:15), writing about the rise of spreadable media, suggest a less active label – user-circulated content – capturing what they see as a decline in individual agency inherent to the production of such content, particularly via social media platforms. Among other suggested alternatives, user-generated content has been called loser-generated content by those positing that the volunteers who contribute content to the public sphere without pay are losing out on the compensation they deserve (Petersen 2008). Yet another label is worker-generated content, which seeks to highlight user contributions as a source of free labour exploited for profit by the media industry (Qiu 2014:385). These different categorizations reflect ongoing debates surrounding the overall purposes and meanings of user-generated content, pitting the argument that it presents a more democratic model of media production against another line of thought according to which user-generated content serves only the interests of corporate commercial structures (Cottle 2014; Mandiberg 2012; Petersen 2008). Further complicating our understanding of this phenomenon is the fact that in many online arenas, such as communities of fans who creatively respond to corporate entertainment media, the rejection of financial compensation is part of the ethos of producing new material based on commercial works (Coppa 2017).

Forms of user-generated content

User-generated content exists in many forms: from fan fiction inspired by corporate entertainment works to citizen witness photographs of destructive hurricanes to racial justice protesters' live video uploads. In some cases, this content challenges existing norms and beliefs, creating alternative fictional or non-fictional narratives, often relating to important social and political issues. In doing so, citizen content potentially "makes visible hidden realities" within social, cultural and/or political life (Stephansen 2016:26). This section highlights several key forms of user-generated content of particular interest to scholars of citizen media.

Fans of popular culture have a long history of reformulating and customizing popular television shows, novels, films and other entertainment materials into new and sometimes vastly different narratives. These derivative works may take the form of short stories, novels, zines and, more recently, gaming content. Such activities have been characterized as resisting corporate hegemony over cultural products and instead supporting independent, grassroots creative expression (Jenkins 2012b). Frequently, in the fan's alteration of often well-known stories, marginalized communities are shifted to the centre of the text, and alternative storylines rarely seen or heard in corporate templates become visible. This has been particularly the case with fan fiction's emphasis on refashioning mainstream characters and plots to focus on LGBTQ relationships that destabilize heteronormativity (Busse 2017). Scholars have further highlighted the inclusive and collaborative culture of the networked communities in which much fan fiction is produced and the extent to which these are often particularly welcoming to women, youth and LGBTQ groups (Busse and Lothian 2018).

More critical observers suggest that fan-produced content can be co-opted by the very media corporations they initially challenged and that recognition by such companies can be less about legitimacy for the fans and more about selling a media product (Busse 2017; Stanfill 2019). Indeed, commercial producers of television shows, music, film, books and other media texts have adopted many practices that originated among fan communities: for example, corporations have created websites where fans can interact, comment and even produce their own content, as they would usually do on their own (Bird 2011; Russo 2009). In addition to performing promotional work for the media product, such sites can allow companies to monitor and sell user activity to advertisers (Stanfill 2019).

Just as fans of entertainment products have worked to collect and share content based on their passion for particular fictional creations, so have citizens with an interest in politics and other areas of public concern sought to republish professionally produced news. Among the earliest digital-era amateur responders were news bloggers who devoted themselves to aggregating professional journalism, which they then redistributed, often providing their own audiences with very different interpretations of the content. In other cases, bloggers who were on the scene as participants in the events that news journalists were covering decided their stories were missing from public narratives. During the American-led bombing of Baghdad in 2003 and the chaotic aftermath of disappearances, violence and survival, a young Iraqi writing under the nom de plume Salam Pax used his blog 'Where is Raed?' to provide firsthand accounts of enduring the invasion and occupation. At the same time, American soldiers sent to Iraq such as Colby Buzzell, who penned the blog 'CBFTW' (Colby Buzzell Fuck the World), revealed their own unofficial day-to-day view of the war their country was waging, often to the US military's chagrin (Wall 2009). While these early producers of journalistic user-generated content often worked alone, some formed collectives, connecting with each other to heighten their impact. Within the field of professional journalism, initial responses to user-generated content were often harsh, rejecting the idea that amateurs could actually produce news of any value; professionals with these views sought to push such content into designated user-generated content hubs or other enclosures (Harrison 2010; Singer 2015). Later, enabled by social media tools, many citizen journalists produced one-off acts of witnessing, as they filmed or photographed often dramatic news events, producing images that they uploaded to their social media accounts (Mortensen 2015b). This form of individualized user-generated content has been more willingly accepted by some professional journalists, in part because it has helped them maintain boundaries between professionals and amateurs.

Finally, social movement activists also produce user-generated content as part of their mobilization and self-documentation efforts. Here, they seek to generate an alternative narrative about their causes and to inspire others to join their work toward social change. While some content is produced using analogue media or may even involve some elements of bodily performance – such as the creation of temporary street temples by Hong Kong's Umbrella Movement in 2014 as part of a demand for greater democracy – much of what is produced by individuals in these resistant contexts is digital. The importance of digital user-generated content within such movements was made particularly evident during the uprisings across the Middle East in the early 2010s collectively known as the Arab Spring, in which citizens challenged their leaders who had often been in power for decades if not life. Massive protests took place against corruption and the lack of freedoms, and demonstrators, often young, turned to social media to post content documenting the action in the streets (Khamis and Vaughn 2011). Around the same time came the emergence of Black Twitter, a social media space in which "black people discuss issues of concern to themselves and their communities" as a means of providing a counterpublic sphere and of challenging racism and police violence in the United States (Clark 2018:38). Black Twitter became an important source of information about a series of protests in the United States (Richardson 2018). As with fan fiction and citizen journalism, at its most effective, such content creates networks among its producers and encourages collaborative peer production practices, operating independently of corporate media.

Yet, some critics suggest that here too the rise of user-generated content has not brought people together but rather individualized media production, lessening the ability of social movements to create lasting networks, which are crucial to the success of activist campaigns. As Poell and van Dijck (2015:532) argue, the social media platforms that have seemingly supported the easy spread of campaign materials also have a downside for activists, as they imputed "their logic onto activist communication practices". This logic favours constant action and reaction as well as highly emotional, rather than more deliberative, content. For those and other reasons, media content created by social movements is usually not described by activists as user-generated content, nor is it presented in this way by those who study them. For example, Ratto and Boler (2014:3) dismiss the term *user-generated content* as a buzzword that limits rather than describes citizen agency. Milan (2013:xiii) similarly suggests that activist media content is actually threatened by what she describes as the global deluge of indiscriminate user-generated content which is frequently intertwined with advertising and which contains no clear political or social meaning.

Disciplining user-generated content

The many professional entities grappling with the rise of user-generated content have often sought to exercise dominion over such activity. The strategies to do so vary but may include attempts by the news and entertainment industry to contain and restrain users' independence by inviting citizens to submit content to specially designed hosting platforms. In other cases, companies may invoke legal or similar means of intimidation to stop the production of citizen content. As Lewis (2012) argues, at the heart of many of these conflicts lies an ongoing battle between open participation and control.

While professional companies may have the capacity to amplify amateur material and to enable citizen voices to reach a much larger audience, in doing so, they set the terms for participation, thus ensuring the user content fits their own purposes and values (Bird 2011; Burgess 2013; Wahl-Jorgensen 2015). Control is often exerted through codes of conduct and terms of agreement which set out what is (and by implication what is not) expected from contributors. This is a typical strategy adopted by both traditional legacy media companies and new media platforms. For example, a television show might hold a contest for fans to create videos using the show's own content but the production company will maintain control over what content may actually be used. Similarly, a news organization might issue calls for eye witnesses during coverage of breaking news stories to submit photographs but then use those images in ways that point out the lack of qualifications or credibility of the submitter (Pantti and Andén-Papadopoulos 2011; Russo 2009; Sjøvaag 2011). As for new media platforms specifically built to host user-generated content, these "do not simply enable user activity, but very much steer" it (Fuchs 2017; Poell and van Dijck 2015:528). Regardless of the field, these companies socialize users to their norms in ways that often go unnoticed (Bird 2011:507). Thus, entering into a relationship with a professional organization is fraught with compromises for amateurs creating their own content.

User-generated content often derives from original materials produced by professional media organizations with a focus on entertainment such as Disney or news from operations such as CNN. Many of these companies continue to seek to protect their content from being altered and/or distributed without permission, while still appearing to welcome participatory practices (van Dijck 2013). They do so for financial reasons and also, as they see it, to protect the integrity of their product. While sometimes making content available for redistribution, entertainment and news companies generally limit or outright oppose user-led appropriation of their creations (Tushnet 2017). Social media sites, which are usually the hosts for such content, collaborate with other corporations to stop such citizen media productions from being made public (Burgess 2013). Thus, in order to maintain control over their content, some media companies deploy the power of copyright laws to demand that unsanctioned usage of their material be removed from the public sphere. Ironically, at the same time, these same media institutions may make use of amateur content without exhibiting similar concerns for the amateur producer's ownership rights (Sjøvaag 2011). Companies too face ownership issues when using such content because the origins of it can be difficult to discern, especially as many creators are anonymous or are sharing someone else's work. Instances of defamation or libel sometimes created by citizen content producers operating in the news realm could damage the professional company's credibility and even put them in legal jeopardy (Cooper, G. 2017).

User-generated content has become engrained in contemporary structures for the production of information, and represents a "force to be reckoned with" across many fields (Wikstrom 2013:233). It can be highly creative and offer innovative reworkings of corporate content. Ordinary people engaged in its production have formed communities of like-minded contributors who can create receptive spaces for those at the margins and possibly expand the public sphere. Unfortunately, however, these tendencies compete against calculated corporate strategies seeking to undermine and control independent citizen participation. In fact, user-generated content has been the object for some time now of efforts to neutralize the agency of ordinary people (Bird 2011; Usher 2011).

This entry ends with the contention that the term user-generated content has reached its limit and that scholars must reject the continued use of it. The expression is all too firmly locked in the commercial dynamics of the contemporary mediascape (van Dijck 2009:53), and, as Cottle (2014:x) notes, "offers a stunted and proprietorial view of 'its' user audience and the corporate utility of 'their' generated content". In other words, the industry view of citizen media as user-generated content fails to capture the civic richness of what such practices might achieve. The label leaves no space for expressions of the very rationale for citizen media practices: the creation of publics not products (Stephansen 2016). In conclusion, user-generated content needs to be allowed to fade into disuse, making room for more visionary terminology which might better acknowledge what public participation in the creation of independent media content can and should achieve.

See also: amateur; authenticity; crowdsourcing and crowdfunding; fandom; immaterial labour; social media

Recommended reading

Dijck, J. van (2009) 'Users Like You? Theorizing agency in user-generated content', *Media, Culture & Society* 31(1): 41–58.

Provides three lenses for thinking about agency and the production of user-generated content – culture, economics and labour – while raising important questions about the ways non-professional media content is produced and consumed. A much-needed critical approach that can be effectively applied across entertainment and news genres.

Jenkins, H., S. Ford and J. Green (2013) *Spreadable Media: Creating value and meaning in a networked culture*, New York: New York University Press.

An essential, highly readable examination of the mechanics and meanings of user-generated content. This volume expands on Jenkins' earlier groundbreaking work on fan cultures to incorporate activism and other forms of user-generated content that become 'spreadable' and 'sticky'.

Stanfill, M. (2019) Exploiting Fandom: How the media industry seeks to manipulate fans, Iowa City, IA: University of Iowa Press.

Critical exploration of fandom that moves away from the conventional case studies approach to provide an overview of the key issues pertaining to audience interactivity with professional entertainment media. Stanfill argues that fan-produced content is a form of free labour that the entertainment industry embraces both for financial benefit and as a means of more effectively managing and controlling audience responses.

VIDEO GAMES

Tonguc Ibrahim Sezen and Digdem Sezen

Before transforming into a multi-billion dollar commercial industry, the design and development of video games was an experimental endeavour practised by computer scientists and pioneering hackers in the 1960s and remaining a popular practice among computer hobbyists from the 1970s onwards (Williams, A. 2017). Since the mid-1990s, the video games industry has been promoting modding, or the alteration or expansion of commercially released video games by amateur video game enthusiasts, leading to the emergence of the modern modding scene (Laukkanen 2005; Champion 2012; Hirvonen 2017). Tools such as Adobe Flash (Adobe 1996), Scratch (Lifelong Kindergarten Group 2003), Twine (Klimas 2009) and Stencyl (Stencyl LLC 2011) have allowed users with little or no programming skills to develop many different types of video games (Burak and Parker 2017). More sophisticated engines such as Unreal Engine (Epic Games 1998) and Unity (Unity Technologies 2005) were made freely available for non-commercial usage and have been seen as responsible in large part for the increasing number of independent games that have emerged since 2010 (Axon 2016). The formation of online game developer communities, facilitating the growth of peer-to-peer learning environments, has also played a significant role in the democratization of game development.

While the outcomes of these advances have mostly included entertainment-oriented mods and games with varying production and design qualities, video games have been developed and circulated within local hobby computing scenes as reactions to and criticisms of current events and political issues at least since the late 1980s (Švelch 2013). Video games gained widespread attention as platforms for advocacy and activism in the mid-2000s when politicians started to use them for campaigning (Bogost 2006). Since then, video game scholars have proposed terms like serious games (Ferdig 2014; Clément 2014; Blumberg et al. 2013), political games (Lerner 2014; Sicart 2014) and activist games (Flanagan 2009) to conceptualize video games which serve expressive and communicative goals beyond entertainment. Harnessing the potential of video games for encouraging social change and enhancing knowledge of civic issues, organizations such as Games for Change (2018) have taken on the role of curating and promoting impactful games to the public. That said, as a quick survey of the Games for Change online archive would show, most civic games continue to be released by other political organizations, NGOs and organized interest groups.

At the same time, following in the footsteps of pioneering code literacy advocates such as Alan Kay and Seymour Papert (Wardrip-Fruin and Montfort 2003), media scholars have been conceptualizing the consumption and production of video games as means of self-expression for citizens and especially youngsters since the mid-2000s. The ability to play, mod and make games was discussed by Jenkins et al. (2006) as an important new media literacy skill which would prepare and empower users for meaningful participation in the new digital culture. Buckingham and Burn (2007) created a pedagogic framework based on game literacy in order to emphasize the potential of video games as an expressive form. Zimmerman (2007) too explored gaming literacy as a valuable tool for helping people to make sense of real-world culture, society, history and politics. Indeed, in many instances, modding, making and playing games can and should be regarded as practices "of affective sociality produced by unaffiliated citizens as they act in public space(s) to effect aesthetic or socio-political change or express personal desires and aspirations, without the involvement of a third party or benefactor" (Baker and Blaagaard 2016a:16), that is, as citizen media practices.

Modding

One of the earliest and most highly organized venues for user-created content within the video game ecology was and still is the modding scene. The different types of mods include corrections of flaws in the game code, the replacement of audiovisual elements by custom content, additions of new playable content such as new levels or scenarios, changes to or expansions of game rules, and large-scale alterations of several of these and other elements at once in ways which potentially transform a modded game into a new one. Mods are built upon the existing structures of video games and require the games they are built upon to function. Due to legal restrictions, in most cases mods cannot be sold or purchased (Laukkanen 2005; Nieborg and van der Graaf 2008) and attempts made by the video game industry to monetize mods for both modders and video game companies were contested by modders due to concerns about creative freedom, collaborative work distribution and the maintenance of the quality of mods (Joseph 2017). On the other hand, by extending the content and thus the lifespan of video games, modding can be seen to provide considerable commercial benefits to game developers (Poretski and Arazy 2017). Due to this asymmetric relationship, modding has been criticized as a form of exploitation (Kücklich 2005). Others, however, have regarded the practice as "a symbiosis, of reciprocal, cultural gift-giving" (Schleiner 2012:36), and a currently limited but potentially rich site of possibility "for [the] co-creation of mainstream games as alternative media" (Prax 2016:81).

The motivations behind modding are diverse but mostly include the desire to engage with the challenge of modifying a complex video game, the enjoyment of social interaction with other members of the modding community and a wish to improve or personalize existing products. Artistic or political motivations are also observed (Sotamaa 2010; Hong 2013; Poor 2014). According to Salen and Zimmerman (2004:560), regardless of the motivations of the modders, "game modifications can act as forms of resistance, affecting the meaning, experience, and cultural identity of a game". They regard early mods – which typically inserted visual elements from shoujo manga or homoerotic characters into action games – as interventions against conventional discourses on masculinity and the traditional heterosexual gaze observed in mainstream video games (Salen and Zimmerman 2004). The introduction of the teenage pregnancy scenario through modding for *The Sims* (Maxis Inc. 2000) was described as a vehicle for the communication of social messages (Sihvonen 2011), while the *FinnWars* (Stimor 2006) mod for *Battlefield 1942* (Electronic Arts 2002) introduced an alternative perspective on the Second World War through the lens of Finland's history (Christiansen 2012). The *Religion and Revolution* (Schmiddie 2017) mod for *Sid Meier's Civilization IV: Colonization* (Firaxis Games 2008) not only expanded the native biosphere, but also added new Native American Nations and controversial concepts like slavery to the game. This had the effect of moving it closer to a version which "would allow players to experience the ugly, authentic colonization that so radically changed and shaped our world" (Mir and Owens 2013:103). Mods adding nudity to games have been described as an expression of players' desire for diversity and a means of enabling alternative gender performances (Sihvonen 2011). Moreover, although the majority of sexual mods can essentially be seen to reinforce the male gaze and depict unrealistic sexual acts, these too can be seen as "a space of resistance for the player looking for sex play in their gameplay" (Wysocki 2015:208).

Another distinct category of mods are so-called art-mods, produced by artists to be exhibited in galleries, art exhibitions, museums and biennials. These combine modding practices with activist performances as politically motivated artistic interventions for social change (Flanagan 2009; Poremba 2010). Anna-Marie Schleiner's *Velvet-Strike* (2002) art-mod, for example, which allows players to spray anti-militarist digital graffiti on surfaces in *Counter-Strike* (Valve 2000), was designed as an anti-war intervention against the War on Terror initiated after the September 11 attacks (Schleiner 2012). The *9–11 Survivor* mod for *Unreal Tournament* by Cole et al. (2003), on the other hand, gives players the role of a survivor stuck in the burning Twin Towers and has been regarded as a "reflection about a traumatic event for the sake of memory rather than decision making" (Bogost et al. 2010).

The political dimensions of video games

Regardless of the motivations behind it and the extent to which it changes games, modding may be considered a counter-hegemonic force, not only against the game industry but also against social and cultural normalities, opening up new avenues for creative expression (Christiansen 2012). At the same time, as Condis (2018) points out, it is also a transformative force impacting professional design decisions by providing hard data on players' preferences not only in relation to gameplay but also in relation to representation and inclusion. According to Condis (2018:71), it is "one way that diverse groups of fans are making their voices heard". In this regard, fostering communication on different levels is one of the key strengths the modding scene offers to modding citizens.

Focusing on the design and content of mods, Galloway (2006:126) criticizes modding and especially art-mods for changing only the audiovisual elements of games in ways that do not advance the gameplay and fail to realize the true potential of games as "a political and cultural avant-garde". In 2006 he called for a counter-gaming movement which would redefine gameplay itself by introducing alternative modes of play (Galloway 2006). While some of the more complex mods and conversions described above may be put forward as responses to Galloway's call, according to Pozo (2015), more significant reactions are to be found in the independent game design of works such as those created by Anna Anthropy. Advocating for a more decentralized and accessible mode of video game production through easy-to-use game development tools, Anthropy proposes to position video games as "transmissions of ideas and culture from person to person, [and] as personal artifacts instead of impersonal creations by teams" (Anthropy 2012b:12). Like zines and many other forms of citizen media, these games may be self-published and self-distributed. A prime example of such games is Anthropy's critically acclaimed autobiographical video game *dys4ia* (2012a), which consists of a series of unchallenging microgames built around her experiences with hormone replacement therapy. Praising the artistic game design behind it, Sharp (2015:114) describes *dys4ia* in glowing terms: "the expression of personal experience, the critical perspectives on gender norms, the choice to view games as a populist outlet – Anthropy's work melds the deft touch of a game designer with a critically engaged artist". Further traces of Anthropy's approach can be also found in so-called vignette games (Ellison 2015): small, poetic, personal, non-commercial games on feelings and moments experienced by their designers, capturing instances of life in a short game format. While vignette games tend to focus on internal emotions, there are also examples which expand this focus to the intersections of personal and social issues, such as Chloe Lister's *Don't Look at Me* (2016) based on her own experience of street harassment, or Hannah Nicklin and George Buckenham's *A Bonfire* (2016) which summarizes their reactions to the stories of Londoners affected negatively by gentrification.

Due to their short format and almost diary-like nature, vignette games bear a strong resemblance to what Bogost et al. (2010) call current event games. A subcategory of news games operating at the intersection of videogames and journalism, current event games are "short, bite-sized works, usually embedded in Web sites, used to convey small bits of news information or opinion" (Bogost et al. 2010:13). These may be released by interest groups or professional organizations such as news corporations and NGOs, but also by individual citizens and activist groups. Gonzalo Frasca's *September 12th* (2003), for example, criticizes the bombing of Afghanistan and Iraq by the US-led coalition as a factor fuelling radicalism in the Middle East: the player is presented with a crowded town and must direct the missiles to kill terrorists without harming civilians; "no matter how carefully you aim", McClellan (2004) notes, "you end up with some collateral damage. When that happens, lots more terrorists appear". *Madrid* (Frasca 2004a), on the other hand, commemorates the 2004 terrorist bombing in Spain's capital by asking the player to help keep candles held by mourning demonstrators alight by clicking on them. Both of these games can be considered early and defining examples of the genre. As one might expect, their release by individuals usually peaks after major social and political events: this was certainly the case after the 2016 coup attempt in Turkey (Sezen and Sezen 2018). Such games may also be released by supporters of both sides in a conflict, as happened for example following the 2008 Gaza Strip bombings (Sezen 2009a, 2009b).

In order to encourage the creation of critical video games by members of disadvantaged groups, Frasca (2004b) has fostered the repurposing of well-known game mechanics within online communal game development platforms called forum videogames. While perhaps not as communal as Frasca envisioned, the practice of socially or politically themed game-jams – 48-hour-long game development marathons – may be considered as an alternative to forum videogames, where unaffiliated game designers may use games to support or criticize specific events or issues, as seen in the Gezi-Jam organized during the 2013 Gezi Park Protests in Turkey (Sezen and Sezen 2016).

Emancipatory in their approach, the development of forum videogames as a technique does not aim to set boundaries, but instead seeks to foster critical reflections and debates among the participants. On the other hand, emphasizing the rhetorical potential of video games, Bogost (2007:ix) uses the term persuasive games to describe video games that mount "the art of persuasion through rule-based representations and interactions". According to Bogost (ibid.:29), "procedural rhetorics afford a new and promising way to make claims about how things work". Relatively large-scale games by the Italian social critic collective Molleindustria can be seen as examples of persuasive games developed by activists. Molleindustria's *The McDonald's Videogame* (2006) for example is a critique of the business practices of the fast-food industry which gives the player the role of an executive who is expected to generate profit by any means necessary.

In evaluating them from their players' position, Holmes (2013:87–88) compares persuasive games to a Socratic dialogue where "the player does not actively participate in the shaping of knowledge as an equal with her interlocutor but enters the terrain as an intellectual inferior". This relationship resembles what Aarseth (2014) calls the positioning of an implied player, that is, the mental construct formulated by game designers based on possible user scenarios and which is expected to be adopted and performed by the player for the game to exercise its effect. In games of progression where players must perform the exact actions the game design dictates (Juul 2005), the domination of the implied player over the real player is most evident. This is seen as a problematic position for a player engaging with games designed to operate as a vehicle for social change since it deprives them "of their capacity to express themselves through play" (Sicart 2011). Sandbox games like *Minecraft* (Mojang Synergies 2011) bestow players with higher levels of creative agency through assemblage and construction, which can also be used for civic purposes as seen in vignettes built to protest current political events (Neltz 2016). According to Cayatte (2014) such instances should be seen more as 'video toys' rather than video games: the combination of a small number of rules allows for a large degree of game variation, and interactions between rules and game objects may lead to occurrences neither players nor designers have foreseen (Juul 2005). Aarseth (2014) uses the term transgressive play to describe such occurrences which are allowed by the game system but which were not part of the game's intended repertoire. "Transgressive play", he writes, "is a symbolic gesture of rebellion against the tyranny of the game, a (perhaps illusory) way for the played subject to regain their sense of identity and uniqueness through the mechanisms of the game itself" (Aarseth 2014:132). In a similar fashion, Flanagan

(2009) proposes the terms subversive play – to describe player actions which defy the authority of game rules – and unplaying – where "players specifically enact 'forbidden' or secret scenes, unfortunate scenarios, or other unanticipated conclusions often in opposition to an acceptable or expected adult-play script" (ibid.:33).

Both transgressive and subversive play hint at the possibility of a type of individualism or even expression in games through actions, scenarios and dynamics unforeseen by game designers, but another concept – queer play – directly aims at critique through play by repurposing game rules (Chang 2017; Scully-Blaker 2018). As an example, we might cite those rare pacifist *World of Warcraft* (Blizzard 2004) players who remain neutral in the dividing war of the game's fiction and who, instead of fighting enemies or engaging quests like most players do, choose to level up in a painstakingly slow manner by picking up flowers and collecting rocks (Goldman 2011; Messner 2018). A more concerted and collective example of queer play can be found in *The Sims* and its later instalments in the form of special scenarios or challenges created and shared by fans within their online communities, which Sihvonen (2011) argues bend and transform the games into cultural products whose uses are impossible to predict. The so-called poverty and immigrant challenges are especially important here, since they not only bend the games but also critique them and the social relationships they represent. Both challenges employ a series of custom settings, such as turning off unrealistic income sources in the poverty challenge or removing charisma skills at the beginning of the game to simulate language barriers and culture shock in the immigrant challenge. Players may additionally upload and share playthroughs via social media, transforming these queer play sessions into commentaries to be followed and discussed among other players (Sezen 2013). Nevertheless, both challenges are played mainly by the official rules of the game. The struggles players face, such as hunger or social isolation, are thus the result of the games' original social model based on a middle-class suburban life scenario.

Future directions

Video games provide diverse means of self-expression and critique for citizens. Modding, game making and play are the foremost modes for these expressive activities. Video games frequently inspire post-play narratives and may be used as tools for creating machinima, that is, appropriating graphics engines from video games to create short animated films. They additionally serve as platforms for socially and politically motivated roleplay and even online protests. A whole subculture has developed around video games which shares many tropes, memes and jargon with other areas of digital media. Studying these intersections would help provide insights on the common and divergent aspects of digital cultures. Future studies of citizen-made video games would also benefit from archival work and comparative approaches which might enhance our understanding of the past and ongoing development of this field. Such research is complicated by the ephemeral nature of video games and mods: over time, technologies become obsolete and storage devices degrade making individual projects harder to track and archive for researchers. The analysis of derivative content such as 'let's play' videos may in some cases be used to overcome this limitation.

See also: remediation

Recommended reading

Bogost, I. (2007) *Persuasive Games: The expressive power of videogames*, Cambridge, MA: The MIT Press.

Provides a theoretical framework and examples useful for exploring how videogames mount arguments and how they are used to make expressive statements about the world. Focuses on their unique persuasive power in terms of their computational properties. A key text regarding the use of video games as a tool for expression.

Flanagan, M. (2009) *Critical Play: Radical game design*, Cambridge, MA: The MIT Press.

A historical and artistic examination of both analogue and digital games designed or used for political and social critique through the subversion of popular gaming tropes. Proposes a theory for alternative game design and for reworking of contemporary game practices.

Lerner, J. A. (2014) *Making Democracy Fun: How game design can empower citizens and transform politics*, Cambridge, MA: The MIT Press.

A study of the use of games and game-like processes to encourage citizen participation in political and civic decision making. It examines multiple examples and offers game-based strategies for attractive, effective and transparent public participation.

WEIBO

Eileen Le Han

Weibo, also known as Sina Weibo, is a Chinese mainstream microblogging service. Launched in 2009 by Sina.com, one of the country's leading technology companies, Weibo has become a prime model for social media platforms in China. Often referred to as Chinese Twitter, Weibo shares a number of features with its non-Chinese microblogging counterpart. For example, users of both platforms have been, at some point, constrained by a 140-character limit per tweet/post; similarly, both services feature the mention function, requiring the use of @ (at) to direct messages to particular users. Users in both platforms can retweet/repost and reply to tweets/posts, and settings can be altered to ensure that tweets/posts can be read by all or part of a user's followers, or by any user of the platform. Finally, connections between individual tweets or posts that revolve around a single topic can be established through the use of hashtags. Since it was first launched, and in keeping with similar developments in other digital media platforms, Weibo has added numerous features to its basic microblogging services, including live streaming and the posting of short videos. In his comparative study of the political economy of major social media platforms in China and the US, including Weibo and Twitter, Fuchs (2016) notes the existence of such similarities and, more importantly, shows that they are all susceptible to corporate and political control.

The similarities between Weibo and other social media platforms outside China have not prevented it from inheriting certain linguistic and cultural traits from its predecessors in the Chinese digital media ecology, or inventing conventions of its own. To date, however, Weibo's idiosyncrasy remains insufficiently explored in the literature. Indeed, Weibo is still often presented as a local variation of Twitter (Sun 2013). While the latter is conceptualized as a "subversive and elite space" (Sullivan 2012:778), Weibo is defined as a "contested force in Chinese politics" (ibid.:779). But as this entry shows, it is the dynamics of the relationship between the platform users (citizens and businesses) and the government that have contributed the most to Weibo's distinctiveness as a Chinese platform, and turned it into a contested space for online activism in China (Yang 2014). Significantly, Weibo has facilitated the reinforcement of rooted Chinese cultural values, including the networks of influence or *guanxi* that lie at the centre of Chinese society (Zhang and Negro 2013).

History

When Weibo was launched in August 2009, access to Twitter in China had been officially blocked for some time. The year 2010, however, should be regarded as the inaugural year of Weibo. Following a number of landmark public events, Weibo became the microblogging site preferred by Chinese citizens. Among other affordances, the platform enabled a new form of online public participation known as *weiguan* – which literally means 'to surround and watch', that is 'to spectate'. Weibo allowed citizens to engage actively in public events as they unfolded, by posting and sharing the latest updates, as well as commenting and calling for collective action. Insofar as most of these landmark events were prompted by the abusive exercise of power by local officials, Weibo empowered grassroots movements, enabling them to publicly voice their grievances, expose social injustices and hold those in power accountable to larger audiences.

As an event-driven platform (Han 2016) capitalizing on the technological affordances of publicness, Weibo was initially marketed and promoted by Sina as a platform for news and information, with a user base consisting primarily of journalists and other media professionals. At that time, journalists were relatively free to utilize the platform for the purposes of news gathering; indeed, they used Weibo to post live updates about a wide range of issues, including topics and events that were the focus of intense government crack-down and intervention. Against this backdrop, journalists and Weibo 'big V' accounts – i.e. influential users such as celebrities, outspoken scholars and business leaders with large numbers of followers whose identity is verified on Weibo – were expected to amplify the voices of the grassroots. However, as Weibo gradually evolved into a popular platform for sharing news and information, it also became highly politicized. Two large and opposing camps within Chinese society – known as public intellectuals and fifty cents – have since used Weibo to conduct (often heated) debate on social and political issues pertaining to the country's evolution towards a modern nation-state. Public intellectuals are liberals supporting constitutionalism and universal values, while the politically conservative members of the fifty cents group advocate that China maintains its unique political system and legitimize the rule of the Communist Party of China (Lei 2017).

As a highly politicized and contentious space, Weibo has always been closely watched by the party-state authority to monitor and weed out potentially subversive content. Since 2012, following Xi Jinping's rise to power, the state's intervention in Weibo through highly sophisticated methods has become a norm. Official media have stepped in and managed to significantly steer public opinion on Weibo. At the same time, a large number of outspoken accounts, mainly from 'big V' users, have been suspended or penalized for their posts on the platform. As these influential users have been silenced, Weibo has tried hard to find new opinion leaders to revitalize the platform, displace news information away from the centre of the platform's activities, and thus minimize political risks in the future. The Chinese state is therefore gradually taking control of Weibo through various strategies: facilitating the official media's occupation of Weibo as a key space within the Chinese public sphere, appropriating fan culture, and capitalizing on digital populism (Guo 2018). The adoption of these strategies has allowed Weibo to turn away from politically sensitive topics towards safer ones, and achieve significant market success and business growth. Although Sina experienced losses between 2009 and 2011 (Fuchs 2016), mainly due to the global financial crisis, Weibo later developed into a global company and improved its financial standing after listing on NASDAQ in 2014. By the second decade of the twenty-first century, Weibo had become an interest-based platform, promoting users with certain areas of expertise who can accumulate a large fan base, and have the potential to monetize their know-how. This turn to monetization amid the tightening of state control over digital media in China has allowed leading tech companies to play a more central role in public life, often blurring the boundary between corporate and public interests.

Following what is widely regarded as Weibo's golden age (2010–2011), the platform's administrators and most influential users have actively contributed to fostering perceptions of and circulating discourses about Weibo's decline as due to two main reasons. First, the changing role of news and information in Weibo. The platform's early years coincided with an explosion of public interest events that attracted the attention of Weibo's users, turning it into an important site of public opinion formation and citizen participation. But it was the centrality of news and information and the platform's potential to mobilize collective action that made Weibo vulnerable to different forms of state supervision and monitoring, including the suspension of active users' accounts. Second, the fast pace of technological innovation in the Chinese media landscape, including the emergence of new platforms. The decline of Weibo, according to Benny and Xu (2017), indicates that the state now has a pervasive presence in the platform, the capacity to manipulate Weibo's content and the means to steer opinions. Although the platform is, as a result, prioritizing individual over public interests, Weibo remains an important barometer of public opinion in China.

Research themes

The study of Weibo should be contextualized within the research tradition of the Chinese Internet and society and the changing sociocultural conditions in China, including the sophisticated mechanisms of state control, the dynamism of the market economy, the collective need to access uncensored global and local information, and the public respect for tradition and remembrance.

Past scholarship on new media (Chan and Qiu 2011) and Internet events (Yang 2009) has paved the way for current research on Weibo as an event-oriented platform whose technological affordances enhance the visibility of topical issues and bestow cultural authority on the platform (Han 2016). Weibo differs from other nonprofit platforms for citizen journalism and participation, as Sina uses it to pursue its commercial and corporate interests, but it remains an important platform for citizen participation and activism. The visibility that such interests and grievances can gain in the public arena through Weibo have led the government to develop and implement various mechanisms to monitor the information flow on the platform. Despite the widely held view that Chinese censorship practices involve an all-powerful state taking down all kinds of subversive or potentially threatening content, the monitoring of Weibo is carried out in a selective and sophisticated manner. King et al. (2013), for example, found that the Chinese state largely censors calls for collective action, but it is generally tolerant of critical views of the government. This is particularly the case with posts relating to specific issues, such as China's territorial disputes with other countries. China's authorities may therefore choose to loosen censorship and allow the public to vent their anger toward the Beijing government online during large-scale nationalist protests – thus signalling its strong determination to defend the national interests on the international scene (Cairns and Carlson 2016).

While monitoring the information flow on Weibo, the state authorities maintain an active presence on the platform to keep citizens informed of new public administration policies (Zhang and Negro 2013). This means that scholars have to study Weibo's role in exposing scandals and mobilizing opinions that challenge the government's abuses of power or its officials' lack of transparency, but they also have to examine how the platform helps the government to advance its own agenda (Sullivan 2014). Ultimately, Weibo facilitates communication between the government and Chinese citizens, enabling user autonomy and timely feedback from the government, and facilitating bilateral supervision and co-governance (Gu 2014).

As a contentious social space (Poell et al. 2014), Weibo has been widely studied for its potential to mobilize activism and collective action (Yang 2014), empower grassroots advocacy (Liu, Y. 2015; Huang and Sun 2014), and boost the impact of public opinion leaders on the community (Zhang and Negro 2013) by challenging and resisting the official propaganda (Nip and Fu 2016). Despite being such a multifaceted platform, Weibo is underpinned by a clear power hierarchy, which favours influential users (such as opinion leaders and celebrities) and reinforces power differentials (Svensson 2014).

Weibo has also been studied as a network of users and meanings. Verified users, who act as key nodes in this network, are essential to the dissemination of information on specific issues or topics (Han and Wang 2015) and the mobilization of collective action (Liu, Y. 2015; Huang and Sun 2014). As a semantic network, Weibo generates public discourses about key social issues that reflect the changing attitudes in a transitional society, attitudes towards issues such as privacy (Yuan et al. 2013) and, in particular, nationalism (Feng and Yuan 2015). The territorial dispute between China and Japan over the Diaoyu/Senkaku Islands, which fuelled nationwide protest actions in China in September 2012, has attracted a significant amount of scholarly interest. In this case, Weibo played a significant role in the formation of nationalist discourses that circulated both online and in street protests. Feng and Yuan (2015) adopted a semantic network approach to analyse Weibo posts during the protest, and found that Weibo enabled diverse expressions of nationalism.

Some of these were not necessarily the result of state manipulation, but emerged spontaneously in a bottom-up fashion. Ng and Han (2018), on the other hand, focused on the comments elicited by two Weibo posts published by a Japanese porn star during the territorial dispute, and found that expressions of nationalism tended to coincide with manifestations of misogyny. More recently, nationalism on Weibo has taken a different shape, as the younger generation is increasingly drawing on an abundance of popular cultural resources to articulate more aggressive and contentious expressions of nationalism on the platform with the backing of official media. As part of these developments, there have been numerous campaigns attacking celebrities and ordinary citizens who show what are regarded as unpatriotic behaviour or views online.

Scholars have also studied Weibo's impact on certain professional fields. Journalism is one of them, due to the extent to which public events drive Weibo's activity. A number of high-profile incidents during the years following the launch of Weibo have had significant practical implications for investigative journalism (Jiao 2013), the fostering of civic engagement and the achievement of social justice (Zhang and Negro 2013). Although the political and institutional controls on the platform are becoming tighter, journalists on Weibo – an extension of the journalistic sphere – enjoy more autonomy and freedom than in their professional news organizations.

Finally, although Weibo appears to be a medium focused on the present, featuring instantaneous updates on ongoing public events, the past also has an important part to play on this platform. And yet, while various attempts have been made to theorize memory practices in the digital age (van Dijck 2007; Hoskins 2009, 2011; Kaun and Stiernstedt 2014), the links between Weibo and the past have not been adequately addressed in the literature. Han (2016), however, has shown that the platform enables various memory practices, in which various symbolic resources are used to narrate the past and present and to make connections between the two during major public events – all of which contributes to supporting activism, citizen journalism and collective expressions of nationalism.

See also: content moderation and volunteer participation; Facebook; social media; Twitter and hashtags; YouTube

Recommended reading

Feng, M. and E. J. Yuan (2015) 'Public Opinion on Weibo: The case of the Diaoyu Islands dispute', in T. A. Hollihan (ed.) *The Dispute over the Diaoyu/Senkaku Islands: How media narratives shape public opinion and challenge the global order*, New York: Palgrave Macmillan, 119–140.

This study is part of a larger project examining the expression of nationalism on Weibo during the territorial dispute between China and Japan over the Diaoyu (Senkaku) Islands. The authors analysed a large set of Weibo posts published during the dispute – as well as the ensuing protests and escalating violence that broke out in multiple Chinese cities. The study concludes that Weibo posts are primarily a reflection of the 'middle stratum' of Chinese society.

Han, E. L. (2016) *Micro-blogging Memories: Weibo and collective remembering in contemporary China*, New York: Palgrave Macmillan.

This is a systematic analysis of Weibo through the lens of collective memory and the deployment of various mnemonic practices during the discussions of major news events. It conceptualizes Weibo as a vehicle that Chinese society uses to push the boundaries of information transparency, citizen participation and individual rights. This study situates Weibo at a critical juncture in terms of China's political and social transformation and contributes to the analysis of social activism through the rise and fall of social media platforms.

Lei, Y.-W. (2017) *The Contentious Public Sphere: Law, media and authoritarian rule in China*, Princeton, NJ: Princeton University Press.

This book takes an institutional approach to media and law in the authoritarian state drawing largely on Weibo data. It delivers a comprehensive analysis of the 'nationwide contentious public sphere, where the centrality of law and the media may pose risks for the state.

WIKIS

Henry Jones

Derived from the Hawaiian word for quick, the term wiki was coined in English in 1995 (Cunningham 2005). Today, it refers to a particular class of networked communication software, designed to enable easy and efficient collaboration between any number of geographically dispersed individuals; the repositories of knowledge that are created and shared online via these digital tools; and the communities of otherwise unaffiliated web users that have formed around and through such technologies and practices of collective enterprise (Shaw and Hill 2014:216). As such, many wikis are often cited as among the most successful examples of citizen media in the digital age (Mittell 2013; Shirky 2008). After briefly introducing the core features of wikis as software tools, this entry provides an overview of the main wiki communities that have captured the attention of citizen media scholars so far, while emphasizing the expanding range of different types of wiki initiative that have appeared on the web since the late 1990s.

Key features

As Leuf and Cunningham (2001:8) discuss in their landmark introduction to wiki technologies, the development of this software has played a key part within a much wider shift which has seen the web transformed "from just a huge collection of static view-only pages of text and graphics to a more interactive model, where users can share and work together in a wide variety of media types as a matter of course". In this regard, it is important to remember that for much of the 1990s the web was generally experienced as "a browsing-only medium" in which most ordinary citizens were effectively restricted to the passive consumption of online content (Reagle 2010:39). Email listservs and Usenet bulletin boards could be used to facilitate a certain level of online collaboration and peer-to-peer knowledge sharing but these were slow, awkward and labour-intensive, often requiring complex hierarchies of roles and requirements among users (Mittell 2013:35; Shirky 2008:111).

With the invention of the wiki, on the other hand, computer engineer Ward Cunningham sought to provide an alternative model for web-based communication and collaboration, and to circumvent the need for formal design and information management processes. He placed the editing functionality of a webpage directly on the server and so made it possible for any visitor to add, remove or otherwise alter content and organizational structures straight from their favourite browser. Instead of requiring these readers-turned-writers to hold prior knowledge of abstract markup coding languages and the underlying mechanisms of a website, Cunningham also stripped the content creation process back to its simplest possible form, targeting speed and ease of use, rather than the construction of highly polished webpages as fixed and final products

(Leuf and Cunningham 2001:15). Indeed, through the inclusion within every wiki interface of an Edit This button, the software explicitly seeks to involve visitors "in an *ongoing* process of creation and collaboration that *constantly changes* the web site landscape" (ibid.:16; emphasis added).

While a minority of wiki platforms impose some form of administrator moderation (e.g. Stay Woke's ResistanceManual.org), most allow users to publish their edits instantly for all the world to see, a practice which stands in marked opposition to the order of top-down editorial control predominant in traditional broadcast media (Shirky 2008:98). Importantly, however, a copy of the previous version of the document is automatically archived within a so-called Revision History and authors are often additionally encouraged – through the presence of a dedicated Talk Page discussion forum – to explain and justify their contributions (ibid.:112). Not only does this built-in transparency permit all contributors to monitor progress but it also makes wikis near-indestructible as online knowledge repositories: low-quality, misleading or destructive contributions can simply be reverted at the click of a mouse, thus reducing the possibility of long-lasting damage by vandalism and encouraging even nontechnical users to "be bold" in their involvement and input (Wikipedia User 'Thespian' 2007).

Applications

Wikis have been widely used in a number of highly institutional settings, from the US intelligence agencies' Intellipedia to NandOpedia, the newsroom wiki of the American regional daily newspaper *The News and Observer* (Jackson 2009; Bradshaw 2009). Of particular interest to citizen media scholars, however, has been the way their ease of use, potential for mass collaboration and technological logic of publish-then-filter have allowed new forms of community and cultural expression to emerge outside of the traditional structures of the media, education and government (Shirky 2008:98). As the foremost example of what Benkler (2002) has described as commons-based peer-production technologies, wikis have been seen to enable individuals to organize themselves in radically different social formations, defined "through voluntary, temporary, and tactical affiliations, reaffirmed through common intellectual enterprises and emotional investments … [and] held together through the mutual production and reciprocal exchange of knowledge" (Jenkins 2006a:27). Rather than operating in strict hierarchies of command and control, wiki users commonly form "loose heterarchies of equapotentiality" or "ad hoc meritocracies" in which the role and power of leaders is much diminished (to that of temporary administrators), and in which all participants have equal ability to make meaningful contributions to the project (Bauwens 2005:1; Bruns 2008:26). These tools have thus shown remarkable capacities for facilitating the production of citizen media, understood as the content, practices and public spaces created by unaffiliated individuals with the aim of effecting aesthetic or sociopolitical change (Baker and Blaagaard 2016a:16).

A particularly high-profile illustration of what the wiki model means for the production and dissemination of knowledge in the twenty-first century is of course the largest and best known wiki, Wikipedia, the "free encyclopedia that anyone can edit" (Wikipedia, n.d.). On the one hand, while it was by no means the primary intended use for this software, there can be no doubt that the application of wiki technology to the encyclopedia-building process has been hugely successful. Despite the near-total absence of editorial oversight and of formal systems for workflow management, by 2019 the platform's 220,000 regular contributors had created just under 50 million articles, published across 295 different language editions, dwarfing previous attempts at collecting the "sum of all human knowledge" (Wales 2004; Feldstein 2011; West et al. 2009; Wikipedia Editor Activity Levels 2019). Perhaps even more strikingly, these Wikipedians have also rapidly succeeded in making Wikipedia sufficiently useful in terms of its quality and comprehensiveness that in 2020 it was the world's most popular information resource, attracting a monthly average of 15 billion page views from a global audience of nearly 500 million individuals (Wikipedia, n.d.). As Reagle (2010:51) writes, the wiki format has facilitated this extraordinary growth and popularity not only by crowdsourcing the creation of such a reference work (i.e. opening the process up to anyone with the time, technical wherewithal and inclination to participate), but also by allowing users to "communicate asynchronously and contribute incrementally". Volunteers need not all be working on the same document at the same time, nor are there fixed deadlines for publication. Thus, "the timing and granularity of a contribution can be as marginal as fixing a typo on a page that hasn't been touched in months" and content is left to emerge organically through an unending process of revision and gradual improvement (ibid.).

As many commentators have observed, Wikipedia's radically democratic "anyone can edit" policy also means that every user's contribution is essentially considered equal, whether distinguished scholar or layperson (Hartelius 2010:506). Participation and adherence to the core values of the community are seen to be valued at least as much as the personal authority and credentialled expertise of any individual user. Consequently, the user-generated encyclopedia has been the focus of widespread debate over the shifting nature and value of expert knowledge in the Internet age. For Wikipedia's critics, the site appears dangerously dismissive of the ideals of scientific truth, professional accuracy and academic rigour that have shaped the construction of encyclopedias from at least the beginning of the nineteenth century onwards, and is symbolic of the populist politics of the post-truth era (Hartelius 2010:509–510; Lovink and Tkacz 2011:9; Reagle 2011:18–19; Sanger 2004). For example, in 2006, US comedian Stephen Colbert famously coined the term wikiality as a satirical attack on Wikipedia's approach to knowledge production, defining his neologism as "a reality where, if enough people agree with a notion, it becomes truth", regardless of any evidence to the contrary (cited in Mittell 2013:41). The site's supporters, on the other hand, argue that Wikipedia's policy of complete openness constitutes a significant and valuable step in helping to dismantle elite control over the processes of determining information as fact. It mounts powerful challenges against top-down, monological and exclusionary structures within science and politics, promoting instead a more transparent, dialogical model of global education in which always partial, always provisional truths emerge through conflict and consensus between multiple points of view (Hartelius 2010). Through this lens, Wikipedia is seen as a prominent new arena for the social construction of knowledge whose founding principles of Neutral Point of View (NPOV) and Assume Good Faith (AGF) promote greater respect for and interaction between a heteroglossic plurality of differing claims about the world (König 2013:163; Reagle 2010:58).

That said, while Wikipedia may now boast "contributors from pretty much every ethnicity, nationality, socioeconomic background, political ideology, religion, sexual orientation and gender" (Gardner 2013), even the project's most dedicated champions are forced to concede that major imbalances continue to persist within this purportedly open knowledge creation community. For example, due to the uneven distribution of digital technologies worldwide, Wikipedia editors are disproportionately based in countries of the Global North and one in five (20 per cent) lives in the USA (Wikimedia Foundation 2011:31; Graham et al. 2015:1160). Still more worryingly, the widespread gender gap affecting many computer-related fields remains unacceptably prominent in the platform's demographics (Reagle and Rhue 2011:1139). The Wikimedia Foundation's (2011:3) most recent survey reports that just 8.5 per cent of all Wikipedians are women, a figure which Massa and Zelenkauskaite (2014:92) have found to drop as low as 3.75 per cent in some individual language editions (e.g. Hindi). Subsequent research (Reagle and Rhue 2011) has highlighted many factors lying behind this disparity: Bear and Collier (2016:255) suggest that the collaborative process of wiki editing appears to involve a culture of "competitive and aggressive behaviour that runs counter to the traditional feminine gender role", while Stephens (2013:984) highlights the influence of broader socioeconomic inequalities, explaining that lower-income women around the world are often tasked with childcare and other household responsibilities, leaving them less opportunity to "tinker with technology". Whatever the reasons, both these geo-technical and gender-based imbalances hold major implications for Wikipedia's representation of knowledge. Graham (2011:273), on the one hand, highlights the astonishing fact that there are more Wikipedia articles written about Antarctica than all but one of the vastly more populous countries of Africa; Reagle and Rhue (2011), on the other hand, uncover a considerable gender bias in the site's coverage of notable persons. It remains a topic of much debate, therefore, as to whether this wiki has truly diminished existing inequalities in the production of knowledge worldwide, or whether it might even have exacerbated them (Graham et al. 2015:1160–1161; Reagle and Rhue 2011:1139).

Despite these controversies, Wikipedia has fuelled the popularity of wikis more generally as technologies for mass adoption and, as Yeomans (2005) puts it, its successes have "spawned something of a Wiki-empire". Toton (2008) suggests this is because "[o]nce one learns a particular platform … it becomes easy to participate in other incarnations of that digital format": many of Wikipedia's key practices and conventions have been transferred directly to other projects, facilitating more fluid interaction between wiki communities across the web. Certainly, following the creation of the nonprofit Wikimedia Foundation hosting platform in June 2003, a whole series of free wiki-based knowledge-gathering initiatives have formed online, including Wiktionary (a multilingual dictionary), Wikiquote (a collection of famous quotations) and Wikivoyage (a travel information site – see Allan 2006:135; Wikimedia Foundation n.d.). The most studied of these sister projects is Wikinews, a free online news site and prominent example of the ways in which wikis are revolutionizing the field of citizen journalism (Bradshaw 2009:251). Stuart Allan (2006:140), for instance, notes how the collection of networked collaboration tools brought by wiki technology offers Wikinews' volunteer reader-writers a radical alternative to conventional news outlets by allowing them "endless scope to pursue stories that matter to them". Indeed, while traditional paid-for news institutions are inevitably subject to influence by market forces and corporate interests when determining what counts as news, 'Wikinewsies' can operate in relative independence of such pressures and, at least in theory, choose to promote additional contestatory perspectives to the prevailing account of world affairs in the mainstream media.

However, much as with Wikipedia (König 2013), the extent to which Wikinews actually achieves this independent point of view is a subject of much discussion among citizen media scholars. For example, Vis (2009) has analysed the process of Wikinews story selection in the immediate aftermath of Hurricane Katrina by tracing the revision history of the seventy-eight articles that were published via the platform in the first twenty days following the disaster. She concludes that, while many of the site's citizen-journalists showed a genuine desire to balance the more sensational claims promoted by many mainstream news outlets with regard to crime and lawlessness in New Orleans, this was often too slow a process to make much of an impact (ibid.:72). Attempts to offer alternative perspectives were frequently thwarted by the practical difficulties of original reporting and the consequent inevitability of having to rely on the established media for much of their information. Similarly, Thorsen (2008:939) commends Wikinews as an innovative attempt to reject the "ready-made truths" of official monologism, encourage citizens to take a more active role in evaluating news content and break away from the traditional debates around journalist objectivity, but he is ultimately forced to admit that the implementation of the community's ideals is variable (ibid.:950).

Another major consequence of the surge of interest in wikis caused by the successes of Wikipedia has been the emergence of so-called fan wikis, i.e. openly editable online repositories collaboratively constructed by otherwise unaffiliated individuals brought together through their shared dedication to a particular television series, computer game or film franchise (Mittell 2013:36). Toton (2008) describes for instance how a fan group associated with the popular science fiction programme *Battlestar Galactica* (*BSG*) first formed within Wikipedia itself, assiduously documenting the key features of the object of their obsession for the purposes of the general encyclopedia, before turning to create their own more detailed database, using the same MediaWiki software, at galactica.wikia.com. The About page for the hugely popular Wookieepedia similarly tells of how this *Star Wars*-themed wiki "was founded when Wikipedia users began to complain of the overabundance of minutiae related to *Star Wars* appearing on Wikipedia" (Wookieepedia, n.d.).

Such wikis serve a variety of different functions within fan praxis (Mittell 2013). Aside from simply providing a virtual meeting place for fan participation (ibid.:39), wikis mainly help satisfy what Toton (2008) calls their collective impulse. If, for fans, shows such as *BSG* constitute "expansive universes with dense histories and sociologies that require archiving and the constant oversight of a fandom's collective intelligence" (Gray 2010:162–163), wikis provide a space for them to assemble, organize and catalogue the ships, characters, geography and vocabulary related to this fictional cosmos, as well as document every plotline, episode by episode. Following Jenkins (2006a), Toton (2008) suggests that this wiki cataloguing activity is driven primarily by the desire to master or conquer cultural objects such as *BSG*, enabling fans to "gain ownership by unhinging parts, rebuilding texts, and re-exhibiting the series online". By opening up endless possibilities for peer-to-peer collaboration, wikis allow fans the tools with which to interact *en masse* with the text itself, to collectively construct meaning and order from and for the narrative world (Mittell 2009).

Through an ethnographic analysis of Lostpedia, a large fan wiki focused on the long-running television drama *Lost*, Mittell (2009) shows how such user-generated sites can additionally function "as a place for the aggregation of fan creativity". Specifically, the flexibility of wiki architecture offers prosumers not only the possibility of documenting and distilling canonical information offered by *Lost*'s creators about the show's characters, setting (the Island) and plotlines, but also a means of collaboratively developing and discussing noncanonical interpretations and speculative analyses, without disturbing the perceived sanctity of the original storyworld. For example, when discussing The Smoke Monster, Lostpedia's users have been able to maintain the distinction between 'fanon' and canon, fiction and truth, simply by creating a subpage connected to the main article where they can include the countless fan-created theories – sometimes serious, sometimes affectionate parody – relating to this mysterious character's origins, role on the island and abilities (Lostpedia contributors, n.d.). Thus, while the wiki's fluid, easy-to-manipulate structure still permits users to uphold the authority of *Lost*'s creators, it also encourages the creation of this supplementary, ludic content which both extends and enriches fans' understanding of and engagement with the source material.

Finally, we must highlight the role of wikis in activist politics. Although largely overlooked in the literature, a growing number of social movements are making use of such technologies to create shared repositories of cause-related information or simply to improve communication between members. Stay Woke's ResistanceManual.org, for example, presents itself as "a Wikipedia for activism" containing "[c]rowd-sourced information and tools to resist the Trump/GOP agenda" (Stay Woke website, n.d.), while many working groups within the alterglobalist Social Forum network also deploy wiki tools to coordinate their actions (e.g. US Social Forum Working Group wiki, n.d.). That said, the best known and most studied wiki in this category is WikiLeaks, the "wiki for whistle-blowers" (Schmidt 2007). Indeed, while the hacktivist group's *modus operandi* has now fundamentally changed, it is important to recognize the extent to which wiki software was integral to WikiLeaks' original aims and methods. Described at its launch in 2006 as "the first intelligence agency of the people" (WikiLeaks, cited in Lynch 2010:312), Julian Assange and his team's initial plan was not only to allow leakers to post documents anonymously to their site, but also to harness the potential of wiki software for massive grassroots communication, collaboration and interaction with these texts (Roberts 2011:18). The wiki interface would, its advocates believed, empower the public to crack the world open by providing them with the means to collectively analyse and expose the government corruption documented in the millions of files to which WikiLeaks had been given access in a way that would be beyond the capacities of most traditional media outlets. In the case of the Somalia cables, for instance, one community member suggested that "feedback will be, like Wikipedia, an act of creation and correction; the [document] will eventually face one hundred thousand enraged Somali refugees, blade and keyboard in hand, cutting apart its pages until all is dancing confetti and the truth" (cited in Lynch 2010:312).

Interestingly, Assange and his colleagues also suggested the wiki model would offer a more reliable method of authentication for the leaked materials than conventional journalism (Lynch 2010:312). In much the same way as has been argued with respect to Wikipedia (Niesyto 2011:145), they asserted that "[p]eddlars of misinformation will find themselves undone by Wikileaks", given the way the platform would be able to profit from the wisdom of the crowds (WikiLeaks cited in Schmidt 2007). Not only would crowdsourcing allow them to analyse and cross-reference a much greater number of texts than otherwise possible, but also to interrogate them from a broader diversity of perspectives. "Instead of a couple of academic specialists", the site's FAQ page argued, "Wikileaks will provide a forum for the entire global community to examine any document relentlessly for credibility, plausibility, veracity and falsifiability" (WikiLeaks, cited in Schmidt 2007). The sheer quantity and democracy of scrutiny, in other words, would more than compensate for the lack of expert quality among analysts.

As WikiLeaks themselves were forced to concede in 2010, this strategy ultimately failed to deliver useable results, forcing the organization to enter into a closed collaboration with a consortium of mainstream newspapers and professional journalists. This fact raises an important point with respect to the promise and pitfalls of wikis as technologies of social and political change. On the one hand, as we have seen with the successes of Wikipedia and its many sister projects, the democratic potential of these collaborative tools is certainly unprecedented in terms of their ability to mobilize individuals into a collective force and destabilize traditional hierarchies determining the relationship between authority, media and citizens (Beckett and Ball 2013). On the other hand, such is the complexity, power and grip of the opaque corporate and governmental structures currently steering the course of world affairs that any belief in wikis and other participatory media forms as a "technological quick fix" for the world's deepest inequalities is inevitably misplaced (Roberts 2011:21). Thus, if we are to advance understandings of wikis as tools for emerging digital forms of engaged citizenship, future analysis must continue to underline the importance of critical perspectives which pay just as much attention to their limits and failings, as to their strengths and possibilities.

See also: citizen journalism; commons; content moderation and volunteer participation; convergence; diversity; user-generated content

Recommended reading

Leuf, B. and W. Cunningham (2001) *The Wiki Way: Quick collaboration on the web*, Boston, MA: Addison-Wesley.

Co-written by Ward Cunningham, the creator of the first wiki site, this widely cited introduction sets out the core principles that inspired the development of this software and that have since shaped the growth of the participatory web.

Lovink, G. and N. Tkacz (eds) (2011) *Critical Point of View: A Wikipedia reader*, Amsterdam: Ten Klein Groep. Available online: www.networkcultures.org/publications.

This edited volume contains a fascinating collection of critical perspectives on Wikipedia by academics based across the humanities and social sciences.

Mittell, J. (2013) 'Wikis and Participatory Fandom', in A. Delwiche and J. Jacobs Henderson (eds) *The Participatory Cultures Handbook*, New York and London: Routledge, 35–42.

Mittell offers a valuable overview of the fan wiki phenomenon, as well as an insightful history of the media context out of which wikis first emerged.

WITNESSING/TESTIMONY

Daniela Mansbach

The twenty-first century is the age of witnessing (Felman and Laub 1992); we are constantly exposed to an immense number of stories, images and videos of human rights violations and other atrocities from around the world. The proliferation of witnessing in this new era has been driven by two main developments. First, since the 1990s, the distinction between mainstream, corporate media and alternative, non-corporate media has weakened (Rodríguez et al. 2014). This means that many more individuals and groups have access to more forms of media and can spread information, challenging the traditional assumption of a fixed and stable distinction between those who produce media and those who consume it (Ashuri and Pinchevski 2009). The second development concerns the increased access to wireless communication, which has given rise to the phenomenon of citizen-camera witnessing: the ability of everyone to constantly document their reality visually and acoustically, and to spread images around the globe easily and quickly, allowing audiences in various locales to watch events as they unfold, or soon after (Andén-Papadopoulos 2014a). These changes have resulted in an increased exposure to human rights violations that take place on the other side of the world.

Unlike eye-witnessing – which happens accidentally and is based on the premise of objective accuracy and truth – witnessing, or bearing witness, is understood as a deliberate political act of telling a story of suffering and oppression from a position of empathy (Oliver 2004; Tait 2011). In this sense, witnessing is not meant to merely reveal a certain truth or specific evidence regarding the events being witnessed. Instead, it aims to expand the circle of those who witness the events, to expose human rights violations and suffering to ensure that others cannot say 'I did not know' (Ellis 2000). Witnessing further aims to define the events in question as disturbing and unjust, using images and stories to encourage the viewer to act in some way to change the situation (Felman and Laub 1992; Schaffer and Smith 2004; Zelizer 2002). This understanding of witnessing as a means of bringing about change blurs the traditional distinction between witnessing and testimony. In theory, it is possible to separate the act of witnessing – watching the events as they unfold – and that of verbally testifying about the events, sometimes long after the events had ended. In practice, however, it is often hard to identify the moment when witnessing ends and testimony begins, especially in the broadened definition of witnessing offered here.

Witnessing, as defined in this entry then, does not merely consist of passive observation but may also constitute a form of action, like testimony. For example, the Israeli movement Checkpoint Watch, established in the early 2000s to oppose the Israeli occupation of the Palestinian Territories, organizes daily shifts at the checkpoints operated by Israeli soldiers. The activists witness the events and write reports which are then publicized in Israel and around the world, in an attempt to draw attention to the way in which Israel controls the Palestinians and limits their freedom. This act of witnessing is not passive: it is part of the movement's activism. Members of the movement hope that as Jewish-Israeli women, their presence at the checkpoints will help prevent some of the most extreme violations of human rights (Mansbach 2016). Witnessing is thus better understood as a performative act of intervention rather than a passive act of gathering information, and it is difficult – as well as unnecessary – to clearly separate it from testimony. This entry uses the two terms interchangeably, except where the distinction is relevant to a specific argument.

This entry focuses on the question of whether increased citizen access to various forms of media and our ability to document human rights violations addresses the two main challenges of witnessing and testimony. The first challenge is that of representation. Witnessing and testimony often reflect existing power relations in society; they give voice to those who already have power, including access to local, national and international media, and are thus in a position to report on events. In addition to access, the individuals and groups bearing witness or giving testimony are often also those who are deemed credible and trustworthy, allowing the audience to take their voices, words or actions seriously. The challenge of witnessing is then to tell the story of suffering without further undermining the subjectivity of marginalized and oppressed populations by a powerful testifier. The second challenge of witnessing and testimony concerns effectiveness: the ability to engage in witnessing and testimony in such a way as to promote a sense of responsibility that can lead the audience to act in an attempt to effect change.

The challenge of representation

Effective witnessing that is able to promote action and effect change sometimes requires a gap between the testifier and the victim. In order to expand the number of people who are exposed to the events and may find the reports credible, testimony needs to be delivered by people who hold positions of power within society. This gap reflects what Fricker (2007) refers to as testimonial injustice; the social position of the subjects and our prejudices against them limit our ability to listen to them and undercut their credibility when they testify. Witnessing and testimony are thus constrained by preconceived ideas about who may be considered a credible and trustworthy individual whose voice, words or actions are worth being taken seriously. If, in order to be effective, testimony must come from individuals who already hold positions of power because of their race, economic status, gender, nationality or any other aspect of their identity, there will be inevitable implications for the construction of subjectivity of those who remain voiceless, and who are often the direct targets of these human rights violations. That is, if the stories of the marginalized can only be heard when they are told by people who enjoy some level of power and influence, there is a risk that testimony will reassert rather than challenge existing power relations.

Testimonial injustice thus means that while witnessing and testimony aim to protect the marginalized and oppressed, these practices risk reinforcing the distinction between those who are voiceless in society and those who are not (Hartman 1997). Since testimony is by definition an act of speaking for someone else, it inevitably re-creates the subject as the Other who cannot speak, promoting an image of the oppressed as unable even to understand their own conditions of oppression (Spivak 1988). By perpetuating the marginalization and oppression of the victim, witnessing and testimony may further undermine subjectivity "by attacking the ability for address and response" (Oliver 2004:80). Ultimately, witnessing and testimony might be able to address a specific case of human rights violations, but without challenging existing power relations they cannot effect durable change in ongoing processes of dehumanization and oppression.

Uneven power relations may also result in more attention being accorded to the witness who is reporting the events, rather than the victim. The 1993 Pulitzer winning photograph by Kevin Carter, which was taken in southern Sudan during a famine that followed political violence and civil war, shows a vulture perching near a dying girl who seems to be so sick and weak that she is unable to move. The photo was used in multiple campaigns by human rights organizations, to collect donations for refugees and mobilize support for social action. At the same time, however, following the suicide of the photographer in 1994, the media moved away from discussing the suffering in Sudan and the fate of the dying girl. Instead, public discussion and media coverage focused on the trauma that witnesses such as Carter experience when they are exposed to these difficult images and events (Kleinman and Kleinman 1996).

The tension between the need to help those who are voiceless on the one hand, and to ensure that their representation by others does not further marginalize and oppress them on the other, has been a central focus of the poststructuralist literature on witnessing and testimony (Agamben 1998/2000; Felman and Laub 1992; Oliver 2004). The question is whether this tension still exists in the age of witnessing, or whether there is a need to rethink the strengths and limitations of witnessing and testimony, especially in light of the growing ability of subjects to represent themselves and their own suffering.

The age of witnessing, with growing access to alternative forms of media and wireless communication, has changed the subject's ability to testify and document, making it easier for everyone – rather than only journalists and human rights activists – to document their everyday reality, including atrocities and human rights violations. Many of the images of human rights violations in the twenty-first century have been taken by ordinary citizens, even as they experience them, including some of the iconic photos of the attack on the World Trade Center in September 2001 (Zelizer 2002). Citizens are thus able to document events while they experience them. In 2009, numerous bystanders took videos that documented the death of the 26-year old Iranian student, Neda Agha-Soltan, during demonstrations in Iran against alleged fraud in the presidential elections. Her death, from a shot by an Iranian paramilitary organization, was broadcast online and around the world, leading to public outrage and further anti-government protest (Andén-Papadopoulos 2014a).

The involvement of ordinary citizens in witnessing and testifying is not new, however; some of the best known photographs of the Holocaust were taken by American soldiers who used their private cameras to document what they saw when they entered the Nazi concentration camps (Zelizer 2002). However, increased access by many around the world to wireless technology means that today, those who witness, document and testify are often not situated outside the events, as was the case of the American soldiers documenting the Holocaust. Instead, they are activists and victims on the ground and often on the receiving end of the violence. This age of witnessing is thus characterized by a proliferation of self-witnessing and self-testimony, even in extreme situations of human rights violations. Many of the images and videos of refugees arriving in Europe, mainly from Africa and the Middle East, since 2015 have been taken by the refugees themselves (Ponzanesi 2016; Weston 2019) and were later published on alternative platforms as well as by mainstream media sources. In 2016, *Frontline*, an American documentary programme, dedicated the episode *Exodus* to refugees and migrants fleeing to Europe to escape war and persecution; much of the episode relies on videos shot and interviews conducted by the refugees themselves during their journey.

Subjectivity may be understood as one's sense of oneself as an *I*, and hence the ability to see oneself as a witness promotes agency, allowing individuals to "speak back to power" (Givoni 2011b:149). Self-witnessing and self-testimony may thus offer one possible corrective to the risk of undermining the subjectivity of victims. Self-representation may allow the oppressed and marginalized to reclaim their subjectivity and autonomy by representing themselves. By claiming their experiences as not only true and valid, but also political, self-witnessing is able to bring moral and political subjects into being (Ashuri and Pinchevski 2009; Felman and Laub 1992; Givoni 2011a; Oliver 2004). Thus, self-witnessing and self-testimony have the potential to liberate the oppressed, albeit momentarily, in the most extreme cases of dehumanization (Agamben 1998/2000; Felman and Laub 1992).

This potential, however, has its own limitations. First, while the spread of wireless communication and alternative forms of media has expanded the range of groups and individuals who can document and amplify their own stories, this does not mean that all people are now able to represent themselves. Instead, many still need others to tell their stories for them, either because of lack of access to media and wireless communication, or because of lack of trust and willingness on the part of society to hear their stories. This is true of the refugees who have been fleeing Syria since the beginning of the civil war in 2012. The refugees who travelled to Europe and whose stories have been told in the mainstream media, including in the *Frontline* episode *Exodus*, are not necessarily representative of the entire population; their socioeconomic class, education and linguistic skills are higher than those of the majority of Syrians, whose chances of leaving the Middle East are low (Betts et al. 2017). Thus, although self-witnessing may be undertaken by the victims, the nature of the identity of the self-witnesses within the larger group of victims often means that some aspects of testimonial injustice endure. Witnesses and testifiers must therefore demonstrate awareness of the limitations of their representation, as Rizk (2013) does in an interview about Mosireen, a collective of filmmakers active in documenting the Egyptian uprising, to which he belonged:

> I don't claim to be able to speak on behalf of a collective that is not uniform. The best I can do is to keep my ears to the ground as much as possible, to spend time with the people that make up this revolution, to listen, to learn and speak in humility. In this act of speaking I do not attempt to 'represent', I try to interpret, but representation is out of the question.

Second, corporate media's use of citizen voices is not undertaken in the context of a partnership (Chouliaraki 2016). Instead, it is an attempt to re-legitimize journalism by prioritizing authenticity as opposed to traditional fact checking and expert analysis. Further, the representation of events and testimonies often relies on a politics of pity, emphasizing the suffering of the victims, who can only be saved through the intervention of the audience. The growing use of self-witnessing thus plays into existing power relations: it gives an impression of self-representation while retaining power in the hands of corporate media.

The challenge of effectiveness

The second challenge of witnessing and testimony concerns the ability to move people to act, without further reinforcing the testimonial injustice perpetrated by issues of representation as discussed above. The oppressed may be able to represent themselves in their own voice, but self-representation does not necessarily fulfil the goal of witnessing and testimony to bring about social and political change, because there remains the difficulty of the oppressed being heard. Even when the audience is exposed to the images and stories of victims, their suffering may still be seen as irrelevant to the audience's own sense of morality, making it harder to effect the shift from witnessing to action. In the case of Checkpoint Watch discussed earlier, for instance, the voices and stories of Palestinians do not reach an Israeli audience not only because of lack of wireless communication or a common language, but because of the longstanding dehumanization of the Palestinians and the image perpetuated of them in Israel as the enemy, making their testimony unreliable and their suffering insignificant in the eyes of most Israelis.

Increased public exposure to human rights violations in the age of witnessing, then, has not necessarily been accompanied by a heightened sense of solidarity and responsibility towards victims. Indeed, as some scholars have argued, constant exposure to images of suffering without the impetus or ability to do something about that suffering desensitizes audiences and discourages them from acting (Cohen 2011; Kleinman 1998). The unwillingness or inability to act are particularly striking given the volume and quality of images now available to audiences around the world, which resolve the thorny question of authenticity and accuracy to a large extent. As audiences become increasingly desensitized and overwhelmed by the amount of suffering they see on their screens, they gradually conclude that there is nothing they can do to change the situation. The age of witnessing, according to this analysis, expands our knowledge about human rights violations, but does not enhance our ability to empathize or act on behalf of others.

Other scholars have argued that there are potential ways in which practices of witnessing might change to help promote a sense of solidarity and action (Ellis 2000; Silverstone 2007). Two characteristics of the age of witnessing and testimony are relevant here. First, unlike the Pulitzer winning photo of the dying girl in Sudan, we are now able to watch the events either as they unfold, or shortly thereafter (Givoni 2011b). This may change the nature of witnessing, as well as the way in which viewers understand their role. In this context, witnessing and testimony are not merely tools that allow us to reconstruct our knowledge and understanding of certain historical events. Instead, in the process of viewing the events as they unfold, the audience itself is transformed into a witness. The audience becomes part of the event itself, and is no longer only "a witness to the witness" (Ashuri and Pinchevski 2009:140). The number of witnesses – on the ground as well as around the globe – subsequently grows, together with the determination of these witnesses to act. This development may promote trust, resulting in more witnesses being seen as credible, irrespective of their background.

Second, new media structures and affordances transform the individual experience of witnessing into a shared experience, as in the case of the hundreds of people who documented the events of 9/11 (Hirsch 2003; Zelizer 2002). Co-witnessing promotes trust, in the testifiers as well as the reality they present. This is especially important in "a social and politically contested terrain in which various agents compete to gain the trust of their designated audience" (Andén-Papadopoulos 2014a:758). Co-witnessing might also help in cases when the victim is considered less than human and his or her suffering is thus dismissed as insignificant; by recasting the event as a shared experience, co-witnessing expands the number of those who see themselves as victims to include, potentially, people and groups whose suffering does matter to the public. Thus, while immediate exposure to human rights violations may reinforce notions of trust and responsibility, co-witnessing further carries the potential to promote among the audience a notion of themselves as the target of these violations.

Ultimately, while images of human rights violations spread more rapidly around the world, often through practices of self-representation, self-witnessing and testimony can limit the risk of undermining autonomy and reinforcing existing power relations if the representations generated are not mediated by powerful individuals or corporate media sources. On the other hand, the increased immediacy and accuracy of testimony offered by new media platforms or attained in the case of shared experience increases its potential to transform the audience into a witness, thus promoting a sense of responsibility and action. Nevertheless, in this inevitably strained dynamic, the transformation of the audience into a witness once again draws power away from the victims and invests it in those in positions of power. The age of witnessing, while powerful in its potential to promote activism as well as self-representation, is still not able to solve the tension between the challenge of representation and that of effectiveness.

See also: citizen journalism; co-optation; documentary filmmaking; film studies and citizen media; migration studies and citizen media; philosophy and citizen media; selfies

Recommended reading

Andén-Papadopoulos, K. (2014) 'Citizen Camera-Witnessing: Embodied political dissent in the age of mediated mass self-communication', *New Media & Society* 16(5): 753–769.

Offers an analysis of the increasing use of mobile camera-phones by political activists and dissidents who risk their lives to provide public testimony of unjust and disastrous events around the world. Examples discussed include recordings of the death of a young woman, Neda Agha-Soltan, during a demonstration in Tehran in 2009, to show how the camera-phone has allowed citizens to create and share their eyewitness records. This new form of bearing witness is shaping demonstrations and resistance today, while also creating a global audience in the attempt to mobilize networks of solidarity worldwide.

Givoni, M. (2011) 'Beyond the Humanitarian/Political Divide: Witnessing and the making of humanitarian ethics', *Journal of Human Rights* 10(1): 55–75.

Provides an in-depth historical analysis of the use of witnessing and testimony in humanitarian work and human rights activism. Through an examination of the humanitarian organization Doctors without Borders, Givoni shows how witnessing has been changing humanitarian work, leading it to focus on moral responsibility while also challenging the distinction between the moral and political spheres.

Tait, S. (2011) 'Bearing Witness, Journalism and Moral Responsibility', *Media, Culture & Society* 33(8): 1220–1235.

Examines the role of witnessing and bearing witness in media and journalism, and the importance of thinking about this practice as a form of action rather than a passive or objective practice. Analysing the columns written by the journalist Nicholas Kristof on Darfur between 2004 and 2009, Tait discusses how journalism may elicit affective responses from readers and advocates a journalism of attachment rather than of impartiality.

WORLD SOCIAL FORUM

Hilde C. Stephansen

The World Social Forum (WSF) is a global gathering of social movements and non-governmental organizations (NGOs) that share a commitment to developing alternatives to neoliberal globalization. First held in 2001 in Porto Alegre, Brazil, the WSF was originally conceptualized as a counterpoint to the World Economic Forum, which gathers the world's economic and political elites every year. Subsequent editions of the WSF, regularly bringing together tens of thousands of participants, have been held in Porto Alegre (2002, 2003, 2005); Mumbai, India (2004); Nairobi, Kenya (2007); Belém, Brazil (2009); Dakar, Senegal (2011); Tunis, Tunisia (2013, 2015); Montréal, Canada (2016) and Salvador, Brazil (2018). The WSF has also spawned regional social forums such as the European Social Forum and the Pan-Amazon Social Forum, national social forums such as the US Social Forum and the India Social Forum, and local social forums, as well as thematic forums in fields such as migration and education. Several editions of the WSF have taken a decentralized format. These include the 'polycentric' WSF 2006, which was held in Bamako (Mali), Caracas (Venezuela) and Karachi (Pakistan), and the WSF 2008, which took the form of a Global Day of Action, with hundreds of local activities taking place simultaneously around the world.

The WSF is governed by a Charter of Principles and an International Council made up of representatives of NGOs and social movements, but is not itself an organization – it was conceived by its founders as a space for reflection and exchange of knowledge and experience among civil society actors. Accounts of its origins typically present the WSF as emerging out of the alterglobalization or global justice movement that burst onto the world stage in the late 1990s and early 2000s with a series of international protest events around summits of the G8, International Monetary Fund, World Trade Organization and World Bank (della Porta et al. 2006; Leite 2005; Smith et al. 2008). While these protest events served to consolidate the emerging 'movement of movements' by bringing together a diverse range of actors in the struggle against neoliberal globalization, the WSF was conceived as a next step in this struggle, as a space in which participants could begin to articulate not only what they stood *against* but what they were *for*. However, the WSF has antecedents in a much broader range of struggles and political traditions (Conway 2013). While many commentators posit the WSF as a continuation of a long history of left movements worldwide (de Sousa Santos 2006), its lineages also include anti-colonial and indigenous peoples' struggles, mass movements against dictatorship and neoliberalism in the Global South, the non-class-based new social movements that emerged (mostly but not exclusively in the Global North) in the 1960s (such as feminism, gay liberation and Black consciousness), as well as environmental and human rights movements (Conway 2013:12–16).

The WSF is, then, a meeting place for a diverse range of movements, political traditions and forms of knowledge. This diversity has been understood as central to the WSF's political novelty and captured by the concept of open space. Described by one of its founders as "only a place, basically a horizontal space" (Whitaker 2008:113), the WSF does not seek to establish consensus or speak on behalf of its participants, is in principle open to all civil society actors that subscribe to the fairly minimal requirement of opposition to neoliberalism and discrimination, and based on the principle of self-organization: those who organize social forums are meant simply to provide a space for participating groups to organize their own activities (Sen 2010). In this way, the WSF was intended to function as an incubator for new initiatives without itself becoming a political actor (Whitaker 2008:113). It has been conceptualized as a pedagogical space, with several commentators such as Andreotti and Dowling (2004), Olivers (2004) and Wright (2005) highlighting the parallels between the ethos of open space and the philosophy of the Brazilian educator Paulo Freire (1972, 1974), whose influential approach to critical pedagogy advocated non-hierarchical, dialogic learning processes aimed at fostering critical resistance to oppression and inequality.

The WSF's politics of open space can be seen as expressive of a broader cultural logic of networking, linked to the rise of new information and communication technologies, which values openness, horizontality and self-organization (Juris 2008). Also central to the notion of open space is a rejection of all forms of *pensamientos únicos* (monolithic forms of thought) and embrace of epistemic plurality. De Sousa Santos (2006) conceptualizes the WSF as expressive of an 'Epistemology of the South': a manifestation of the plurality of knowledges and epistemic practices that exist in the world and the possibility of non-hierarchical, horizontal exchange between them. De Sousa Santos understands domination as profoundly epistemic in character: neoliberal globalization asserts its hegemony by discrediting other available forms of knowledge and social experience while denying the possibility of future alternatives. The WSF resists this dynamic by affirming the existence and validity of such alternatives (de Sousa Santos 2006).

Debate and critique

The WSF has, since its inception, been the subject of a rich critical debate (Sen and Waterman 2008). While de Sousa Santos and proponents of the open space concept have painted a rather optimistic picture of the WSF, critics have highlighted several ways in which it falls short of its own ideals of openness and inclusion. These include structural barriers to participation such as travel costs and visa restrictions, as well as more subtle mechanisms of exclusion arising from cultural norms and discursive practices (Andretta and Doerr 2007; Doerr 2007; Vinthagen 2009; Wright 2005; Ylä-Anttila 2005). The WSF has suffered from hierarchies along lines of race (Conway 2013; Naidoo 2010), gender (Conway 2013; Desai 2016; Eschle and Maiguashca 2010; Karides 2013), class (de Sousa Santos 2006; Smith et al. 2014), age (Jabberi and Laine 2015; Laine 2013), education (Pleyers 2008; Worth and Buckley 2009) and religion (Caruso 2012). Conway (2013) conceptualizes the WSF from a postcolonial, anti-racist feminist perspective as a site of tension between modern emancipatory traditions (socialism, liberalism, anarchism, feminism) and the political praxis and visions of various subaltern 'others', such as indigenous and poor peoples' movements in the Global South. While the WSF has enabled such subaltern actors to enter the global stage, it remains dominated by movements grounded in the political traditions of European modernity (Conway 2013). Far from simply an open, horizontal space, the WSF is not immune to the hierarchies and exclusions that structure the world at large.

Another strand of critique has focused on the WSF's lack of political efficacy. In the so-called space versus movement debate (Conway 2005, 2013; Kohler 2005; Marcuse 2005; Patomäki and Teivainen 2004; Ponniah 2005, 2008; de Sousa Santos 2006; Teivainen 2004; Wallerstein 2004), critics of the open space concept have argued that it has resulted in nothing more than a talking shop (Worth and Buckley 2009) and called for the WSF to become more active in formulating and acting on collective proposals, while its defenders have argued that such a move would destroy the WSF's capacity to attract a diversity of actors (Whitaker 2008). Some scholars have conceptualized this debate as a tension between verticals and horizontals (Kavada 2009; Smith and Smythe 2009), the former consisting of 'older' political actors such as trade unions, NGOs and organized movements that favour more hierarchical forms of organization, and the latter represented by a new generation of activists who operate with a political logic of openness and horizontal networking. This new political culture has perhaps found its most coherent expression in the WSF Youth Camps, a staple feature of WSF events, where young activists gather to experiment with non-hierarchical, participatory modes of organization.

Tensions between different political cultures have been further highlighted by the emergence in the 2010s of social media-facilitated movements of the squares, which operate according to a logic of connective action (Bennett and Segerberg 2013), based on individualized and personalized modes of political participation. At the WSF 2013 and 2015 in Tunis and the WSF 2016 in Montreal, activists linked to these movements mobilized under the moniker GlobalSquare to create their own open space, many adopting a critical position towards what they perceived as informal hierarchies and lack of transparency within the WSF. Much of this criticism has centred on the undemocratic nature of the International Council – which is not elected and is only open to organizations – and what many see as the dominance of NGOs within the WSF. This critique of the WSF can be linked to a tension between the logic of aggregation characteristic of the newest movements, which involves the assembly of unaffiliated individuals within public spaces, and the logic of networking widespread within the global justice movement, which involves the networking of already formed collectives (Juris 2012; Smith et al. 2014).

Another area of debate about the WSF has focused on its status as a global phenomenon. The Charter of Principles states that the World Social Forum is "a world process", and it is frequently referred to in such terms by both activists and scholars. However, the WSF is perhaps more appropriately conceptualized as a multi-scalar process bringing together place-based movements and actors that operate on a more self-evidently global scale (Conway 2008; Osterweil 2005; Stephansen 2013a, 2013b, 2019). As it has travelled around the world, the WSF has become a site for claims by various 'local subalterns' – such as urban slum-dwellers in Nairobi and indigenous peoples in the Amazon – who have come to make their voices heard and assert their right to be present in the spaces of global civil society (Conway 2004, 2008). However, the WSF has frequently been criticized for excluding the local resident population, and it has been dominated by a highly mobile cosmopolitan elite of scholar-activists who have the resources and inclination to travel to international events (Conway 2008; Pleyers 2008). This raises questions about how local or global the forum should be (Conway 2008) and how its globality is to be defined. As Osterweil has argued, the WSF can be conceptualized as a site of tension between a "universalizing globalist" perspective, which involves moving beyond place-based and local struggles to create a united global movement, and "place-based globalism", which sees true globality as "comprised of many nodes, places, interconnections and relations that at no point are totally consolidated into a singular global entity" (2005:26).

Questions of scale are also brought to the fore by efforts to conceptualize the WSF as a manifestation of global civil society, or, more specifically, a global public sphere (Kaldor et al. 2003; Glasius 2005; Wright 2005; Smith 2004, 2008; Smith et al. 2008, 2014). As Conway (2013:67–90) explains, such theorizations rely on an understanding of global civil society as a stakeholder within global governance networks – as a counterpart to state power – and see the WSF as an emergent global public sphere. Based on the Habermasian concept of the public sphere as an open and inclusive communication space located within civil society, in which citizens can come together to debate issues of common concern, such accounts conceptualize the WSF as a force for global democratization. However, although the WSF's politics of open space on the face of it has many similarities with this understanding of the public sphere, it cannot straightforwardly be conceptualized in such terms. While the public sphere is understood within the Habermasian tradition as a counterpart to state power, the WSF has no obvious counterpart in the form of a global state authority, and while the concept of the public sphere is rooted in the assumption that it is possible (and desirable) to arrive at consensus about the common good through rational deliberation, the WSF's open space is deliberately structured so as not to produce consensus (Conway and Singh 2009; Conway 2013). Moreover, the oppositional character of the WSF means that it is perhaps more appropriately conceptualized as a counterpublic (Fraser 1990) or a formation of multiple counterpublics (Conway 2004).

Citizen media in the WSF

Although new communication technologies and diverse forms of cultural and creative expression have been central to the global justice movement, there has been surprisingly little consideration of the role of citizen media – or media, culture and communications more generally – in the academic literature on the WSF. Even those who analyse the WSF from the perspective of public sphere theory focus overwhelmingly on the physical space provided by forum events, neglecting the role of media and communications. Forum organizers, along with many of the major movements that participate in the WSF, have also been criticized for not taking media, culture and communication seriously (Waterman 2005; Couture et al. 2016).

The global gatherings of the WSF provide occasions for a diverse range of citizen media practices such as theatrical protest, music concerts and impromptu street performances, and the forum site is usually overflowing with leaflets, posters and other print media. More specifically, however, communication media such as radio, video and online technologies have played an important role within the WSF, as a means for activists to circulate and share coverage of the issues and debates that social movements bring to the forum. Given that the WSF has struggled to achieve visibility and standing within mainstream media, citizen and alternative media have arguably been the main sources of information about the forum for participants and interested publics. The WSF has also provided an important site for exchange and collaboration among citizen media initiatives from different parts of the world. Since its inception, citizen journalists and alternative media producers have used the WSF as a space for network-building and experimentation with new communication practices (Stephansen 2013b, 2016, 2019). Brazilian media activists, who have played an important role in this process, developed the concept and practice of 'shared communication' (*comunicação compartilhada*), characterized by an explicit focus on creating collaborative production processes that bring media activists together. The idea of shared communication emerged with the first WSF in 2001, when a copyleft-based web publication system named Ciranda was created to enable media activists to share coverage of the forum. Having initially emerged from a need to enable sharing of content at a time before Web 2.0 technologies were widely available, the idea of shared communication soon also came to signify collaborative media production and exchange of knowledge and experience among media activists (Stephansen 2016). Dedicated spaces for citizen and alternative media (including community radios, independent journalists, video producers and tech activists) have since become a staple feature of social forum events, with the explicit aim of facilitating collaboration and mutual learning among participants.

As the WSF has travelled around the world, shared communication activists have sought to establish links with citizen media producers in the places where the forum has been held (Stephansen 2013b, 2019). At the WSF 2009 in Belém, members of the Ciranda network worked closely with local media activists, including community radios and an organization that used audio-visual media as a tool for community engagement, to organize independent coverage of the WSF and in the process create networks and build capacity among grassroots activists (Stephansen 2013b). The WSF 2011 in Dakar provided an occasion for Latin American media activists to connect with their counterparts in the Indymedia Africa network. When the WSF moved to Tunis in 2013 and 2015, the network expanded further to include community radios, bloggers and other media activists in the Maghreb-Mashreq region. The WSF 2016 in Montreal, meanwhile, provided an opportunity to connect with North American activists (Couture et al. 2016). Such efforts to engage with and mobilize local media activists form part of a movement-building approach to media activism. Shared communication activists understand the mobilizing function of citizen and alternative media not just in terms of their role in disseminating convincing messages, but also in mobilizing people to participate in media production and build solidarity-based networks among media activists

(Stephansen 2016). This movement-building approach has enabled grassroots citizen media initiatives in different parts of the world to connect with a growing transnational network of media activists and feel part of a global struggle. For example, Stephansen (2013b) details how community radios and video activists in Belém used the WSF 2009 as an opportunity to build capacity among local media activists and strengthen links between social movements in the Amazon, while simultaneously connecting with transnational networks. In this way, citizen media practices concerned with capacity-building, networking and movement-building support a place-based globalism that allows marginalized actors to connect to global networks while remaining committed to place-based struggles (Stephansen 2019).

As well as enabling collaboration and networking among citizen media producers, the WSF has also provided a site for political discussions about media and technology. Media and communication first appeared as a thematic axis at the WSF 2003 (Milan 2013:36), and since then activists have organized seminars and workshops at every WSF to discuss issues ranging from censorship and repression to community media and Internet governance. Since 2009, these discussions have been brought together under the banner of the World Forum of Free Media (FMML, for *Fórum Mundial de Mídia Livre* and its French/ Spanish equivalents), a thematic forum linked to the WSF that brings together a wide range of actors, including community radios, free software developers, citizen journalists, bloggers and NGOs, that support access to information and communication. Following the first FMML, which was held alongside the WSF 2009 in Belém, subsequent editions have been organized in conjunction with the Rio +20 summit in 2012, the WSF 2013 and 2015 in Tunis, and the WSF 2016 in Montreal. For media activists who have been involved in the WSF process since the beginning, the development of the FMML has been accompanied by a shift in collective identity as they have gradually come to see themselves not only as producers of citizen and alternative media but as participants in a social movement that takes media and communication as a subject of contention in its own right (Stephansen 2017).

Future directions

Citizen media have played a central role in documenting the ideas and proposals emerging from the WSF and making these more widely known. The significance of citizen media within the WSF process, however, is not limited to their capacity to disseminate information but extends to a range of practices such as capacity-building, networking and movement-building. The WSF and FMML have provided important sites for the development of transnational networks of media activists and for enabling political discussions about media and technology issues – including, *inter alia*, the challenges that citizen media face due to repression, censorship and the growing power of media corporations. Thus, while much recent literature on citizen media has focused on individuals' use of Web 2.0 technologies to make their voices heard, citizen media practices in the context of the WSF are notable for their collective dimension and activists' efforts to politicize media and technology issues.

As already noted, media and communication have not received much attention in the academic literature about the WSF, and have not always been high on the agenda of forum organizers and participating movements. However, given the role that citizen and alternative media play both in terms of disseminating information and mobilizing new actors, they should be considered fundamental to the WSF process. Citizen media contribute to the open space of the WSF by disseminating information about participants' ideas and proposals, thus facilitating exchange of knowledge and experience beyond forum events. The emphasis that many activists place on supporting grassroots movements to produce their own media is evidence of a commitment to epistemic plurality that recognizes the importance of giving voice to marginalized groups. Citizen media can thus be seen as central to the WSF's pedagogical praxis and to the idea of the WSF as 'Epistemology of the South' (de Sousa Santos 2006). Citizen and alternative media should furthermore be central to theorizations of the WSF as a global public sphere (Conway and Singh 2009; Smith et al. 2014; Fraser 2005) – most obviously, because such media disseminate information about the WSF and thereby contribute to making it more public and global, but also because citizen media practices like those discussed here contribute to the making and interlinking of publics at different scales (Stephansen 2016, 2019). The practices of media activists within the WSF thus point towards an understanding of the global public sphere as de-centred, multi-scalar and constituted through diverse communication practices (Stephansen 2019). Finally, future research on global-scale media activism, within and beyond the WSF, needs to recognize that activists operate on an unequal playing field, crisscrossed by hierarchies of race, gender and class (among others), and that this makes a difference to how they are positioned within global communication networks.

See also: activism; autonomous movements; citizen journalism; diversity; Indymedia; prefiguration

Recommended reading

Conway, J. (2013) *Edges of Global Justice: The World Social Forum and its 'others'*, London: Routledge.

Analyses the World Social Forum from a postcolonial, anti-racist feminist perspective, focusing on core political currents within the Forum, including liberalism, feminism and autonomism, as well as emerging movements of 'subaltern others'. Argues that the WSF is at once at the forefront of a fundamental transition towards a new politics beyond modernist traditions *and* a site where their exclusions and hierarchies are reproduced.

Santos, B. de Sousa (2006). *The Rise of the Global Left: The World Social Forum and beyond*, London: Zed Books.

Theorizes the WSF as the manifestation of an alternative, counter-hegemonic globalization and as epistemology of the South – an affirmation of epistemic pluralism and the validity of knowledges that have been suppressed by colonialism and neoliberal capitalism. Provides an overview of the WSF's organizational structure and main political cleavages, and conceptualizes its key challenge in terms of constructing mutual intelligibility among the movements that participate in it.

Sen, J. and P. Waterman (eds) (2008) *World Social Forum: Challenging empires*, second edition, Montreal: Black Rose Books.

An edited collection bringing together key essays and documents from the early years of the WSF. An excellent introduction to the main controversies surrounding the Forum; gives a clear sense of the richness of scholarly and activist debates.

Stephansen, H. C. (2019) 'Conceptualizing a Distributed, Multi-scalar Global Public Sphere through Activist Communication Practices in the World Social Forum', *Global Media and Communication* 15(3): 345–360.

Draws on ethnographic research on activist communication practices in the WSF to develop an understanding of the global public sphere as an emergent formation made up of multiple, interlinked publics at different scales. Challenges contemporary theorizations of global publics as personalized networks and emphasizes the continued importance of collective communication spaces for actors at the margins of the global network society.

YOUTUBE

Abigail Keating

While widely promoted as a grassroots community with an emphasis on 'you' – the user, the content creator, the viewer –YouTube has, since its inception, fallen under the rubrics of two traditionally opposing logics: it is at once both an amateur space for individual expression and personal exhibition, and a commercial platform for professional, corporate media. Thus, to consider YouTube as a public space to which the citizen has access, and whose content 'you' can shape, we must evaluate the conditions under which this user-led activity plays out. As such, it is necessary to reflect on the site's mechanics, the ways in which it has democratized media space, the incentives it offers (explicitly or indirectly) to active citizens, and its disruption of traditional definitions of the amateur and the professional, but also of consumers and producers.

The first major scholarly works on YouTube, produced in 2009, addressed questions relating to its importance in the context of broader shifts in media and participatory culture. Burgess and Green (2009:15), for instance, argued that "YouTube clearly represent[s] a disruption to existing media business models and … a new site of media power". The website was also discussed through a series of metaphors widely used to stress YouTube's growing prominence and influence in society (Snickars and Vonderau 2009:13): it was explored as a tool of cultural memory, as an archive, as a laboratory and as a medium. Indeed, the various ways in which YouTube bears a resemblance to a number of established cultural institutions provided the main critical frameworks for early scholarly attempts to theorize this phenomenon (Snickars and Vonderau 2009:13). As the website has developed as part of new trends in digital culture, research has continued to explore the nature of the evolving relationship between 'you' and the platform, between the user and the broader media landscape, and between the micro and the macro of contemporary screen media and society.

This entry maps four major themes within research on the social media ecology of which YouTube is a significant part. Firstly, the question of anarchy and YouTube as a site of disruption is addressed, before we turn in the second section to consider YouTube as a site of conformity in which platform-specific trends and attributes have led to the standardization (or professionalization) of the amateur moving image. These themes are interrelated and do not always stand in contrast to one another. Indeed, to argue that YouTube is a site of disruption and/or conformity, we must view the cultural, aesthetic and thematic specificities of YouTube – its hosting and fostering of particular genres and categories of videos, of YouTubers and their followings, of viral videos and global phenomena – within the contexts of its broader media ecology. Thirdly, then, YouTube is considered within the realm of the private sphere, alongside contemporary definitions of personal media space. And finally, YouTube is addressed as a public sphere and as a space for promoting social justice, social change and citizen activism. Here, it is also important to keep in mind traditional distinctions between the private and the public spheres: the former associated with privacy and intimacy, the home, familial life, personal finances and emotions; the latter grounded in public discussion, debate, politics, society and mostly facilitated by mass media. We may ask to what extent YouTube contributes to the wider digital revision of these distinctions. The influencer and the vlogger (or video diarist) are two of the most significant figures to have emerged from this landscape and to have contributed to the erosion of the boundaries between private and public; similarly, we often see the blending and collapsing of traditional private–public distinctions in the context of social media activism and movements. As such, to theorize YouTube and its relevance for the study of citizen media, one must not only highlight its relationship with already established institutions (Snickars and Vonderau 2009:13), but also explore how it interacts with these traditional boundaries.

YouTube and disruption

To explore YouTube as a site of disruption, three areas must be considered: its beginnings and the ways in which its early development differed from the trajectories of other forms of screen media, its continuation of older practices within and beyond screen media, and its definitional revision of the spaces in which these practices transpire. In October 2006, 'A Message from Chad and Steve' was posted on YouTube to announce the website's acquisition by Google (YouTube Spotlight 2006). In the short video we see Chad Hurley and Steve Chen, two of the platform's three founders, inform viewers of the exciting news and express their commitment to the growth of YouTube as an innovative service. Here, the language used in the video reflects the somewhat utopian notion of YouTube as a community (Nielsen 2006), while the video's aesthetic mirrors YouTube's status as a user-generated, predominantly amateur platform: the lighting and sound are of poor quality, the camera is shaky, the dialogue appears unscripted. It is interesting to reflect on this video as a moment that was to foreshadow the website's dual identity as both a community and a business (Snickars and Vonderau 2009:11), but also as part of YouTube's broader function as a social media platform, in which the role of the citizen was to become a lot more multifaceted than merely that of content creator or consumer.

Of Jenkins' many insightful contributions to contemporary discourses on screen media culture, one especially pertinent point, when considering YouTube's early development, is his discussion on what came before October 2006. Here, Jenkins (2009:109) notes the novelty of "everyday people [being] able to seize the means of cultural production and distribution", thus reiterating the innovative qualities that have been attributed to YouTube within both academic and media debate. Yet, he also argues that "the emergence of participatory cultures of all kinds over the past several decades paved the way for the early embrace, quick adoption, and diverse use of such platforms" (ibid.). While YouTube may have appeared as an overnight success story, many of the communities creating and sharing content via this site – Jenkins (ibid.:110) suggests – were already in existence and practising some form of media output before the website's inception. For many decades prior to 2006, fan communities were publishing zines, activists and other citizen groups were engaged in television and radio, and amateurs were producing home videos. As Jenkins concludes, "YouTube may represent the epicenter of today's participatory culture but it doesn't represent its origin point for any of the cultural practices people associate with it" (ibid.). In light of these arguments, we might legitimately question to what extent we can truly think about YouTube as a site of innovation and disruption.

Scholarly work has already addressed many aspects of this issue by considering YouTube as a space of anarchy in which what might once have been considered throwaway pieces of culture may be broadcast: more than any previous technology, the platform has popularized the production of short viral videos where 'you' are at the centre of the frame and which potentially garner more views than any box office hit of its respective year (Keating 2013). Research has also explored YouTube as a site in which both bottom-up and top-down models for media production converge (Keating 2014). On the one hand, by engaging with YouTube, traditional television programmes and networks have been able to embrace amateur culture by disseminating for free already-broadcast clips, extras and behind-the-scenes footage via dedicated channels, encouraging fans to subscribe, like and comment on the content. Put simply, these are professional media's responses to the increasing popularity of participatory culture among the masses, and are reflective of contemporary shifts in (free) advertising and market research. At the same time, the aesthetics of amateur content have become increasingly standardized and professionalized through the popularity of YouTube How-To videos teaching citizens how to enhance their media productions according to established norms (Keating 2014). However, alongside Jenkins' (2009:110) argument concerning how YouTube facilitates already-existing cultural practices, it is also important to keep in mind that throughout film history the amateur has been encouraged to professionalize their output (Zimmermann 1995). Thus, in order to conceptualize YouTube as a site of disruption, one needs to go beyond the content of the platform itself and to consider it within the broader context of Internet media and the ways in which digital networked communications technologies facilitate disorder (or at least a reordering of the status quo) more generally.

YouTube and conformity

Paradoxically, the theme of disruption is fundamental to an understanding of YouTube as a site of conformity. Indeed, in the first edition of their foundational study of YouTube, Burgess and Green (2009:1) explore the societal impact of YouTube from a range of perspectives, underlining its place within media history and its role in a new era of digital innovation stemming from the work of "youthful visionaries working outside of established enterprises". Yet, like Jenkins (2009), they also remain cautious of defining the platform as a radical historical break from all that has come before it (Burgess and Green 2009:14). While they seek to foreground YouTube as a site of cultural and economic disruption, they simultaneously present the site as part of an ongoing process of co-evolution and negotiation between old and new media, between old and new practices (ibid.). The value and validity of this theoretical framework becomes especially clear when viewing YouTube videos through the prism of category or genre: from this perspective, YouTube videos are seen as artefacts created by "amateurs working outside the institutional structures of the television and movie industry" (Strangelove 2010:3), but which conform to broader trends in video production, and thus to older practices of genre within the screen media industries. For example, in another of the first major anthologies on YouTube (Snickars and Vonderau 2009), Peters and Seier's (2009) chapter explores YouTube dance videos within the broader context of relations between music and the moving image, before considering whether the actors in these videos are attempting to adhere to professional aesthetic and generic standards. While a certain degree of conformity to professional conventions is observed, Peters and Seier also suggest that it is just as arguable that the aesthetic standards of YouTube itself and its social media environment play an important, determining role in shaping the output of the site's citizen users. This influence is made clear through the interaction that follows the dissemination of a video (likes, comments, video responses), and through the presence of the digital device. The latter's influence can be exerted both implicitly, given that the digital device is the essential tool enabling the video to be broadcast to the world, and diegetically, through various references to the YouTube community or to digital video culture more broadly. This can be seen, for example, in the form of the standard, jump-cut vlog when the vlogger interacts with the webcam that captures the action, or when we hear dialogical references to other videos or other YouTubers. Here, as Peters and Seier suggest, the digital device serves "as a medium of reference to existing texts, poses and videos already in circulation" (ibid.:193). Thus, an awareness of the culture of the platform that hosts these videos is made explicit. This argument plays out more tangibly when we consider the generic codes and aesthetics of broader, collective video trends: for instance, where videos are posted in response to other videos or to the general trend itself. This was the case with Harlem Shake, a video trend that went viral in 2013, where people were filmed in everyday settings suddenly starting to dance and gyrate at the beat-drop of the extra-diegetic song playing on the soundtrack (which featured the lyrics 'then do the Harlem Shake'). Further examples include the Ice Bucket Challenge, a video 'challenge' that went viral in 2014 in which participants have someone pour a bucket of ice water over their heads as part of a campaign to raise awareness for the disease ALS (amyotrophic lateral sclerosis); as well as the more general trend of 'reaction' videos, wherein a viewer's immediate reaction to a film trailer, music video, sporting event or any other piece of media is recorded and uploaded, quite often contributing to a fandom or online community.

It is, however, limiting to argue that YouTube represents a site of conformity solely on the basis of textual coding across videos and trends, particularly when considering the evolution of that commerce–community negotiation that we see in and since 'A Message from Chad and Steve' (YouTube Spotlight 2006). We also see another of YouTube's similarities with television, in both its traditional and evolved guises, through its use of mandatory and skippable advertisements. Moreover, through the rise of the YouTube celebrity and social media influencer, we see the website's conformity to broader neoliberal trends driving the commodification of the self.

YouTube as private sphere

The individualization of society brought about by neoliberalism is a particularly interesting starting point when considering YouTube's relation to the private and public spheres, its contribution to the evolution of both, and to their increasingly blurred boundaries. In their article on self-branding and the rise of the social media influencer, Khamis et al. (2017) argue that there are three main reasons why the culture of influencing has become so significant a phenomenon: firstly, they contend that influencing has emerged as a result of social media platforms like YouTube and the promises they bring of fame for ordinary citizens; secondly, they suggest that, within a climate of political neoliberalism, the potential rewards of self-branding are promoted; and, lastly, they highlight that the formulae according to which already established social media influencers have attained a degree of micro-celebrity appear to be replicable (ibid.:194). Ultimately, they suggest, "the increased ease of projecting one's image through social media coupled with the rise of individualism has made the notion of self-branding more popular" (ibid.:194). Perhaps there is no greater symbol of this within the digital landscape than that which aims first and foremost to exhibit 'you'.

A cross-disciplinary consideration of contemporary individualism is worth pursuing when reflecting on YouTube as an interstitial space not only between community and commerce, but also between the private and the public spheres. Bauman, who has written extensively on modern and postmodern (or liquid modern) identity, society, media and culture, describes how the meaning of individualization has changed drastically over the last century: "[t]o cut a long story short", he summarizes, "there is a growing gap between individuality as fate and individuality as practical capacity for self-assertion" (2001:xvi). Over a decade later, Bauman (2014) further elaborates the root of this point in a treatise on what he terms the confessional society. In this confessional society, the private, Bauman argues, has colonized the public sphere. Whereas for much of the modern era it was the public that posed a threat to the private through attacks on free will and on the privacy of the citizen, it is now the private which saturates the public arena: personal secrets and security are currency to spend in return for the joys that the Internet can offer (ibid.). Amid this paradox is another layer of negotiation between the individual and community. Bauman (ibid.) describes the correlation between belonging and autonomy in this regard, where social media platforms are known for their facilitation of communication, the making of friends and the building of communities, while presenting the user with the choice of opting in whenever and however they want – through, for instance, anonymous connections, the manipulated presentation of self or fragmented communication.

The vlogger, video diarist and influencer on YouTube exist within these paradoxes. This transmedial ecology is no longer unchartered ground for researchers, in that already successful models exist with which to demonstrate how platforms like YouTube "accelerate and accentuate the means by which users can package, perform and sell a lucrative personal brand across several online sites" (Khamis et al. 2017:195). For Bauman, the mechanics of this ecology, and the ways in which it encourages the fast-paced flow of information and bite-size portions of media, and thus the fragmentation of identity, communication and information, present one of the main issues: "attention tends to be trained to skate over the surface much too fast for getting a glimpse of what hides beneath" (2014). Herein, it could be argued, lies one of the reasons for YouTube's success as a platform, which has, since its beginnings, fallen under the rubrics of conventionally opposing logics, without explicitly drawing attention to them. The site is consistently presented as 'your' platform, 'your' community, but capitalizes on much more than self-commodification.

This interstitiality is considered under a more optimistic framework in Papacharissi's *A Private Sphere: Democracy in a Digital Age* (2010), in which the plurality of spaces within the digital environment of contemporary democracies, both private and public, are explored. Here, the political sphere is examined in light of new civic habits adapting and responding to new realms of digital communication. Posting, liking or commenting on a political video on YouTube may be seen as a public act of expression, for instance, but doing so "stands as a private, digitally enabled intrusion on a public agenda determined by others" (Papacharissi 2010:131). In this context, the citizen is engaged, digitally connected and must negotiate the boundary between private and public for themselves.

YouTube as public sphere

In 2006, Jenkins et al. produced a white paper that sought from a pedagogical perspective to address the cultural competencies and social skills needed for full involvement in what they called the participatory culture of contemporary media. This participatory culture has "relatively low barriers to artistic expression and civic engagement" (Jenkins et al. 2006:3). The potential benefits of this new media context are vast, particularly in relation to the spread of human knowledge, the diversification of cultural expression and citizen empowerment (ibid.). The latter point is highlighted through the notion that politics, as usually facilitated by traditional media, has historically been a spectator sport. In the digital era, by contrast, social media, along with increased access to both information and media tools, have offered citizens opportunities for further engagement in political debates. As Jenkins et al. (ibid.:10) suggest, "we learn the skills of citizenship by becoming political actors". This point was fleshed out in more practical terms in 2011 when Jenkins commented on the visual rhetoric of the Occupy Wall Street movement. He notes that the movement's use of pop cultural imagery to spread its message did not merely serve as a way to present a more light-hearted, twenty-first-century strategy for political activism, but more significantly as a digital version of protest media that circulates across social platforms (Jenkins 2011). YouTube, as the world's largest moving image depository, has played a crucial role in contemporary iterations of both protest media and traditional political campaigning.

On the latter, one of the most notable milestones has been former US President Barack Obama's 2008 campaign. "Throughout the world", Losh (2012:256) writes, "government agencies have adopted YouTube as a mode for broadcasting state-sanctioned video messages" (ibid.:256), and many of them have emulated the Obama 2008 strategy. While the production of media content as part of a political campaign is not in itself new, Losh suggests that "[w]hat is new about Obama and his YouTube performances … could be the way his rhetoric draws attention to what could be called 'mediated transparency' or 'transparent mediation'" (ibid.:257). Here, the diegetic presence of the camera, lights, the computer and other digital tools in the images accompanying Obama's videos is significant, in that such features emphasize "the existence of an authentic rhetorical moment not mediated by the technological apparatus foregrounded in the scene" (ibid.:257). In this sense, a level of realism, and indeed familiarity among YouTube users, is achieved through the overt presence of mediation.

It is useful to consider this technique within the broader context of YouTube videos and mediation. If we reflect on the tendency towards transparent mediation in an incalculable number of YouTube home videos in which the digital apparatus or landscape of which the video is a part is diegetically or thematically present, it is worth recalling early theorizations of this aesthetic and its association with the concept of the collective. As Peters and Seier have proposed, "the actors are engaged with symbolic structures and mediation … establishing referential chains or a spreading network among video clips and commentaries" (2009:192–193). Many aesthetic, generic and thematic features of YouTube have now been established, and the theme of collectivity is present both within individual vlogs and in the website's own emphasis on community in its promotional materials. Collectivity is similarly evident in video productions like the YouTube Rewind series released at the end of each year, in which some of the most popular YouTubers and YouTube trends of the year are celebrated.

Yet, importantly, under the framework of the public sphere, the referential chains of YouTube go beyond aesthetic and generic coding within the website itself, and stretch across media sites whose own micro-ecologies add and accentuate meaning. Just as we have seen with the Occupy Wall Street movement (2011), as well as during earlier uprisings in the Middle East and North Africa between 2009 and 2011, an especially significant phenomenon of transmediality can be found in online protest media, coinciding with a paradigm shift in social justice activism brought about by social media platforms and the ease with which content and information can be shared between them. While the Internet is often a free-for-all in both progressive and regressive ways, human rights issues and the social discourses around them have benefitted greatly from the interventionist capabilities of social media: such platforms have facilitated attempts to inform, educate and mobilize citizens cross-demographically, and to circumvent state-run media traditions and regulations, along with mass media's potential ties to specific sociopolitical ideologies, corporations or institutions. Moreover, strategies of corporeal protest associated with civil rights demonstrations and protests of the past take on new meaning when the body is (trans)mediated and exhibited globally, from the documentation of sociopolitical rallies to witness footage of human mistreatment. The rise of hashtivism, or hashtag activism, is a particularly powerful example in this context, and often strikingly more powerful when embedded with YouTube content providing social commentary and/or direct evidence of brutality, harassment or inequality. A significant embodiment of all three of these examples can be seen across Black Lives Matter, a powerful, cross-medial, social justice movement that, since 2013, has highlighted the mistreatment of black people within the US and beyond. This online movement has its roots in the use of the hashtag #BlackLivesMatter on social media in 2013, in response to the acquittal of George Zimmerman in the trial of the fatal shooting of African American teenager Trayvon Martin. The hashtag has continued to be used across social media platforms in response to injustices. While street demonstrations and marches have stemmed from the activism online, it is important to underline the movement's powerful use of social media as a tool of education, organization and evidence, such as the use of online video to document and circulate the mistreatment of and violence against African Americans in contemporary US society. Arguably, there is no greater YouTube paradox than when it is used as a political tool: on the one hand, because such practices represent a continuation of older forms of political campaigning, and on the other, because they offer an outlet for radical strategies in the arena of citizen-driven social justice.

Future directions

Since the website's inception, we have seen a notable diversification in the labels that can be attributed to the YouTube user: activist, campaigner, witness, diarist, celebrity, entertainer, social commentator, influencer. While the platform continues to expand beyond the social, an awareness of the constant negotiations between commerce and community, producer and consumer, local and global, public and private, remain fundamental to its mechanics and identity. Its contributions to the digital citizen media landscape are incomparable, yet they function under the framework of old and new, already established and perennially evolving practices.

Going forward, the increasingly political use of online platforms and media trends will continue to present fertile ground for the exploration of the paradoxes of YouTube, as discussed in this entry. We need only turn to the US general election of 2016, and beyond, to see how saturated with media content political campaigning has become (in both official and unofficial capacities). We see how memes and gifs – media trends conventionally featuring humorous content – have been used to spread political messages, how the darker problematics of social media have come to light in their facilitation of the spread of fake news, how the circulation of video has never had such a broad, global reach and how audiovisual content has played such a powerful role in the promotion of ideologies. In turn, future research on YouTube and the citizen must continue to look beyond the surface, as Bauman (2014) has warned, beyond the presentation of self, beyond 'you', and deeper into the mechanics of the platform, its role in the broader ecology of user-generated media and, most significantly, its role in democracy.

See also: content moderation and volunteer participation; media ecologies; public sphere; social media

Recommended reading

Jenkins, H. (2006) *Convergence Culture: Where old and new media collide*, New York: New York University Press.

Blends discussions of an array of topics and significant events under the rubric of media convergence in the twenty-first century. The notion of participatory culture is teased out in all its plurality and transmediality. An important text that makes broad sense of the cultural changes that are key to an understanding of a digital media ecology.

Snickars, P. and P. Vonderau (eds) (2009) *The YouTube Reader*, Stockholm: National Library of Sweden.

An important starting point for any research on YouTube. This edited collection is rich in content and theoretical scope, reading YouTube in aesthetic, political and cultural capacities and drawing on the works of a number of key screen media scholars.

Tufekci, Z. (2017) *Twitter and Tear Gas: The power and fragility of networked protest*, New Haven, CT: Yale University Press.

An analysis of contemporary protest with, crucially, a transnational focus. It tracks the history of online political activism through a number of globally important events and uprisings, and across the most powerful online platforms, including YouTube, Facebook and Twitter.

BIBLIOGRAPHY

15Mpedia (2016) 'Acampada Sol. Organización'. Available online: https://15mpedia.org/wiki/Acampada_Sol#Organizaci.C3.B3n.

18 Days in Egypt http://beta.18daysinegypt.com/.

858: An Archive of Resistance. https://858.ma.

A Wall Is a Screen (n.d.) 'Project Homepage'. Available online: www.awallisascreen.com/.

Aarseth, E. (2014) 'I Fought the Law: Transgressive play and the implied player', in N. Segal and D. Koleva (eds) *From Literature to Cultural Literacy*, Basingstoke, UK: Palgrave Macmillan, 180–188.

Abah, A. L. (2009) 'Popular Culture and Social Change in Africa: The case of the Nigerian video industry', *Media, Culture & Society* 31(5): 731–748.

Abaza, M. (2011) 'Cairo's Downtown Imagined: Dubaisation or nostalgia', *Urban Studies Journal* 48(6): 1075–1087.

Abaza, M. (2013a) 'Mourning, Narratives and Interactions with the Martyrs through Cairo's Graffiti', E-International Relations, 7 October. Available online: www.e-ir.info/2013/10/07/mourning-narratives-and-interactions-with-the-martyrs-through-cairos-graffiti.

Abaza, M. (2013b) 'Walls, Segregating Downtown Cairo and the Mohammed Mahmud Street Graffiti', *Theory, Culture & Society* 30(1): 122–139.

Abaza, M. (2017) 'Repetitive Repertoires: How writing about Cairene graffiti has turned into a serial monotony', in K. Avramidis and M. Tsilimpounidi (eds) *Graffiti and Street Art: Reading, writing and representing the city*, London: Routledge, 177–194.

Abbas, B. and R. Abou-Rahme (2013) 'The Archival Multitude', *Journal of Visual Culture* 12(3): 345–363.

Abdelrahman, M. (2014) *Egypt's Long Revolution: Protest movements and uprisings*, London and New York: Routledge.

Abdulhadi, R. (1998) 'The Palestinian Women's Autonomous Movement: Emergence, dynamics, and challenges', *Gender and Society* 12(6): 649–673.

Abercrombie, N. and B. Longhurst (1998) *Audiences: A sociological theory of performance and imagination*, London: Sage.

Abidin, C. (2016) '"Aren't These Just Young, Rich Women Doing Vain Things Online?": Influencer selfies as subversive frivolity', *Social Media + Society* 2(2): 1–17.

Abrams, J. J. (dir.) (2015) Star Wars: The force awakens. IMDb entry: www.imdb.com/title/tt2488496/?ref_=fn_al_tt_1.

Abu-Lughod, L. (1986) *Veiled Sentiments: Honor and poetry in a Bedouin society*, Berkeley, CA: University of California Press.

Abu-Lughod, L. (1990) 'The Romance of Resistance: Tracing transformations of power through Bedouin women', *American Ethnologist* 17(1): 41–55.

Actipedia (2017) 'Homepage'. Available online: https://actipedia.org/.

Actipedia (n.d.) 'Create Project'. Available online: https://actipedia.org/node/add/action.

Adam, A. (2005) *Gender, Ethics and Information Technology*, London: Palgrave Macmillan.

Adams, P. C., J. Cupples, K. Glynn, A. Jansson and S. Moores (2017) *Communication/Media/Geographies*, New York and London: Routledge.

Adams, T. and D. Fuller (2006) 'The Words Have Changed but the Ideology Remains the Same: Misogynistic lyrics in rap music', *Journal of Black Studies* 36(6): 938–957.

Aday, S., H. Farrell, D. Freelon, M. Lynch, J. Sides and M. Dewar (2013) 'Watching from Afar: Media consumption patterns around the Arab Spring', *American Behavioral Scientist* 57: 899–919.

Adbusters (2011) '#OCCUPYWALLSTREET', Adbusters.org, 13 July. Available online: https://web.archive.org/web/20120317105025/http://www.adbusters.org/blogs/adbusters-blog/occupywallstreet.html.

Adbusters Media Foundation (2019) *Adbusters: Journal of the Mental Environment*. www.adbusters.org/.

Addams, J. (1907) 'Introduction', in J. Addams (ed.) *Newer Ideals of Peace*, New York: Macmillan, 3–30. Available online: https://brocku.ca/MeadProject/Addams/Addams_1907/Addams_1907_1.html.

Adkins, L. and C. Lury (eds) (2012) *Measure and Value*, Malden, MA: Wiley-Blackwell.

Adobe (1996) Adobe Flash. Available online: www.adobe.com/products/flashruntimes.html

Adorno, T. (1941) 'On Popular Music', *Studies in Philosophy and Social Sciences* 9(1): 17–48.

Adorno, T. (1951/2005) *Minima Moralia: Reflections from a damaged life*, translated by E. F. N. Jephcott, London: Verso.

Adorno, T. and M. Horkheimer (1944/1979) *Dialectic of Enlightenment*, translated by J. Cummings, London: Verso.

Adorno, T. and M. Horkheimer (1944/1997) *Dialectic of Enlightenment*, translated by J. Cummings, second edition, London: Verso.

Agamben, G. (1996/2000) *Means without End: Notes on politics*, translated by V. Binetti and C. Casarino, Minneapolis, MN: University of Minnesota Press.

Agamben, G. (1998/2000) *Remnants of Auschwitz: The witness and the archive*, translated by D. Heller-Roazen, New York: Zone Books.

Agathagnelou, A. and L. Ling (2004) 'Power, Borders, Security, Wealth: Lessons of violence and desire from September 11', *International Studies Quarterly* 48: 517–538

Agier, M. (2013/2016) *Borderlands:* Towards and anthropology of the cosmopolitan condition, translated by M. Fernbach, Cambridge: Polity Press.

Aguayo, A. J. (2006) 'Activist Filmmaking', in I. Aitken (ed.) *Encyclopedia of the Documentary Film*, Volume 1, London and New York: Routledge, 6–9.

Aguayo, A. J. (2019) *Documentary Resistance: Social change and participatory media*, New York: Oxford University Press.

Ahmed, S. (2011) *On Being Included: Racism and diversity in institutional life*, Durham, NC: Duke University Press.

Aiello, L. M., A. Barrat, R. Schifanella, C. Cattuto, B. Markines and F. Menczer (2012) 'Friendship Prediction and Homophily in Social Media', *ACM Transactions on the Web* 6(2): 1–33.

Aites, A., A. Ewell, N. Krstic and L. Read (dir.) (2013) 99%: The Occupy Wall Street Collaborative Film. IMDb entry: www.imdb.com/title/tt2201488/.

Akomfrah, J. (dir.) (2013) The Stuart Hall Project, British Film Institute (BFI) Films.

Alabed, B. (2016) Twitter account. Available online: https://twitter.com/alabedbana?lang=en.

Alacovska, A. (2017) 'The History of Participatory Practices: Rethinking media genres in the history of user-generated content in 19th-century travel guidebooks', *Media, Culture & Society* 39(5): 661–679.

Alaimo, K. (2015) 'How the Facebook Arabic Page "We are all Khaled Said" Helped Promote the Egyptian Revolution', *Social Media + Society*, July–December: 1–10.

Albury, K. and P. Byron (2016) 'Safe on My Phone? Same-sex attracted young people's negotiations of intimacy, visibility, and risk on digital hook-up apps', *Social Media + Society* 2(4). Available online: https://journals.sagepub.com/doi/epub/10.1177/2056305116672887.

Alcoff, L. (1991) 'The Problem of Speaking for Others', *Cultural Critique* 20(5): 5–32.

Alderman, L. (2014) 'Across Athens, Graffiti Worth a Thousand Worlds of Malaise', *The New York Times*, 16 April. Available online: www.nytimes.com/2014/04/16/world/europe/across-athens-graffiti-worth-a-thousand-words-of-malaise.html?_r=0.

Alessandrin, P. (dir.) (2009) District 13: Ultimatum. IMDb entry: www.imdb.com/title/tt1247640.

Alexa (2018a) 'The Top 500 Sites on the Web'. Available online: www.alexa.com/topsites.

Alexa (2018b) 'Top Sites in United States'. Available online: www.alexa.com/topsites/countries/US.

Alexander, B. (2011) *The New Digital Storytelling: Creating narratives with new media*, Santa Barbara, CA: ABC-CLIO.

Alexander, C., R. Edwards and B. Temple (2007) 'Contesting Cultural Communities: Language, ethnicity and citizenship in Britain', *Journal of Ethnic and Migration Studies* 33(5): 783–800.

Alexander, J. C. and R. N. Jacobs (1998) 'Mass Communication, Ritual and Civil Society', in T. Liebes and J. Curran (eds) *Media, Ritual and Identity*, London: Routledge, 23–41.

Alexander, W. (1981) *Film on the Left: American documentary film from 1931 to 1942*, Princeton, NJ: Princeton University Press.

Alexander, W. (2013) 'A Chat with Some Immoral Hackers Who Don't Care about Your Feelings', *Vice*, 24 June. Available online: www.vice.com/en_us/article/znqpj5/i-spoke-to-some-black-hat-hackers-about-internet-ethics.

Alfaro, R. M. (1985) *Los altoparlantes. Recuperación de la palabra popular*, Mimeo, Calandria.

Alfaro, R. M. (1986) 'Talleres de dramatización popular: Educar desde los sujetos y en los procesos', *Revista Tarea* 15. Available online: www.ts.ucr.ac.cr/binarios/accioncritica/ac-cr-019-06.pdf.

Alfaro, R. M. (1988) *De la conquista de la ciudad a la apropiación de la palabra*, second edition, Lima: Calandria Tarea.

Ali, C. (2017) *Media Localism: The policies of place*, Urbana, IL: University of Illinois Press.

Ali, S. R. and S. Fahmy (2013) 'Gatekeeping and Citizen Journalism: The use of social media during the recent uprisings in Iran, Egypt, and Libya', *Media, War and Conflict* 6(1): 55–69.

Alim, S. H., A. Ibrahim and A. Pennycook (2009) *Global Linguistic Flows: Hip hop cultures, youth identities and the politics of language*, London: Routledge.

Alinejad, D. (2011) 'Mapping Homelands through Virtual Spaces: Transnational embodiment and Iranian diaspora bloggers', *Global Networks* 11(1): 43–62.

Allagui, I. (2014) 'Waiting for Spring: Arab resistance and change', *International Journal of Communication* 8: 983–1007.

Allan, S. (2006) *Online News: Journalism and the Internet*, Maidenhead, UK: Open University Press.

Allan, S. (2009) 'Histories of Citizen Journalism', in S. Allan and E. Thorsen (eds) *Citizen Journalism: Global perspectives*, Volume 1, New York: Peter Lang, 17–32.

Allan, S. (2010) *News Culture*, third edition, Buckingham, UK: Open University Press.

Allan, S. (2013) *Citizen Witnessing: Revisioning journalism in times of crisis*, Cambridge: Polity Press.

Allan, S. (2014) 'Witnessing in Crisis: Photo-reportage of terror attacks in Boston and London', *Media, War & Conflict* 7(2): 133–151.

Allan, S. and E. Thorsen (eds) (2009) *Citizen Journalism: Global perspectives*, Volume 1, New York: Peter Lang.

Allen, B. L. (2000) 'The Popular Geography of Illness in the Industrial Corridor', in C. E. Colten (ed.) *Transforming New Orleans and Its Environs: Centuries of change*, Pittsburgh, PA: University of Pittsburgh Press, 178–201.

Alloco, K. (2018) *Videocracy: How YouTube is changing the world*, New York: Bloomsbury.

Alooshe, N. (2011) 'Zenga Zenga Song, Libya'. Available online: www.youtube.com/watch?v=cBY-0n4esNY.

Alphen, E. van (2014) *Staging the Archive: Art and photography in the age of new media*, London: Reaktion Books.

Alridge, D. (2005) 'From Civil Rights to Hip Hop: Toward a nexus of ideas', *The Journal of African American History* 90(3): 226–252.

Altheide, D. L. and R. P. Snow (1979) *Media Logic*, Beverly Hills, CA: Sage.

Althusser, L. (1965/1969) *For Marx*, translated by B. Brewster, London: Allen Lane.

Althusser, L. (1971) *Lenin and Philosophy and Other Essays*, translated by B. Brewster, London: New Left Books.

Ameel, L. and S. Tani (2012) 'Parkour: Creating loose spaces?', *Geografiska Annaler: Series B, Human Geography* 94(1): 17–30.

Amerika, M. (2011) 'Overview: Source material everywhere', Remixthebook. Available online: www.remixthebook.com/the-course/collagereadymadescombines.

Amin, A. (2008) 'Collective Culture and Urban Public Space', *City* 12(1): 5–24.

Amnesty International (2016) *The Human Cost of 'Crushing' the Market: Criminalization of sex work in Norway*, London: Amnesty International.

Amsler, S. S. (2015) *The Education of Radical Democracy*, New York: Routledge.

Andén-Papadopoulos, K. (2014a) 'Citizen Camera-Witnessing: Embodied political dissent in the age of mediated mass self-communication', *New Media & Society* 16(5): 753–769.

Andén-Papadopoulos, K. (2014b) 'Citizen Camera-Witnessing: Crisis testimony in the age of mediated mass self-communication', *New Media & Society* 16(5): 117–133.

Andén-Papadopoulos, K. and M. Pantti (eds) (2011) *Amateur Images and Global News*, Bristol, UK: Intellect.

Anderson, B. (1991) *Imagined Communities: Reflections on the origin and spread of nationalism*, London: Verso.

Anderson, B. (2010) 'Preemption, Precaution, Preparedness: Anticipatory action and future geographies', *Progress in Human Geography* 34(6): 777–798.

Anderson, C. W. (2011) 'From Indymedia to Demand Media: Journalism's visions of its audience and the horizons of democracy', in M. Mandiberg (ed.) *The Social Media Reader*, New York: New York University Press, 77–99.

Anderson, C. W. (2018) *Apostles of Certainty: Data journalism and the politics of doubt*, Oxford: Oxford University Press.

Anderson, J., M. Stender, S. M. West and J. C. York (2016) 'Unfriending Censorship: Insights from four months of crowdsourced data on social media censorship'. Available online: https://onlinecensorship.org/news-and-analysis/onlinecensorship-org-launches-first-report-download.

Anderson, K. M. and B. Wallace (2015) 'Digital Storytelling as a Trauma Narrative Intervention for Children Exposed to Domestic Violence', in J. L. Cohen, J. L. Johnson and P. Orr (eds) *Film and Video-based Therapy*, New York: Routledge, 95–107.

Anderson, M. and A. Perrin (2017) *Disabled Americans Are Less Likely to Use Technology*, Washington, DC: Pew Research Center. Available online: www.pewresearch.org/fact-tank/2017/04/07/disabled-americans-are-less-likely-to-use-technology/.

Anderson, P. (2011) 'Lucio Magri', *New Left Review* 72: 111–121.

Anderson, P.T. (dir.) (2007) There Will Be Blood. IMDb entry: www.imdb.com/title/tt0469494/?ref_=fn_al_tt_1.

Andrejevic, M. (2010) 'Social Network Exploitation', in Z. Papacharissi (ed.) *A Networked Self: Identity, community, and culture on social network sites*, New York and London: Routledge, 82–101.

Andrejevic, M. (2017) 'To Preempt a Thief', *International Journal of Communication* 11: 879–896.

Andreotti, V. and E. Dowling (2004) 'WSF, Ethics and Pedagogy', *International Social Science Journal* 56(182): 605–613.

Andretta, M. and N. Doerr (2007) 'Imagining Europe: Internal and external non-state actors at the European crossroads', *European Foreign Affairs Review* 12: 385–400.

Andrew X (1999/2005) 'Give Up Activism', libcom.org. Available online: http://libcom.org/library/give-up-activism.

Andrijasevic, R., C. Aradau, J. Huysmans and V. Squire (2012) 'European Citizenship Unbound: Sex work, mobility, mobilisation', *Environment and Planning D: Society and Space* 30: 497–514.

Andro, M. (2018) *Digital Libraries and Crowdsourcing*, London: Wiley.

Androutsopoulos, J. and A. Scholz (2003) 'Spaghetti Funk: Appropriations of hip-hop culture and rap music in Europe', *Popular Music and Society* 26(4): 463–479.

Ang, I. (1996) *Living Room Wars*, London: Routledge.

Angel, J. (2011) *Ciné Parkour*, London: CreateSpace Independent Publishing Platform.

Angel, J. (2016) *Breaking the Jump: The secret story of parkour's high-flying rebellion*, London: Aurum Press.

Angelis, M. De (2005) 'PR Like Process! Strategy from the bottom-up', *Ephemera* 5(2): 193–204. Available online: www.ephemerajournal.org/sites/default/files/5-2deangelis.pdf.

Angelis, M. De (2012) 'Crises, Capital, and Co-optation: Does capital need a commons fix?', in D. Bollier and S. Helfrich (eds) *The Wealth of the Commons: A world beyond market and state*, Amherst, MA: Levellers Press, 184–191. Available online: http://wealthofthecommons.org/essay/crises-capital-and-co-optation-does-capital-need-commons-fix.

Angwin, J., T. Parris and S. Mattu (2016) 'Breaking the Black Box: What Facebook knows about you', *ProPublica*, 28 September. Available online: www.propublica.org/article/breaking-the-black-box-what-facebook-knows-about-you.

Anstead, N. and A. Chadwick (2010) 'Parties, Election Campaigning, and the Internet: Towards a comparative institutional approach', in A. Chadwick (ed.) *The Routledge Handbook of Internet Politics*, London and New York: Routledge, 56–71.

Anthropy, A. (2012a) Dys4ia. IGDB entry: www.igdb.com/games/dys4ia.

Anthropy, A. (2012b) *Rise of the Videogame Zinesters: How freaks, normals, amateurs, artists, dreamers, drop-outs, queers, housewives, and people like you are taking back an art form*, New York: Seven Stories Press.

Antony, M. G. and R. J. Thomas (2010) '"This Is Citizen Journalism at Its Finest": YouTube and the public sphere in the Oscar Grant shooting incident', *New Media & Society* 12(8): 1280–1296.

Anzaldúa, G. (1990) 'Haciendo caras, una entrada / An Introduction', in G. Anzaldúa (ed.) *Making Face, Making Soul / Haciendo Caras*, San Francisco, CA: Aunte Lute Books, xv–xxviii.

Aouragh, M., S. Gürses, J. Rocha and F. Snelting (2015) 'Let's First Get Things Done! On division of labour and techno-political practices of delegation in times of crisis', *The Fibreculture Journal* 26: 208–235.

AP (2017) 'Protests Erupt at Airports Nationwide over Immigration Action', *CBS News*, 29 January. Available online: www.cbsnews.com/news/protests-airports-immigration-action-president-trump/.

Appadurai, A. (1990) 'Disjuncture and Difference in the Global Cultural Economy', *Theory, Culture, Society* 7: 295–310.

Appadurai, A. (2003) 'Archive and Aspiration', in J. Brouwer and A. Mulder (eds) *Information Is Alive*, Rotterdam: V2 Publishing/NAI Publishers, 14–25.

Appalshop. www.appalshop.org/.

Appel, V., H. Boulanger and L. Massou (2010) *Les dispositifs de l'information et de la communication: Concept, usages et objets*, Brussels: De Boeck.

Appelt, N. (2017) 'You Haven't Seen Anything' / 'I've Seen It All': Post 2011 Syrian documentaries and their frontiers', Bidayyat. Available online: http://bidayyat.org/opinions_article.php?id=175#.Wy1AixIzaHo.

Aradau, C., J. Huysmans, P. G. Macioti and V. Squire (2013) 'Mobility Interrogating Free Movement: Roma Acts of European Citizenship', in E. Isin (ed.) *Enacting European Citizenship*, Cambridge: Cambridge University Press, 132–154.

Aradau, C., J. Huysmans and V. Squire (2010) 'Acts of European Citizenship: A political sociology of mobility', *Journal of Common Market Studies* 48: 945–965.

Araujo, T., P. Neijens and R. Vliegenthart (2017) 'Getting the Word out on Twitter: The role of influentials, information brokers and strong ties in building word-of-mouth for brands', *International Journal of Advertising* 36(3): 496–513.

Ardévol, E. and E. Gómez Cruz (2013) 'Digital Ethnography and Media Practices', *The International Encyclopedia of Media Studies*, Wiley-Blackwell.

Ardévol, E., A. Roig, G. San Cornelio, R. Pagès and P. Alsina (2010). 'Playful Practices: Theorising "new media" cultural production', in B. Bräuchler and J. Postill (eds) *Theorising Media and Practice*, New York: Berghahn Books, 259–279.

Arendt, H. (1958/1998) *The Human Condition,* second edition, Chicago, IL: University of Chicago Press.

Arendt, H. (1972) *Crises of the Republic: Lying in politics; Civil disobedience; On violence; Thoughts on politics and revolution,* Orlando, FL: Mariner Books.

Armbrust, W. (1998) 'When the Lights Go Down in Cairo: Cinema as secular ritual', *Visual Anthropology* 10(2–4): 413–442.

Arnold, M. (1960) *Culture and Anarchy,* London: Cambridge University Press.

Arnold, M. (1960–77) *Complete Prose Works,* Volume 3, Ann Arbor, MI: University of Michigan Press.

Article 19 (2013) *The Right to Blog,* Policy Brief, London: Article 19.

Artikişler Collective (eds) (2016) *Autonomous Archiving,* Barcelona: dpr-barcelona.

Arvidsson, A. (2006) *Brands: Meaning and value in media culture,* London and New York: Routledge.

Arvidsson, A. (2016) 'Facebook and Finance: On the social logic of the derivative', *Theory, Culture & Society* 33(6): 3–23.

Arvidsson, A. and E. Colleoni (2012) 'Value in Informational Capitalism and on the Internet', *Information Society* 28(3): 135–150.

Asad, T. (1994) 'From the History of Colonial Anthropology to the Anthropology of Western Hegemony', in G. W. Stocking, Jr. (ed.) *Colonial Situations: Essays on the contextualization of ethnographic knowledge,* Madison, WI: University of Wisconsin Press, 314–324.

Asenbaum, H. (2017) 'Cyborg Activism: Exploring the reconfigurations of democratic subjectivity in Anonymous', *New Media & Society* 20(4): 1543–1563.

Ashuri, T. and A. Pinchevski (2009) 'Witnessing as a Field', in P. Frosh and A. Pinchevski (eds) *Media Witnessing,* London: Palgrave Macmillan, 133–157.

Askanius, T. (2013) 'Online Activism and Political Mashup Genres', *JOMEC Journal of Journalism, Media and Cultural Studies* 4. Available online: https://jomec.cardiff universitypress.org/articles/abstract/10.18573/j.2013.10257/.

Askanius, T. (2014) 'Video for Change', in K. G. Wilkins, T. Tufte and R. Obregon (eds) *Handbook on Development Communication and Social Change,* Malden, MA: Wiley-Blackwell, 453–470.

Askanius, T. (2018) '(Social) Media Time, Connective Memory and Activist Television Histories: The case of *TV Stop*', in M. Mortensen, C. Neumeyer and T. Poell (eds) *Social Media Materialities and Protest: Critical reflections,* London: Routledge, 59–71.

Askanius, T. and N. Gustafsson (2010) 'Mainstreaming the Alternative: The changing media practices of protest movements', *Interface* 2(2): 23–41.

Askanius, T. and J. Uldam (2011) 'Online Social Media for Radical Politics: Climate change activism on YouTube', *International Journal of Electronic Governance* 4(1/2): 69–84.

Aslama, M. and M. Pantti (2006) 'Talking Alone: Reality TV, emotions and authenticity', *European Journal of Cultural Studies* 9(2): 167–184

Assange, J., J. Appelbaum, A. Muller Maguhn and J. Zimmermann (2012) *Cypherpunks: Freedom and the future of the Internet,* New York: OR Books.

Assman, A. (2010) 'Canon and Archive', in A. Erll and A. Nunning (eds) *A Companion to Cultural Memory Studies,* New York: de Gruyter, 97–108.

Association for Historical Dialogue and Research (2011) *Thinking Historically about Missing Persons: A guide for teachers, 4. Missing persons in Cyprus,* AHDR. Available online: www.ahdr.info/ckfinder/userfiles/files/MISSING%20PPL_S4.pdf.

Association of Internet Researchers (2012) 'Ethical Decision-Making and Internet Research'. Available online: http://aoir.org/reports/ethics2.pdf.

Aston, J., S. Gaudenzi and M. Rose (2017) *i-Docs: The evolving practices of interactive documentary,* London and New York: Wallflower Press.

Atkinson, J. D. (2017) *Journey into Social Activism: Qualitative approaches,* New York: Fordham University Press.

Atkinson, M. (2009) 'Parkour, Anarcho-environmentalism and Poiesis', *Journal of Sport and Social Issues* 33(2): 169–194.

Atkinson, M. (2012) 'Free Running: Post-sport liminality in a neoliberal world', in D. Andrews and M. Silk (eds), *Sport and Neo-Liberalism,* Philadelphia, PA: Temple University Press, 193–208.

Atkinson, M. (2013a) 'Heidegger, Free Running and the Essence of Being', in B. Carrington and D. Andrews (eds) *Blackwell Companion to Sport,* Hoboken, NJ: Blackwell, 359–374.

Atkinson, M. (2013b) 'Norbert Elias: The quest for excitement in parkour', in E. Pike and S. Beames (eds) *Outdoor Adventure and Social Theory,* London: Taylor & Francis, 69–81.

Atton, C. (2002) *Alternative Media,* London: Sage.

Atton, C. (2004) *An Alternative Internet: Radical media, politics and creativity,* Edinburgh: Edinburgh University Press.

Atton, C. (2006) 'Far-Right Media on the Internet: Culture, discourse and power', *New Media & Society* 8(4): 573–587.

Atton, C. (2008/2015) 'Citizen Journalism', in W. Donsbach (ed.) *The International Encyclopedia of Communication,* Wiley Online Library. Available online: https://onlinelibrary.wiley.com/doi/10.1002/9781405186407.wbiecc027.pub3.

Atton, C. (2009) 'Alternative and Citizen Journalism', in K. Wahl-Jorgensen and T. Hanitzsch (eds) *Handbook of Journalism Studies*, New York: Routledge, 265–278.

Atton, C. (2015) *The Routledge Companion to Alternative and Community Media*, New York: Routledge.

Atton, C. and J. F. Hamilton (2008) *Alternative Journalism*, London: Sage.

Au, W. (2005) 'Fresh out of School: Rap music's discursive battle with education', *The Journal of Negro Education* 74(3): 210–220.

Aufderheide, P. (1992) 'Grassroots Video in Latin America', unpublished paper presented at the *International Communication Association Annual Convention*, Miami, Florida.

Aufderheide, P. (1993) 'Film and Video in the Cultural Struggle of Latin America', *Media Development* 40(1): 30–31.

Aufderheide, P. (1995) 'The Video in the Villages Project: Videomaking with and by Brazilian Indians', *Visual Anthropology Review* 11(2): 83–92.

Augé, M. (1995) *Non-Places: Introduction to an anthropology of supermodernity*, London: Verso.

Aukta, S. (2018) '10 Types of Hackers You Should Know', MalwareFox, 13 June. Available online: www.malwarefox.com/types-of-hackers/.

Auletta, K. (2011) 'You've Got News', *The New Yorker*. Available online: www.newyorker.com/magazine/2011/01/24/youve-got-news.

Austin, B. (2017) 'Wireless in the Boer War', *IEEE Electromagnetic Compatibility Magazine* 6(1): 30–35.

Austin, J. (2001) *Taking the Train: How graffiti art became an urban crisis in New York City*, New York: Columbia University Press.

Austin, J. L. (1962) *How to Do Things with Words*, Oxford: Oxford University Press.

Austin, J. L. (1970) 'Performative Utterances', in J. O. Urmson and G. J. Warnock (eds) *Philosophical Papers*, second edition, Oxford: Oxford University Press, 233–252.

Avaaz.org (2017) 'Privacy Policy'. Available online: www.avaaz.org/en/privacy/.

Aviram, A. (2016) 'Revealed: Bristol's police and mass mobile phone surveillance', *Bristol Cable*, 10 October. Available online: https://thebristolcable.org/2016/10/imsi/.

Avramidis, K. (2012) '"Live Your Greece in Myths": Reading the crisis on Athens' walls', Professional Dreamers, Working paper no.8. Available online: www.professionaldreamers.net/_prowp/wp-content/uploads/Avramides-Reading-the-Crisison-Athens-walls-fld.pdf.

Avramidis, K. (2014) 'Mapping the Geographical and Spatial Characteristics of Politicized Urban Art in the Athens of Crisis', in M. Tsilimpounidi and A. Walsh (eds) *Remapping 'Crisis': A Guide to Athens*, London: Zero Books, 183–203.

Axford, B. (2011) 'Talk about a Revolution: Social media and the MENA uprisings', *Globalizations* 8(5): 681–686. Available online: www.tandfonline.com/doi/full/10.1080/14747731.2011.621281#.Uteqw_RdWed.

Axon, S. (2016) 'Unity at 10: For better – or worse – game development has never been easier'. Available online: https://arstechnica.com/gaming/2016/09/unity-at-10-for-better-or-worse-game-development-has-never-been-easier/.

Azoulay, A. (2007/2008) *The Civil Contract of Photography*, translated by R. Mazali and R. Danieli, New York: Zone Books.

Baack, S. (2018) 'Practically Engaged: The entanglements between data journalism and civic tech', *Digital Journalism* 6(6): 673–692.

Baca, M. (ed.) (2016) *Introduction to Metadata*, third edition, Los Angeles, CA: The Getty Research Institute.

Bacallao-Pino, L. M. (2016) 'Transmedia Events: Media coverage of global transactional repertoires of collective action', in B. Mitu and S. Poulakidakos (eds) *Media Events: A critical contemporary approach*, London: Palgrave Macmillan, 189–206.

Bachelard, G. (1938/1964) *The Psychoanalysis of Fire*, translated by A. Ross, Boston, MA: Beacon Press.

Bachelard, G. (1961) *La flamme d'une chandelle*, Paris: Presses Universitaires de France.

Backaler, J. (2018) 'Working with Influencers: Potential paths to take', in *Digital Influence*, Cham, Switzerland: Springer, 119–135.

Bacon-Smith, C. (1992) *Enterprising Women: Television fandom and the creation of popular myth*, Philadelphia, PA: University of Pennsylvania Press.

Bailard, C. S. and S. Livingston (2014) 'Crowdsourcing Accountability in a Nigerian Election', *Journal of Information Technology & Politics* (11)4: 349–367.

Bailey, O., B. Cammaerts and N. Carpentier (2008) *Understanding Alternative Media Issues in Cultural and Media Studies*, Maidenhead, UK and New York: McGraw-Hill and Open University Press.

Bailey, O., M. Georgiou and R. Harindranath (eds) (2007) *Transnational Lives and the Media, Re-imagining Diaspora*, Basingstoke, UK: Palgrave Macmillan.

Baines, D. (2012) 'Hyper-local News: A glue to hold rural communities together?', *Local Economy* 27(2): 152–166.

Baishya, A. (2015) '#NaMo: The political work of the selfie in the 2014 Indian general elections', *International Journal of Communications* 9: 1686–1700.

Bakardjieva, M. (2005) *Internet Society: The internet in everyday life*, London: Sage.

Bakardjieva, M. (2009) 'Subactivism: Lifeworld and politics in the age of the internet', *The Information Society* 25(2): 91–104.

Bakardjieva, M. (2012) 'Reconfiguring the Mediapolis: New media and civic agency', *New Media & Society* 14(1): 63–79.

Baker, G. (1999) 'The Taming of the Idea of Civil Society', *Democratization* 6: 1–29.

Baker, G. (2005) 'Hip Hop Revolucion! Nationalizing rap in Cuba', *Ethnomusicology* 49(3): 368–402.

Baker, M. (2016) 'The Prefigurative Politics of Translation in Place-Based Movements of Protest: Subtitling in the Egyptian Revolution', *The Translator* 22(1): 1–21.

Baker, M. and B. B. Blaagaard (2016a) 'Reconceptualizing Citizen Media: A preliminary charting of a complex domain', in M. Baker and B. B. Blaagaard (eds) *Citizen Media and Public Spaces: Diverse expressions of citizenship and dissent*, London and New York: Routledge, 1–22.

Baker, M. and B. B. Blaagaard (eds) (2016b) *Citizen Media and Public Spaces*, London: Routledge.

Bakhtin, M. (1965/1984) *Rabelais and His World*, translated by H. Iswolsky, Bloomington, IN: Indiana University Press.

Bakir, V. (2010) *Sousveillance, Media and Strategic Political Communication: Iraq, USA, UK*, New York: Continuum.

Bakir, V. (2015) '"Veillant Panoptic Assemblage": Mutual watching and resistance to mass surveillance after Snowden', *Media and Communication* 3(3): 12–25.

Bakker, P. and Y. Matras (eds) (2013) *Contact Languages: A comprehensive guide*, Boston, MA and Berlin: De Gruyter Mouton.

Baladi, L. (2016) 'Archiving a Revolution in the Digital Age: Archiving as an act of resistance', *Ibraaz*, 28 July. Available online: www.ibraaz.org/essays/163.

Balbi, G. (2010) 'Radio before Radio: Araldo Telefonico and the invention of Italian broadcasting', *Technology and Culture* 51(4): 786–808.

Balbi, G. (2017) 'Deconstructing "Media Convergence": A cultural history of the buzzword, 1980s–2010s', in S. Sparviero, C. Peil and G. Balbi (eds) *Media Convergence and Deconvergence*, Cham, Switzerland: Palgrave Macmillan, 31–51.

Baldé, C., F. Wang, R. Kuehr, and J. Huisman (2015) *The Global E-Waste Monitor 2014: Quantities, flows and resources*, Bonn, Germany: United Nations University, IAS- SCYCLE.

Balibar, É. (2012/2015) *Citizenship*, translated by T. Scott-Railton, Cambridge: Polity Press.

Balkin, J. M. (2016) 'Information Fiduciaries and the First Amendment', *UC Davis Law Review* 49(4): 1183–1234. Available online: https://ssrn.com/abstract=2675270.

Ball, J. (2009) 'FreeMix Radio: The original mixtape radio show: A case study in mixtape "radio" and emancipatory journalism', *Journal of Black Studies* 39(4): 614–634.

Ball, J. (2015) 'State-Media-Citizen Relations in the Surveillance Society', *Surveillance and Citizenship* conference, Cardiff University, June 18–19.

Banerjee, S. (2010) *Becoming Imperial Citizens: Indians in the late-Victorian empire*, Durham, NC: Duke University Press.

Banet-Weiser, S. (2012) *Authentic*™, New York and London: New York University Press.

Bank of the Commons (2019a) 'Vision and Mission'. Available online: https://bankofthecommons.coop/vision-and-mission/.

Bank of the Commons (2019b) 'Who We Are'. Available online: https://bankofthecommons.coop/who-we-are/.

Barabási, A. L. (2002) *Linked: The new science of networks*, Cambridge, MA: Perseus Publishing.

Barassi, V. (2013) 'When Materiality Counts: The social and political importance of activist magazines in Europe', *Global Media and Communication* 9(2): 135–151.

Barassi, V. (2015a) *Activism on the Web: Everyday struggles against digital capitalism*, New York: Routledge.

Barassi, V. (2015b) 'Social Media, Immediacy and the Time for Democracy: Critical reflections on social media as "temporalizing practices"', in L. Dencik and O. Leistert (eds) *Critical Perspectives on Social Media and Protest: Between control and emancipation*, London and New York: Rowman & Littlefield International, 73–90.

Barbero, I. (2012) 'Expanding Acts of Citizenship: The struggles of sinpapeles migrants', *Social and Legal Studies* 21: 529–547.

Barker, C. (2000) *Cultural Studies: Theory and practice*, London: Sage.

Barnard, S. R. (2018) 'Tweeting #Ferguson: Mediatized fields and the new activist journalist', *New Media & Society* 20(7): 2252–2271.

Barnes, B. (2016) *The Minority Body: A theory of disability*, Oxford: Oxford University Press.

Barry, A. (2001) *Political Machines: Governing a technological society*, London: The Athlone Press.

Bartesaghi, S. (dir.) (2013) Run. IMDb entry: www.imdb.com/title/tt2215457.

Barthes, R. (1957/1973) *Mythologies*, translated by A. Lavers, London: Paladin.

Bartlett, J. (2014) *The Dark Net: Inside the digital underworld*, London: William Heinemann.

Bartlett, R. A. (2007) *The World of Ham Radio, 1901–1950*, Jefferson, NC and London: McFarland & Company.

Bartlett, R. D. (2017a) 'How a Global Network of #FearlessCities is Making Racist Colonial Nation States Obsolete'. Available online: https://blog.p2pfoundation.net/global-network-fearlesscities-making-racist-colonial-nation-states-obsolete/2017/06/15.

Bartlett, R. D. (2017b) 'Minimum Viable Structure: Organisational scaffolding to get out of emergency mode'. Available online: https://medium.com/enspiral-tales/minimum-viable-structure-92e91048ff66.

Baruch, A., A. May and D. Yu (2016) 'The Motivations, Enablers and Barriers for Voluntary Participation in an Online Crowdsourcing Platform', *Computers in Human Behavior* 64: 923–931.

Bassel, L. (2008) 'Citizenship as Interpellation: Refugee women and the state', *Government and Opposition* 43: 293–314.

Bassel, L. and C. Lloyd (2011) 'Rupture or Reproduction? "New" citizenship in France', *French Politics* 9: 21–49.

Bastian, J. and B. Alexander (eds) (2009) *Community Archives: The shaping of memory*, London: Facet Publishing.

Bateman, M. (2012) 'Microcredit and Marikana: How they are linked', *The Star*, 18 September. Available online: www.iol.co.za/the-star/microcredit-and-marikana-how-they-are-linked-1385126.

Batiste, P. D. (2013) '0 Feet Away': The queer cartography of French gay men's geosocial media use', *Anthropological Journal of European Cultures* 22(2): 111–132.

Batty, M. (1993) 'The Geography of Cyberspace', *Environment and Planning B: Planning and Design* (20): 615–661.

Baú, V. (2015) 'Participatory Photography for Peace: Using images to open up dialogue after violence', *Journal of Peacebuilding & Development* 10(3): 74–88.

Baú, V. (2018) 'Media, Participation and Healing: Community-led content to overcome the trauma of conflict', *Online Journal of Art and Design* 6(2): 61–81.

Baudrillard, J. (1981/1983) *Simulations*, translated by P. Foss, P. Patton and P. Beitchman, New York: Semiotext(e).

Bauman, Z. (2001) 'Foreword: Individually together', in U. Beck and E. Beck-Gernsheim (eds) *Individualization: Institutionalized individualism and its social and political consequences*, London: Sage, xiv–xix.

Bauman, Z (2014) 'Privacy, Secrecy, Intimacy, Human Bonds—and Other Collateral Casualties of Liquid Modernity', *The Hedgehog Review*. Available online: https://iasc-culture.org/THR/THR_article_2011_Spring_Bauman.php.

Bauman, Z., D. Bigo, P. Esteves, E. Guild, V. Jabri, D. Lyon and R. B. J. Walker (2014) 'After Snowden: Rethinking the impact of surveillance', *International Political Sociology* 8(2): 121–144.

Bauman, Z. and D. Lyon (2012) *Liquid Surveillance: A conversation*, Cambridge: Polity Press.

Bauwens, M. (2005) 'Peer to Peer and Human Evolution', *Integral Visioning*, 15 June. Available online: http://z.agoravox.fr/IMG/P2PandHumanEvolV2.pdf.

Bayat, A. (2010) *Life as Politics: How ordinary people change the Middle East*, Amsterdam: Amsterdam University Press.

Bayerl, P. S. and L. Stoynov (2016) 'Revenge by Photoshop: Memefying police acts in the public dialogue about injustice', *New Media & Society* 18(6): 1006–1026.

Baym, N. K. (2000) *Tune In, Log On: Soaps, fandom, and online community*, Thousand Oaks, CA: Sage.

Baym, N. K. (2010) *Personal Connections in the Digital Age*, Cambridge, UK and Malden, MA: Polity Press.

Baym, N. K. (2015) *Personal Connections in the Digital Age*, second edition, Cambridge: Polity Press.

BBC (2013) 'Twitter Translates Tweets from Leading Egyptians', *BBC News*, 4 July. Available online: www.bbc.co.uk/news/technology-23179269

BBC News (2017) 'Fake Obama Created Using AI Tool to Make Phoney Speeches', *BBC Click*. Available online: www.bbc.com/news/av/technology-40598465/fake-obama-created-using-ai-tool-to-make-phoney-speeches.

Bear, J. and B. Collier (2016) 'Where Are the Women in Wikipedia? Understanding the different psychological experiences of men and women in Wikipedia', *Sex Roles* 74(5–6): 254–265.

Beck, S. (2016) *Appropriating Narrative of Conflict in Contemporary Verbatim Theatre: A practice-as-research investigation into the role of the playwright*, Unpublished PhD thesis, London: Goldsmiths University of London.

Beck, U. (2000) *The Brave New World of Work*, Cambridge: Polity Press.

Beck, U. and E. Beck-Gernsheim (2001) *Individualization*, London: Sage.

Beck, U. and M. Ritter (1992) *Risk Society: Towards a new modernity*, London: Sage.

Beck, U., J. Van Loon and B. Adam (2000) *The Risk Society and Beyond: Critical issues for social theory*, London: Sage.

Beckers, K. (2017) 'How Ordinary is the Ordinary (Wo)man on the Street? An analysis of vox pop characteristics in television news', *Journalism Practice* 11(8): 1026–1041.

Beckers, K., S. Walgrave and H. van den Bulck (2018) 'Opinion Balance in Vox Pop Television News', *Journalism Studies* 19(2): 284–296.

Beckett, C. and J. Ball (2013) *WikiLeaks: News in the networked era*, Cambridge, UK and Malden, MA: Polity Press.

Beer, D. (2016) *Metric Power*, London: Palgrave.

Beier, J. M. (2007) 'Grave Misgivings: Allegory, catharsis, composition', *Security Dialogue* 38(2): 251–269.

Bélair-Gagnon, V. and C. W. Anderson (2015) 'Citizen Media and Journalism', *The International Encyclopedia of Digital Communication and Society*, Wiley Online Library. Available online: https://onlinelibrary.wiley.com/doi/10.1002/9781118767771.wbiedcs028.

Bell, D. (1973) *The Coming of Post-Industrial Society*, New York: Basic Books.

Bell, D. (1976) *The Coming of Post-Industrial Society: A venture in social forecasting*, New York: Basic Books.

Bell, J. and D. Hartmann (2007) 'Diversity in Everyday Discourse: The cultural ambiguities and consequences of "happy talk"', *American Sociological Review* 72: 895–914.

Bell¿ngcat (n.d.) 'Bell¿ngcat: The home of online investigations'. Available online: www.bellingcat.com/.

Benecchi, E. (2015) 'Online Italian Fandoms of American TV shows', *Transformative Works and Cultures* 19.

Benedikt, M. (1991) 'Introduction', in M. Benedikt (ed.) *Cyberspace: First steps*, Cambridge, MA: The MIT Press, 1–26.

Bengtsson, S. (2006) 'Framing Space: Media and the intersection of work and leisure', in A. Jansson and J. Falkheimer (eds) *Geographies of Communication: The spatial turn in media studies*, Gothenburg: Nordicom, 189–204.

Ben Hassine, S. (2011) 'Tunisia's Youth Finally Has Revolution on Its Mind', *The Guardian*, 13 January. Available online: www.theguardian.com/commentisfree/2011/jan/13/tunisia-youth-revolution.

Beninger, K. (2017) 'Social Media Users' Views on the Ethics of Social Media Research', in L. Sloan and A. Quan-Haase (eds) *The Sage Handbook of Social Media Methods*, London: Sage, 57–73.

Benjamin, R. (2019) *Race after Technology*, Cambridge: Polity Press.

Benjamin, W. (1931/1972) 'A Short History of Photography', *Screen* 13(1): 5–26.

Benjamin, W. (1936/1969) 'The Work of Art in the Age of Mechanical Reproduction', in H. Arendt (ed.) *Illuminations*, translated by H. Zohn, New York: Schocken Books. Available online: http://web.mit.edu/allanmc/www/benjamin.pdf.

Benjamin, W. (1940/1999) 'Thesis on the Philosophy of History', in H. Arendt (ed.) *Illuminations*, translated by H. Zohn, London: Random House, 245–255.

Benkler, Y. (2002) 'Coase's Penguin, or, Linux and the Nature of the Firm', *The Yale Law Journal* 112: 369–446.

Benkler, Y. (2006) *The Wealth of Networks: How social production transforms markets and freedom*, New Haven, CT and London: Yale University Press.

Benkler, Y. (2011) 'A Free Irresponsible Press: Wikileaks and the battle over the soul of the networked fourth estate', *Harvard CR-CLL Review* 46: 311–397.

Benmayor, D. (dir.) (2015) Tracers. IMDb entry: www.imdb.com/title/tt2401097.

Bennett, A. (1999) 'Hip Hop Am Main: The localization of rap music and hip hop culture', *Media, Culture & Society* 21(1): 77–91.

Bennett, L. (2012) 'Fan Activism for Social Mobilization: A critical review of the literature', *Transformative Works and Cultures* 10.

Bennett, L. (2014a) '"If We Stick Together We Can Do Anything": Lady Gaga fandom, philanthropy and activism through social media', *Celebrity Studies* 5(1–2): 138–152.

Bennett, L. (2014b) 'Tracing *Textual Poachers*: Reflections on the development of fan studies and digital fandom', *The Journal of Fandom Studies* 2(1): 5–20.

Bennett, L., B. Jones and B. Chin (2015) *Crowdfunding the Future: Media industries, ethics and digital society*, New York: Peter Lang.

Bennett, W. L. (2013) 'Researching Online Fandom', *Cinema Journal* 52(4): 129–134.

Bennett, W. L. and A. Segerberg (2011) 'The Ecologies of Two Climate Change Protests', *Communication Review* 14: 197–215.

Bennett, W. L. and A. Segerberg (2012) 'The Logic of Connective Action: Digital media and the personalization of contentious politics', *Information, Communication and Society* 15(5): 739–768.

Bennett, W. L. and A. Segerberg (2013) *The Logic of Connective Action: Digital media and the personalization of contentious politics*, Cambridge: Cambridge University Press.

Benny, J. and J. Xu (2017) 'The Decline of Sina Weibo: A technological, political and market analysis', in M. Kent, K. Ellis and J. Xu (eds) *Chinese Social Media: Social, cultural and political implications*, London and New York: Routledge, 221–235.

Benski, T., L. Langman, I. Perugorría and B. Tejerina (2013) 'From the Streets and Squares to Social Movement Studies: What have we learned?', *Current Sociology* 61(4): 541–561.

Bentham, J. (1791) *Panopticon; Or, the Inspection House*, Volume 1, London: Mews Gate.

Benzaquen, M., D. Cave and R. Oliver (2018) 'The Raw Videos that Have Sparked Outrage over Police Treatment of Blacks', *The New York Times*, 19 April. Available online: www.nytimes.com/interactive/2017/08/19/us/police-videos-race.html.

Berardi, F. (2003) *La fábrica de la infelicidad, Nuevas formas de trabajo y movimiento global*, Madrid: Traficantes de Sueños. Available online: http://biblioweb.sindominio.net/pensamiento/fabrica_infelicidad.pdf.

Berardi, F. (2007) *Generación post-alfa, Patologías e imaginarios en el semiocapitalismo*. Madrid: Tinta Limon. Available online: http://tintalimon.com.ar/archivos/file/pdfs/pdf_978-987-23140-4-0.pdf.

Berardi, F. (2009) *The Soul at Work: From alienation to autonomy*, New York: Semiotext(e).

Berardi, F. (2011) *After the Future*, Oakland, CA: AK Press.

Berenger, R. (2013) *Social Media Go to War: Rage, rebellion and revolution in the age of Twitter*, Spokane, WA: Marquette Books.

Beretta, S. (2016) 'Media Participation and Desiring Subjects', in M. Baker and B. B. Blaagaard (eds) *Citizen Media and Public Spaces: Diverse expressions of citizenship and dissent*, London and New York: Routledge, 139–154.

Berger, P. L. and T. Luckmann (1966) *The Social Construction of Reality: A treatise in the sociology of knowledge*, London: Penguin.

Berger, R. (2015) 'Now I See It, Now I Don't: Researcher's position and reflexivity in qualitative research', *Qualitative Research* 15(2): 219–234.

Berger, W. (2000) 'Commercial Rebellion: Advertising's voracious appetite for underground culture swallows another victim: Culture Jamming', *Metropolis* (October). Available online: https://groups.google.com/forum/#!topic/misc.activism.progressive/1KjZ2hKvvWE.

Berlant, L. (2011) *Cruel Optimism*, Durham, NC: Duke University Press.

Berlet, C. and C. Mason (2015) 'Swastikas in Cyberspace: How hate went online', in A. P. Simpson and H. Druxes (eds) *Digital Media Strategies of the Far Right in Europe and the United States*, Lanham, MD: Lexington Books, 21–36.

Berman, S. (1997) 'Civil Society and the Collapse of the Weimar Republic', *World Politics* 49: 401–429.

Bernstein, A. (2013) 'An Inadvertent Sacrifice: Body politics and sovereign power in the Pussy Riot affair', *Critical Inquiry* 40(1): 220–241.

Berntzen, L. (2019) *Liberal Roots of Far Right Activism: The anti-Islamic movement in the 21st century*, London: Routledge.

Berntzen, L. and M. Weisskircher (2016) 'Anti-Islamic PEGIDA beyond Germany: Explaining differences in mobilisation', *Journal of Intercultural Studies* 37(6): 556–573.

Berrigan, F. J. (1979) *Community Communications: The role of community media in development*, Paris: UNESCO.

Bessant, J. (2017) 'Right-Wing Populism and Young "Stormers": Conflict in democratic politics', in S. Pickard and J. Bessant (eds) *Young People Re-Generating Politics in Times of Crises*, London: Palgrave, 139–159.

Betts, A., A. Ali and F. Memişoğlu (2017) *Talent Displaced: The economic lives of Syrian refugees in Europe*, Oxford: Deloitte and Oxford Refugee Studies Centre. Available online: www2.deloitte.com/global/en/pages/about-deloitte/articles/talent-displaced.html.

Bevington, D. and D. Dixon (2005) 'Movement-relevant Theory: Rethinking social movement scholarship and activism', *Social Movement Studies* 4(3): 185–208.

Bey, H. (1985) *T.A.Z.: The temporary autonomous zone, ontological anarchy, poetic terrorism*, Brooklyn, NY: Autonomedia.

Beyerle, M. and C. N. Brinckmann (eds) (1991) *Der amerikanische Dokumentarfilm der 60er Jahre. Direct Cinema und Radical Cinema*, Frankfurt a.M. and New York: Campus.

Bhabha, H. (1990) 'The Third Space. Interview with Homi Bhabha', in J. Rutherford (ed.) *Identity: Community, culture, difference*, London: Routledge, 207–221

Bhabha, H. (1994) *The Location of Culture*, New York and London: Routledge.

Bhabha, H. (2004) 'Of Mimicry and Man: The ambivalence of colonial discourse', in *The Location of Culture*, London: Routledge Classics, 85–92.

Bhambra, G. K. (2014) 'Postcolonial and Decolonial Dialogues', *Postcolonial Studies* 17(2): 115–121.

Bhatia, R. (2016) 'The Inside Story of Facebook's Biggest Setback', *The Guardian*, 12 May. Available online: www.theguardian.com/technology/2016/may/12/facebook-free-basics-india-zuckerberg.

Bichlbaum, A., M. Bonanno and K. Engfehr (dir.) (2009) The Yes Men Fix the World. IMDb entry: www.imdb.com/title/tt1352852/?ref_=nv_sr_1?ref_=nv_sr_1.

Bielsa, E. and S. Bassnett (2009) *Translation in Global News*, London: Routledge.

Bird, S. E. (2003) *The Audience in Everyday Life: Living in a media world*, New York: Routledge.

Bird, S. E. (2010) 'From Fan Practice to Mediated Moments: The value of practice theory in the understanding of media audiences', in B. Bräuchler and J. Postill (eds) *Theorising Media and Practice*, New York: Berghahn Books, 85–104.

Bird, S. E. (2011) 'Are We All Produsers Now? Convergence and media audience practices', *Cultural Studies* 25(4–5): 502–516.

Birks, J. (2014) *News and Civil Society*, Farnham, UK and Burlington, VT: Ashgate.

Bishop, C. (2012) *Artificial Hells: Participatory art and the politics of spectatorship*, Brooklyn, NY: Verso.

Bishop, L. and D. Gray (2017) 'Ethical Challenges of Publishing and Sharing Social Media Research Data', in K. Woodfield (ed.) *The Ethics of Online Research*, Bingley, UK: Emerald Publishing, 159–187.

Blaagaard, B. B. (2013a) 'Shifting Boundaries: Objectivity, citizen journalism and tomorrow's journalists', *Journalism* 14(8): 1076–1090.

Blaagaard, B. B. (2013b) 'Situated, Embodied and Political: Expressions of citizen journalism', *Journalism Studies* 14(2): 187–200.

Blaagaard, B. B. (2016a) 'Cosmopolitan Relations in the Colonies: Redefining citizen journalism through a historical reading of *The Herald*, St. Croix 1915–25', *European Journal of Cultural Studies* 21(4): 418–434.

Blaagaard, B. B. (2016b) 'Reading *The Herald* Today: Notes on journalism and citizen media', in S. Ponzanesi and G. Colpani (eds) *Postcolonial Transitions in Europe*, London and New York: Rowman and Littlefield, 231–250.

Blaagaard, B. B. and S. Allan (2016) 'Citizen Mediations of Connectivity: Narrowing the "culture of distance" in television news', in M. Baker and B. B. Blaagaard (eds) *Citizen Media and Public Spaces: Diverse expressions of citizenship and dissent*, London and New York: Routledge, 58–75.

Blaagaard, B. B., M. Mortensen and C. Neumayer (2017) 'Digital Images and Globalized Conflict', *Media, Culture & Society* 39: 1111–1121.

Black Panther Party Research Project, 'Black Panther Community Programs'. Available online: https://web.stanford.edu/group/blackpanthers/programs.shtml.

Blackledge, A. (2004) 'Constructions of Identity in Political Discourse in Multilingual Britain', in A. Pavlenko and A. Blackledge (eds) *Negotiation of Identities in Multilingual Contexts*, Clevedon, UK: Multilingual Matters, 68–92.

Blackledge, A. and A. Creese (2010) *Multilingualism*, New York: Continuum.

Blackwell, M. (2011) *¡Chicana Power! Contested histories of feminism in the Chicano movement*, Austin, TX: University of Texas Press.

Blair, M. (2004) 'Commercialization of the Rap Music Youth Subculture', in M. Forman and M. Neal (eds) *That's the Joint! The hip hop studies reader*, London: Routledge, 497–504.

Blee, K. and K. Creasap (2010) 'Conservative and Right-Wing Movements', *Annual Review of Sociology* 36: 269–286.

Blizzard (2004) World of Warcraft. Available online: https://worldofwarcraft.com.

Block, H. (2018) Personal conversation with Hans Block, co-director of the film *The Cleaners*.

Bloemraad, I. (2017) 'Theorising the Power of Citizenship as Claims-Making', *Journal of Ethnic and Migration Studies* 44: 4–26.

Blood, R. (2017) 'Weblogs: A history and perspective'. Available online: www.rebeccablood.net/essays/weblog_history.html.

Bloom, T. (2017) *Noncitizenism: Recognising noncitizen capabilities in a world of citizens*, London: Routledge.

Bloore, C. and G. Seiberling (1986) *Amateurs, Photography and the Mid-Victorian Imagination*, London and Chicago, IL: University of Chicago Press.

Blumberg, F. C., D. E. Almonte, J. S. Anthony and N. Hashimoto (2013) 'Serious Games: What are they? What do they do? Why should we play them?', in K. Dill (ed.) *Oxford Handbook of Media Psychology*, New York, Oxford University Press, 334–351.

Blumer, H. (1954) 'What Is Wrong with Social Theory?', *American Sociological Review* 19: 3–10.

Blunt, A. and G. Rose (1994) 'Introduction: Women's colonial and postcolonial geographies', in A. Blunt and G. Rose (eds) *Writing Women and Space: Colonial and postcolonial geographies*, New York: Gilford Press, 1–29.

Boaden, H. (2008) 'The Role of Citizen Journalism in Modern Democracy', *BBC News*, Editors' Blog, Thursday, 13 November. Available online: www.bbc.co.uk/blogs/theeditors/2008/11/the_role_of_citizen_journalism.html.

Boal, A. (1974/1979) *Theatre of the Oppressed*, translated by A. Charles, M-O. Leal McBride and E. Fryer, second edition, London: Pluto Press.

Bob, C. (2011) 'Civil and Uncivil Society', in M. Edwards (ed.) *The Oxford Handbook of Civil Society*, Oxford: Oxford University Press, 209–219.

Bobel, C. (2007) '"I'm Not an Activist, Though I've Done a Lot of It": Doing activism, being activist and the "perfect standard" in a contemporary movement', *Social Movement Studies* 6(2): 147–159.

Boeder, P. (2005) 'Habermas' Heritage: The future of the public sphere in the network society', *First Monday* 10(9), 5 September. Available online: https://firstmonday.org/article/view/1280/1200

Boéri, J. (2009) *Babels, the Social Forum and the Conference Interpreting Community: Overlapping and competing narratives on activism and interpreting in the era of globalisation*, Manchester: University of Manchester.

Boéri, J. (2010) 'Emerging Narratives of Conference Interpreters' Training: A case study of ad hoc training in Babels and the Social Forum', *Puentes* 9: 61–70. Available online: www.ugr.es/~greti/puentes/puentes9/08-Julie-Boeri.pdf.

Boéri, J. (2012) 'Translation/Interpreting Politics and Praxis: The impact of political principles on Babels' interpreting practice', *The Translator* 18(2): 269–290.

Boéri, J. (2013) 'Le défi de la traduction dans les forums sociaux: Plusieurs langues pour un autre monde', *Le monde diplomatique* 715(October) : 9. Available online: www.monde-diplomatique.fr/2013/10/BOERI/49730.

Boéri, J. (2014) 'Vers une approche communicationnelle de l'engagement: Les récits des traducteurs-interprètes du réseau Babels dans le mouvement altermondialiste', *Revue française des sciences de l'information et de la communication*, 5. Available online: http://rfsic.revues.org/1172.

Boéri, J. (2015) 'Émergence d'une communauté virtuelle de pratique: Regards croisés sur les dispositifs et les récits', in S. Parrini-Alemanno (ed.) *Communications organisationnelles, management et numérique*. Paris: L'Harmattan.

Boéri, J. and S. Hodkinson (2004) 'Social Forums after London: The politics of language', *Red Pepper*. Available online: www.redpepper.org.uk/Social-forums-after-London-The/.

Boggs, C. (1977a) 'Marxism, Prefigurative Communism, and the Problem of Workers' Control', *Radical America* 11: 99–122.

Boggs, C. (1977b) 'Revolutionary Process, Political Strategy and the Dilemma of Power', *Theory and Society* 4(3): 359–393.

Bogost, I. (2006) 'Playing Politics: Videogames for politics, activism, and advocacy', *First Monday* 7. Available online: http://firstmonday.org/issues/special11_9/bogost/index.html.

Bogost, I. (2007) *Persuasive Games*, Cambridge, MA: The MIT Press.

Bogost, I., S. Ferrari and B. Schweizer (2010) *Newsgames: Journalism at play*, Boston, MA: The MIT Press.

Böhm, S., A. C. Dinerstein and A. Spicer (2010) '(Im)possibilities of Autonomy: Social movements in and beyond capital, the state and development', *Social Movement Studies* 9(1): 17–32.

Böhm, S., S. Sullivan and O. Reyes (2005) 'The Organization and Politics of Social Forums', *Ephemera* 5(2). Available online: www.ephemerajournal.org/issue/organisation-and-politics-social-forums.

Boido, P. (2003) 'Combining Digital Technology and Traditional Media to Create Cooperatives and Participatory Forms of Communication', *Our Media/NuestrosMedios 3*, Barranquilla, Colombia.

Boland, G. and D. Cameron (2005) 'Newspaper Theatre: Applying performance-based learning to journalism education', *Conference Proceedings, Journalism Education Association Conference*. Available online: https://pdfs.semanticscholar.org/0b18/20cfea28ae64fc98d9b6e139e39efad6b4fc.pdf?_ga=2.94949845.833841611.1577880208-1174748008.1577880208.

Bollier, D. and S. Helfrich (2012) 'Introduction: The commons as a transformative vision', in D. Bollier and S. Helfrich (eds) *The Wealth of the Commons: A world beyond market and state*, Amherst, MA: Levellers Press, xi. Available online: http://wealthofthecommons.org/essay/introduction-commons-transformative-vision.

Boltanski, L. and E. Chiapello (2005) *The New Spirit of Capitalism*, London: Verso.

Bolter, J. D. and R. A. Grusin (1999) *Remediation: Understanding new media*, Cambridge, MA: The MIT Press.

Bond, J.-M. (2018) 'Understanding Reddit: A beginner's guide to the front page of the internet', *The Daily Dot*, 2 September. Available online: www.dailydot.com/debug/what-is-reddit/.

Bond, P. and S. Mottiar (2013) 'Movements, Protests and a Massacre in South Africa', *Journal of Contemporary African Studies* 31(2): 283–302.

Bondebjerg, I. (2016) 'Documentary, Multi-platform Production, and Cosmopolitan Dialogues', in D. Marcus and S. Kara (eds) *Contemporary Documentary*, London and New York: Routledge, 171–186.

Bonechi, M. (2012) 'Interview with Simon Rogers', *Editor Guardian Data Blog*. Available online: www.youtube.com/watch?v=bU8DiD2J-tw.

Bonilla, Y. (2010) 'Guadeloupe is Ours: The prefigurative politics of the mass strike in the French Antilles', *Interventions* 12(1): 125–137.

Bonilla, Y. and J. Rosa (2015) '#Ferguson: Digital protest, hashtag ethnography and the racial politics of social media in the United States', *American Ethnologist* 42(1): 4–16.

Bonini, T. (2017) 'Twitter or Radio Revolutions? The central role of Açık Radyo in the Gezi protests of 2013', *Westminster Papers in Communication and Culture* 12(2): 1–17.

Bonter, D. N. (2012) 'From Backyard Observations to Continent-wide Trends: Lessons from the first twenty-two years of Project FeederWatch', in J. L. Dickinson and R. Bonney (eds) *Citizen Science: Public participation in environmental research*, Ithaca, NY: Cornell University Press, 27–34.

Boochani, B. (2016a) 'Mehdi Savari: Actor, prisoner, and improbable star of Manus Island', translated by O. Tofighian, *New Matilda*. Available online: https://newmatilda.com/2016/06/23/mehdi-savari-actor-prisoner-and-improbable-star-of-manus-island/.

Boochani, B. (2016b) 'Cruel and Violent Political Strategy', translated by O. Tofighian, *The Huffington Post*. Available online: www.huffingtonpost.com.au/behrouz-boochani/a-cruel-and-violent-political-strategy/.

Boochani, B. (2017a) '"This is Hell Out Here": How Behrouz Boochani's diaries expose Australia's refugee shame', translated by O. Tofighian, *The Guardian*. Available online: www.theguardian.com/world/2017/dec/04/this-is-hell-behrouz-boochani-diaries-expose-australia-refugee-shame.

Boochani, B. (2017b) 'A Letter from Manus', translated by O. Tofighian, *The Saturday Paper*. Available online: www.thesaturdaypaper.com.au/news/politics/2017/12/09/letter-manus-island/15127380005617.

Boochani, B. (2018) *No Friend but the Mountains: Writing from Manus prison*, translated by O. Tofighian, with translator's note and essay, Sydney: Picador.

Boorstin, D. J. (1961/1992) *The Image: A guide to pseudo-events in America*, New York: Vintage Books.

Booth, N. and J. A. Matic (2011) 'Mapping and Leveraging Influencers in Social Media to Shape Corporate Brand Perceptions', *Corporate Communications: An International Journal* 16(3): 184–191.

Booth, P. (2015) *Playing Fans: Negotiating fandom and media in the digital age*, Iowa City, IA: University of Iowa Press.

Bordwell, M. (2002) 'Jamming Culture: Adbusters' hip media campaign against consumerism', in T. Princen, M. Maniates and K. Conca (eds) *Confronting Consumption*, Cambridge, MA: The MIT Press, 237–253.

Bordwell, T. and K. Thompson (2010) *Film Art: An Introduction*, ninth edition, New York: McGraw-Hill.

Boser, S. (2007) 'Power, Ethics, and the IRB: Dissonance over human participant review of participatory research', *Qualitative Inquiry* 13(8): 1060–1074.

Bosi, L. and H. Reiter (2014) 'Historical Methodologies: Archival research and oral history in social movement research', in D. della Porta (ed.) *Methodological Practices in Social Movement Research*, Oxford: Oxford University Press, 117–143.

Bosk, C. and R. de Vries (2004) 'Bureaucracies of Mass Deception: Institutional review boards and the ethics of ethnographic research', *Annals of the American Academy of Political and Social Science* 595: 249–263.

Botman, S. (1991) *Egypt from Independence to Revolution, 1919–1952*, Syracuse, NY: Syracuse University Press.

Boulianne, S. (2015) 'Social Media Use and Participation: A meta-analysis of current research', *Information, Communication and Society* 18(5): 524–538.

Boullier, D. (1989) 'Du bon usage d'une critique du modèle diffusionniste: Discussion prétexte des concepts de Everett M. Rogers', *Reseaux* 7(36): 31–51.

Bouncken, R., M. Komorek and S. Kraus (2015) 'Crowdfunding: The current state of research', *International Business and Economics Research Journal* 14(3): 407–415.

Bourdieu, P. (1972/1977) *Outline of a Theory of Practice*, translated by R. Nice, Cambridge: Cambridge University Press.

Bourdieu, P. (1977/1993) 'The Production of Belief: Contribution to an economy of symbolic goods', in R. Johnson (ed.) *The Field of Cultural Production: Essays on art and literature*, translated by R. Nice, New York: Columbia University Press, 74–111.

Bourdieu, P. (1980/1990) *The Logic of Practice*, translated by R. Nice, Stanford, CA: Stanford University Press.

Bourdieu, P. (1982/1991) *Language and Symbolic Power*, translated by G. Raymond and M. Adamson, Cambridge, MA: Harvard University Press.

Bourdieu, P. (1979/2010) *Distinction: A social critique of the judgement of taste*, translated by R. Nice, London and New York: Routledge.

Bourdieu, P. (1983/2011) 'The Forms of Capital', in I. Szeman and T. Kaposy (eds) *Cultural Theory: An anthology*, translated by R. Nice, Hoboken, NJ: Wiley, 81–93.

Bowen, J. and J. Purkis (2004) 'Introduction: Why anarchism still matters', in J. Purkis and J. Bowen (eds) *Changing Anarchism*, Manchester: Manchester University Press.

Bowker, G. and S. Leigh Star (1999) *Sorting Things Out: Classification and its consequences*, Cambridge: The MIT Press.

Bowman, S. and C. Willis (2003) *We Media: How audiences are shaping the future of news and information*, Reston, VA: The Media Center at The American Press Institute. Available online: www. hypergene.net/wemedia/.

boyd, d. (2014) *It's Complicated: The social lives of networked teens*, New Haven, CT: Yale University Press.

boyd, d. (2017a) 'The Information War Has Begun', Zephora.org. Available online: www.zephoria.org/thoughts/archives/2017/01/27/the-information-war-has-begun.html.

boyd, d. (2017b) 'The Radicalization of Utopian Dreams', *Data & Society*. Available online: https://points.datasociety.net/the-radicalization-of-utopian-dreams-e1b785a0cb5d.

boyd, d. and K. Crawford (2012) 'Critical Questions for Big Data: Provocations for a cultural, technological, and scholarly phenomenon', *Information, Communication & Society* 15(5): 662–679.

boyd, d. and N. Ellison (2007) 'Social Network Sites: Definition, history, and scholarship', *Computer-Mediated Communication* 13(1): 210–230.

boyd, d. and A. Marwick (2011) 'Social Privacy in Networked Publics: Teens' attitudes, practices, and strategies'. Available online: www.danah.org/papers/2011/SocialPrivacyPLSC-Draft.pdf.

Boyle, D. (1997) *Subject to Change: Guerilla television revisited*, Bloomington, IN: Indiana University Press.

Brabham, D. C. (2012) 'Motivations for Participation in a Crowdsourcing Application to Improve Public Engagement in Transit Planning', *Journal of Applied Communication Research* 40(3): 307–328.

Brabham, D. C. (2013a) *Crowdsourcing*, Cambridge, MA: The MIT Press.

Brabham, D. C. (2013b) 'Crowdsourcing: A model for leveraging online communities', in A. Delwiche and J. Henderson (eds) *The Participatory Cultures Handbook*, New York: Routledge, 120–130.

Bradshaw, P. (2009) 'Wiki Journalism', in S. Allan and E. Thorsen (eds) *Citizen Journalism: Global perspectives*, Volume 1, New York: Peter Lang, 243–254.

Braman, S. (2006) *Change of State: Information, policy, and power*, Cambridge, MA: The MIT Press.

Brandzel, A. L. (2016) *Against Citizenship: The violence of the normative*, Chicago, IL: University of Illinois Press.

Brantner, C. and K. Lobinger (2015) 'In the Eye of the Beholder: Subjective views on the authenticity of selfies', *International Journal of Communications* 9: 1848–1860.

Bratteteig, T. (2008) 'Does It Matter that It's Digital?', in K. Lundby (ed.) *Digital Storytelling, Mediatized Stories*, New York: Peter Lang, 271–284.

Bräuchler, B. and J. Postill (2010). *Theorising Media and Practice*, New York: Berghahn Books.

Braun, J. (2015) 'Social Media and Distribution Studies', *Social Media + Society*, April–June, 1–2. Available online: http://doi.org/10.1177/2056305115580483.

Bravo, D. (2014) 'Occupy Gezi', Public Space. Available online: www.publicspace.org/works/-/project/h312-occupy-gezi.

Breech, R., C. Cox, E. Crowe, J. Hendricks and D. Larson (2014) *Warning Signs: Toxic air pollution identified at oil and gas development sites*, Brattleboro, VT: Coming Clean and Global Community Monitor.

Breines, W. (1981) *Community and Organization in the New Left 1962–1968: The great refusal*, New Brunswick, NJ: Rutgers University Press.

Brejzek, T. (2010) 'From Social Network to Urban Intervention: On the scenographies of flash mobs and urban swarms', *International Journal of Performance Arts and Digital Media* 6(1): 109–122.

Breslow, J. (2018) 'Moderating the "Worst of Humanity": Sexuality, witnessing, and the digital life of coloniality', *Porn Studies* 5(3): 225–240.

Briant, E., N. Watson and G. Philo (2013) 'Reporting Disability in the Age of Austerity: The changing face of media representation of disability and disabled people in the United Kingdom and the creation of "folk devils"', *Disability and Society* 28(6): 874–889.

Brighenti, A. M. (2010) 'At the Wall: Graffiti writers, urban territoriality and the public domain', *Space and Culture* 13(3): 315–322.

Brighenti, A. M. (ed.) (2016) *Urban Interstices: The aesthetics and the politics of the in-between*, New York: Routledge.

Britton, C. M. (1999) *Edouard Glissant and Postcolonial Theory: Strategies of language and resistance*, Charlottesville, VA: University Press of Virginia.

Broersma, M. (2010) 'Journalism as Performative Discourse: The importance of form and style in journalism', in V. Rupar (ed.) *Journalism and Meaning-making: Reading the newspaper*, Cresskill, NJ: Hampton Press, 15–35.

Broersma, M. and T. Graham (2012) 'Social Media as Beat', *Journalism Practice* 6(3): 403–419.

Broersma, M. and T. Graham (2013) 'Twitter as a News Source', *Journalism Practice* 7(4): 446–464.

Bromley, R. (2011) 'Beast, Vermin, Insect – Hate media and the construction of the enemy: The case of Rwanda, 1990–1994', *At the Interface/Probing the Boundaries* 75: 39–59.

Brookes, R. A., J. Lewis and K. Wahl-Jorgensen (2004) 'The Media Representation of Public Opinion: British television news coverage of the 2001 general election', *Media, Culture & Society* 26(2): 63–80.

Brooten, L. and G. Hadl (2008) 'Gender and Hierarchy: A case study of the Independent Media Center Network', in C. Rodríguez, D. Kidd and L. Stein (eds) *Making Our Media*, New York: Hampton Press, 203–222.

Brophy, E. and G. de Peuter (2008) 'Immaterial Labor, Precarity, and Recomposition', in C. McKercher and V. Mosco (eds) *Knowledge Workers in the Information Society*, Lanham, MD: Lexington Books, 177–191.

Brough, M. M. and S. Shresthova (2011) 'Fandom Meets Activism: Rethinking civic and political participation', *Transformative Works and Cultures* 10.

Brown, K. (2013) 'Agitprop in Soviet Russia', *Constructing the Past* 14(1). Available online: https://digitalcommons.iwu.edu/constructing/vol14/iss1/4.

Brown, P. (1992) 'Popular Epidemiology and Toxic Waste Contamination: Lay and professional ways of knowing', *Journal of Health and Social Behavior* 33(3): 267–281.

Brown, P. and S. Zavestocki (2004) 'Social Movements in Health: An introduction', *Sociology of Health and Illness* 26(6): 679–694.

Brown, W. (2003) 'Neo-Liberalism and the End of Liberal Democracy', *Theory and Event* 7(1): 1–19.

Brown, W. (2015) *Undoing the Demos: Neoliberalism's stealth revolution*, New York: Zone Books.

Browne, D. R. (1990) 'Aboriginal Radio in Australia: From dreamtime to prime time?', *Journal of Communication* 40(Winter): 111–120.

Browne, D. R. (1996) *Electronic Media and Indigenous Peoples: A voice of our own?* Ames, IO: Iowa State University Press.

Bruns, A. (2005) *Gatewatching: Collaborative online news production*, New York: Peter Lang.

Bruns, A. (2006) 'Towards Produsage: Futures for user-led content production', in F. Sudweeks, H. Hrachovec and C. Ess (eds) *Proceedings Cultural Attitudes towards Communication and Technology*, 28 June – 1 July, Tartu, Estonia, 275–284. Available online: https://eprints.qut.edu.au/4863/1/4863_1.pdf.

Bruns, A. (2008) *Blogs, Wikipedia, Second Life, and Beyond: From production to produsage*, New York: Peter Lang.

Bruns, A. (2010a) 'Citizen Journalism and Everyday Life: A case study of Germany's myHeimat.de', in B. Franklin and M. Carlson (eds) *Journalists, Sources, and Credibility: New perspectives*, London and New York: Routledge, 182–194.

Bruns, A. (2010b) 'From Reader to Writer: Citizen journalism as news produsage', in J. Hunsinger, L. Klastrup and M. Allen (eds) *International Handbook of Internet Research*, Dordrecht: Springer, 119–133.

Bruns, A. (2010c) 'News Produsage in a Pro-Am Mediasphere: Why citizen journalism matters', in G. Meikle and G. Redden (eds) *News Online: Transformations and Continuities*, London: Palgrave Macmillan, 1–22.

Bruns, A. (2018) *Gatewatching and News Curation: Journalism, social media, and the public sphere*, New York: Peter Lang.

Bruns, A. and J. Jacobs (2007) *Uses of Blogs*, New York: Peter Lang.

Bryant-Bertail, S. (2010) 'Erwin Piscator', *Oxford Reference*. Available online: www.oxfordreference.com/view/10.1093/acref/9780199574193.001.0001/acref-9780199574193-e-3098.

Brym, R., M. Godbout, A. Hoffbauer, G. Menard and T. Huiquan Zhang (2014) 'Social Media in the 2011 Egyptian Uprising', *British Journal of Sociology* 65(2): 266–292.

Bucher, T. and A. Helmond (2018) 'The Affordances of Social Media Platforms', in J. Burgess, T. Poell and A. Marwick (eds) *The Sage Handbook of Social Media*, London and New York: Sage, 233–253.

Buckingham, D. and A. Burn (2007) 'Game Literacy in Theory and Practice', *Journal of Educational Multimedia and Hypermedia* 16(3): 323–349.

Buck-Morss, S. (2000) 'Hegel and Haiti', *Critical Inquiry* 26(4): 821–865.

Buck-Morss, S. (2009) *Hegel, Haiti, and Universal History*, Pittsburgh, PA: University of Pittsburgh Press.

Budka, P. (2015) 'From Marginalization to Self-Determined Participation: Indigenous digital infrastructures and technology appropriation in Northwestern Ontario's remote communities', *Journal des Anthropologues* 142–143(3): 127–153.

Buettner, E. (2016) *Europe after Empire: Decolonization, society, and culture*, Cambridge: Cambridge University Press.

Buhle, P. and N. Schulman (eds) (2005) *Wobblies!: A graphic history of the Industrial Workers of the World*, London and New York: Verso.

Bulliet, R. (1998) 'Popular Culture', in R. Bulliet (ed.) *The Columbia History of the 20th Century*, New York: Columbia University Press, 31–52.

Bulut, E. and E. Yörük (2017) 'Digital Populism: Trolls and political polarization of Twitter in Turkey', *International Journal of Communication* 11(1): 4093–4117.

Buni, C. and S. Chemaly (2016) 'The Secret Rules of the Internet', *The Verge*, 13 April. Available online: www.theverge.com/2016/4/13/11387934/internet-moderator-history-youtube-facebook-reddit-censorship-free-speech.

Burak, A. and L. Parker (2017) *Power Play: How video games can save the world*, New York: St. Martin's Press.

Burchell, J. and J. Cook (2013) 'CSR, Co-optation and Resistance: The emergence of new agonistic relations between business and civil society', *Journal of Business Ethics* 115(4): 741–754.

Burgess, J. (2006) 'Hearing Ordinary Voices: Cultural studies, vernacular creativity and digital storytelling', *Continuum: Journal of Media & Culture Studies* 20(2): 201–214.

Burgess, J. (2013) 'YouTube and the Formalization of Amateur Media', in D. Hunter, R. Lobato, M. Richardson and J. Thomas (eds) *Amateur Media: Social, cultural and legal perspectives*, New York: Routledge, 53–58.

Burgess, J. and J. Green (eds) (2009) *YouTube: Online video and participatory culture*, Cambridge: Polity Press.

Burns, S. and B. Haller (2015) 'The Politics of Representing Disability: Exploring news coverage of the Americans with Disabilities Act and the National Disability Insurance Scheme', *Asia Pacific Media Educator* 25(2): 262–277.

Burrell, I. (2019) 'Are Google and Facebook killing advertising?', *The Raconteur*, 13 March. Available online: www.raconteur.net/business-innovation/google-facebook-duopoly.

Burstein, P. (1999) 'Social Movements and Public Policy', in M. Giugni, D. McAdam and C. Tilly (eds) *How Social Movements Matter*, Minneapolis, MN: University of Minnesota Press, 3–21.

Burton, J. (ed.) (1986) *Cinema and Social Change in Latin America: Conversations with filmmakers*, Austin, TX: University of Texas Press.

Bury, R. (2005) *Cyberspaces of Their Own: Female fandoms online*, New York: Peter Lang.

Bush, S. (2017) 'Putting on the Red Boots: Role-play as "coping work" and "creative work" in the theatrical representation of prostitution', *Studies in Theatre and Performance* 39(1): 54–69.

Busse, K. (2017) *Framing Fan Fiction: Literary and social practices in fan fiction communities*, Ames, IA: University of Iowa Press.

Busse, K. and K. Hellekson (2006) 'Introduction: Work in progress', in K. Hellekson, and K. Busse (eds) *Fan Fiction and Fan Communities in the Age of the Internet: New essays*, Jefferson, NC: McFarland and Company, 5–32.

Busse, K. and A. Lothian (2018) 'A History of Slash Sexualities', in C. Smith, F. Attwood and B. McNair (eds) *The Routledge Companion to Media, Sex and Sexuality*, New York: Routledge, 117–129.

Butler, B. (2009) 'Palestinian Heritage "to the Moment": Archival memory and the representation of heritage', *Conflict, Conservation and Management of Archaeological Sites* 11(3–4): 236–261.

Butler, J. (1990) *Gender Trouble: Feminism and the subversion of identity*, New York: Routledge.

Butler, J. (1993) *Bodies that Matter: On the discursive limits of 'sex'*, London: Routledge.
Butler, J. (1997) *Excitable Speech: A politics of the performative*, London: Routledge.
Butler, J. (2004) *Precarious Life: The powers of mourning and violence*, London: Verso.
Butler, J. (2009) *Frames of War: When Is Life Grievable?*, London: Verso.
Butler, J. (2012) 'Precarity Talk: A virtual roundtable with Lauren Berlant, Judith Butler, Bojana Cvejić, Isabell Lorey, Jasbir Puar, and Ana Vujanović', *TDR: The Drama Review* 56(4): 163–177.
Butler, J. (2015a) *Notes Towards a Performative Theory of Assembly*, Cambridge, MA: Harvard University Press.
Butler, J. (2015b) *Senses of the Subject*, New York: Fordham University Press.
Butler, J. and A. Athanasiou (2013) *Dispossession: The performative in the political*, Cambridge: Polity Press.
Butsch, R. (2011) 'Audiences and Publics, Media and Public Spheres', in V. Nightingale (ed.) *The Handbook of Media Audiences*, Malden, MA: Blackwell, 149–167.
Byrd, A. (2014) 'How TuneCore Is Making Record Labels Unnecessary'. Available online: www.fastcompany.com/3034888/innovation-agents/how-tunecore-is-making-record-labels-unncessary/.
Byrd Phillips, L. (2014) 'The Role of Open Authority in a Collaborative Web', in M. Ridge (ed.) *Crowdsourcing Our Cultural Heritage*, Burlington, VT: Ashgate, 247–269.
Cable, S., T. Mix and D. Hastings (2005) 'Mission Impossible? Environmental justice activists' collaborations with professional environmentalists and with academics', in D. N. Pellow and R. J. Brulle (eds) *Power, Justice, and the Environment: A critical appraisal of the environmental justice movement*, Cambridge, MA: The MIT Press, 55–76.
Caffentzis, G. (2005) 'Immeasurable Value? An essay on Marx's legacy', *The Commoner* 10(Spring–Summer): 87–104.
Caffentzis, G. (2016) 'Commons', in K. Fritsch, C. O'Connor and A. K. Thompson (eds) *Keywords for Radicals: The contested vocabulary of late-capitalist struggle*, Oakland, CA: AK Press, 95–101.
Çağlar, A. (2016) 'Displacement of European Citizen Roma in Berlin: Acts of citizenship and sites of contentious politics', *Citizenship Studies* 20: 647–663.
Cai, W. and S. Landon (2019) 'Attacks by White Extremists Are Growing: So are their connections', *The New York Times*, 3 April. Available online: www.nytimes.com/interactive/2019/04/03/world/white-extremist-terrorism-christchurch.html.
Caiani, M. and P. Kröll (2015) 'The Transnationalization of the Extreme Right and the Use of the Internet', *International Journal of Comparative and Applied Criminal Justice* 39: 331–351.
Cain, N. (2011) *Video Days and What We Saw through the Viewfinder*, Palm Springs, CA: Event Horizon Press.
Cairns, C. and A. Carlson (2016) 'Real-World Islands in a Social Media Sea: Nationalism and censorship on Weibo during the 2012 Diaoyu/Senkaku Crisis', *The China Quarterly* 225: 23–49.
Caldwell, G. (2005) 'Using Video for Advocacy', in S. Gregory, G. Caldwell, R. Avni and T. Harding (eds) *Video for Change: A guide for advocacy and activism*, London: Pluto Press, 1–19.
Calhoun, C. (1992) 'Introduction: Habermas and the public sphere', in C. Calhoun (ed.) *Habermas and the Public Sphere: Studies in contemporary German social thought*, Cambridge, MA: The MIT Press, 1–50.
Calhoun, C. (1998) 'The Public Good as a Social and Cultural Project', in W. Powell and E. S. Clemens (eds) *Private Action and the Public Good*, New Haven, CT: Yale University Press, 20–35.
Cambridge English Dictionary (2018) 'Remediation'. Available online: https://dictionary.cambridge.org/dictionary/english/remediation.
Cambridge English Dictionary (2019) 'Co-opt, verb'. Available online: https://dictionary.cambridge.org/dictionary/english/co-opt
Cammaerts, B. (2007a) 'Activism and Media', in B. Cammaerts and N. Carpentier (eds) *Reclaiming the Media: Communication rights and democratic media roles*, Bristol: Intellect, 217–224.
Cammaerts, B. (2007b) 'Jamming the Political: Beyond counter-hegemonic practices', *Continuum: Journal of Media & Cultural Studies*, 21(1): 71–90.
Cammaerts, B. (2009) 'Radical Pluralism and Free Speech in Online Public Spaces: The case of North Belgian extreme right discourses', *International Journal of Cultural Studies* 12(6): 555–575.
Cammaerts, B. (2012) 'Protest Logics and the Mediation Opportunity Structure', *European Journal of Communication* 27(2): 117–134.
Cammaerts, B. (2015) 'Movement Media as Technologies of Self-mediation', in C. Atton (ed.) *The Routledge Companion to Alternative and Community Media*, New York: Routledge, 445–456.
Cammaerts, B. and N. Couldry (2016) 'Digital Journalism as Practice', in T. Witschge, C. W. Anderson, D. Domingo and A. Hermida (eds) *The Sage Handbook of Digital Journalism*, London: Sage, 326–340.
Cammaerts, B., A. Mattoni and P. McCurdy (eds) (2013) *Mediation and Protest Movements*, Chicago, IL: Intellect.
Campbell, M. (dir.) (2006) Casino Royale. IMDb entry: www.imdb.com/title/tt0381061.
Canalp, A. (2017) 'Breaking Barriers', *Cyprus Today*, 3 March. Available online: www.cyprustodayonline.com/cyprus/breaking-barriers/93.

Canetti, E. (1960/1984) *Crowds and Power*, translated by C. Stewart, New York: Noonday Press.
Caplan, R. (2018) *Content or Context Moderation? Artisanal, community-reliant, and industrial approaches*, New York: Data and Society. Available online: https://datasociety.net/wp-content/uploads/2018/11/DS_Content_or_Context_Moderation.pdf.
Caplan, R. and d. boyd (2018) 'Isomorphism through Algorithms: Institutional dependencies in the case of Facebook', *Big Data & Society*, January–June: 1–12.
Cardon, D. and F. Granjon (2010) *Médiactivistes*, Paris: Presses de la Fondation Nationale des Sciences Politiques.
Cardoso, G. (2006) *The Media in the Network Society: Browsing, news, filters and citizenship*, Lisbon: Centre for Research and Studies in Sociology.
Carey, J. (1987) 'The Press and Public Discourse', *The Center Magazine* 20: 4–32.
Carlson, M. (2003) *The Haunted Stage: The theatre as memory machine*, second edition, Ann Arbor, MI: The University of Michigan Press.
Carlson, M. (2015) *Boundaries of Journalism: Professionalism, practices and participation*, Oxon: Routledge.
Carney, N. (2016) 'All Lives Matter, but So Does Race', *Humanity and Society* 40(2): 180–199.
Carpentier, N. (2011) *Media and Participation: A site of ideological-democratic struggle*, Bristol, UK and Chicago, IL: Intellect.
Carpentier, N. (2012) 'The Concept of Participation. If They Have Access and Interact, Do They Really Participate?', *Fronteiras – estudos midiáticos* 14(2): 164–177.
Carpentier, N. (2014) 'The Cypriot Web Radio Mycyradio as a Participatory Mélange: Overcoming dichotomies in the era of web 2.0', *Sociologia e Politiche Sociali* 17(2): 91–108.
Carpentier, N. (2015) 'Articulating Participation and Agonism: A case study on the agonistic re-articulations of the Cyprus problem in the broadcasts of the community broadcaster MYCYradio', *The Cyprus Review* 27(1): 129–153.
Carpentier, N. (2017) *The Discursive-Material Knot: Cyprus in conflict and community media participation*, New York: Peter Lang.
Carpentier, N. (ed.) (2019) *Respublika! Experiments in the performance of participation and democracy*, Limassol, Cyprus: NeMe.
Carpentier, N. and B. De Cleen (2008) 'Introduction: Blurring participations and convergences', in N. Carpentier and B. De Cleen (eds) *Participation and Media Production: Critical reflections on content creation*, Newcastle: Cambridge Scholars Publishing, 1–12.
Carpentier, N. and V. Doudaki (2013) 'Community Media for Reconciliation: A Cypriot case study', *Communication, Culture & Critique* 7(4): 415–434.
Carter, A. (1983) *Direct Action*, third edition, London: Housmans.
Caruso, G. (2012) 'Toward an Emancipatory Cosmopolitan Project: The World Social Forum and the transformation of conflicts', *Globalizations* 9(2): 211–224.
Casati, R. and A. Varzi (2015) 'Events', in N. Zalta (ed.) *The Stanford Encyclopedia of Philosophy*, Stanford, CA: The Metaphysics Research Lab, Stanford University, 1–14.
Castelli, P. and C. Froio (2018) 'Getting "Right" into the News: Grassroots far-right mobilization and media coverage in Italy and France', *Comparative European Politics* 17(5): 738–758.
Castelli Gattinara, P. and S. Bouron (2019) 'Extreme-Right Communication in Italy and France: Political culture and media practices in CasaPound Italia and Les Identitaires', *Information, Communication & Society*. Available online: https://doi.org/10.1080/1369118X.2019.1631370.
Castells, M. (1996) *The Rise of the Network Society*, Cambridge, MA and Oxford: Blackwell.
Castells, M. (2000) *The Rise of the Network Society*, Volume 1, second edition, Oxford and Malden, MA: Blackwell.
Castells, M. (2004) 'Informationalism, Networks, and the Network Society: A theoretical blueprint', in M. Castells (ed.) *The Network Society: A cross-cultural perspective*, Northampton, MA: Edward Elgar Publishing, 3–45.
Castells, M. (2009) *Communication Power*, first edition, Oxford: Oxford University Press.
Castells, M. (2010) *The Information Age*, Volume 1, *The Rise of the Network Society*, second edition, Malden, MA: Wiley-Blackwell.
Castells, M. (2011) *Communication Power*, second edition, Oxford: Oxford University Press.
Castells, M. (2012) *Networks of Outrage and Hope: Social movements in the Internet age*, Cambridge: Polity Press.
Castells, M. and M. Kumar (2014) 'A Conversation with Manuel Castells', *Berkeley Planning Journal* 27(1): 93–99. Available online: http://escholarship.org/uc/item/2ns059h3.
Castells-Talens, A. (2010) 'When Our Media Belong to the State: Policy and negotiation in indigenous-language radio in Mexico', in C. Rodríguez, D. Kidd and L. Stein (eds) *Making Our Media: Global initiatives toward a democratic public sphere*, Cresskill, NJ: Hampton, 249–270.
Castells-Talens, A. (2011) 'Indigenous Radio Stations (Mexico)', in J. D. H. Downing (ed.) *Encyclopedia of Social Movement Media*, Thousand Oaks, CA: Sage, 263–266.
Castells-Talens, A. (2012) '¿Ni indígena ni comunitaria? La radio indigenista en tiempos neoindigenistas' [Not indigenous nor community-based? Indigenist radio during neoindigenist times], in E. Rashkin and N. E. García Meza

(eds) *Escenarios de la cultura y la comunicación en México: De la memoria al devenir cultural*, Xalapa, Veracruz: Universidad Veracruzana, 253–277.

Castells-Talens, A. and J. M. Ramos Rodríguez (2013) 'Technology for Cultural Survival: Indigenous-language radio at the end of the twentieth century', in A. Tinajero and J. B. Freeman (eds) *Technology and Culture in Twentieth Century Mexico*, Tuscaloosa, AL: University of Alabama, 178–193.

Castells-Talens, A., J. M. Ramos Rodríguez and M. Chan Concha (2009) 'Radio, Control, and Indigenous Peoples: The failure of state-invented citizens' media in Mexico', *Development in Practice* 19(4): 525–537.

Caswell, M. (2014) 'Seeing Yourself in History: Community archives in the fight against symbolic annihilation', *The Public Historian: A Journal of Public History* 36(4): 26–37.

Caswell, M., H. Cole and Z. Griffith (2018) 'Images, Silences, and the Archival Record: An interview with Michelle Caswell', *disClosure: A Journal of Social Theory* 27: 21–27.

Cattaneo, C. and E. Tudela (2014) '¡El Carrer Es Nostre! The autonomous movement in Barcelona, 1980–2012', in B. van der Steen, A. Katzeff and L. van Hoogenhuijze (eds) *The City Is Ours: Squatting and autonomous movements in Europe from the 1970s to the present*, Oakland, CA: PM Press, 95–130.

Causer, T., K. Grint, A.-M. Sichani and M. Terras (2018) '"Making Such Bargain": *Transcribe Bentham* and the quality and cost-effectiveness of crowdsourced transcription', *Digital Scholarship in the Humanities* 33(3): 1–21.

Cayatte, R. (2014) 'Where, Game, Play and Art Collide', in N. Garrelts (ed.) *Understanding Minecraft: Essays on play, community and possibilities*, Jefferson, NC: McFarland and Company, 203–213.

cDa Communications and Hacktivismo (2001) 'A Special Message of Hope', CultDeadCow.com. Available online: https://web.archive.org/web/20180131075946/http://www.cultdeadcow.com/cDc_files/declaration.html.

Center for Digital Storytelling. https://elmcip.net/organization/center-digital-storytelling.

Centivany, A. (2016) 'Values, Ethics and Participatory Policymaking in Online Communities', *Proceedings of the Association for Information Science and Technology* 53(1): 1–10.

Centivany, A. and B. Glushko (2016) '"Popcorn Tastes Good": Participatory policymaking and Reddit's "AMAgeddon"', *Proceedings of the SIGCHI Conference on Human Factors in Computing Systems, CHI'16, May 07–12, 2016, San Jose, CA*, New York: ACM.

Cere, R. (2011) 'Postcolonial and Media Studies: A cognitive map', in R. Cere and R. Brunt (eds) *Postcolonial Media Culture in Britain*, Basingstoke, UK: Palgrave Macmillan, 1–13.

Certeau, M. de (1980/1984) *The Practice of Everyday Life*, translated by S. F. Rendall, Berkeley, CA: University of California Press.

Chadwick, A. (2013) *The Hybrid Media System: Politics and power*, first edition, Oxford: Oxford University Press.

Chadwick, A. (2017) *The Hybrid Media System: Politics and power*, second edition, Oxford: Oxford University Press.

Chaffee, L. (1993) *Political Protest and Street Art: Popular tools for democratization in Hispanic countries*, Westport, CT: Greenwood Press.

Chambat, P. (1994) 'Usages des technologies de l'information et de la communication (TIC): Évolution des problématiques', *TIS* 6(3): 249–270.

Champion, E. (ed.) (2012) *Game Mods: Design, theory and criticism*, Pittsburgh, PA: ETC Press.

Chan, J. (2014) 'Hong Kong's Umbrella Movement', *The Round Table* 103(6): 571–580.

Chan, J. M. and J. L. Qiu (2011) 'Toward the Research on New Media Events", in J. M. Chan and J. L. Qiu (eds) *Research on New Media Events*, Beijing: Renmin University Press.

Chanan, M. (2008) *The Politics of Documentary*, London: British Film Institute.

Chanan, M. (2012) 'Video, Activism and the Art of Small Media', *Transnational Cinemas* 2(2): 217–226.

Chandhoke, N. (2007) 'Civil Society', *Development in Practice* 17: 607–614.

Chang, E.Y. (2017) 'Queergaming', in B. Ruberg, and A. Shaw (eds) *Queer Game Studies*, Minneapolis, MN: Minnesota University Press, 15–23.

Chang, J. (2005) *Can't Stop Won't Stop: A history of hip-hop generation*, New York: St. Martin's Press.

Change.org (2017) 'Business Model'. Available online: www.change.org/about/business-model.

Chatman, D. (2017) 'Black Twitter and the Politics of Viewing Scandal', in J. Gray, C. Sandvoss and C. L. Harrington (eds) *Fandom: Identities and communities in a mediated world*, second edition, New York: New York University Press, 299–314.

Chatterton, P. and J. Pickerill (2010) 'Everyday Activism and Transitions towards Post-capitalist Worlds', *Transactions of the Institute of British Geographers* 35(4): 475–490.

Chen, A. (2010) 'What's Really behind those Philadelphia "Flash Mobs": A breakdancing crew'. Available online: http://gawker.com/5504131/whats-really-behind-those-philadelphia-flash-mobs-a-breakdancing-crew.

Chen, A. (2012a) 'Facebook Releases New Content Guidelines, Now Allows Bodily Fluids', *Gawker*, 16 February. Available online: http://gawker.com/5885836/facebook-releases-new-content-guidelines-now-allows-bodily-fluids.

Chen, A. (2012b) 'Inside Facebook's Outsourced Anti-Porn and Gore Brigade, Where "Camel Toes" Are More Offensive Than "Crushed Heads"', *Gawker*, 16 February. Available online: https://web.archive.org/web/20140301083111/http://gawker.com/5885714/inside-facebooks-outsourced-anti+porn-and-gore-brigade-where-camel-toes-are-more-offensive-than-crushed-heads.

Chen, A. (2014) 'The Laborers Who Keep Dick Pics and Beheadings out of Your Facebook Feed', *Wired*, 23 October. Available online: www.wired.com/2014/10/content-moderation/.

Chen, A. (2017) 'The Human Toll of Protecting the Internet from the Worst of Humanity', *The New Yorker*, 28 January. Available online: www.newyorker.com/tech/elements/the-human-toll-of-protecting-the-internet-from-the-worst-of-humanity.

Chen, L. (2016) 'The Emergence of the Anti-Hallyu Movement in China', *Media, Culture & Society* 39(3): 374–390.

Chen, L. (2017) *Chinese Fans of Japanese and Korean Pop Culture: Nationalistic narratives and international fandom*, Oxon: Routledge.

Cheney-Lippold, J. (2017) *We Are Data: Algorithms and the making of our digital selves*, New York: New York University Press.

Chesters, G. and I. Welsh (2004) 'Rebel Colours: "Framing" in global social movements', *Sociological Review* 53(3): 314–335.

Cheyfitz, E. (1997) *The Poetics of Imperialism: Translation and colonization from The Tempest to Tarzan*, expanded edition, Philadelphia, PA: University of Pennsylvania.

Chikamori, T. (2009) 'Between the "Media City" and the "City as a Medium"', *Theory, Culture & Society* 26(4): 147–154.

Chin, B. and L. H. Morimoto (2013) 'Towards a Theory of Transcultural Fandom', *Participations: Journal of Audience Research* 10(1): 92–108.

Chin, B., A. Punathambekar and S. Shresthova (2018) 'Advancing Transcultural Fandom: A conversation', in M. A. Click and S. Scott (eds) *The Routledge Companion to Media Fandom*, New York: Routledge, 298–306.

Chmielewska, E. (2007) 'Graffiti and Place', *Space and Culture* 10(2): 145–169.

Chouliaraki, L. (2010a) 'Ordinary Witnessing in Post-television News: Towards a new moral imagination', *Critical Discourse Studies* 7(4): 305–319.

Chouliaraki, L. (2010b) 'Self-mediation: New media and citizenship', *Critical Discourse Studies* 7(4): 227–232.

Chouliaraki, L. (2013a) 'Re-Mediation, Inter-Mediation, Trans-Mediation: The cosmopolitan trajectories of convergent journalism', *Journalism Studies* 14(2): 267–283.

Chouliaraki, L. (2013b) *The Ironic Spectator: Solidarity in the age of post-humanitarianism*, Cambridge: Polity Press.

Chouliaraki, L. (2015) 'Digital Witnessing in Conflict Zones: The politics of remediation', *Information, Communication & Society* 18(11): 1362–1377.

Chouliaraki, L. (2016) 'The Securitization of Citizen Reporting in Post-Arab Spring Conflicts', in M. Baker and B. Blaagaard (eds) *Citizen Media and Public Spaces*, London: Routledge, 189–207.

Chouliaraki, L. (2017) 'Symbolic Bordering: The self-representation of migrants and refugees in digital news', *Popular Communication: The International Journal of Media and Culture* 15: 78–94.

Chouliaraki, L. and B. B. Blaagaard (2013) 'Cosmopolitanism and the New News Media', *Journalism Studies* 14(2): 150–155

Christensen, C. (2011) 'Twitter Revolutions? Addressing social media and dissent', *The Communication Review* 14(3): 155–157.

Christensen, C. (2017) 'Images That Last? Iraq videos from YouTube to WikiLeaks', in J. Eder and C. Klonk (eds) *Image Operations: Visual media and political conflict*, Manchester: Manchester University Press, 118–127.

Christensen, H. S. (2011) 'Political Activities on the Internet: Slacktivism or political participation by other means?', *First Monday* 16(2).

Christensen, T. H. and I. Røpke (2010) 'Can Practice Theory Inspire Studies of ICTs in Everyday Life', in B. Bräuchler and J. Postill (eds) *Theorising Media and Practice*, New York: Berghahn Books, 233–256.

Christiansen, P. (2012) 'Between a Mod and a Hard Place', in E. Champion (eds) *Game Mods: Design, Theory and Criticism*, Pittsburgh, PA: ETC Press, 27–50.

Christie, M. (dir.) (2003) Jump London, Channel 4. IMDb entry: www.imdb.com/title/tt0380472/.

Christl, W. (2017) 'Corporate Surveillance in Everyday Life', Cracked Labs. Available online: http://crackedlabs.org/en/corporate-surveillance.

Christophers, B. (2018) *The New Enclosure: The appropriation of public land in neoliberal Britain*, London: Verso.

Ciranda. www.ciranda.net.

CKAN (n.d.) 'Homepage'. Available online: https://ckan.org/.

Clapp, J. T., J. A. Roberts, B. Dahlberg, L. S. Berry, L. M. Jacobs, E. A. Emmett and F. K. Barg (2016) 'Realities of Environmental Toxicity and Their Ramifications for Community Engagement', *Social Science and Medicine* 170: 143–151.

Clark, J. A. (2006) 'Field Research Methods in the Middle East', *PS: Political Science & Politics*, 39(3): 417–424.

Clark, L. S. (2016) 'Participants on the Margins: #BlackLivesMatter and the role that shared artifacts of engagement played among minoritized political newcomers on Snapchat, Facebook, and Twitter', *International Journal of Communication* 10(26): 235–253.

Clark, M. (2018) 'Black Twitter', in D. Freelon, L. Lopez, M. D. Clark, and S. J. Jackson (eds) *How Black Twitter and Other Social Media Communities Interact with Mainstream News*, Miami, FL: John and James Knight Foundation, 36–46.

Clarke, J., K. M. Coll, E. Dagnino and C. Neveu (2014) *Disputing Citizenship*, Bristol, UK: Policy Press.

Class War (2017) 'Belgravia: The anarchist present', *Class War*, 1 February. Available online: www.classwarparty.org.uk/belgravia-anarchist-present/.

Cleaver, H. (1994) 'The Chiapas Uprising', *Studies in Political Economy* 44(1): 141–157.

Cleaver, H. (1995) 'The Zapatistas and the Electronic Fabric of Struggle'. Available online: https://la.utexas.edu/users/hcleaver/zaps.html.

Cleen, B. De (2015) '"Flemish Friends, Let Us Separate!": The discursive struggle for Flemish Nationalist Civil Society in the media', *Javnost – The Public* 22(1): 37–54.

Clegg, J. and T. Butryn (2012) 'An Existential Phenomenological Examination of Parkour and Freerunning', *Qualitative Research in Sport, Exercise and Health* 4(4): 320–340.

Clément, F. (2014). 'Players/gamers', in M. J. P. Wolf and B. Perron (eds) *The Routledge Companion to Video Game Studies*, New York: Routledge, 197–203

Clery, D. (2011) 'Galaxy Zoo Volunteers Share the Pain and Glory of Research', *Science* 333(6039): 173–175.

Click, M. A. and S. Scott (2018) 'Introduction', in M. A. Click, and S. Scott (eds) *The Routledge Companion to Media Fandom*, New York: Routledge, 1–5.

Clogston, J. (1994) 'Disability Coverage in American Newspapers', in J. Nelson (ed.) *The Disabled, the Media, and the Information Age*, Westport, CT: Greenwood Press, 45–57.

Cloke, P., P. Cooke, J. Cursons, P. Milbourne and R. Widdowfield (2000) 'Ethics, Reflexivity and Research: Encounters with homeless people', *Ethics Place and Environment* 3(2): 133–154.

Coban, B. and B. Yanikkaya (2013) 'Barış Medyası Olarak Alternatif Medyalar ya da Alternatif Medyanın Barış ile İmtihanı' [Alternative Media as Peace Media or Alternative Media's Test with Peace], in T. İlter, H. Aliefendioğlu, P. Behçetoğulları and N. Kara (eds) *(Re)Making and Undoing of Peace/Conflict*, Famagusta, Cyprus: Eastern Mediterranean University Press, 17–22.

Cockburn, A., J. St. Clair and A. Sekula (eds) (2000) *Five Days that Shook the World: Seattle and beyond*, London: Verso.

Coëgnarts, M. and P. Kravanja (2012) 'From Thought to Modality: A Theoretical framework for analysing structural-conceptual metaphors and image metaphors in film', *Image & Narrative* 13(1): 96–113.

Cohen, S. (2011) *States of Denial: Knowing about atrocities and suffering*, Cambridge: Polity Press.

Cohn, J. P. (2008) 'Citizen Science: Can volunteers do real research?', *BioScience* 58(3): 192–197.

Cole, J., M. Caloud and J. Brennon (2003) *9–11 Survivor*. Available online: http://ljudmila.org/~selectparks/archive/911survivor/911game.html.

Cole, P., D. Struthers and K. Zimmer (eds) (2017) *Wobblies of the World: A global history of the IWW*, London: Pluto Press.

Coleman, E. G. (2013) *Coding Freedom: The ethics and aesthetics of hacking*, Princeton, NJ: Princeton University Press.

Coleman, G. (2014) *Hacker, Hoaxer, Whistleblower, Spy: The many faces of Anonymous*, London: Verso.

Coleman, J. S. (1988) 'Social Capital in the Creation of Human Capital', *American Journal of Sociology* 94: S95–S120.

Coleman, J. S. (1999) 'Cutting out the Middle Man: From virtual representation to direct deliberation', in B. N. Hague and B. D. Loader (eds) *Digital Democracy: Discourse and decision-making in the information age*, London: Routledge, 195–210.

Collier, G. (1999) *Basta!: Land and the Zapatista rebellion in Chiapas*, Oakland, CA: Food First Books.

Conant, J. (2010) *A Poetics of Resistance: The revolutionary public relations of the Zapatista insurgency*. Oakland, CA: AK Press.

Condis, M. (2018) *Gaming Masculinity: Trolls, fake geeks, and the gendered battle for online culture*, Iowa City, IA: Iowa University Press.

Condry, I. (2001) 'A History of Japanese Hip-Hop: Street dance, club scene, pop market', in T. Mitchell (ed.) *Global Noise: Rap and hip-hop outside the USA*, Middletown, CT: Wesleyan University Press, 222–247.

Confessore, N., G. J. X. Dance, R. Harris and M. Hansen (2018) 'The Follower Factory', *The New York Times*, 27 January. Available online: www.nytimes.com/interactive/2018/01/27/technology/social-media-bots.html.

Conklin, B. A. (1997) 'Body Paint, Feathers & VCRs: Aesthetics and authenticity in Amazonian activism', *American Ethnologist* 24(4): 711–737.

Connolly, K., A. Chrisafis, P. McPherson, S. Kirchgaessner, B. Haas, D. Phillips and M. Safi (2016) 'Fake News: An insidious trend that's fast becoming a global problem', *The Guardian*, 2 December.

Conover, M., J. Ratkiewicz and M. Francisco (2011) 'Political Polarization on Twitter', *ICWSM* 133(26): 89–96.

ControlaTuGobierno (n.d.) 'ControlaTuGobierno. Herramientas para el control social de la acción gubernamental'. Available online: https://controlatugobierno.com/.

Conway, J. (2003) 'Civil Resistance and the Diversity of Tactics in the Anti-Globalization Movement: Problems of violence, silence, and solidarity in activist politics', *Osgoode Hall Law Journal* 41(2): 505–530.

Conway, J. (2004) 'Citizenship in a Time of Empire: The World Social Forum as a new public space', *Citizenship Studies* 8(4): 367–381.

Conway, J. (2005) 'Social Forums, Social Movements and Social Change: A response to Peter Marcuse on the subject of the World Social Forum', *International Journal of Urban and Regional Research* 29(2). Available online: https://onlinelibrary.wiley.com/doi/full/10.1111/j.1468-2427.2005.00594.x.

Conway, J. (2008) 'Reading Nairobi: Place, space, and difference at the 2007 World Social Forum', *Societies without Borders* 3: 48–70.

Conway, J. (2013) *Edges of Global Justice: The World Social Forum and its 'others'*, London: Routledge.

Conway, J. and J. Singh (2009) 'Is the World Social Forum a Transnational Public Sphere? Nancy Fraser, critical theory and the containment of radical possibility', *Theory, Culture & Society* 26(5): 61–84.

Cook, C. E., K. Geels and P. Bakker (2016) *Hyperlocal Revenues in the UK and Europe: Mapping the road to sustainability and resilience*, London: Nesta.

Cooke, T. (2018) 'A Cybernetic Epistemology of iOS: Or, how jailbreaking (re)establishes time as a crucial dimension of user-first privacy', *Academia.edu*. Available online: www.academia.edu/36775519/A_Cybernetic_Epistemology_of_iOS._Or_How_Jailbreaking_re_Establishes_Time_as_a_Crucial_Dimension_of_User-first_Privacy.

Coombs, M. (2011) 'Lobstering and Common Pool Resource Management in Maine', *Grassroots Economic Organizing (GEO) Newsletter* 2(9). Available online: http://geo.coop/node/654.

Cooper, C. B. and B. V. Lewenstein (2016) 'Two Meanings of Citizen Science', in D. Cavalier and E. B. Kennedy (eds) *The Rightful Place of Science: Citizen science*, Tempe, AZ: Consortium for Science, Policy, and Outcomes, 51–62.

Cooper, C. B., J. Shirk and B. Zuckerberg (2014) 'The Invisible Prevalence of Citizen Science in Global Research: Migratory birds and climate change', *PLoS ONE* 9(9). Available online: https://doi.org/10.1371/journal.pone.0106508.

Cooper, D. (2017) 'Prefiguring the State', *Antipode* 49(2): 335–356.

Cooper, F. (2014) *Citizenship between Empire and Nation: Remaking France and French Africa, 1945–1960*, Princeton, NJ: Princeton University Press.

Cooper, G. (2008) 'Tawhaki and Māui: Critical literacy in indigenous epistemologies', *Critical Literacy* 2(1): 37–42.

Cooper, G. (2012) 'Kaupapa Maori Research: Epistemic wilderness as freedom?', *New Zealand Journal of Educational Studies* 47(2): 64–73.

Cooper, G. (2017) 'Rights and Responsibilities When Using User-Generated Content to Report Crisis Events', in H. Tumber and S. Waisbord (eds) *The Routledge Companion to Media and Human Rights*, New York and London: Routledge, 257–267. Available online: https://openaccess.city.ac.uk/id/eprint/19100/.

Coover, R. (2009) 'On Verité to Virtual: Conversations on the frontier of film and anthropology', *Visual Studies* 24(3): 235–249.

Coppa, F. (2008) 'Women, Star Trek, and the Early Development of Fannish Vidding', *Transformative Works and Cultures* 1.

Coppa, F. (2017) 'Five Things Fan Fiction Is and One It Isn't', in F. Coppa (ed.) *The Fanfiction Reader: Folk tales for the digital age*, Lansing, MI: University of Michigan Press, 1–17.

Coppa, F., A. Lothian and T. Turk (2018) 'Vidding and Identity: A conversation', in M. A. Click and S. Scott (eds) *The Routledge Companion to Media Fandom*, New York: Routledge, 230–240.

Corburn, J. (2005) *Street Science: Community knowledge and environmental health justice*, Cambridge, MA: The MIT Press.

Corens, C. (2016) 'Dislocation and Record-Keeping: The counter archives of the Catholic diaspora', *Past and Present* 230(11): 269–287.

Costanza-Chock, S. (2012) 'Mic Check! Media cultures and the Occupy Movement', *Social Movement Studies* 11(3–4): 375–385.

Costanza-Chock, S. (2014) *Out of the Shadows, Into the Streets! Transmedia organizing in the immigrant rights movement*, Cambridge, MA: The MIT Press.

Cottle, S. (2006) 'Mediatized Rituals: Beyond manufacturing consent', *Media, Culture & Society* 28(3): 411–432.

Cottle, S. (2014) 'Series Editor's Preface', in E. Thorsen and S. Allan (eds) *Citizen Journalism: Global perspectives*, Volume 2, New York: Peter Lang, ix–xii.

Cottle, S. and L. Lester (2011) *Transnational Protests and the Media*, New York: Peter Lang.

Couldry, N. (2004) 'Theorising Media as Practice', *Social Semiotics* 14(2): 115–132.

Couldry, N. (2008a) 'Mediatization or Mediation? Alternative understandings of the emergent space of digital storytelling', *New Media & Society* 10(3): 373–391.

Couldry, N. (2008b) 'Digital Storytelling, Media Research and Democracy: Conceptual choices and alternative futures', in K. Lundby (ed.) *Mediatized Stories: Self-representation in new media*, New York: Peter Lang, 41–60.
Couldry, N. (2010) *Why Voice Matters: Culture and politics after neoliberalism*, London: Sage.
Couldry, N. (2011) 'More Sociology, More Culture, More Politics', *Cultural Studies* 25(4–5): 487–501.
Couldry, N. (2012) *Media, Society, World: Social theory and digital media practice*, Cambridge: Polity Press.
Couldry, N. (2015) 'The Myth of "Us": Digital networks, political change and the production of collectivity', *Information, Communication & Society* 18(6): 608–626.
Couldry, N. (2017) 'Surveillance Democracy', *Journal of Information Technology and Politics* 18(2): 182–188.
Couldry, N. and J. Curran (2003) *Contesting Media Power: Alternative media in a networked world*, Lanham, MD: Rowman and Littlefield.
Couldry, N. and A. Hepp (2013) 'Conceptualizing Mediatization: Contexts, traditions, arguments', *Communication Theory* 23: 191–202.
Couldry, N. and A. Hepp (2017) *The Mediated Construction of Reality*, Cambridge: Polity Press.
Couldry, N., A. Hepp and F. Krotz (eds) (2010) *Media Events in a Global Age*, Abingdon, UK: Routledge.
Couldry, N. and A. McCarthy (2004) *MediaSpace: Place, scale and culture in a media age*, New York: Routledge.
Couldry, N. and A. Powell (2014) 'Big Data from the Bottom Up', *Big Data & Society* 1(2): 1–5.
Couture, S., G. King, S. Toupin and B. Lentz (2016) 'Can Media and Technology Help Make "Another World Possible"? Reflections from the Media@ McGill Research Delegation to the 2015 World Social Forum in Tunisia', *Canadian Journal of Communication* 41(1): 157–167.
Cox, T. E., J. Philippoff, E. Baumgartner and C. M. Smith (2012) 'Expert Variability Provides Perspective on the Strengths and Weaknesses of Citizen-driven Intertidal Monitoring Program', *Ecological Applications* 22(4): 1201–1212.
Coy, P. G. and T. Hedeen (2005) 'A Stage Model of Social Movement Co-optation: Community mediation in the United States', *The Sociological Quarterly* 46(3): 405–435.
Coyer, K. and A. Hintz (2010) 'Developing the "Third Sector": Community media policies in Europe', in B. Klimkiewicz (ed.) *Media Freedom and Pluralism: Media policy challenges in the enlarged Europe*, Budapest: Central European University Press. Available online: http://books.openedition.org/ceup/2186.
Craig, L. S. and L. McInroy (2014) 'You Can Form a Part of Yourself Online: The influence of new media on identity development and coming out for LGBTQ youth', *Journal of Gay & Lesbian Mental Health* (18): 95–109.
Crawford, K. and T. Gillespie (2016) 'What Is a Flag for? Social media reporting tools and the vocabulary of complaint', *New Media & Society* 18(3): 410–428.
Crenshaw, K. (1989) 'Demarginalizing the Intersection of Race and Sex: A black feminist critique of antidiscrimination doctrine, feminist theory, and antiracist politics', *Feminist Legal Theory: Readings in Law and Gender* (1): 57–80.
Cresswell, T. (1992) 'The Crucial "Where" of Graffiti: A geographical analysis of reactions to graffiti in New York', *Environment and Planning D: Society and Space* 10(3), 329–344.
Cresswell, T. (1996) *In Place/Out of Place: Geography, ideology, and transgression*, Minneapolis, MN: University of Minnesota Press.
Cresswell, T. (2009) 'Place', in N. Thrift and R. Kitchen (eds) *International Encyclopedia of Human Geography*, Volume 8, Oxford: Elsevier, 169–177.
Critical Art Ensemble (1994) 'Electronic Civil Disobedience', Critical-Art.net. Available online: www.critical-art.net/books/ecd/ecd2.pdf.
Critical Art Ensemble (1996) *Electronic Civil Disobedience and Other Unpopular Ideas*, Brooklyn, NY: Autonomedia.
Croeser, S. and T. Highfield (2014) 'Occupy Oakland and #oo: Uses of Twitter within the Occupy movement', *First Monday* 19(3). Available online: https://firstmonday.org/ojs/index.php/fm/article/view/4827.
Cronin, M. (2003) *Translation and Globalization*, London: Routledge.
Cross, K. (2016) 'Memory, Guardianship and the Witnessing Amateur in the Emergence of Citizen Journalism', in M. Baker and B. B. Blaagaard (eds) *Citizen Media and Public Spaces: Diverse expressions of citizenship and dissent*, London: Routledge, 225–238.
Crossley, N., and J. M. Roberts (2004) *After Habermas: New perspectives on the public Sphere*, Malden, MA: Blackwell
Cruz, E. G. and H. Thornham (2015) 'Selfies beyond Self-Representation: The (theoretical) f(r)ictions of a practice', *Journal of Aesthetics and Culture* 7: 1–10.
Cubitt, S. (2016) 'Ecopolitics of Cinema', in Y. Tzioumakis and C. Molloy (eds) *The Routledge Companion to Cinema and Politics*, New York and London: Routledge, 40–49.
Cucco, I. and M. Fonte (2015) 'Local Food and Civic Food Networks as a Real Utopias Project', *Socio.hu Social Science Review*, Special issue in English No. 3.
Culp, A. C. (2013) 'Dispute or Disrupt? Desire and violence in protests against the Iraq War', *Affinities: A Journal of Radical Theory, Culture, and Action* 6(1): 16–47.

Cultural Survival Quarterly Magazine (1987) 'Native Rice, Native Hands: The Ikwe Marketing Collective'. Available online: www.culturalsurvival.org/publications/cultural-survival-quarterly/native-rice-native-hands-ikwe-marketing-collective.

Cuninghame, P. (2002) Autonomia: A movement of refusal. Social movements and social conflict in Italy in the 1970s, Unpublished PhD thesis, London: Middlesex University. Available online: http://eprints.mdx.ac.uk/6688/.

Cunningham, S. and T. Nguyen (2003) 'Actually Existing Hybridity: Vietnamese diasporic music video', in H. K. Karim (ed.) *The Media of Diaspora*, London and New York: Routledge, 119–134.

Cunningham, W. (2005) 'Correspondence on the Etymology of Wiki', Cunningham and Cunningham, Inc. Available online: http://c2.com/doc/etymology.html.

Curle, A. (1971) *Making Peace*, London: Tavistock.

Curle, A. (1990) *Tools for Transformation: A personal study*, London: Hawthorn Press.

Curran, J. (1991a) 'Rethinking the Media as a Public Sphere', in P. Dahlgren and C. Sparks (eds) *Communication and Citizenship*, London: Routledge, 27–57.

Curran, J. (1991b) 'Mass Media and Democracy: A reappraisal', in J. Curran and M. Gurevitch (eds) *Mass Media and Society*, London: Arnold, 82–117.

Curran, J., N. Fenton and D. Freedman (eds) (2012) *Misunderstanding the Internet*, London: Routledge.

Custódio, L. (2017) *Favela Media Activism: Counterpublics for human rights in Brazil*, Lanham, MD, New York and London: Lexington Books.

Cutler, C. and U. Royneland (2015) 'Where the Fuck Am I From? Hip-Hop youth and the (re)negotiation of language and identity in Norway and the US', in J. Nortier and B. A. Svendsen (eds) *Language, Youth and Identity in the 21st Century*, Cambridge: Cambridge University Press, 139–164.

Cvetkovich, A. (2012) 'Queer Archival Futures: Case Study Los Angeles', *e-misférica* 9(1). Available online: www.hemisphericinstitute.org/hemi/en/e-misferica-91/cvetkovich.

Dahl, R. A. (1961) *Who Governs? Democracy and power in an American city*, New Haven, CT: Yale University Press.

Dahlberg, L. and S. Phelan (eds) (2011) *Discourse Theory and Critical Media Politics*, Basingstoke, UK: Palgrave Macmillan.

Dahlgren, P. (2005) 'The Internet, Public Spheres, and Political Communication: Dispersion and deliberation', *Political communication* 22(2): 147–162.

Dahlgren, P. (2006) 'Doing Citizenship: The cultural origins of civic agency in the public sphere', *European Journal of Cultural Studies* 9(3): 267–286.

Dahlgren, P. (2009) *Media and Political Engagement*, Cambridge: Cambridge University Press.

Dahlgren, P. (2013) *The Political Web: Media, participation and alternative democracy*, London: Palgrave Macmillan.

Daisuke, O. and M. Ito (2003) 'Camera Phones Changing the Definition of Picture-worthy', *Japan Media Review*. Available online: www.dourish.com/classes/ics234cw04/ito3.pdf.

Danesi, M. (ed.) (2013) *Encyclopedia of Media and Communication*, Toronto, Buffalo, NY and London: University of Toronto Press.

Dangl, B. (2007) *The Price of Fire: Resource Wars and Social Movements in Bolivia*, Oakland, CA: AK Press.

Darling, J. and V. Squire (2013) 'Everyday Enactments of Sanctuary: The UK City of Sanctuary Movement', in R. K. Lippert and S. Rehaag (eds) *Sanctuary Practices in International Perspectives: Migration, citizenship and social movements*, London: Routledge, 191–204.

Darmon, K. (2013) 'Introduction: Protest in the new media ecology', *Networking Knowledge: Journal of the MeCCSA Postgraduate Network* 6(3): 1–2.

Das Commons Institut (2018) 'Was sind Commons?', *Das Commons Institut*. Available online: https://commons-institut.org/was-sind-commons.

Davies, K. (2014) 'Citizen Journalism in Indonesia's Disputed Territories', in E. Thorsen and S. Allan (eds) *Citizen Journalism: Global Perspectives*, Volume 2, New York: Peter Lang, 243–258.

Davies, S. R. (2017) *Hackerspaces: Making the maker movement*, Cambridge: Polity Press.

Davies, W. (2018) 'Short Cuts', *London Review of Books* 40(7), 5 April: 20–21. Available online: www.lrb.co.uk/v40/n07/william-davies/short-cuts.

Davis, A. Y. (1971) 'Political Prisoners, Prisons, and Black Liberation'. Available online: www.historyisaweapon.com/defcon1/davispoprprblli.html.

Davis, S. (2000) 'Public Journalism: The case against', *Journalism Studies* 1(4): 686–686.

Davison, P. (2012) 'The Language of Internet Memes', in M. Mandiberg (ed.) *The Social Media Reader*, New York: NYU Press, 120–134.

Day, B. (2011) 'In Defense of Copyright: Record labels, creativity, and the future of music', *Seton Hall Journal of Sports and Entertainment Law* 21(1): 61–103.

Day, J. (2005) 'We Had Fifty Images an Hour', *The Guardian*, 11 July. Available online: www.theguardian.com/technology/2005/jul/11/mondaymediasection.attackonlondon.

Dayan, D. and E. Katz (1992) *Media Events: The live broadcasting of history*, Cambridge, MA: Harvard University Press.
Dayter, D. (2015) 'Small Stories and Extended Narratives on Twitter', *Discourse, Context and Media* 10: 19–26.
DCMS (2019) 'Online Harms White Paper', presented to Parliament by the Secretary of State for Digital, Culture, Media and Sport and the Secretary of State for the Home Department by Command of Her Majesty, April. Available online: https://assets.publishing.service.gov.uk/government/uploads/system/uploads/attachment_data/file/793360/Online_Harms_White_Paper.pdf.
Dean, J. (2005) 'Communicative Capitalism: Circulation and the foreclosure of politics', *Cultural Politics* 1(1): 51–74..
Dean, J. (2011) 'Claiming Division, Naming a Wrong', *Theory and Event* 14(4).
Dearing, J. W. and E. M. Rodgers (1996) *Agenda-Setting*, Thousand Oaks, CA: Sage.
Debord, G. and G. Wolman (1956/2006) 'A User's Guide to Détournement', in K. Knabb (ed.) *Situationist International Anthology*, Bureau of Public Secrets. Available online: www.bopsecrets.org/SI/detourn.htm.
Decolonize Oakland (2012) 'Communiqué from Decolonize Oakland 3.18.12'. Available online: https://occupyoakland.org/2012/03/communique-from-decolonize-oakland-3-18-12/.
Deger, J. (2011) 'Global Indigenous Media: Cultures, poetics and politics', *The Australian Journal of Anthropology* 22(1): 142–143.
DeHaan, S., E. Kuper, C. J. Magee, L. Bigelow and S. B. Mustanski (2013) 'The Interplay between Online and Offline Explorations of Identity, Relationships, and Sex: A mixed-methods study with LGBT Youth', *The Journal of Sex Research* 50(5): 421–434.
Delamarre, C. (dir.) (2014) Brick Mansions. IMDb entry: www.imdb.com/title/tt1430612.
DeLaure, M., M. Fink and M. Dery (eds) (2017) *Culture Jamming: Activism and the art of cultural resistance*, New York: NYU Press.
Deleuze, G. and F. Guattari (1980) *Mille Plateaux: Capitalisme et schiziphrénie*, Volume 2, Paris: Les Éditions de Minuit.
Deleuze, G. and F. Guattari (1980/1987) *A Thousand Plateaus: Capitalism and Schizophrenia*, translated by B. Massumi, Minneapolis: University of Minnesota Press.
Deleuze, G. and F. Guattari (1980/1988) *A Thousand Plateaus: Capitalism and schizophrenia*, translated by B. Massumi, London and New York: The Athlone Press.
Delhaye, C. (2008) 'Immigrants' Artistic Practices in Amsterdam, 1970–2007: A political issue of inclusion and exclusion', *Journal of Ethnic and Migration Studies* 34(8): 1301–1321.
della Porta, D. (2005) 'Deliberation in Movement: Why and how to study deliberative democracy and social movements', *Acta Politica* 40(3): 336–350.
della Porta, D., M. Andretta, L. Mosca and H. Reiter (2006) *Globalization from Below: Transnational activists and protest networks*, Minneapolis, MN: University of Minnesota Press.
della Porta, D. and M. Diani (2011) 'Social Movements', in M. Edwards (ed.) *The Oxford Handbook of Civil Society*, Oxford: Oxford University Press, 68–79.
della Porta, D. and A. Mattoni (2015) 'Social Movements', in G. Mazzoleni, K. Barnhurst, R. C. M Maia and H. Weddler (eds) *The International Encyclopedia of Political Communication*, Malden, MA: John Wiley & Sons, 1496–1502.
della Porta, D. and H. Reiter (1998) *Policing Protest: The control of mass demonstrations in Western democracies*, Minneapolis, MN: University of Minnesota Press.
della Porta, D. and S. G. Tarrow (eds) (2004) *Transnational Protest and Global Activism*, Lanham, MD: Rowman & Littlefield.
della Ratta, D. (2018) *Shooting a Revolution: Visual media and warfare in Syria*, London: Pluto Press.
Dencik, L. (2015) 'Social Media and the "New Authenticity" of Protest', in L. Dencik and O. Leistert (eds) *Critical Perspectives on Social Media and Protest: Between control and emancipation*, London: Rowman & Littlefield, 203–218.
Dencik, L. and J. Cable (2017) 'The Advent of Surveillance Realism: Public opinion and activist responses to the Snowden leaks', *International Journal of Communication* 11: 763–781.
Dencik, L. and O. Leistert (eds) (2015) *Critical Perspectives on Social Media and Protest: Between control and emancipation*, New York: Rowman & Littlefield.
Dencik, L. and P. Wilkin (2015) *Worker Resistance and Media: Challenging global corporate power in the 21st century*, New York: Peter Lang.
Dennis, C. (2006) 'Afro-Columbian Hip-Hop: Globalization, popular music and ethnic identities', *Studies in Latin American Popular Culture* 25: 271–295.
Denson, S. and J. Leyda (eds) (2016) *Post-Cinema: Theorizing 21st-century film*, Falmer, UK: Reframe Books.
Denzin, N. K. and Y. S. Lincoln (2014) 'Introduction: Critical methodologies and critical inquiry', in N. K. Denzin, Y. S. Lincoln and L. T. Smith (eds) *Handbook of Critical and Indigenous Methodologies*, London: Sage, 1–20.
Derrida, J. (1995/1998) *Archive Fever: A Freudian impression*, translated by E. Prenowitz, Chicago, IL: University of Chicago Press.
Dery, M. (1993a) 'Culture Jamming: Hacking, slashing, and sniping in the empire of the signs'. Available online: http://project.cyberpunk.ru/idb/culture_jamming.html.

Dery, M. (1993b) *Culture Jamming: Hacking, slashing and sniping in the empire of signs*, Westfield, NJ: Open Magazine.

Des Verney Sinnette, E. (1989) *Arthur Alfonso Schomburg, Black Bibliophile and Collector: A biography*, Detroit, MI: New York Public Library and Wayne State University Press.

Desai, M. (2016) 'SWS 2015 Feminist Lecture: The gendered geographies of struggle: The World Social Forum and its sometimes overlapping other worlds', *Gender & Society* 30(6): 869–889.

Desai, R. (dir.) (2014) Miners Shot Down, Uhuru Productions, South Africa. IMDb entry: www.imdb.com/title/tt3677896/.

Deseriis, M. (2015) *Improper Names: Collective pseudonyms from the Luddites to Anonymous*, Minneapolis, MN: University of Minnesota Press.

De-Shalit, A. (2001) 'Ten Commandments of How to Fail in an Environmental Campaign', *Environmental Politics* 10(1): 111–137.

Deutch, J. and H. Habal (2018) 'The Syrian Archive: A methodological case study of open-source investigation of state crime using video evidence from social media platforms', State Crime Journal 7(1): 46–76.

Deuze, M. (2006a) 'Ethnic Media, Community Media and Participatory Culture', *Journalism* 7: 262–280.

Deuze, M. (2006b) 'Participation, Remediation, Bricolage: Considering principal components of a digital culture', *The Information Society* 22(2): 63–75.

Deuze, M. (2012) *Medialife*, Cambridge, UK and Malden, MA: Polity Press.

Deuze, M. and The Janissary Collective (2012) 'Mobile Media Life', in P. Snickars and P. Vonderau (eds) *Moving Data: The iPhone and the future of media*, New York: Columbia University Press, 296–308.

Dewey, C. (2015) 'Wikipedia Has a Ton of Money. So why is it begging you to donate yours?', *Washington Post*, 2 December. Available online: www.washingtonpost.com/news/the-intersect/wp/2015/12/02/wikipedia-has-a-ton-of-money-so-why-is-it-begging-you-to-donate-yours/?utm_term=.9508f89d7ba1.

Dewey, J. (1927) *The Public and Its Problems*, Athens, OH: Ohio University Press.

Diamond, L. (2010) 'Liberation Technology', *Journal of Democracy* 21(3): 69–83.

Dibashram: A Space for the Migrant Worker in Singapore (n.d.) 'About'. https://dibashrammock.wordpress.com/about/.

Dickinson, E. (2011) 'The First WikiLeaks Revolution?', *Foreign Policy*, 13 January. Available online: http://wikileaks.foreignpolicy.com/posts/2011/01/13/wikileaks_and_the_tunisia_protests.

Dickinson, J. L. and R. Bonney (2012) 'Introduction: Why citizen science', in J. L. Dickinson and R. Bonney (eds) *Citizen Science: Public participation in environmental research*, Ithaca, NY: Cornell University Press, 1–14.

Dickinson, M. (ed.) (1999) *Rogue Reels: Oppositional Film in Britain, 1945–90*, London: BFI.

Diehl, P. (2016) 'Exploring Peace: Looking beyond war and negative peace', *International Studies Quarterly* 60(1): 1–10.

DiFilippo, D. (2009) 'Flash Mobs Draw S. Phila. Ire: Community calls for police oversight of online networking', *Philadelphia Daily News*, 5 June: 3.

Dijck, J. van (2006/2012) *The Network Society: Social aspects of new media*, third edition, London: Sage.

Dijck, J. van (2007) *Mediated Memories in the Digital Age*, Stanford, CA: Stanford University Press.

Dijck, J. van (2009) 'Users Like You? Theorizing agency in user-generated content', *Media, Culture & Society* 31(1): 41–58.

Dijck, J. van (2013) *The Culture of Connectivity: A critical history of social media*, Oxford: Oxford University Press.

Dijck, J. van (2014) 'Datafication, Dataism and Dataveillance: Big data between scientific paradigm and ideology', *Surveillance & Society* 12(2): 197–208.

Dijck, J. van (2017) 'Flickr: Photo sharing sites between collective and connective memory', in O. Shevchenko (ed.) *Double Exposure: Memory and photography*, Abingdon, UK: Routledge, 211–231.

Dijck, J. van and Poell, T. (2013) 'Understanding Social Media Logic', *Media and Communication* 1(1): 2–14.

Dijck, J. van, T. Poell and M. De Waal (2018) *The Platform Society: Public values in a connective world*, Oxford: Oxford University Press.

Dimitrov, S. and M. Chow (2013) 'Privacy and Security Benefits of Jailbreaking iOS'. Available online: www.cs.tufts.edu/comp/116/archive/fall2013/sdimitrov.pdf.

Dinerstein, A. C. (2015) *The Politics of Autonomy in Latin America: The art of organizing hope*, London: Palgrave Macmillan.

Dirksmeier, P. and I. Helbrecht (2008) 'Time, Non-representational Theory and the "Performative Turn": Towards a new methodology in qualitative social research', *Forum Qualitative Sozialforschung/Forum: Qualitative Social Research* 9(2). Available online: http://nbn-resolving.de/urn:nbn:de:0114-fqs0802558.

Diverlus, R. (2016) 'Re/imagining Artivism', in D. Elliot, M. Silverman and W. Bowman (eds) *Artistic Citizenship: Artistry, social responsibility, and ethical praxis*, New York: Oxford University Press, 189–212.

Dixon, C. (2014) *Another Politics: Talking across today's transformative movements*, Berkeley, CA: University of California Press.

Dobson, M. (2011) Shakespeare and Amateur Performance: A cultural history, Cambridge: Cambridge University Press.

Doerr, N. (2007) 'Is "Another" Public Sphere Actually Possible? The case of "women without" in the European Social Forum process as a critical test for deliberative democracy', *Journal of International Women's Studies* 8(3): 71–87.

Doerr, N. (2008) 'Deliberative Discussion, Language and Efficiency in the World Social Forum Process', *Mobilization: An International Journal* 13(4): 395–410.

Doerr, N. (2012) 'Translating Democracy: How activists in the European Social Forum practice multilingual deliberation', *European Political Science Review* 4(3): 361–384.

Doerr, N. (2014) 'Memory and Culture in Social Movements', in B. Baumgarten, P. Daphi and P. Ullrich (eds) *Conceptualizing Culture in Social Movement Research*, London: Palgrave Macmillan, 206–226.

Doerr, N. (2018) *Political Translation: How social movement democracies survive*, Cambridge: Cambridge University Press.

Doerr, N., A. Mattoni and S. Teune (2014) 'Visuals in Social Movements', in D. della Porta and M. Diani (eds) *The Oxford Handbook for Social Movements*, Oxford: Oxford University Press, 557–566.

Doherty, B., M. Paterson and B. Seel (2000) 'Direct Action in British Environmentalism', in B. Seel, M. Paterson and B. Doherty (eds) *Direct Action in British Environmentalism*, London: Routledge, 15–38.

Domingo, D. and A. Heinonen (2008) 'Weblogs and Journalism', *Nordicom Review* 29(1): 3–15.

Dominguez, R. (2009) 'Electronic Civil Disobedience: Inventing the future of online Agitprop theater', *PMLA* (124)5: 1806–1812.

Donk, W. van de, B. Loader, P. Nixon and D. Rucht (2004) *Cyberprotest: New media, citizens and social movements*, London: Routledge.

Donovan, J. (2013) 'Occupy 3.0: A slow network movement'. Available online: http://wagingnonviolence.org/feature/occupy-3-0-slow-network-movement/.

Donsbach, W. (ed.) (2008) *The International Encyclopedia of Communication*, 12 volumes, Malden, MA: Wiley-Blackwell.

Doran, M. (2004) 'Negotiating between Bourg and Racaille: Verlan as youth identity practice in suburban Paris', in A. Pavlenko and A. Blackledge (eds) *Negotiation of Identities in Multilingual Contexts*, Clevedon: Cromwell Press, 68–92.

Dotson, K. (2012) 'A Cautionary Tale: On limiting epistemic oppression', *Frontiers: A Journal of Women Studies* 33(1): 24–47.

Dotson, K. (2015) 'Inheriting Patricia Hill Collins's *Black Feminist* Epistemology', *Ethnic and Racial Studies* 38(13): 2322–2328.

Dotson, K. (2016) 'Between Rocks and Hard Places: Introducing black feminist professional philosophy', *Journal of Black Studies and Research* 46 (2): 46–56.

Doty, A. (1993) *Making Things Perfectly Queer: Interpreting mass culture*, Minneapolis, MN: University of Minnesota Press.

Doumani, B. (2009) 'Archiving Palestine and the Palestinians: The patrimony of Ihsan Nimr', *Jerusalem Quarterly* 36: 3–12.

Dowling, E., R. Nunes and B. Trott (eds) (2007) 'Immaterial and Affective Labour: Explored', in E. Dowling, R. Nunes and B. Trott (eds) *Immaterial and Affective Labour: Explored*. Special issue of *Ephemera: Theory and Politics in Organization* 7(1): 1–7. Available online: www.ephemerajournal.org/issue/immaterial-and-affective-labour-explored.

Dowmunt, T. (2002) 'Access: Television at the margins', in P. Holland (ed.), *The Television Handbook*, London and New York: Routledge, 188–193.

Downey, A. (ed.) (2015) *Dissonant Archives: Contemporary visual culture and contested narratives in the Middle East*, London: I. B. Tauris.

Downey, J. and N. Fenton (2003) 'New Media, Counter Publicity and the Public Sphere', *New Media & Society* 5(2): 185–202.

Downing, J. D. H. (1984) *Radical Media: The political experience of alternative communication*, Boston, MA: South End Press.

Downing, J. D. H. (2001) *Radical Media: Rebellious communication and social movements*, Thousand Oaks, CA: Sage.

Downing, J. D. H. (2003) 'Audiences and Readers of Alternative Media: The absent lure of the virtually unknown', *Media, Culture & Society* 25(5): 625–645.

Downing, J. D. H. (2008) 'Social Movement Theories and Alternative Media: An evaluation and critique', *Communication, Culture & Critique* 1(1): 40–50.

Downing, J. D. H. (2011) *Encyclopedia of Social Movement Media*. Thousand Oaks, CA: Sage.

Downing, J. D. H., T. Villarreal Ford, G. Gil and L. Stein (2001) *Radical Media: Rebellious communication and social movements*, Thousand Oaks, CA: Sage.

Downing, J. D. H. and C. Husband (2005) *Representing Race: Racisms, ethnicities and media*, London: Sage.

Dragomir, M. and M. Thompson (2014) 'Mapping the World's Digital Media', in Open Society Foundations, *Mapping Digital Media: Global Findings*, New York: Open Society Foundations, 11–19.

Drake, J. (2016) '#ArchivesForBlackLives: Building a community archives of police violence in Cleveland', Medium.com. Available online: https://medium.com/on-archivy/archivesforblacklives-building-a-community-archives-of-police-violence-in-cleveland-93615d777289.

Draper, N. and J. Turow (2019) 'The Corporate Cultivation of Digital Resignation', *New Media and Society* 21(8): 1824–1839.

Dreher, T. (2009) 'Listening Across Difference: Media and multiculturalism beyond the politics of voice', *Continuum: Journal of Media & Cultural Studies* 23(4): 445–458.

Drew, J. (2013) *A Social History of Contemporary Democratic Media*, London and New York: Routledge.

Drissel, D. (2009) 'Hip-Hop Hybridity for a Glocalized World', *The Global Studies Journal* 2(3): 121–142.

DuComb, C. and J. Benmen (2014) 'Flash Mobs, Violence, and the Turbulent Crowd', *Performance Research* 19(5): 34–40.

Duffett, M. (2013) *Understanding Fandom: An introduction to the study of media fan culture*, New York: Bloomsbury.

Duffett, M. (2017) 'I Scream Therefore I Fan? Music audiences and affective citizenship', in J. Gray, C. Sandvoss and C. L. Harrington (eds) *Fandom: Identities and communities in a mediated world*, second edition, New York: New York University Press, 143–156.

Duffy, B. E. (2017) *(Not) Getting Paid to Do What You Love: Gender, social media, and aspirational work*, New Haven, CT: Yale University Press.

Dugan, M. (1996) 'A Nested Theory of Conflict', *Women in Leadership: Sharing the Vision* 1: 9–20.

Duguay, S. (2016) 'Dressing up Tinderella: Interrogating authenticity claims on the mobile dating app Tinder', *Information, Communication & Society* 20(3): 351–367.

Duguay, S., J. Burgess and N. Suzor (2018) 'Queer women's experiences of patchwork platform governance on Tinder, Instagram, and Vine', *Convergence: The International Journal of Research into New Media Technologies*. Available at: https://doi.org/10.1177/1354856518781530.

Dunbar-Hester, C. (2009) '"Free the Spectrum!" Activist encounters with old and new media technology', *New Media & Society* 11(1–2): 221–240.

Dunbar-Hester, C. (2010) 'Beyond "Dudecore"? Challenging gendered and "raced" technologies through media activism', *Journal of Broadcasting and Electronic Media* 54(1): 121–135.

Dunbar-Hester, C. (2014) *Low Power to the People! Pirates, protest, and politics in FM radio activism*, Cambridge, MA: The MIT Press.

Dunbar-Hester, C. (2017) 'Feminists, Geeks and Geek Feminists: Understanding gender and power in technological activism', in V. Pickard and G. Yang (eds) *Media Activism in the Digital Age*, London: Routledge, 187–204.

Duncan, J. (2013) 'South African Journalism and the Marikana Massacre: A case study of an editorial failure', *The Political Economy of Communication* 1(2): 65–88.

Durham, M. (2011) 'Body Matters', *Feminist Media Studies* 11(1): 53–60.

Dutton, W. H. (2007) 'Through the Network (of Networks): The Fifth Estate', Inaugural Lecture, University of Oxford, 15 October.

Dwyer, T. (2010) *Media convergence*, Maidenhead, UK: McGraw Hill.

Dyer, R. (2004) *Heavenly Bodies: Film stars and society*, second edition, New York: Routledge.

Dyer-Witheford, N. (2015) *Cyber-Proletariat: Global labour in the digital vortex*, London: Pluto Press.

Dyer-Witheford, N. and G. de Peuter (2005) 'A Playful Multitude? Mobilising and counter-mobilising immaterial game labour', *The Fibreculture Journal* (5). Available online: http://five.fibreculturejournal.org/fcj-024-a-playful-multitude-mobilising-and-counter-mobilising-immaterial-game-labour/.

Dyer-Witheford, N. and G. de Peuter (2009) *Games of Empire: Global capitalism and video games*, Minneapolis, MN: University of Minnesota Press.

Dyson, M. E. (2010) *Know What I Mean: Reflections on hip-hop*, New York: Basic Civitas.

Eagleton, T. (1984) *The Function of Criticism*, London: Verso.

Earl, J. and K. R. Garrett (2017) 'The New Information Frontier: Toward a more nuanced view of social movement communication', *Social Movement Studies* 16(4): 479–493.

Earl, J. and K. Kimport (2009) 'Movement Societies and Digital Protest: Fan activism and other nonpolitical protest online', *Sociological Theory* 27(3): 220–243.

Earl, J. and K. Kimport (2011) *Digitally Enabled Social Change: Activism in the internet age*, Cambridge, MA: The MIT Press.

Earl, J. and D. A. Rohlinger (eds) (2010) *Media, Movements, and Political Change*, Bingley, UK: Emerald Publishing.

Earth First! (2015) *The Earth First! Direct Action Manual*, third edition.

Eder, J. (2010) 'Understanding Characters', *Projections: The Journal for Movies and Mind* 4(1): 16–40.

Eder, J. (2017) 'Affective Image Operations', in J. Eder and C. Klonk (eds) *Image Operations: Visual media and political conflict*, Manchester: Manchester University Press, 63–78.

Eder, J. (2018) 'Collateral Emotions: Political web videos and divergent audience responses', in C. Brylla and M. Kramer (eds) *Cognitive Theory and Documentary Film*, London: Palgrave Macmillan, 183–203.

Eder, J. and C. Klonk (eds) (2017) 'Introduction', in J. Eder and C. Klonk (eds) *Image Operations: Visual media and political conflict*, Manchester: Manchester University Press, 1–22.

Eder, J., B. Hartmann and C. Tedjasukmana (eds) (2015) *Videoactivism.net*. Available online: http://videoactivism.net/en/.

Eder, J., B. Hartmann and C. Tedjasukmana (2020) *Bewegungsbilder. Politische Videos in Sozialen Medien*, Berlin: Bertz + Fischer.

Edgerton, K. (2013) 'How "The Occupy Wall Street Collaborative Film" Demonstrates Crowdsourcing in Action', Open Documentary Lab at MIT, 13 February. Available online: http://opendoclab.mit.edu/how-the-occupy-wall-street-collaborative-film-demonstrates-crowdsourcing-in-action.

Edkins, J., N. Persram and V. Pin-Fat (1999) *Sovereignty and Subjectivity*, London: Lynne Rienner.

Edwards, D. (2009) *The Parkour and Freerunning Handbook*, New York: Dey Street Books.

Edwards, E. (2012) *The Camera as Historian: Amateur photographers and historical imagination, 1885–1918*, Durham, NC and London: Duke University Press.

Edwards, M. (ed.) (2011) *The Oxford Handbook of Civil Society*, Oxford: Oxford University Press.

Edwards, M. (2014) *Civil Society*, Cambridge: Polity Press.

Edwards, P. N. (2003) 'Infrastructure and Modernity: Force, time, and social organization in the history of sociotechnical systems', in T. Misa, P. Brey and A. Feenberg (eds) *Modernity and Technology*, Cambridge, MA: The MIT Press, 185–226.

Edwards, R. and Tryon, C. (2009) 'Political Video Mashups as Allegories of Citizen Empowerment', *First Monday* 14(10): 1. Available online: http://firstmonday.org/article/view/2617/2305.

Effendi, R. (2017) 'I Posted a Graphic Selfie of My Beating to Tell the World #MeToo', *Time*, 24 October. Available online: https://time.com/4991494/metoo-beating-graphic-selfie-rena-effendi/.

Eglinton, K., A. Gubrium and L. Wexler (2017) 'Digital Storytelling as Arts-Inspired Inquiry for Engaging, Understanding and Supporting Indigenous Youth', *International Journal of Education & the Arts* 18(5): 1–28.

Ehrenberg, J. (1999) *Civil Society: The critical history of an idea*, New York: NYU Press.

Eisendrath, C. R. (1979) 'Back to the People with the Mom-and-Pop Press', *Columbia Journalism Review* 18(4): 72.

Ekman, M. (2014) 'The Dark Side of Online Activism: Swedish right-wing extremist video activism on YouTube', *MedienKultur* 30(56): 79–99.

El Zein, R. and A. Ortiz (2011) 'Signs of the Times: The popular literature of Tahrir', *ArteEast* (Spring). Available online: https://issuu.com/arteeast/docs/shahadat_january25_final.

Elder, S. (2016) 'Core Story 1: Trusting the process – Sarah Elder and Leonard Kamerling's community-collaborative approach to filmmaking', in A. Gubrium and K. Harper (eds) *Participatory Visual and Digital Methods*, London and New York: Routledge, 99–110.

Eldridge II, S. (2018) *Online Journalism from the Periphery: Interloper media and the journalistic field*, Oxon: Routledge.

Electronic Arts (2002) Battlefield 1942. IGDB entry: www.igdb.com/games/battlefield-1942.

Electronic Frontier Foundation (2018) 'Section 230 of the Communications Decency Act'. Available online: www.eff.org/issues/cda230.

Ellcessor, E. (2017) 'Kickstarting Community: Disability, access, and participation in My Gimpy Life', in E. Ellcessor and B. Kirkpatrick (eds) *Disability Media Studies*, New York: New York University Press, 31–51.

Elliott, K. C. (2011) *Is a Little Pollution Good for You?: Incorporating societal values in environmental research*, Oxford: Oxford University Press.

Ellis, J. (2000) *Seeing Things: Television in the age of uncertainty*, London: I. B. Tauris.

Ellis, K. and G. Goggin (2015a) *Disability and the Media*, Basingstoke, UK: Palgrave.

Ellis, K. and G. Goggin (2015b) 'Disability Media Participation: Opportunities, obstacles and politics', *Media International Australia* 154(1): 78–88.

Ellis, K., G. Goggin, B. Haller and R. Curtis (eds) (2019) *The Routledge Companion to Disability and Media*, New York: Routledge.

Ellis, K. and M. Kent (eds) (2017) *Disability and Social Media: Global perspectives*, New York: Routledge.

Ellison, C. (2015) *Embed with Games: A year on couch with game developers*, Edinburgh, UK: Birlinn.

Elmer, G. (2015) 'Preempting Dissent: From participatory policing to collaborative filmmaking', in L. Dencik. and O. Leistert (eds) *Critical Perspectives on Social Media and Protest: Between control and emancipation*, London: Rowman & Littlefield, 145–161.

Emirbayer, M. and M. Sheller (1999) 'Publics in History', *Theory and Society* 28(1): 143–197.

Engelhardt, J. and J. Jansz (2014) 'Challenging Humanitarian Communication: An empirical exploration of Kony 2012', *The International Communication Gazette* 76(6): 464–484.

Enli, G. (2015) *Mediated Authenticity: How the media construct reality*, New York: Peter Lang.

Enli, G. (2016) '"Trust Me, I Am Authentic!" Authenticity illusions in social media politics', in A. Bruns, G. Enli, E. Skogerbo, A. O. Larson and C. Christensen (eds) *The Routledge Companion to Social Media and Politics*, New York and London: Routledge, 121–137.

Enli, G. (2017) 'Twitter as Arena for the Authentic Outsider: Exploring the social media campaigns of Trump and Clinton in the 2016 US presidential election', *European Journal of Communication* 31(1): 50–61.

Epic Games (1998) Unreal Engine. Available online: www.unrealengine.com.

Ernesto (2012) 'Netflix Caught Using "Pirated" Subtitles in Finland', *TorrentFreak*, 19 October. Available online: https://torrentfreak.com/netflix-caught-using-pirated-subtitles-in-finland-121019/.

Eschle, C. and B. Maiguashca (2010) *Making Feminist Sense of the Global Justice Movement*, Lanham, MD: Rowman & Littlefield.

Essed, P. (1991) *Understanding Everyday Racism: An interdisciplinary theory*, London: Sage.

Estellés-Arolas, E. (2018) 'The Need of Co-utility for Successful Crowdsourcing', in J. Domingo-Ferrer and D. Sánchez (eds) *Co-utility: Studies in systems, decision and control*, Volume 110, Cham, Switzerland: Springer, 189–200.

Estellés-Arolas, E. and F. González Ladrón-de-Guevara (2012) 'Towards an Integrated Crowdsourcing Definition', *Journal of Information Science* 38(2): 1–14.

Ettlinger, N. (2007) 'Precarity Unbound', *Alternatives: Global, Local, Political* 32(3): 319–340.

Ettlinger, N. (2016) 'The Governance of Crowdsourcing: Rationalities of the new exploitation', *Environment and Planning A: Economy and Space* 48(11): 2162–2180.

Evans, S. (1980) *Personal Politics: The roots of women's liberation in the civil rights movement and the New Left*, New York: Vintage Books.

Eveleigh, A. (2014) 'Crowding Out the Archivist? Locating crowdsourcing within the broader landscape of participatory archives', in M. Ridge (ed.) *Crowdsourcing our Cultural Heritage*, Burlington, VT: Ashgate, 233–252.

Ezra, E. (ed.) (2006) *Transnational Cinema: The Film Reader*. London and New York: Routledge.

Facebook (n.d.) 'Mission Statement'. Available online: www.facebook.com/notes/mark-zuckerberg/bringing-the-world-closer-together/10154944663901634/.

Facebook (2018) 'Facebook Statement of Rights and Responsibilities'. Available online: www.facebook.com/legal/terms.

Fahlenbrach, K. (2017) 'Images and Imagery of Protest', in K. Fahlenbrach, M. Klimke and J. Scharloth (eds) *Protest Cultures: A companion*, Oxford and New York: Berghahn Books, 243–258.

Fahmy, Z. (2011) *Ordinary Egyptians: Creating the modern nation through popular culture*, Palo Alto, CA: Stanford University Press.

Fairchild, C. (2001) *Community Radio and Public Culture*, Cresskill, NJ: Hampton Press.

FakeApp (n.d.) Available online: www.malavida.com/en/soft/fakeapp/.

Falkheimer, J. and A. Jansson (eds) (2006) *Geographies of Communication: The spatial turn in media studies*, Gothenburg: Nordicom.

Fanon, F. (1986) *Black Skin, White Masks*, London: Pluto.

Farman, J. (2012) *Mobile Interface Theory: Embodied space and locative media*, London and New York: Routledge.

Farnel, M. (2015) 'Kickstarting Trans*: The crowdfunding of gender/sexual reassignment surgeries', *New Media & Society* 17(2): 215–230.

Faust III, A. T. (2011) 'Occupy App Gives You Even More News on Growing Global Phenomenon', *AppAdvice*, 20 November. Available online: http://appadvice.com/appnn/2011/11/occupy-app-gives-you-even-more-news-on-growing-global-phenomenon.

Featherstone, A. (2000) 'Peacekeeping, Conflict Resolution and Peacebuilding: A reconsideration of theoretical frameworks', *International Peacekeeping* 7(1): 190–218.

Featherstone, D. (2012) *Solidarity: Hidden histories and geographies of internationalism*, London and New York: Zed Books.

Featherstone, M. (2006) 'Archive', *Theory, Culture & Society* 23(2–3): 591–596.

Federici, S. (2004) *Caliban and the Witch: Women, the body and primitive accumulation*, New York: Autonomedia.

Federici, S. and G. Caffentzis (2019) 'Commons against and beyond Capitalism', in S. Federici (ed.) Re-enchanting the World: Feminism and the politics of the commons, Oakland, CA and New York: PM Press and Autonomedia, 85–98.

Feenberg, A. (2002) *Transforming Technology: A critical theory revisited*, New York: Oxford University Press.

Feenberg, A. and M. Bakardjieva (2004) 'Virtual Community: No "killer implication"', *New Media & Society* 6(1): 37–43.

Feeney, T. (2015) 'Greek Bailout Fund', *Indiegogo*. Available online: www.indiegogo.com/projects/greek-bailout-fund.

Feenstra, R., S. Tormey, A. Casero-Ripollés and J. Keane (2017) *Refiguring Democracy: The Spanish political laboratory*, London: Routledge.

Feigenbaum, A. (2014) 'Resistant Matters: Tents, tear gas and the "other media" of Occupy', *Communication and Critical/Cultural Studies* 11(1): 15–24.

Feigenbaum, A., F. Frenzel and P. McCurdy (2013) *Protest Camps*, London: Zed Books.

Feldstein, A. (2011) 'Deconstructing Wikipedia: Collaborative content creation in an open process platform', *Procedia. Social and Behavioral Sciences* 26: 76–84.

Felman, S. and D. Laub (1992) *Testimony: Crises of witnessing in literature, psychoanalysis, and history*, London and New York: Taylor & Francis.

Feng, M. and E. J. Yuan (2015) 'Public Opinion on Weibo: The case of the Diaoyu Islands dispute', in T. A. Hollihan (ed.) *The Dispute over the Diaoyu/Senkaku Islands: How media narratives shape public opinion and challenge the global order*, New York: Palgrave Macmillan, 119–140.
Fenton, N. (2008) 'Mediating Hope: New media, politics and resistance', *International Journal of Cultural Studies* 11(2): 230–248.
Fenton, N. (2011) 'Multiplicity, Autonomy, New Media, and the Networked Politics of New Social Movements', in L. Dahlberg and S. Phelan (eds) *Discourse Theory and Critical Media Politics*, Basingstoke, UK: Palgrave Macmillan, 178–200.
Fenton, N. (2012a) 'De-democratizing the News? New media and the structural practices of journalism', in E. Siapera and A. Veglis (eds) *The Handbook of Global Online Journalism*, London: John Wiley and Sons, 119–134.
Fenton, N. (2012b) 'The Internet and Radical Politics', in J. Curran, N. Fenton and D. Freedman (eds) *Misunderstanding the Internet*, London: Routledge, 149–176.
Fenton, N. and V. Barassi (2011) 'Alternative Media and Social Networking Sites: The politics of individuation and political participation', *The Communication Review* 14(3): 179–196.
Fenton, N. and D. Freedman (2017) 'Fake Democracy, Bad News', in L. Panitch and G. Albo (eds) *Socialist Register* 54, London: Merlin, 130–149.
Ferdig, R. E. (2014) 'Education', in M. J. P. Wolf and B. Perron (eds) *The Routledge Companion to Video Game Studies*, New York: Routledge, 317–323.
Ferguson, M. (ed.) (1986) *New Communication Technologies and the Public Interest*, London: Sage.
Ferguson, P. P. (1998) 'A Cultural Field in the Making: Gastronomy in 19th-century France', *American Journal of Sociology* 104(3): 597–641.
Fernández, F. (forthcoming) *Translating the Crisis: Culture and politics in Spain after the 15M*, Abingdon: Routledge.
Fernández-Savater, A., C. Flesher Fominaya, L. Carvalho, C. Hoda et al. (2016) 'Life after the Squares: Reflections on the consequences of the Occupy movements', *Social Movement Studies* 16(1): 119–151.
Ferrari, E. (2017) 'Fake Accounts, Real Activism: Political faking and user-generated satire as activist intervention', *New Media & Society* 20(6): 2208–2223.
Ferrell, J. (1993) *Crimes of Style: Urban graffiti and the politics of criminality*, New York: Garland.
Ferrell, J. (1995) 'The World Politics of Wall Painting', in J. Ferrell and C. R. Sanders (eds) *Cultural Criminology*, Boston, MA: Northeastern University Press, 277–294.
Fidler, D. P. (ed.) (2015) *The Snowden Reader*, Bloomington, IN: Indiana University Press.
Fielding, M. and P. Moss (2011) *Radical Education and the Common School: A democratic alternative*, New York: Routledge.
Fielitz, M. and N. Thurston (eds) (2019) *Post-Digital Cultures of the Far-Right: Online actions and offline consequences in Europe and the US*, Bielefeld: Transcript.
Figueiras, R. and P. do Espírito Santo (2015) *Beyond the Internet: Unplugging the protest movement wave*, London: Routledge.
Filar, R. (2016) 'Meet the Sex Workers Using Art to Expose Truths about the Sex Industry', *Open Democracy*, 16 April. Available online: www.opendemocracy.net/transformation/ray-filar/meet-sex-workers-using-art-to-expose-truths-about-sex-industry.
Finnemann, N. O. (2014) 'Digitalization: New trajectories of mediatization?', in K. Lundby (ed.) *Mediatization of Communication*, Berlin and New York: de Gruyter, 297–322.
Finneran, P. (2014) 'Documentary Impact: Social change through storytelling', *Hotdocs. Outspoken understanding.* Available online: http://s3.amazonaws.com/assets.hotdocs.ca/doc/HD14_Documentary_Impact_Report.pdf.
Firaxis Games (2008) Sid Meier's Civilization IV: Colonization. IGDB entry: www.igdb.com/games/sid-meier-s-civilization-iv-colonization.
Firmstone, J. and S. Coleman (2015) 'Rethinking Local Communicative Spaces: Implications of digital media and citizen journalism for the role of local journalism in engaging citizens', in R. K. Nielsen (ed.) *Local Journalism: The decline of newspapers and the rise of digital media*, London: Reuters Institute for the Study of Journalism, 117–140.
Fischer, M. and K. Mohrman (2016) 'Black Deaths Matter? Sousveillance and the invisibility of black life', *Ada: A Journal of Gender, New Media, and Technology* 10.
Fisher, D. (2009) 'Mediating Kinship: Country, family, and radio', *Cultural Anthropology* 24(2): 280–312.
Fisher, E. (2010) *Media and New Capitalism in the Digital Age: The spirit of networks*, New York: Palgrave Macmillan.
Fiske, J. (1987) *Television Culture*, New York and London: Routledge.
Fiske, J. (1989) *Understanding Popular Culture*, London: Unwin Hyman.
Fiske, J. (1994) *Media Matters: Everyday culture and political change*, Minneapolis, MN: University of Minnesota Press.
Fitzpatrick, J. (2018) *Digital Civil Society: Wie zivilgesellschaftliche Organisationen im Web 2.0 politische Ziele verfolgen*, Weisbaden: Springer.
Flaherty, R. J. (dir.) (1922) Nanook of the North. IMDb entry: www.imdb.com/title/tt0013427/?ref_=fn_al_tt_1.
Flanagan, M. (2009) *Critical Play: Radical game design*, Cambridge, MA: The MIT Press.

Flesher Eguiarte, C. (2005) The Logic of Autonomy: Principles, praxis, and challenges of autonomous anti-capitalist movement; Three case studies from Madrid, Unpublished PhD thesis, Berkeley, CA: University of California.

Flesher Fominaya, C. (2007) 'Autonomous Movements and the Institutional Left: Two approaches in tension in Madrid's anti-globalization network', *South European Society & Politics* 12(3): 335–358.

Flesher Fominaya, C. (2010) 'Creating Cohesion from Diversity: The challenge of collective identity formation in the global justice movement', *Sociological Inquiry* 80(3): 377–404.

Flesher Fominaya, C. (2013) 'Movement Culture Continuity: The British anti-roads movement as precursor to the global justice movement', in C. Flesher Fominaya and L. Cox (eds) *Understanding European Movements: New social movements, global justice struggles, anti-austerity protests*, London: Routledge, 109–124.

Flesher Fominaya, C. (2015a) 'Autonomous Social Movements and the Paradox of Anti-identitarian Collective Identity', in A. McGarry and J. M. Jasper (eds) *The Identity Dilemma: Social movements and collective identity*, Philadelphia, PA: Temple University Press, 65–84.

Flesher Fominaya, C. (2015b) 'Debunking Spontaneity: Spain's 15-M/indignados as autonomous movement', *Social Movement Studies* 14(2): 142–163.

Flesher Fominaya, C. (2016a) 'Cultural Barriers to Activist Networking: Habitus (in)action in three European transnational encounters', *Antipode* 48(1): 151–171.

Flesher Fominaya, C. (2016b) 'Unintended Consequences: The negative impact of e-mail use on participation and collective identity in two "horizontal" social movement groups', *European Political Science Review* 8(1): 95–122.

Flesher Fominaya, C. (2017) 'European Anti-austerity and Pro-democracy Protests in the Wake of the Global Financial Crisis', *Social Movement Studies* 16(1): 1–20.

Flesher Fominaya, C. (2020a) *Social Movements in a Globalized World*, London: Macmillan and Red Globe Press.

Flesher Fominaya, C. (2020b) *Democracy Reloaded: Inside Spain's political laboratory from 15-M to Podemos*, Oxford: Oxford University Press.

Flesher Fominaya, C. and K. Gillan (2017) 'Navigating the Technology-Media-Movements Complex', *Social Movement Studies* 16(4): 383–402.

Fletcher, R. and R. K. Nielsen (2017) 'Are People Incidentally Exposed to News on Social Media? A comparative analysis', *New Media & Society* 20(7): 1–19.

Flinn, A. (2008) 'Other Ways of Thinking, Other Ways of Being. Documenting the margins and the transitory: What to preserve, how to collect', in L. Craven (ed.) *What Are Archives? Cultural and theoretical perspectives*, Aldershot: Ashgate, 109–128.

Flinn, A., M. Stevens and E. Shepherd (2009) 'Whose Memories, Whose Archives? Independent community archives, autonomy and the mainstream', *Archival Science* 9: 71–86.

Flores, R. and G. Tanaka (2001) 'Autonomy and Participatory Democracy: An ongoing discussion on the application of Zapatista autonomy in the United States', Interview by P. McLaren, *International Journal of Educational Reform* 10(2): 130–144. Republished at www.inmotionmagazine.com/auto/ijer.html.

Flores-Márquez, D. (2010) 'Estar con los otros: Presencias, proximidades y sentidos de vínculo en las redes de bloggers', Mediaciones Sociales 6: 145–161.

Flores-Márquez, D. (2013a) 'From Internet to the Streets: An approach to activism in Aguascalientes, Mexico', *Selected Papers of Internet Research – IR 14.0, Resistance + appropriation, AoIR*.

Flores-Márquez, D. (2013b) 'Formas distintas de vivir la ciudad: La experiencia urbana en las prácticas de comunicación digital en Aguascalientes', *Caleidoscopio* 28: 79–94.

Flores-Márquez, D. (2019) *Imaginar un mundo mejor: La expresión pública de los activistas en Internet*, Guadalajara: ITESO.

Florida, R. (2003) *The Rise of the Creative Class: And how it's transforming work, leisure, community, and everyday life*, North Melbourne, Vic: Basic Books.

Florini, S. (2015) 'This Week in Blackness, the George Zimmerman Acquittal, and the Production of a Networked Collective Identity', *New Media & Society* 19(3): 439–454.

Flouch, H. and K. Harris (2010) *London's Digital Neighbourhoods Study: Typology of local websites*, London: Connected London.

Flynn, E. G. (1916/2014) 'Sabotage: The conscious withdrawal of the workers' industrial efficiency', in S. Salerno (ed.) *Direct Action and Sabotage: Three classic IWW pamphlets from the 1910s*, Chicago, IL: H. Kerr and Company, 91–115.

Foellmer, S., M. Lünenborg and C. Raetzsch (2018) *Media Practices, Social Movements, and Performativity: Transdisciplinary approaches*, New York: Routledge.

Forde, S., K. Foxwell and M. Meadows (2009) *Developing Dialogues: Indigenous and ethnic community broadcasting in Australia*, London: Intellect and Chicago: University of Chicago Press.

Forensic Architecture (n.d.) 'The Left-to-die Boat: The deadly drift of a migrants' boat in the Central Mediterranean'. Available online: www.forensic-architecture.org/case/left-die-boat/.

Forman, M. (2000) 'Represent: Race, place, and space in rap music', *Popular Music* 19(1): 65–90.

Forman, M. and M. A. Neal (2004) *That's The Joint! The hip-hop studies reader*, New York: Routledge.

Fortunati, L. (2002) 'The Mobile Phone: Towards new categories and social relations', *Information, Communication and Society* 5(4): 513–528.

Fortunati, L. (2007) 'Immaterial Labor and Its Machinization', in E. Dowling, R. Nunes and B. Trott (eds) *Immaterial and Affective Labour: Explored*. Special issue of *Ephemera: Theory and Politics in Organization* 7(1): 139–157. Available online: www.ephemerajournal.org/sites/default/files/7-1fortunati.pdf.

Foster, H. (2004) 'An Archival Impulse', *October* 110: 3–22.

Fotopoulou, A. (2017) *Feminist Activism and Digital Networks: Between empowerment and vulnerability*, London: Palgrave Macmillan.

Foucault, M. (1969/1972) *The Archaeology of Knowledge and the Discourse on Language*, translated by A. M. S. Smith, New York: Pantheon.

Foucault, M. (1975/1977) *Discipline and Punish: The birth of the prison*, translated by A. Sheridan, London: Penguin Books.

Foucault, M. (1976/1990) *The Will to Knowledge: The history of sexuality*, Volume 1, translated by R. Hurley, London: Penguin.

Foucault, M. (1989) 'Rituals of Exclusion', in S. Lotringer (ed.) *Foucault Live: Collective interviews, 1961–1984*, translated by L. Hochroth and J. Johnston, New York: Semiotexte.

Foucault, M. (2008) *The Birth of Biopolitics: Lectures at the Collège de France, 1978–1979*, edited by A. I. Davidson, translated by G. Burchell, Basingstoke, UK and New York: Palgrave Macmillan.

Fouché, C. B. and L. A. Chubb (2017) 'Action Researchers Encountering Ethical Review: A literature synthesis on challenges and strategies', *Educational Action Research* 25(1): 23–34.

Fountain, A. (2008) 'Alternative Film, Video, and Television, 1965–2005', in K. Coyer, T. Dowmunt and A. Fountain (eds) *The Alternative Media Handbook*, London and New York: Routledge, 29–46.

Foust, C. R. and K. D. Hoyt (2018) 'Social Movement 2.0: Integrating and assessing scholarship on social media and movement', *Review of Communication* 18(1): 37–55.

Fox, A. (ed.) (2016) *Global Perspectives on Media Events in Contemporary Society*, Hershey, PA: IGI Global.

Frabetti, F. (2015) *Software Theory*, London: Rowman & Littlefield.

Francis, D. (2011) 'New Thoughts on Power: Closing the gaps between theory and action', in B. Austin, M. Fischer and H. J. Giessmann (eds), *Advancing Conflict Transformation: The Berghof handbook II*, Opladen and Farmington Hills, MI: Barbara Budrich Publishers, 506–525.

Frank, T. (1997) *The Conquest of Cool*, Chicago, IL: University of Chicago Press.

Franklin, B., M. Hamer, M. Hanna, M. Kinsey and J. Richardson (2005) *Key Concepts in Journalism Studies*, London: Sage.

Franklin, K. (n.d.) Benefit Scrounging Scum blog. Available online: www.BenefitsScroungingScum.blogspot.com.

Franks, B. (2003) 'The Direct Action Ethic from 59 Upwards', *Anarchist Studies* 11(1): 13–41.

Franks, B. (2014) 'Anti-Fascism and Prefigurative Ethics', *Affinities: A Journal of Radical Theory, Culture, and Action* 8(1): 44–72.

Frasca, G. (2003) *September 12th*. Available online: www.ludology.org/my_games.html.

Frasca, G. (2004a) *Madrid*. Available online: www.ludology.org/my_games.html.

Frasca, G. (2004b) 'Videogames of the Oppressed: Critical thinking, education, tolerance, and other trivial issues', in N. Wardrip-Fruin and P. Harrigan (eds) *First Person: New media as story, performance, and game*, Cambridge, MA: The MIT Press, 85–94.

Fraser, A. and E. C. Y. Li (2017) 'The Second Life of Kowloon Walled City: Crime, media and cultural memory', *Crime, Media, Culture* 13(2): 217–234.

Fraser, N. (1990) 'Rethinking the Public Sphere: A contribution to the critique of actually existing democracy', *Social Text* 25–26: 56–80.

Fraser, N. (1997) *Justice Interruptus: Critical reflections on the 'postsocialist' condition*, London: Routledge.

Fraser, N. (2005) 'Reframing Justice in a Globalizing World', *New Left Review* 36: 69–88.

Fraser, N. (2007) 'Transnationalizing the Public Sphere: On the legitimacy and efficacy of public opinion in a post-Westphalian world', *Theory, Culture & Society* 24(4): 7–30.

Free Dictionary (2003–2018) 'Citizen Media'. Available online: www.thefreedictionary.com/Citizen+media.

Free Software Foundation (2018) 'Free Software Definition', GNU.org. Available online: www.gnu.org/philosophy/free-sw.en.html.

Freeden, M. (1996) *Ideologies and Political Theory: A conceptual approach*, Oxford: Clarendon.

Freelon, D., C. D. McIlwain and M. D. Clark (2016a) 'Quantifying the Power and Consequence of Social Media Protest', *New Media & Society* 20(3): 990–1011.

Freelon, D., C. D. McIlwain, and M. D. Clark (2016b) *Beyond the Hashtags: #Ferguson, #Blacklivesmatter, and the online struggle for offline justice*, Washington DC: Center for Media and Social Impact, American University.

Freeman, J. (1971/2013) 'The Tyranny of Structurelessness', *WSQ: Women's Studies Quarterly* 41(3): 231–246.

Freidberg, J. and R. Rowley (dir.) (2000) This Is What Democracy Looks Like. IMDb entry: www.imdb.com/title/tt0265871/.

Freire, P. (1968) *Pedagogía del oprimido*, Rio de Janeiro: Paz e Terra.

Freire, P. (1972) *Pedagogy of the Oppressed*, Harmondsworth, UK: Penguin.

Freire, P. (1974) *Education for Critical Consciousness*, London: Sheed & Ward.

Frickel, S. (2004) 'Scientist Activism in Environmental Justice Conflicts: An argument for synergy', *Society and Natural Resources* 17(4): 359–366.

Fricker, M. (2007) *Epistemic Injustice: Power and the ethics of knowing*, Oxford: Oxford University Press.

Frith, S. (1997a) 'Popular Culture', in M. Payne (ed.) *A Dictionary of Cultural and Critical Theory*, Oxford: Blackwell, 415–417.

Frith, S. (1997b) 'Folk Culture', in M. Payne (ed.) *A Dictionary of Cultural and Critical Theory*, Oxford: Blackwell, 199–200.

Fritzsche, K. and C. Löser (eds) (1996) *Gegenbilder – Filmische Subversion in der DDR 1976–1989*, Berlin: Janus Press.

Fröhlich, R., O. Quiring and S. Engesser (2012) 'Between Idiosyncratic Self-interests and Professional Standards: A contribution to the understanding of participatory journalism in Web 2.0. Results from an online survey in Germany', *Journalism* (13)8: 1041–1063.

Frosh, P. (2015) 'The Gestural Image: The selfie, photography theory, and kinesthetic sociability', *International Journal of Communication* 9: 1607–1628.

Fuad-Luke, A. (2013) *Design Activism: Beautiful strangeness for a sustainable world*, London: Routledge.

Fuchs, C. (2009) 'Some Theoretical Foundations of Critical Media Studies', *International Journal of Communication* 3: 369–402.

Fuchs, C. (2010) 'Alternative Media as Critical Media', *European Journal of Social Theory* 13(2): 173–192.

Fuchs, C. (2012a) 'Social Media, Riots, and Revolutions', *Capital & Class* 36(3): 383–391.

Fuchs, C. (2012b) 'The Political Economy of Privacy on Facebook', *Television and New Media* 13(2): 139–159.

Fuchs, C. (2012c) 'Dallas Smythe Today: The audience commodity, the digital labour debate, Marxist political economy and critical theory. Prolegomena to a digital labour theory of value', *tripleC: Communication, Capitalism & Critique: Journal for a Global Sustainable Information Society* 10(2): 692–740. Available online: www.triple-c.at/index.php/tripleC/article/view/443/414.

Fuchs, C. (2013) 'Class and Exploitation on the Internet', in T. Scholtz (ed.) *Digital Labor: The internet as playground and factory*, New York: Routledge, 211–224.

Fuchs, C. (2014a) *Digital Labour and Karl Marx*, London and New York: Routledge.

Fuchs, C. (2014b) 'Digital Prosumption Labour on Social Media in the Context of the Capitalist Regime of Time', *Time and Society* 23(1): 97–123.

Fuchs, C. (2014c) *Social Media: A critical introduction*, first edition, Los Angeles, CA: Sage.

Fuchs, C. (2015) 'Dallas Smythe and Digital Labor', in R. Maxwell (ed.) *The Routledge Companion to Labor and Media*, London and New York: Routledge, 51–62.

Fuchs, C. (2016) 'Baidu, Weibo and Renren: The global political economy of social media in China', *Asian Journal of Communication* 26(1): 14–41.

Fuchs, C. (2017) *Social Media: A critical introduction*, second edition, Thousand Oaks, CA: Sage.

Fuchs, C. and M. Sandoval (2014) 'Digital Workers of the World Unite! A framework for critically theorising and analysing digital labour', *tripleC* 12(2): 486–563.

Fuentes-Bautista, M. and G. Gil-Egui (2011) 'Community Media and the Rearticulation of State: Civil society relations in Venezuela', *Communication, Culture and Critique* 4: 250–274.

Fukuyama, F. (2001) 'Social Capital, Civil Society and Development', *Third World Quarterly* 22: 7–20.

Fuller, M. (2005) Media Ecologies: Materialist energies in art and technoculture, Cambridge, MA: The MIT Press.

Fusco, C. (2001) *The Bodies that Were Not Ours*, London and New York: Routledge.

Fuster Morell, M. (2011) 'The Unethics of Sharing: Wikiwashing', *International Review of Information* 15: 9–16.

Futrell, R. and P. Simi (2004) 'Free Spaces, Collective Identity, and the Persistence of U.S. White Power Activism', *Social Problems* 51(1): 16–42.

Gabriel, D. (2016) 'Blogging while Black, British and Female: A Critical study on discursive activism', *Information, Communication and Society* 19(11): 1622–1635.

Gagyi, Á. (2013) 'The Shifting Meaning of "Autonomy" in the East European Diffusion of the Alterglobalization Movement', in C. Flesher Fominaya and L. Cox (eds) *Understanding European Movements: New social movements, global justice struggles, anti-austerity protests*, London: Routledge, 143–153.

Gaines, J. (1999) 'Political Mimesis', in J. Gaines and M. Renov (eds) *Collecting Visible Evidence*, Minneapolis, MN: University of Minnesota Press, 84–102.

Galis, V. and C. Neumayer (2016) 'Laying Claim to Social Media by Activists: A cyber-material détournement', *Social Media + Society* 2(3): 1–14.

Gallagher, O. (2018) *Reclaiming Critical Remix Video: The role of sampling in transformative works*, London and New York: Routledge.
Galloway, A. R. (2006) *Gaming Essays on Algorithmic Culture*, Minneapolis, MN: University of Minnesota Press.
Galtung, J. (1969) 'Violence, Peace and Peace Research', *Journal of Peace Research* 6(3): 167–191.
Galtung, J. (1986) 'On the Role of the Media in Worldwide Security and Peace', in T. Varis (ed.) *Peace and Communication*, San Jose, Costa Rica: Universidad para La Paz, 249–266.
Galtung, J. (1996) *Peace by Peaceful Means: Peace and conflict, development and civilisation*, Thousand Oaks, CA: Sage Publications.
Games for Change (2018) Available online: www.gamesforchange.org.
Gamson, W. A. (1975) *The Strategy of Social Protest*, Homewood, IL: Dorsey Press.
Gamson, W. A. (1992) 'The Social Psychology of Collective Action', in J. Goodwin and J. M. Jasper (eds) *Social Movements*, London: Routledge, 168–187.
Gamson, W. (1995) 'Constructing Social Protest', in H. Johnston and B. Klandermans (eds) *Social Movements and Culture*, London: UCL, 85–106.
Gandhi, M. (1987) *The Moral and Political Writings of Mahatma Gandhi: Non-violent resistance and social transformation*, New York: Oxford University Press.
Gandy, O. (1993) *The Panoptic Sort: A political economy of personal information*, Boulder, CO: Westview Press.
Gani, A. (2015) 'Russian Lesbians Stage Selfie Kiss Plane Protest', *The Guardian*, 3 February. Available online: www.theguardian.com/world/2015/feb/03/russian-lesbians-selfie-kiss-plane-protest-vitaly-milonov.
García, D. and G. Lovink (1997) 'The ABC of Tactical Media', *NetTime.org*, 16 May. Available online: www.nettime.org/Lists-Archives/nettime-l-9705/msg00096.html.
García-Canclini, N. (1998) 'Monuments, Billboards and Graffiti', in H. Escobedo (ed.) *Mexican Monuments: Strange encounters*, New York: Abbeville Press, 215–251.
García Márquez, G. (1986) *Clandestine in Chile: The adventures of Miguel Littín*, New York: Henry Holt and Company.
Gardner, S. (2013) 'Wikipedia, the People's Encyclopedia', *Los Angeles Times*, 13 January. Available online: http://articles.latimes.com/2013/jan/13/opinion/la-oe-gardner-Wikipedia-20130113.
Garfinkel, H. (1967) *Studies in Ethnomethodology*, Upper Saddle River, NJ: Prentice-Hall.
Garnham, N. (1986) 'The Media and the Public Sphere', *Intermedia* 14(1): 28–33.
Garrahan, M. (2017) 'Google and Facebook Dominance Forecast to Rise', *Financial Times*, 4 December. Available online: www.ft.com/content/cf362186-d840-11e7-a039-c64b1c09b482.
Gaudeul, A. and C. Peroni (2010) 'Reciprocal Attention and Norm of Reciprocity in Blogging Networks', *Economics Bulletin* 3(30): 2230–2248.
Gehl, R. W. (2011) 'The Archive and the Processor: The internal logic of Web 2.0', *New Media & Society* 13(8): 1228–1244.
Gehl, R. W. (2015) 'The Case for Alternative Social Media', *Social Media + Society*, July–December: 1–12. Available online: http://sms.sagepub.com/content/1/2/2056305115604338.full.pdf+html.
Gehring, D. and D. E. Wittkower (2015) 'On the Sale of Community in Crowdfunding: Questions of power, inclusion, and value', in L. Bennett, B. Chin and B. Jones (eds) *Crowdfunding the Future: Media industries, ethics, and digital societies*, New York: Peter Lang, 65–79.
Gennep, A. van (1909/2010) *The Rites of Passage*, Abingdon, UK and New York: Routledge.
Genova, N. De (2018) 'Europe's Racial Borders', Monitor Racism Project, Robert Schuman Centre for Advanced Studies. Available online: http://monitoracism.eu/europes-racial-borders.
George, N. (2005) *Hip-Hop America*, New York: Viking.
Georgeon, D. (2012) 'Revolutionary Graffiti: Street art and revolution in Tunisia', *Wasafiri* 27(4): 70–75.
Gerbaudo, P. (2012) *Tweets and the Streets: Social media and contemporary activism*, London: Pluto Press.
Gerbaudo, P. (2014) 'Populism 2.0: Social media activism, the generic internet user and interactive direct democracy', in D. Trottier and C. Fuchs (eds) *Social Media, Politics and the State: Protests, revolutions, riots, crime and policing in the age of Facebook, Twitter and YouTube*, London and New York: Routledge, 67–87.
Gerbaudo, P. (2016) 'Rousing the Facebook Crowd: Digital enthusiasm and emotional contagion in the 2011 protests in Egypt and Spain, *International Journal of Communication* 10: 254–273. Available online: http://ijoc.org/index.php/ijoc/article/view/3963.
Gerbaudo, P. (2017a) 'From Cyber-Autonomism to Cyber-Populism: An ideological history of digital activism', *tripleC* 15(2): 477–489.
Gerbaudo, P. (2017b) *The Mask and the Flag: Populism, citizenism, and global protest*, London: Hurst and Co.
Gerbaudo, P. and E. Treré (2015) 'In Search of the "We" of Social Media Activism: Introduction to the special issue on social media and protest identities', *Information, Communication & Society* 18(8): 865–871.
Gerecke, A. and L. Levin (2018) 'Moving Together in an Era of Assembly', *Canadian Theatre Review* 176: 5–10.
Geyh, P. (2006) 'Urban Free Flow: A poetics of parkour', *Journal of Media and Culture* 9(3).

Ghosh, B. (2011) *Global Icons: Apertures to the popular*, Durham, NC: Duke University Press.
Ghoshal, A. (2017) 'Microsoft Sued by Employees Who Developed PTSD after Reviewing Disturbing Content', *The Next Web*, 12 January. Available online: https://thenextweb.com/microsoft/2017/01/12/microsoft-sued-by-employees-who-developed-ptsd-after-reviewing-disturbing-content/#.tnw_RL7zXFcI.
Gibson, J. (2011) 'Saying It Right: Creating ethical verbatim theatre', *NEO: Journal for Higher Degree Research Students in the Social Sciences and Humanities* 4: 1–18. Available online: www.arts.mq.edu.au/documents/hdr_journal_neo/neoJanet2011_2.pdf.
Gibson, R. K. (2015) 'Party Change, Social Media and the Rise of "Citizen-initiated" Campaigning', *Party Politics* 21(2): 183–197
Giddens, A. (1984) *The Constitution of Society: Outline of the theory of structuration*, Berkeley and Los Angeles, CA: University of California Press.
Giddens, A. (1990) *The Consequences of Modernity*, Cambridge: Polity Press.
Giddens, A. (1991) *Modernity and Self-identity: Self and society in the late modern age*, Cambridge: Polity Press.
Gilbert, A. (2017) 'Live from Hall H: Fan/producer symbiosis at San Diego Comic-Con', in J. Gray, C. Sandvoss and C. L. Harrington (eds) *Fandom: Identities and communities in a mediated world*, second edition, New York: New York University Press, 354–368.
Gilbert, J. (2013) *Common Ground: Democracy and collectivity in an age of individualism*, London, Pluto Press.
Gilchrist, P. and B. Wheaton (2011) 'Lifestyle Sport, Public Policy and Youth Engagement: Examining the emergence of parkour', *International Journal of Sports Policy and Politics* 3(1): 109–131.
Gilchrist, P. and B. Wheaton (2013) 'New Media Technologies and Lifestyle Sport' in B. Hutchins and D. Rowe (eds) *Digital Media Sport: Technology, Power and Culture in the Network Society*, New York: Routledge, 169–185.
Gill, K. (1993) 'On the Metaphysical Distinction between Processes and Events', *Canadian Journal of Philosophy* 23(3): 365–384.
Gill, R. and A. Pratt (2008) 'Precarity and Cultural Work: In the social factory? Immaterial labour, precariousness and cultural work', *Theory, Culture & Society* 25(7–8): 1–30.
Gillan, K. and J. Pickerill (2012) 'The Difficult and Hopeful Ethics of Research on, and with, Social Movements', *Social Movement Studies* 11(2): 133–143.
Gillespie, N. (2018) 'FakeApp: Finally, a program that will let you seamlessly put Donald Trump in porn', *Reason*. Available online: http://reason.com/blog/2018/01/25/fakeapp-finally-a-program-that-will-let.
Gillespie, T. (2010) 'The Politics of Platforms', *New Media & Society* 12(3): 347–364.
Gillespie, T. (2014) 'The Relevance of Algorithms', in T. Gillespie, P. J. Boczkowski and K. A. Foot (eds) *Media Technologies: Essays on communication, materiality, and society*, Cambridge, MA and London: The MIT Press, 167–194.
Gillespie, T. (2018a) *Custodians of the Internet: Platforms, content moderation, and the hidden decisions that shape social media*, New Haven, CT: Yale University Press.
Gillespie, T. (2018b) 'Regulation of and by Platforms', in J. Burgess, T. Poell and A. Marwick (eds) *Sage Handbook of Social Media*, Los Angeles, CA: Sage, 254–278.
Gillespie, T., P. J. Boczkowski and K. A. Foot (eds) (2014) *Media Technologies: Essays on communication, materiality, and society*, Cambridge, MA and London: The MIT Press.
Gillespie, T. and N. Seaver (2016) 'Critical Algorithm Studies: A reading list', *Social Media Collective*. Available online: https://socialmediacollective.org/reading-lists/critical-algorithm-studies/.
Gilliland, A. and A. Flinn (2013) 'Community Archives: What are we really talking about?', *CIRN Prato Community Informatics – Conference Proceedings*, Monash Centre, Prato, Italy.
Gillmor, D. (2004) 'We the Media: The rise of citizen journalists', *National Civic Review* 93(3): 58–63.
Gillmor, D. (2006) *We the Media: Grassroots journalism by the people, for the people*, Beijing and Sebastopol, CA: O'Reilly Media.
Gilroy, P. (1993a) *The Black Atlantic*, London: Verso Books.
Gilroy, P. (1993b) *Small Acts: Thoughts on the politics of black cultures*, London: Serpent's Tail.
Gilroy, P. (2004) *After Empire*, London: Routledge.
Ginsburg, F. (1991) 'Indigenous Media: Faustian contract or global village?', *Cultural Anthropology* 6(1): 92–111.
Ginsburg, F. (1994) 'Culture/Media: A (mild) polemic', *Anthropology Today* 10(2): 5–15.
Ginsburg, F. (1997) 'From Little Things, Big Things Grow!', in R. G. Fox and O. Starn (eds) *Between Resistance and Revolution: Cultural politics and social protest*, New Brunswick, NJ: Rutgers University Press, 118–144.
Ginsburg, F. (1999) 'Shooting Back: From ethnographic film to indigenous production/ethnography of media', in T. Miller and R. Stam (eds) *A Companion to Film Theory*, Oxford: Blackwell, 295–322.
Ginsburg, F. (2003) 'Indigenous Media: Negotiating control over images in image ethics', in L. Gross, J. Katz and J. Ruby (eds) *The Digital Age*, Minneapolis, MN: University of Minnesota Press, 295–312.

Ginsburg, F. (2004) 'Atanarjuat Off-Screen: From "media reservations" to the world stage', *American Anthropologist* 105(4): 827–831.

Ginsburg, F. (2007) 'Shooting Back: From ethnographic film to indigenous production/ethnography of media', in T. Miller and R. Stam (eds) *A Companion to Film Theory*, Oxford: Blackwell, 295–322.

Ginsburg, F. (2012) 'Indigenous Counter-Publics: A foreshortened history', in Y. McKee and M. McLagan (eds) *Sensible Politics: The visual culture of nongovernmental activism*, New York: Zone Books, 87–108.

Ginsburg, F. and F. Myers (2006) 'A History of Indigenous Futures: Accounting for Indigenous art and media', *Aboriginal History* 30: 95–110.

Giraud, E. (2014) 'Has Radical Participatory Online Media Really "Failed"? Indymedia and its legacies', *Convergence* 20(4): 419–437.

Giraud, E. (2017) 'Displacement, "Failure" and Friction', in T. Schneider, K. Eli, C. Dolan and S. Ulijaszek (eds) *Digital Food Activism*, London and New York: Routledge, 130–150.

Gitlin, T. (1980) *The Whole World Is Watching: Mass media in the making and unmaking of the New Left*, Berkeley, CA: University of California Press.

Gitlin, T. (1987/1993) *The Sixties: Years of hope, days of rage*, New York: Banthan Books.

Gitlin, T. (2003a) *The Whole World Is Watching: Mass media in the making and unmaking of the New Left*, Oakland, CA: University of California Press.

Gitlin, T. (2003b) *Letters to a Young Activist*, New York: Basic Books.

Givoni, M. (2011a) 'Beyond the Humanitarian/Political Divide: Witnessing and the making of humanitarian ethics', *Journal of Human Rights* 10(1): 55–75.

Givoni, M. (2011b) 'Witnessing/Testimony', Mafte'akh. Lexical Review of Political Thought, Issue 2e-Winter: 147–169. Available online: http://mafteakh.tau.ac.il/en/wp-content/uploads/2011/01/Witnessing_Testimony-.pdf.

Gladwell, M. (2010) 'Small Change: Why the revolution will not be tweeted', *The New Yorker*, 4 October. Available online: www.newyorker.com/magazine/2010/10/04/small-change-malcolm-gladwell.

Glaser, M. (2010) 'Citizen Journalism: Widening world democracy', in S. Allan (ed.) *The Routledge Companion to News and Journalism*, London and New York: Routledge, 578–590.

Glasius, M. (2005) 'Deliberation or Struggle? Civil society traditions behind the social forums', *Ephemera* 5(2): 240–252.

Glasser, T. L. (2000) 'The Politics of Public Journalism', *Journalism Studies* 1(4): 683–685.

Gleasure, J. (2016) 'Emerging Technologies and the Democratisation of Financial Services: A metatriangulation of crowdfunding research', *Information and Organization* 26(4): 101–115.

Gleditsch, N. P., J. Nordkvelle and H. Strand (2014) 'Peace Research: Just the study of war?', *Journal of Peace Research* 51(2): 145–158.

Glenn, E. N. (1992) 'From Servitude to Service Work: Historical continuities in the racial division of paid reproductive labour', *Signs* 18(1): 1–43.

Glenny, M. (2011) *Dark Markets: Cyberthieves, cybercops and you*, New York: Random House.

Global Revolution (2011) 'Globalrevolution: Live streaming video powered by Livestream', Livestream.com. Available online: www.livestream.com/globalrevolution.

Gloeckle, M. and K. Royal (2017) 'Cover Your Assets', Association of Corporate Counsel. Available online: www.acc.com/sites/default/files/resources/vl/public/ACCDocketArticle/1465495_1.pdf.

Godin, M. and G. Doná (2016) '"Refugee Voices", New Social Media and Politics of Representation: Young Congolese in the diaspora and beyond', *Refuge: Canada's Journal on Refugees* 32(1): 60–71.

Goffman, E. (1959/1990) *The Presentation of Self in Everyday Life*, London: Penguin.

Goffman, E. (1961) *Asylums*, New York: Anchor Books.

Goggin, G. (2006) *Cell Phone Culture: Mobile technology in everyday life*, New York: Routledge.

Goggin, G. (2011) 'Ubiquitous Apps: Politics of openness in global mobile cultures', *Digital Creativity* 22(3): 148–159.

Goggin, G. and J. Clark (2009) 'Mobile Phones and Community Development: A contact zone between media and citizenship', *Development in Practice* 19(4–5): 585–597.

Goggin, G. and T. Noonan (2006) 'Blogging Disability: The interface between new cultural movements and internet technology', in A. Bruns and J. Jacobs (eds) *Uses of Blogs*, New York: Peter Lang, 161–172.

Goldman, E. (1923) *My Disillusionment with Russia*, New York: Double Day.

Goldman, T. (2011) 'Peaceful World of Warcraft Player Hits Max Level without Kills', *The Escapist*, 10 April. Available online: www.escapistmagazine.com/news/view/109134-Peaceful-World-of-Warcraft-Player-Hits-Max-Level-Without-Kills.

Gollatz, K., M. J. Riedl and J. Pohlmann (2018) 'Removals of online hate speech in numbers', HIIG Science Blog, 9 August. Available online: www.hiig.de/en/removals-of-online-hate-speech-numbers/.

Gómez Cruz, E. and E. Ardévol (2013a) 'Ethnography and the Field in Media(ted) Studies: A practice theory approach', *Westminster Papers in Communication and Culture* 9(3): 29–46.

Gómez Cruz, E. and E. Ardévol (2013b) 'Performing Photography Practices in Everyday Life: Some ethnographic notes on a Flickr group', *Photographies* 6(1): 35–44.
González, J. and J. Torres (2012) *News for All the People: The epic story of race and the American media*, London: Verso.
González-Bailón, S., J. Borge-Holthoefer, A. Rivero and Y. Moreno (2011) 'The Dynamics of Protest Recruitment through an Online Network', *Scientific Reports* 1. Available online: www.nature.com/articles/srep00197.
Goodchild, M. F. and J. A. Glennon (2010) 'Crowdsourcing Geographic Information for Disaster Response: A research frontier', *International Journal of Digital Earth* 3(3): 231–241.
Goodman, M. (2015) *Future Crimes: Inside the digital underground and the battle for our connected world*, New York: Penguin Random House.
Goodwin, J. and J. M. Jasper (2004) 'Caught in a Winding, Snarling Vine: The structural bias of political process theory', in J. Goodwin, and J. M. Jasper (eds) *Rethinking Social Movements: Structure, meaning, and emotion*, Lanham, MD: Rowman & Littlefield, 3–30.
Goodyear-Smith, F., C. Jackson and T. Greenhalgh (2015) 'Co-design and Implementation Research: Challenges and solutions for ethics committees', *BMC Medical Ethics* 16(1): 1–5.
Gordon, E. and P. Mihailidis (eds) (2016) *Civic Media: Technology, design, practice*, Cambridge, MA: The MIT Press.
Gordon, R. (2003) 'The Meanings and Implications of Convergence', in K. Kawamoto (ed.) *Digital Journalism: Emerging media and the changing horizons of journalism*, Lanham, MD: Rowman & Littlefield, 57–73.
Gordon, U. (2018) 'Prefigurative Politics between Ethical Practice and Absent Promise', *Political Studies* 66(2): 521–537.
Gorman, D. (2006) *Imperial Citizenship: Empire and the question of belonging*, Manchester: Manchester University Press.
Gotto, L. (2011) 'Life in a Day/One Day on Earth: Visibility and visuality in the digital arena', *The International Journal of the Image* 1. Available online: http://otheimage.com/journal.
Gould, D. (2010) 'On Affect and Protest', in J. Staiger and A. Cvetkovich (eds) *Political Emotions*, New York and London: Routledge, 18–44.
Grabe, M. E. and E. P. Bucy (2009) *Image Bite Politics*, Oxford: Oxford University Press.
Graeber, D. (2002) 'The New Anarchists', *New Left Review* 13: 61–73.
Graeber, D. (2008) 'The Sadness of Post-Workerism or "Art and Immaterial Labour" Conference: A sort of review', *The Commoner*, 1 April, 1–16.
Graeber, D. (2009) *Direct Action: An ethnography*, Edinburgh: AK Press.
Graeber, D. (2012) *Debt: The first 5,000 years*, Brooklyn, NY: Melville House.
Graeber, D. (2013) *The Democracy Project: A history, a crisis, a movement*, New York: Spiegel and Grau.
Graeff, E., M. Stempeck and E. Zuckerman (2014) 'The Battle for "Trayvon Martin": Mapping a media controversy online and off-line', *First Monday* 19(2). Available online: https://firstmonday.org/article/view/4947/3821.
Graham, M. (2011) 'Wiki Space: Palimpsests and the politics of exclusion', in G. Lovink and N. Tkacz (eds) *Critical Point of View: A Wikipedia reader*, Amsterdam: Ten Klein Groep, 269–282. Available online: www.networkcultures.org/publications.
Graham, M., R. Straumann and B. Hogan (2015) 'Digital Divisions of Labor and Informational Magnetism: Mapping participation in Wikipedia', *Annals of the Association of American Geographers* 105(6): 1158–1178.
Grahm, L. (2016) 'Representing Marikana', *Interventions: International Journal of Postcolonial Studies* 18(6): 834–851.
Grainge, P. (2011) 'A Song and Dance: Branded entertainment and mobile promotion', *International Journal of Cultural Studies* 15(2): 165–180.
Granjon, F., B. Lelong and J.-L. Metzger (2009) *Inégalités numériques: Clivages sociaux et modes d'appropriation des TIC*, Cachan, France: Hermes-Lavoisier.
Grant, A. (2009) 'Dimensions of Media Convergence', in A. Grant and J. Wilkinson (eds) *Media Convergence: The state of the field*, New York: Oxford University Press, 3–17.
Grant, D. E. (2015) 'Fire and the Body of Yahweh', *Journal of the Study of the Old Testament* 40(2): 139–161.
Grant, P. and E. Boon (2013) 'When the Persuasion Attempt Fails: An examination of consumers' perception of branded flash mobs', *Journal of Public Affairs* 13(2): 190–201.
Grauwacke, A. G. (2004) *Autonome in Bewegung: Aus den Ersten 23 Jahren*, Berlin: Assoziation A.
Gray, J. (2010) *Show Sold Separately: Promos, spoilers, and other media paratexts*, New York: New York University Press.
Gray, M. L. (2007) 'Face Value', *Contexts: A Publication of the American Sociological Association* 6(2): 73–75.
Gray, M. L. (2009) 'Negotiating Identities/Queering Desires: Coming out online and the remediation of the coming-out story', *Journal of Computer-Mediated Communication* (14): 1162–1189.
Green, S., C. Jenkins and H. Jenkins (2006) '"Normal Female Interest in Men Bonking": Selections from the terra nostra underground and strange bedfellows', in H. Jenkins (ed.) *Fans, Bloggers, and Gamers: Exploring participatory culture*, New York: New York University Press, 61–88.
Green Anarchist (1990) *Urban Attack: A Primer for Direct Action*.

Greenburg, Z. (2013) *Tech N9ne: Hip-hop's secret mogul*. Available online: www.forbes.com/sites/zackomalleygreenburg/2013/09/24/tech-n9ne-hip-hops-secret-mogul/.

Greenfield, P. (2018) 'The Cambridge Analytica Files: The story so far', *The Guardian*, 26 March. Available online: www.theguardian.com/news/2018/mar/26/the-cambridge-analytica-files-the-story-so-far.

Greengrass, P. (dir.) (2007) The Bourne Ultimatum. IMDb entry: www.imdb.com/title/tt0440963.

Greenpeace (2010) 'Lobbying', *Greenpeace*, 9 December. Available online: https://storage.googleapis.com/gpuk-archive/about/lobbying.html.

Greenpeace (2019) 'Volunteer to Take Direct Action', *Greenpeace*. Available online: www.greenpeace.org.uk/volunteering/direct-action/.

Greenwald, G. (2014) *No Place to Hide: Edward Snowden, the NSA and the surveillance state*, London: Hamish Hamilton.

Greenwald, G. (2015) 'With Power of Social Media Growing, Police Now Monitoring and Criminalizing Online Speech', *The Intercept*, 6 January. Available online: https://firstlook.org/theintercept/2015/01/06/police-increasingly-monitoring-criminalizing-online-speech/.

Greenwald, G. and M. Lee (2019) 'The U.S. Government's Indictment of Julian Assange Poses Grave Threats to Press Freedom', *The Intercept*, 12 April. Available online: https://theintercept.com/2019/04/11/the-u-s-governments-indictment-of-julian-assange-poses-grave-threats-to-press-freedoms/.

Gregg, M. (2011) *Work's Intimacy*, Cambridge: Polity Press.

Gregory, S. (2005) 'Introduction', in S. Gregory, G. Caldwell, R. Avni and T. Harding (eds) *Video for Change: A guide for advocacy and activism*, London: Pluto Press, xii–xvii.

Gregory, S. (2012) 'The Participatory Panopticon and Human Rights: WITNESS's experience supporting video advocacy and future possibilities', in M. McLagan and Y. McKee (eds) *Sensible Politics: The visual culture of nongovernmental activism*, New York: Zone Books, 517–549.

Gregory, S. (2017) 'Human Rights in an Age of Distant Witnesses: Remixed lives, incarnated images and live-streamed co-presence', in J. Eder and C. Klonk (eds) *Image Operations: Visual media and political conflict*, Manchester: Manchester University Press, 184–196.

Gregory, S., G. Caldwell, R. Avni and T. Harding (eds) (2005) *Video for Change: A guide for advocacy and activism*, London: Pluto Press.

Greskovits, B. (2019) 'Rebuilding the Hungarian Right through Conquering Civil Society: The Civic Circles Movement', *East European Politics* 36(2): 247–266.

Grieve, J., A. Nini and D. Guo (2017) 'Analyzing Lexical Emergence in American English Online', *English Language and Linguistics* 21: 99–127.

Grimmelmann, J. (2014) 'Anarchy, Status Updates, and Utopia', *Pace Law Review* 35(1): 135–153.

Grosfoguel, R. (2011) 'Decolonizing Post-Colonial Studies and Paradigms of Political Economy: Transmodernity, decolonial thinking, and global coloniality', *TRANSMODERNITY: Journal of Peripheral Cultural Production of the Luso-Hispanic World* 1(1): 1–36.

Grossberg, L. (2002) 'Postscript', *Communication Theory* 12(3): 367–370.

Grue, J. (2016) 'The Problem with Inspiration Porn: A tentative definition and a provisional critique', *Disability and Society* 31(6): 838–849.

Gruzd, A., B. Wellman and Y. Takhteyev (2011) 'Imagining Twitter as an Imagined Community', *American Behavioral Scientist* 55(10): 1294–1318.

Gu, Q. (2014) 'Sina Weibo: A mutual communication apparatus between the Chinese government and Chinese citizens', *China Media Research* 10(2): 72–86.

Guarnieri, C. (2017) 'What Is to Be Hacked?', *limn 8*. Available online: https://limn.it/articles/what-is-to-be-hacked/.

Gubrium, A. (2009) 'Digital Storytelling: An emergent method for health promotion research and practice', *Health Promotion Practice* 10(2): 186–191.

Gubrium, A. and K. Harper (2016) *Participatory Visual and Digital Methods*, London and New York: Routledge.

Gumucio Dagron, A. (2001) *Making Waves: Participatory communication for social change*, New York: The Communication for Social Change Consortium.

Gumucio Dagron, A. and L. Cajías (1989) *Las radios mineras de Bolivia*, CIMCA-UNESCO.

Gunning, J. and L. Z. Baron (2014) *Why Occupy a Square? People, protests and movements in the Egyptian Revolution*, Oxford: Oxford University Press.

Guo, S. (2018) '"Occupying" the Internet: State media and the reinvention of official culture online', *Communication and the Public* 3(1): 19–33.

Gurumurthy, A. and N. Chami (2016) 'Internet Governance as "Ideology in Practice": India's "free basics" controversy', *Internet Policy Review* 5(3): 2–17. Available online: https://policyreview.info/articles/analysis/internet-governance-ideology-practice-indias-free-basics-controversy.

Guss, N. (2011) 'Parkour and the Multitude: Politics of a dangerous art', *French Cultural Studies* 22(1): 73–85.

Gutiérrez, M. (2018) *Data Activism and Social Change*, Cham, Switzerland: Palgrave Macmillan.

Gutiérrez, M. and S. Milan (2019) 'Playing with Data and Its Consequences', *First Monday* 24(1). Available online: https://journals.uic.edu/ojs/index.php/fm/article/view/9554.

Haan, F. De (2017) 'Archive Fever, Resistance and Loss: A rereading of the IAV's early history', in R. Römkens (ed.) *Gender and Archiving: Past, present, future, 37th volume of the Yearbook of Women's History*, Amsterdam and Hilversum: Verloren, 21–38.

Habermas, J. (1962/1989) *The Structural Transformation of the Public Sphere: An inquiry into a category of bourgeois society*, translated by T. Burger and F. Lawrence, Cambridge, MA: The MIT Press.

Habermas, J. (1964/1974) 'The Public Sphere: An encyclopaedia article', translated by S. Lennox and F. Lennox, *New German Critique* 3: 49–55.

Habermas, J. (1964/1989) 'The Public Sphere: An encyclopaedia article', translated by S. Lennox and F. Lennox, in S. E. Bronner and D. Kellner (eds) *Critical Theory and Society: A reader*, New York: Routledge, 136–142.

Habermas, J. (1983/1990) 'Discourse Ethics: Notes on a program of philosophical justification', in J. Habermas (ed.) *Moral Consciousness and Communicative Action*, translated by C. Lenhart and S. W. Nicholson, Cambridge. MA: The MIT Press, 43–115

Habermas, J. (1992/1996) *Between Facts and Norms: Contributions to a discourse theory of law and democracy*, Cambridge: Polity Press.

Habermas, J. (2008/2009) 'Media, Markets and Consumers: The quality press as the backbone of the political public sphere', in J. Habermas (ed.) *Europe: The faltering project*, translated by C. Cronin, Cambridge: Polity Press, 131–137.

Hackett, R. and W. Carroll (2006) *Remaking Media: The struggle to democratize communication*, New York: Routledge.

Hader, B. and F. Armisen (2015) Documentary Now! Broadway Video. Available online: www.ifc.com/shows/documentary-now.

Hadden, J. (2015) *Networks of Contention: The divisive power of climate change*, Cambridge: Cambridge University Press.

Hagener, M., V. Hediger and A. Strohmeier (eds) (2016) *The State of Post-Cinema: Tracing the moving image in the age of digital dissemination*, Basingstoke, UK: Palgrave Macmillan.

Haggerty, K. D. and R. V. Ericson (2000) 'The Surveillant Assemblage', *British Journal of Sociology* 51(4): 605–622.

Haiven, M. and A. Khasnabish (2014) *The Radical Imagination: Social movement research in the age of austerity*, London: Zed Books.

Häkkinen, A. and S. Leppänen (2014) 'YouTube Meme Warriors: Mashup videos as political critique', *Varieng: Texts and Discourses of New Media* 15(1): 1. Available online: www.helsinki.fi/varieng/series/volumes/15/hakkinen_leppanen/.

Hale, C. (2006) 'Activist Research v. Cultural Critique: Indigenous land rights and the contradictions of politically engaged anthropology', *Cultural Anthropology* 21(1): 96–120.

Halilovich, H. (2013) 'Trans-local and Cyber Spaces', *Journal of Refugee Studies* 26(4): 524–540.

Hall, R. (2011) 'Police Hurt in Violent Anti-Tesco Protests', BBC News, 22 April. Available online: http://bbc.co.uk/news/uk-england-bristol-13167041.

Hall, S. (1980) 'Encoding/Decoding', in S. Hall, D. Hobson, A. Lowe and P. Tillis (eds) *Culture, Media, Language*, London: Hutchinson, 128–138.

Hall, S. (1990) 'Cultural Identity and Diaspora', in J. Rutherford (ed.) *Identity: Community, Culture, Difference*, London: Lawrence & Wishart, 222–237.

Hall, S. (1992) 'The Question of Cultural Identity', in S. Hall, D. Held and T. McGrew (eds) *Modernity and Its Futures*, Cambridge: Polity Press, 274–323.

Hall, S. (1995) 'When Was the "Post-Colonial"? Thinking at the limit', in I. Chambers and L. Curti (eds) *The Postcolonial Question: Common skies, divided horizons*, New York and London: Routledge, 242–260.

Hall, S. (1996) 'Race, Articulation, and Societies Structured in Dominance', in R. A. Lindeborg, R. H. Lindeborg, H. A. Baker Jr. and M. Diawara (eds) *Black British Cultural Studies: A reader*, Chicago, IL: University of Chicago Press, 15–33.

Hall, S. (1997a) *Representation*, London, Thousand Oaks, CA and New Delhi: Sage.

Hall, S. (1997b) 'The Spectacle of the Other', in S. Hall (ed.) *Representation: Cultural representations and signifying practices*, London: Sage, 223–290.

Hall, S. (2001) 'Constituting an Archive', *Third Text* 54: 89–92.

Hall, S. (2006) 'Notes on Deconstructing the "popular"', in J. Storey (ed.) *Cultural Theory and Popular Culture: A reader*, fourth edition, Harlow, UK: Pearson Education, 43–4.

Hall, S., C. Critcher, T. Jefferson, J. Clarke and B. Roberts (1978) *Policing the Crisis: Mugging, the state and law and order*, London: Macmillan.

Halleck, D. (2002) *Hand-held Visions: The uses of community media*, New York: Fordham University Press.

Haller, B. (2006) 'Journalism', in G. Albrecht (ed.) *Encyclopedia of Disability*, London: Sage, 992–996.

Haller, B. (2010) *Representing Disability in an Ableist World: Essays on mass media*, Louisville, KY: Avocado Press.
Haller, B. and J. Preston (2017) 'Confirming Normalcy: "Inspiration porn" and the construction of the disabled subject', in K. Ellis and M. Kent (eds) *Disability and Social Media: Global perspectives*, New York: Routledge, 41–56.
Haller, B. and L. Zhang (2014) 'Stigma or Empowerment?: What do disabled people say about their representations in the news and entertainment media?', *Review of Disability Studies* 9(4): 19–33.
Halloran, J. D., P. Elliott and G. Murdock (1970) *Demonstrations and Communication: A case study*, London: Penguin.
Hamilton, C. and L. Shepherd (eds) (2016) *Understanding Popular Culture and World Politics in the Digital Age*, London and New York: Routledge.
Hampton, K. N. and B. Wellman (2003) 'Neighboring in Netville: How the Internet supports community and social capital in a wired suburb', *City & Community* 2(4): 277–311.
Hampton, K. N., L. Rainie, W. Lu, M. Dwyer, I. Shin and K. Purcell (2014) *Social Media and the 'Spiral of Silence'*, Washington, DC: Pew Research Center. Available online: www.pewinternet.org/files/2014/08/PI_Social-networks-and-debate_082614.pdf.
Han, E. L. (2016) *Micro-Blogging Memories: Weibo and collective remembering in contemporary China*, New York: Palgrave Macmillan.
Han, G. and W. Wang (2015) 'Mapping User Relationships for Health Information Diffusion on Microblogging in China: A social network analysis of Sina Weibo', *Asian Journal of Communication* 25(1): 65–83.
Hanifan, L. J. (1916) 'The Rural School Community Center', *Annals of the American Academy of Political and Social Science* 67(1): 130–138.
Hannerz, U. (1991) 'Scenarios for Peripheral Cultures', in A. D. King (ed.) *Culture, Globalization and the World-System: Contemporary conditions for the representation of identity*, Minneapolis, MN: University of Minnesota Press, 107–128.
Hanusch, F. (2017) 'Transformations of Journalism Culture', in B. Franklin and S. Eldridge II (eds) *The Routledge Companion to Digital Journalism Studies*, London: Taylor & Francis, 387–395.
Harcup, T. and D. O'Neill (2016) 'What Is News? News values revisited (again)', *Journalism Studies* 18(12): 1470–1488.
Hardt, M. and A. Negri (2001) *Empire*, Cambridge, MA: Harvard University Press.
Hardt, M. and A. Negri (2004) *Multitude: War and democracy in the age of empire*, New York: Penguin.
Hardt, M. and A. Negri (2011) 'The Fight for "Real Democracy" at the Heart of Occupy Wall Street', *Foreign Affairs*, 11 October. Available online: www.foreignaffairs.com/articles/north-america/2011-10-11/fight-real-democracy-heart-occupy-wall-street.
Hare, D. (2005) 'On Factual Theatre', in R. Soans, *Talking to Terrorists*, London: Oberon Books, 111–113.
Hari, D. (2009) *The Translator*, New York: Random House.
Harkness, G. (2012) 'True School: Situational authenticity in Chicago's hip-hop underground', *Cultural Sociology* 6(3): 283–298.
Harlow, S. (2016) 'Reconfiguring and Remediating Social Media as Alternative Media: Exploring youth activists' digital media ecology in El Salvador', *Palabra Clave* 19(4): 997–1026.
Harmon, A. (2003) 'Ideas & Trends: Flash mobs; Guess some people don't have anything better to do'. Available online: www.nytimes.com/2003/08/17/weekinreview/ideas-trends-flash-mobs-guess-some-people-don-t-have-anything-better-to-do.html.
Harney, N. (2013) 'Precarity, Affect and Problem Solving with Mobile Phones by Asylum Seekers, Refugees and Migrants in Naples, Italy', *Journal of Refugee Studies* 26(4): 541–557.
Haro Barba, C. and V. F. Sampedro Blanco (2011) *Activismo político en Red: del Movimiento por la Vivienda Digna al 15M*, *Teknokultura*, Madrid: Grupo de Investigación Cibersomosaguas, Universidad Complutense de Madrid.
Harold, C. (2004) 'Pranking Rhetoric: "Culture jamming" as media activism', *Critical Studies in Media Communication* 21(3): 189–211.
Harrington, S., T. Highfield and A. Bruns (2013) 'More than a Backchannel: Twitter and Television', *Participations: Journal of Audience and Reception Studies* 10(1): 405–409.
Harris, J. (2011) 'About', Cowbird. Available online: http://cowbird.com/about/.
Harris, V. (2012) 'Genres of the Trace: Memory, archives and trouble', *Archives and Manuscripts* 40(3): 147–157.
Harrison, A. K. (2006) 'Cheaper than a CD, Plus We Really Mean It: Bay area underground hip hop tapes as subcultural artefacts', *Popular Music* 25: 283–301.
Harrison, J. (2010) 'User-Generated Content and Gatekeeping at the BBC Hub', *Journalism Studies* 11(2): 243–256.
Harte, D. (2013) '"One Every Two Minutes": Assessing the scale of hyperlocal publishing in the UK', *JOMEC Journal*. Available online: https://jomec.cardiffuniversitypress.org/articles/abstract/10.18573/j.2013.10240/.
Harte, D., R. Howells and A. Williams (2018) *Hyperlocal Journalism: The decline of local newspapers and the rise of online community news*, London and New York: Routledge.
Harte, D., J. Turner and A. Williams (2016) 'Discourses of Enterprise in Hyperlocal Community News in the UK', *Journalism Practice* 10(2): 233–250.

Harte, D., A. Williams and J. Turner (2017) 'Reciprocity and the Hyperlocal Journalist', *Journalism Practice* 11(2–3): 160–176.
Hartelius, E. J. (2010) 'Wikipedia and the Emergence of Dialogic Expertise', *Southern Communication Journal* 75(5): 505–526.
Hartley, J. (1982) *Understanding News*, London and New York: Routledge.
Hartley, J. and K. McWilliam (2009a) 'Computational Power Meets Human Contact', in J. Hartley and K. McWilliam (eds) *Story Circle: Digital storytelling around the world*, Malden, MA: Wiley-Blackwell, 3–15.
Hartley, J. and K. McWilliam (eds) (2009b) *Story Circle: Digital storytelling around the world*, Malden, MA: Wiley-Blackwell.
Hartman, S. (1997) *Scenes of Subjection: Terror, slavery, and self-making in nineteenth-century America*, New York: Oxford University Press.
Harvey, D. (1989) *The Condition of Postmodernity: An enquiry into the origins of cultural change*, London: Basil Blackwell.
Harvey, D. (2007) *A Brief History of Neoliberalism*, Oxford: Oxford University Press.
Harvey, D. (2012) *Rebel Cities: From the right to the city to the urban revolution*, London and New York: Verso Books.
Haunss, S. (2015) 'Promise and Practice in Studies of Social Media and Movements', in L. Dencik and O. Leistert (eds) *Critical Perspectives on Social Media and Protest: Between control and emancipation*, Lanham, MD and London: Rowman & Littlefield, 13–34.
Hawdon, J., A. Oksanen and P. Räsänen (2014) 'Victims of Online Hate Groups: American youth's exposure to online hate speech', in J. Hawdon, M. Lucht and J. Ryan (eds) *The Causes and Consequences of Group Violence: From bullies to terrorists*, Lanham, MD: Lexington Books, 165–182.
Hay, J. and N. Couldry (eds) (2011) 'Rethinking Convergence/Culture'. Special Issue of *Cultural Studies* 25(4–5).
Hazard, M. (2015) 'Happy Collaborator: George Stoney', DER. Available online: www.der.org/films/happy-collaborator.html.
Hazou, R. (2009) 'Refugee Advocacy and the Theatre of Inclusion', *About Performance* 9: 67–85.
Heaney, C. D., S. M. Wilson and O. R. Wilson (2007) 'The West End Revitalization Organization's Community-owned and -managed Research Model: Development, implementation, and action', *Progress in Community Health Partnerships: Research, Education, and Action* 1(4): 339–349.
Hearn, A. (2017) 'Verified: Self-presentation, identity management, and selfhood in the age of big data', *Popular Communication* 15(2): 62–77.
Hearn, A. and S. Schoenhoff (2016) 'From Celebrity to Influencer: Tracing the diffusion of celebrity value across the data stream', in P. D. Marshall and S. Redmond (eds) *A Companion to Celebrity*, Hoboken, NJ: John Wiley and Sons, 194–211.
Hearn, G. N. and M. Foth (2007) 'Communicative Ecologies', *The Electronic Journal of Communication* 17(1–2). Available at http://eprints.qut.edu.au/8171/1/8171.pdf.
Heath, J. and A. Potter (2004) *The Rebel Sell: Why the culture can't be jammed*, Toronto: HarperCollins Canada.
Heathershaw, J. (2008) 'Unpacking the Liberal Peace: The dividing and merging of peacebuilding discourses', Millennium: *Journal of International Studies* 36(3): 597–621.
Hebdige, D. (1979) *Subculture: The meaning of style*, London: Routledge.
Heidegger, M. (1954/1993) 'Building, Dwelling, Thinking', in D. F. Krell (ed.) *Martin Heidegger: Basic writings*, London: Routledge, 361–362.
Heidenreich, T., J. Eberl, F. Lind and H. Boomgarden (2019) 'Political Migration Discourses on Social Media: A comparative perspective on visibility and sentiment across political Facebook accounts in Europe', *Journal of Ethnic and Migration Studies*. Available online: https://doi.org/10.1080/1369183X.2019.1665990.
Heikka, T. (2015) 'The Rise of the Mediating Citizen: Time, space, and citizenship in the crowdsourcing of Finnish legislation', *Policy and Internet* 7(3): 268–291.
Heimo, A. (2017) 'The Italian Hall Tragedy, 1913: A hundred years of remediated memories', in T. Andersen and B. Törnquist-Plewa (eds) *The Twentieth Century in European Memory: Transcultural mediation and reception*, Leiden and Boston, MA: Brill, 240–267.
Heiner, B. T. (2008) 'Foucault and the Black Panthers', *City* 11(3): 313–356.
Held, D. (2006) *Models of Democracy*, third edition, Cambridge: Polity Press.
Heldt, A. (2019) 'Reading between the Lines and the Numbers: An analysis of the first NetzDG reports', *Internet Policy Review* 8(2).
Helle-Valle, J. (2010) 'Language-games, In/dividuals and Media Uses: What a practice perspective should imply for media studies', in B. Bräuchler and J. Postill (eds) *Theorising Media and Practice*, New York: Berghahn Books, 191–212.
Helmond, A. (2015) 'The Platformization of the Web: Making web data platform ready', *Social Media + Society* 15: 1–11. Available online: https://journals.sagepub.com/doi/full/10.1177/2056305115603080.
Helms, E. (2017) *The Fake News with Ted Nelms*, USA: Pacific Electric Picture Company. Available online: www.amazon.com/Fake-News-Ted-Nelms-Season/dp/B0788HG6PD.

Hemer, O. and T. Tufte (eds) (2016) *Voice and Matter: Communication, development and the cultural return*, Stockholm: Nordicom.

Hemmi, A. and I. Graham (2014) 'Hacker Science versus Closed Science: Building environmental monitoring infrastructure', *Information, Communication, and Society* 17(7): 830–842.

Hemsley, B., S. Dann, S. Palmer, S. Allan and S. Balandin (2015) 'We Definitely Need an Audience: Experiences of Twitter, Twitter networks and tweet content in adults with severe communication disabilities who use augmentative and alternative communication (AAC)', *Disability and Rehabilitation* 37(17): 1531–1542.

Henley, P. (2010) *The Adventure of the Real: Jean Rouch and the craft of ethnographic cinema*, Chicago, IL: University of Chicago Press.

Hepp, A. (2013) *Cultures of Mediatization*, Cambridge: Polity Press.

Hepp, A. (2016) 'Pioneer Communities: Collective actors of deep mediatisation', *Media, Culture & Society* 38(6): 918–933.

Hepp, A. (2020) *Deep Mediatization*, London: Routledge.

Hepp, A. and N. Couldry (2010) 'Introduction: Media events in globalized media cultures', in N. Couldry, A. Hepp and F. Krotz (eds) *Media Events in a Global Age*, Abingdon, UK: Routledge, 1–20.

Hepp, A. and U. Hasebrink (2017) 'Researching Transforming Communications in Times of Deep Mediatization: A figurational approach', in A. Hepp, A. Breiter and U. Hasebrink (eds) *Communicative Figurations: Transforming communications in times of deep mediatization*, London: Palgrave Macmillan, 51–80.

Hepp, A., S. Hjarvard and K. Lundby (2015) 'Mediatization: Theorizing the interplay between media, culture and society', *Media, Culture & Society* 37(2): 314–324.

Hepp, A. and W. Vogelgesang (eds) (2003) *Populäre Events: Medienevents, Spielevents und Spaßevents*, Opladen: Leske + Budrich.

Herman, E. S. (2000) 'The Propaganda Model: A retrospective', *Journalism Studies* 1(1): 101–112.

Herman, E. S. and N. Chomsky (1988) *Manufacturing Consent*, New York: Pantheon Books.

Hermida, A. (2010) 'Twittering the News: The emergence of ambient journalism', *Journalism Practice* 4(3): 297–308.

Hermida, A. (2013) '#Journalism', *Digital Journalism* 1(3): 295–313.

Hermida, A., F. Fletcher, D. Korell and D. Logan (2012) 'SHARE, LIKE, RECOMMEND: Decoding the social media news consumer', *Journalism Studies* 13(5–6): 815–824.

Herndon, S. (2003) 'Panel Presentation', OurMedia Conference, Barranquilla, Colombia.

Herrera, L. (2014) *Revolution in the Age of Social Media: The Egyptian popular insurrection and the internet*, London and New York: Verso.

Hertzberg, B. and K. Lundby (2008) 'Autobiography and Assumed Authenticity in Digital Storytelling', in K. Lundby (ed.) *Mediatized Stories: Self-representation in new media*, New York: Peter Lang, 105–222.

Hesling, W. (1989) 'Documentary Film and Rhetorical Analysis', in W. de Greef and W. Hesling (eds) *Image–Reality–Spectator*, Leuven: Acco, 101–131.

Hesmondhalgh, D. (2010) 'User-Generated Content: Free labour and the cultural industries', *Ephemera: Theory and Politics in Organization* 10(3–4): 267–284.

Hesmondhalgh, D. and S. Baker (2011) *Creative Labour: Media work in three cultural industries*, London and New York: Routledge.

Hesmondhalgh, D. and C. Melville (2002) 'Urban Breakbeat Culture: Repercussions of hip-hop in the United Kingdom', in T. Mitchell (eds) *Global Noise: Rap and hip hop outside the USA*, Middletown, CT: Wesleyan University Press, 86–110.

Hess, A. (2015) 'The Selfie Assemblage', *International Journal of Communication* 9: 1629–1646.

Hess, K. and L. Waller (2016) 'Hip to Be Hyper: The subculture of excessively local news', *Digital Journalism* 4(2): 193–210.

Hesse, B. (2004) 'Im/plausible Deniability: Racism's conceptual double bind', *Social Identities* 10(1): 9–29.

Hickman, P. (2013) '"Third Places" and Social Interaction in Deprived Neighbourhoods in Great Britain', *Journal of Housing and the Built Environment* 28(2): 221–236.

Higgins, J. W. (2011) 'Peace-Building through Listening, Digital Storytelling, and Community Media in Cyprus', *GMJ: Mediterranean Edition* 6(1), Spring. Available online: http://connection.ebscohost.com/c/articles/71847751/peace-building-through-listening-digital-storytelling-community-media-cyprus.

Higgins, J. W. (2014) 'Changing the World, One Story at a Time: Oral histories and self-reflexivity in Cypriot university classrooms', in H. Briel (ed.) *Challenging History: Oral History Work in Cyprus*, Nicosia: University of Nicosia Press, 103–123.

Hight, C. (2014) 'Shoot, Edit, Share: Cultural software and user-generated documentary practice', in K. Nash, C. Hight and C. Summerhayes (eds) *New Documentary Ecologies: Emerging platforms, practices and discourses*, Basingstoke, UK and New York, AIAA, 219–236.

Hildebrandt, P., K. Evert, S. Peters, M. Schaub, K. Wildner and G. Ziemer (eds) (2018) *Performing Citizenship: Bodies, agencies, limitations*, New York: Springer.

Hill, B. M. (2012) 'Freedom for Users, Not for Software', in D. Bollier and S. Helfrich (eds) *The Wealth of the Commons: A world beyond market and state*, Amherst, MA: Levellers Press, 305–308. Available online: http://wealthofthecommons.org/essay/freedom-users-not-software.

Hill, M. (2009) *Beats, Rhymes, and Classroom Life: Hip-hop pedagogy and the politics of identity*, New York: Teachers College Press.

Hill Collins, P. and J. Solomos (eds) (2009) *The Sage Handbook of Race and Ethnicity Studies*, London: Sage.

Hillier, L. and L. Harrison (2007) 'Building Realities Less Limited than Their Own: Young people practicing same sex attraction on the Internet', *Sexualities: Studies in Culture and Society* 10(1): 82–100.

Hillis, K., S. Paasonen and M. Petit (2015) *Networked Affect*, Cambridge, MA: The MIT Press.

Hills, M. (2002) *Fan Cultures*, Oxon: Routledge.

Himanen, P., L. Torvalds and M. Castells (2001) *The Hacker Ethic and the Spirit of the Information Age*, New York: Random House.

Himmelweit, S. (2007) 'The Prospects for Caring: Economic theory and policy analysis', *Cambridge Journal of Economics* 31(4): 581–599.

Hinck, A. (2011) 'Theorizing a Public Engagement Keystone: Seeing fandom's integral connection to civic engagement through the case of the Harry Potter Alliance', *Transformative Works and Cultures* 10.

Hinegardner, L. (2009) 'Action, Organization, and Documentary Film: Beyond a communications model of human rights videos', *Visual Anthropology Review* 25(2): 172–185.

Hintz, A. (2003) 'Indymedia Germany: A local node of the global Indymedia network', *Media Development* XLX(4): 21–25.

Hintz, A., L. Dencik and K. Wahl-Jorgensen (2018) *Digital Citizenship in a Datafied Society*, Cambridge: Polity Press.

Hintz, A. and S. Milan (2009) 'At the Margins of Internet Governance: Grassroots Tech groups and communication policy', *International Journal of Media and Cultural Politics* 5(1): 23–38.

Hirsch, M. (2003) 'I Took Pictures: September 2001 and beyond', in J. Greenberg (ed.) *Trauma at Home: After 9/11*, Lincoln, NE and London: University of Nebraska Press, 69–86.

Hirschman, A. O. (1970) *Exit, Voice, and Loyalty: Responses to decline in firms, organizations, and states*, Cambridge, MA: Harvard University Press.

Hirvonen, E. (2017) Improving the Game with User Generated Content: An overview of Skyrim mod users, Unpublished Masters thesis, Jyväskylä, Finland: University of Jyväskylä Digital Repository. Available online: https://jyx.jyu.fi/handle/123456789/56234.

Hjarvard, S. (2017) 'Mediatization', in C. A. Hoffner and L. V. Zoonen (eds) *International Encyclopedia of Media Effects*, New York: Wiley. Available online: https://onlinelibrary.wiley.com/doi/abs/10.1002/9781118783764.wbieme0107.

Hjorth, L. and N. Hendry (2015) 'A Snapshot of Social Media: Camera phone practices', *Social Media + Society* 1(1): 1–3.

Hjorth, L. and S. Pink (2012) 'Emplaced Cartographies: Reconceptualising camera phone', *MIA (Media International Australia)* 145(1): 145–155.

Ho, M. (2010) 'Co-opting Social Ties: How the Taiwanese petrochemical industry neutralized environmental opposition', *Mobilization: An International Quarterly* 15(4): 447–463.

Hoechsmann, M. (1997) 'Benetton Culture: Marketing difference to the new global consumer', in S. H. Riggins (ed.) *The Language and Politics of Exclusion: Others in discourse*, London: Sage, 183–202.

Hoffman, A. L., N. Proferes and M. Zimmer (2018) '"Making the World More Open and Connected": Mark Zuckerberg and the discursive construction of Facebook and its users', *New Media & Society* 20(1): 199–218.

Hoffman, D. (2013) 'Citizens: The new fourth estate'. Available online: www.huffingtonpost.com/david-hoffman/citizens-the-new-fourth-e_b_3894819.html.

Hoffmann, J. (2013) 'Conceptualizing "Communication for Peace"', UPEACE Open Knowledge Network Occasional Working Papers 1 (January 2013), University for Peace.

Hoffmann, K. and E. Wottrich (eds) (2015) *Protest – Film – Bewegung. Neue Wege im Dokumentarischen*, Munich: Edition Text + Kritik.

Hogan, B. (2012) 'Pseudonyms and the Rise of the Real-Name Web', in J. Hartley, J. Burgess and A. Bruns (eds) *A Companion to New Media Dynamics*, Chichester, UK: Blackwell, 290–308.

Hogan, M. (2015) 'The Archive as Dumpster', *Pivot* 4(1).

Hoggart, R. (1957) *The Uses of Literacy*, Harmondsworth, UK: Penguin.

Hohendahl, P. U. (1979) 'Critical Theory, Public Sphere and Culture: Jurgen Habermas and his critics', *New German Critique* (16): 89–118.

Hollander, E., J. Stappers and N. W. Jankowski (2002) 'Community Media and Community Communication', in N. W. Jankowski and O. Prehn (eds) *Community Media in the Information Age: Perspectives and prospects*, Cresskill, NJ: Hampton Press, 19–30.

Holloway, J. (2004) '"Walking We Ask Questions": An interview with John Holloway', interviewed by M. Sitrin. Available online: www.leftturn.org/"walking-we-ask-questions"-interview-john-holloway.

Holmes, S. K. (2013) 'From Social Realism to Allegorithms of Assemblage as Gaming Activism in Minecraft', in Z. Waggoner (ed.) *Terms of Play: Essays on words that matter in videogame theory*, Jefferson, NC: McFarland and Company, 85–116.

Holt, D. B. (2002) 'Why Do Brands Cause Trouble? A dialectical theory of consumer culture and branding', *Journal of Consumer Research* 29(1): 70–90.

Hong, R. (2013) 'Game Modding, Prosumerism and Neoliberal Labor Practices', *International Journal of Communication* 7: 984–1002.

Hoofd, I. M. (2012) *Ambiguities of Activism: Alter-globalism and the imperatives of speed*, New York: Routledge.

hooks, b. (1989) *Talking Back: Thinking feminist, thinking black*, Boston, MA: Ellen Herman and South End Press.

Horn, G.-R. (2007) *The Spirit of '68: Rebellion in Western Europe and North America 1956–1976*, Oxford: Oxford University Press.

Horsti, K., G. Hultén and G. Titley (2013) *National Conversations? Public service media and cultural diversity*, Bristol, UK: Intellect.

Hoskins, A. (2009) 'The Mediatization of Memory', in J. Garde-Hansen, A. Hoskins and A. Reading (eds) *Save as … Digital Memories*, New York: Palgrave Macmillan, 27–43.

Hoskins, A. (2011) '7/7 and Connective Memory: Interactional trajectories of remembering in post-scarcity culture', *Memory Studies* 4(3): 269–280.

Hoskins, A. (2014) 'A New Memory of War', in B. Zelizer and K. Tenenboim-Weinblatt (eds) *Journalism and Memory*, Basingstoke, UK: Palgrave Macmillan, 179–191.

Hossain, M. and I. Kauranen (2015) 'Crowdsourcing: A comprehensive literature review', *Strategic Outsourcing: An International Journal* 8(1): 2–22.

House, N. Van, M. Davis, M. Ames, M. Finn and V. Viswanathan (2005) 'The Uses of Personal Networked Digital Imaging: An empirical study of cameraphone photos and sharing', *CHI '05 Extended Abstracts on Human Factors in Computing Systems (CHI EA '05)*, New York: Association for Computing Machinery, 1853–1856.

Howard, P. N. (2010) *The Digital Origins of Dictatorship and Democracy: Information technology and political Islam*, Oxford: Oxford University Press.

Howe, J. (2006) 'The Rise of Crowdsourcing', *Wired*, 1 June. Available online: http://wrd.cm/1nkRGsc.

Howley, K. (2005) *Community Media: People, places, and communication technologies*, Cambridge: Cambridge University Press.

Howley, K. (2009) *Understanding Community Media*, Los Angeles, CA: Sage.

Howley, K. (2010) 'Introduction', in K. Howley (ed.) *Understanding Community Media*, Thousand Oaks, CA: Sage, 1–14.

Huang, R. and X. Sun (2014) 'Weibo Network, Information Diffusion and Implications for Collective Action in China', *Information, Communication & Society* 17(1): 86–104.

Huang, Y., D. Guo, A. Kasakoff and J. Grieve (2016) 'Understanding US Regional Linguistic Variation with Twitter Data Analysis', *Computers, Environment and Urban Systems* 59: 244–255.

Hudson, B., J. Rosenbloom and D. Cole (eds) (2019) *Routledge Handbook of the Study of the Commons*, Abingdon, UK: Routledge.

Hudson, D. and P. R. Zimmermann (2015) *Thinking through Digital Media: Transnational environments and locative places*, New York: Palgrave Macmillan.

Huesca, R. (1995) 'Subject-Authored Theories of Media Practice: The case of Bolivian tin miners' radio', *Communication Studies* 46 (Fall/Winter): 149–168.

Huesca, R. (1996) 'Participation for Development in Radio: An ethnography of the *Reporteros populares* of Bolivia', *Gazette* 57(1): 29–52.

Huesca, R. and B. Dervin (1994) 'Theory and Practice in Latin American Alternative Communication Research', *Journal of Communication* 44(4): 53–73.

Hughes, J. and S. Parry (2016) 'Theatricality and Gesture as Citizen Media: Composure on a precipice', in M. Baker and B. B. Blaagaard (eds) *Citizen Media and Public Spaces: Diverse expressions of citizenship and dissent*, London and New York: Routledge, 79–95.

Huguet, F. (2013) 'Tahrir et ses écritures dans le miroir', in F. Daghmi, F. Toumi and A. Amsidder (eds) *Les médias font-ils les révolutions? Regards critiques sur les soulèvements Arabes*, Paris: L'Harmattan, 101–117.

Hui, A., T. Schatzki and E. Shove (2017) *The Nexus of Practices: Connections, constellations, practitioners*, London: Routledge.

Hunter, A. (2016) 'Monetizing the Mommy: Mommy blogs and the audience commodity', *Information, Communication & Society* 19(9): 1306–1320.

Hunter, D., R. Lobato, M. Richardson and J. Thomas (eds) (2013) *Amateur Media: Social, cultural and legal perspectives*, London: Routledge.

Hurl, C. (2009) 'Antiglobalization and "Diversity of Tactics"', *Upping the Anti* 1. Available online: https://uppingtheanti.org/journal/article/01-anti-globalization-and-diversity-of-tactics.

Hutchinson, P. (2015) 'Riot Girls on Film: How silent cinema lent the suffragettes a voice', *The Guardian*, 19 October. Available online: www.theguardian.com/film/filmblog/2015/oct/19/silent-film-suffragettes-bfi-movie.

Hutter, S. (2014) *Protesting Culture and Economics in Western Europe*, Minneapolis, MN: University of Minnesota Press.

Huurdeman, A. A. (2003) *The Worldwide History of Telecommunications*, New York: John Wiley and Sons.

Hyde, L. (1998) *Trickster Makes this World: Mischief, myth and art*, New York: Farrar, Straus and Giroux.

Iannelli, L. and C. M. Marelli (2017) 'Cultures in Action: Public art and political participation', in L. Iannelli and P. Musarò (eds) *Performance Citizenship: Public art, urban design, and political participation*, Milan: Mimesis International, 51–71.

Iannelli, L. and P. Muscarò (eds) (2017) *Performative Citizenship: Public art, urban design, and political participation*, Milan: Mimesis International.

ICRSE (2005a) 'The Declaration of the Rights of Sex Workers in Europe', International Committee on the Rights of Sex Workers in Europe. Available online: www.sexworkeurope.org.

ICRSE (2005b) 'Sex Workers in Europe Manifesto', International Committee on the Rights of Sex Workers in Europe. Available online: www.sexworkeurope.org.

Iglesias Turrión, P. (2003) 'Ciclos de movimiento: Conversando con Luca Casarini', *El Viejo topo, Iniciativas Editoriales* 175: 6–15.

ILO (2012) 'South Africa Could Do More for Miners, Says ILO Mining Specialist', International Labour Organization, 24 August. Available online: www.ilo.org/global/about-the-ilo/newsroom/news/WCMS_187783/lang--en/index.htm.

Improv Everywhere (2008) *Frozen Grand Central*. Available online: https://improveverywhere.com/2008/01/31/frozen-grand-central/.

Ince, A. (2010) Organising Anarchy: Spatial strategy prefiguration and the politics of everyday life, Unpublished PhD thesis, London: Queen Mary University of London. Available online: https://qmro.qmul.ac.uk/xmlui/bitstream/handle/123456789/496/INCEOrganisingAnarchy2010.pdf?sequence=1&isAllowed=y.

Ince, J., F. Rojas and C. A. Davis (2017) 'The Social Media Response to Black Lives Matter: How Twitter users interact with Black Lives Matter through hashtag use', *Ethnic and Racial Studies* 40(11): 1814–1830.

Indiegogo (2017) 'Indiegogo Pricing and Fees'. Available online: https://learn.indiegogo.com/pricing-and-fees/.

Industrial Workers of the World (1905) 2019 Preamble, Constitution, & General Bylaws of the Industrial Workers of the World. Available online: www.iww.org/PDF/Constitutions/CurrentIWWConstitution.pdf.

Indymedia (2017a) 'Homepage'. Available online: www.indymedia.org.uk/.

Indymedia (2017b) 'Repression of Independent Media Won't Silence Us: Not in Germany, not anywhere else', 4 September. Available online: www.indymedia.org/el/2017/09/988285.shtml.

Infoamazonia (n.d.) 'Homepage'. Available online: https://infoamazonia.org/en/.

Inghilleri, M. (2017) *Translation and Migration*, London and New York: Routledge.

Ingold, T. (2007) 'Anthropology Is Not Ethnography (Radcliffe-Brown Lecture in Social Anthropology, March 14, 2007)', *Proceedings of the British Academy*, Edinburgh: The British Academy, 69–92.

Ingram, D. (2018) 'Facebook Critics Want Regulation, Investigation after Data Misuse', *Reuters, Technology News*, 18 March. Available online: www.reuters.com/article/us-facebook-cambridge-analytica-regulati/facebook-critics-want-regulation-investigation-after-data-misuse-idUSKCN1GU01A.

Innes, C. D. (2010) *Erwin Piscator's Political Theatre: The development of modern German drama*, Cambridge: Cambridge University Press.

Interference Archive (2012) 'Who Will Maintain the Archives of Occupy Wall Street?'. Available online: http://interferencearchive.org/28773516602/.

International State Crime Initiative (2010) 'Genocide and the Contextual Evidence of Children'. Available online: http://statecrime.org/state-crime-research/genocide-and-the-contextual-evidence-of-children/.

Internet Live Stats (2019) 'Twitter Usage Statistics'. Available online: www.internetlivestats.com/twitter-statistics/.

Iordanova, D. and L. Torchin (eds) (2012) *Film Festivals and Activism*, St. Andrews, Scotland: St. Andrews Film Studies.

Iosifidis, P. (2014) 'Social Media, Democracy and Public Service Media', *Online Journal of Communication and Media Technologies* 4: 71–88.

Iosifidis, P. and M. Wheeler (2016) *Public Spheres and Mediated Social Networks in the Western Context and Beyond*, London: Palgrave Macmillan.

Ippolita, Lovink, G. and N. Rossiter (2009) 'The Digital Given: 10 Web 2.0 theses', *The Fibreculture Journal* 14. Available online: http://fourteen.fibreculturejournal.org/fcj-096-the-digital-given-10-web-2-0-theses.

Irani, L. (2015) 'Hackathons and the Making of Entrepreneurial Citizenship', *Science, Technology, and Human Values* 40(5): 799–824.
Irwin, A. (1995) *Citizen Science: A study of people, expertise, and sustainable development*, London: Routledge.
Isacat, B. (2015) 'How to Do Animal Rights … Legally, with Confidence', AnimalEthics.org. Available online: www.animalethics.org.uk/How-to-Do-Animal-Rights-2015.pdf.
Isin, E. (2002) *Being Political: Genealogies of citizenship*, London: Routledge.
Isin, E. (2017) 'Performative Citizenship', in A. Shachar, R. Bauböck, I. Bloemraad and M. Vinck (eds) *The Oxford Handbook of Citizenship*, Oxford: Oxford University Press, 500–523.
Isin, E. and E. Ruppert (2015) *Being Digital Citizens*, London: Rowman & Littlefield.
Iskandar, A. (2014) "The Meme-ing of Revolution: Creativity, folklore, and the dislocation of power in Egypt", *Jadaliyya*, September. Available online: http://interviews.jadaliyya.com/pages/index/19122/the-meme-ing-of-revolution_creativity-folklore-and.
Ito, M. (2012) 'Contributors versus Leechers: Fansubbing ethics and a hybrid public culture', in M. Ito, D. Okabe and I. Tsuji (eds) *Fandom Unbound: Otaku culture in a connected world*, New Haven, CT: Yale University Press, 179–204.
Ivakhiv, A. (2016) 'The Art of Morphogenesis: Cinema in and beyond the Capitalocene', in S. Denson and J. Leyda (eds) *Post-Cinema: Theorizing 21st-century film*, Falmer, UK: Reframe Books. Available online: https://reframe.sussex.ac.uk/post-cinema/6-1-ivakhiv/.
Iveson, K. (2007) *Publics and the City*, Oxford: Blackwell.
Iveson, K. (2010) 'Social or Spatial Justice? Marcuse and Soja on the right to the city', *City* 15(2): 250–259.
Iyengar, S. (2010) 'The State of Media-effects Research', in J. Curran (ed.) *Media and Society*, fifth edition, London: Bloomsbury, 273–289.
Jabberi, F. and S. Laine (2015) 'Spaces of Dialogue? The case of the World Social Forum Tunis 2013 from the perspective of young, local volunteers', *Global Studies of Childhood* 5(2): 178–190.
Jabri, V. (1996) *Discourses on Violence: Conflict analysis reconsidered*, Manchester: Manchester University Press.
Jackson, J. (2009) 'Intellipedia Suffers Midlife Crisis', *GCN Magazine*, 18 February. Available online: https://gcn.com/Articles/2009/02/18/Intellipedia.aspx.
Jackson, S. and B. Foucault Welles (2015) 'Hijacking #myNYPD: Social media dissent and networked counterpublics', *Journal of Communication* 65(6): 932–952.
Jackson, S. and B. Foucault Welles (2016) '#Ferguson Is Everywhere: Initiators in emerging counterpublic networks', *Information, Communication & Society* 19(3): 397–418.
Jaffee, D. (2012) 'Weak Coffee: Certification and co-optation in the Fair Trade Movement', *Social Problems* 59(1): 94–116.
Jahangard, R. and K. Duffy (2016) 'Unheard Voices, Unseen Faces: Staging stories of male refugee youth', *Critical Stages/Scènes Critiques, The IATC journal/Revue de i'AICT* (December/14). Available online: www.critical-stages.org/14/unseen-faces-and-unheard-voices-how-a-group-of-refugee-teenage-boys-living-in-the-uk-brought-their-hidden-stories-to-the-stage-in-dear-home-office/.
Janoschka, M. and E. Mateos (2015) 'Cracking Capitalism through Acts of Citizenship and the Use of Politics of Place: Geographies of the #Spanishrevolution', *Acme: An International E-Journal for Critical Geographies* 14: 75–89.
Jasper, J. M. and J. Poulsen (1993) 'Fighting Back: Vulnerabilities, blunders, and countermobilization by the targets in three animal rights campaigns', *Sociological Forum* 8: 639–657.
Jasper, J. M., M. Young and E. Zuern (2018) 'Character Work in Social Movements', *Theory and Society* 47: 113–131.
Jawad, R. (2011) 'Staging Resistance in Bil'in: The performance of violence in a Palestinian village', *The Drama Review* 55(4): 128–143.
Jean-Marie, V. (2013) 'Kant and Trouillot on the Unthinkability of the Haitian Revolution', *Souls* 15(3): 241–257.
Jeffares, S. (2014) *Interpreting Hashtag Politics*, London: Palgrave Macmillan.
Jeffers, A. (2006) 'Refugee Perspectives: The practice and ethics of verbatim theatre and refugee stories', *Platform* 1(1): 1–17.
Jeffreys, E. and J. Xu (2017) 'Celebrity-inspired, Fan-driven: Doing philanthropy through social media in mainland China', *Asian Studies Review* 41(2): 244–262.
Jenkins, H. (1992) *Textual Poachers: Television fans and participatory culture*, New York: Routledge.
Jenkins, H. (1995) 'Out of the Closet and into the Universe', in H. Jenkins and J. Tulloch J (eds) *Science Fiction Audiences: Watching Star Trek and Doctor Who*, London: Routledge, 237–266.
Jenkins, H. (2001) 'Convergence? I Diverge', *MIT Technology Review*, 1 June. Available online: www.technologyreview.com/s/401042/convergence-i-diverge/.
Jenkins, H. (2004) 'The Cultural Logic of Media Convergence', *International Journal of Cultural Studies* 7(1): 33–43.
Jenkins, H. (2006a) *Convergence Culture: Where old and new media collide*, New York: New York University Press.
Jenkins, H. (2006b) *Fans, Bloggers, and Gamers: Exploring participatory culture*, New York: New York University Press.

Jenkins, H. (2008) *Convergence Culture: Where old and new media collide*, second edition, New York: New York University Press.

Jenkins, H. (2009) 'What Happened before YouTube', in J. Burgess and J. Green (eds) *YouTube: Online video and participatory culture*, Cambridge: Polity Press, 109–125.

Jenkins, H. (2011) '"The Revolution Will Be Hashtagged": The visual culture of the Occupy Movement', Confessions of an Aca-fan blog, 28 October. Available online: http://henryjenkins.org/blog/2011/10/the_revolution_will_be_hashtag.html.

Jenkins, H. (2012a) '"Cultural Acupuncture": Fan activism and the Harry Potter Alliance', *Transformative Works and Cultures* 10.

Jenkins, H. (2012b) *Textual Poachers: Television fans and participatory culture*, New York: Routledge.

Jenkins, H. (2014) 'Fan Activism as Participatory Politics: The case of the Harry Potter Alliance', in M. Ratto, and M. Boler (eds) *DIY Citizenship: Critical Making and Social Media*, Cambridge, MA: The MIT Press, 65–73.

Jenkins, H., K. Clinton, R. Purushotma, A. J. Robinson and M. Weigel (2006) *Confronting the Challenges of Participatory Culture: Media education for the 21st century*, Chicago, IL: The MacArthur Foundation. Available online: www.macfound.org/media/article_pdfs/JENKINS_WHITE_PAPER.PDF.

Jenkins, H., S. Ford and J. Green (2013) *Spreadable Media: Creating value and meaning in a networked culture*, New York: New York University Press.

Jenkins, H., M. Ito and d. boyd (2015) *Participatory Culture in a Networked Era*, Cambridge: Polity Press.

Jenkins, H., R. Purushotma, M. Weigel, K. Clinton and A. J. Robison (2005) *Confronting Challenges of Participatory Culture: Media education for the 21st century*, Cambridge, MA: The MIT Press.

Jenkins, H. and S. Shresthova (eds) (2012) 'Special Issue: Transformative Works and Fan Activism', Transformative Works and Cultures 10.

Jenkins, H., S. Shresthova, L. Gamber-Thompson, N. Kligler-Vilenchik and A. Zimmerman (2016) *By Any Media Necessary: The new youth activism. Connected youth and digital futures*, New York: New York University Press.

Jensen, J. (1992) 'Fandom as Pathology: The consequence of characterization', in L. A. Lewis (ed.) *The Adoring Audience*, London: Routledge, 9–29.

Jensen, K. B. (2013) 'Definitive and Sensitizing Conceptualizations of Mediatization', *Communication Theory* 23(3): 203–222.

Jensen, K. B. (2016) 'Metamedium', in K. B. Jensen, E. W. Rothenbuhler, J. D. Pooley and R. T. Craig (eds) *The International Encyclopedia of Communication Theory and Philosophy*, Chichester, UK: Wiley-Blackwell.

Jeppesen, S., A. Kruzynski, A. Lakoff and R. Sarrasin (2014) 'Grassroots Autonomous Media Practices: A diversity of tactics', *Journal of Media Practice* 15(1): 21–38.

Jerslev, A. and M. Mortensen (2016) 'What Is the Self in the Celebrity Selfie? Celebrification, phatic communication and performativity', *Celebrity Studies* 7: 1–10.

Jethwany, J. (2016) *Social Sector Communication in India: Concepts, practices, and case studies*, New Delhi: Sage.

Jiao, B. (2013) 'How Chinese Journalists Use Weibo Microblogging for Investigative Reporting', Reuters Institute Fellowship Paper, University of Oxford. Available online: https://reutersinstitute.politics.ox.ac.uk/our-research/how-chinese-journalists-use-weibo-microblogging-investigative-reporting.

Joerges, B. (1999) 'Do Politics Have Artefacts?', *Social Studies of Science* 29(3): 411–431.

Johnson, C. K. (2008) 'Danceable Capitalism: Hip-hop's link to corporate space', *The Journal of Pan African Studies* 2(4): 80–92.

Johnson, D. (2007) 'Martin Luther King Jr.'s 1963 Birmingham Campaign as Image Event', *Rhetoric and Public Affairs* 10(1): 1–25.

Johnson, D. (2013) *Media Franchising: Creative license and collaboration in the culture industries*, New York: NYU Press.

Johnson, R. (dir.) (2017) Star Wars: The last Jedi. IMDb entry: www.imdb.com/title/tt2527336/?ref_=fn_al_tt_1.

Johnston, C. (2005) *House of Games: Making theatre from everyday life*, London: Nick Hern.

Johnston, C. (2017) *Disobedient Theatre: Alternative ways to inspire, animate and play*, London and New York: Bloomsbury.

Johnston, D. (ed.) (2003) *Encyclopedia of International Media and Communications*, Amsterdam: Academic Press.

Jones, C. and A. Murie (2006) *The Right to Buy: Analysis and evaluation of a housing policy*, Oxford: Blackwell.

Jones, D. (2012) *Who Cares Wins: Why good business is better business*, Harlow, UK: Pearson.

Jones, E. (2016) 'Five Reasons Why We Don't Have a Free and Independent Press in the UK and What Can We Do about It', *Open Democracy*, 18 April. Available online: www.opendemocracy.net/uk/ed-jones/five-reasons-why-we-don-t-have-free-and-independent-press-in-uk-and-what-we-can-do-about.

Jones, S. G. (2000) 'Histories, Fictions, and Xena: Warrior Princess', *Television and New Media* 1(4): 403–418.

Jonge, L. de (2018) 'The Populist Radical Right and the Media in the Benlux: Friend or foe?', *The International Journal of Press/Politics* 24(2): 189–209.

Joppke, C. (2010) *Citizenship and Immigration*, Cambridge: Polity Press.

Jorba, L. and B. Bimber (2012) 'The Impact of Digital Media on Citizenship from a Global Perspective', in E. Anduiza, M. J. Jensen and L. Jorba (eds) *Digital Media and Political Engagement Worldwide: A comparative study*, Cambridge: Cambridge University Press, 16–38.

Jordan, N. (dir.) (1996) Michael Collins, Ireland and USA: The Geffen Film Company/Warner Bros.

Jordan, T. (2002a) *Activism! Direct action, hactivism and the future of society*, London: Reaktion Books.

Jordan, T. (2002b) *Cyberpower: An introduction to the politics of cyberspace*, London: Routledge.

Jordan, T. (2016) 'A Genealogy of Hacking', *Convergence: The International Journal of Research into New Technologies* 23(5): 528–544.

Jordan, T. and P. Taylor (2004) *Hacktivism and Cyberwars: Rebels with a cause?*, London: Routledge.

Joseph, D. (2017) Distributing Productive Play: A materialist analysis of steam, Unpublished PhD thesis, Toronto: Ryerson University. Available online: https://digital.library.ryerson.ca/islandora/object/RULA%3A6820.

Jouët, J. (2000) 'Retour critique sur la sociologie des usages', *Reseaux* 18: 487–521.

Jourdan, B. and M. Maeckelbergh (2012) 'Occupy Wall Street: One year later'. Available online: www.globaluprisings.org/occupy-wall-street-one-year-later/.

Juhasz, A. (1995) *Aids TV: Identity, community, and alternative video*, Durham, NC: Duke University Press.

Juhasz, A. (1999) 'The Politics of Realist Feminist Documentaries', in J. Gaines and M. Renov (eds) *Collecting Visible Evidence*, Minneapolis, MN and London: University of Minnesota Press, 190–215.

Juhasz, A. and C. Gund (1995) *AIDS TV: Identity, community, and alternative video*, Durham, NC: Duke University Press.

Juhasz, A. and A. Lebow (eds) (2015) *A Companion to Contemporary Documentary Film*, Hoboken, NJ: Wiley-Blackwell.

Jung, S. (2012) 'Fan activism, Cybervigilantism, and Othering Mechanisms in K-pop Fandom', *Transformative Works and Cultures* 10.

Junka-Aikio, L. (2018) 'Indigenous Culture Jamming: *Suohpanterror* and the articulation of Sami political community', *Journal of Aesthetics & Culture* 10(4). DOI: 10.1080/20004214.2017.1379849.

Jürgenmeyer, J. and K. Krenn (2016) 'Classification Situations: A new field of research for valuation studies?', *Valuation Studies* 4(2): 177–189.

Juris, J. (2004) 'Networked Social Movements: Global movements for global justice', in M. Castells (ed.) *The Network Society: A cross-cultural perspective*, Northampton, MA: Edward Elgar Publishing, 341–362.

Juris, J. (2008) *Networking Futures*, Durham, NC: Duke University Press.

Juris, J. (2012) 'Reflections on #Occupy Everywhere: Social media, public space, and emerging logics of aggregation', *American Ethnologist* 39(2): 259–279.

Juris, J., G. Caruso, S. Couture and L. Mosca (2013) 'The Cultural Politics of Free Software and Technology within the Social Forum Process', in J. Juris and A. Khasnabish (eds) *Insurgent Encounters: Transnational activism, ethnography, and the political*, Durham, NC: Duke University Press, 342–366.

Jusić, T. (2009) 'Media Discourse and the Politics of Ethnic Conflict: The case of Yugoslavia', in P. Kolstø (ed.) *Media Discourse and the Yugoslav Conflicts: Representations of self and other*, Farnham, UK: Ashgate Publishing, 1–21.

Juul, J. (2005) *Half-Real: Videogames between real rules and fictional worlds*, Cambridge, MA: The MIT Press.

Kadet, A. (2011) 'The Occupy Economy', *Wall Street Journal*, 15 October. Available online: http://online.wsj.com/article/SB10001424052970204002304576631084250433462.html.

Kaganof, A. (dir.) (2014) Night *Is* Coming: A threnody for the victims of Marikana, South Africa. IMDb entry: www.imdb.com/title/tt3719338/.

Kagumire, R. (2018) 'Artists Rise up as Ugandan MP "Bobi Wine" Faces Fresh Charges of Treason', *Global Voices*, 24 August. Available online: https://globalvoices.org/2018/08/24/artists-rise-up-as-ugandan-mp-bobi-wine-faces-fresh-charges-of-treason/.

Kahn, R. and D. Kellner (2004) 'New Media and Internet Activism: From the "Battle of Seattle" to blogging', *New Media & Society* 6(1): 87–95.

Kahn, R. and D. Kellner (2008) 'Technopolitics, Blogs, and Emergent Media Ecologies: A critical/reconstructive approach', in B. Hawk, D. Riedler and O. Oviedo (eds) *Small Tech: The culture of digital tools*, Minneapolis, MN: University of Minnesota Press, 22–37.

Kalathil, S., J. Langlois and A. Kaplan (2008) 'Towards a New Model: Media and communication in post-conflict and fragile states', Communication for Governance and Accountability Program Brief for Policy Makers, World Bank. Available online: http://documents.worldbank.org/curated/en/621951468328558714/pdf/454140WP0REPLA1AP0NewModel01PUBLIC1.pdf.

Kaldor, M. (2003) *Global Civil Society: An answer to war*, Cambridge: Polity Press.

Kaldor, M., H. Anheier and M. Glasius (2003) 'Global Civil Society in an Era of Regressive Globalization', in H. Anheier, M. Glasius and M. Kaldor (eds) *Global Civil Society 2003*. Oxford: Oxford University Press, 3–33.

Kannengießer, S. (2012) 'Digital Storytelling to Empower Sex Workers: Warning, relieving and liberating', in E. Zobel and R. Drücke (eds) *Feminist Media: Participatory spaces, networks and cultural citizenship*, Bielefeld: Transcript, 238–249.

Kannengießer, S. and S. Kubitschko (2017) 'Acting on Media: Influencing, shaping and (re)configuring the fabric of everyday life', *Media and Communication* 5(3): 1–4.

Kant, I. (1784/1996) 'An Answer to the Question: What is enlightenment?', in M. J. Gregor (ed. and trans.) *Practical Philosophy: The Cambridge edition of the works of Immanuel Kant*, Cambridge: Cambridge University Press, 11–22.

Kant, I. (1793/1991) 'On the Common Saying: "This may be true in theory, but it does not apply in practice"', in H. Reiss (ed.) *Political Writings*, Cambridge: Cambridge University Press.

Kapoor, I. (2013) *Celebrity Humanitarianism: The ideology of global charity*, London: Routledge.

Kaplún, M. (1983) 'La comunicación popular. ¿Alternativa válida?', *Chasqui* 7: 40–43.

Kaplún, M. (1986) 'Uruguay: Participación, praxis, problema. La experiencia del casete-foro', in M. S. Grinberg (ed.) *Comunicación alternativa y cambio social: América Latina*, second edition, Mexico DF: Premià Editora, 266–283.

Kara, H. (2018) 'Qualitative Research Ethics by Technique', in R. Iphofen and M. Tolich (eds) *The Sage Handbook of Qualitative Research Ethics*, London: Sage, 103–113.

Karatzogianni, A. and A. Kuntsman (eds) (2012) *Digital Cultures and the Politics of Emotion: Feelings, affect and technological change*, London: Palgrave Macmillan.

Karides, M. (2013) 'Women, Gender, Feminism: Marginalization at the inception of the World Social Forum', *International and Multidisciplinary Journal of Social Sciences* 2(2): 166–197.

Karim, H. K. (2003) 'Mapping Diasporic Mediascapes', in H. K. Karim (ed.) *The Media of Diaspora*, London and New York: Routledge, 2–17.

Karra, M. (ed.) (2014) *No Respect*, Athens: Onassis Cultural Centre.

Karubian, S. (2009) '360 Deals: An industry reaction to the devaluation of recorded music', *Southern California Interdisciplinary Law Journal* 18: 395–462.

Katsiaficas, G. (2006) *The Subversion of Politics: European autonomous social movements and the decolonization of everyday life*, Oakland, CA: AK Press.

Katz, E. and P. E. Lazarsfeld (1995) 'The TwoStep Flow of Communication', in O. Boyd-Barrett and C. Newbold (eds) *Approaches to Media: A reader*, London: Arnold, 124–134.

Katz, E. and T. Liebes (1989) 'On the Critical Abilities of Television Viewers', in E. Seiter, H. Borchers, G. Kreitzner and E.-M. Warth (eds) *Remote Control: Television, audiences, and cultural power*, London: Routledge, 204–222.

Katz, E. and T. Liebes (2007) 'No More Peace! How disaster, terror and war have upstaged media events', *International Journal of Communication* 1: 157–166.

Kauffman, L. (2017) *Direct Action: Protest and the reinvention of American radicalism*, London: Verso.

Kaulingfreks, R. and S. Warren (2010) 'SWARM: Flash mobs, mobile clubbing and the city', *Culture and Organization* 16(3): 211–227.

Kaun, A. (2015) 'Regimes of Time: Media practices of the dispossessed', *Time & Society* 24(2): 221–243.

Kaun, A. (2016a) *Crisis and Critique: A history of media participation in times of crisis*, London: Zed Books.

Kaun, A. (2016b) 'Archiving Protest Digitally: The temporal regime of immediation', *International Journal of Communication* 10: 5395–5408. Available online: https://ijoc.org/index.php/ijoc/article/view/5079.

Kaun, A. and F. Stiernstedt (2014) 'Facebook Time: Technological and institutional affordances for media memories', *New Media & Society* 16(7): 1154–1168.

Kaun A. and J. Uldam (2017) 'Digital Activism: After the hype', in A. Kaun and J. Uldam (eds) *Digital Activism*. Special issue of *New Media & Society* 20(6): 2099–2106.

Kaun, A. and J. Uldam (2019) '"It Only Takes Two Minutes": The so-called migration crisis and Facebook as civic infrastructure', in T. Thomas, M. Kruse and M. Stehling (eds) *Media and Participation in Post-migrant Societies*, London: Rowman & Littlefield, 117–131.

Kavada, A. (2009) 'Email Lists and the Construction of an Open and Multifaceted Identity: The case of the London 2004 European Social Forum', *Information, Communication and Society* 12(6): 817–839.

Kavada, A. (2015) 'Creating the Collective: Social media, the Occupy Movement and its constitution as a collective actor', *Information, Communication & Society* 18(8): 872–886.

Kavada, A. (2016) 'Social Movements and Political Agency in the Digital Age: A communication approach', *Media and Communication* 4(4): 8–12.

Kaye, D. (2018) 'Report of the Special Rapporteur on the Promotion and Protection of the Right to Freedom of Opinion and Expression', Freedex.org. Available online: https://freedex.org/a-human-rights-approach-to-platform-content-regulation/.

Kaye, J. and S. Quinn (2010) *Funding Journalism in the Digital Age: Business models, strategies, issues and trends*, New York: Peter Lang.

Keane, J. (1988) 'Despotism and Democracy, the Origins and Development of the Distinction between Civil Society and the State, 1750–1850', in J. Keane (ed.) *Civil Society and the State: New European perspectives*, London: Verso, 35–72.

Keane, J. (2009) 'Civil Society, Definitions and Approaches', in H. Anheier and S. Toepler (eds) *International Encyclopedia of Civil Society*, New York: Springer, 461–464.

Keating, A. (2013) 'Video-making, Harlem Shaking: Theorizing the interactive amateur', *New Cinemas: Journal of Contemporary Film* 11(2–3): 99–110.

Keating, A. (2014) 'Home Movies in the Age of Web 2.0: The case of "*Star Wars* Kid"', in L. Rascaroli, G. Young and B. Monahan (eds) *Amateur Filmmaking: The home movie, the archive, the web*, London: Bloomsbury, 301–314.

Keen, A. (2007) *The Cult of the Amateur: How today's internet is killing our culture*, London: Doubleday.

Keightley, E. (2012) 'Introduction: Time, media, modernity', in E. Keightley (ed.) *Time, Media and Modernity*, Basingstoke, UK: Palgrave Macmillan, 1–24.

Keightley, E. (2013) 'From Immediacy to Intermediacy: The mediation of lived time', *Time & Society* 22(1): 55–75.

Keirstead, P. O. and S.-K. Keirstead (1987) 'The Many Faces of Community Radio', *International Communication Bulletin* 22(1–2): 6–12.

Kelion, L. (2012) 'Hackers Retaliate over Megaupload Website Shutdown', *BBC News*, 8 March. Available online: www.bbc.co.uk/news/technology-16646023.

Keller, J. (2010) 'Evaluating Iran's Twitter Revolution', *The Atlantic*. Available online: www.theatlantic.com/technology/archive/2010/06/evaluating-irans-twitter-revolution/58337/.

Kellner, D. (2010) 'Media Spectacle and Media Events', in N. Couldry, A. Hepp and F. Krotz (eds) *Media Events in a Global Age*, Abingdon, UK: Routledge, 76–91.

Kellner, M. (1975) 'Democracy and Civil Disobedience', *The Journal of Politics* 37(4): 899–911.

Kelty, C. M. (2010) 'Theorising the Practices of Free Software: The movement', in B. Bräuchler and J. Postill (eds) *Theorising Media and Practice*, New York: Berghahn Books, 281–302.

Kennedy, H., R. L. Hill, W. Allen and A. Kirk (2016) 'Engaging with (Big) Data Visualizations: Factors that affect engagement and resulting new definitions of effectiveness', *First Monday* 21(11). Available online: http://firstmonday.org/ojs/index.php/fm/rt/printerFriendly/6389/0.

Kerkhoven, M. van and P. Bakker (2014) 'The Hyperlocal in Practice', *Digital Journalism* 2(3): 296–309.

Kessler, F. (2007) 'Notes on *dispositif*'. Available online: www.frankkessler.nl/wp-content/uploads/2010/05/Dispositif-Notes.pdf.

Keyes, C. L. (2002) *Rap Music and Street Consciousness*, Chicago, IL: University of Illinois Press.

Khalaf, A. (2013) 'The Many Afterlives of Lulu', *Ibraaz* 56: 1–18.

Khamis, S., L. Ang and R. Welling (2017) 'Self-Branding, "Micro-Celebrity" and the Rise of Social Media Influencers', *Celebrity Studies* 8(2): 191–208.

Khamis, S. and K. Vaughn (2011) 'Cyberactivism in the Egyptian Revolution: How civic engagement and citizen journalism tilted the balance', *Arab Media and Society* 14(3): 1–25.

Khasnabish, A. (2008) *Zapatismo beyond Borders: New imaginations of political possibility*, Toronto: University of Toronto Press.

Khasnabish, A. (2010) *Zapatistas: Rebellion from the grassroots to the global*, London: Zed Books.

Khasnabish, A. (2015) 'Outside but Along-Side: Stumbling with social movements as academic activists', Studies in Social Justice 9(1): 18–33.

Khatib, L. (2012) *Image Politics in the Middle East: The role of the visual in political struggle*, London and New York: I.B. Tauris.

Khosravi, S. (2013) 'Graffiti in Tehran', *Anthropology Now* 5(1): 1–17.

Kidd, D. (1999) 'The Value of Alternative Media', *Peace Review* 11(1): 113–119.

Kidd, D. (2003a) 'Indymedia.org: A new communications commons', in M. McGaughey and M. Ayers (eds) *Cyberactivism: Critical theories and practices of on-line activism*, New York: Routledge, 47–69.

Kidd, D. (2003b) 'From Carnival to Commons: The global IMC network', in E. Yuen, D. Burton-Rose and G. Katsiaficas (eds) *Confronting Capitalism: Dispatches from a global movement*, New York: Soft Skull Press, 330–340.

Kidd, D. (2015) 'Occupy and Social Movement Communication', in C. Atton (ed.) *The Routledge Companion to Alternative and Community Media*, New York: Routledge, 457–468.

Kidd, D., C. Rodríguez and L. Stein (2009) *Making Our Media: Global initiatives toward a democratic public sphere*, Cresskill, NJ: Hampton Press.

Kidder, J. (2012) 'Parkour, the Affective Appropriation of Urban Space, and the Real/Virtual Dialectic', *City and Community* 11(3): 229–253.

Kidder, J. (2017) *Parkour and the City: Risk, masculinity, and meaning in a postmodern sport*, New Brunswick, NJ: Rutgers University Press.

Killick, A. (2016) 'Introduction to Part II: Cinema, activism, and opposition', in Y. Tzioumakis and C. Molloy (eds) *The Routledge Companion to Cinema and Politics*, London and New York: Routledge, 77–78.

Kim, Y. and W. Lowrey (2015) 'Who Are Citizen Journalists in the Social Media Environment? Personal and social determinants of citizen journalism activities', *Digital Journalism* 3(2): 298–314.

Kim, Y., S. Russo and E. Amnå (2017) 'The Longitudinal Relation between Online and Offline Political Participation among Youth at Two Different Developmental Stages', *New Media & Society* 19(6): 899–917.
Kimura, A. H. and A. Kinchy (2016) 'Citizen Science: Probing the virtues and contexts of participatory research', *Engaging Science, Technology, and Society* 2: 331–361.
King, G., J. Pan and M. Roberts (2013) 'How Censorship in China Allows Government Criticism but Silences Collective Expression', *American Political Science Review* 107(2): 326–343.
King, L. (2015) *Innovators in Digital News*, London and New York: I. B. Tauris.
King, L. and J. Busa (2017) 'When Corporate Actors Take Over the Game: The corporatization of organic, recycling and breast cancer activism', *Social Movement Studies* 16(5): 549–563.
King, M. L., Jr. (1963) 'Letter from Birmingham Jail', *Liberation: An Independent Monthly* 8(4): 10–16.
King, M. L., Jr. (1989) *The Trumpet of Conscience*, San Francisco, CA: Harper.
Kingkade, T. (2011) 'New Protest Apps Crowd-Sourced from Occupy Wall Street Hackers', *Huffington Post*, 26 October. Available online: www.huffingtonpost.com/2011/10/26/occupy-wall-street-a-diy-tech-tools-protest_n_1032518.html.
Kingsley, P. (2011) 'Inside the Anti-Kettling HQ', *The Guardian*, 2 February. Available online: www.theguardian.com/uk/2011/feb/02/inside-anti-kettling-hq.
Kinna, R. E. (2016) 'Utopianism and Prefiguration', in S. D. Chrostowska and J. D. Ingram (eds) *Political Uses of Utopia: New Marxist, anarchist, and radical democratic perspectives*, New York: Columbia University Press, 198–218.
Kip, M. (2016) 'Solidarity', in A. K. Thompson, K. Fritsch and C. O'Connor (eds) *Keywords for Radicals*, Oakland, CA: AK Press, 391–398.
Kirby, M. (1965) *Happenings*, New York: Dutton.
Kitchin, R. (2017) 'Big Data: Hype or revolution?', in L. Sloan, A. Quan-Haase and K. Metzler (eds) *The Sage Handbook of Social Media Research Methods*, London: Sage, 27–39.
Kittler, F. A. (1996) 'The City Is a Medium', *New Literary History* 27(4): 717–729.
Klein, N. (2000) *No Logo: Taking aim at the brand bullies*, New York: Picador.
Kleinman, A. (1998) 'Experience and Its Moral Modes: Culture, human conditions, and disorder', *Tanner Lectures on Human Values* 20: 355–420. Available online: https://tannerlectures.utah.edu/_documents/a-to-z/k/Kleinman 99.pdf.
Kleinman, A. and J. Kleinman (1996) 'The Appeal of Experience: The dismay of image: Cultural appropriations of suffering in our times', *Daedalus* 125(1): 1–23.
Klimas, C. (2009) *Twine*. Available online: http://twinery.org/.
Klonick, K. (2017) 'The New Governors: The people, rules, and processes governing online speech', *Harvard Law Review* 131: 1598–1670.
Knight, J. and P. Thomas (2011) *Reaching Audiences: Distribution and promotion of alternative moving image*, Bristol, UK and Chicago, IL: Intellect.
Knight, M. (2012) 'Journalism as Usual: The use of social media as a newsgathering tool in the coverage of the Iranian elections in 2009', *Journal of Media Practice* 13(1): 61–74.
Knoblauch, H. (2013) 'Communicative Constructivism and Mediatization', *Communication Theory* 23(3): 297–315.
Koch, G., J. Füller and S. Brunswicker (2011) 'Online Crowdsourcing in the Public Sector: How to design open government platforms', in A. Ozok and P. Zaphiris (eds) *Online Communities and Social Computing*, Berlin: Springer, 203–212.
Kohler, B. (2005) 'Social Forums as Space: A response to Peter Marcuse', *International Journal of Urban and Regional Research* 29(2): 429–432.
Köhler, M. (ed.) (1980) *Alternative Medienarbeit. Videogruppen in der Bundesrepublik*, Opladen: Leske und Budrich.
Kokkinidis, G. (2015) 'Spaces of Possibilities: Workers' self-management in Greece', *Organization* 22(6): 847–871.
Koliska, M. and J. Roberts (2015) 'Selfies: Witnessing and participatory journalism with a point of view', *International Journal of Communication* 9(1): 1672–1685.
Kollanyi, B., P. N. Howard and S. C. Woolley (2016) 'Bots and Automation over Twitter during the US Election', COMPROP DATA MEMO 2016.4. Available online: http://comprop.oii.ox.ac.uk/wp-content/uploads/sites/89/2016/11/Data-Memo-US-Election.pdf.
König, R. (2013) 'Wikipedia: Between lay participation and elite knowledge representation', *Information, Communication and Society* 16(2): 160–177.
Kopecký, P. and C. Mudde (2003) 'Rethinking Civil Society', *Democratization* 10: 1–14.
Koser, K. (2007) 'Refugees, Transnationalism and the State', *Journal of Ethnic and Migration Studies* 33(2): 233–254.
Kosnik, A. De (2016) *Rogue Archives: Digital cultural memory and fandom*, Cambridge, MA: The MIT Press.
Kosnik, A. De (2018) 'Filipinos' Forced Fandom of US Media: Protests against *The Daily Show* and *Desperate Housewives* as bids for cultural citizenship', in M. A. Click, and S. Scott (eds) *The Routledge Companion to Media Fandom*, New York: Routledge, 262–270.

Koss, S. (1981, 1984) *The Rise and Fall of the Political Press in Britain*, Volumes 1 and 2, London: Hamish Hamilton.

Kovic, M., A. Rauchfleisch, M. Sele and C. Caspar (2018) 'Digital Astroturfing in Politics: Definition, typology, and countermeasures', *Studies in Communication Sciences* 18(1): 69–85.

Kperogi, F. (2011) 'Cooperation with the Corporation? CNN and the hegemonic cooptation of citizen journalism through iReport.com', *New Media and Society* 13: 314–329.

Kraidy, M. M. (2010) *Reality Television and Arab Politics: Contention in public life*, New York: Cambridge University Press.

Kraidy, M. M. (2013) 'The Body as Medium in the Digital Age: Challenges and opportunities', *Communication and Critical-Cultural Studies* 10(2–3): 285–290.

Kraidy, M. M. (2016) *The Naked Blogger of Cairo: Creative insurgency in the Arab world*, Cambridge, MA: Harvard University Press.

Kramer, A. D. I., J. E. Guillory, J. Hancock and T. Jeffrey (2014) 'Experimental Evidence of Massive-Scale Emotional Contagion through Social Networks', *PNAS* 111(24): 8788–8790. Available online: www.pnas.org/content/111/24/8788.abstract?tab=author-info.

Kristofferson, K., K. White and J. Peloza (2014) 'The Nature of Slacktivism: How the social observability of an initial act of token support affects subsequent prosocial action', *Journal of Consumer Research* 40(6): 1149–1166.

Krivitsky, P., M. Handcock, A. Raftery and P. Hoff (2009) 'Representing Degree Distributions, Clustering, and Homophily in Social Networks with Latent Cluster Random Effects Models', *Social Networks* 31(3): 204–213.

Krotz, F. (2014) 'Mediatization as a Mover in Modernity', in K. Lundby (ed.) *Mediatization of Communication*, Berlin and New York: de Gruyter, 131–161.

Kubitschko, S. (2015) 'Hackers' Media Practices: Demonstrating and articulating expertise as interlocking arrangements', *Convergence: The International Journal of Research into New Media Technologies* 21(3): 388–402.

Kubitschko, S. (2017) 'Acting on Media Technologies and Infrastructures: Expanding the media as practice approach', *Media, Culture & Society* 40(4): 629–635.

Kubrin, C. (2005) 'Gangstas, Thugs, and Hustlas: Identity and the code of the street in rap music', *Social Problems* 52(3): 360–378.

Kücklich, J. (2005) 'Precarious Playbour: Modders and the digital games industry', *The Fibreculture Journal* 5. Available online: http://five.fibreculturejournal.org/fcj-025-precarious-playbour-modders-and-the-digital-games-industry/.

Kulick, R. (2014) 'Making Media for Themselves: Strategic dilemmas of prefigurative work in independent media outlets', *Social Movement Studies* 13(3): 365–380.

Kulik, D. (2003) 'Sex in the New Europe: The criminalization of clients and Swedish fear of penetration', *Anthropological Theory* 3(2): 199–218.

Kumar, S. (2014) 'Media, Communication, and Postcolonial Theory', in R. S. Fortner and P. M. Fackler (eds) *The Handbook of Media and Mass Communication Theory*, New York: Wiley, 380–399.

Kunelius, R., H. Heikkilä, A. Russell and E. Eide (2017) 'The NSA Revelations as a Prism', in R. Kunelius, H. Heikkilä, A. Russell and D. Yagodin (eds) *Journalism and the NSA Revelations: Privacy, security and the press*, London: Reuters Institute/I. B. Tauris, 1–22.

Kuntsman, A. (ed.) (2017a) *Selfie Citizenship*, Cham, Switzerland: Palgrave Macmillan.

Kuntsman, A. (2017b) 'Introduction: Whose selfie citizenship?', in A. Kuntsman (ed.) *Selfie Citizenship*, Cham, Switzerland: Palgrave Macmillan.

Kuntsman, A. and R. Stein (2015) *Digital Militarism: Israel's occupation in the social media age*, Palo Alto, CA: Stanford University Press.

Kuo, R. (2018) 'Racial Justice Activist Hashtags: Counterpublics and discourse circulation', *New Media & Society* 20(2): 495–514.

Kuper, L. (1957) *Passive Resistance in South Africa*, New Haven, CT: Yale University Press.

Kurpius, D. D., E. T. Metzgar and K. M. Rowley (2010) 'Sustaining Hyperlocal Media: In search of funding models', *Journalism Studies* 11(3): 359–376.

Kuttab, D. (1993) 'Grass Roots TV Production in the Occupied Territories', in T. Downmunt (ed.) *Channels of Resistance: Global television and local empowerment*, London: British Film Institute, 138–145.

Kwak, H., C. Lee, H. Park and S. Moon (2010) 'What Is Twitter, a Social Network or a News Media?', *The International World Wide Web Conference Committee (IW3C2)*, New York: ACM, 1–10.

Kymlicka, W. (1995) *Multicultural Citizenship*, Oxford: Oxford University Press.

La Ingobernable (2019) 'Nosotras', *La Ingobernable*. Available online: https://ingobernable.net/nosotras/.

Labahn, A. (2006) 'Fire from Above: Metaphors and images of God's actions in Lamentations 2.1–9', *Journal of the Study of the Old Testament* 31(2): 239–256.

Lacan, J. (1977/1997) 'The Mirror Stage as Formative of the Function of the I as Revealed in Psychoanalytic Experience', in J. Lacan (ed.) *Écrits: A selection*, translated by B. Fink, London: Routledge, 1–7.

Laclau, E. and C. Mouffe (1985) *Hegemony and Socialist Strategy: Towards a radical democratic politics*, London: Verso.

Laclau, E. and C. Mouffe (2001) *Hegemony and Socialist Strategy: Towards a radical democratic politics*, second edition, London: Verso.

Lacol and La Ciutat Invisible (2018) *Habitar en comunidad. La vivienda cooperativa en cesión de uso*, Madrid: La Catarata.

LaFlamme, M. (2017) 'A Sky Full of Signal: Aviation media in the age of the drone', *Media, Culture & Society* 40(5): 689–706.

Laine, S. (2013) 'Possibilities for Cosmopolitan Emancipation among Local Poor and Minority Youth in Transnational Political Events', *Global Society* 27(1): 73–90.

LaLone, N., A. Tapia, C. Zobel, C. Caraega, V. Kishore Neppalli and S. Halse (2017) 'Embracing Human Noise as Resilience Indicator: Twitter as power grid correlate', *Sustainable and Resilient Infrastructure* 2(4): 169–178.

Lamazares, A. (2017) 'São Paulo's Pixação and Street Art: Representations of or responses to Brazilian Modernism?', in K. Avramidis and M. Tsilimpounidi (eds) *Graffiti and Street Art: Reading, writing and representing the city*, London: Routledge, 198–215.

Lamb, P. F. and D. Veith (1986) 'Romantic Myth, Transcendence, and Star Trek Zines', in D. Palumbo (ed.) *Erotic Universe: Sexuality and fantastic literature*, Westport, CT: Greenwood Press, 236–255.

Lambert, A. (2013) *Intimacy and Friendship on Facebook*, Basingstoke, UK and New York: Palgrave Macmillan.

Lambert, J. (2009) 'Where It All Started: The Center for Digital Storytelling in California', in J. Hartley and K. McWilliam (eds) *Story Circle: Digital storytelling around the world*, Malden, MA: Wiley-Blackwell, 79–90.

Lamerichs, N. (2011) 'Stranger than Fiction: Fan identity in cosplay', *Transformative Works and Cultures* 7.

Lampe, C. and P. Resnick (2004) 'Slash (Dot) and Burn: Distributed moderation in a large online conversation space', CHI '04: Proceedings of the SIGCHI conference on human factors in computing systems, New York: ACM, 543–550.

Lapegna, P. (2014) 'The Problem with "Cooptation"', *States, Power and Society* 20(1): 7–10.

Laplante, L. J. and K. Phenicie (2009) 'Mediating Post-Conflict Dialogue: The media's role in transitional justice processes', *Marquette Law Review* 93(1): 251–283.

Larkin, B. (2018) 'Promising Forms: The political aesthetics of infrastructure', in N. Anand, A. Gupta and H. Appel (eds) *The Promise of Infrastructure*, Durham, NC: Duke University Press, 175–202.

Laroussi, F. and F. Lienard (2013) 'Révolution Tunisienne et communautés de parole virtuelles: Les écrilectes pour communiquer', in F. Daghmi, F. Toumi and A. Amsidder (eds) *Les médias font-ils les révolutions? Regards critiques sur les soulèvements Arabes*, Paris: L'Harmattan, 71–86.

Lash, S. and J. Urry (1994) *Economies of Signs and Space*, London: Sage.

Lasica, J. D. (2003) 'Random Acts of Journalism', *Inside Social Media*, 12 March. Available online: http://insidesocialmedia.com/2003/03/12/random-acts-of-journalism/.

Lasn, K. (1999) *Culture Jam: The uncooling of America*, New York: Eagle Brook.

Latour, B. (1987) *Science in Action: How to follow scientists and engineers through society*, Cambridge, MA: Harvard University Press.

Latour, B. (1999) *Pandora's Hope: Essays on the reality of science studies*, Cambridge, MA: Harvard University Press.

Latour, B. (2006) *Changer de société: Refaire de la sociologie*, Paris: La Découverte.

Latzko-Toth, G. and F. Millerand (2012) 'Sociologie des usages et science & technology studies: Un dialogue à poursuivre', in G. Vidal (ed.) *La sociologie des usages: Continuités et transformations*, Cachan, France: Hermès-Lavoisier, 119–150.

Laukkanen, T. (2005) *Modding Scenes: Introduction to user-created content in computer gaming*, Tampere, Finland: University of Tampere Hypermedia Laboratory. Available online: http://tampub.uta.fi/handle/10024/65431.

Lavin, M., L. Yang and J. J. Zhao (2017) *Boys' Love, Cosplay, and Androgynous Idols: Queer fan cultures in mainland China, Hong Kong, and Taiwan*, Hong Kong: Hong Kong University Press.

Lavin, M. and X. Zhu (2014) 'Alexter: Boys' Love meets Hong Kong activism', *FNewsMagazine*, 17 November. Available online: http://fnewsmagazine.com/2014/11/alexter-boys-love-meets-hong-kong-activism/.

Lazzarato, M. (1997) *Lavoro immateriale. Forme di vita e produzione di soggettività* [Immaterial Labour: Forms of life and the production of subjectivity], Verona: Ombre Corte.

Lazzarato, M. (2006) 'Immaterial Labor', in P. Virno and M. Hardt (eds) *Radical Thought in Italy: A potential politics*, Minneapolis, MN: University of Minnesota Press, 131–147.

Lazzarato, M. (2007) 'Strategies of the Political Entrepreneur', *SubStance* 36(1): 87–97.

Lazzarato, M. (2018) *Experimental Politics: Work, welfare and creativity in the neoliberal age*, edited by J. Gilbert, Cambridge, MA: The MIT Press.

Lazzarato, M. and A. Negri (1993) *Benetton-Sentier: Des entreprises pas comme les autres* [Benetton-Sentier: Companies like no other], Paris: Publisud.

Lazzarato, M. and A. Negri (1994) *Du service à la relation de service* [From service to the relation of service], Paris: M.I.R.E.

Lazzarato, M. and A. Negri (1997) 'Lavoro immateriale e soggettivitá', in *Lavoro immateriale. Forme di vita e produzione di soggettività* [Immaterial Labour: Forms of life and the production of subjectivity], Verona: Ombre Corte, 23–38.
Le Bon, G. (1895/1960) *The Crowd: A study of the popular mind*, New York: Viking Press.
Leavis, F. R. and D. Thompson (1933) *Culture and Environment*, Westport, CT: Greenwood Press.
Leavitt, A. and J. Clark (2014) 'Upvoting Hurricane Sandy: Event-based news production processes on a social news site', *Proceedings of CHI 2014, One of a CHInd, Toronto, ON, Canada*, New York: ACM, 1495–1504.
Lebow, A. (2018) *Filming Revolution*, a Stanford Digital Project, Stanford, CA: Stanford University Press. Available online: www.filmingrevolution.org/.
Lederach, J. P. (1996) *Preparing for Peace: Conflict transformation across cultures*, New York: Syracuse University Press.
Lederach, J. P. (2003) *The Little Book of Conflict Transformation*, New York: Good Books.
Lee, C. K. and Y. Kofman (2012) 'The Politics of Precarity: Views beyond the United States', *Work and Occupations* 39(4): 388–408.
Lee, C. T. (2008) 'Undocumented Workers' Subversive Citizenship Acts', *Peace Review* 20: 330–338.
Lee, F. and J. Chan (2016) 'Digital Media Activities and Mode of Participation in a Protest Campaign: A study of the Umbrella Movement', *Information, Communication and Society* 19: 4–22.
Lee, H. K. (2011) 'Participatory Media Fandom: A case study of anime fansubbing', *Media, Culture & Society* 33(8): 1131–1147.
Lefebvre, H. (1974/1991) *The Production of Space*, translated by D. Nicholson-Smith, Malden, MA and Oxford: Wiley-Blackwell.
Lefkowitz, D. (2007) 'On a Moral Right to Civil Disobedience', *Ethics* 117: 202–233.
Lei, Y.-W. (2017) *The Contentious Public Sphere: Law, media and authoritarian rule in China*, Princeton, NJ: Princeton University Press.
Leigh Star, S. and G. Bowker (2002) 'How to Infrastructure', in L. Lievrouw and S. Livingstone (eds) *Handbook of New Media: Social shaping and consequences of ICTs*, London: Sage, 151–162.
Leistert, O. (2015) 'The Revolution Will Not Be Liked: On the systematic constraints of corporate social media platforms for protest', in L. Dencik and O. Leistert (eds) *Critical Perspectives on Social Media and Protest: Between control and emancipation*, London: Rowman & Littlefield, 35–52.
Leite, J. C. (2005) *The World Social Forum: Strategies of resistance*, Chicago, IL: Haymarket Books.
Lekakis, E. J. (2017) 'Culture Jamming and Brandalism for the Environment: The logic of appropriation', *Popular Communication* 15(4): 311–327.
Lena, J. (2006) 'Social Context and Musical Content of Rap Music. 1979–1995', *Social Forces* 85(1): 479–496.
Lena, J. (2013) 'Authenticity and Independence in Rap Music and Other Genre Communities', in S. Horsfall, J. M. Meij and M. Probstfield (eds) *Explorations in Music Sociology: Examining the role of music in social life*, Boulder, CO: Paradigm Publishing, 232–240.
Lena, J. and R. Peterson (2008) 'Classification of Culture: Types and trajectories of music genres', *American Sociological Review* 73: 697–718.
Lentin, A. (2014) 'Postracial Silences: The othering of race in Europe', in W. D. Hund and A. Lentin (eds) *Racism and Sociology*, Berlin: Lit Verlag, 1268–1685.
Lentin, A. and G. Titley (2011) *The Crises of Multiculturalism: Racism in a neo-liberal age*, London and New York: Zed Books.
Lentz, R. (2011) 'Civil Society in the Digital Age', in M. Edwards (ed.) *The Oxford Handbook of Civil Society*, Oxford: Oxford University Press, 337–348.
Lerner, J. (2014) *Making Democracy Fun*, Cambridge, MA: The MIT Press.
Lesbian Herstory Archives (2018). Available online: www.lesbianherstoryarchives.org/history.html.
Lessig, L. (2008) *Remix: Making art and commerce thrive in the hybrid economy*, New York: Penguin Press.
Leuf, B. and W. Cunningham (2001) *The Wiki Way: Quick collaboration on the Web*, Boston, MA: Addison-Wesley.
Leurs, K. (2015) *Digital Passages: Migrant youth 2.0*, Amsterdam: Amsterdam University Press.
Leurs, K. (2016) 'Social Media as Contact Zones: Young Londoners remapping the metropolis through digital media', in S. Ponzanesi and G. Colpani (eds) *Postcolonial Transitions in Europe*, New York and London: Rowman & Littlefield, 251–272.
Leurs, K. (2017) 'Communication Rights from the Margins: Politicizing young refugees' smartphone pocket archives', *The International Communication Gazette* 79(6–7): 674–698.
Levi, A., M. McHarry and D. Pagliassotti (2010) *Boys' Love Manga: Essays on the sexual ambiguity and cross-cultural fandom of the genre*, Jefferson, NC: McFarland.
Levinson, P. (2001) *Digital McLuhan: A guide to the information millennium*, London and New York: Routledge.
Levy, S. (1984) *Hackers: Heroes of the computer revolution*, Sebastopol, CA: O'Reilly.
Levy, S. (2002) *Crypto: How the code rebels beat the government*, London: Penguin Random House.

Lewicki, A. (2017) '"The Dead Are Coming": Acts of citizenship at Europe's borders', *Citizenship Studies* 21: 275–290.
Lewicki, A. and T. O'Toole (2017) 'Acts and Practices of Citizenship: Muslim women's activism in the UK', *Ethnic and Racial Studies* 40: 152–171.
Lewis, J., S. Inthorn and K. Wahl-Jorgensen (2005) *Citizens or Consumers? What the media tell us about political participation*, Buckingham, UK: Open University Press.
Lewis, L. A. (ed.) (1992) *The Adoring Audience: Fan culture and popular media*, London: Routledge.
Lewis, P. (2018) '"Fiction Is Outperforming Reality": How YouTube's algorithm distorts truth', *The Guardian*, 2 February.
Lewis, P. M. (1978) *Community Television and Cable in Britain*, London: British Film Institute.
Lewis, P. M. (1984a) 'Europe', in P. M. Lewis (ed.) *Media for People in Cities: A study of community media in the urban context*, Paris: UNESCO, 89–107.
Lewis, P. M. (1984b) 'Community Radio: The Montreal conference and after', *Media, Culture & Society* 6: 137–150.
Lewis, S. C. (2012) 'The Tension between Professional Control and Open Participation: Journalism and its boundaries', *Information, Communication and Society* 15(6): 836–866.
Li, D. (2018) 'Critical Media Literacy: A social semiotic analysis and multimodal discourse of corpocracy', *International Journal of Education & the Arts* 19(16).
Li, E. C. Y. (2012) 'The Absence of Fan Activism in the Queer Fandom of Ho Denise Wan See (HOCC) in Hong Kong', *Transformative Works and Cultures* 10.
Li, E. C. Y. (2017) 'Desiring Queer, Negotiating Normal: Denise Ho (HOCC) fandom before and after the coming-out', in M. Lavin, L. Yang and J. J. Zhao (eds) *Boys' Love, Cosplay, and Androgynous Idols: Queer fan cultures in mainland China, Hong Kong, and Taiwan*, Hong Kong: Hong Kong University Press, 131–156.
Li, Y., C. Lai and C. Chen (2011) 'Discovering Influencers for Marketing in the Blogosphere', *Information Sciences* 181(23): 5143–5157.
Liamputtong, P. (2010) *Performing Qualitative Cross-Cultural Research*, Cambridge: Cambridge University Press.
Lievrouw, L. A. (2011) *Alternative and Activist New Media*, Cambridge: Polity Press.
Lifelong Kindergarten Group (2003) *Scratch*. Available online: https://scratch.mit.edu/.
Light, B. (2016) 'Producing Sexual Cultures and Pseudonymous Publics with Digital Networks', in R. A. Lind (ed.) *Race and Gender in Electronic Media: Challenges and opportunities*, London: Routledge, 231–246.
Lim, M. (2012) 'Clicks, Cabs, and Coffee Houses: Social media and oppositional movements in Egypt, 2004–2011', *Journal of Communication* 62: 231–248.
Lim, M. (2015) 'A CyberUrban Space Odyssey: The spatiality of contemporary social movements', *New Geographies* 7: 117–123.
Lin, A. (2006) 'Independent Hip-Hop Artists in Hong-Kong: Cultural capitalism, youth subculture resistance, and alternative modes of cultural production', *Mobile and Popular Culture* 1: 1–18.
Lin, C. S, A. A. Pykett, C. Flanagan and K. R. Chávez (2016) 'Engendering the Prefigurative: Feminist praxes that bridge a politics of prefigurement and survival', in *Rethinking Prefigurative Politics*, special thematic section of *Journal of Social and Political Psychology* 4(1): 302–317.
Lin, Z. and S.-D. Liu (2016) 'Occupation as Prefiguration? The emergence of a new political form in the Occupy Central movement', *Contemporary Chinese Political Economy and Strategic Relations: An International Journal* 2(2): 775–794.
Linebaugh, P. (2008) *The Magna Carta Manifesto: Liberties and commons for all*, Berkeley, CA: University of California Press.
Linebaugh, P. and M. Rediker (2000) *The Many-headed Hydra: Sailors, slaves, commoners, and the hidden history of the revolutionary Atlantic*, Boston, MA: Beacon Press.
Lingel, J. (2017) 'Fight for Your Platform to Party: Brooklyn drag and the battle for a queerer Facebook', *Digital Countercultures and the Struggle for Community*, Cambridge, MA and London: The MIT Press, 99–122.
Linksvayer, M. (2012) 'Creative Commons: Governing the intellectual commons from below', in D. Bollier and S. Helfrich (eds) *The Wealth of the Commons: A world beyond market and state*, Amherst, MA: Levellers Press, 299–304. Available online: http://wealthofthecommons.org/essay/creative-commons-governing-intellectual-commons-below.
Lister, C. (2016) *Don't Look at Me*. Available online: https://napstack.itch.io/.
Liu, J. (2015) 'Mobile Activism and Contentious Politics in Contemporary China', in W. Chen and S. D. Reese (eds) *Networked China: Global dynamics of digital media and civic engagement*, London: Routledge, 97–113.
Liu, J. (2016) 'Digital Media, Cycle of Contention, and Sustainability of Environmental Activism: The case of anti-PX protests in China', *Mass Communication and Society* 19(5): 604–625.
Liu, Y. (2015) 'Tweeting, Re-tweeting and Commenting: Microblogging and social movements in China', *Asian Journal of Communication* 25(6): 567–583.
Livingstone, S. (2005) 'On the Relation between Audiences and Publics', in S. Livingstone (ed.) *Audiences and Publics: When cultural engagement matters for the public sphere*, Changing Media – Changing Europe series, Volume 2, Bristol, UK: Intellect, 17–41.

Livingstone, S. and R. Das (2009) 'The End of Audiences?: Theoretical echoes of reception amidst the uncertainties of use', Paper presented at Transforming Audiences, University of Westminster 3–4 September.

Livingstone, S. M. (2009) 'On the Mediation of Everything', *Journal of Communication* 59(1): 1–18.

Loader, B. D. and D. Mercea (2012) 'Networking Democracy? Social media innovations in participatory politics', in B. D. Loader and D. Mercea (eds) *Social Media and Democracy: Innovations in participatory politics*, London and New York: Routledge, 1–10.

Lochman, Z. and J. Beinin (eds) (1989) *Intifada: The Palestinian uprising against Israeli occupation*, Toronto: Between the Lines.

Longhi, V. (2013) *The Immigrant War*, translated by J. Eastwood, London: Polity Press.

Longman, C. and K. de Graeve (2014) 'From Happy to Critical Diversity: Intersectionality as a paradigm for gender and diversity research', *Journal of Diversity and Gender Studies* 1(1): 33–39.

Lopes, P. (1992) 'Innovation and Diversity in the Popular Music Industry: 1969–1990', *American Sociological Review* 57(1): 56–71.

López-Bech, L. and R. Zúñiga (2017) 'Digital Storytelling: Putting young asylum seekers at the heart of the story', *Intercultural Education* 28(2): 224–228.

Lorey, I. (2012) 'Precarity Talk: A virtual roundtable with Lauren Berlant, Judith Butler, Bojana Cvejić, Isabell Lorey, Jasbir Puar, and Ana Vujanović', *TDR: The Drama Review* 56(4): 163–177.

Lorr, M. (2005) 'Skateboarding and the X-Gamer Phenomenon: A case of subcultural cooptation', *Humanity and Society* 29(2): 140–147.

Löser, C. (2011) *Strategien der Verweigerung/Untersuchungen zum politisch-ästhetischen Gestus unangepasster filmischer Artikulationen in der Spätphase der DDR*, Berlin: DEFA-Stiftung.

Losh, E. (2012) 'Channelling Obama: YouTube, Flickr, and the social media President', *Comparative American Studies: An International Journal* 10(2–3): 255–268.

Lostpedia contributors (n.d.) 'The Monster/Theories'. Available online: http://lostpedia.wikia.com/wiki/The_Monster/Theories?oldid=1107919.

Lotan, G. (2014) 'Israel, Gaza, War and Data', Medium.com, 4 August. Available online: https://medium.com/i-data/israel-gaza-war-data-a54969aeb23e.

Lotringer, S. and C. Marazzi (2007) *Autonomia: Post-political politics*, Los Angeles, CA: Semiotext(e).

Lott, T. (1994) 'Black Cultural Politics: An interview with Paul Gilroy', *Found Object* 4: 46–81.

Lotz, A. (2014) *The Television Will Be Revolutionized*, second edition, New York: New York University Press.

Louisiana Bucket Brigade (2001) *Land Sharks: Orion's predatory property purchases*, New Sarpy, LA: Louisiana Bucket Brigade and the New Sarpy Concerned Citizens.

Love, N. (2017) 'Back to the Future: Trendy fascism, the Trump effect, and the alt-right', *New Political Science* 39(2): 263–268.

Loviglio, J. (2002) 'Vox Pop: Network radio and the voice of the people', in M. Hilmes and J. Loviglio (eds) *Radio Reader: Essays in the cultural history of radio*, New York and London: Routledge, 89–112.

Lovink, G. (2012) *Networks without a Cause: A critique of social media*, Cambridge: Polity Press.

Lovink, G. and M. Rasch (eds) (2013) *Unlike Us Reader: Social media monopolies and their alternatives*, Amsterdam: Institute of Network Cultures.

Lovink, G. and N. Tkacz (2011) 'The "C" in CPOV: Introduction to the CPOV reader', in G. Lovink and N. Tkacz (eds) *Critical Point of View: A Wikipedia reader*, Amsterdam: Ten Klein Groep, 9–13. Available online: www.networkcultures.org/publications.

Lowenthal, A. (2004/2006) 'Nodes of Conflict: Weaving a web of radical communications'. Available online: www.undergrowth.org/nodes_of_conflict_by_andrew_lowenthal.

Loxley, J. (2007) *Performativity*, London: Routledge.

Lubanga, D. and T. Ominde (2018) 'Shocking Video Showing Brutality of Uasin Gishu "Kanjo" Goes Viral', *Nairobi News*. Available online: http://nairobinews.nation.co.ke/news/shocking-video-showing-brutality-uasin-gishu-kanjo-goes-viral-video/.

Lucas, G. (dir.) (1999) Star Wars: Episode I – The phantom menace. IMDb entry: www.imdb.com/title/tt0120915/?ref_=fn_al_tt_1.

Luce, A., D. Jackson and E. Thorsen (2017) 'Citizen Journalism at the Margins', *Journalism Practice* 11(2–3): 266–284.

Luchies, T. (2015) 'Towards an Insurrectionary Power/Knowledge: Movement-relevance, anti-oppression, prefiguration', *Social Movement Studies* 14(5): 523–538.

Lüders, M. (2015) 'Researching Social Media: Confidentiality, anonymity and reconstructing online practices', in H. Fossheim and H. Ingierd (eds) *Internet Research Ethics*, Oslo: Cappelen Damm Akademisk, 77–97.

Luiselli, V. (2017) *Tell Me How It Ends: An essay in 40 questions*, Minneapolis MN: Coffee House Press.

Luke, T. (2017) 'Overtures for the Triumph of the Tweet: White power music and the alt-right in 2016', *New Political Science* 39(2): 277–282.

Lull, J. (1997) *Media Scandals: Morality and desire in the popular culture marketplace*, Oxford: Polity Press.

Lundby, K. (2008a) 'Digital Storytelling, Mediatized Stories', in K. Lundby (ed.) *Digital Storytelling, Mediatized Stories: Self-representation in new media*, New York: Peter Lang, 1–20.

Lundby, K. (ed.) (2008b) *Digital Storytelling, Mediatized Stories: Self-representations in new media*, New York: Peter Lang.

Lundby, K. (ed.) (2014) *Mediatization of Communication*, Berlin and New York: de Gruyter.

Lunnay, B., J. Borlagdan, D. McNaughton and P. Ward (2015) 'Ethical Use of Social Media to Facilitate Qualitative Research', *Qualitative Health Research* 25(1): 99–109.

Lunt, P. and S. Livingstone (2016) 'Is "Mediatization" the New Paradigm for Our Field? A commentary on Deacon and Stanyer (2014, 2015) and Hepp, Harvard and Lundby (2015)', *Media, Culture & Society* 38(3): 462–470.

Lury, C. (2004) *Brands: The logos of the global economy*, London and New York: Routledge.

Lusthaus, J. and F. Varese (2017) 'Offline and Local: The hidden face of cybercrime', *Policing: A Journal of Policy and Practice*, pax042: 1–11.

Lynch, L. (2010) 'We're Going to Crack the World Open', *Journalism Practice* 4(3): 309–318.

Lynch, L. (2019) 'Wikileaks and Investigative Journalism: The organization's effects and unfinished legacy', in S. Price (ed.) *Journalism, Power and Investigation: Global and activist perspectives*, London: Routledge, 265–287.

Lynd, S. and A. Grubačić (2008) *Wobblies & Zapatistas: Conversations on anarchism, Marxism and radical history*, Oakland, CA: PM Press.

Lyon, D. (2003) *Surveillance as Social Sorting*, London and New York: Routledge.

Lyon, D. (2007) *Surveillance Studies: An overview*, Cambridge: Polity Press.

Lyon, D. (2014) 'Surveillance, Snowden, and Big Data: Capacities, consequences, critique', *Big Data and Society* 1(2): 1–13.

Lyon, D. (2015) *Surveillance after Snowden*, Cambridge: Polity Press.

Lyon, D. (2017) 'Surveillance Culture, Ethics and Digital Citizenship', *International Journal of Communication* 11: 824–842.

Lyon, W. (2014) 'Client Criminalisation and Sex Workers' Right to Health', *Hibernian Law Journal* 13(58).

Macey, G. P., R. Breech, M. Chernaik, C. Cox, D. Larson, D. Thomas and D. O. Carpenter (2014) 'Air Concentrations of Volatile Compounds Near Oil and Gas Production: A community-based exploratory study', *Environmental Health* 13(1): 82. Available online: https://ehjournal.biomedcentral.com/articles/10.1186/1476-069X-13-82.

Macguire, E. (2017) 'How Graffiti Artists Turned Former Warring Areas into a Giant Artwork for Peace', *CNN.com*, 24 November. Available online: www.cnn.com/style/article/peace-graffiti-lebanon-operation-salam/index.html.

MacKenzie, D. and J. Wajcman (1999) *The Social Shaping of Technology*, second edition, Buckingham, UK: Open University Press.

Mackenzie, I. (2009) *Politics: Key concepts in philosophy*, New York: Continuum.

Macková, A. and J. Macek (2014) '"Žít Brno": Czech online political activism from jokes and tactics to politics and strategies', *Cyberpsychology* 8(3). Available online: https://cyberpsychology.eu/article/view/4313/3363.

Madani, D. (2017) '7-Year-Old Refugee Who Tweeted during Aleppo Siege Is Writing a Memoir'. Huffington Post, 12 April. Available online: www.huffingtonpost.co.uk/entry/7-year-old-refugee-who-tweeted-during-aleppo-siege-is-writing-a-memoir_n_58ee4779e4b0f392747482b8.

Madden, K. (1992) 'Video and Cultural Identity: The Inuit broadcasting corporation experience', in F. Korzenny, S. Ting-Toomey and E. Schiff (eds) *Mass Media Effects across Cultures*, Newbury Park, CA: Sage, 130–149.

Madden, S., M. Janoske, R. B. Winkler and Z. Harpole (2018) 'Who Loves Consent? Social media and the culture jamming of Victoria's Secret', Public Relations Inquiry 7(2): 171–186.

Madianou, M. (2012) 'Migration and the Accentuated Ambivalence of Motherhood: The role of ICTs in Filipino transnational families', *Global Networks* 12(3): 277–295.

Madianou, M. (2014) 'Polymedia Communication and Mediatized Migration: An ethnographic approach', in K. Lundby (ed.) *Mediatization of Communication*, Berlin and New York: de Gruyter, 323–348.

Madianou, M. and D. Miller (2012) 'Polymedia: Towards a new theory of digital media in interpersonal communication', *International Journal of Cultural Studies* 16(2): 169–182.

Madison, M., B. Frischmann and K. Strandburg (2019) 'Knowledge Commons', in B. Hudson, J. Rosenbloom and D. Cole (eds) *Routledge Handbook of the Study of the Commons*, Abingdon, UK: Routledge, 76–90.

Maeckelbergh, M. (2007) Decentralized Network Democracy: Prefiguring horizontality and diversity in the alterglobalization movement, Unpublished PhD thesis, Brighton, UK: University of Sussex.

Maeckelbergh, M. (2009) *The Will of the Many: How the alterglobalisation movement is changing the face of democracy*, London: Pluto Press.

Maeckelbergh, M. (2011) 'Doing Is Believing: Prefiguration as strategic practice in the alterglobalization movement', *Social Movement Studies* 10(1): 1–20.

Maeckelbergh, M. (2012a) 'Horizontal Democracy Now: From alterglobalization to occupation', *Interface: The Journal of and for Social Movements* 4(1): 207–234.

Maeckelbergh, M. (2012b) 'Occupy the US: Musings on horizontal decision-making and bureaucracy', *Stir* (Spring). Available online: www.kosmosjournal.org/article/occupy-the-us-musings-on-horizontal-decision-making-and-bureaucracy/.

Maeckelbergh, M. (2014) 'Activism: Integration', *Cultural Anthropology*, 26 May. Available online: https://culanth.org/fieldsights/535-activism-integration.

Maeckelbergh, M. and B. Jourdan (2013) *Fighting Sexual Assault in Egypt: Interview with Mariam Kirollos*. Available online: www.globaluprisings.org/global-uprising-conference-interview-1-fighting-sexual-assault-in-egypt/.

Magallanes Blanco, C. (2008) 'Video as a Tool for Change: Gender discourse in Zapatista indigenous communities', *Agenda* 22: 106–114.

Magallanes Blanco, C. (2014) 'Talking about our Mother: Indigenous videos on nature and the environment', *Communication Culture & Critique* 8(2): 199–216.

Magallanes Blanco, C. and L. Rodríguez Medina (2016) 'Give Me a Mobile and I Will Raise a Community', *Studies in Media and Communications* 12: 315–343.

Magaudda, P. (2011) 'When Materiality "Bites Back": Digital music consumption practices in the age of dematerialization', *Journal of Consumer Culture* 11(1): 15–36.

Magaudda, P. (2012) 'How to Make a "Hackintosh": A journey into the "consumerization" of hacking practices and culture', *Journal of Peer Production* (2). Available online: http://peerproduction.net/issues/issue-2/peer-reviewed-papers/how-to-make-a-hackintosh/.

Magno, F. (2017) 'The Influence of Cultural Blogs on Their Readers' Cultural Product Choices', *International Journal of Information Management* 37(3): 142–149.

Maher, G. C. (2005) 'Brechtian Hip-Hop: Didactics and self-production in post-gangsta political mix-tapes', *Journal of Black Studies* 26(1): 129–160.

Maira, S. (2005) 'The Intimate and the Imperial: South Asian Muslim immigrant youth after 9/11', in S. Maira and E. Soep (eds) *Youthscapes: The popular, the national, the global*, Philadelphia, PA: University of Pennsylvania Press, 63–81.

Maireder, A. and C. Schwarzenegger (2012) 'A Movement of Connected Individuals', *Information, Communication & Society* 15(2): 171–195.

Malik, S. el. (2013) '"Creative Diversity": UK public service broadcasting after multiculturalism', *Popular Communication* 11(3): 227–241.

Malik, S. el. (2016) 'Subjectivity', in A. Ní Mhurchú and S. Shindo (eds) *Critical Imaginations in International Relations*, London: Routledge, 212–227.

Malkowski, J. (2017) *Dying in Full Detail: Mortality and digital documentary*, Durham, NC: Duke University Press.

Mandiberg, M. (ed.) (2012) *Social Media Reader*, New York: New Yok University Press.

Mann, S. (2004) '"Sousveillance": Inverse surveillance in multimedia imaging', *ACM Multimedia 2004 – Proceedings of the 12th ACM International Conference on Multimedia*, New York: ACM, 620–627

Mann, S. (2013) 'Veillance and Reciprocal Transparency: Surveillance versus sousveillance, AR glass, lifeglogging, and wearable computing', *2013 IEEE International Symposium on Technology and Society (ISTAS): Social implications of wearable computing and augmediated reality in everyday life*, Toronto: IEEE, 1–12.

Mann, S. (2017) 'Big Data Is a Big Lie without Little Data: Humanistic intelligence as a human right', *Big Data and Society* 4(1): 1–10.

Mann, S. and J. Ferenbok (2013) 'New Media and the Power Politics of Sousveillance in a Surveillance Dominated World', *Surveillance and Society* 11(2): 18–34.

Mann, S., J. Nolan, and B. Wellman (2003) 'Sousveillance: Inventing and using wearable computing devices for data collection in surveillance environments', *Surveillance and Society* 1(3): 331–355.

Manning, C. (2010) '"My Memory's Back!" Inclusive learning disability research using ethics, oral history and digital storytelling', *British Journal of Learning Disabilities* 38(3): 160–167.

Manoff, M. (2004) 'Theories of the Archive from across the Disciplines', *Libraries and the Academy* 4(1): 9–25.

Manovich, L. (2001) *The Language of New Media*, Cambridge, MA and London: The MIT Press.

Manovich, L. (2013) *Software Takes Command*, London: Bloomsbury.

Mansbach, D. (2016) 'Witnessing as Activism: Watching the other at the Israeli checkpoints', *Journal of Human Rights* 15(4): 496–508.

Mansbridge, J. (1983) *Beyond Adversary Democracy*, Chicago, IL: University of Chicago Press.

Manyozo, L. (2010) 'Mobilizing Rural and Community Radio in Africa', *Ecquid Novi* 30(1): 1–23.

Manzerolle, V. (2013) 'Brave New Wireless World: Mapping the rise of ubiquitous connectivity from myth to market', Faculty of Information and Media Studies, London, Ontario: The University of Western Ontario.

Marazzi, C. (2011) *Capital and Affects: The politics of the language economy*, Los Angeles, CA and Cambridge, MA: Semiotext(e).

Marback, R. and M.W. Kruman (2015) *The Meaning of Citizenship*, Detroit, MI: Wayne State University Press.

Marchart, O. (2011) 'From Media to Mediality: Mediatic (counter-)apparatuses and the concept of the political in communication studies', in L. Dahlberg and S. Phelan (eds) *Discourse Theory and Critical Media Politics*, Basingstoke, UK: Palgrave Macmillan, 64–81.

Marchetti, G. (2016) 'Activism', *Journal of Chinese Cinemas* 10(1): 4–7.

Marchevska, E. (2017a) 'Performing the Present, Saving the Future: Digital media presence in the work of Raspeani Skopjani', *Contemporary Theatre Review* 27(3): 391–397.

Marchevska, E. (2017b) 'Fluidity and Friendship: The choir that surprised the city. Elena Marchevska in conversation with Nita Çavolli, Jana Jakimovska, and Katerina Mojanchevska', *Contemporary Theatre Review (Interventions)*, December. Available online: www.contemporarytheatrereview.org/2017/raspeani-skopjani/.

Marcos, S. I. (2002) *Our Word Is Our Weapon: Selected writings*, New York: Seven Stories Press.

Marcos, S. I. (2004) 'A Death Has Been Decided', in Ž. Vodovnik (ed.) *¡Ya Basta!: Ten years of the Zapatista uprising*, Oakland, CA: AK Press, 593–601.

Marcus, D. (2016) 'Documentary and Video Activism', in D. Marcus and S. Kara (eds) *Contemporary Documentary*, London and New York: Routledge, 187–203.

Marcus, G. (1996) 'Introduction', in G. Marcus (ed.) *Connected: Engagements with media*, Chicago, IL: University of Chicago Press, 1–18.

Marcuse, P. (2005) 'Are Social Forums the Future of Social Movements?', *International Journal of Urban and Regional Research* 29(2). Available online: https://onlinelibrary.wiley.com/doi/full/10.1111/j.1468-2427.2005.00593.x.

Marino, A. (2015) *Acts of Angry Writing: On citizenship and orientalism in postcolonial India*, Detroit, MI: Wayne State University Press.

Markham, A. and N. Baym (2009) *Internet Inquiry: Conversations about method*, Thousand Oaks, CA: Sage.

Markham, A. and E. Buchanan (2012) *Ethical Decision-Making and Internet Research: Recommendations from the AoIR Ethics Working Committee (version 2.0)*. Available online: http://aoir.org/reports/ethics2.pdf.

Marres, N. (2007) 'The Issues Deserve More Credit: Pragmatist contributions to the study of public involvement in "controversy"', *Social Studies of Science* 37(5): 759–780.

Marsden, G., T. Reitmaier, N. J. Bidwell and E. Blake (2010) 'Digital Storytelling in Africa', *International Journal of Computing* 9(3): 257–265.

Marsh, Sue (n.d.) Diary of a Benefit Scrounger website. Available online: www.DiaryOfABenefitsScrounger.blogspot.com.

Marshall, L. (2013) 'The 360 Deal and the "New" Music Industry', *European Journal of Cultural Studies* 16(1): 77–99.

Marshall, P. (1992) *Demanding the Impossible: A history of anarchism*, London: Fontana Press.

Marshall, T. H. (1992) *Citizenship and Social Class*, London: Pluto Press.

Marthews, A. and C. Tucker (2015) 'Government Surveillance and Internet Search Behaviour', *SSRN*. Available online: http://ssrn.com/abstract=2412564.

Martin, C. (2003) *Framed: Labor and the Corporate Media*, Ithaca, NY: Cornell University Press.

Martin, M. T. (ed.) (1997) *New Latin American Cinema*, Detroit, MI: Wayne State University Press.

Martínez, M. and Á. García (2015) 'Ocupar las plazas, liberar edificios (The Occupation of Squares and the Squatting of Buildings)', *ACME: An International Journal for Critical Geographies* 14(1): 157–184.

Martínez López, M. A. (2016) 'Between Autonomy and Hybridity: Urban struggles within the 15M movement in Spain', in M. Meyer, K. Thörn and H. Thörn (eds) *Urban Uprisings*, London: Palgrave, 253–281.

Martínez Martínez, M. J. (2017) 'Prácticas mediáticas y movimientos sociales: El activismo transnacional de Marea Granate', *index.comunicación* 7(3): 31–50. Available online: http://journals.sfu.ca/indexcomunicacion/index.php/indexcomunicacion/article/view/340/305.

Marty, E., N. Pignard-Cheynel and B. Sebbah (2017) 'Internet Users' Participation and News Framing: The Strauss-Kahn case-related live blog at Le Monde.fr', *New Media & Society* 19(12): 1964–1982.

Marwick, A. (2013) *Status Update: Celebrity, publicity, and branding in the social media age*, New Haven, CT and London: Yale University Press.

Marwick, A. (2015) 'Instafame: Luxury selfies in the attention economy', *Public Culture* 27(1): 137–160.

Marwick, A. (2017) 'Are There Limits to Online Free Speech?', *Data and Society*, 5 January. Available online: https://points.datasociety.net/are-there-limits-to-online-free-speech-14dbb7069aec.

Marx, K. (1993) *Grundrisse: Foundations of the critique of political economy*, reprint edition, translated by M. Nicolaus, London: Penguin Books.

Masden, C. A., C. Grevet, R. E. Grinter, E. Gilbert and W. K. Edwards (2014) 'Tensions in Scaling-up Community Social Media: A multi-neighborhood study of nextdoor', *Proceedings of the SIGCHI Conference on Human Factors in Computing Systems*, New York: ACM, 3239–3248.

Maslin Nir, S. (2011) 'Wall Street Protesters Broadcast Arrests on Social Media', City Room Blog, 24 September. Available online: http://cityroom.blogs.nytimes.com/2011/09/24/wall-street-protesters-broadcast-arrests-on-social-media.

Mason, G. (1977) 'Documentary Drama from the Revue to the Tribunal', *Modern Drama* 20(3): 263–277.

Massa, P. and A. Zelenkauskaite (2014) 'Gender Gap in Wikipedia Editing: A cross-language comparison', in P. Fichman and N. Hara (eds) *Global Wikipedia: International and cross-cultural issues in online collaboration*, Washington DC: Scarecrow Press, 85–96.

Massanari, A. L. (2015) *Participatory Culture, Community, and Play: Learning from Reddit*, New York: Peter Lang.

Massanari, A. L. (2017) '#Gamergate and The Fappening: How Reddit's algorithm, governance, and culture support toxic technocultures', *New Media & Society* 19(3): 329–346.

Massey, J. and B. Snyder (2012) 'Occupying Wall Street: Places and spaces of political action', *Places Journal*. Available online: https://doi.org/10.22269/120917.

Massumi, B. (2002) *Parables for the Virtual: Movement, affect, sensation*, Durham, NC and London: Duke University Press.

Massumi, B. (2015) *The Politics of Affect*, Malden, MA: Polity Press.

Massumi, B. (2016) 'The Autonomy of Affect', *Cultural Critique* 31: 83–109.

Massung, E., D. Coyle, K. Cater, M. Jay and C. Preist (2013) 'Using Crowdsourcing to Support Pro-Environmental Community Activism', *Proceedings of the SIGCHI Conference on Human Factors in Computing Systems*, New York: ACM, 371–380.

Matamoros-Fernández, A. (2017) 'Platformed Racism: The mediation and circulation of an Australian race-based controversy on Twitter, Facebook and YouTube', *Information, Communication and Society* 20(6): 930–946.

Mateos, C. and C. Gaona (2015) 'Video Activism. A descriptive typology', *Global Media Journal*. Available online: www.globalmediajournal.com/open-access/video-activism-a-descriptive-typology.php?aid=62532.

Matheson, D. and S. Allan (2009) *Digital War Reporting*, Malden, MA: Polity Press.

Matias, J. N. (2016) 'The Civic Labor of Online Moderators', IPP2016: The Platform Society Conference, University of Oxford, 22–23 September 2016.

Mattelart, T. (2008) 'Les théories de la mondialisation culturelle', *Hermès* 2: 17–22.

Mattoni, A. (2012) *Media Practices and Protest Politics: How precarious workers mobilise*, London: Routledge.

Mattoni, A. (2013a) 'Repertoires of Communication in Social Movement Processes', in B. Cammaerts, A. Mattoni and P. McCurdy (eds) *Mediation and Protest Movements*, Bristol, UK: Intellect, 39–56.

Mattoni, A. (2013b) 'Media Activism', in D. A. Snow, D. della Porta, B. Klandermans and D. McAdam (eds) *The Wiley-Blackwell Encyclopedia of Social and Political Movements*, Malden, MA: Blackwell.

Mattoni, A. (2017) 'A Situated Understanding of Digital Technologies in Social Movements: Media ecology and media practice approaches', *Social Movement Studies* 16(4): 494–505.

Mattoni, A. and N. Doerr (2007) 'Images within the Precarity Movement in Italy', *Feminist Review* 87: 130–135.

Mattoni, A. and E. Treré (2014) 'Media Practices, Mediation Processes, and Mediatization in the Study of Social Movements', *Communication Theory* 24(3): 252–271.

Mattoni, A. and M. Vogiatzoglou (2014) '"Today, We Are Precarious. Tomorrow, We Will Be Unbeatable": Early struggles of precarious workers in Italy and Greece', in D. Chabanet and F. Royall (eds) *From Silence to Protest: International perspectives on weakly resourced groups*, Burlington, VT: Ashgate, 67–82.

Maxis Inc. (2000) *The Sims*. IGDB entry: www.igdb.com/games/the-sims.

Maxwell, I. (1994) 'True to the Music: Authenticity, articulation and authorship in Sydney hip hop culture', *Social Semiotics* 4(1–2): 117–137.

Maxwell, R. and T. Miller (2011) 'Old, New and Middle-aged Media Convergence', *Cultural Studies* 25(4–5): 585–603.

Maxwell, R. and T. Miller (2012) *Greening the Media*, Oxford: Oxford University Press.

May, H. and G. N. Hearn (2005) 'The Mobile Phone as Media', *Cultural Studies* 8(2): 195–211.

May, T. (1994) *The Political Philosophy of Poststructuralist Anarchism*, University Park, PA: Pennsylvania State University Press.

Mayer-Schönberger, V. and K. Cukier (2013) *Big Data: A revolution that will transform how we live, work, and think*, Boston, MA: Houghton Mifflin Harcourt.

Mayo, B. (1961) 'Objects, Events, and Complementarity', *Mind* 70: 340–361.

Mazierska. E. and L. Kristensen (eds) (2015) *Marxism and Film Activism: Screening alternative worlds*, New York and Oxford: Berghahn, 186–212.

McCall, J. C. (2002) 'Madness, Money, and Movies: Watching a Nigerian popular video with the guidance of a native doctor', *Africa Today* 49(3): 79–94.

McCarthy, J. D. and M. Wolfson (1992) 'Consensus Movements, Conflict Movements, and the Cooptation of Civic and State Infrastructures', in A. D. Morris and C. M. Mueller (eds) *Frontiers in Social Movement Theory*, New Haven, CT: Yale University Press, 273–297.

McCaughey, M. and M. D. Ayers (eds) (2013) *Cyberactivism: Online activism in theory and practice*, London: Routledge.

McClellan, J. (2004) 'The Role of Play', *The Guardian*, 13 May. Available online: www.theguardian.com/technology/2004/may/13/games.onlinesupplement.

McCormack, D. and S. Salmenniemi (2016) 'The Biopolitics of Precarity and the Self', *European Journal of Cultural Studies* 19(1): 3–15.

McCormack, D. P. (2017) 'Elemental Infrastructures for Atmospheric Media: On stratospheric variations, value and the commons', *Environment and Planning D: Society and Space* 35(3): 418–437.

McCosker, A. and A. Johns (2014) 'Contested Publics: Racist rants, bystander action and social media acts of citizenship', *Media International Australia*: 66–72.

McCosker, A., S. Vivienne and A. Johns (eds) (2016) *Negotiating Digital Citizenship: Control, contest and culture*, Lanham, MD: Rowman & Littlefield.

McCurdy, P. (2011) 'Theorizing "Lay Theories of Media": A case study of the Dissent! Network at the 2005 Gleneagles G8 Summit', *International Journal of Communication* 5: 619–638.

McDonald, K. (2015) 'From Indymedia to Anonymous: Rethinking action and identity in digital cultures', *Information, Communication & Society* 18(8): 968–982.

McIntosh, H. (2016) 'Producing the Crowdsourced Documentary: The implications of storytelling and technology', in D. Marcus and S. Kara (eds) *Contemporary Documentary*, London and New York: Routledge, 57–71.

McKay, G. (1998) *DIY Culture: Party and protest in nineties Britain*, London: Verso.

McKay, I. (2012) *An Anarchist FAQ*, Volume 2, Edinburgh: AK Press.

McKenzie, P., J. Burkell, L. Wong, C. Whippey, S. Trosow and M. McNally (2012) 'User-Generated Online Content 1: Overview, current state and context', *First Monday* 17(6).

McLagan, M. (1996) 'Computing for Tibet: Virtual politics in the post-Cold War era', in G. Marcus (ed.) *Connected: Engagements with media*, Chicago, IL: University of Chicago Press, 159–194.

McLagan, M. (2012) 'Imagining Impact: Documentary film and the production of political effects', in M. McLagan and Y. McKee (eds) *Sensible Politics: The visual culture of nongovernmental activism*, New York and Cambridge, MA: Zone Books, 305–320.

McLagan, M. and Y. McKee (eds) (2012) *Sensible Politics: The visual culture of nongovernmental activism*, New York and Cambridge, MA: Zone Books.

McLuhan, E. and F. Zingrone (1995) *Essential McLuhan*, New York: Basic Books.

McLuhan, M. (1964) *Understanding Media: The extensions of man*, New York and Toronto: McGraw Hill.

McLuhan, M. (1964/1994) *Understanding Media: The extensions of man*, reissued edition, Cambridge, MA: The MIT Press.

McNevin, A. (2011) *Contesting Citizenship: Irregular migrants and new frontiers of the political*, New York: Columbia University Press.

McPherson, M., L. Smith-lovin and J. M. Cook (2001) 'Homophily in Social Networks', *Annual Review of Sociology* 27: 415–444.

McQuail, D. (1992) *Media Performance: Mass communication and the public interest*, London: Sage.

McQuire, S. (2006) 'The Politics of Public Space in the Media City', First Monday 4. Available online: http://firstmonday.org/article/view/1544/1459.

McRobbie, A. (2011) 'Reflections on Feminism, Immaterial Labour and the Post-Fordist Regime', *New Formations* 70: 60–76.

McRobbie, A. (2016) *Be Creative: Making a living in the new culture industries*, Cambridge, UK and Malden, MA: Polity Press.

Meadows, M. (1992) 'Broadcasting in Aboriginal Australia: One mob, one voice, one land', in S. H. Riggins (ed.) *Ethnic Minority Media: An international perspective*, Newbury Park, CA: Sage, 82–101.

Mecchia, G. and M. Henninger (eds) (2007) *Italian Post-Workerist Thought*. Special issue of SubStance 36(1).

Medeiros, B. (2017) 'Platform (Non-)Intervention and the "Marketplace" Paradigm for Speech Regulation', *Social Media + Society* 3(1).

Medina, J. (2012) *Epistemologies of Resistance: Gender and racial oppression, epistemic injustice, and resistant imaginations*, Oxford: Oxford University Press.

Medina, J. (2017) 'Varieties of Hermeneutical Injustice', in I. J. Kidd, J. Medina and G. Pohlhaus Jr. (eds) *The Routledge Handbook of Epistemic Injustice*, New York: Routledge, 41–52.

Meeker, M. (2016) *Internet Trends 2016: Code Conference*. Available online: www.kpcb.com/blog/2016-internet-trends-report.

Megwa, E. R. (2007) 'Bridging the Digital Divide: Community radio's potential for extending information and communication technology benefits to poor rural communities in South Africa', *The Howard Journal of Communications* 18(4): 335–352.

Meikle, G. (2016) *Social Media: Communication, sharing and visibility*, London and New York: Routledge.

Meikle, G. and S. Young (2012) *Media Convergence: Networked digital media in everyday life*, Basingstoke, UK: Palgrave Macmillan.

Meiksins Wood , E. (2017) *The Origin of Capitalism*, London: Verso.

Melucci, A. (1986) *Nomads of the Present: Social movements and individual needs in contemporary society*, Philadelphia, PA: Temple University Press.

Melucci, A. (1994) 'A Strange Kind of Newness: What's "new" in new social movements?', in E. Laraña, H. Johnston and J. R. Gusfield (eds) *New Social Movements: From ideology to identity*, Philadelphia, PA: Temple University Press, 103–130.

Melucci, A. (1996) *Challenging Codes: Collective action in the information age*, Cambridge: Cambridge University Press.

Membretti, A. and P. Mudu (2013) 'Where Global Meets Local', in C. Flesher Fominaya and L. Cox (eds) *Understanding European Movements: New social movements, global justice struggles, anti-austerity protests*, London: Routledge, 76–93.

Meng, B. (2011) 'From Steamed Bun to Grass Mud Horse: E Gao as alternative political discourse on the Chinese Internet', *Global Media and Communication* 7(1), 33–51.

Meraz, S. and Z. Papacharissi (2013) 'Networked Gatekeeping and Networked Framing on #Egypt', *The International Journal of Press/Politics* 18(2): 138–166.

Mercea, D. (2013) 'Probing the Implications of Facebook Use for the Organizational Form of Social Movement Organizations', *Information Communication and Society* 16(8): 1306–1327.

Mercea, D., L. Iannelli and B. Loader (2016) 'Protest Communication Ecologies', *Information, Communication & Society* 19(3): 279–289.

Merleau-Ponty, M. (1962/1997) *Phenomenology of Perception*, London: Routledge.

Merriam-Webster (2018). Available online: www.merriam-webster.com/dictionary/.

Merrifield, A. (2017) *The Amateur: The pleasures of doing what you love*, London: Verso.

Mesch, G. S. and Y. Levanon (2003) 'Community Networking and Locally-Based Social Ties in Two Suburban Localities', *City & Community* 2(4): 335–351.

Messner, S. (2018) 'World of Warcraft's Pacifist Panda Has Reached Level 120 by Only Picking Flowers', *PCGamer*, 29 October. Available online: www.pcgamer.com/world-of-warcrafts-pacifist-panda-has-reached-level-120-by-only-picking-flowers/.

Mestrovic, S. (1997) *Postemotional Society*, London: Sage.

Metzgar, E. T., D. Kurpius and K. Rowley (2011) 'Defining Hyperlocal Media: Proposing a framework for discussion', *New Media & Society* 13(5): 772–787.

Meyer, R. (2014) 'Everything We Know about Facebook's Mood Manipulation Study', *The Atlantic Monthly*, 28 June. Available online: www.theatlantic.com/technology/archive/2014/06/everything-we-know-about-facebooks-secret-mood-manipulation-experiment/373648/.

Meyrowitz, J. (1985) *No Sense of Place: The impact of electronic media on social behavior*, New York: Oxford University Press.

Mezzadra, S. (2009) 'Italy, Operaism and Post-Operaism', in I. Ness (ed.) *International Encyclopedia of Revolution and Protest*, Oxford: Blackwell, 1841–1845.

Miall, H. (2004) 'Conflict Transformation: A multi-dimensional task', in B. Austin, M. Fischer and H. J. Giessmann (eds) *Berghof Handbook for Conflict Transformation*, Berlin: Berghof Foundation, 1–20.

Michaels, E. (1994) 'For a Cultural Future: Francis Jupurrurla makes TV at Yuendumu', in E. Michaels (ed.) *Bad Aboriginal Art: Tradition, media, and technological horizons*, Minneapolis, MN: University of Minnesota Press, 99–125.

Midnight Notes (ed.) (2001) *Auroras of the Zapatistas: Local and global struggles of the fourth world war*, Brooklyn, NY: Autonomedia.

Mignolo, W. (2011) 'Geopolitics of Sensing and Knowing: On (de)colniality, border thinking and epistemic disobedience', *European Institute for Progressive Cultural Politics* 9. Available online: http://eipcp.net/transversal/0112/mignolo/en.

Mignolo, W. (2013) 'Geo-politics of Sensing and Knowing: On (de)coloniality, border thinking, and epistemic disobedience', *Confero* 1(1): 129–150.

Migrant Worker Poetry Competition. www.singaporeworkerpoetry.com/.

Mikkelsen, F. and R. Karpantschof (2001) 'Youth as a Political Movement: Development of the squatters' and autonomous movement in Copenhagen', *International Journal of Urban and Regional Research* 25(3): 593–608.

Milan, S. (2013) *Social Movements and Their Technologies: Wiring social change*, New York: Palgrave Macmillan.

Milan, S. (2015) 'From Social Movements to Cloud Protesting: The evolution of collective identity', *Information, Communication & Society* 18: 887–900.

Milan, S. (2017) 'Data Activism as the New Frontier of Media Activism', in V. Pickard and G. Yang (eds) *Media Activism in a Digital Age*, New York: Routledge, 151–163.
Milan, S. and M. Gutiérrez (2015) 'Citizens' Media Meets Big Data: The emergence of data activism', *Mediaciones* 14: 120–133.
Milan, S. and M. Gutiérrez (2017) 'Technopolitics in the Age of Big Data', in F. Sierra Caballero and T. Gravante (eds) *Networks, Movements and Technopolitics in Latin America: Critical analysis and current challenges*, London: Palgrave Macmillan, 95–109.
Milan, S. and L. van der Velden (2016) 'The Alternative Epistemologies of Data Activism', *Digital Culture and Society* 2(2): 57–74.
Milan, S. and L. van der Velden (eds) (2018) *Data Activism*. Special issue of Krisis: Journal for Contemporary Philosophy 1. Available online: https://krisis.eu/issue-1-2018-data-activism/.
Milgram, S., L. Bickman and L. Berkowitz (1969) 'Note on the Drawing Power of Crowds of Different Size', *Journal of Personality and Social Psychology* 13(2): 79–82.
Miller, D. (2011) *Tales from Facebook*, Cambridge, UK and Malden, MA: Polity Press.
Miller, D. and D. Slater (2000) *The Internet: An ethnographic approach*, New York: Berg Press.
Miller, J. (1994) *Democracy Is in the Streets*, Cambridge, MA: Harvard University Press.
Miller, J. (2014) 'Intensifying Mediatization: Everyware media', in A. Hepp and F. Krotz (eds) *Mediatized worlds: Culture and society in a media age*, Basingstoke, UK: Palgrave Macmillan, 107–122.
Miller, T. and M. M. Kraidy (2016) *Global Media Studies*, London: Polity Press.
Miller, V. (2011) *Understanding Digital Culture*, London: Sage.
Miller-Rushing, A., R. Primack and R. Bonney (2012) 'The History of Public Participation in Ecological Research', *Frontiers in Ecology and the Environment* 10(6): 285–290.
Milligan, T. (2013) *Civil Disobedience: Protest, justification and the law*, New York: Bloomsbury Academic.
Milošević-Đorđević, J. S. and I. L. Žeželj (2017) 'Civic Activism Online: Making young people dormant or more active in real life?', *Computers in Human Behavior* 70(C): 113–118.
Milstein, C. (2015a) 'Solidarity, as Weapon and Practice, versus Killer Cops and White Supremacy', in C. Milstein (ed.) *Taking Sides: Revolutionary solidarity and the poverty of liberalism*, Oakland, CA: AK Press, 138–156.
Milstein, C. (ed.) (2015b) *Taking Sides: Revolutionary solidarity and the poverty of liberalism*, Oakland, CA: AK Press.
Minghella, A. (dir.) (2006) Breaking and Entering. IMDb entry: www.imdb.com/title/tt0443456.
Mir, R. and T. Owens (2013) 'Modeling Indigenous Peoples: Unpacking ideology in Sid Meier's *Colonization*', in M. Kapell and A. B. R. Elliott (eds) *Playing with the Past: Digital games and the simulation of history*, New York: Bloomsbury Academic, 91–106.
Mirowski, P. (2017) 'Against Citizen Science', *Aeon*. Available online: https://aeon.co/essays/is-grassroots-citizen-science-a-front-for-big-business.
Mirzoeff, N. (2011) 'The Right to Look', *Critical Inquiry* 37(3): 473–496.
Mirzoeff, N. (2017) *The Appearance of Black Lives Matter*, New York: NAME Publications.
Misera, T. (2018) 'About Meta-Metareddit', *Metareddit*. Available online: https://web.archive.org/web/20180724222526/http://www.metareddit.com/about/.
Mitchell, C. (2002) 'Beyond Resolution: What does conflict transformation actually transform?', *Peace and Conflict Studies* 9(1). Available online: http://nsuworks.nova.edu/pcs/vol9/iss1/1.
Mitchell, D. (2016) 'Tent Cities: Interstitial spaces of survival', in A. M. Brighenti (ed.) *Urban Interstices: The Aesthetics and the Politics of the In-between*, New York: Routledge, 65–86.
Mitchell, T. (1996) *Popular Music and Local Identity: Rock, pop and rap in Europe and Oceania*, Leicester, UK: Leicester University Press.
Mitchell, T. (2000) 'Doin' Damage in my Native Language: The use of "resistance vernaculars" in hip hop in France, Italy, and Aotearoa/New Zealand', *Popular Music and Society* 24(3): 41–54.
Mitchell, W. (1994) *The Reconfigured Eye: Visual truth in the post-photographic era*, Cambridge, MA: The MIT Press.
Mittell, J. (2009) 'Sites of Participation: Wiki fandom and the case of Lostpedia', *Transformative Works and Cultures* 3. Available online: http://journal.transformativeworks.org/index.php/twc/article/view/118.
Mittell, J. (2013) 'Wikis and Participatory Fandom', in A. Delwiche and J. Jacobs Henderson (eds) *The Participatory Cultures Handbook*, New York and London: Routledge, 35–42.
Mitu, B. (2016) 'Web 2.0 Media Events: Barack Obama's inauguration', in B. Mitu and S. Poulakidakos (eds) *Media Events: A critical contemporary approach*, London: Palgrave Macmillan, 230–242.
Mitu, B. and S. Poulakidakos (eds) (2016) *Media Events: A critical contemporary approach*, London: Palgrave Macmillan.
Mogoş, A. A. (2016) 'The Self (Re)presentation: Selfie as cultural artefact and social practice', *Studia Universitatis Babes-Bolyai, Ephemerides* 61(1): 51–67.
Mojang Synergies (2011) Minecraft. IGDB entry: www.igdb.com/games/minecraft.

Molleindustria (2006) The McDonald's Videogame. Available online: www.mcvideogame.com/index-deu.html.

Mollerup, N. G. (2015) Media and Place in Revolutionary Egypt: An anthropological exploration of information activism and journalism, Unpublished PhD thesis, Roskilde, Denmark: Roskilde University.

Mollerup, N. G. and S. Gaber (2015) 'Making Media Public: On revolutionary street screenings in Egypt', *International Journal of Communication* 9: 2903–2921.

Molnar, H. (1994) 'Remote Indigenous Community Broadcasting in Australia: An extension of Indigenous cultural expression', paper presented at the *International Communication Association Conference*, Sydney, Australia.

Molnar, H. and M. Meadows (2001) *Songlines to Satellites: Indigenous communication in Australia, the South Pacific and Canada*, Leichhardt, NSW: Pluto Press.

Mols, A. and S. Janssen (2017) 'Not Interesting Enough to Be Followed by the NSA: An analysis of Dutch privacy attitudes', *Digital Journalism* 5(3): 277–298.

Momeni Roochi, E. (2014) Adaptive Moderation of User-Generated Content on Web, Unpublished PhD thesis, Vienna: University of Vienna, Austria. Available online: http://othes.univie.ac.at/33946/.

Monani, S. (2014) 'Evoking Sympathy and Empathy: The ecological Indian and indigenous eco-activism', in A. Weik von Mossner (ed.) *Moving Environments: Affect, emotion, ecology and film*, Waterloo, Ontario: Wilfrid Laurier University Press, 225–247.

Monterroso, I., P. Cronkleton and A. Larson (2019) 'Commons, Indigenous Rights, and Governance', in B. Hudson, J. Rosenbloom and D. Cole (eds) *Routledge Handbook of the Study of the Commons*, Abingdon, UK: Routledge, 376–391.

Montgomery, M. (2001) 'The Uses of Authenticity: "Speaking from experience" in a UK election broadcast', *The Communication Review* 4(4): 447–462.

Moore, J. (2013) 'How to Start an Independent Record Label'. Available online: www.pigeonsandplanes.com/2013/03/how-to-start-an-independent-record-label/s/be-careful-with-your-cash/.

Moore, R. (2010) *Cybercrime: Investigating high-technology computer crime*, Amsterdam: Elsevier.

Moore, S. and S. Pell (2010) 'Autonomous Archives', *International Journal of Heritage Studies* 16(4–5): 255–268.

Moores, S. (2004) 'The Doubling of Place: Electronic media, time–space arrangements and social relationships', in N. Couldry and A. McCarthy (eds) *Mediaspace: Place, scale and culture in a media age*, London and New York: Routledge, 21–36.

Moores, S. (2012) *Media, Place and Mobility*, Basingstoke, UK: Palgrave Macmillan.

Moores, S. (2017) *Digital Orientations: Non-media-centric media studies and non-representational theories of practice*, New York: Peter Lang.

Morayef, S. (2016) 'Pharaonic Street Art: The challenge of translation', in M. Baker (ed.) *Translating Dissent: Voices from and with the Egyptian revolution*, London: Routledge, 194–207.

Morel, P. (dir.) (2004) *District 13*. IMDb entry: www.imdb.com/title/tt0414852.

Moreton-Robinson, A. M. (2003) 'I Still Call Australia Home: Indigenous belonging and place in a white postcolonising society', in S. Ahmed (ed.) *Uprootings/Regroundings: Questions of home and migration*, Oxford: Berg Publishing, 23–40.

Morgan, M. and D. Bennett. (2011) 'Hip-Hop and the Global Imprint of a Black Cultural Form', *Dædalus: The Journal of the American Academy of Arts & Sciences* 140(2): 176–196.

Morimoto, L. H. and B. Chin (2017) 'Reimagining the Imagined Community: Online media fandoms in the age of global convergence', in J. Gray, C. Sandvoss and C. L. Harrington (eds) *Fandom: Identities and communities in a mediated world*, second edition, New York: New York University Press, 174–188.

Morini, C. (2007) 'The Feminisation of Labour in Cognitive Capitalism', *Feminist Review* 87: 40–59.

Moritz, A. and J. Block (2016) 'Crowdfunding: A literature review and research directions', in D. Brüntje and O. Gajda (eds) *Crowdfunding in Europe*, Cham, Switzerland: Springer, 25–55.

Morozov, E. (2009) 'The Brave New World of Slacktivism', *Foreign Policy*. Available online: http://foreignpolicy.com/2009/05/19/the-brave-new-world-of-slacktivism/.

Morozov, E. (2011) *The Net Delusion: How not to liberate the world*, London and New York: Allen Lane.

Morozov, E. (2012) *The Net Delusion: The dark side of Internet freedom*, New York: PublicAffairs.

Morris, P. (ed.) (1994) *The Bakhtin Reader*, London: Edward Arnold.

Morrow, O., R. Hawkins and L. Kern (2015) 'Feminist Research in Online Spaces', *Gender, Place & Culture* 22(4): 526–543.

Mortensen, F. (1977) 'The Bourgeois Public Sphere: A Danish mass communications research project', in M. Berg, P. Hemanus and J. Ekecrantz (eds) *Current Theories in Scandinavian Mass Communications Research*, Grenaa, Denmark: GMT, 293–354.

Mortensen, M. (2015a) 'Conflictual Media Events, Eyewitness Images, and the Boston Marathon Bombing (2013)', *Journalism Practice* 9(4): 536–551.

Mortensen, M. (2015b) *Journalism and Eyewitness Images: Digital media, participation, and conflict*, New York: Routledge.

Mortensen, M., C. R. Andersen and G. With (eds) (2003) *Geometry of the Face: Photographic portraits*, Copenhagen: The National Museum of Photography.

Mortensen, M. and S. Lomborg (2019) 'The Mobile Archive of the Self: On the interplay between aesthetic and metric modes of communication', *International Journal of Communication* 13: 6006–6020.
Mortensen, M., C. Neumayer and T. Poell (eds) (2018) *Social Media Materialities and Protest: Critical reflections*, London: Routledge.
Mosco, V. (2014) *To the Cloud: Big data in a turbulent world*, New York: Paradigm Publishers.
Mosco, V. (2017) *Becoming Digital: Toward a post-internet society*, Bingley, UK: Emerald Publishing.
Mosco, V. (2019) *The Smart City in a Digital World*, Bingley, UK: Emerald Publishing.
Mosireen (2018) 'What Is This?'. Available online: https://858.ma.
Mossberger, K., C. Tolbert and R. S. McNeal (2007) *Digital Citizenship: The internet, society, and participation*, Cambridge, MA: The MIT Press.
Mouffe, C. (1992) 'Democratic Citizenship and Political Community', in C. Mouffe (ed.) *Dimensions of Radical Democracy: Pluralism, citizenship, community*, London: Verso, 225–239.
Mouffe, C. (2000a) *The Democratic Paradox*, London: Verso.
Mouffe, C. (2000b) 'Deliberative Democracy or Agonistic Pluralism', *Political Science Series* 72, Vienna: Institute for Advanced Studies, 1–17. Available online: www.ihs.ac.at/publications/pol/pw_72.pdf.
Mouffe, C. (2005) *On the Political*, London and New York: Routledge.
Mouffe, C. (2007) 'Artistic Activism and Agonistic Spaces', *Art & Research: A Journal of Ideas, Contexts and Methods* 1(2): 1–5.
Mouffe, C. (2018) *For a Left Populism*, London: Verso.
Mould, O. (2009) 'Parkour, the City, the Event', *Environment and Planning D: Society and Space* 27(4): 738–750.
Mudu, P. and S. Chattopadhyay (2017) *Migration, Squatting and Radical Autonomy: Resistance and destabilization of racist regulatory policies and b/ordering mechanisms*, London: Routledge.
Mufson, B. (2015) '600 Fake Ads Flood Paris to Protest Climate Talks', *Vice*, 2 December. Available online: https://creators.vice.com/en_us/article/ez5q3m/600-fake-ads-flood-paris-to-protest-climate-talks.
Mulholland, R. (2011) 'The Libyan Artists Driving Gaddaffi to the Wall', *The Guardian*, 5 June. Available online: www.theguardian.com/world/2011/jun/05/gaddafi-rebels-art-graffiti-benghazi.
Munck, R. (2013) 'The Precariat: A view from the South', *Third World Quarterly* 34(5): 747–762.
Muñoz Ramírez, G. (2008) *The Fire & the Word: A history of the Zapatista movement*, San Francisco, CA: City Lights Books.
Murakami Wood, D. (2009) 'The "Surveillance Society": Questions of history, place and culture', *European Journal of Criminology* 6(2): 179–194.
Murdock, G. (1999) 'Rights and Representations: Public discourse and cultural citizenship', in J. Gripsrud (ed.) *Television and Common Knowledge*, London: Routledge, 7–17.
Murdock, G. (2000) 'Digital Futures: European television in the age of convergence', in J. Wieten, G. Murdock and P. Dahlgren (eds) *Television Across Europe*, London: Sage, 35–57.
Murji, K. and J. Solomos (2015) 'Introduction: Situating the present', in K. Murji and J. Solomos (eds) *Theories of Race and Ethnicity: Contemporary debates and perspectives*, Cambridge: Cambridge University Press, 1–22.
Murray, D. C. (2015) 'Notes to Self: The visual culture of selfies in the age of social media', *Consumption Markets and Culture* 18: 490–516.
Murthy, D. (2018a) *Twitter: Social communication in the Twitter age*, second edition, Cambridge, UK and Malden, MA: Polity Press.
Murthy, D. (ed.) (2018b) 'Special Issue: Social media, activism, and organizations', *Social Media + Society* 4(1).
Muse, J. (2010) 'Flash Mobs and the Diffusion of Audience', *Theater* 40(3): 9–23.
Myer, L. and C. Kleck (2007) 'From Independent to Corporate: A political economic analysis of rap billboard toppers', *Popular Music and Society* 30(2): 137–148.
Myers West, S. (2018) 'Censored, Suspended, Shadowbanned: User interpretations of content moderation on social media platforms', *New Media & Society* 20(11): 4366–4383.
Myrick, J. G., A. E. Holton, I. Himelboim and B. Love (2015) '#Stupidcancer: Exploring a typology of social support and the role of emotional expression in a social media community', *Health Communication* 31(5): 596–605.
Naficy, H. (2001) *An Accented Cinema: Exilic and diasporic filmmaking*, Princeton, NJ: Princeton University Press.
Nagel, E. van der (2017a) 'From Usernames to Profiles: The development of pseudonymity in Internet communications', *Internet Histories* 1(4): 312–331.
Nagel, E. van der (2017b) 'Scan', *Journal of Media Arts Culture* 10(2). Available online: http://scan.net.au/scn/journal/vol10number2/Emily van-der-Nagel.html.
Nagel, E. van der and J. Frith (2015) 'Anonymity, Pseudonymity, and the Agency of Online Identity: Examining the social practices of r/Gonewild', *First Monday* 20(3).
Nagle, A. (2017) *Kill All Normies: Online culture wars from 4Chan and Tumblr to Trump and the alt-right*, Croydon, UK: Zero Books.

Naidoo, P. (2010) 'Subaltern Sexiness: From a politics of representation to a politics of difference', *African Studies* 69(3): 439–456.
Napoli, P. M. and M. Aslama (eds) (2010) *Communications Research in Action: Scholar–Activist collaborations for a democratic public sphere*, New York: Fordham University Press.
Napoli, P. M. and R. Caplan (2017) 'Why Media Companies Insist They're Not Media Companies, Why They're Wrong, and Why It Matters', *First Monday* 5(5).
Narayanan, A. (2013) 'What Happened to the Crypto Dream? Part 1', *IEEE Security and Privacy* 11(2): 75–76.
Nardi, B. and V. O'Day (1999) *Information Ecologies: Using technology with heart*, Boston, MA: The MIT Press.
Nash, K. and J. Corner (2016) 'Strategic Impact Documentary: Contexts of production and social intervention', *European Journal of Communication* 31(3): 227–242.
Nashmi, E. A. (2017) 'From Selfies to Media Events', *Digital Journalism* 6(1): 98–117.
National Advisory Council for Environmental Policy and Technology (2016) Environmental Protection Belongs to the Public: A vision for citizen science at EPA. Available online: www.epa.gov/faca/nacept-2016-report-environmental-protection-belongs-public-vision-citizen-science-epa.
Navas, E., O. Gallagher and x. burrough (2015) *The Routledge Companion to Remix Studies*, London and New York: Routledge.
Navas , E., O. Gallagher, and x. burrough (2018) *Keywords in Remix Studies*, London and New York: Routledge.
Negri, A. (1980) 'Domination and Sabotage', in S. Lotringer and C. Marazzi (eds) *Autonomia: Post-political politics*, New York: Semiotexte, 62–71.
Negus, K. (2002) 'The Cultural Work of Intermediaries and the Enduring Distance between Production and Consumption', *Cultural Studies* 16(4): 501–515.
Nelson, A., D. Cordova, A. S. Walters and E. Szecsy (2016) 'Storytelling for Empowerment for Latino Teens: Increasing HIV prevention knowledge and attitudes', *Journal of Adolescent Research* 31(2): 202–231.
Nelson, J. (2000) 'The Media Role in Building Disability Community', *Journal of Mass Media Ethics* 15(3): 180–193.
Neltz, A. (2016) *Minecraft as American Political Commentary*. Available online: https://kotaku.com/minecraft-as-american-political-commentary-1771749115.
Nesbitt, N. (2008) *Universal Emancipation: The Haitian Revolution and the radical Enlightenment*, Charlottesville, VA: University of Virginia Press.
Neumayer, C. and L. Rossi (2016) '15 Years of Protest and Media Technologies Scholarship: A sociotechnical timeline', *Social Media + Society* 2(3): 1–13.
Neumayer, C. and G. Stald (2014) 'The Mobile Phone in Street Protest: Texting, tweeting, tracking, and tracing', *Mobile Media & Communication* 2(2): 117–133.
Neumayer, C. and D. M. Struthers (2018) 'Social Media as Activists Archives', in M. Mortensen, C. Neumayer and T. Poell (eds) *Social Media Materialities and Protest: Critical reflections*, London: Routledge, 86–98.
Neumayer, C. and J. Svensson (2016) 'Activism and Radical Politics in the Digital Age: Towards a typology', *Convergence* 22(2): 131–146.
Neverson, N. and C. T. Adeyanju (2017) 'Worth a Thousand Words: Tasers, new media events, and narrative struggle', *Journalism Studies* 19(11): 1633–1651. Available online: https://doi.org/10.1080/1461670X.2017.1289113.
Newbold, C. (1995) 'The Media Effects Tradition', in O. Boyd-Barrett and C. Newbold (eds) *Approaches to Media: A reader*, London: Arnold, 118–124.
Newell, M. (dir.) (2010) Prince of Persia: The Sands of Time. IMDb entry: www.imdb.com/title/tt0473075.
Newman, N., W. H. Dutton and G. Blank (2012) 'Social Media in the Changing Ecology of News: The Fourth and Fifth Estates in Britain', *International Journal of Internet Science* 7(1): 6–22.
Ng, J. Q. and E. L. Han (2018) 'Slogans and Slurs, Misogyny and Nationalism: A case study of anti-Japanese sentiment by Chinese netizens in contentious social media comments', *International Journal of Communication* 12: 1988–2009.
Nguyen, T. (2014) Spirit of Life Episode 438. Available online: www.youtube.com/watch?v=gHfGmx8O0eE&t=171s.
Ní Mhurchú, A. (2014) *Ambiguous Citizenship in an Age of Global Migration*, Edinburgh: Edinburgh University Press.
Ní Mhurchú, A. (2016) 'Unfamiliar Acts of Citizenship: Enacting citizenship in vernacular music and language from the space of marginalised intergenerational migration', *Citizenship Studies* 20(2): 156–172.
Nicey, J. (2016) 'Semi-Professional Amateurs', in T. Witschge, C. W. Anderson, D. Domingo and A. Hermida (eds) *The Sage Handbook of Digital Journalism*, Thousand Oaks, CA: Sage, 222–235.
Nichols, B. (1980) *Newsreel: Documentary Filmmaking on the American Left, 1969–1974*, New York: Arno Press.
Nichols, B. (2001) *Introduction to Documentary*, Bloomington, IN: Indiana University Press.
Nichols, B. (2016) *Speaking Truth with Film: Evidence, ethics, politics in documentary*, Oakland, CA: University of California Press.
Nicholson, J. (2005) 'Flash! Mobs in the Age of Mobile Connectivity'. Available online: http://six.fibreculturejournal.org/fcj-030-flash-mobs-in-the-age-of-mobile-connectivity/.

Nicklin, H. and G. Buckenham (2016) 'A Bonfire'. Available online: www.hannahnicklin.com/portfolio-items/a-bonfire/.

Nicolini, D. (2017) 'Practice Theory as a Package of Theory, Method and Vocabulary: Affordances and limitations', in M. Jonas, B. Littig and A. Wroblewski (eds) *Methodological Reflections on Practice Oriented Theories*, Berlin: Springer, 19–34.

Nieborg, D. B. and S. van der Graaf (2008) 'The Mod Industries? The industrial logic of non-market game production', *European Journal of Cultural Studies* 11(2): 177–195. Available online: https://journals.sagepub.com/doi/10.1177/1367549407088331.

Nielsen, J. (2006) 'The 90-9-1 Rule for Participation Inequality in Social Media and Online Communities', Nielsen Norman Group, 8 October. Available online: www.nngroup.com/articles/participation-inequality/.

Nielsen, R. K. (2015) *Local Journalism: The decline of newspapers and the rise of digital media*, London: Reuters Institute for the Study of Journalism.

Nielsen, R. K. (2016) 'The Many Crises of Western Journalism: A comparative analysis of economic crises, professional crises, and crises of confidence', in J. C. Alexander, E. B. Breese, and M. Luengo (eds) *The Crisis of Journalism Reconsidered: Democratic culture, professional codes, digital future*, Cambridge: Cambridge University Press, 77–97.

Nieminen, H. (2009) 'Media in Crisis? Social, economic and epistemic dimensions', in N. Carpentier, P. Pruulmann-Vengerfeldt, R. Kilborn, T. Olsson, H. Nieminen, E. Sundin and K. Nordenstreng (eds) *Communicative Approaches to Politics and Ethics in Europe*, Tartu, Estonia: Tartu University Press, 31–43.

Niesyto, J. (2011) 'A Journey from Rough Consensus to Political Creativity: Insights from the English and German language Wikipedias', in G. Lovink and N. Tkacz (eds) *Critical Point of View: A Wikipedia reader*, Amsterdam: Ten Klein Groep, 139–158. Available online: www.networkcultures.org/publications.

Nigg, H. (ed.) (2017) *Rebel Video. Die Videobewegung der 1970er- und 1980er-Jahre*, London: Scheidegger & Spiess.

Nikunen, K. (2018) *Media Solidarities: Emotions, power and justice in the digital age*, London: Sage.

Nim, E. (2016) 'Nanodemonstrations as Media Events: Networked forms of the Russian protest movement', in M. Baker and B. B. Blaagaard (eds) *Citizen Media and Public Spaces: Diverse expressions of citizenship and dissent*, London and New York: Routledge, 96–117.

Nip, J. Y. and K. W. Fu (2016) 'Challenging Official Propaganda? Public opinion leaders on Sina Weibo', *The China Quarterly* 225: 122–144.

Nirmala, Y. (2015) 'The Role of Community Radio in Empowering Women in India', *Media Asia* 42(1–2): 41–46.

Nix, L., J. Servin and I. Vamos (dir.) (2014) The Yes Men are Revolting. IMDb entry: www.imdb.com/title/tt2531282/?ref_=nv_sr_1?ref_=nv_sr_1.

Noll, J. M. (2003) 'The Myth of Convergence', *International Journal on Media Management* 5(1): 12–13.

Nölleke, D. and A. M. Scheu (2018) 'Perceived Media Logic: A point of reference for mediatization', in C. Thimm, M. Anastasiadis and J. Einspänner-Pflock (eds) *Media Logic(s) Revisited: Modelling the interplay between media institutions, media technology and societal change*, London: Palgrave Macmillan, 195–216.

Nomai, A. (2008) Culture Jamming: Ideological struggle and the possibilities for social change, Unpublished PhD thesis, Austin, TX: University of Texas at Austin, USA. Available online: https://repositories.lib.utexas.edu/handle/2152/3971.

Noppe, N. (2011) 'Why We Should Talk about Commodifying Fan Work', *Transformative Works and Cultures* 8.

Nord, D. P. (2001) *Communities of Journalism: A history of American newspapers and their readers*, Champaign, IL: University of Illinois Press.

Norris, W. (2017) 'Digital Humanitarians: Citizen journalists on the virtual front line of natural and human-caused disasters', *Journalism Practice* 11(2–3): 213–228.

Nortier, J. and M. Dorleijn (2013) 'Multi-Ethnolects', in P. Bakker and Y. Matras (eds) *Contact Languages*, Boston, MA: De Gruyter Mouton, 229–272.

Nortier, J. and B. A. Svendsen (eds) (2015) *Language, Youth and Identity in the 21st Century: Linguistic practices across urban spaces*, Cambridge: Cambridge University Press.

Notes from Nowhere (ed.) (2003) *We Are Everywhere: The irresistible rise of global anticapitalism*, New York: Verso.

Nunes, M. (2017) 'Selfies, Self-Witnessing and the "Out-of-Place" Digital Citizen', in A. Kuntsman (ed.) *Selfie Citizenship*, Cham, Switzerland: Palgrave Macmillan, 109–116.

Nunes, R. (2007) 'Forward How? Forward Where? (Post-)Operaismo beyond the immaterial labour thesis', in E. Dowling, R. Nunes and B. Trott (eds) *Immaterial and Affective Labour: Explored*. Special issue of *Ephemera: Theory and Politics in Organization* 7(1): 178–202. Available online: www.ephemerajournal.org/contribution/forward-how-forward-where-i-post-operaismo-beyond-immaterial-labour-thesis.

Nunns, A. and N. Idle (2011) *Tweets from Tahrir*, New York: OR Books.

Nyers, P. (2003) 'Abject Cosmopolitanism: The politics of protection in the anti-deportation movement', *Third World Quarterly* 24: 1069–1093.

Nyers, P. (2008) 'In Solitary, in Solidarity: Detainees, hostages and contesting the anti-policy of detention', *European Journal of Cultural Studies* 11: 333–349.

O'Brien, R. (1992) *Global Financial Integration: The end of geography*, London: Pinter.

O'Connor, A. (1990) 'Between Culture and Organization: The radio studios of Cotopaxi, Ecuador', *Gazette* 46: 81–91.

O'Rourke, S. (2011) 'Empowering Protest through Social Media', *Proceedings of the 2nd International Cyber Resilience Conference*, Perth, Australia: Edith Cowan University, 47–54.

O'Sullivan, D. (2017) 'Myanmar: Video prompts investigation into police brutality against Rohingya', *CNN Edition*. Available online: https://edition.cnn.com/2017/01/02/world/myanmar-investigates-brutality-against-rohingya-muslims/index.html.

Obar, J. A. and S. Wildman (2015) 'Social Media Definition and the Governance Challenge: An introduction to the special issue', *Telecommunications Policy* 39(9): 745–750.

Occupy Portland (2011) 'Occupy Portland Oregon: GA structures and process proposal to change and mix meeting styles and types passes GA 12/14/2011'. Available online: http://interoccupy.net/blog/occga-structures-and-process-proposal-passed-portland/.

Occupy Wall Street (2011) 'Principles of Solidarity'. Available online: www.declarationproject.org/?p=487.

Odin, R. (ed.) (1995) *Le film de famille: Usage privé, usage public*, Paris: Mérridiens Klincksieck.

Ofcom (2012) *The Communications Market Report*, London: Ofcom.

Ogbar, J. (2007) *Hip-Hop Revolution: The culture and politics of rap*, Lawrence, KS: University Press of Kansas.

Ogola, G. (2015) 'Social Media as a Heteroglossic Discursive Space and Kenya's Emergent Alternative/Citizen Experiment', *African Journalism Studies* 36(4): 66–81.

Ohler, J. B. (2013) *Digital Storytelling in the Classroom: New media pathways to literacy, learning and creativity*, second edition, Thousand Oaks, CA: Corwin Press.

Ohmae, K. (1995) *The End of the Nation: The end of geography*, London: Pinter.

Okabe, D. and K. Ishida (2012) 'Making *Fujoshi* Identity Visible and Invisible', in M. Ito, D. Okabe and I. Tsuji (eds) *Fandom Unbound: Otaku culture in a connected world*, New Haven, CT: Yale University Press, 207–224.

Oldenburg, R. (1997) *The Great Good Place: Cafes, coffee shops, bookstores, bars, hair salons, and other hangouts at the heart of a community*, Boston, MA: Da Capo Press.

Oldenburg, R. (ed.) (2001) *Celebrating the Third Place: Inspiring stories about the great good places at the heart of our communities*, Boston, MA: Da Capo Press.

Oliver, K. (2004) 'Witnessing and Testimony', *Parallax* 10(1): 79–88.

Olivers, D. (2004) 'Counter Hegemonic Dispersions: The World Social Forum model', *Antipode* 36(2): 175–182.

Ollman, D., S. Price and C. Smith (dir.) (2003) The Yes Men. IMDb entry: www.imdb.com/title/tt0379593/?ref_=ttco_co_tt.

Olszanowski, M. (2014) 'Feminist Self-Imaging and Instagram: Tactics of circumventing sensorship', *Visual Communication Quarterly* 21: 83–95.

Omar, B. (2017) 'Online News Production, Consumption and Immediacy: The remediation perspective', *Jurnal Komunikasi Malaysian Journal of Communication* 33(3): 250–266.

Omoniyi, T. (2006) 'Hip-Hop through the World Englishes Lens: A response to globalization', *World Englishes* 25(2): 195–208.

Oomen, J. and L. Aroyo (2011) 'Crowdsourcing in the Cultural Heritage Domain: Opportunities and challenges', *5th International Conference on Communities and Technologies*, New York: ACM, 138–149.

Openspending (n.d.) 'Homepage'. Available online: https://openspending.org/.

Ordóñez, V. (2018) 'Direct Action', in B. Franks, N. Jun and L. Williams (eds) *Anarchism: A conceptual approach*, London: Routledge, 74–85.

Orgad, S. (2012) *Media Representation and the Global Imagination*, Cambridge: Polity Press.

Ortuzar, J. (2009) 'Parkour or L'art du deplacement', *The Drama Review* 53(3): 54–66.

Osterweil, M. (2005) 'Place-based Globalism: Theorizing the global justice movement', *Development* 48(2): 23–28.

Osterweil, M. (2013) 'The Italian Anomaly', in C. Flesher Fominaya and L. Cox (eds) *Understanding European Movements: New social movements, global justice struggles, anti-austerity protests*, London: Routledge, 33–46.

Östman, S. and R. Turtiainen (2016) 'From Research Ethics to Researching Ethics in an Online Specific Context', *Media and Communication* 4(4): 66–74.

Ostrom, E. (1990) *Governing the Commons: The evolution of institutions for collective action*, Cambridge: Cambridge University Press.

Ostrove, G. (2014) 'The Political Economy of Financially Successful Independent Artists', *Class, Race and Corporate Power* 2(1): 1–22.

Ottinger, G. (2010) 'Buckets of Resistance: Standards and the Effectiveness of Citizen Science', *Science, Technology, and Human Values* 35(2): 244–270.

Ottinger, G. (2017) 'Reconstructing or Reproducing? Scientific authority and models of change in two traditions of citizen science', in D. Tyfield, R. Lave, S. Randalls and C. Thorpe (eds) *The Routledge Handbook of the Political Economy of Science*, London and New York: Routledge, 351–363.

Overseas Development Institute (n.d.) 'Western Africa's Missing Fish: The impacts of illegal, unreported and unregulated fishing and under-reporting catches by foreign fleets'. Available online: www.odi.org/sites/odi.org.uk/files/resource-documents/10665.pdf.

Oware, M. (2018) *I Got Something to Say: Gender, race, and social consciousness in rap music*, Cham, Switzerland: Palgrave Macmillan.

Oxford English Dictionary (n.d.). 'Solidarity'. Available online: https://en.oxforddictionaries.com/definition/solidarity.

Oxford Living Dictionaries (2018) 'Citizen Media'. Available online: https://en.oxforddictionaries.com/definition/citizen_media.

Özdemir, C. (2017) 'What's Left of Turkey's Gezi Protest Movement?', *Deutsche Welle*, 31 May. Available online: www.dw.com/en/whats-left-of-turkeys-gezi-protest-movement/a-39049440.

Paffenholz, T. (2009) 'Understanding Peacebuilding Theory: Management, resolution and transformation', in The Life and Peace Institute (ed.) *Conflict Transformation: Three lenses in one frame. New Routes* 14 (2): 3–6.

Page, R. (2012) 'The Linguistics of Self-Branding and Micro-Celebrity in Twitter: The role of hashtags', *Discourse and Communication* 6(2): 181–201.

Paget, D. (1987) '"Verbatim Theatre": Oral history and documentary techniques', *New Theatre Quarterly* 3(12): 317–336.

Pain, P. (2017) 'Framing Citizen Activism: A comparative study of the CGNET Swara and MobileVoices projects', *Media Asia* 44(2): 107–120.

PalestineRemembered.com (2002) 'Mission Statement'. Available online: www.palestineremembered.com:80/MissionStatement.htm.

Palys, T. and C. Atchison (2008) *Research Decisions: Quantitative and qualitative perspectives*, Canada: Nelson College Indigenous.

Pan, Y. and E. Blevis (2011) 'A Survey of Crowdsourcing as a Means of Collaboration and the Implications of Crowdsourcing for Interaction Design', *International Conference on Collaboration Technologies and Systems*, Philadelphia, PA: IEEE, 397–403.

Pande, R. (2016) 'Squee from the Margins: Racial/cultural/ethnic identity in global media fandom', in P. Booth and L. Bennett (eds) *Seeing Fans: Representations of fandom in media and popular culture*, New York: Bloomsbury, 209–220.

Pantti, M. and K. Andén-Papadopoulos (2011) 'Transparency and Trustworthiness: Strategies for incorporating amateur photography into news discourse', in K. Andén-Papadopoulos and M. Pantti (eds) *Amateur Images and Global News*, Chicago, IL: University of Chicago Press, 97–112.

Pantti, M. and K. Andén-Papadopoulos (2012) *Amateur Images and Global News*, Bristol, UK: Intellect.

Pantti, M. and P. Bakker (2009) 'Misfortunes, Memories and Sunsets: Non-professional images in Dutch news media', *International Journal of Cultural Studies* 15(5): 471–489.

Papacharissi, Z. (2010) *A Private Sphere: Democracy in a digital age*, Malden, MA: Polity Press.

Papacharissi, Z. (2011a) *A Networked Self: Identity, community and culture on social network sites*, London and New York: Routledge.

Papacharissi, Z. (2011b) 'Conclusion. A networked self', in Z. Papacharissi (ed.) *A Networked Self: Identity, community, and culture on social network sites*, London and New York: Routledge, 304–318.

Papacharissi, Z. (2015a) *Affective Publics: Sentiment, technology and politics*, Oxford: Oxford University Press.

Papacharissi, Z. (2015b) 'Affective Publics and Structures of Storytelling: Sentiment, events and mediality', *Information, Communication and Society*: 1–18.

Papacharissi, Z. and M. De Fatima Oliveira (2012) 'Affective News and Networked Publics: The rhythms of news storytelling on #Egypt', *Journal of Communication* 62(2): 266–282.

Papadopoulos, D. (2011) 'Alter-ontologies: Towards a constituent politics in technoscience', *Social Studies of Science* 41(2): 177–201.

Paper Tiger Television. https://papertiger.org/.

Paré, D. and S. Smeltzer (2013) 'ICTs as a Catalyst for Social Justice? A capabilities perspective', in S. Ilcan (ed.) *Mobilities, Knowledge, and Social Justice*, Montreal: McGill-Queens University Press, 320–339.

Parikha, J. (2010) *Insect Media: An archaeology of animals and technology*, Minneapolis, MN: University of Minnesota Press.

Pariser, E. (2011) *The Filter Bubble: What the internet is hiding from you*, London: Penguin.

Parkour.com (n.d.) 'Homepage'. Available online: www.parkour.com.

Parreñas, R. (2005) 'Long Distance Intimacy: Class, gender and intergenerational relations between mothers and children in Filipino transnational families', *Global Networks* 5(4): 317–336.

Patomäki, H. and T. Teivainen (2004) 'The World Social Forum: An open space or a movement of movements?', *Theory, Culture & Society* 21(6): 145–154.

Pavlenko, A. and A. Blackledge (eds) (2004) *Negotiation of Identities in Multilingual Contexts*, Clevedon, UK: Cromwell Press.
Pearson, C. and F. Trevisan (2015) 'Disability Activism in the New Media Ecology: Campaigning strategies in the digital era', *Disability and Society* 30(6): 924–940.
Pedroni, M. (2015) '"Stumbling on the Heels of My Blog": Career, forms of capital, and strategies in the (sub) field of fashion blogging', *Fashion Theory* 19(2): 179–199.
Peeples, J. and B. Mitchell (2007) 'No Mobs – No Confusions – No Tumult: Networking civil disobedience', *Electronic Journal of Communication* 17(1–2). Available online: www.cios.org/EJCPUBLIC/017/1/01714.HTML.
Pegg, D. and R. Evans (2016) 'Controversial Snooping Technology "Used by at Least Seven Police Forces"', *The Guardian*, 10 October. Available online: www.theguardian.com/world/2016/oct/10/controversial-phone-snooping-technology-imsi-catcher-seven-police-forces.
PEN (2013) 'Chilling Effects: NSA surveillance drives U.S. writers to self-censor', New York: PEN American Center. Available online: www.pen.org/sites/default/files/Chilling%20Effects_PEN%20American.pdf.
Penney, J. (2016) 'Chilling Effects: Online surveillance and Wikipedia use', *Berkeley Technology Law Journal* 31(1): 117.
Penney, J. and C. Dadas (2014) '(Re)Tweeting in the Service of Protest: Digital composition and circulation in the Occupy Wall Street movement', *New Media & Society* 16(1): 74–90.
Pennycook, A. (2007) 'Language, Localization, and the Real: Hip-hop and the global spread of authenticity', *Journal of Language, Identity, and Education* 6(2): 101–115.
Pentley, J. (2015) 'Which Side Were They On? Rundfunk, Filmmacher und der Bergarbeiterstreik 1984/85', in K. Hoffmann and E. Wottrich (eds) *Protest – Film – Bewegung. Neue Wege im Dokumentarischen*, Munich: Text + Kritik, 93–106.
People's Republic of Stokes Croft (2010) 'No Tesco in Stokes Croft Campaign Update', PRSC.org.uk, 17 April. Available online: https://prsc.org.uk/no-tesco-in-stokes-croft-campaign-update/.
Pérez-González, L. (2016) 'The Politics of Affect in Activist Amateur Subtitling: A biopolitical perspective', in M. Baker and B. B. Blaagaard (eds) *Citizen Media and Public Spaces: Diverse Expressions of Citizenship and Dissent*, London and New York: Routledge, 118–135.
Perin, C. (2009) 'Hacker vs Cracker', *TechRepublic*, 17 April. Available online: www.techrepublic.com/blog/it-security/hacker-vs-cracker/.
Perkins, W. (1996) *Droppin' Science: Critical essays on rap music and hip-hop culture*, Philadelphia, PA: Temple University Press.
Perreault, G. P. and T. P. Vos (2016) 'The GamerGate Controversy and Journalistic Paradigm Maintenance', *Journalism* 19(4): 553–569.
Perry, H. (1950) 'The Living Newspaper', *The English Journal* 39(1): 11–15.
Perry, I. (2004) *Prophets of the Hood: Politics and poetics in hip hop*, Durham, NC: Duke University Press.
Perry, K. H. (2011) 'Ethics, Vulnerability, and Speakers of Other Languages: How university IRBs (do not) speak to research involving refugee participants', *Qualitative Inquiry* 17(10): 899–912.
Perucci, T. (2018) 'Irritational Aesthetics: Reality friction and indecidable theatre', *Theatre Journal* 70(4): 473–498.
Pesce, S. (2016) 'Short Shelf-Life Media: Ephemeral digital practices and the contemporary dream of permanence', in S. Pesce and P. Noto (eds) *The Politics of Ephemeral Digital Media: Permanence and obsolescence in paratexts*, New York: Routledge, 13–31.
Peteet, J. (1996) 'The Writing's on the Wall', *Cultural Anthropology* 11(2): 139–159.
Peters, J. D. (2011) 'Witnessing', in P. Frosh and A. Pinchevski (eds) *Media Witnessing*, Basingstoke, UK: Palgrave Macmillan, 23–41.
Peters, J. D. (2015) *The Marvellous Clouds: Towards a philosophy of elemental media*, Chicago, IL: University of Chicago Press.
Peters, K. and A. Seier (2009) 'Home Dance: Mediacy and aesthetics of the self on YouTube', in P. Snickars and P. Vonderau (eds) *The YouTube Reader*, Stockholm: National Library of Sweden, 187–203.
Peters, S. (2016) 'Performing Citizenship', in D. Elliot, M. Silverman and W. Bowman (eds) *Artistic Citizenship: Artistry, social responsibility, and ethical praxis*, New York: Oxford University Press, 469–479.
Petersen, S. M. (2008) 'Loser Generated Content: From participation to exploitation', *First Monday* 13(3).
Peterson, M. A. (2010) 'But It Is My Habit to Read the Times': Metaculture and practice in the reading of Indian newspapers', in B. Bräuchler and J. Postill (eds) *Theorising Media and Practice*, New York: Berghahn Books, 127–145.
Pettit, J., J. F. Salazar and A. G. Dagron (2009) 'Citizens' Media and Communication', *Development in Practice* 19(4–5): 443–452.
Pew (2015) 'Investigative Journalists and Digital Security', Pew Research Center in Association with Columbia University's Tow Center for Digital Journalism, 5 February. Available online: www.journalism.org/2015/02/05/investigative-journalists-and-digital-security/.
Pickard, V. (2006) 'United yet Autonomous: Indymedia and the struggle to sustain a radical democratic network', *Media, Culture & Society* 28(3): 315–336.

Pickard, V. (2013) 'Indymedia', in G. Ritzer (ed.) *The Wiley-Blackwell Encyclopedia of Globalization*, first edition, Hoboken, NJ: Blackwell, 1–3.
Pickard, V. and G. Yang (eds) (2017) *Media Activism in the Digital Age*, London and New York: Routledge.
Pickerill, J. and P. Chatterton (2006) 'Notes towards Autonomous Geographies: Creation, resistance and self-management as survival tactics', *Progress in Human Geography* 30(6): 730–746.
Peil, C. and S. Sparviero (2017) 'Media Convergence Meets Deconvergence', in S. Sparviero, C. Peil and G. Balbi (eds) *Media Convergence and Deconvergence*, Cham, Switzerland: Palgrave Macmillan, 3–30.
Pink, S. (2008) 'Sense and Sustainability: The case of the Slow City movement', *Local Environment* 13(2): 95–106.
Pink, S. (2012) *Situating Everyday Life: Practices and places*, London: Sage.
Pink, S. and K. Leder Mackley (2012) 'Video and a Sense of the Invisible: Approaching domestic energy consumption through the sensory home', *Sociological Research Online* 17(1): 1–19.
Pink, S. and K. Leder Mackley (2013) 'Saturated and Situated: Expanding the meaning of media in the routines of everyday life', *Media, Culture & Society* 35(6): 677–691.
Pink, S. and K. Leder Mackley (2016) 'Moving, Making and Atmosphere: Routines of home as sites for mundane improvisation', *Mobilities* 11(2): 171–187.
Pinxten, H. and C. Longman (2008) 'Culturele diversiteit: van een essentialistisch naar een kosmopolitisch perspectief', *Volkskunde* 109(3–4): 229–237.
Piscator, E. (1925) *Trotz Alledem!* [In Spite of Everything], directed by Erwin Piscator. Grosses Schauspielhaus, Berlin.
Pitch Interactive (n.d.) 'Out of Sight, Out of Mind Homepage'. Available online: http://drones.pitchinteractive.com/.
Pizzorno, A. (1978) 'Political Change and Collective Identity in Industrial Conflict', in C. Crouch and A. Pizzorno (eds) *The Resurgence of Class Conflict in Western Europe since 1968*, London: Macmillan, 277–298.
Plantin, J. C., C. Lagoze, P. N. Edwards and C. Sandvig (2016) 'Infrastructure Studies Meet Platform Studies in the Age of Google and Facebook', *New Media & Society* 20(1): 293–310.
Plantinga, C. (2013) 'The Affective Power of Movies', in Arthur Shimamura (ed.) *Psychocinematics: Exploring cognition at the movies*, Oxford: Oxford University Press, 94–111.
Plantinga, C. (2018) *Screen Stories: Emotion and the ethics of engagement*, Oxford: Oxford University Press.
Platon, S. and M. Deuze (2003) 'Indymedia Journalism: A radical way of making, selecting and sharing news?', *Journalism* 4(3): 336–355.
Pleyers, G. (2008) 'The World Social Forum, a Globalisation from Below?', *Societies without Borders* 3(1): 71–89.
Pleyers, G. (2010) *Alter-globalization: Becoming actors in the global age*, Cambridge: Polity Press.
Plows, A. (1998) 'Earth First! Defending mother earth, direct-style', in G. McKay (ed.) *DIY Culture: Party and protest in nineties Britain*, London: Verso, 152–172.
Plows, A. (2008) 'Towards an Analysis of the "Success" of UK Green Protests', *British Politics* 3(1): 92–109.
Poell, T., R. Abdullah, B. Rieder, R. Woltering and L. Zack (2016) 'Protest Leadership in the Age of Social Media', *Information, Communication and Society* 19(7): 994–1014.
Poell, T. and E. Borra (2012) 'Twitter, YouTube, and Flickr as Platforms of Alternative Journalism: The social media account of the 2010 Toronto G20 protests', *Journalism* 13(6): 695–713.
Poell, T., J. de Kloet and G. Zeng (2014) 'Will the Real Weibo Please Stand Up? Chinese online contention and actor-network theory', *Chinese Journal of Communication* 7(1): 1–18.
Poell, T. and J. van Dijck (2015) 'Social Media and Activist Communication', in C. Atton (ed.) *The Routledge Companion to Alternative and Community Media*, London: Routledge, 527–537.
Pohlhaus, G. (2012) 'Relational Knowing and Epistemic Injustice: Toward a theory of willful hermeneutical ignorance', *Hypatia: A Journal of Feminist Philosophy* 27(4): 715–735.
Polletta, F. (1998) '"It Was Like a Fever …": Narrative and identity in social protest', *Social Problems* 45(2): 137–159.
Polletta, F. (2002) *Freedom Is an Endless Meeting*, Chicago, IL: University of Chicago Press.
Polletta, F. (2006) *It Was Like a Fever: Storytelling in protest and politics*, Chicago, IL: University of Chicago Press.
Polletta, F. and K. Hoban (2016) 'Why Consensus?: Prefiguration in three activist eras', in *Rethinking Prefigurative Politics*, Special thematic section of *Journal of Social and Political Psychology* 4(1): 286–301.
Ponniah, T. (2005) 'Autonomy and Political Strategy: Building the other superpower', *International Journal of Urban and Regional Research* 29(2): 441–443.
Ponniah, T. (2008) 'The Meaning of the US Social Forum: A reply to Whitaker and Bello's debate on the open space', *Societies without Borders* 3(1): 187–195.
Ponzanesi, S. (2014) *The Postcolonial Cultural Industry: Icons, markets, mythologies*, Basingstoke, UK: Palgrave Macmillan.
Ponzanesi, S. (2016) 'Frontiers of the Political: "Closed Sea" and the cinema of discontent', in M. Baker and B. Blaagaard (eds) *Citizen Media and Public Spaces: Diverse expressions of citizenship and dissent*, London and New York: Routledge, 42–57.
Pool, I. de Sola (1983) *Technologies of Freedom*, Cambridge, MA: Harvard University Press.

Poor, N. (2014) 'Computer Game Modders' Motivations and Sense of Community: A mixed-methods approach', *New Media & Society* 16(8): 1249–1267. Available online: https://journals.sagepub.com/doi/abs/10.1177/1461444813504266.

Poremba, C. (2010) 'Discourse Engines for Art Mods', *Eludamos Journal for Computer Game Culture* 4(1): 41–56. Available online: www.eludamos.org/index.php/eludamos/article/view/vol4no1-4/155.

Poretski, L. and O. Arazy (2017) 'Placing a Value on Community's Co-creation: A study of video game "modding" community', *Proceedings of the 2017 ACM Conference on Computer Supported Cooperative Work and Social Computing*, New York: ACM, 480–491.

Porritt, J. (1997) 'Environmental Politics: The old and the new', *The Political Quarterly* 68: 62–73.

Porto, M. P. (2005) 'The Principle of Diversity in Journalism: Jornal Nacional and political deliberation in Brazil', *Brazilian Journalism Research* 1(1): 135–153.

Postill, J. (2008) 'Localizing the Internet beyond Communities and Networks', *New Media & Society* 10(3): 413–431.

Postill, J. (2010) 'Introduction: Theorising media and practice', in B. Bräuchler and J. Postill (eds) *Theorising Media and Practice*, New York: Berghahn Books, 1–32.

Postill, J. (2011) *Localizing the Internet: An anthropological account*, New York: Berghahn Books.

Postill, J. (2013) 'The Multilinearity of Protest: Understanding new social movements through their events, trends and routines'. Available online: http://johnpostill.com/2013/10/19/the-multilinearity-of-protest/.

Postill, J. (2014) 'Spain's Indignados and the Mediated Aesthetics of Nonviolence', in P. Werbner, K. Spellman-Poots and M. Webb (eds) *The Political Aesthetics of Global Protest: Beyond the Arab Spring*, Edinburgh: Edinburgh University Press, 341–367.

Postill, J. (2018) *The Rise of Nerd Politics: Digital activism and political change*, London: Pluto Press.

Postill, J. and S. Pink (2012) 'Social Media Ethnography: The digital researcher in a messy web', *Media International Australia* 145(2): 123–134.

Potter, W. J. (2011) 'Conceptualizing Mass Media Effects', *Journal of Communication* 61(5): 896–915.

Potts, L., J. Seitzinger, D. Jones and A. Harrison (2011) 'Tweeting Disaster: Hashtag constructions and collisions', *Proceedings of the 29th Annual International ACM SIGIR Conference on Research and Development in Information Retrieval*, New York: ACM, 235–40.

Pough, G. (2004) *Check It While I Wreck It: Black womanhood, hip-hop culture, and the public sphere*, Lebanon, NH: Northeastern University Press.

Power, C. (2016) 'The Integrity of Process: Is inner transition sufficient?', in *Rethinking Prefigurative Politics*, Special thematic section of *Journal of Social and Political Psychology* 4(1): 347–363.

Pozo, D. (2015), 'Countergaming's Porn Parodies, Hard Core and Soft', in M. Wysocki and E. W. Lauteria (eds) *Rated M for Mature: Sex and sexuality in video games*, New York: Bloomsbury Academic, 133–146.

Prashad, V. (2008) *The Darker Nations: A people's history of the Third World*, London and New York: New Press.

Prax, P. (2016) *Co-creative Game Design as Participatory Alternative Media*, Uppsala: Uppsala University Publications. Available online: http://uu.diva-portal.org/smash/record.jsf?pid=diva2%3A923235&dswid=9803.

Precarias a la Deriva (2004) *A la deriva: por los circuitos de la precariedad femenina*, Madrid: Traficantes de Sueños.

Prentoulis, M. and L. Thomassen (2013) 'Political Theory in the Square: Protest, representation and subjectification', *Contemporary Political Theory* 12(3): 166–184.

Presence, S. (2015) 'The Contemporary Landscape of Video-Activism in Britain', in E. Mazierska and L. Kristensen (eds) *Marxism and Film Activism: Screening alternative worlds*, New York and Oxford: Berghahn, 186–212.

Presence, S., M. Wayne and J. Newsinger (eds) (2020) *Contemporary Radical Film Culture: Networks, organisations and activists*, London and New York: Routledge.

Price, A. (2010) 'Local Voices: The regional press and user-generated content', in G. Monaghan and S. Tunney (eds) *Web Journalism: A new form of citizenship?*, Eastbourne, UK: Sussex Academic Press, 137–149.

Price, V. (1992) *Public Opinion*, Volume 4, London: Sage.

Proulx, S. (2005) 'Penser les usages des technologies de l'information et de la communication aujourd'hui: Enjeux – modèles – tendances', in L. Viera and N. Pinède-Wojciechowski (eds) *Enjeux et usages des TIC: Aspects sociaux et culturels*, Volume 1, Bordeaux, France: Presses Universitaire de Bordeaux, 7–20.

Proulx, S., L. Poissant and M. Sénécal (2006) *Communautés virtuelles: Penser et agir en réseau*, Laval, Quebec: Presses de l'Université Laval.

Pugh, S. (2005) *The Democratic Genre: Fan fiction in a literary context*, Bridgend, UK: Seren.

Putnam, R. (2000) *Bowling Alone: The collapse and revival of American community*, New York: Simon and Schuster.

Qiu, J. L. (2014) '"Power To the People!": Mobiles, migrants, and social movements in Asia', *International Journal of Communication* 8: 376–391.

Quijano, A. (2000) 'Coloniality of Power, Eurocentrism and Latin America', *Nepantla* 1(3): 533–580.

Rabinowitz, P. (1994) *They Must Be Represented: The politics of documentary*, London: Verso.

Radcliffe, D. (2012) *Here and Now: UK hyperlocal media today*, London: Nesta. Available online: www.nesta.org.uk/sites/default/files/here_and_now_uk_hyperlocal_media_today.pdf.

Radhhani, N. (2016) 'Language Still a Barrier in the Banlieu', *The Guardian*, 16 December.

Radical Film Network (RFN) (n.d.) Available online: https://radicalfilmnetwork.com.

Radway, J. A. (1984/1991) *Reading the Romance: Women, patriarchy, and popular literature*, Chapel Hill, NC: University of North Carolina Press.

Rainie, L. and B. Wellman (2012) *Networked: The new social operating system*, Cambridge, MA and London: The MIT Press.

Raley, R. (2009) *Tactical Media*, Minneapolis, MN: University of Minnesota Press.

Ramos Rodríguez, J. M. (2005) 'Indigenous Radio Stations in Mexico: A catalyst for social cohesion and cultural strength. The radio journal', *Studies in Broadcast and Audio Media* 3: 155–169.

Ramsbotham, O. (2010) *Transforming Violent Conflict: Radical disagreement, dialogue and survival*, London and New York: Routledge.

Rancière, J. (2004) *The Politics of Aesthetics: The distribution of the sensible*, London: Continuum.

RAND Corporation (2005) *Aptitude for Destruction: Case studies of organizational learning in five terrorist groups*, Volume 2, Santa Monica, CA: RAND Corporation.

Ranieri, M. and I. Bruni (2012) 'Mobile Storytelling and Informal Education in a Suburban Area: A qualitative study on the potential of digital narratives for young second-generation immigrants', *Learning, Media and Technology* 38(2): 217–235.

Ransom, L. (1996) Disability Magazine and Newsletter Editors: Perceptions of the disability press, community, advocacy, and mainstreaming diversity, Unpublished PhD thesis, College Park, MD: University of Maryland. Available online: https://drum.lib.umd.edu/bitstream/handle/1903/20184/1160354.pdf?sequence=1&isAllowed=y.

Rao, U. (2010) 'Embedded/Embedding Media Practices and Cultural Production', in B. Bräuchler and J. Postill (eds) *Theorising Media and Practice*, New York: Berghahn Books, 147–168.

Raspeani Skopjani, *Plostad Sloboda (City Square Freedom)*. Available online: www.youtube.com/user/PlostadSloboda/videos.

Ratto, M. and M. Boler (2014) 'Introduction', in M. Ratto and M. Boler (eds) *DIY Citizenship: Critical making and social media*, Boston, MA: The MIT Press, 1–22.

Rawls, J. (1972) *A Theory of Justice*, Oxford: Oxford University Press.

Rawls, J. (1999) *A Theory of Justice*, revised edition, Cambridge, MA: Harvard University Press.

Raymond, E. (1999) *The Cathedral and the Bazaar: Musings on Linux and Open Source by an accidental revolutionary*, Sebastopol, CA: O'Reilly Media.

Razsa, M. J. (2014) 'Beyond "Riot Porn": Protest video and the production of unruly subjects', *Ethnos* 79(4): 496–524.

Razsa, M. J. and A. Kurnik (2012) 'The Occupy Movement in Žižek's Hometown: Direct democracy and a politics of becoming', *American Ethnologist* 39(2): 238–258.

Reagle Jr., J. M. (2010) *Good Faith Collaboration: The culture of Wikipedia*, Cambridge, MA: The MIT Press.

Reagle Jr., J. M. (2011) 'The Argument Engine', in G. Lovink and N. Tkacz (eds) *Critical Point of View: A Wikipedia reader*, Amsterdam: Ten Klein Groep, 14–34. Available online: www.networkcultures.org/publications.

Reagle Jr., J. M. and L. Rhue (2011) 'Gender Bias in Wikipedia and Britannica', *International Journal of Communication* 5: 1138–1158.

Reckwitz, A. (2002) 'Toward a Theory of Social Practices: A development in culturalist theorizing', *European Journal of Social Theory* 5(2): 243–263.

Rediker, M. (2004) *Villains of All Nations: Atlantic pirates in the golden age*, Boston, MA: Beacon Press,.

Rediker, M. (2007) *The Slave Ship: A human history*, New York: Viking.

Reed, A. (2010) '"Don't Keep It to Yourself!" Digital storytelling with South African Youth', *International Journal of Media, Technology and Lifelong Learning* 6(2). Available online: http://seminar.net/index.php/home/75-current-issue/146-dont-keep-it-toyourself-digital-storytelling-with-south-african-youth.

Reestorff, C. M. (2014) 'Mediatised Affective Activism: The activist imaginary and the topless body in the Femen movement', *Convergence: The International Journal of Research into New Media Technologies* 20(4): 478–495.

Reich, H. (2006) 'Local Ownership in Conflict Transformation Projects: Partnership, participation or patronage?', Berghof Occasional Papers 27, Berlin: Berghof Foundation.

Reichert, R. (2012) 'Uploading Politics: Herausbildung einer zivilgesellschaftlichen Medienpolitik', in A. Dörner and L. Vogt (eds) *Unterhaltungskultur als politische Kultur. Vorstellungswelten zwischen Politik und Entertainment*, Bonn: Bundeszentrale für politische Bildung, 203–216.

Reilly, D. (2016) 'Quichael Collins'. Available online: www.youtube.com/watch?v=rp6DFsXPCN8.

Reilly, P. (2014) 'The "Battle of Stokes Croft" on YouTube: The development of an ethical stance for the study of online comments', *Sage Research Methods Cases*. Available online: http://dx.doi.org/10.4135/978144627305013509209.

Reilly, P. (2015) 'Every Little Helps? YouTube, sousveillance and the "anti-Tesco" riot in Bristol', *New Media & Society* 17(5): 755–771.

Reilly, P. and F. Trevisan (2016) 'Researching Protest on Facebook: Developing an ethical stance for the study of Northern Irish flag protest pages', *Information, Communication & Society* 19(3): 419–435.

Reis, L. A. (1995) 'The Rodney King Beating: Beyond fair use: A broadcaster's right to air copyrighted videotape as part of a newscast', *John Marshall Journal of Computer and Information Law* 13: 269–311.

Relph, E. (1976) *Place and Placelessness*, London: Pion.

Rennie, E. (2006) *Community Media: A global introduction*, Lanham, MD and Oxford: Rowman & Littlefield.

Res Publica (n.d.) 'Homepage'. Available online: https://web.archive.org/web/20110728100532/http://www.therespublica.org/.

Resnikoff, P. (2014) 'Why Major Labels Are the Best Thing that Happened to Artists', *Digital Music News*, 19 September. Available online: www.digitalmusicnews.com/2014/09/19/major-labels-best-thing-happened-artists/.

Rettberg, J. W. (2016) *Seeing Ourselves through Technology*, London: Palgrave Macmillan.

Rettberg, J. W. (2017) 'Biometric Citizens: Adapting our selfies to machine vision', in A. Kuntsman (ed.) *Selfie Citizenship*, Cham, Switzerland: Palgrave Macmillan, 89–96.

Reuters (2017) 'Why Google and Facebook Prove the Digital Ad Market Is a Duopoly', *Fortune*, 28 July. Available online: http://fortune.com/2017/07/28/google-facebook-digital-advertising/.

Reuters (2018) 'Facebook Not Protecting Content Moderators from Mental Trauma: Lawsuit', *Reuters.com*, 24 September. Available online: www.reuters.com/article/us-facebook-lawsuit/facebook-not-protecting-content-moderators-from-mental-trauma-lawsuit-idUSKCN1M423Q.

Reynolds, A. (2017) 'From Crowdsourcing to the Commons: Towards critical and meaningful digital collaboration in museums', *The International Journal of the Inclusive Museum* 10(2): 49–68.

Rheingold, H. (1993) *The Virtual Community: Homesteading on the electronic frontier*, Reading, MA: Addison-Wesley.

Rheingold, H. (2002) *Smart Mobs: The next social revolution*, Cambridge, MA: Basic Books.

Rheingold, H. (2008) 'Virtual Communities: Exchanging ideas through computer bulletin boards', *Journal of Virtual Worlds Research* 1(1): 1–5.

Rheingold, H. (2012) *Net Smart: How to thrive online*, Cambridge, MA and London: The MIT Press.

Rhizome (2017) *FloodNet: Electronic Disturbance Theater, 1998*. Available online: https://anthology.rhizome.org/floodnet.

Ribeiro, S. (2016) 'Developing Intercultural Awareness Using Digital Storytelling', *Language and Intercultural Communication* 16(1): 69–82.

Richardson, A. V. (2016) 'Bearing Witness While Black: Theorizing African American mobile journalism after Ferguson', *Digital Journalism* 5(6): 673–698.

Richardson, A.V. (2018) 'A Phenomenology of Black Lives Matter Journalist-Activists', in S. Eldridge II and B. Franklin (eds) *The Routledge Handbook of Developments in Digital Journalism Studies*, Abingdon, UK: Routledge, 387–400.

Richmond, O. (2005) *The Transformation of Peace*, London: Palgrave Macmillan.

Richmond, O. (2008) *Peace in International Relations*, London and New York: Routledge.

Ridge, M. (ed.) (2014) *Crowdsourcing Our Cultural Heritage*, Burlington, VT: Ashgate.

Riedlmayer, A. and S. Naron, (2009) 'From Yizkor Books to Weblogs: Genocide, grassroots documentation and new technologies', in J. Bastion and B. Alexander (eds) *Community Archives: The shaping of memory*, London: Facet Publishing, 151–163.

Rigoni, I. and E. Saitta (eds) (2012) *Mediating Cultural Diversity in a Globalized Public Space*, Basingstoke, UK: Palgrave Macmillan.

Riismandel, P. (2002) 'Radio by and for the Public: The death and resurrection of low-power radio', in M. Hilmes and J. Loviglio (eds) *Radio Reader: Essays in the cultural history of radio*, New York and London: Routledge, 423–451.

Riley, C. (2005) *Disability and the Media: Prescriptions for change*, Lebanon, NH: University Press of New England.

Rinke, E. M. and M. Röder (2011) 'The Arab Spring Media Ecologies, Communication Culture, and Temporal–Spatial Unfolding: Three components in a communication model of the Egyptian regime change', *International Journal of Communication* 5: 1273–1285.

Risam, R. and A. Koh (n.d.) *Postcolonial Digital Humanities*. Available online: http://dhpoco.org/.

Riseup (1999) 'Radical Servers'. Available online: https://riseup.net/en/security/resources/radical-servers.

Ritchin, F. (2013) *Bending the Frame: Photojournalism, documentary and the citizen*, New York: Aperture.

Ritzer, G. (2011) *Sociological Theory*, New York: McGraw-Hill.

Rizk, P. (2013) 'Interview with Philip Rizk by Shuruq Harb', *ArtTerritories*, 20 May. Available online: www.monabaker.org/2015/09/20/interview-with-philip-rizk-by-shuruq-harb/.

Robé, C. (2016) 'Criminalizing Dissent: Western state repression, video activism, and counter-summit protests', *Framework: The Journal of Cinema and Media*, 57(2): 161–188.

Roberts, A. (2011) 'The Wikileaks Illusion', *The Wilson Quarterly* 35(3): 16–21.

Roberts, J. (1998) *The Art of Interruption: Realism, photography, and the everyday*, Manchester: Manchester University Press.

Roberts, J. (2008) 'The Amateurs Retort', in R. Rugoff (ed.) *Amateurs*, San Francisco, CA: California College of the Arts, The Wattis Institute for Contemporary Art.

Roberts, J. and M. Koliska (2015) 'Selfies: Witnessing and participatory journalism with a point of view', *International Journal of Communications* 9: 1672–1685.

Roberts, S. T. (2014) Behind the Screen: The hidden digital labor of commercial content moderation, Unpublished PhD thesis, Urbana-Champaign, IL: University of Illinois at Urbana-Champaign.

Roberts, S. T. (2016) 'Commercial Content Moderation: Digital laborers' dirty work', in S. U. Noble and B. M. Tynes (eds) *The Intersectional Internet: Race, sex, class and culture online*, New York: Peter Lang, 147–160.

Roberts, S. T. (2017) 'Content Moderation', in L. A. Schintler and C. L. McNeely (eds) *Encyclopedia of Big Data*, Cham, Switzerland: Springer.

Roberts, S. T. (2018) 'Digital Detritus: "Error" and the logic of opacity in social media content moderation', *First Monday* 23(3–5).

Roberts, S. T. (2019) *Behind the Screen: Content moderation in the shadows of social media*, New Haven, CT: Yale University Press.

Robertson, A. (2013) 'Connecting in Crisis: "Old" and "new" media and the Arab Spring', *The International Journal of Press/Politics* 18(3): 325–341.

Robin, B. R. (2008) 'Digital Storytelling: A powerful technology tool for the 21st century classroom', *Theory into Practice* 47(3): 220–228.

Robinson, S. (2009) '"If You Had Been with Us": Mainstream press and citizen journalists jockey for authority over the collective memory of Hurricane Katrina', *New Media & Society* 11(5): 795–814.

Robinson, S. and M. L. Schwartz (2014) 'The Activist as Citizen Journalist', in E. Thorsen and S. Allan (eds) *Citizen Journalism: Global perspectives*, Volume 2, New York: Peter Lang, 372–390.

Rodino-Colocino, M. (2012) 'Participant Activism: Exploring a methodology for scholar-activists through lessons learned as a precarious labor organizer', *Communication, Culture & Critique* 5(4): 541–562.

Rodino-Colocino, M. (2016) 'Critical-Cultural Communication Activism Research Calls for Academic Solidarity', *International Journal of Communication* 10(10): 4017–4026.

Rodríguez, C. (2000) 'Civil Society and Citizens' Media: Peace architects for the new millennium', in K. Wilkins (ed.) *Redeveloping Communication for Social Change: Theory, practice, power*, Boulder, CO: Rowman & Littlefield, 147–160.

Rodríguez, C. (2001) *Fissures in the Mediascape: An international study of citizens' media*, Cresskill, NJ: Hampton Press.

Rodríguez, C. (2003) 'The Bishop and His Star: Citizens' communication in Southern Chile', in N. Couldry and J. Curran (eds) *Contesting Media Power: Alternative media in a networked world*, Boulder, CO: Rowman & Littlefield, 177–194.

Rodríguez, C. (2010a) 'Citizens' Media', in J. Downing (ed.) *The Encyclopedia of Social Movement Media*, Thousand Oaks, CA: Sage, 98–103.

Rodríguez, C. (2010b) 'Knowledges in Dialogue: A participatory evaluation study of citizens' radio stations in Magdalena Medio, Colombia', in C. Rodríguez, D. Kidd and L. Stein (eds) *Making Our Media: Global initiatives toward a democratic public sphere*, Volume 1, *Creating New Communication Spaces*, Cresskill, NJ: Hampton Press, 131–154.

Rodríguez, C. (2011) *Citizens' Media against Armed Conflict: Disrupting violence in Colombia*, Minneapolis, MN and London: University of Minnesota Press.

Rodríguez, C. (2017) 'Studying Media at the Margins: Learning from the field', in V. Pickard and G. Yang (eds) *Media Activism in the Digital Age*, New York: Routledge, 49–62.

Rodríguez, C., B. Ferron and K. Shamas (2014) 'Four Challenges in the Field of Alternative, Radical and Citizens' Media Research', *Media, Culture & Society* 36(2): 150–166.

Rodríguez, C., D . Kidd and L. Stein (eds) (2010) *Making Our Media: Global initiatives toward a democratic public sphere*, Volume 1, *Creating New Communication Spaces*, Cresskill, NJ: Hampton Press.

Roediger, D. R. (2017) 'Making Solidarity Uneasy: Cautions on a keyword from Black Lives Matter to the past', in *Class, Race, and Marxism*, London and New York: Verso, 157–188.

Roggeband, C. and B. Klandermans (2007) 'Introduction', in C. Roggeband and B. Klandermans (eds) *Handbook of Social Movements across Disciplines*, New York: Springer, 1–13.

Rohlinger, D. and C. Vaccaro (2013) 'Media and Social Movements', *The Wiley-Blackwell Encyclopedia of Social and Political Movements*, Malden, MA: Blackwell, 736–739.

Romanos, E. (2016) 'Immigrants as Brokers: Dialogical diffusion from Spanish Indignados to Occupy Wall Street', *Social Movement Studies* 15(3): 247–262.

Romanos, E. and I. Sádaba (2015) 'La evolución de los marcos (tecno) discursivos del movimiento 15M y sus consecuencias', *Empiria. Revista de metodología de ciencias sociales* 32: 15–36.

Roncagliolo, R. (1986) 'New Information Order in Latin America: A taxonomy of national communication policies', in J. Becker (ed.) *Communication and Domination*, Norwood, NJ: Ablex, 168–185.
Rone, J. (2014) 'Anonymous Bulgaria', *Internet Law and Politics. Proceedings of the 10th International Conference on Internet Law and Politics Universitat Oberta de Catalunya, Barcelona, 3–4 July, 2014*, Barcelona: Huygens Editorial.
Rone, J. (2016) 'The People Formerly Known as the Oligarchy: The co-option of citizen journalism', in M. Baker and B. Blaagaard (eds) *Citizen Media and Public Spaces: Diverse expressions of citizenship and dissent*, London: Routledge, 208–224.
Rone, J. (2017) 'Anyone Can Do It', *ARestlessArt.com*, 12 July. Available online: https://arestlessart.com/2017/07/12/anyone-can-do-it/.
Rone, J. (2018) 'Don't Worry, We Are from the Internet': The diffusion of protest against the anti-counterfeiting trade agreement in the age of austerity, Unpublished PhD thesis, Fiesole, Italy: European University Institute.
Rone, J. (2019) 'Fake Profiles, Trolls, and Digital Paranoia: Digital media practices in breaking the Indignados movement', *Social Movement Studies*.
Roos, J. and L. Oikonomakis (2014) 'They Don't Represent Us! The global resonance of the Real Democracy Movement from the Indignados to Occupy', in D. della Porta and A. Mattoni (eds) *Spreading Protest: Social movements in times of crisis*, Colchester, UK: ECPR Press, 117–136.
Rooted in Rights (n.d.) 'Homepage'. Available online: www.rootedinrights.org.
Rootes, C. and C. Saunders (2007) 'The Global Justice Movement in Great Britain', in D. della Porta (ed.) *Global Justice Movement: Cross-national and transnational perspectives*, London: Routledge, 128–156.
Rosa, H. (2013) *Social Acceleration: A new theory of modernity*, New York: Columbia University Press.
Rose, T. (2008) *The Hip Hop Wars: What we talk about when we talk about hip hop – And why it matters*, New York: BasicCivitas.
Rosen, G., A. Stamos, S. Chakrabarti, T. Lyons and R. Leathern (2018) 'Hard Questions: What is Facebook doing to protect election security?', *Facebook*, 29 March. Available online: https://newsroom.fb.com/news/2018/03/hard-questions-election-security/.
Rosen, J. (1999) 'The Action of the Idea: Public journalism in built form', in T. L. Glasser (ed.) *The Idea of Public Journalism*, New York: Guilford Press, 21–48.
Rosen, J. (2000) 'Questions and Answers about Public Journalism', *Journalism Studies* 1(4): 679–682.
Rosen, J. (2006) 'The People Formerly Known as the Audience', *PressThink*, 27 June. Available online: http://archive.pressthink.org/2006/06/27/ppl_frmr.html.
Rosen, J. (2010) 'The Afghanistan War Logs Released by Wikileaks, the World's First Stateless News Organization', *Pressthink*, 26 July. Available online: http://pressthink.org/2010/07/the-afghanistan-war-logs-released-by-wikileaks-the-worlds-first-stateless-news-organization/.
Rosenberg, E. (2016) 'Manhattan Subway Passage Becomes Emotional Outlet after Election', *The New York Times*, 10 November. Available online: http://nyti.ms/2uK33UN.
Rosenberg, E. (2017) 'Protest Grows "Out of Nowhere" at Kennedy Airport after Iraqis Are Detained', *The New York Times*, 28 January. Available online: www.nytimes.com/2017/01/28/nyregion/jfk-protests-trump-refugee-ban.html.
Rosenberg, M. (2018) 'How Trump Consultants Exploited the Facebook Data of Millions', *The New York Times*, 17 March. Available online: www.nytimes.com/2018/03/17/us/politics/cambridge-analytica-trump-campaign.html.
Roslyng, M. M. and B. B. Blaagaard (2018) 'Networking the Political: On the dynamic interrelations that create publics in the digital age', *International Journal of Cultural Studies* 21(2): 124–138.
Ross, A. (1989) *No Respect: Intellectuals and popular culture*, London and New York: Routledge.
Ross, A. (2012) 'In Search of the Lost Paycheck', in T. Scholz (ed.) *Digital Labor: The internet as playground and factory*, New York: Routledge, 13–32.
Roth, L. (2005) *Something New in the Air: The story of first peoples television in Canada*, Montreal: McGill-Queens University Press.
Rousseau, J. J. (1762/1992) *Du contrat social*, Paris: GF-Flammarion.
Routledge, P. (2004) 'Relational Ethics of Struggle', in D. Fuller and R. Kitchin (eds) *Radical Theory/Critical Praxis: Making a difference beyond the academy*, Vernon and Victoria, British Columbia: Praxis (e)Press, 79–91.
Roy, A. (1999) *The Cost of Living*, London: Flamingo.
Roy, A. (2011) *Walking with the Comrades*, London and New York: Penguin.
Rubdy, R. and S. L. McKay (2013) 'Foreign Workers' in Singapore: Conflicting discourses, language politics and the negotiation of immigrant identities', *International Journal of the Sociology of Language* 222: 157–185.
Rubinstein, D. and K. Sluis (2008) 'A Life More Photographic: Mapping the networked image', *Photographies* 1(1): 9–28.
Rucht, D. (2004) 'The Quadruple "A": Media strategies of protest movements since the 1960s', in W. van de Donk, B. D. Loader and D. Rucht (eds) *Cyberprotest: New media, citizens and social movements*, London: Routledge, 29–56.

Rupesinghe, K. (1995) 'Conflict Transformation', in K. Rupesinghe (ed.) *Conflict Transformation*, London: Macmillan, 65–92.
Russell, A. and N. Echchaibi (eds) (2009) *International Blogging: Identity, politics, and networked publics*, New York: Peter Lang.
Russell, A. and S. Waisbord (2017) 'News Flashpoints and the Snowden Revelations', in R. Kunelius, H. Heikkila, A. Russell and D. Yagodin (eds) *Journalism and the NSA Revelations: Privacy, security and the press*, London: Reuters Institute/I. B. Tauris, 69–90.
Russo, J. L. (2009) 'User-Penetrated Content: Fan video in the age of convergence', *Cinema Journal* 48(4): 125–130.
Russo, J. L. (2014) 'Textual Orientation: Queer female fandom online', in C. Carter, L. Steiner and L. McLaughlin (eds) *The Routledge Companion to Media and Gender*, London: Routledge, 450–460.
Russo, J. L. (2018) 'The Queer Politics of Femslash', in M. A. Click and S. Scott (eds) *The Routledge Companion to Media Fandom*, New York: Routledge, 155–164.
Rustomji, N. (2009) *The Garden and the Fore: Heaven and hell in Islamic culture*, New York: Columbia University Press.
Rygiel, K. (2010) *Globalizing Citizenship*, Vancouver: UBC Press.
Sadik, A. (2008) 'Digital Storytelling: A meaningful technology-integrated approach for engaged student learning', *Educational Technology Research and Development* 56(4): 487–506.
Sadler, N. (2017) 'Narrative and Interpretation on Twitter: Reading tweets by telling stories', *New Media & Society* 20(9): 3266–3282.
Safran, W. (1992) 'Language, Ideology and the State in French Nation-Building', *History of European Ideas* 15: 793–800.
Saha, A. (2017) *Race and the Cultural Industries*, Cambridge: Polity Press.
Said, E. (1978) *Orientalism*, London and New York: Routledge.
Said, E. (1994) *Representations of the Intellectual: The 1993 Reith Lecture*, London: Vintage.
Said, E. (1997) *Covering Islam: How the media and the experts determine how we see the rest of the world*, London and New York: Vintage.
Sajed, A. (2010) 'Postcolonial Strangers in a Cosmopolitan World: Hybridity and citizenship in the Franco-Maghrebian borderland', *Citizenship Studies* 14(4): 363–380.
Salazar, E. (2017) 'Hashtags 2.0: An annotated history of the hashtag and a window to its future', *Icono* 15(2): 16–54.
Salazar, J. F. (2007) 'Indigenous Peoples and the Cultural Constructions of Information and Communication Technology in Latin America', in L. E. Dyson, M. Hendriks and S. Grant (eds) *Indigenous People and Information Technology*, Hershey, PA: Idea Book Publishing, 14–26.
Salazar, J. F. (2010) 'Making Culture Visible: The mediated construction of a Mapuche nation in Chile', in C. Rodríguez, D. Kidd and L. Stein (eds) *Making Our Media: Global initiatives toward a democratic public sphere*, Volume 1, *Creating New Communication Spaces*, Cresskill, NJ: Hampton Press, 29–46.
Salen, K. and E. Zimmerman (2004) *Rules of Play: Game Design Fundamentals*, Cambridge, MA: The MIT Press.
Salerno, S. (ed.) (2014) *Direct Action and Sabotage: Three classic IWW pamphlets from the 1910s*, Chicago, IL: H. Kerr and Company.
Salisbury, M. and J. Pooley (2017) 'The #nofilter Self: The contest for authenticity among social networking sites, 2002–2016', *Social Sciences* 6(1): 10.
Sampedro Blanco, V. (2005) *Multitudes on Line*. Available online: www.nodo50.org/multitudesonline/cap9.pdf.
Sande, M. van de (2015) 'Fighting with Tools: Prefiguration and radical politics in the twenty-first century', *Rethinking Marxism* 27(2): 177–194.
Sanders, L. (2012) 'Reclaiming the City: Street art of the revolution', in S. Mehrez (ed.) *Translating Egypt's Revolution: The language of Tahrir*, Cairo: American University in Cairo Press, 143–182.
Sandoval, M. (2014) *From Corporate to Social Media: Critical perspectives on corporate social responsibility in media and communications industries*, London and New York: Routledge.
Sandvoss, C. (2005) *Fans: The mirror of consumption*, Cambridge: Polity Press.
Sandvoss, C., J. Gray and C. L. Harrington (2017) 'Introduction: Why still study fans?', in J. Gray, C. Sandvoss and C. L. Harrington (eds) *Fandom: Identities and communities in a mediated world*, second edition, New York: New York University Press, 1–26.
Sanger, L. (2004) 'Why *Wikipedia* Must Jettison its Anti-elitism', blog post at LarrySanger.org, 31 December. Available online: http://larrysanger.org/2004/12/why-wikipedia-must-jettison-its-anti-elitism.
Santos, B. de Sousa (2006) *The Rise of the Global Left: The World Social Forum and beyond*, London: Zed Books.
Saragas, A. (dir.) (2017) *Strike a Rock*, Elafos Productions, South Africa. IMDb entry: www.imdb.com/title/tt7199700.
Saunders, B., J. Kitzinger and C. Kitzinger (2015) 'Participant Anonymity in the Internet Age: From theory to practice', *Qualitative Research in Psychology* 12(2): 125–137.
Saunders, C. (2008) 'Double-edged Swords? Collective identity and solidarity in the environment movement', *British Journal of Sociology* 59: 227–253.
Saunders, D. (2007) *Direct Cinema: Observational documentary and the politics of the sixties*, London: Wallflower Press.

Sauter, T. (2014) '"What's on Your Mind?" Writing on Facebook as a tool for self-formation', *New Media & Society* 16(5): 823–839.

Savage, S. (2003) 'Flash Mobs Take Manhattan'. Available online: www.cheesebikini.com/?p=67.

Saward, M. (2015) 'Afterword: Sovereign and critical grammars', in S. M. Rai and J. Reinelt (eds) *The Grammar of Politics and Performance*, Abingdon, UK and New York: Routledge, 217–225.

Scannell, P. (1996) 'For a Phenomenology of Radio and Television', *Journal of Communication* 45(3): 4–19.

Scannell, P. (2014) *Television and the Meaning of Live*, Cambridge, UK and Malden, MA: Polity Press.

Scardaville, M. C. (2005) 'Accidental Activists: Fan activism in the soap opera community', *American Behavioral Scientist* 48(7): 881–901.

Schaefer, T. and T. Birkland (eds) (2006) *Encyclopedia of Media and Politics*, Washington, DC: CQ Press.

Schaffer, K. and S. Smith (2004) *Human Rights and Narrated Lives*, New York: Palgrave Macmillan.

Schatzki, T. R. (1996) *Social Practices: A Wittgensteinian approach to human activity and the social*, Cambridge: Cambridge University Press.

Schatzki, T. R. (2001) 'Introduction: Practice theory', in T. R. Schatzki, K. K. Cetina and E. von Savigny (eds) *The Practice Turn in Contemporary Theory*, London: Routledge, 1–14.

Schatzki, T. R. (2012) ' A Primer on Practices', in J. Higgs, R. Barnett, S. Billett, M. Hutchings and F. Trede (eds) *Practice-based Education*, Rotterdam: Sense Publishers, 13–26.

Schatzki, T. R., K. K. Cetina and E. von Savigny (eds) (2001) *The Practice Turn in Contemporary Theory*, London: Routledge.

Schechner, R. (2006) *Performance Studies: An introduction*, second edition, New York and Abingdon, UK: Routledge.

Schielke, S. and J. Winegar (2012) 'The Writing on the Walls of Egypt', *Middle East Report* 42(265).

Schilk, F. (2017) *Souveränität statt Komplexität. Wie das Querfront-Magazin ›Compact‹ die politische Legitimationskrise der Gegenwart bearbeitet*, Münster: Unrast.

Schleiner, A. M. (2002) 'Velvet-Strike: War times and reality games', OpenSorcery.net, 2 March. Available online: www.opensorcery.net/velvet-strike/about.html.

Schleiner, A. M. (2012) Ludic Mutation: The player's power to change the game, Unpublished PhD thesis, Amsterdam: University of Amsterdam. Available online: https://dare.uva.nl/search?identifier=a67537a6-61e7-4423-acbb-fd990cdddbb1.

Schlesinger, P. (2009) 'Cultural and Communications Policy and the Stateless Nation', *Catalan Journal of Communication and Cultural Studies* 1(1): 9–14. Available online: http://eprints.gla.ac.uk/8800/1/8800.pdf.

Schmiddie (2017) Religion and Revolution Mod. ModDB entry: www.moddb.com/games/civilization-iv-colonization/downloads/religion-and-revolution-27.

Schmidt, T. S. (2007) 'A Wiki for Whistle-Blowers', *Time* Magazine, 22nd January. Available online: http://content.time.com/time/printout/0,8816,1581189,00.html.

Schmitt, I. (2005) 'Germany Speaking? Rap and Kanak Attak and dominant discourses on language', in D. Hoerder, Y. Herbert and I Schmitt (eds) *Negotiating Transcultural Lives*, Toronto: University of Toronto Press, 213–234.

Schneider, A. (2004) *'Die Stars sind wir'. Heimkino als filmische Praxis*, Marburg: Schüren.

Scholz, T. (ed.) (2012a) *Digital Labour: The internet as playground and factory*, London and New York: Routledge.

Scholz, T. (2012b) 'Cheaper by the Dozen: An introduction to "crowdsourcing"', in x. burrough (ed.) *Net Works: Case studies in web art and design*, New York and London: Routledge, 47–54.

Schrag, Z. M. (2011) 'The Case against Ethics Review in the Social Sciences', *Research Ethics* 7(4): 120–131.

Schrock, A. R. (2016) 'Civic Hacking as Data Activism and Advocacy: A history from publicity to open government data', *New Media & Society* 18(4): 581–599.

Schroeder, R. (2014) 'Big Data and the Brave New World of Social Media Research', *Big Data & Society* 1(2): 1–11.

Schwartz, J. M. and T. Cook (2002) 'Archives, Records, and Power: The making of modern memory', *Archival Science* 2: 1–19.

Schwartz, R. and G. R. Halegoua (2015) 'The Spatial Self: Location-based identity performance on social media', *New Media & Society* 17(10): 1643–1660.

Schwarzmantel, J. (2015) *The Routledge Guidebook to Gramsci's Prison Notebooks*, Abingdon, UK: Routledge.

Scifo, S. (2014) 'Communication Rights as a Networking Reality: Community radio in Europe', in C. Padovani and A. Calabrese (eds) *Communication Rights and Social Justice: Global transformations in media and communication research*, London: Palgrave Macmillan, 164–179.

Scolari, C. (2009) 'Mapping Conversations about New Media: The theoretical field of digital communication', *New Media & Society* 11(6): 943–964.

SCOT-PEP (2015) 'The Swedish Model: A briefing'. Available online: www.scot-pep.org.uk/sites/default/files/reports/the_swedish_model_summary.pdf.

Scott, S. (2015) 'The Moral Economy of Crowdfunding and the Transformative Capacity of Fan-ancing', *New Media & Society* 17(2): 167–182.

Scribe Video Center. https://scribe.org/.

Scully-Blaker, R. (2018) 'Make Tents, Not War: Queer play as play-that-critiques', *DiGRA 2018*. Available online: www.digra.org/wp-content/uploads/digital-library/DIGRA_2018_Paper_15.pdf.

Seamon, D. (1979) *A Geography of the Lifeworld: Movement, rest and encounter*, New York: Routledge.

Seaver, N. (2017) 'Algorithms as Culture: Some tactics for the ethnography of algorithmic systems', *Big Data and Society* 4(2): 1–12.

Segerberg, A. and W. L. Bennett (2011) 'Social Media and the Organization of Collective Action: Using Twitter to explore the ecologies of two climate change protests', *The Communication Review* 14(3): 197–215.

Seidman, G. (2001) 'Guerrillas in Their Midst: Armed struggle in the South African anti-apartheid movement', *Mobilization: An International Quarterly* 6(2): 111–127.

Sekula, A. (1986) 'The Body and the Archive', *October* 39: 3–64.

Seltzer, E. and D. Mahmoudi (2012) 'Citizen Participation, Open Innovation, and Crowdsourcing: Challenges and opportunities for planning', *Journal of Planning Literature* 28(1): 3–18.

Sen, J. (2010) 'On Open Space: Explorations towards a vocabulary of a more open politics', *Antipode* 42(4): 994–1018.

Sen, J. and P. Waterman (eds) (2008) *World Social Forum: Challenging empires*, second edition, Montreal: Black Rose Books.

Senft, T. (2008) *Camgirls: Celebrity and community in the age of social networks*, New York: Peter Lang.

Senft, T. (2015) 'The Skin of the Selfie', in S. Zarinbal and S. Michalskaja (eds) *Ego Update: The future of digital identity*, Dusseldorf: NRW Forum, 134–160.

Sennett, R. (2006) *The Culture of the New Capitalism*, New Haven, CT and London: Yale University Press.

Seri, J. (dir.) (2004) Les *fils* du *vent*. IMDb entry: www.imdb.com/title/tt0345235.

Servaes, J. (1999) *Communication for Development: One world, multiple cultures*, Cresskill, NJ: Hampton Press.

Servaes, J., T. L. Jacobson, S. A. White (eds) (1996) *Participatory Communication for Social Change*, New Delhi: Sage.

Sewell, W. (2005) *Logics of History: Social theory and social transformation*, Chicago, IL: University of Chicago Press.

Seymour, R. (2013) 'Turkey's "Standing Man" Shows How Passive Resistance Can Shake a State', *The Guardian*, 18 June. Available online: www.theguardian.com/commentisfree/2013/jun/18/turkey-standing-man.

Sezen, D. and T. I. Sezen (2018) 'Press Start to Remember the Martyrs: On video games commemorating the 2016 coup attempt in Turkey', in F. Çiçekoğlu and Ö. Turan (eds) *The Dubious Case of a Failed Coup*, Singapore: Palgrave Macmillan, 123–139.

Sezen, T. I. (2009a) 'Raid Gaza – Elestirel ve rahatsiz edici bir Gazze oyunu', Tongucs.Blogspot.com. Available online: http://tongucs.blogspot.com/2009/01/raid-gaza-elestirel-ve-rahatsiz-edici.html.

Sezen, T. I. (2009b) 'Krizin Oteki Yuzu: Israil yanlisi Gazze oyunlari', Tongucs.Blogspot.com. Available online: http://tongucs.blogspot.com/2009/01/krizin-oteki-yuzu.html.

Sezen, T. I. (2013) 'Representation of Poverty in Digital Games', in V. Bernard and S. Oksay (eds) *IMAGES (II) – Images of the Poor*, Berlin: LIT Verlag, 47–61.

Sezen, T. I. and D. Sezen (2016) 'Designing and Playing to Protest: Looking back to Gezi Games', in B. Bostan (ed.) *Gamer Psychology and Behavior*, Cham, Switzerland: Springer International, 73–89.

Shalson, L. (2017) *Theatre and Protest*, Basingstoke, UK: Palgrave Macmillan.

Shantz, J. and J. Tomblin (2014) *Cyber Disobedience: Re://Presenting online anarchy*, Croydon, UK: Zero Books.

Shapiro, M. (2013) *Studies in Transdisciplinary Methods*, London: Routledge.

Shapiro, N., N. Zakariya and J. A. Roberts (2017) 'A Wary Alliance: From enumerating the environment to inviting apprehension', *Engaging Science, Technology, and Society* 3: 575–602.

Sharma, A. and S. K. Kashyap (2015) 'Community Radio: A community media module for empowerment of rural women', *Media Asia* 42(3–4): 180–191.

Sharma, S. (2013) 'Because the Night Belongs to Lovers: Occupying the time of precarity', *Communication and Critical/Cultural Studies* 11(1): 5–14.

Sharma, S. (2014) *In the Meantime: Temporality and cultural politics*, Durham, NC: Duke University Press.

Sharp, J. (2015) *Works of Game: On the Aesthetics of Games and Art*, Cambridge, MA: The MIT Press.

Shaw, A. and B. M. Hill (2014) 'Laboratories of Oligarchy? How the iron law extends to peer production', *Journal of Communication* 64(2): 215–238.

Shaw, B. (ed.) (1994) *Ragged Edge: The disability experience from the pages of the first fifteen years of the Disability Rag*, Louisville, KY: Avocado Press.

Shearer, E. and K. E. Matsa (2018) 'News Use across Social Media Platforms 2018', Pew Research Center: Journalism and media, 10 September. Available online: www.journalism.org/2018/09/10/news-use-across-social-media-platforms-2018/.

Shepard, A. (1994) 'The Gospel of Public Journalism', *American Journalism Review* September: 28–34.
Shepherd, S. and M. Wallis (2004) *Drama / Theatre / Performance*, London and New York: Routledge.
Sheridan, G. (1982) 'Introduction', in P. Cohen and C. Gardner (eds) *It Ain't Half Racist, Mum: Fighting racism in the media*, London: Comedia, 1–5.
Shields, P. (ed.) (2017) 'Jane Addams: Pioneer in American sociology, social work and public administration', in P. Shields (ed.) *Jane Addams: Progressive pioneer of peace, philosophy, sociology, social work and public administration*. Pioneers in Arts, Humanities, Science, Engineering, Practice Volume 10, Cham, Switzerland: Springer International, 43–67.
Shilton, K and S. Sayles (2016) '"We Aren't All Going to Be on the Same Page about Ethics": Ethical practices and challenges in research on digital and social media', *Conference Proceedings, 49th Hawaii International Conference on System Sciences (HICSS 2016), Kauai*, 1909–1918.
Shim, E. (2014) 'Selfies with Soldiers Is the Coolest New Thing in Crimea', mic.com, 3 March. Available online: www.mic.com/articles/84089/selfies-with-soldiers-is-the-coolest-new-thing-in-crimea.
Shirk, J., H. L. Ballard, C. C. Wilderman, T. Phillips, A. Wiggins, R. Jordan, et al. (2012) 'Public Participation in Scientific Research: A framework for deliberate design', *Ecology and Society* 17(2): 29–48. Available online: www.ecologyandsociety.org/vol17/iss2/art29/.
Shirky, C. (2008) *Here Comes Everybody: The power of organizing without organizations*, New York: Penguin.
Shirky, C. (2011) 'The Political Power of Social Media: Technology, the public sphere, and political change', *Foreign Affairs* 90(1): 28–41.
Shnayerson, M. and L. Goldstein (2003) 'At Least They Don't Do the Wave'. Available online: http://content.time.com/time/magazine/article/0,9171,474568,00.html.
Shohat, E. and R. Stam (2012) *Race in Translation: Culture wars around the postcolonial Atlantic*, New York: NYU Press.
Shome, R. (2016) 'When the Postcolonial Meets Media Studies', *Critical Studies in Media Communication* 33(3): 245–263.
Shotwell, A. (2016) *Against Purity: Living ethically in compromised times*, Minneapolis, MN: University of Minnesota Press.
Shoup, J. (2013) 'Rai: North Africa's music of the working class', in W. El Hamamsy and M. Soliman (eds) *Popular Culture in the Middle East and North Africa: A postcolonial outlook*, London a New York: Routledge, 46–63.
Shove, E. and M. Pantzar (2005) 'Consumers, Producers and Practices: Understanding the invention and reinvention of Nordic Walking', *Journal of Consumer Culture* 5(1): 43–64.
Shove, E., M. Pantzar and M. Watson (2012) *Dynamics of Social Practice: Everyday life and how it changes*, London: Sage.
Shove, E., M. Watson, M. Hand and J. Ingram (eds) (2007) *The Design of Everyday Life*, Oxford: Berg.
Shresthova, S. (2013) 'Bollywood Dance as Political Participation? On flash mobs, new media, and political potential', *Convergence* 19(3): 311–317.
Shresthova, S. (2016) 'Dance It, Film It, Share It', in D. Elliot, M. Silverman and W. Bowman (eds) *Artistic Citizenship: Artistry, social responsibility, and ethical praxis*, New York: Oxford University Press, 146–162.
Siapera, E. and M. Theodosiadis (2017) '(Digital) Activism at the Interstices: Anarchist and self-organizing movements in Greece', *tripleC* 15(2): 505–523.
Sicart, M. (2011) 'Against Procedurality', *Game Studies International Journal of Computer Game Research* 11(3). Available online: http://gamestudies.org/1103/articles/sicart_ap.
Sicart, M. (2014) *Play Matters*, Cambridge, MA: The MIT Press.
Sienkiewicz, M. (2014) 'Start Making Sense: A three-tier approach to citizen journalism', *Media, Culture & Society* 36(5): 691–701.
Sihvonen, T. (2011) *Players Unleashed!: Modding The Sims and the culture of gaming*, Amsterdam: Amsterdam University Press.
Siltanen, J., F. Klodawsky and C. Andrew (2015) '"This Is How I Want to Live My Life": An experiment in prefigurative feminist organizing for a more equitable and inclusive city', *Antipode* 47(1): 260–279.
Silva, B., F. Vegetti and L. Littvay (2017) 'The Elite Is up to Something: Exploring the relation between populism and belief in conspiracy theories', *Swiss Political Science Review* 23 (4): 423–443.
Silverstein, L. (dir.) (2011) Freerunner. IMDb entry: www.imdb.com/title/tt1579232.
Silverstone, R. (1995) 'Convergence Is a Dangerous Word', *Convergence* 1(1): 11–13.
Silverstone, R. (2002) 'Complicity and Collusion in the Mediation of Everyday Life', *New Literary History* 33(4): 761–780.
Silverstone, R. (2005) 'The Sociology of Mediation and Communication', in C. Calhoun, C. Rojek and B. Turner (eds) *Sage Handbook of Sociology*, London: Sage, 188–207.
Silverstone, R. (2007) *Media and Morality: On the rise of the mediapolis*, Cambridge: Polity Press.
Simons, S. (2012) 'Ich bin Neda. Zur Authentizität ästhetischer Brüche und ihrer sozialen Resonanz im Web 2.0', in A. Weixler (ed.) *Authentisches Erzählen. Produktion, Narration, Rezeption*, Berlin and Boston MA: Walter de Gruyter, 293–320.
Simpson, A. (2014) *Mohawk Interruptus: Political life across the borders of settler states*, Durham, NC: Duke University Press.
Simpson, A. P. and H. Druxes (eds) (2015) *Digital Media Strategies of the Far Right in Europe and the United States*, Lanham, MD: Lexington Books.

Simsek, B. (2012) Using Digital Storytelling as a Change Agent for Women's Participation in the Turkish Public Sphere, Unpublished PhD thesis, Brisbane: Queensland University of Technology. Available online: http://eprints.qut.edu.au/50894/1/Burcu_Simsek_Thesis.pdf.

Singer, J. B. (2015) 'Out of Bounds: Professional norms as boundary markers', in M. Carlson and S. C. Lewis (eds) *Boundaries of Journalism: Professionalism, practices and participation*, New York: Routledge, 21–36.

Singer, P. (2011) *Practical Ethics*, third edition, Cambridge: Cambridge University Press.

Sitrin, M. (2006) *Horizontalism: Voices of popular power in Argentina*, Oakland, CA: AK Press.

Sivanandan, A. (1990) *Communities of Resistance: Writing on black struggles for socialism*, London: Verso.

Sjøvaag, H. (2011) 'Amateur Images and Journalistic Authority', in K. Andén-Papadopoulos, and M. Pantti (eds) *Amateur Images and Global News*, Chicago, IL: University of Chicago Press, 79–95.

Skilling, G. (1989) *Samizdat and an Independent Society in Central and Eastern Europe*, Basingstoke, UK: Macmillan.

Sloan, L. and A. Quan-Haase (eds) (2017) *The Sage Handbook of Social Media Research Methods*, London: Sage.

Smeltzer, S. (2012) 'Asking Tough Questions: The ethics of studying activism in democratically restricted environments', *Social Movement Studies* 11(2): 255–271.

Smeltzer, S. and S. Cantillon (2015) 'Guest Editor's Introduction: Scholar-activist terrain in Canada and Ireland', *Studies in Social Justice* 9(1): 7–17.

Smith, A. (1989) 'The Public Interest', *Intermedia* 17(2): 10–24.

Smith, A. M. (1998) *Laclau and Mouffe: The radical democratic imaginary*, London and New York: Routledge.

Smith, C. (1997) 'Method in the Madness: Exploring the boundaries of identity in hip-hop performativity, social identities', *Journal for the Study of Race, Nation and Culture* 3(3): 345–374.

Smith, C. (2011) 'Egypt's Facebook Revolution: Wael Ghonim thanks the social network', *Huffpost*. Available online: www.huffingtonpost.com/2011/02/11/egypt-facebook-revolution-wael-ghonim_n_822078.html.

Smith, J. (2004) 'The World Social Forum and the Challenges of Global Democracy', *Global Networks: A Journal of Transnational Affairs* 4(4): 413–421.

Smith, J. (2008) *Social Movements for Global Democracy*, Baltimore, MD: Johns Hopkins University Press.

Smith, J. (2015) 'BBC Reports on Royal Family Celebrations'. Available online: www.youtube.com/watch?=107&v=9pnt_uT5YzQ.

Smith, J., M. Karides, M. Becker, D. Brunelle, C. Chase-Dunn et al. (2008) *Global Democracy and the World Social Forums*, Boulder, CO and London: Paradigm Publishers.

Smith, J., M. Karides, M. Becker, D. Brunelle, C. Chase-Dunn, D. Della Porta et al. (2014) *Global Democracy and the World Social Forums*, second edition, New York: Routledge.

Smith, L. T. (1999) *Decolonizing Methodologies: Research and indigenous peoples*, London: Zed Books.

Smith, M. (1995) *Engaging Characters: Fiction, emotion, and the cinema*, Oxford: Clarendon Press.

Smith, P. J. and E. Smythe (2009) 'Open Spaces, Open Sources: The World Social Forum and international communication rights in a digital world', *Information, Communication & Society* 12(6): 793–816.

Smith, R. (2003) *Stories of Peoplehood: The Politics and Morals of Political Membership*, Cambridge: Cambridge University Press.

Smith, S., M. Choueiti and K. Pieper (2016) *Inequality in 800 Popular Films: Examining portrayals of gender, race/ethnicity, LGBT, and disability from 2007–2015*, Los Angeles, CA: USC Annenberg School for Communication and Journalism. Available online: https://annenberg.usc.edu/sites/default/files/2017/04/10/MDSCI_Inequality_in_800_Films_FINAL.pdf.

Smythe, D. W. (1977) 'Communications: Blindspot of Western Marxism', *Canadian Journal of Political and Social Theory* 1(3): 1–27.

Smythe, D. W. (2009) 'On the Audience Commodity and Its Work', in M. G. Durham and D. M. Kellner (eds) *Media and Cultural Studies: Key works*, Malden, MA: Blackwell, 230–256.

Snickars, P. (2009) 'The Archival Cloud', in P. Snickars and P. Vonderau (eds) *The YouTube Reader*, London: Wallflower Press, 292–313.

Snickars, P. and P. Vonderau (eds) (2009) *The YouTube Reader*, Stockholm: National Library of Sweden.

Snickars, P. and P. Vonderau (eds) (2009) *The YouTube Reader*, London: Wallflower Press.

Snow, D. A. (2004) 'Framing Processes, Ideology, and Discursive Fields', in D. A. Snow, S. A. Soule and H. Kriesi (eds) *The Blackwell Companion to Social Movements*, Oxford: Blackwell, 380–412.

Snow, D. A. and R. D. Benford (1988) 'Ideology, Frame Resonance, and Participant Mobilization', *International Social Movement Research* 1: 197–217.

Snowdon, P. (2014) 'The Revolution Will Be Uploaded: Vernacular video and the Arab Spring', *Culture Unbound: Journal of Current Cultural Research*, 6: 401–429. Available online: www.cultureunbound.ep.liu.se/v6/a21/cu14v6a21.pdf.

Sobchack, V. (1992) *The Address of the Eye: A phenomenology of film experience*, Princeton, NJ: Princeton University Press.

Solanas, F. and O. Getino (dirs.) (1969) *The Hour of the Furnaces*. IMDb entry: www.imdb.com/title/tt0063084/.

Solanas, F. and O. Getino (2000) 'Towards a Third Cinema', in R. Stam and T. Miller (eds) *Film and Theory: An anthology*, Malden, MA: Blackwell, 298–316.

Somers, M. R. (2008) *Genealogies of Citizenship: Markets, statelessness, and the right to have rights*, Cambridge: Cambridge University Press.

Sonnevend, J. (2016) *Stories without Borders: The Berlin Wall and the making of a global iconic event*, Oxford: Oxford University Press.

Sonnevend, J. (ed.) (2018) 'Special Section: Media events', *Media, Culture & Society* 40(1): 110–157.

Sotamaa, O. (2010) 'When the Game Is Not Enough: Motivations and practices among computer game modding culture', Games and Culture 5(3): 239–255.

Soukup, C. (2006) 'Computer-mediated Communication as a Virtual Third Place: Building Oldenburg's great good places on the world wide web', *New Media & Society* 8(3): 421–440.

Soysal, Y. (1994) Limits of Citizenship: Migrants and postnational membership in Europe, Chicago, IL: University of Chicago Press.

Soysal, Y. (1997) 'Changing Parameters of Citizenship and Claims-Making: Organized Islam in European public spheres', *Theory and Society* 26: 509–527.

Sparviero, S., C. Peil and G. Balbi (eds) (2017) *Media Convergence and Deconvergence*, Cham, Switzerland: Palgrave Macmillan.

Spence, J. and P. Holland (eds) (1991) *Family Snaps: The meanings of domestic photography*, London: Virago.

Spigel, L. (2014) 'Housing Television: Architecture of the archive', in C. Robertson (ed.) *Media History and the Archive*, London: Routledge, 55–77.

Spivak, G. (1985) 'The Rani of Sirmur: An essay in reading the archives', *History and Theory* 24(3): 247–272.

Spivak, G. (1988) 'Can the Subaltern Speak?', in C. Nelson and L. Grossberg (eds) *Marxism and the Interpretation of Culture*, Chicago, IL: University of Illinois Press, 271–313.

Spivak, G. (2005) 'Scattered Speculations on the Subaltern and the Popular', *Postcolonial Studies* 8(4): 475–486.

Splichal, S. (2006) 'In Search of a Strong European Public Sphere: Some critical observations on conceptualizations of publicness and the (European) public sphere', *Media, Culture & Society* 28(5): 695–714.

Sprouse, M. (ed.) (1992) *Sabotage in the American Workplace: Anecdotes of dissatisfaction, mischief and revenge*, San Francisco, CA and Edinburgh, UK: Pressure Drop and AK Press.

Squire, V. (2009) *The Exclusionary Politics of Asylum*, Basingstoke, UK: Palgrave Macmillan.

Squires, C. R. (2002) 'Rethinking the Black Public Sphere: An alternative vocabulary for multiple public spheres', *Communication Theory* 12(4): 446–468.

Sreberny, A. (2005) '"Not Only, But Also": Mixedness and media', *Journal of Ethnic and Migration Studies* 31(3): 443–459.

Srinivasan, R. and A. Fish (2011) 'Revolutionary Tactics, Media Ecologies, and Repressive States', *Public Culture* 23(365): 505–510.

Srnicek, N. (2017) *Platform Capitalism*, Cambridge, UK and Malden, MA: Polity Press.

SSL Nagbot (2016) 'Feminist Hacking/Making: Exploring new gender horizons of possibility', *The Journal of Peer Production: New Perspectives on the Implications of Peer Production for Social Change* 8. Available online: http://peerproduction.net/issues/issue-8-feminism-and-unhacking/.

St. John, B., K. Johnson and S. Nah (2014) 'Patch.com: The challenge of connective community journalism in the digital sphere', *Journalism Practice* 8(2): 197–212.

Stahl, M. and L. Meier (2012) 'The Firm Foundation of Organizational Flexibility: The 360 contract in the digitalizing music industry', *Canadian Journal of Communication* 37: 441–458.

Stallman, R. (1987) 'GNU Manifesto', GNU.org. Available online: www.gnu.org/gnu/manifesto.en.html.

Stallman, R. (2002) *Free Software, Free Society: Selected essays of Richard Stallman*, Boston, MA: Free Software Society.

Stam, R. (1998) 'The Two Avant-gardes', in B. K. Grant and J. Sloniowski (eds) *Documenting the Documentary: Close readings of documentary film and video*, Detroit, MI: Wayne State University Press, 254–268.

Stanfill, M. (2019) *Exploiting Fandom: How the media industry seeks to manipulate fans*, Iowa City, IA: University of Iowa Press.

Stanley, J. and B. Steinhardt (2003) 'Bigger Monster, Weaker Chains: The growth of an American surveillance society', Washington, DC: Technology and Liberty Program, American Civil Liberties Union.

Stapleton, K. (1998) 'From the Margins to Mainstream: The political power of hip-hop', *Media, Culture & Society* 20: 219–234.

Stapleton, M. and T. Froese (2015) 'Is Collective Agency a Coherent Idea? Considerations from the enactive theory of agency', in C. Misselhorn (ed.) *Collective Agency and Cooperation in Natural and Artificial Systems*, Cham, Switzerland: Springer International, 219–236.

Stapleton, S. and S. Terrio (2010) 'Le Parkour: Urban street culture and the commoditization of male youth expression', *International Migration* 50(6): 18–27.

Statista (2019) 'Number of Monthly Active Twitter Users Worldwide from 1st Quarter 2010 to 1st Quarter 2019'. Available online: www.statista.com/statistics/282087/number-of-monthly-active-twitter-users/.

Stavrides, S. (2012) 'Squares in Movement', *South Atlantic Quarterly* 111(3): 585–596.

Stay Woke website (n.d.) Available online: www.staywoke.org/.

Steavenson, W. (2011) 'Revolution in Cairo: A graffiti story', *The New Yorker*, 27 July. Available online: www.newyorker.com/news/wendell-steavenson/revolution-in-cairo-a-graffiti-story.

Stebbins, R. A. (1992) *Amateurs, Professionals and Serious Leisure*, Montreal: McGill-Queen's University Press.

Stebbins, R. A. (2014) *Between Work and Leisure: The common ground of two separate worlds*, New Brunswick, NJ and London: Transaction Publishers.

Steen, B. van der, A. Katzeff and L. van Hoogenhuijze (eds) (2014) *The City Is Ours: Squatting and autonomous movements in Europe from the 1970s to the present*, Oakland, CA: PM Press.

Stein, L. and K. Busse (2017) 'Limit Play: Fan authorship between source text, intertext, and context', *Popular Communication* 7(4): 192–207.

Steinberg, M. (2019) *The Platform Economy: How Japan transformed the consumer internet*, Minneapolis, MN: University of Minnesota Press.

Steinberg, G. and B. Wertman (2018) 'Value Clash: Civil society, foreign funding, and national sovereignty', *Global Governance* 24: 1–10.

Stencyl LLC (2011) Stencyl. Available online: www.stencyl.com/.

Stephansen, H. C. (2013a) 'Connecting the Peripheries: Networks, place and scale in the World Social Forum process', *Journal of Postcolonial Writing* 49(5): 506–518.

Stephansen, H. C. (2013b) 'Starting from the Amazon: Communication, knowledge and politics of place in the World Social Forum', *Interface: A Journal for and about Social Movements* 5(1): 102–127.

Stephansen, H. C. (2016) 'Understanding Citizen Media as Practice: Agents, processes, publics', in M. Baker and B. Blaagaard (eds) *Citizen Media and Public Spaces: Diverse expressions of citizenship and dissent*, London: Routledge, 25–41.

Stephansen, H. C. (2017) 'Media Activism as Movement? Collective identity formation in the World Forum of Free Media', *Media and Communication* 5(3): 59–66.

Stephansen, H. C. (2019) 'Conceptualizing a Distributed, Multi-scalar Global Public Sphere through Activist Communication Practices in the World Social Forum', *Global Media and Communication* 15(3): 345–360.

Stephansen, H. C. and E. Treré (eds) (2019) *Citizen Media and Practice: Connections, currents, challenges*, London: Routledge.

Stephens, M. (2013) 'Gender and the GeoWeb: Divisions in the production of user-generated cartographic information', *GeoJournal* 78(6): 981–996.

Stevens, M. (2017) 'Firm behind "Fearless Girl" Statue Underpaid Women, U.S. Says', *The New York Times*, 6 October. Available online: www.nytimes.com/2017/10/06/business/fearless-girl-settlement.html.

Stevenson, N. (2009) 'Talking to Bowie Fans: Masculinity, ambivalence and cultural citizenship', *European Journal of Cultural Studies* 12(1): 79–98.

Steward, H. (1997) *The Ontology of Mind: Events, processes, and states*, Oxford: Oxford University Press.

Stimor (2006) FinnWars. ModDB entry: www.moddb.com/mods/finnwars.

Stoler, A. L. (2002) 'Colonial Archives and the Arts of Governance', *Archival Science* 2(1–2): 87–109.

Stone, L. K. (2015) 'Suffering Bodies and Scenes of Confrontation: The art and politics of representing structural violence', *Visual Anthropology Review* 31(2): 177–189.

Storey, J. (2003) *Inventing Popular Culture: From folklore to globalization*, Oxford: Blackwell.

Storey, J. (2012) *Cultural Theory and Popular Culture: An introduction*, sixth edition, Harlow, UK: Pearson Education Limited.

Strachan, R. (2007) 'Micro-Independent Record Labels in the UK', *European Journal of Cultural Studies* 10(2): 245–265.

Strangelove, M. (2010) *Watching YouTube: Extraordinary videos by ordinary people*, Toronto: University of Toronto Press.

Stringer, V. (2013) 'This Is What Democracy Looked Like', in J. Juris and A. Khasnabish (eds) *Insurgent Encounters: Transnational activism, ethnography, and the political*, Durham, NC: Duke University Press, 318–341.

Stromae (2013a) 'Bâtard', *Racine Carreé*, 16 August, Universal Music France.

Stromae (2013b) 'Racine Carreé Tour', Mosaert Website. Available online: https://mosaert.com/projects/racince-carree-tour.

Strömbäck, J. (2005) 'In Search of a Standard: Four models of democracy and their normative implications for journalism', *Journalism Studies* 6(3): 331–345.

Strömbäck, J. and F. Esser (2014) 'Mediatization of Politics: Towards a theoretical framework', in F. Esser and J. Strömbäck (eds) *Mediatization of Politics: Understanding the transformation of Western democracies*, Basingstoke, UK: Palgrave Macmillan, 3–28.

Stryker, S., T. J. Owens and R. W. White (2000) 'Social Psychology and Social Movements: Cloudy past and bright future', in S. Stryker, T. J. Owens and R. W. White (eds) *Self, Identity and Social Movements*, Minneapolis, MN: University of Minnesota Press, 1–17.

Stuart Fisher, A. (2011) 'Trauma, Authenticity, and the Limits of Verbatim', *Performance Research* 16(1): 112–122.
Styan, L. J. (1981) *Modern Drama in Theory and Practice*, Volume 3, *Expressionism and epic theatre*, Cambridge: Cambridge University Press.
Subirats, J. and C. Rendueles (2016) *Los (bienes) comunes. ¿Oportunidad o espejismo?*, Barcelona: Icaria.
Sullivan, J. (2012) 'A Tale of Two Microblogs in China', *Media, Culture & Society* 34(6): 773–783.
Sullivan, J. (2014) 'China's Weibo: Is faster different?', *New Media & Society* 16(1): 24–37.
Sullivan, M. (2001) *African-American Music as Rebellion: From slavesong to hip-hop*, Ithaca, NY: Cornell University Press.
Sullivan, S., A. Spicer and S. Böhm (2011) 'Becoming Global (Un)Civil Society: Counter-hegemonic struggle and the Indymedia network', *Globalizations* 8(5): 703–717.
Sumiala, J., M. Tikka, J. Huhtamäki and K. Valaskivi (2016) '#JeSuisCharlie: Towards a multi-method study of hybrid media events', *Media and Communication* 4(4): 97–108.
Sun, H. (2013) 'Sina Weibo of China: From a copycat to local uptake of a global technology assemblage', *International Journal of Sociotechnology and Knowledge Development* 5(4): 24–37.
Sun, W. (2002) *Leaving China: Media, migration, and transnational imagination*, Lanham, MD: Rowman & Littlefield.
Sundararajan, A. (2016) *The Sharing Economy: The end of employment and the rise of crowd-based capitalism*, Cambridge, MA: The MIT Press.
Sundén, J. (2003) *Material Virtualities: Approaching online textual embodiment*, New York: Peter Lang.
Sunstein, C. (2018) *#Republic: Divided democracy in the age of social media*, Princeton, NJ and Woodstock: Princeton University Press.
Suran, M. and D. K. Kilgo (2017) 'Freedom from the Press? How anonymous gatekeepers on Reddit covered the Boston Marathon bombing', *Journalism Studies* 18(8): 1035–1051.
Sussman, L. R. (2001) *Press Freedom in Our Genes*, Reston, VA: World Press Freedom Committee.
Sutko, D. M. and A. de Souza e Silva (2011) 'Location-Aware Mobile Media and Urban Sociability', *New Media & Society* 13(5): 807–823.
Suzor, N. (2018) 'Digital Constitutionalism: Using the rule of law to evaluate the legitimacy of governance by platforms', *Social Media + Society* 4(3): 1–11.
Suzor, N., T. Van Geelen and S. M. West (2018) 'Evaluating the Legitimacy of Platform Governance: A review of research and a shared research agenda', *International Communication Gazette* 80(4): 385–400.
Švelch, J. (2013) 'Say It with a Computer Game: Hobby computer culture and the non-entertainment uses of homebrew games in the 1980s Czechoslovakia', *Game Studies: The International Journal of Computer Game Research* 13(2). Available online: http://gamestudies.org/1302/articles/svelch.
Svensson, M. (2014) 'Voice, Power and Connectivity in China's Microblogosphere: Digital divides on SinaWeibo', *China Information* 28(2): 168–188.
Svensson, J., C. Neumayer, A. Banfield-Mumb and J. Schossböck (2014) 'Identity Negotiation in Activist Participation', *Communication, Culture & Critique* 8(1): 144–162.
Swatuk, L. A. and P. Vale (2016) '"A Better Life for All": Prefigurative and strategic politics in Southern Africa', in *Rethinking Prefigurative Politics*, Special thematic section of *Journal of Social and Political Psychology* 4(1): 332–346.
Swedenburg, T. (2004) 'Homies in the 'Hood: Rap's commodification of insubordination', in M. Forman and M. Neal (eds) *That's the Joint! The hip hop studies reader*, London: Routledge, 579–592.
Sweney, M. (2010) 'T-Mobile Flash Mob Wins TV Ad of the Year'. Available online: www.theguardian.com/media/2010/mar/11/tmobile-flashmob-ad-of-year.
Swidler, A. (2001) 'What Anchors Cultural Practices', in T. R. Schatzki, K. K. Cetina and E. von Savigny (eds) *The Practice Turn in Contemporary Theory*, London: Routledge, 83–101.
Syrian Archive (n.d.) 'Syrian Archive: Curating visual documentation for justice and human rights'. Available online: https://syrianarchive.org.
Szolucha, A. (2013b) 'Learning Consensus Decision-Making in Occupy: Uncertainty, responsibility, commitment', in P. G. Coy (ed.) *Research in Social Movements, Conflicts and Change*, Bingley, UK: Emerald Publishing, 205–233.
Tacchi, J. (2009) 'Finding a Voice: Participatory development in Southeastern Asia', in J. Hartley and K. McWilliam (eds) *Story Circle: Digital storytelling around the world*, Malden, MA: Wiley-Blackwell, 167–175.
Tacchi, J., D. Slater and G. N. Hearn (2003) *Ethnographic Action Research: A user's handbook*, New Delhi: UNESCO. Available online: http://eprints.qut.edu.au/archive/00004399/.
Tactical Tech (2018) Security in a Box [Key Project]. Available online: https://tacticaltech.org/themes/digital-security/.
Tadros, M. (2015) 'Contentious and Prefigurative Politics: Vigilante groups' struggle against sexual violence in Egypt (2011–2013)', *Development and Change* 46(6): 1345–1368.
Tagg, J. (1988) *The Burden of Representation: Essays on photographies and histories*, Houndmills, UK: Macmillan.

Taher, M. (2011) 'Egyptian Cinema in 2011 between Activism and Art', *Ahram Online* (December). Available online: http://english.ahram.org.eg/NewsContent/1/114/30336/Egypt/-January-Revolution-continues/Egyptian-Cinema-in--between-activism-and-art.aspx.

Tait, S. (2011) 'Bearing Witness, Journalism and Moral Responsibility', *Media, Culture & Society* 33(8): 1220–1235.

Tanchak, P. (2016) 'The Invisible Front: Russia, trolls, and the information war against Ukraine', in O. Bertelsen (ed.) *Revolution and War in Contemporary Ukraine: The challenge of change*, Stuttgart: Ibidem, 253–282.

Tanczer, L. M. (2015) 'Hacktivism and the Male-Only Stereotype', *New Media & Society* 18(8): 1599–1615.

Tandoc Jr., E., Z. W. Lim and R. Ling (2018) 'Defining "Fake News"', *Digital Journalism* 6(2): 137–153.

Tari, M. and I. Vani (2005) 'On the Life and Seeds of San Precario: Patron saint of precarious workers and lives', *Fibreculture Journal* 5.

Tarifa, A. (2012) 'Hip Hop as Empowerment: Voices in El Alto, Bolivia', *International Journal of Qualitative Studies in Education* 25(4): 397–415.

Tarlau, R. (2014) 'From a Language to a Theory of Resistance: Critical pedagogy, the limits of "framing", and social change', *Educational Theory* 64(4): 369–392.

Tarrow, S. G. (1994) *Power in Movement: Social movements and contentious politics*, Cambridge: Cambridge University Press.

Tascón, S. and T. Wils (2017) *Activist Film Festivals: Towards a political subject*, Chicago, IL: University of Chicago Press.

Taylor, A. (2016) 'In Aleppo's Misinformation War, a 7-Year-Old Girl Prompts a Fact Check', *The Washington Post*, 14 December. Available online: www.washingtonpost.com/news/worldviews/wp/2016/12/14/in-aleppos-misinformation-war-a-7-year-old-girl-prompts-a-fact-check/.

Taylor, C. (1990) 'Modes of Civil Society', *Public Culture* 3(1): 95–118.

Taylor, C. and V. Taylor (2007) 'Hip-Hop Is Now: An evolving youth culture', *Reclaiming Children and Youth* 15(4): 210–213.

Taylor, K.-Y. (2016) *From #BlackLivesMatter to Black Liberation*, Chicago, IL: Haymarket Books.

Taylor, P. (2003) 'Maestros or Misogynists? Gender and the social construction of hacking', in Y. Jewkes (ed.) *Dot.cons: Crime, deviance and identity on the internet*, Cullompton, UK: Willan, 126–146.

Taylor, P. A. (2005) 'From Hackers to Hacktivists: Speed bumps on the global superhighway?', *New Media & Society* 7(5): 625–646.

Taylor, S. and S. Luckman (eds) (2018) *The New Normal of Working Lives: Critical studies in contemporary work and employment*, Basingstoke, UK: Palgrave Macmillan.

Tazzioli, M. (2016) 'Border Displacements: Challenging the politics of rescue between Mare Nostrum and Triton', *Migration Studies* 4(1): 1–19.

Team Nike (2010) Team Nike Presented South Street (The Takeover). Available online: https://youtu.be/EQ-4SgQPTHM.

Tech N9ne (2008) 'Crybaby', *Killer*. Strange Music.

Tedjasukmana, C. (2018) 'Das Netzvideo als Flaschenpost. Sylvia Rivera und queere Filmgeschichte', in S. Deckel and D. Brunow (eds) *Queer Cinema*, Mainz: Ventil, 1–22.

Teivainen, T. (2004) 'The World Social Forum: Arena or actor?', in J. Sen, A. Anand, A. Escobar and P. Waterman (eds) *World Social Forum: Challenging empires*, New Delhi: The Viveka Foundation, 122–129.

ten Brink, J. (ed.) (2007) *Building Bridges: The cinema of Jean Rouch*, London: Wallflower Press.

Tenhunen, S. (2018) *A Village Goes Mobile: Telephony, mediation, and social change in rural India*, New York: Oxford University Press.

Terranova, T. (2000) 'Free Labor: Producing culture for the digital economy', *Social Text* 18(2): 33–58.

Terranova, T. (2004) *Network Culture: Politics for the information age*, London and Ann Arbor, MI: Pluto Press.

Terranova, T. and J. Donovan (2013) 'Occupy Social Networks: Paradoxes of corporate social media for networked social movements', in G. Lovink and M. Rasch (eds) *Unlike Us Reader: Social media monopolies and their alternatives*, Amsterdam: Institute of Network Cultures, 296–311.

Terras, M. (2016) 'Crowdsourcing in the Digital Humanities', in S. Schreibman, R. Siemens and J. Unsworth (eds) *A New Companion to Digital Humanities*, second edition, Chichester, UK: Wiley-Blackwell, 420–439.

The Blackout (2017) 'Official #BlackoutDay Masterpost', *Tumblr*, 6 March. Available online: https://tumblr.theblackout.org/post/114966275331/official-blackoutday-masterpost-created-march.

The Daily Moth (n.d.) 'Homepage'. Available online: www.thedailymoth.com.

The Gift of Refuge (n.d.) Facebook page: www.facebook.com/giftofrefuge.

The Guardian (2010) 'WikiLeaks Embassy Cables: The key points at a glance', *The Guardian*, 7 December. Available online: www.theguardian.com/world/2010/nov/29/wikileaks-embassy-cables-key-points.

The Guardian (2014) '"I Can't Breathe": Eric Garner put in chokehold by NYPD officer – video', *The Guardian*, 4 December. Available online: www.theguardian.com/us-news/video/2014/dec/04/i-cant-breathe-eric-garner-chokehold-death-video.

The Mighty (n.d.) 'Disability'. Available online: www.themighty.com/topic/disability.
The Online Citizen (2017) 'Migrant Worker Competition 2017', 30 November. Available online: www.theonlinecitizen.com/2017/11/30/migrant-worker-poetry-competition-2017-2/.
Theocaris, Y., W. Lowe, J. van Deth and G. García-Albacete (2014) 'Using Twitter to Mobilize Protest Action: Online mobilization patterns and action repertoires in the Occupy Wall Street, Indignados, and Aganaktismenoi movements', *Information, Communication and Society* 18(2): 202–220.
Thibault, V. (2013) *Parkour and the Art du Deplacement: Strength, dignity, community*, Montreal: Baraka Books.
Thiéblemont-Dollet, S. and A. Koukoutsaki-Monnier (2010) *Médias, dispositifs, médiation*, Nancy: Presses Universitaires de Nancy.
Thimm, C., M. Anastasiadis and J. Einspänner-Pflock (eds) (2018) *Media Logic(s) Revisited: Modelling the interplay between media institutions, media technology and societal change*, London: Palgrave Macmillan.
Thompson, E. (1963) *The Making of the English Working Class*, New York: Vintage.
Thompson, E. C. (2009) 'Mobile Phones, Communities and Social Networks among Foreign Workers in Singapore', *Global Networks* 9(3): 359–380.
Thompson, J. B. (2000) Political Scandal: Power and visibility in the media age, Cambridge: Polity Press.
Thompson, J. B. (2005) 'The New Visibility', *Theory, Culture & Society* 22: 31–51.
Thorburn, D. (2014) 'Social Media, Subjectivity, and Surveillance: Moving on from Occupy, the rise of live streaming video', *Communication and Critical/Cultural Studies* 11(1): 52–63.
Thorburn, D. and H. Jenkins (2004) 'Introduction: Toward an aesthetics of transition', in D. Thorburn and H. Jenkins (eds) *Rethinking Media Change: The aesthetics of transition*, Cambridge, MA: The MIT Press, 1–16.
Thoreau, E. (2006) 'Ouch!: An examination of the self-representation of disabled people on the internet', *Journal of Computer-Mediated Communication* 11(2): 442–468.
Thoreau, H. D. (1849) 'The Resistance to Civil Government', in E. Peabody (ed.) *Æsthetic Papers*, New York: G. P. Putnam, 189–211.
Thorpe, H. and N. Ahmed (2015) 'Youth, Action Sports and Political Agency in the Middle East: Lessons from a grassroots parkour group in Gaza', *International Review for the Sociology of Sport* 50(5): 678–704.
Thorpe, K. (2017) 'Aboriginal Community Archives: A case study in ethical community research', in A. Gilliland, S. McKemmish and A. Lau (eds) *Research in the Archival Multiverse*, Clayton, Victoria: Monash University Publishing, 900–934.
Thorsen, E. (2008) 'Journalistic Objectivity Redefined? Wikinews and the neutral point of view', *New Media & Society* 10(6): 935–954.
Thorsen, E. (2014) 'Live Reporting Terror: Remediating citizen crisis communication', in E. Thorsen and S. Allan (eds) *Citizen Journalism: Global perspectives*, Volume 2, New York: Peter Lang.
Thorsen, E. (2016) 'Citizen Participation and Remediation across Media', in J. Linaa Jensen, M. Mortensen and J. Ørmen (eds) *News across Media Production, Distribution and Consumption*, Abingdon, UK and New York: Routledge.
Thorsen, E. and S. Allan (2014a) 'Introduction', in E. Thorsen and S. Allan (eds) *Citizen Journalism: Global perspectives*, Volume 2, New York: Peter Lang, 1–13.
Thorsen, E. and S. Allan (eds) (2014b) *Citizen Journalism: Global perspectives*, Volume 2, New York: Peter Lang.
Thorsen, E., D. Jackson and A. Luce (2015) '"I Wouldn't Be a Victim When It Comes to Being Heard": Citizen journalism and civic inclusion', in H. Savigny, E. Thorsen and D. Jackson (eds) *Media, Margins and Civic Agency*, Basingstoke, UK: Palgrave Macmillan, 43–61.
Thorson, K., K. Driscoll, B. Ekdale, S. Edgerly, L. G. Thompson, A. Schrock, L. Swartz, E. K. Vraga and C. Wells (2013) 'YouTube, Twitter and the Occupy Movement: Connecting content and circulation practices', *Information, Communication & Society* 16(3): 421–451.
Thumim, N. (2012) *Self-Representation and Digital Culture*, London: Palgrave Macmillan.
Thurman, N., J.-C. Pascal and P. Bradshaw (2011) 'Can Big Media Do "Big Society"?: A critical case study of commercial, convergent hyperlocal news', *International Journal of Media and Cultural Politics* 8(2): 269–285.
Tickner, A. (2008) 'Aqui en el Ghetto: Hip-hop in Colombia, Cuba, and Mexico', *Latin American Politics and Society* 50(3): 121–146.
Tifentale, A. and L. Manovich, (2015) 'Selfiecity: Exploring photography and self-fashioning in social media', in D. Berry and M. Dieter (eds) *Postdigital Aesthetics: Art, computation and design*, London: Palgrave Macmillan, 109–122.
Tilly, C. (1997) 'A Primer on Citizenship', *Theory and Society* 26: 599–602.
Tilly, C. (2004) *Social Movements: 1768–2004*, London: Paradigm Publications.
Tilly, C. (2005) *Identities, Boundaries and Social Ties*, Boulder, CO: Paradigm Publishers.
Times Now Bureau (2018) 'Delhi Cops Thrash Biker for Riding without Helmet, Claim He Slapped Them First', *Times Now News*, 1 March. Available online: www.timesnownews.com/india/video/delhi-police-viral-video-assault-biker-thrashed-new-delhi-patel-nagar-biker-helmet/203754.

TIS (2007) 'J'ai besoin', *La Petite Voiture*, 31 December, Bandcamp. Available online: https://tisdasupaemceeee.bandcamp.com/track/jai-besoin-prod-kronos.

Titley, G. (2012) 'Budgetjam! A communications intervention in the political-economic crisis in Ireland', *Journalism* 14(2): 292–306.

T-Mobile (2009) *The T-Mobile Dance*. Available online: https://youtu.be/VQ3d3KigPQM.

Tocqueville, A. de (1835–1840/2003) *Democracy in America, and Two Essays on America*, translated by G. Bevan, London: Penguin Books.

Toffler, A. (1970) *Future Shock*, New York: Bantam Books.

Toffler, A. (1980) *The Third Wave*, New York: Bantam Books.

Tofighian, O. (2018a) 'Translator's Tale: A window to the mountains', in B. Boochani, *No Friend but the Mountains: Writing from Manus prison*, translated by O. Tofighian, Sydney: Picador.

Tofighian, O. (2018b) 'No Friend but the Mountains: Translator's reflections in B. Boochani, *No Friend but the Mountains: Writing from Manus prison*, translated by O. Tofighian, Sydney: Picador.

Tomlinson, J. (1999) *Globalization and Culture*, Cambridge: Polity Press.

Tong, S. and J. B. Walther (2011) 'Relational Maintenance and CMC', in K. B. Wright and L. M. Webb (eds) *Computer-mediated Communication in Personal Relationships*, New York: Peter Lang, 98–118.

Torchin, L. (2016) 'KONY 2012. Anatomy of a campaign video and a video campaign', in Y. Tzioumakis and C. Molloy (eds) *The Routledge Companion to Cinema and Politics*, London: Routledge, 123–134.

Torgerson, D. (2010) 'Policy Discourse and Public Spheres: The Habermas paradox', *Critical Policy Studies* 4(1): 1–17.

Toton, S. (2008) 'Cataloguing Knowledge: Gender, generative fandom, and the Battlestar Wiki', *Flow Journal* 1. Available online: www.flowjournal.org/2008/01/cataloging-knowledge-gender-generative-fandom-and-the-battlestar-wiki/.

Touraine, A. (1985) 'An Introduction to the Study of Social Movements', *Social Research* 52(4): 749–787.

Townsend, L. and C. Wallace (2016) Social Media Research: A guide to ethics. Available online: www.gla.ac.uk/media/media_487729_en.pdf.

Trautmann, W. (1912/2014) 'Direct Action and Sabotage', in S. Salerno (ed.) *Direct Action and Sabotage: Three classic IWW pamphlets from the 1910s*, Chicago, IL: H. Kerr and Company: 29–56.

Trejo Delarbre, R. (2015) 'Ser visibles para ser ciudadanos. Política y redes sociodigitales en América Latina', *Revista Latinoamericana de Ciencias de la Comunicación* 22: 56–69

Treré, E. (2011) *Social Movements and Alternative Media*: The "anomalous wave" movement and the ambivalences of the online protest ecology. Unpublished Doctoral dissertation, Udine, Italy: University of Udine. Available online: http://dspace-uniud.cilea.it/handle/10990/306.

Treré, E. (2012) 'Social Movements as Information Ecologies: Exploring the coevolution of multiple Internet technologies for activism', *International Journal of Communication* 6: 2359–2377.

Treré, E. (2015) 'The Struggle Within: Discord, conflict and paranoia in social media protest', in L. Dencik and O. Leistert (eds) *Critical Perspectives on Social Media and Protest: Between control and emancipation*, Lanham, MD: Rowman & Littlefield, 163–180.

Treré, E. (2016) 'The Dark Side of Digital Politics: Understanding the algorithmic manufacturing of consent and the hindering of online dissidence', *IDS Bulletin* 47(1). Available online: http://bulletin.ids.ac.uk/index.php/idsbo/article/view/41.

Treré, E. (2018) 'Nomads of Cyber-Urban Space: Media hybridity as resistance', in M. Mortensen, C. Neumayer and T. Poell (eds) *Social Media Materialities and Protest: Critical reflections*, London: Routledge, 42–55.

Treré, E. (2019) *Hybrid Media Activism: Ecologies, imaginaries, algorithms*, London: Routledge, 31–103.

Treré, E., S. Jeppesen and A. Mattoni (2017) 'Comparing Digital Protest Media Imaginaries: Anti-austerity movements in Greece, Italy and Spain', *tripleC* 15(2): 404–422.

Treré, E. and A. Mattoni (2016) 'Media Ecologies and Protest Movements: Main perspectives and key lessons', *Information, Communication and Society* 19(3): 290–306.

Trevisan, F. (2016) *Disability Rights Advocacy Online: Voice, empowerment and global connectivity*, New York: Routledge.

Trevisan, F. (2017) 'Crowd-sourced Advocacy: Promoting disability rights through online storytelling', *Public Relations Inquiry* 6(2): 191–208.

Trevisan, F. (2018) 'Connective Action Mechanisms in a Time of Political Turmoil: Virtual disability rights protest at Donald Trump's inauguration', *Australian Journal of Political Science* 53(1): 103–115.

Trillò, T. (2018) 'Can the Subaltern Tweet? Reflections on Twitter as a space of appearance and inequality in accessing visibility', *Studies on Home and Community Science* 11(2): 116–124.

Trott, C. D. (2016) 'Constructing Alternatives: Envisioning a critical psychology of prefigurative politics', in *Rethinking Prefigurative Politics*, Special thematic section of *Journal of Social and Political Psychology* 4(1): 266–285.

Trottier, D. and C. Fuchs (eds) (2014) *Social Media, Politics and the State: Protests, revolutions, riots, crime and policing in the age of Facebook, Twitter and YouTube*, London: Routledge.

Trottier, D. and D. Lyon (2012) 'Key Features of Social Media Surveillance', in C. Fuchs, K. Boersma, A. Albrechtslund and M. Sandoval (eds) *Internet and Surveillance: The challenges of Web 2.0 and social media*, Abingdon, UK: Routledge, 89–105.

Trouillot, M. (1995) *Silencing the Past: Power and the production of history*, Boston, MA: Beacon.

Trouillot, M. (2013) 'An Unthinkable History: The Haitian Revolution as a non-event', in A. Goldstein Sepinwall (ed.) *Haitian History: New perspectives*, New York: Routledge, 33–54.

Troullinou, P. (2016) Exploring the Seduced Surveilled Subject: The case of smartphone devices as means of seductive surveillance, Unpublished PhD thesis, Milton Keynes, UK: Open University.

Trumpy, A. J. (2008) 'Subject to Negotiation: The Mechanisms behind co-optation and corporate reform', *Social Problems* 55(4): 480–500.

Tsilimpounidi, M. (2012) 'Athens 2012: Performances in crisis, or what happens when a city goes soft?', *City: Analysis of Urban Trends, Culture, Theory, Policy, Action* 16(5): 546–556.

Tsilimpounidi, M. (2017) *Sociology of Crisis: Visualising urban austerity*, London: Routledge.

Tsilimpounidi, M. and A. Walsh, (2010) 'Painting Human Rights: Mapping street art in Athens', *Journal of Arts and Communities* 2(2): 111–122.

Tsui, L. (2015) 'The Coming Colonization of Hong Kong Cyberspace: Government responses to the use of new technologies by the umbrella movement', *Chinese Journal of Communication* 8(4): 1–9.

Tuan Yi-Fu (1977) *Space and Place: The perspective of experience*, Minneapolis, MN: University of Minnesota Press.

Tudela, E. and C. Cattaneo (2016) 'Beyond Desencanto: The slow emergence of new social youth movements in Spain during the early 1980s', in K. Andresen and B. van der Steen (eds) *A European Youth Revolt*, London: Palgrave Macmillan, 127–141.

Tudor, M. (2018) *Desire Lines: Towards a queer digital media phenomenology*, Stockholm: Södertörn Doctoral Dissertations.

Tufekci, Z. (2013) 'Not This One: Social movements, the attention economy, and microcelebrity networked activism', *American Behavioral Scientist* 57(7): 848–870.

Tufekci, Z. (2014) 'Engineering the Public: Big data, surveillance and computational politics', *First Monday*, 19 July. Available online: http://firstmonday.org/article/view/4901/4097.

Tufekci, Z. (2017a) *Twitter and Tear Gas: The power and fragility of networked protest*, New Haven, CT and London: Yale University Press.

Tufekci, Z. (2017b) 'Facebook's Ad Scandal Isn't a "Fail", It's a Feature', *The New York Times, Sunday Review*, 23 September. Available online: www.nytimes.com/2017/09/23/opinion/sunday/facebook-ad-scandal.html.

Tufekci, Z. (2018) 'Why Zuckerberg's 15-Year Apology Tour Hasn't Fixed Facebook', *Wired*, 6 April. Available online: www.wired.com/story/why-zuckerberg-15-year-apology-tour-hasnt-fixed-facebook/.

Tufekci, Z. and C. Wilson (2012) 'Social Media and the Decision to Participate in Political Protest: Observations from Tahrir Square', *Journal of Communication* 62(2): 363–379.

Tufte, T. and P. Mefalopulos (2009) 'Participatory Communication: A practical guide', World Bank Working Paper 170, Washington, DC: The World Bank.

Tully, J. (2009) 'Lineages of Contemporary Imperialism', in D. Kelly (ed.) *Lineages of Empire: The historical roots of British imperial thought*, Oxford: Oxford University Press, 3–29.

Tully, J. (2014) *On Global Citizenship: James Tully in dialogue*, London: Bloomsbury.

Turkle, S. (1995) *Life on the Screen: Identity in the age of the Internet*, New York: Simon & Schuster.

Turner, B. S. (2001) 'The Erosion of Citizenship', *British Journal of Sociology* 52: 189–209.

Turner, J. (2017) Hyperlocal Community Media Audiences: An ethnographic study of local social media spaces and their place in everyday life, Unpublished PhD thesis, Birmingham, UK: Birmingham City University.

Turner, V. (1969/2008) *The Ritual Process: Structure and antistructure*, New Brunswick, NJ and London: Aldine Transaction.

Tushnet, R. (2017) 'Copyright Law, Fan Practices and the Rights of the Author', in J. Gray, S. Sandvoss, and C. L. Harrington (eds) *Fandom: Identities and communities in a mediated world*, New York: New York University Press, 77–90.

Twitter (2018) 'Tweet Data Dictionaries: Tweet metadata timeline'. Available online: https://developer.twitter.com/en/docs/tweets/data-dictionary/guides/tweet-timeline.

Tzioumakis, Y. and C. Molloy (eds) (2016) *The Routledge Companion to Cinema and Politics*, London and New York: Routledge.

Ugor, P. (2009) 'Small Media, Popular Culture, and New Youth Spaces in Nigeria', *Review of Education, Pedagogy, and Cultural Studies* 31(4): 387–408.

UK Parliament (2019) Disinformation and 'Fake News': Final report, February. Available online: https://publications.parliament.uk/pa/cm201719/cmselect/cmcumeds/1791/179102.htm.

Uldam, J. (2013) 'Activism and the Online Mediation Opportunity Structure: Attempts to impact global climate change policies?', *Policy & Internet* 5(1): 56–71.
Uldam, J. (2016) 'Corporate Management ofVisibility and the Fantasy of the Post-political: Social media and surveillance', *New Media & Society* 18(2): 201–219.
Uldam, J. (2018) 'Social Media Visibility: Challenges to activism', *Media, Culture & Society* 40(1): 41–58.
Uldam, J., and A. Vestergaard (eds) (2015) *Civic Engagement and Social Media: Political participation beyond protest*, London: Palgrave Macmillan.
ULEX (2008) *Autonomía y metrópolis. Del movimiento okupa a los centros sociales de segunda generación*, Malaga, Spain: Cedma.
Uludag, S. (2005a) *Oysters with the Missing Pearls*, Nicosia: IKME and Bilban.
Uludag, S. (2005b) *Cyprus: The untold stories*, Mannheim: Bibliopolis.
UN General Assembly (2013) 'Report of the Special Rapporteur on the promotion and protection of the right to freedom of expression', OHCHR.org. Available online: www.ohchr.org/Documents/HRBodies/HRCouncil/RegularSession/Session23/A.HRC.23.40_EN.pdf.
UNESCO MacBride Report (1980) 'Many Voices One World: Towards a new more just and efficient world information and communication order'. Available online: https://unesdoc.unesco.org/ark:/48223/pf0000040066.
United States Securities and Exchange Commission (2017) 'State Street Paying Penalties to Settle Fraud Charges and Disclosure Failures', 7 September. Available online: www.sec.gov/news/press-release/2017–159.
Unity Technologies (2005) Unity. Available online: https://unity3d.com/.
UrbanFreeflow.com (n.d.) 'Homepage'. Available online: https://web.archive.org/web/20171101105116/http://urbanfreeflow.com/.
Urla, J. (2012) *Reclaiming Basque: Language, nation, and cultural activism*, Reno, NV: University of Nevada Press.
Urla, J. and J. Helepololei (2014) 'The Ethnography of Resistance Then and Now: On thickness and activist engagement in the twenty-first century', *History and Anthropology* 25(4): 431–451.
US Social Forum Working Group wiki (n.d.) Available online: http://wiki.ussocialforum.net/wiki/USSF_Planning_Wiki.
Usher, N. (2011) 'Professional Journalists, Hands Off! Citizen journalism as civic responsibility', in R. W. McChesney and V. Pickard (eds) *Will the Last Reporter Please Turn out the Lights*, New York: The New Press, 264–276.
Usher, N. (2017) 'The Appropriation/Amplification Model of Citizen Journalism', *Journalism Practice* 11(2–3): 247–265.
Uzelman, S. (2011) 'Dangerous Practices: "Determinism of technique" in alternative media and their literature', *International Journal of Media and Cultural Politics* 7(1): 21–35.
Uzunoğlu, E. and S. Kip (2014) 'Brand Communication through Digital Influencers: Leveraging blogger engagement', *International Journal of Information Management* 34(5): 592–602.
Väätäjä, H. (2012) 'Readers' Motivations to Participate in Hyperlocal News Content Creation', *Proceedings of the 17th ACM International Conference on Supporting Group Work*, ACM, 309–312.
Vacha, J. E. (1986) 'The Federal Theater's Living Newspapers: New York's docudramas of the thirties', *New York History* 67(1): 66–88.
Vaidhyanathan, S. (2018a) *Antisocial Media: How Facebook disconnects us and undermines democracy*, Oxford: Oxford University Press.
Vaidhyanathan, S. (2018b) 'The Incomplete Political Economy of Social Media', in J. Burgess, T. Poell and A. Marwick (eds) *The Sage Handbook of Social Media*, London and New York: Sage, 213–230.
Valaskakis, G. (1983) 'Communication and Control in the Canadian North: The Inuit experience', in B. D. Singer (ed.) *Communications in Canadian Society*, Menlo Park, CA: Addison-Wesley, 237–247.
Valck, M. de (ed.) (2016) *Film Festivals*, London and New York: Routledge.
Valck, M. de, B. Kredell and S. Loist (eds) (2016) *Film Festivals: History, Theory, Method, Practice*. London: Routledge.
Valdez, A. (2010) 'There Will Be Oil'. Available online: www.youtube.com/watch?v=g3b-KMv8NO0.
Valdivia, A. N. (ed.) (2013) *The International Encyclopedia of Media Studies*, Malden, MA: Wiley-Blackwell.
Valle, F. S. (2014) 'Getting into the Mainstream: The digital/media strategies of a feminist coalition in Puerto Rico', in E. Thorsen and S. Allan (eds) *Citizen Journalism: Global perspectives*, Volume 2, New York: Peter Lang, 211–266.
Valve (2000) Counter-Strike. IGDB entry: www.igdb.com/games/counter-strike-1-6.
Vandevoordt, R. (2016) 'The Politics of Food and Hospitality: How Syrian refugees in Belgium create a home in hostile environments', *Journal of Refugee Studies* 30(4): 605–621.
Vargas, L. (1995) *Social Uses and Radio Practices: The use of participatory radio by ethnic minorities in Mexico*, Boulder, CO: Westview Press.
Vermeersch, P. (2014) 'The Romani Perspective: Experiences and acts of citizenship across Europe', in E. Isin and P. Nyers (eds) *The Routledge Handbook of Global Citizenship Studies*, London: Routledge, 477–486.

Vickery, G. and S. Wunsch-Vincent (2007) 'Participative Web and User-Created Content: Web 2.0, wikis and social networking', Organization for Economic Cooperation and Development, Paris: OECD. Available online: www.oecd.org/internet/ieconomy/38393115.pdf.

Vierkant, A. (2009) 'Surveillance vs. Sousveillance', *Hyperallergic*, 4 November. Available online: https://hyperallergic.com/867/surveillance-vs-sousveillance/.

Villard, F. and P. Yan-Sayegh (2013) 'Redefining a (Mono)cultural Nation: Political discourse against multiculturalism in contemporary France', in R. Taras (ed.) *Challenging Multiculturalism*, Edinburgh: Edinburgh University Press, 236–254.

Villeneuve, N. (2012) 'Fake Skype Encryption Software Cloaks Darkcomet Trojan', Trend Micro Malware Blog, 20 April. Available online: http://blog.trendmicro.com/fake-skype-encryption-software-cloaks-darkcomet-trojan/.

Vimercati, G. (2016) 'Blu v Bologna: New shades of grey in the street art debate', *The Guardian*, 17 March. Available online: www.theguardian.com/artanddesign/2016/mar/17/street-artist-blu-destroys-murals-in-bologna.

Vinthagen, S. (2009) 'Is the World Social Forum a Democratic Global Civil Society?', in J. Blau and M. Karides (eds) *The World and US Social Forums: A better world is possible and necessary*, Plymouth: Lexington Books, 131–148.

Virilio, P. (1977/1986) *Speed and Politics: An essay on dromology*, translated by M. Polizzotti, New York: Semiotext(e).

Virilio, P. (1995) 'Speed and Information: Cyberspace alarm!', translated by P. Riemens, *CTheory*. Available online: www.ctheory.net/articles.aspx?id=72.

Virno, P. (2004) *A Grammar of the Multitude: For an analysis of contemporary forms of life*, Los Angeles, CA and Cambridge, MA: Semiotext(e).

Virno, P. and M. Hardt (eds) (2006) *Radical Thought in Italy: A potential politics*, Minneapolis, MN: University of Minnesota Press.

Vis, F. (2009) 'Wikinews Reporting of Hurricane Katrina', in S. Allan and E. Thorsen (eds) *Citizen Journalism: Global perspectives*, Volume 1, New York: Peter Lang, 65–74.

Visualising Information for Advocacy (n.d.) 'Homepage'. Available online: https://visualisingadvocacy.org/.

Vitak, J., K. Shilton and Z. Ashktorab (2016) 'Beyond the Belmont Principles: Ethical challenges, practices, and beliefs in the online data research community', *Conference Proceedings, 19th ACM Conference on Computer-Supported Cooperative Work & Social Computing, San Francisco*, 941–953.

Vito, C. (2015) 'Who Said Hip-Hop Was Dead? The politics of hip-hop culture in Immortal Technique's lyrics', *International Journal of Cultural Studies* 18(4): 395–411.

Vittet-Philippe, P. and P. Crookes (1983) *Local Radio and Regional Development in Europe*, Media Monograph no. 7. The European Institute for the Media and The European Centre for Political Studies.

Vivienne, S. and J. Burgess (2013) 'The Remediation of the Personal Photograph and the Politics of Self-Representation in Digital Story-telling', *Journal of Material Culture* 18: 279–298.

Vladeck, D. C. (2018) 'Facebook, Cambridge Analytica, and the Regulator's Dilemma: Clueless or venal?', Harvard Law Review Blog, 4 April. Available online: https://blog.harvardlawreview.org/facebook-cambridge-analytica-and-the-regulators-dilemma-clueless-or-venal/.

Voniati, C., V. Doudaki and N. Carpentier (2018) 'Mapping Community Media Organisations in Cyprus: A methodological reflection', *Journal of Alternative and Community Media* 3(1): 17–32.

Vysotsky, S. (2015) 'The Anarchy Police: Militant anti-fascism as alternative policing practice', *Critical Criminology* 23(3): 235–253.

Wahl-Jorgensen, K. (2002a) 'The Normative-Economic Justification for Public Discourse: Letters to the editor as a "wide open" forum', *Journalism & Mass Communication Quarterly* 79(1): 121–133.

Wahl-Jorgensen, K. (2002b) 'Understanding the Conditions for Public Discourse: Four rules for selecting letters to the editor', *Journalism Studies* 3(1): 69–81.

Wahl-Jorgensen, K. (2007) *Journalists and the public: Newsroom culture, letters to the editor, and democracy*, Cresskill, NJ: Hampton Press.

Wahl-Jorgensen, K. (2015) 'Resisting Epistemologies of User-Generated Content? Cooptation, segregation and the boundaries of journalism', in M. Carlson and S. C. Lewis (eds) *Boundaries of Journalism: Professionalism, practices and participation*, New York: Routledge, 169–185.

Wahl-Jorgensen, K. (2018) *Emotions, media and politics*, Cambridge: Polity Press.

Wahl-Jorgensen, K., L. Bennett and G. Taylor (2017) 'The Normalisation of Surveillance: The invisibility of citizens and their rights in media coverage of the Snowden revelations', *International Journal of Communication* 11: 740–762.

Wainwright, L. S. (2008/2011) 'Happening (Art Event)', *Encyclopedia Britannica*. Available online: www.britannica.com/art/Happening.

Wajcman, J. (2015) *Pressed for Time: The acceleration of life in digital capitalism*, Chicago, IL: University of Chicago Press.

Waldstreicher, D. (1997) *In the Midst of Perpetual Fetes: The making of American nationalism, 1776–1820*, Chapel Hill, NC: University of North Carolina Press.

Wales, J. (2004) 'Wikipedia Founder Jimmy Wales Responds', *Slashdot*, 28 July. Available online: https://slashdot.org/story/04/07/28/1351230/wikipedia-founder-jimmy-wales-responds.

Walia, H. (2013) *Undoing Border Imperialism: Anarchist Interventions*, Oakland, CA: AK Press.

Walker, R. (2013) 'Fill/Flash/Memory: A history of flash mobs', *Text and Performance Quarterly* 33(2): 115–132.

Walker, S. (2015) 'Salutin' Putin: Inside a Russian troll house', *The Guardian*, 2 April. Available online: www.theguardian.com/world/2015/apr/02/putin-kremlin-inside-russian-troll-house.

Wall, D. (1999) *Earth First! and the Anti-roads Movement: Radical environmentalism and comparative social movements*, London: Routledge.

Wall, D. (2014) *The Commons in History: Culture, conflict, and ecology*, Cambridge, MA: The MIT Press.

Wall, M. (2009) 'Taming the Warblog', in S. Allan and E. Thorsen (eds) *Citizen Journalism: Global perspectives*, Volume 1, New York: Peter Lang, 33–42.

Wall, M. (2012) *Citizen Journalism: Valuable, Useless, or Dangerous?*, New York: International Debate Education Association.

Wall, M. (2015) 'Citizen Journalism: A retrospective on what we know, an agenda for what we don't', *Digital Journalism* 3(6): 797–813.

Wall, M. (2019) *Citizen Journalism: Practices, propaganda, pedagogy*, London: Routledge.

Wallerstein, I. (1989) '1968, Revolution in the World-System: Theses and queries', *Theory and Society* 18: 431–449.

Wallerstein, I. (2004) 'The Dilemmas of Open Space: The future of the WSF', *International Social Science Journal* 56(4): 629–637.

Walzer, M. (1991) 'The Idea of Civil Society', *Dissent* 39: 293–304.

Wang, O. (2014) *Legions of Boom: Filipino American mobile DJ crews in the San Francisco Bay area*, Durham, NC: Duke University Press.

Wanzo, R. (2015) 'African American Acafandom and Other Strangers: New genealogies of fan studies', *Transformative Works and Cultures* 20.

Wardle, C. (2017) 'Fake News: It's complicated', First Draft. Available online: https://medium.com/1st-draft/fake-news-its-complicated-d0f773766c79.

Wardle, C. and S. Dubberly (2014) 'Amateur Footage: A global study of user-generated content', *Tow Center for Digital Journalism*. Available online: https://towcenter.org/research/amateur-footage-a-global-study-of-user-generated-content/.

Wardrip-Fruin, N and N. Montfort (2003) *The New Media Reader*, Cambridge: The MIT Press.

Wargo, J. (2015) 'Spatial Stories and Nomadic Narrators: Affect, snapchat, and feelings embodied in youth mobile composing', *Journal of Language and Literacy Education* 11(1): 47–64

Wark, M. (2004) *A Hacker Manifesto*, Cambridge, MA: Harvard University Press.

Wark, M. (2008) *Fifty Years of Recuperation of the Situationist International*, Princeton, NJ: Princeton Architectural Press.

Wark, M. (2012) 'Considerations on a Hacker Manifesto', in T. Scholz (ed.) *Digital Labor: The internet as playground and factory*, London and New York: Routledge, 69–75.

Wark, M. (2019) *Capital Is Dead: Is this something worse?* London and New York: Verso.

Warner, K. J. (2018) '(Black Female) Fans Strike Back: The emergence of the Iris West Defense Squad', in M. A. Click and S. Scott (eds) *The Routledge Companion to Media Fandom*, New York: Routledge, 253–261.

Warner, M. (2002) 'Publics and Counter Publics', *Public Culture* 14(1): 49–90.

Warr, D., J. Waycott, M. Guillemin and S. Cox (2016) 'Ethical Issues in Visual Research and the Value of Stories from the Field', in D. Warr, M. Guillemin, S. Cox and J. Waycott (eds) *Ethics and Visual Research Methods*, New York: Palgrave Macmillan, 1–16.

Wasik, B. (2006) 'My Crowd; or Phase 5: A report from the invertor of the flash mob', *Harper's Magazine* (March): 56–66.

Wasik, B. (2011) '"Flash Robs": Trying to stop a meme gone wrong'. Available online: www.wired.com/2011/11/flash-robs/.

Wasserman, H. (2015) 'Marikana and the Media: Acts of citizenship and a faith in democracy-to-come', *Social Dynamics: A Journal of African Studies* 41: 375–386.

Waterman, P. (2005) 'Between a Political-Institutional Past and a Communicational-Networked Future? Reflections on the Third World Social Forum, 2003', in W. de Jong, M. Shaw and N. Stammers (eds) *Global Activism, Global Media*, London and Ann Arbor, MI: Pluto Press, 68–91.

Watkins, S. C. (2005) *Hip Hop Matters: Politics, pop culture, and the struggle for the soul of a movement*, Boston, MA: Beacon Press.

Watts, E. (2004) 'An Exploration of Spectacular Consumption: Gangsta rap as cultural commodity', in M. Forman and M. Neal (eds) *That's the Joint! The hip hop studies reader*, London: Routledge, 593–609.

Waugh, T. (ed.) (1984) *'Show Us Life': Towards a history and aesthetics of the committed documentary*, Metuchen, NJ and London: Scarecrow Press.

Waugh, T. (2011) *The Right to Play Oneself: Looking back on documentary film*, Minneapolis, MN and London: University of Minnesota Press.

Waugh, T. (2016) *The Conscience of Cinema: The works of Joris Ivens 1912–1989*, Amsterdam: Amsterdam University Press.

Wayne, M. (2016) 'The Dialectics of Third Cinema', in Y. Tzioumakis and C. Molloy (eds) *The Routledge Companion to Cinema and Politics*, London and New York: Routledge, 17–28.

We Are Social (2019) *We Are Social – Digital Report 2019*. Available online: https://wearesocial.com/global-digital-report-2019/.

We are the 99% (2013) 'Anonymous Asked: How can you claim to speak for 99% of people? We don't claim to speak for anyone, we merely present stories', *Tumblr*, 14 October. Available online: https://wearethe99percent.tumblr.com/post/64064302796/how-can-you-claim-to-speak-for-99-of-people.

Weaver, C. K. (2018) 'The Slow Conflation of Public Relations and Activism: Understanding trajectories in public relations theorising', in A. Adi (ed.) *Protest Public Relations: Communicating dissent and activism*, London: Routledge, 12–28.

Weber, C. (2011) *I Am an American: Filming the fear of difference*', Bristol, UK: Intellect.

Webster, F. (2006) *Theories of the Information Society*, third edition, London: Routledge.

Wei, H. (2015) '#Me, #I, #Selfie: The seduction of protest selfies', onlineopen.org, 30 August. www.onlineopen.org/download.php?id=478.

Weigel, M. (2016) 'Political Correctness: How the right invented a phantom enemy', *The Guardian*, 30 November. Available online: www.theguardian.com/us-news/2016/nov/30/political-correctness-how-the-right-invented-phantom-enemy-donald-trump.

Weintraub, J. (1997) 'The Theory and Politics of the Public/Private Distinction', in J. Weintraub and K. Kumar (eds) *Public and Private in Thought and Practice: Perspectives on a grand dichotomy*, Chicago, IL: University of Chicago Press, 1–7.

Weintraub, J. and K. Kumar (eds) (1997) *Public and Private in Thought and Practice: Perspectives on a grand dichotomy*, Chicago, IL: University of Chicago Press.

Weisskircher, M. (2019) 'New Technologies as a Neglected Social Movement Outcome: The case of activism against animal experimentation', *Sociological Perspectives* 62(1): 59–76.

Weisskircher, M. and L. Berntzen (2019) 'Remaining on the Streets: Anti-Islamic PEGIDA mobilization and its relationship to far-right party politics', in M. Caiani and O. Císař (eds) *Radical Right Movement Parties in Europe*, Abingdon, UK: Routledge, 114–130.

Weller, K. and K. E. Kinder-Kurlanda (2015) 'Uncovering the Challenges in Collection, Sharing and Documentation: The hidden data of social media research?', *Standards and Practices in Large-Scale Social Media Research*, International Conference on Web and Social Media, Oxford, 28–37.

Welles, O. (1973) 'F for Fake'. Available online: www.youtube.com/watch?v=gIVgUjj6RxU.

Wellman, B. (1999) 'Preface', in B. Wellman (ed.) *Networks in the Global Village: Life in contemporary communities*, Boulder, CO: Westview Press, xi–xxii.

Wellman, B., A. Quan-Haase, J. Boase, W. Chen, K. Hampton, I. Díaz and K. Miyata (2003) 'The Social Affordances of the Internet for Networked Individualism', *Journal of Computer-Mediated Communication* 8(3).

Wells-Wilbon, R., N. Jackson and J. Schiele (2010) 'Lessons from the Maafa: Rethinking the legacy of slain hip-hop icon Tupac Amaru Shakur', *Journal of Black Studies* 40(4): 504–526.

West, R., D. Precup and J. Pineau (2009) 'Completing Wikipedia's Hyperlink Structure through Dimensionality Reduction', *Proceedings of 18th International Conference on Information and Knowledge Management*, New York: ACM, 1097–1106.

Westmoreland, M. (2016) 'Street Scenes: The politics of revolutionary video in Egypt', *Visual Anthropology* 29(3): 243–262.

Weston, K. (2019) 'Midnight Traveler', *Film Comment* 55(5): 71–71.

Whitaker, C. (2008) 'The World Social Forum as Open Space', in J. Sen and P. Waterman (eds) *World Social Forum: Challenging empires*, second edition, Montreal: Black Rose Books, 81–93.

White, J. (2017) 'Twitter Strips Prominent White Supremacists and Neo-Nazis of Blue Ticks', *The Independent*, 16 November. Available online: www.independent.co.uk/life-style/gadgets-and-tech/news/twitter-white-supremacists-blue-ticks-neo-nazis-richard-spencer-tommy-robinson-a8057771.html.

White, M. (2010) 'Clicktivism Is Ruining Leftist Activism', *The Guardian*. Available online: www.guardian.co.uk/commentisfree/2010/aug/12/clicktivism-ruining-leftist-activism.

White, R. A. (1994) 'The "Public Sphere" as an Integrating Concept for Development Communication', in C. Hamelink and O. Linne (eds) *Mass Communication Research: On problems and policies*, Norwood, NJ: Ablex, 249–265.

Wiese, H. (2013) 'The Potential of Multi-ethnic Speakers/ Das Potential Multiethnischer Sprechergemeinschaften', in D. Arnulf (ed.) *Das Deutsch der Migranten*, Berlin and New York: de Gruyter.

Wiese, H. (2014) 'Voices of Linguistic Outrage: Standard language constructs and the discourse on new urban dialects', *Working Papers in Urban Language and Literacies* 120: 1–25.
WikiLeaks (2015) 'What Is WikiLeaks?'. Available online: https://wikileaks.org/What-is-Wikileaks.html.
Wikimedia Foundation (n.d.) 'Frequently Asked Questions'. Available online: https://wikimediafoundation.org/wiki/FAQ/en.
Wikimedia Foundation (2011) 'Wikipedia Editors Study: Results from the editor survey, April 2011'. Available online: https://meta.wikimedia.org/wiki/Research:Wikipedia_Editors_Survey_2011_April.
Wikipedia (n.d.) 'Main page'. Available online: https://en.wikipedia.org/w/index.php?title=Main_Pageandoldid=807996266.
Wikipedia (2019) 'Wikipedia', 25 February. Available online: https://en.wikipedia.org/wiki/Wikipedia.
Wikipedia Editor Activity Levels (2019) Available online: https://stats.wikimedia.org/EN/TablesWikipediaZZ.htm#editor_activity_levels.
Wikipedia User 'Thespian' (2007) 'Wikipedia: Can't break it', 25 June. Available online: https://en.wikipedia.org/w/index.php?title=Wikipedia:Can%27t_break_itandoldid=140566552.
Wikstrom, P. (2013) 'The Dynamics of Multi-Sided Media Content', in J. Hartley, J. Burgess and A. Bruns (eds) *A Companion to New Media Dynamics*, Malden, MA: Blackwell, 231–245.
Wiles, R. (2013) *What Are Qualitative Research Ethics?*, London: Bloomsbury Academic.
Wilhelm, A. G. (1999) 'Virtual Sounding Boards: How deliberative is online political discussion?', in B. N. Hague and B. D. Loader (eds) *Digital Democracy: Discourse and decision-making in the information age*, London: Routledge, 154–178.
Williams, A. (2017) *History of Digital Games: Developments in art, design and interaction*, Florida: CRC Press.
Williams, A., D. Harte and J. Turner (2015) 'The Value of UK Hyperlocal Community News', *Digital Journalism* 3(5): 680–703.
Williams, A., K. Wahl-Jorgensen and C. Wardle (2011) '"More Real and Less Packaged": Audience discourse on amateur news content and its effects on journalism practice', in K. Andén-Papadopoulous and M. Pantti (eds) *Amateur Images and Global News*, Bristol, UK: Intellect, 193–210.
Williams, A. A. and B. A. Marquez (2015) 'The Lonely Selfie King: Selfies and the conspicuous prosumption of gender and race, *International Journal of Communication* 9: 1775–1787.
Williams, Q. (2017) *Remix Multilingualism: Hip-hop ethnography and performing marginalised voice*, London: Bloomsbury.
Williams, R. (1965) *The Long Revolution*, London: Penguin.
Williams, R. (1974) *Television: Technology and Cultural Form*, London: Fontana.
Williams, R. (1976) *Keywords: A vocabulary of culture and society*, London: Fontana.
Williams, R. (1989) *Resources of Hope*, London: Verso.
Williams, R. (1991) 'Base and Superstructure in Marxist Cultural Theory', in C. Mukerji and M. Schudson (eds) *Rethinking Popular Culture: Contemporary perspectives in cultural studies*, Berkeley, CA: University of California Press, 407–423.
Williams, R. (2009) 'The Analysis of Culture', in J. Storey (ed.) *Cultural Theory and Popular Culture: A reader*, fourth edition, Harlow, UK: Pearson Education, 48–56.
Williams, R. and D. Edge (1996) 'The Social Shaping of Technology', *Research Policy* 26: 856–899.
Williams, S. (2015) 'Digital Defence: Black feminists resist violence with hostage activism', Feminist Media Studies 15(2): 341–344.
Willis, J. (2014) 'The Social Media as a Megaphone for Exposing Global Oppression and Urging Help', in B. A. Musa and J. Willis (eds) *From Twitter to Tahrir Square: Ethics in social and new media communication*, Santa Barbara, CA, Denver, CO and Oxford: Praeger, 301–315.
Wilson, C. and A. Dunn (2011) 'The Arab Spring: Digital media in the Egyptian revolution: Descriptive analysis from the Tahrir data set', *International Journal of Communication* 5: 1248–1272.
Wilson, D. and S. Quinton (2012) 'Let's Talk about Wine: Does Twitter have value?', *International Journal of Wine Business Research* 24(4): 271–286.
Wilson, P. (2016) 'Indigenous Documentary Media', in D. Marcus and S. Kara (eds) *Contemporary Documentary*, London and New York: Routledge, 87–104.
Wilson, P. and M. Stewart (eds) (2008) *Global Indigenous Media: Cultures, poetics, and politics*, Durham, NC: Duke University Press.
Wingfield, N. (2014) 'Feminist Critics of Video Games Facing Threats in "Gamergate" Campaign', *The New York Times*, 15 October. Available online: www.nytimes.com/2014/10/16/technology/gamergate-women-video-game-threats-anita-sarkeesian.html.
Winner, L. (1980) 'Do Artefacts Have Politics?', *Daedalus* 109(1): 121–136.
Winner, L. (1986/2009) 'Do Artifacts Have Politics?', in D. G. Johnson and J. M. Wetmore (eds) *Technology and Society: Building our sociotechnical future*, Cambridge, MA: The MIT Press, 209–226.

Wiseman, L. (dir.) (2007) *Live Free or Die Hard*. IMDb entry: www.imdb.com/title/tt0337978.
Wittkower, D. E. (2016) 'Principles of Anti-discriminatory Design', Philosophy Faculty Publications 28. Available online: http://digitalcommons.odu.edu/philosophy_fac_pubs/28.
Wolfsfeld, G., E. Segev and T. Sheafer (2013) 'Social Media and the Arab Spring: Politics comes first', *The International Journal of Press/Politics* 18: 115–137.
Wolfson, T. (2012) 'From the Zapatistas to Indymedia: Dialectics and orthodoxy in contemporary social movements', *Communication, Culture & Critique* 5: 149–170.
Wolfson, T. (2013) 'Democracy or Autonomy? Indymedia and the contradictions of global social movement networks', *Global Networks* 13(3): 410–424.
Wolfson, T. (2014) *Digital Rebellion: The birth of the cyber left*, Urbana, IL: University of Illinois Press.
Womack, J. (1999) *Rebellion in Chiapas: An historical reader*, New York: New Press.
Wong, A. (n.d.) Disability Visibility Project. Available online: www.disabilityvisibilityproject.com.
Wong, C. H.-Y. (2011) *Film Festivals: Culture, people and power on the global screen*, New Brunswick, NJ: Rutgers University Press.
Wood, A. (2006) '"Straight" Women, Queer Texts: Boy-love manga and the rise of a global counterpublic', *Women's Studies Quarterly* 34(1/2): 394–414.
Woodcock, J. and M. Graham (2020) *The Gig Economy: A critical introduction*, Cambridge and Medford, MA: Polity Press.
Wookieepedia (n.d.) 'About Wookieepedia'. Available online: http://starwars.wikia.com/wiki/Wookieepedia.
Woolley, J. P., M. L. McGowan, H. J. A. Teare, V. Coathup, J. R. Fishman, R. A. Settersten Jr., S. Sterckx, J. Kaye and E. T. Juengst (2016) 'Citizen Science or Scientific Citizenship? Disentangling the uses of public engagement rhetoric in national research initiatives', *BMC Medical Ethics* 17(33). Available online: https://doi.org/10.1186/s12910-016-0117-1.
Woolley, S. C. and P. N. Howard (2016) 'Political Communication, Computational Propaganda, and Autonomous Agents – Introduction', *International Journal of Communication*, 10(0): 4882–4890.
Woolley, S. C. and P. N. Howard (2017) 'Computational Propaganda Worldwide: Executive summary', Working paper no. 2017.11, Computational Propaganda Research Project. Available online: http://comprop.oii.ox.ac.uk/wp-content/uploads/sites/89/2017/06/Casestudies-ExecutiveSummary.pdf.
World Health Organization (2011) *World Report on Disability*, Geneva: World Health Organization.
World Wide Web Foundation (2019) 'A Contract for the Web'. Available online: https://contractfortheweb.org/.
Worrell, T. R. (2018) *Disability in the Media*, Lanham, MD: Lexington Books.
Worth, O. and K. Buckley (2009) 'The World Social Forum: Postmodern prince or court jester?', *Third World Quarterly* 30(4): 649–661.
Worth, S. and J. Adair (1972) *Through Navajo Eyes: An exploration in film communication and anthropology*, Bloomington, IN and London: Indiana University Press.
Wright, C. (2005) 'Opening Spaces: Power, participation and plural democracy at the World Social Forum', *Ephemera: Theory, Politics and Organization* 5(2): 409–422.
Wright, D. and R. Kreissl (2013) 'European Responses to the Snowden Revelations: A discussion paper', Increasing Resilience in Surveillance Societies project. Available online: http://irissproject.eu/wp-content/uploads/2013/12/IRISS_European-responses-to-the-Snowden-revelations_18-Dec-2013_Final.pdf.
Wright, E. O. (2010) *Envisioning Real Utopias*, London and New York: Verso Books.
Wright, M. (2006) *Disposable Women and Other Myths of Global Capitalism*, New York: Routledge.
Wright, S. (2007) 'Back to the Future: Italian workerists reflect upon the Operaista project', in E. Dowling, R. Nunes and B. Trott (eds) *Immaterial and Affective Labour: Explored*. Special issue of *Ephemera: Theory and Politics in Organization* 7(1): 270–281. Available online: www.ephemerajournal.org/contribution/back-future-italian-workerists-reflect-upon-operaista-project.
Wu, S., J. M. Hofman, W. Mason and D. Watts (2011) 'Who Says What to Whom on Twitter', *Proceedings of the 20th International Conference on World Wide Web*, New York: ACM, 705–714.
WWF (2019) 'Deforestation in the Amazon', *WWF*. Available online: https://wwf.panda.org/our_work/forests/deforestation_fronts/deforestation_in_the_amazon/.
Wylie, S. A., K. Jalbert, S. Dosemagen and M. Ratto (2014) 'Institutions for Civic Technoscience: How critical making is transforming environmental research', *The Information Society* 30(2): 116–126.
Wynne, B. (1996) 'Misunderstood Misunderstandings: Social identities and public uptake of science', in A. Irwin and B. Wynne (eds) *Misunderstanding Science? The public reconstruction of science and technology*, Cambridge: Cambridge University Press, 19–46.
Wysocki, M. (2015) 'It's Not Just the Coffee That's Hot: Modding sexual content in video games', in M. Wysocki and E. W. Lauteria (eds) *Rated M for Mature: Sex and sexuality in video games*, New York: Bloomsbury Academic, 194–209.

X-Net (2018) 'Homepage'. Available online: https://xnet-x.net/en/.

Xu, K. (2013) 'Framing Occupy Wall Street: A content analysis of *The New York Times* and *USA Today*', *International Journal of Communication* 7: 2412–2432.

Yagoda, B. (2014) 'A Short History of "Hack"', *The New Yorker*, 6 March. Available online: www.newyorker.com/tech/elements/a-short-history-of-hack.

Yang, G. (2009) *The Power of the Internet in China: Citizen activism online*, New York: Columbia University Press.

Yang, G. (2014) 'Internet Activism and the Party-State in China', *Daedalus* 143(2): 110–123.

Yang, S., A. Quan-Haase, A. D. Nevin and Y. Chen (2017) 'The Role of Online Reputation Management, Trolling, and Personality Traits in the Crafting of the Virtual Self on Social Media', in L. Sloan and A. Quan-Haase (eds) *The Sage Handbook of Social Media Research Methods*, London: Sage, 74–89.

Yardi, S. and d. boyd (2010) 'Dynamic Debates: An analysis of group polarization over time on Twitter', *Bulletin of Science, Technology and Society* 30(5): 316–327.

Yates, L. (2015a) 'Rethinking Prefiguration: Alternatives, micropolitics and goals in social movements', *Social Movement Studies* 14(1): 1–21.

Yates, L. (2015b) 'Everyday Politics, Social Practices and Movement Networks: Daily life in Barcelona's social centres', *The British Journal of Sociology* 66(2): 236–258.

Yefimova, K., M. Neils, B. C. Newell and R. Gomez (2015) 'Fotohistorias: Participatory photography as a methodology to elicit the life experiences of migrants', *48th Hawaii International Conference on System Sciences*, Kauai, HI: IEEE, 3672–3681.

Yeginsu, C. (2018) 'Benetton "Migrants" Ads Draw Outrage for Using Photos of Real Migrants', *The New York Times*, 21 June. Available online: www.nytimes.com/2018/06/21/world/europe/benetton-ad-migrants.html.

Yeomans, M. (2005) 'The Birth of Wikinews', *Citizen's Kane*, 28 April. Available online: http://citizenskane.blogspot.com/.

Ylä-Anttila, T. (2005) 'The World Social Forum and the Globalization of Social Movements and Public Spheres', *Ephemera: Theory, Politics and Organization* 5(2): 423–442.

Youmans, W. L. and J. York (2012) 'Social Media and the Activist Toolkit: User agreements, corporate interests, and the information infrastructure of modern social movements', *Journal of Communication* 62(2): 315–329.

Young, A. (2014) *Street Art, Public City: Law, crime and the urban imagination*, London: Routledge.

Young, I. M. (1980) 'Throwing Like a Girl: A phenomenology of feminine body comportment motility and spatiality', *Human Studies* 3(2): 137–156.

Young, I. M. (1986) 'The Ideal of Community and the Politics of Difference', *Social Theory and Practice* 12(1): 1–26.

Young, I. M. (1989) 'Polity and Group Difference: A critique of the ideal of universal citizenship', *Ethics* 99(2): 250–274.

YouTube Spotlight (2006) 'A Message from Chad and Steve'. Available online: www.youtube.com/watch?v=QCVxQ_3Ejkg.

Ytreberg, E. (2011) 'Convergence: Essentially confused?', *New Media & Society* 13(3): 502–508.

Yuan, E. J., M. Feng and J. A. Danowski (2013) '"Privacy" in Semantic Networks on Chinese Social Media: The case of Sina Weibo', *Journal of Communication* 63(6): 1011–1031.

Zamponi, L. (2015) Memory in Action: Mediatised public memory and the symbolic construction of conflict in student movements, Unpublished PhD thesis, Florence, Italy: European University Institute.

Zamponi, L. (2018) *Social Movements, Memory and Media: Narrative in action in the Italian and Spanish student movements*, London: Palgrave Macmillan.

Zappavigna, M. (2012) *Discourse of Twitter and Social Media: How we use language to create affiliation on the web*, London, New Delhi, New York and Sydney: Bloomsbury.

Zappavigna, M. (2015) 'Searchable Talk: The linguistic functions of hashtags', *Social Semiotics* 25(3): 274–291.

Zarrilli, P. (1986) 'Review: Toward a definition of performance studies, part I', *Theatre Journal* 38(3): 372–376.

Zavia, M. (2017) 'Pillan a Telefónica usando subtítulos descargados de Internet para sus series de Movistar+', *Gizmodo*, 1 October. Available online: https://es.gizmodo.com/pillan-a-telefonica-usando-subtitulos-descargados-de-in-1791035537.

Zayani, M. (2015) *Networked Publics and Digital Contention: The politics of everyday life in Tunisia*, London: Oxford University Press.

Zeitoun, A. and J. Seri (dir.) (2001) Yamakasi. IMDb entry: www.imdb.com/title/tt0267129.

Zelizer, B. (2002) 'Finding Aids to the Past: Bearing personal witness to traumatic public events', *Media, Culture & Society* 24(5): 697–714.

Zembylas, M. (2010) 'Negotiating Co-Existence in Divided Societies: Teachers, students and parents' perspectives at a shared school in Cyprus', *Research Papers in Education* 25(4): 433–455.

Zetter, K. (2015) *Countdown to Zero Day: Stuxnet and the launch of the world's first digital weapon*, New York: Broadway Books.

Zhang, Z. and G. Negro (2013) 'Weibo in China: Understanding its development through communication analysis and cultural studies', *Communication, Politics & Culture* 46 (2): 199–216.

Zhao, S. (2006) 'The Internet and the Transformation of the Reality of Everyday Life: Toward a new analytic stance in sociology', *Sociological Inquiry* 76(4): 458–474.

Ziegele, M., T. K. Naab and P. Jost (2019) 'Lonely Together? Identifying the determinants of collective corrective action against uncivil comments', *New Media & Society*. Available online: https://doi.org/10.1177/1461444819870130.

Zimmerman, E. (2007) 'Gaming Literacy: Game design as a model for literacy in the twenty-first century', in T. Binder, J. Löwgren and L. Malmborg (eds) *(Re)Searching the Digital Bauhaus*, London: Springer Verlag, 179–190.

Zimmermann, P. R. (1995) *Reel Families: A social history of amateur film*, Bloomington, IN: Indiana University Press.

Zimmermann, P. R. and H. de Michiel (2017) *Open Space New Media Documentary: A toolkit for theory and practice*, New York: Routledge.

Zittrain, J. (2018) 'Mark Zuckerberg Can Fix this Mess', *The New York Times*, 7 April. Available online: www.nytimes.com/2018/04/07/opinion/sunday/zuckerberg-facebook-privacy-congress.html?smid=tw-nytopinion&smtyp=cur.

Zivi, K. (2012) *Making Rights Claims: A practice of democratic citizenship*, Oxford: Oxford University Press.

Zoo Project (2011) 'Overview of His Works'. Available online: www.zoo-project.com/hommage/zoo-project/apercu-de-ses-oeuvres/.

Zoonen, L. van (2004) *Entertaining the Citizen: When politics and popular culture converge*, Lanham, MD: Rowman & Littlefield.

Zuckerberg, M. (2018) 'I Want to Share an Update on the Cambridge Analytica Situation …', Facebook, 21 March. Available online: www.facebook.com/zuck/posts/10104712037900071.

Zuckerman, E. (2013) 'Cute Cats to the Rescue? Participatory media and political expression', in D. Allen and J. Light (eds) *Youth, New Media and Political Participation*, Cambridge, MA and London: The MIT Press.

Züger, T. (2013) 'Re-thinking Civil Disobedience', *Internet Policy Review* 2(4). Available online: http://policyreview.info/articles/analysis/re-thinking-civil-disobedience.

Zukin, S. and M. Papadantonakis (2017) 'Hackathons as Co-optation Ritual: Socializing workers and institutionalizing innovation in the "new" economy', in A. L. Kalleberg and S. P. Vallas (eds) *Precarious Work*, Bingley, UK: Emerald Publishing, 157–181.

Zutavern, J. (2015) *Politik des Bewegungsfilms*, Marburg: Schüren.

AUTHOR INDEX

SUBJECT INDEX

For Product Safety Concerns and Information please contact our EU
representative GPSR@taylorandfrancis.com
Taylor & Francis Verlag GmbH, Kaufingerstraße 24, 80331 München, Germany

www.ingramcontent.com/pod-product-compliance
Ingram Content Group UK Ltd.
Pitfield, Milton Keynes, MK11 3LW, UK
UKHW050233200926
475702UK00020B/3010
* 9 7 8 0 3 6 7 5 4 4 1 6 4 *